Accelerating MATLAB® Performance

1001 tips to speed up MATLAB programs

Accelerating MATLAB® Performance

1001 tips to speed up MATLAB programs

Yair Altman

CRC Press
Taylor & Francis Group
Boca Raton   London   New York

CRC Press is an imprint of the
Taylor & Francis Group, an **informa** business

A CHAPMAN & HALL BOOK

CRC Press
Taylor & Francis Group
6000 Broken Sound Parkway NW, Suite 300
Boca Raton, FL 33487-2742

© 2015 by Taylor & Francis Group, LLC
CRC Press is an imprint of Taylor & Francis Group, an Informa business

No claim to original U.S. Government works

Printed on acid-free paper
Version Date: 20141031

International Standard Book Number-13: 978-1-4822-1129-0 (Hardback)

Visit the Taylor & Francis Web site at
http://www.taylorandfrancis.com

and the CRC Press Web site at
http://www.crcpress.com

To Tovi, Gali, Liat, and Lavi

Contents at a Glance

Contents

Preface

The MATLAB® programming environment is often perceived as a platform suitable for prototyping and modeling but not for actual real-life applications. One of the reasons that I constantly hear when consulting with clients is that *"MATLAB is slow"*.

This book aims to help reduce this perception and shows that MATLAB programs can in fact be made to run extremely fast, in a wide variety of different ways.

MathWorks, who develop MATLAB, invests a significant amount of R&D effort in constantly improving MATLAB's performance and advocating best practices for improved performance.[1] Postings for performance-related R&D jobs are periodically posted[2] and the engine's performance improves with almost every semi-annual MATLAB release. In fact, the same MATLAB programs that might have been slow 10 or more years ago may now be blazingly fast when run using the latest MATLAB release, on the very same platform.

Using programming techniques presented in this book, MATLAB applications can be made even faster, fast enough for most uses. This enables significant reduction of the development time and cost, since we can use MATLAB from end to end, from prototyping to deployment, without having to maintain a mirror code–base using a different programming language and environment.

So, the perception of MATLAB as a slow environment may at least partly be due to a combination of factors, ranging from negative experience from past releases, to application code that does not follow good programming practices.

Some people say that when performance is really important, we should use a different programming language such as C/C++. While using C/C++ can certainly improve performance if done correctly, it is not a general panacea. MATLAB provides many benefits that could be very important during both development (e.g., rapid application development, short development cycles, ease of use, lenient environment, gentle learning curve) and run time (e.g., built-in vectorization, simple parallelization, automated memory management, and JIT optimizations). This is the reason we use High-Level Languages in the first place, trading some performance for functionality, development time, and so on. Otherwise, we would program in Assembly, use FPGAs, or even develop custom ASIC chips for top performance....

Using MATLAB does not however mean that we should abandon performance altogether. This book shows that using some very easy-to-follow techniques we can significantly improve MATLAB code speed without sacrificing MATLAB's benefits.

Some authors who write about performance like to demonstrate ideas with colorful graphs that show the performance behavior as a function of one or more parameters. Such comparisons are often academic in nature. This book tries to take a more practical approach with the presented recommendations. The ideas are explained verbally and short code snippets are often included to illustrate the point, usually without a rigorous comparison of all the parameter variants. This is not a PhD thesis, but rather a practical hands-on book intended for day-to-day use by engineers.

On one hand, enough information is provided to enable engineers to immediately apply the suggestions to their MATLAB programs. On the other hand, many references are also provided to enable readers who wish to expand the treatment of a particular topic to easily do so. There is always a delicate line between providing too much and too little information in the main text, so I hope my choices were adequate.

Performance is a term that can refer to many things, from functional (*"Does it perform well enough?"*) to speed (*"Does it perform fast enough?"*). In this book we are interested in only the latter aspect: speed. Other aspects of performance (accuracy, stability, robustness, etc.) are not less important, but are outside the scope of this text.

If this book may seem verbose at times, this is because I have tried to explain the reasons behind the recommendations, in the hope that users will gain insight. After all, this book cannot cover every possible aspect in all possible situations; gaining insight will enable readers to search for other ways to tune their specific program. Naturally, not all the numerous individual speedup suggestions can be remembered. But in my experience, once we understand and internalize the reasoning, we naturally "rediscover" these techniques whenever we come across a situation that merits them.

This book contains a wide variety of suggestions. Some may not be relevant for a specific application, or we may decide not to use them for some reason (e.g., due to their extra development and/or maintenance cost). Do not despair—plenty of other suggestions are available that could be helpful. There are many ways to achieve our target performance goals, so even if one technique fails, there are alternatives that we could try. In fact, there are so many different ways to achieve these goals that we can take a pick based on aesthetic preferences and subjective experience: Some people use vectorization, others like parallelization, some others prefer to invest in smarter algorithms, others trade memory for performance, still others display a GUI that gives the impression of being fast.

All of these routes and more are valid alternatives for making a program answer user expectations of speed and responsiveness. Moreover, it is expected that readers will become more proficient in efficient programming techniques, such that their programs will run faster in the first place, even before any tuning is actually done.

The book is meant as a generic reference for MATLAB performance tuning. As such, it does not include detailed discussion of domain-specific topics such as numerical analysis, optimization, statistics, algorithms, or image processing. These topics are well worth discussing for performance aspects, but they too are outside the scope of this book. Some discussion is included, but is not intended to be comprehensive nor detailed. Interested readers are referred to dedicated works on these specific topics.

Book Organization

This book is organized in chapters grouped by related functionality/usage. It is not necessary to read the book in order: the chapters and sections are mostly independent and stand alone. You can safely skip parts that you find difficult or uninteresting.

We begin with a theoretical description of performance tuning in Chapter 1. The discussion includes typical pitfalls, tradeoffs, and considerations that need to be kept in mind before and during any tuning process. Chapter 1 is not meant to be a comprehensive discussion of the theory of performance tuning; there are other books fully devoted to this subject. In contrast, I attempted to describe the essence of the major practical issues as I see them. It should be noted that this is not an exact science, and my subjective opinions on tradeoff considerations may well be disputed by others. Still, I hope that by reading this chapter, readers will at least be exposed to the underlying questions and considerations that relate to performance tuning, even if they disagree with my analysis or recommendations.

As long as you keep the underlying questions in mind when you tune an application, you should be okay.

Chapter 2 provides an overview of tools that are available in MATLAB in order to diagnose an application to determine the locations of, and reasons for, its performance hotspots. There are several different manners by which we can profile application run time in MATLAB, and different situations may dictate different tools.

Chapters 3 through 11 discuss specific speedup techniques that can be used in MATLAB:

- Chapter 3 explains standard techniques adapted from non-MATLAB programming languages.

- Chapter 4 discusses techniques that are unique to MATLAB code.

- Chapter 5 discusses implicit parallelization, with indexing and vectorization.

- Chapters 6 and 7 discuss explicit parallelization using a variety of means (CPU, GPU, and multi-threading).

- Chapter 8 discusses techniques for using compiled (binary) code.

- Chapter 9 discusses specific techniques that are memory-related. The nontrivial relationship between memory and performance is explained, and a variety of tuning techniques are presented in light of these explanations.

- Chapter 10 discusses techniques related to graphics, GUI, and user interaction.

- Chapter 11 concludes the list of specific tuning techniques with a discussion of techniques related to I/O, particularly reading and writing files.

Chapters 3 through 11 are intended for use as a random-access reference. The sections and techniques can typically be used independently of each other. You can directly use any section or technique, without reading or using any other.

Appendix A presents online and offline resources that expand the information presented in the text and enable further research. Appendix B concludes the text by providing a non-comprehensive general checklist for performance tuning.

Throughout the text, references are provided to enable interested readers to expand their knowledge of specific issues. Footnotes are used to clarify some points and to provide cross-references to other sections within this book; endnotes are used to provide references to related online resources. Most online references are provided in both full and shortened format, to enable easy usage when transcribed from hardcopy.

Conventions Used in This Book

The following special text formatting conventions are used within this book:

- `Fixed-width` font is used for MATLAB code segments. The Command-Line prompt (>>) is provided only where it would help to distinguish between user-entered text and MATLAB's response in the console:

```
>> version
ans =
8.3.0.532 (R2014a)
```

In other places, the console output is indicated using an arrow sign:
```
tic, pause(2), toc
⇨ Elapsed time is 2.001925 seconds.
```

- **Regular bold** font is used for object property names (e.g., **UserData**), as well as for occasional emphasis.
- *Bold italic* font is used for MATLAB function names (e.g., *max* or *ismember*)
- *Regular italic* font is used for file names (e.g., *data.mat*), utility names (e.g., *grep*), introduction of new terms, as well as for occasional emphasis.

The duration terms "minutes", "seconds", and "milliseconds" are used extensively throughout the text and are usually shortened to "min", "s", and "ms", respectively (e.g., 3 min, 5 s or 45 ms).

Icons are sometimes placed next to the text, as follows:

- The lightning icon[3] indicates a suggestion with potentially high impact on the program's performance. These suggestions should typically be considered first, before trying other alternatives.

- The warning icon indicates a suggestion that relies on undocumented or unsupported features that may not be available in future MATLAB releases and may not work correctly (or at all) on some platforms, releases, or situations. Use of such suggestions should be done only after careful testing and at the user's own risk. Neither MathWorks nor this book's author or publisher can take any responsibility for possible consequences due to using unsupported functionality. It is the author's explicit suggestion that such features should be considered last, only after all the other (supported) venues have been tried.

MATLAB and Simulink are registered trademarks of The MathWorks, Inc. For product information, please contact:

The MathWorks, Inc.
3 Apple Hill Drive
Natick, MA 01760-2098 USA
Tel: 508 647 7000
Fax: 508-647-7001
E-mail: info@mathworks.com
Web: www.mathworks.com

Acknowledgments

I owe a debt of gratitude to several MATLAB experts who have helped me prepare this manuscript. A few experts were most gracious to provide significant contributions in their field of expertise: Pavel Emeliyanenko (parallelization); James Tursa (MEX), and Igal Yaroslavski (FPGA). I received invaluable assistance from MathWorks, both in the Books Program (Naomi Fernandes) and R&D (Michelle Hirsch, Ken Atwell, Gaurav Sharma, Pat

Quillen, and others).* Systematics Ltd. were very helpful: Michael Donnefeld helped me wade through MATLAB Coder issues, Igal provided the FPGA section, and David Gochman assisted behind-the-scenes. John D'Errico and Bruno Luong are MATLAB giants—I have learned countless techniques from them over the years. Jim Hokanson, Bill McKeeman, Malcolm Lidierth, Mike Croucher, Eric Sampson, Kadin Tseng, and Oliver Woodford reviewed and provided valuable feedback. At CRC Press I found a truly caring publisher: Bob Stern and Bob Ross for helping with issues large and small, always with patience, understanding, and a kind word; and Kyle Meyer and Syed Shajahan for careful editing.

This book is much better thanks to the contributions by these people, and yet I take responsibility for any errors, inaccuracies, and omissions that might have inadvertently entered the text. In such a large project, as in any large program, some "bugs" have surely escaped scrutiny. I hope that you accept them with the understanding that such work is never perfect. I have tried to generalize my findings for improved performance, and yet it is quite possible that the situation may be different or even diametrically opposite under a different set of parameters such as hardware, OS, MATLAB release, data size, data type, or any of myriad other possible factors. Readers should therefore not rely on the suggestions within the text before carefully testing them on their specific system. Neither the author nor publisher can accept any responsibility for possible consequences due to this book. Please report any issue that you discover. My direct email is altmany (at) gmail.com. A detailed list of the major errata will be posted on http://UndocumentedMatlab.com/books/matlab-performance.

Finally, this book is devoted to my family. I said it in my first book and find no better words to say it again—this book would never have seen the light of day without your loving support and understanding. I owe you more than words can express.

---

* MathWorks employees are affectionately called *MathWorkers* by the MATLAB community. I use this term in this book to refer to MathWorks engineers, either past or current, who contributed some useful utility, suggestion, or insight.

Author

Yair Altman, author of the extremely popular UndocumentedMatlab.com website, is well respected by the MATLAB® community as a leading expert on advanced MATLAB programming.

Yair's first book, *Undocumented Secrets of MATLAB-Java Programming*, was published in 2011 to rave reviews and became the standard textbook on the subject. His many years of public contribution on MATLAB performance, plus a multitude of useful tips never before published, are now available in this highly readable volume.

Yair holds a BSc in physics and an MSc in computer science, both with high honors. Yair has over 20 years of professional software development experience at various levels of organizational responsibility, from programmer to VP R&D. He has developed systems using two dozen programming languages, on a dozen different platforms, half a dozen databases, and countless MATLAB releases.

Yair became an independent MATLAB consultant several years ago, and has never looked back. He currently assists clients world-wide in various MATLAB-related aspects: consulting, training, and programming.

Yair can be reached at altmany@gmail.com.

1

Introduction to Performance Tuning

Performance tuning, as used in the context of this book, is the process of improving the run speed of an application, function, or algorithm.[4] Tuning is often repetitive, each tuning pass improving the application's speed compared to the previous version. It is much more of an art than a science, and a wide variety of different techniques can often be employed to achieve the required speed boost.

Before jumping into the thicket of techniques for profiling and improving program performance, we should understand the larger picture. Why should we bother to performance tune? How should we handle the tuning process? When should we stop tuning? What are the tradeoffs that we should consider when tuning? These are all questions of a theoretical nature that have very practical effects.

I have seen many engineers who jump straight into tuning without considering these questions. Quite often this leads to frustrations at the cost-effectiveness of the tuning. On the other hand, by giving these questions some thought before tuning, we could reach unexpected conclusions that will affect the manner by which we would tune our specific application. In simple words, we will spend more time tuning what is most effective, and our overall investment will be more cost-effective.

1.1 Why Should We Bother?

This is not a trivial question. Performance tuning takes time and effort. Significant speedups are sometimes achieved with a single tuning pass, but are more often than not a result of accumulated small speedups in several passes, employing different tuning techniques. The development effort is certainly not negligible.

So why indeed should we bother to performance tune?

One quick answer is that we usually hope that investing some of *our* time in tuning an application, will make it faster and more usable to its users, saving time for *them*. We will invest the tuning development time once, but the users will enjoy its benefits multiple times, whenever they use the tuned function.

A variant of this answer is that not all times are equal, some times are simply worth more than others. A developer's workhour may cost less than a client's. Or a workhour in the development phase may be worth less than an hour under tight operational run-time constraints. Or maybe the client's perception of the application as fast and reactive is worth an extra workhour by the developer.

Another reason for tuning is that very often the speed difference could be the deciding factor for the users whether or not to use the application.

For example, let's say that we have an algorithm that processes real-time stock data and decides whether to buy or sell on the stock exchange. In order to be effective, the algorithm must run within (say) 10 seconds. If the algorithm is slower, then it would not be effective, and we might end up losing a lot of money by missing the narrow window of opportunity for trading a stock. The difference between an algorithm that runs in 12 s versus 8 s, by itself not such a major performance boost, could well be the deciding factor between a usable and a nonusable algorithm.

As another example, an algorithm needs to run within 1 s in order to process real-time data from some hardware. If the algorithm takes even a bit longer than that to run, it will quickly lose sync with the real-time data and the program will fail due to input buffer overflow, dropped input events, or CPU churning causing a MATLAB freeze-up. The simple fix of improving performance so that the algorithm runs faster than a second, is the deciding factor between a usable and a nonusable algorithm.

Similarly, if a GUI takes too long to process user inputs, the entire application may become practically unusable. Countless studies have shown that websites generate significantly less user-generated revenue when they slow down, and that users quickly stop browsing if a webpage fails to load within a few seconds.[5] One might think that browsing behavior is unrelated to MATLAB, but the underlying human psychology is very similar in web browsing and desktop applications. The nonintuitive result is that users often prefer imperfect immediate answers, than a perfect but delayed response.

Finally, applications often need to run within specified performance constraints in order to be formally accepted by a client. In fact, such constraints are often specified in the project's specifications (SRS or SOW). Performance constraints are often considered critical acceptance criteria. In such cases, we may have to performance-tune our code, otherwise the entire application might not be accepted by the client.

Even when we develop for ourselves, increasing application speed allows us to run multiple times in short succession, waiting a shorter time for each run to finish. This is important for parameter sweeps — tuning application parameters by scanning various combinations. Alternatively, we could run larger or more accurate models at the same time. *Performance optimization* thus enables better *functional optimization*.

This highlights the dual meaning of the term *"optimization"*: in the context of performance it means improving speed, throughput, or latency, without ever achieving an optimal result; in the context of functionality, optimization means finding an optimal set of parameters for best functional outcome. In this book, we usually refer to optimization in the performance context, rarely the functional one (see §4.5.5).

In a recent poll of Java programmers,[6] 72% felt that performance was important or critical for their application's success, and only 7% believed that it did not matter that much for their applications. Although the survey may not be fully representative, the bottom line is pretty clear: Performance is a very important aspect of application functionality, one which we should not ignore.

1.2 When to Performance-Tune and When Not to Bother

Efficient code that is used once may only be good for our self-respect. The run-time savings often do not justify the extra development time investment. If brute force solves a problem slowly while we work on something more important, then brute force may be just right for our needs. Our time is valuable, at least to us, whereas computers are relatively cheap: If our brute force code runs for a few minutes or even hours while we are able to produce other useful work, then there is no reason to be more proactive. Having said that, some problems, when solved using a brute-force approach may take the lifetime of the universe to solve, while an intelligently devised algorithm might solve the problem in seconds.* It is always good to understand such methods to optimize code, as we will surely benefit from them one day.

Sometimes we can simply postpone a long-running task until we get faster hardware. Same-cost hardware constantly performs faster over time. As Eric Raymond said:[7]

> *"Don't just do something – stand there!"*

We can distinguish between several levels of code:

- Code that we will use once, and then dispose
- Code that we will use often
- Code to be distributed, used by many others

If we need to use a program often or immediately, it makes sense to amortize the time that we will spend optimizing the code over the number of uses it will see. If an hour of time invested now will save many hours later, it may be worth the investment.

If the code is used by many others, it can be a good investment to save even a few seconds per run, multiplied by thousands or millions of uses into a real benefit.

We need to consider the total potential savings when deciding whether to tune: after all, it would not make much sense to spend an hour to tune an algorithm that would save a single noncritical second in run time, if the function would only run 60 times (a total savings of only 1 min). However, if the same function would be run 15000 times, the total run-time saving would be over 4 hours, so tuning makes sense.

Do not optimize any code before it is functionally stable. Tuning buggy code is useless. We should ensure the code works properly, before considering any tuning.

Moreover, we should not tune our code before the entire functionality is implemented, the run-time performance has been profiled, and the performance bottlenecks (if any) have been identified. The reason is that when we would later add the missing functionality, the code could change to such a degree that much of our tuning effort might be irrelevant and would need to be redone. Moreover, it may well turn out that the code that we prematurely tuned is in fact not a performance bottleneck of our program, so we have wasted time tuning a piece of code that has no real effect on the overall performance. Tuning before these steps are done is often called *premature tuning/optimization*, on which computer-science guru Donald Knuth once said:[8]

---

* Project Euler (http://projecteuler.net) is a splendid online repository of problems that teaches methods in problem solving.

"The real problem is that programmers have spent far too much time worrying about efficiency in the wrong places and at the wrong times; premature optimization is the root of all evil (or at least most of it) in programming."

On the other hand, we should also avoid the very common reverse practice, of leaving performance tuning to the very end of the development. At this time, real-world pressures typically force developers to release a product that was not properly tuned, if at all. It is also sometimes difficult to make anything but minor code changes at this stage, reducing the effectiveness of the tuning process.

One way to reconcile this conflict is to monitor performance early and frequently during the development process, but without actually spending time to improve performance, unless we discover a critical show-stopper performance bottleneck. We can then redesign our code to fix this bottleneck, but still leave the major tuning effort to a later part of the development. We would only invest the major tuning effort near the end of development, when the requirements are stable and the program performs as requested at a functional level. This is sometimes termed:[9]

"Optimize late, benchmark early"

As with other things, tuning should not be taken to an extreme: It should not be started in full-force at the very beginning, but also not left to the very end. We should keep performance in mind throughout the development, but remember that it is only one aspect of the program goals. A good place to plan for a dedicated performance-tuning effort is around the 80%–85% mark of the development. By this time, the application should be stable enough to prevent the drawbacks of premature optimization, but also far enough away from the end of the project to shield us from client deadlines and other exogenic project constraints.

When considering the question of whether or not to performance-tune code, we should take into consideration not just the development time required for the tuning. Whenever we modify our code, there is a good chance that we will inadvertently introduce bugs into the code. A well-known rule-of-thumb is that a single bug is introduced for every 10 new or modified lines of code. Therefore, we should also take into consideration additional time required for debugging, verification, and fixing bugs introduced during the tuning process, and the possibility that some additional bugs will not be detected. Because of this, and as a general rule in all engineering aspects,

"Don't fix it if it ain't broke!"

So, if our code runs fast *enough*, leave it alone. Do not spend time optimizing something that does not really require optimization. It would just waste time and possibly introduce bugs. As engineers, it is often very tempting to optimize our code, even when this is not strictly needed. We should be aware of this tendency and avoid it. One of the development manager's responsibilities is to ensure that premature and/or excessive tuning does not occur. When a project starts, we should allocate 3%–5% for the tuning phase. If it takes longer, this could indicate that something is wrong.

1.3 The Iterative Performance-Tuning Cycle

Performance tuning is a repetitive development cycle task that is typically performed following the first complete pass of development and testing. This ensures that we tune a stable program that works well in all respects excluding speed/responsiveness, rather than a buggy program. Performance tuning includes the following sub-tasks:

- We first **measure** the overall code performance in order to determine whether tuning is at all necessary. If not, performance tuning stops and we proceed to deploy the application. Never tune a program that meets its requirements.

- If we determine that tuning is in fact required, we **profile** the program to determine the location of the performance hotspots (see Chapter 2).

- We then **modify** the code to fix these hotspots, focusing our attentions only at a few top hotspots, using a wide variety of techniques (see Chapters 3 through 11).

- We now **test** the program to ensure that we did not introduce any bugs. This is very important, since we often inadvertently introduce new bugs in the tuning process.

- Finally, we **loop** back to the first tuning stage, to measure the performance again and determine whether or not we should continue tuning.[10] Moreover, we should test that the program is actually faster. More often than we care to admit, would-be optimizations degrade performance and should be reverted.*

No performance improvement is really possible without measurement. Remember the well-known adage that says (somewhat paraphrased[11]):

> *"If you don't know where you are and what your goal is, then it is nearly impossible to reach that goal."*

It is important to measure the correct things, and to do this correctly, without external artifacts that may affect the measurements. Moreover, it is important to constantly compare the results to quantifiable (numeric), measurable, and comparable performance goals. These metrics should be comprehensive enough to enable hotspot localization.

Performance measurement and profiling are often done and discussed together, although their purpose is different: *Measuring* performance is meant to test the overall program's run time and check whether or not it is within the performance goals. This check is the main criteria for deciding whether or not any tuning is at all needed. On the other hand, *profiling* is performed once we have decided that some tuning is needed, in order to isolate the location and reasons of the performance hotspots.

Measuring and profiling might use the same or different tools. For example, MATLAB's *tic* and *toc* functions are usually used for measuring but sometimes also for pinpoint profiling, while MATLAB's Profiler is primarily used for profiling.†

---

* Code changes that degrade performance are sometimes called *pessimizations* (in contrast to optimizations).

† Measurement and profiling tools available in MATLAB, and techniques for using them, are discussed in Chapter 2 and §9.2.

1.3.1 Pareto's Principle and the Law of Diminishing Returns

In each step of the tuning cycle, we should use Pareto's principle[12] (also known as the *"80–20 rule"**) to concentrate our energy on the 20% of the code that accounts for 80% of the application's time. Within that 20% of code, only about 20% (i.e., altogether just 4%–5% of the total program code) accounts for 80% of the code that can actually be tuned to improve performance. In practice this means that we should concentrate on the top 3–5 time hogging functions (as reported by the profiling — see Chapter 2), and within them only on the top 3–5 hogging lines or code segments. These are called performance *bottlenecks*, or *hotspots*.[13]

Note that every rule has an exception: The top time-hogging items may possibly be less attractive for tuning than others. For example, they may require more time to tune than others ("low-hanging fruits") that would yield immediate improvements.

We should also be aware of cases where several items, which are individually relatively low in the Pareto list, are slow due to the same underlying reason. Therefore, solving the underlying reason will automatically improve all these items together, in one fell swoop, making this an ideal tuning candidate. For instance, in the following simplified example, blindly following Pareto's principle might lead us to tune mainAlgo, whereas in fact it would make more sense to optimize the legend first, since the three legend-related functions together outweigh mainAlgo's run time:[†]

Function	Relative Run Time (%)
mainAlgo	35
createLegend	25
initLegend	15
updateLegend	10
(all others)	15

When deciding what to tune, we should NEVER rely on guesses or intuition. More often than not in performance tuning, we are completely wrong in our assumptions of the program's hotspots, and surprised to see the actual profiling results. Therefore, always profile using one of MATLAB's available profiling tools (see Chapter 2).

Do not apply optimizations across-the-board to the entire application code — only to the top few bottlenecks in each tuning iteration. There are several reasons for this:

1. Tuning non-top hotspots presents smaller benefits and cost-effectiveness.

2. The more we disperse our efforts, the higher the risk of introducing bugs.

3. A corollary of the previous point is that we would need to retest the entire application, rather than just the few isolated sections that were modified.

4. We increase the risk of hurting performance (!), due to resource contention, memory allocations, and other similar factors that are hard to foretell before we profile the modified application. Making small modifications at each tuning cycle facilitates finding performance anomalies and reverting adverse fixes.

---

* Many computer scientists claim that in practice, software actually follows a 90-10 rule, rather than 80-20.
† The astute reader will note that this does not in fact break Pareto's principle, since we can consider all the legend-related functions as being a single factor worth 50% of the run time, making it the top Pareto tuning candidate.

5. Modifying multiple code segments hampers our ability to track the specific speedup impact of each of our fixes. It could be that one of the fixes is bad for performance while others are good — by making all the changes in one tuning round, the bad fix may get lost in the overall improvements of the others.

It is very important to re-profile the application in each tuning cycle, because the bottlenecks may well move to other code sections after the cycle's fixes. Also, it is possible that our changes did not fix the hotspot but actually made performance worse! This is in fact quite common when trying to optimize code using vectorization on a relatively recent MATLAB release: sometimes MATLAB's internal JIT (*just-in-time compiler*,[14] see §3.1.15) is more effective than hand-crafted vectorization. By re-profiling after each fix, we can easily roll-back ineffective modifications and try a different approach. A source-code versioning system might be handy for this.

When attempting to use any specific technique to solve a performance bottleneck, we should be ready for occasions when our attempts will fail to produce the required speedup and, as said, might even make the situation worse. Do not be discouraged: There are numerous different techniques that can be applied and we can always try to use one of the others. This book lists hundreds of small suggestions, grouped into dozens of technique types — surely one or more of the other techniques will work.

Performance bottlenecks are typically caused by one of three possible limiting resources: CPU, memory, or I/O. Different code sections may be limited by a different resource. The ways to tackle each resource limitation is different:

- CPU — see Chapters 3 through 8, and 10
- Memory — see Chapter 9
- I/O — see §3.2 through §3.8, §10.4, and Chapter 11

Do not tune MATLAB's built-in functions except as last resort. Instead check whether:

- The number of calls to these functions can be reduced or even eliminated
- The functions can be replaced with equivalent faster variants

The *law of diminishing returns* stipulates that the initial tuning rounds will be much more effective in terms of achieved speedups than subsequent rounds. For example, the first tuning round may provide a 3× speedup while the fifth round only 10%. Such behavior is typical and is to be expected. In fact, if it does not happen, then you have probably not correctly identified the top bottlenecks according to Pareto's principle.

It is partly due to the diminishing speedups that it is tempting to continue tuning, in the hope that "just a few more rounds" will give us another 50% speedup. But at some point, the cost–benefit ratio of the tuning round simply becomes too high.

1.3.2 When to Stop Tuning

Following each tuning re-run, we should recheck whether the application is fast enough, and continue tuning only if it is not. When the code is fast enough, we say that the application is *performant*. We should avoid the natural temptation to tune beyond what is really needed. The decision of when to stop tuning is obviously application dependent. The easy cases are when the application needs to fulfill strict and precise performance requirements. Unfortunately, such precise requirements are usually unavailable. We often find that the target performance is a moving goal, and the tuning process a never-ending task. Obsessive optimization can be a search for perfection, a state that we can never achieve, and that we should actively avoid.

To avoid moving performance goals I suggest to define clear measurable performance target(s), under a pre-defined set of parameters (e.g., mean/median/max), *before* we start tuning. Keeping this goal in focus during the tuning cycles will help us achieve it. For example, in GUI applications a typical criterion is that graphical updates complete in 1.0 s or less (optimally within 0.3 s, which humans perceive as "instant").[15]

A few years ago I developed an application (IB-Matlab[16]) that connects MATLAB to Interactive Brokers (IB)[17] for online analysis and trading of stocks, bonds, and other securities. The target here was to process 200 IB events per second (i.e., 5 ms per event) in MATLAB on a "standard" laptop to enable effective streaming quotes. It took many hours and dozens of tuning iterations, some improving by only a few percent, to improve the program's response time from the original 45 ms per event, down to the required 5 ms.

Deciding the target speedup may affect the tuning path. For example, if we need 100× speedup, then no simple code improvement will suffice — we need a huge algorithm breakthrough, removal of I/O, massive hardware upgrades, and/or parallel processing.

When we reach a point where most of the time is spent on calls to a small number of built-in functions (e.g., core arithmetic operators/functions or internal indexing), we have probably optimized the code as much as can be expected from a source-code perspective. This does not mean that we cannot improve the performance any further: Additional speedups can still be achieved by other means, such as algorithm changes, or hardware aspects (I/O, GPU, parallelization, memory, etc.). We should also use perceived performance as a complement to actual performance, where the actual performance cannot be improved further.*

A variant of this tuning-stop criteria, is when the incremental speed improvements are significantly less (e.g., 5%–10% or less) than the initial improvement. We can keep track of these values by recording the measurement results in each tuning cycle. Examples can be found of 20× or 100× speedups, but in real life there is often no need for such speedups — 3–10× is more than enough for most practical cases.

Another indication that source-code tuning has reached its usefulness limit, is when the top time-hoggers only use 5%–10% of the total program time. This indicates a breakup of the Pareto principle upon which effective tuning is based. Further tuning will probably not be very cost-effective from this point onward.

Performance speedup goals need not (and in fact should not) be a single number, but rather a function of the system load and other external considerations. We should decide on acceptable performance degradation in system load, prepare the test harness and profile. We might well find that having *graceful degradation* means that the system is suboptimal (a bit slower) for the common scenario (no load) — this is definitely a tradeoff worth considering, so that the common scenario achieves its higher performance goal, at the expense of uncommon load scenarios. Alternatively, we could decide to sacrifice a bit of functionality rather than performance in case of load, to achieve graceful degradation by other means. As long as we are aware of all these factors, we can make an informed decision suited for the specific case.

Remember that tuning is not a standalone aim by itself. It depends on other factors such as development time/cost, maintainability, robustness, and other tradeoff factors that shall be discussed in §1.6. It is quite possible that in any particular case, one or another of these tradeoffs will dictate a different tuning-limit criterion.

---

* Perceived performance is discussed in §1.8.

Finally, we should stop tuning when we reach the limit in which platform differences could come into play. We might well over-optimize our code so that it runs extremely fast on our specific development workstation, but fails miserably on other computers, due to targeted use of hardware-specific techniques such as CPU cache size.

1.3.3 Periodic Performance Maintenance

Every now and then, we should recheck our application's performance and decide whether the conditions suggest re-tuning the application. Performance degradations tend to accumulate over time due to increased data and usage, and due to changing usage patterns. By performing proactive performance maintenance we can detect and fix these problems before they become a crisis. Moreover, it is possible that even if there is no performance *degradation*, new capabilities could *improve* performance:

A colleague once wrote a tool that took roughly an hour to solve relatively small problems, running on a corporate mainframe. This code was useful enough for small problems. At the time, while it was state of the art, it was relatively slow, and scaled poorly. In fact, the time required was $O(n^9)$, so simply doubling the problem size would take 512 times as much time to solve. However, my colleague waited and watched. Every time the capabilities of the mathematical algorithms improved, he rewrote the code, improving performance by an order of magnitude each time. In some cases, this was based on algorithmic improvements in MATLAB. In other cases it was simply a reflection of improved algorithmic understanding, since he too was learning new methods over the years. Finally, computers themselves have gotten vastly more capable: what was once a super-computer now sits on every desktop, and modern phones are smarter than the computers on the Apollo lunar modules.

1.4 What to Tune

So we have decided that we need to improve our program's performance. Good for us. But what exactly is it that we wish to improve? In other words, which "performance" exactly are we trying to tune?

It seems that the answer to this seemingly simple question is not trivial at all. There are several separate aspects of performance that we could tune. Some of the important performance aspects to consider include:[18]

- **Availability** — Over-clocking the CPU will increase its speed but reduce the computer's life span and overall availability to future re-runs.
- **Latency** — The time delay between user request and program response.
- **Throughput** — The rate at which a program can accept and transmit data.*
- **Data capacity** — The amount of data that a program can process at once.
- **Processing speed** — The speed at which a program processes the data.
- **Initial response** — The ability of a program to provide a rough initial estimate response, before delivering the more accurate final response later on.†
- **Scalability** — The ability of a program to easily scale up or scale out.‡

---

* The common tradeoff between latency and throughput is discussed in §3.8.
† See §1.8.4. Also see §3.9.2 for a related discussion of the performance-accuracy tradeoff.
‡ See §1.7.

- **Resource usage** — The ability of a program to minimize external resource (memory, I/O, CPU, power, cost, etc.) usage while doing its work.
- **User perception** — The ability of a program to convey to the user a feeling that it is fast enough.*
- **Graceful degradation** — The ability of a program to remain reasonably performant in the face of deteriorating external conditions (e.g., load or data size).

In many cases, improving one aspect comes at the expense of another aspect.† We can rarely optimize more than a handful of performance aspects of a program. In fact, tuning cycles typically focus on only a single aspect, or two in rare cases.

We need to decide the relative priorities of the tunable aspects as they pertain to each and every specific application. This needs to be done in advance, before the tuning cycle begins. For different applications, different aspects may be more important than others, resulting in radically different profiling methodologies and tuning paths.

1.5 Performance Tuning Pitfalls

Performance tuning is an art, in which one gets better with time. Getting better in performance-tuning involves avoiding some common mistakes and pitfalls, no less than gaining experience in a variety of optimization techniques. Here are some of the pitfalls (also called *performance antipatterns*[19]) that are often encountered:

1.5.1 When to Tune

- **Unnecessary optimization** — This is probably the most common pitfall: If an application is fast enough, leave it alone; *don't fix it if it ain't broke*.
- **Premature optimization** — Tuning the application too early in the development process (see §1.2).
- **Belated optimization** — This refers to the reverse practice, of leaving performance tuning to the very end of the development (§1.2 again).
- **Rushed optimization** — This happens when performance is not treated as an important application feature, and is not allocated reasonable resources and priority by the project manager. It can also occur in teams where it is unclear that programmers have a personal responsibility for performance.[20] The result is often ineffective off-hand optimization at the last moment before delivery.

1.5.2 Performance Goals

- **Missing metrics** — Without clear quantifiable measurable goals, we will likely spend too much time tuning yet still fall short of our targets.
- **Irrelevant or unrealistic metrics** — Using incorrect performance goals is worse than not having any goals at all. It could lead to irrelevant tuning that does not

---

* See §1.6.
† See §1.6 for a discussion of common performance-tuning tradeoffs.

apply to the end user, and lull us into a false sense of security that the performance is suitable. The goals have to be relevant to the end user, and use realistic work-flow, hardware specifications and load scenarios.[21]

- **Partial metrics** — Performance goals that cover only part of the application can also be worse than not having any goal at all. Such *"coin under the lamp"* goals tend to focus our attention on the measured code sections, not necessarily the real hotspots. This could mislead us to tune only part of the application, and to believe that (following the tuning) it is fully optimized.

- **Moving targets** — A management pitfall no less than an engineering one, this is the mistake of modifying the performance goals during the iterative development cycle. The goals should be set independently of the performance tuning cycle, preferably by other persons, outside the R&D group.

- **Ignoring performance goals** — The purpose of the performance targets is to be a criterion for performance tuning. Ignoring them easily leads to a disproportion-ate, unnecessary investment in over-tuning.

1.5.3 Profiling

- **Using intuition** — It is a very common tendency to rely on intuition and "gut feel-ing" regarding what and where are the application's hotspots. Such intuition is often incorrect. So do not assume, profile.

- **Using profiling to avoid optimization** — It is very easy to blame performance hotspots on components outside our control.[22] For example, it is easy to blame Microsoft Excel for the long time it takes to update an Excel file using *xlswrite*. We can even "prove" it using profiling. Correct tuning should not stop there but ask whether the I/O is in fact needed, or whether MATLAB's *xlswrite* can be improved.*

- **Using inappropriate tools** — A surprising number of MATLAB users have never used the built-in Profiler tool, basing their performance tuning on *tic/toc* or debug printouts. This is inaccurate at best, misleading in the common case, and nearly always highly inefficient.†

- **Ignoring the differences between CPU and wall-clock profiling** — It makes no sense to tune CPU-intensive segments of I/O-bound applications. If we rely on CPU profiling we might miss this aspect and optimize the wrong thing.‡

- **Non-representative profiling** — Profiling only a small part of the application (*micro-benchmarking*), or a non-representative small sample of data or user interac-tion, or a non-representative platform.

- **Not repeating timing measurements** following each tuning iteration.

1.5.4 Optimization

- **Optimizing irrelevant code** — Developers often optimize code that is easier to fix rather than code that the profiling session determines is the real performance bottleneck. Maintaining optimization discipline is hard…

---

* See §11.5.
† See §2.1.
‡ See §2.1.5.

- **Optimizing code, not data** — We often approach tuning with the notion that we should optimize the speed that the program processes our data. In fact, we should also question whether we can optimize the data itself, by possibly reducing (e.g., uniquify or sample),* or simplifying it (e.g., convert text to binary). Also, since most programs are data-driven, optimizing data and/or data structures is often more cost-effective than optimizing algorithms.[23]

- **Ignoring tradeoffs** — Performance is not an end by itself. It is part of a complex eco-system of factors and constraints that should be monitored, in order to achieve an overall optimum.† Tuning an application while ignoring tradeoffs causes more harm than benefit.

- **Algorithmic antipathy** — Some engineers have a natural aversion to tuning core algorithms and data structures. In fact, fixing the design, core algorithms and data structures can significantly speed-up some CPU-bound applications.

- **Algorithmic frenzy** — This is the exact opposite of the previous pitfall: We replace a simple $O(n^2)$ algorithm with a complex $O(n)$ one, only to discover that in real-world situations where n is small, overall run time has degraded. Brute force is often underrated…‡ Moreover, I/O and other non-algorithmic factors are the real performance bottlenecks in most applications.[24] Finally, modifying algorithms is often more difficult and bug-prone than simpler techniques.[25]

- **Using a single tuning strategy** — Performance goals can often be achieved in many different ways. Do not bang your head against an impossible task of improving the algorithm (i.e., CPU) if you can more easily improve I/O or memory to achieve the required goals. Similarly, do not focus on just one tuning technique, use a variety in combination.

- **Optimizing infrequent use cases** — By tuning infrequently used code flows, we are spending time and energy optimizing irrelevant code. As an extension of Pareto's principle, tune the common use cases first.§

- **Undocumented assumptions** — We often make implicit assumptions on the nature of the data or the environment. If we do not document these assumptions in the code, it is harder to revisit and retune the application when the underlying factors change in the future, as they invariably do. Self-optimizing code¶ is much harder to develop than simple documentation.

- **Speeding up unused code** — More often than we might care to admit, applications do stuff that is not strictly required. Speeding up such code, rather than removing it altogether, is a natural tendency that should be avoided. The first question we should ask ourselves when tuning a hotspot, is whether the hotspot can be avoided in certain cases, or eliminated altogether.**

- **Multiple intermediate updates** — Instead of computing, saving, and displaying intermediate results, it is much faster to skip the intermediate steps, saving and

---

* See §3.6.3.
† See §1.6.
‡ Computer scientist Ken Thompson's famous quote: *"When in doubt, use brute force"*.
§ See the end of §3.3 for one specific utilization of this concept.
¶ See §3.10.3.2.
** See §3.3, a special case of this. Ken Thompson again: *"One of my most productive days was throwing away 1000 lines of code"*.

displaying only the end results.* Instead of spending time optimizing the intermediate steps, spend the effort to avoid or merge them.

- **Bottom-up optimization** — Many performance hotspots are caused by underlying factors that are not reported in the profiling result. For example, if the Profiler may report much time spent in some I/O function, we might be tempted to tune that function. Instead, check whether the functional flow can be modified to prevent or reduce I/O.

1.6 Performance Tuning Tradeoffs

Performance tuning is not an end by itself. The tuning cycle contains explicit and implicit costs that are sacrificed in order to achieve the requested speedups. These costs should constantly be monitored during the tuning process, to determine whether tuning should be stopped or continued. In fact, the tradeoff ratios vary during the tuning process, due to the diminishing speedup returns.

At one time or another all of the tradeoffs would indicate that it is no longer cost-effective to tune the application any further. Even if we are not concerned about some of these tradeoffs for a specific application, it is still wise to at least be aware of them. This enables us to take informed decisions about the tuning. For example, investing in hardware upgrades rather than software development to achieve the required speedup.

Some of the common tradeoffs include:

- **Development (tuning) time/cost** versus **total expected run-time savings** — This is probably the most obvious tradeoff: It makes sense to spend a minute in order to save one hour, but not the reverse (spending an hour to save a minute).

- **Total run time** versus **total expected run-time savings** — If the savings are very small compared to the overall run time, tuning might not be reasonable.

- **Functionality** versus **performance** — This is one of the most well-known tradeoffs. Performance can very often be improved by removing some hotspot functionality, possibly enabling the user to run that functionality only by special demand.[26] For example, by using a dedicated <Save to Excel> button rather than an automated save whenever the data changes.

- **Accuracy** versus **performance** — This is a variant of the functionality tradeoff. After all, accuracy is one aspect of the functionality: Performance can often be achieved by reducing the algorithm's target accuracy. Sometimes the lowered accuracy is still sufficient, but we often have a true tradeoff.† Just remember that a program that is faster but produces incorrect results is useless. Computer scientist John Ousterhout is quoted as saying: *"A program that produces incorrect results twice as fast is infinitely slower"*.

- **Portability** versus **performance** — Software reusability and portability often dictates using layers of abstraction (e.g., using polymorphism and multiple class

---

* See §10.4.2.7.
† See §3.9.2.

inheritances). Unfortunately, abstraction layers very often have a devastating effect on run-time performance.

In a related matter, some performance optimizations can be platform dependent. For example, vectorization may be faster than a simple JIT-ted loop on Windows, but slower on a Mac.[27] We should decide whether we optimize for a specific platform, or for a variety of different systems.

- **Maintainability** versus **performance** — Performance-tuned code is sometimes harder to understand and maintain. For example, mex C-code is harder to maintain than the equivalent m-file; complex vectorization is sometimes harder to maintain than simple loops. The future maintenance cost may not be worth the current run-time saving. As a reader on Loren's blog* commented:[28]

> *"Pay a LOT of attention to code clarity and code design. In 97% of the code you write, tiny improvements in execution speed don't matter, and yet you (and your colleagues/coworkers) will spend 10-20 times as much time reading your code as you spent typing it… Programmer time is vastly more valuable than machine time in almost all cases."*

Comments have zero performance impact, yet are incredibly valuable. They are easily written as you write code, with everything fresh in your mind. Next year, you may forget how the code works, but you may still need to debug and maintain it. So take a few seconds to insert comments that describe how the vectorization works, or to explain why you selected algorithm B over A.

When tuning, keep the old (verbose) code in a comment, if you feel that the old code variant would help explain the new (faster) code. However, if the older code does not enhance understanding, remove it so that your files do not become littered with old, unused code, which reduces readability.

- **Robustness** versus **performance** — Some performance gains can be achieved at the expense of robustness. For example, by avoiding checks of uncommon edge cases, input arguments validity, data underflow/overflow or results sanity. Another example is reduced sensitivity to exceptional conditions such as singularity, instability, and ill-formed/degenerate data. The built-in functions *polyfit* and *robustfit* illustrate this tradeoff. The performance benefits should be weighed against the risk of reduced program robustness:[29]

> *"More computing sins are committed in the name of efficiency (without necessarily achieving it) than for any other single reason — including blind stupidity."*

- **Repeatability** versus **performance** — Speedup is sometimes achieved by adapting the order of execution of code sections to external conditions, sacrificing results repeatability at the expense of performance. In fact, sometimes the *order* of execution is more important than the *speed* of execution.

- **Variability** versus **performance** — As we shall see in §1.8.6, variability of the performance results directly affects the perceived performance. It is often better to reduce the performance variability, at the expense of slower *average* execution time, for the benefit of a faster *maximal* run time. The variability can be between

---

* Loren Shure was one of the first MathWorkers and is currently principal MATLAB developer at MathWorks. She has helped design and develop much of the core MATLAB functionality.

different runs (which affects the perceived performance), and different system loads (which affects the application's graceful degradation).

- **System resources** versus **performance** — Performance can often be improved at the expense of global system resources: increased memory (e.g., caching*); higher CPU load (e.g., pre-calculation); or increased network traffic (e.g., pre-fetching). In some cases, this may be impossible due to external restrictions.

- **Generality** versus **specialized** — Some performance improvements can be gained by specializing the code to only handle the specific data at hand, rather than making the code generic so that it would handle different kinds of future data.

- **Scalability** versus **specialized** — This is a variant of the specialization tradeoff, but from a different angle: It sometimes improves performance to use a nonscalable algorithm if we know that it is only expected to work on a small subset of data values. For example, an $O(n^2)$ algorithm may actually be faster for small values of n than a more scalable $O(n)$ algorithm. On the other hand, if the algorithm would need to support a higher n, the $O(n^2)$ algorithm's non-linear run time might become intolerable.[30]

- **Hand-crafted** versus **automated optimization** — MATLAB employs a wide variety of internal performance optimizations, often mistakenly classified as part of its JIT (JIT has a large, but not unique, part in these optimizations). Some extra drops of performance can sometimes be squeezed in MATLAB by hand-crafting the code. However, this technique (sometimes called *"coding against the JIT"*) could backfire if the program is used on a different platform or MATLAB release, where the modification can actually hurt performance.

- **Selective** versus **overall tuning** — In some cases, the performance of a specific functionality can be improved at the expense of some other functionality(ies). Relative expenses vary during the optimization cycles; an expense that is acceptable in cycle 1 may become unacceptable in cycle 5 (or vice versa).

- **Hardware** versus **software** — Extra hardware power is often cheaper than software development for isolated computers, but more expensive for deployment onto multiple stations; closely related to the following tradeoff.

- **Vertical** versus **horizontal scaling** — This relates to the hardware/software tradeoff, but is not limited to it. It is often better to invest in making scalable parallel code, than making a single chunk faster.[†]

- **Actual** versus **perceived performance** — Are we really interested in speeding up the program? Perhaps, we just need to make the user more comfortable?[‡]

- **Interactive** versus **batch** — The performance emphases for offline overnight jobs may be drastically different than for interactive user GUI.

- **Throughput** versus **latency** — Which is more important: the program's initial response or its ongoing performance? The answer affects performance tuning focus and techniques such as lazy evaluation and pre-fetching.[§]

---

* See §3.2.
† See §1.7.
‡ See §1.8 on this very *non*-intuitive tradeoff.
§ See §3.8.

1.7 Vertical versus Horizontal Scaling

Many of our programs start small and later grow to handle larger data sets, or expanded functionality, using additional or improved hardware. The ability of a program to handle such growth is called *scalability*.[31] Naturally, we wish all of our programs to *scale well* and not to crash with increased workload (refer to the discussion of graceful degradation in the preceding sections), nor demand a disproportionate investment in hardware.

We generally distinguish between two different types of scaling:

- **Vertical scaling** (also called *scaling up*) means that the application can use additional or better resources on the same computer. For example, by adding memory, or switching to a faster disk and CPU. In practice, this means upgrading the computer hardware to become a more high-end system.

  Vertical scaling is extremely easy to manage from a software standpoint. In fact, in most cases the scaling is entirely transparent to the software, and minimal (if any) source-code changes need to be done. It also enables simpler virtualization, meaning that multiple users could share the now-more-powerful computer resources at the same time.

  On the other hand, vertical scaling typically involves higher-than-linear costs per performance. This means that in general (and yes, there are always exceptions to this rule) in order to achieve a 2× speedup we would need to more-than-double our existing hardware value, typically by a factor much larger than 2. The larger the requested speedup, the (much) larger the cost.

  Vertical scaling is typically used when programming labor costs are higher than the expected hardware costs; when the amount of required scaling is small (no more than 2–3×); and/or when parallelization is impractical.

- **Horizontal scaling** (also called *scaling out*, or *parallelization*) means that we are adding computing nodes: multi-core CPUs, GPUs, FPGAs, or cluster computers. Such a solution typically involves multiple identical, simple, low-cost mass-produced (*commodity*) nodes. Cloud and cluster computing are typical usage examples of horizontal scaling.

  Horizontal scaling is harder to develop and manage. The application needs to take care to distribute the program sub-tasks (*jobs*) among the computing nodes, wait for them to complete and then collect and assemble them in the correct manner. The application needs delicate coordination to prevent task contention between nodes. This is certainly not a trivial programming task.*

  In addition, horizontal scaling involves significant overhead in terms of hardware resources: We usually need a central node to run the main program and distribute/assemble jobs; and we often need each of the nodes to have similar memory/disks/CPU, even if they do not make full utilizations of these resources (that can sometimes be shared). We also need to invest in fast network connections and efficient hardware oversight monitoring. The nodes may be low-cost commodity computers, but the overall cluster cost may be quite large.

  On the other hand, horizontal scaling enables use of clusters that have dozens, hundreds, or many thousands of computer nodes. This enables us to achieve performance speedups that would be impossible with vertical scaling.

---

* See Chapters 6 and 7 for solutions available within MATLAB.

An important aspect of application design is *capacity planning*, in which we analyze the requirements and the expected usage and data growth, and ensure that our system can scale (up or out) to handle this growth as it is encountered. Different scaling can be applied to different system components or levels. For example, we might scale up individual computer nodes, while retaining an overall horizontal scaling (cluster/cloud) approach to the system as a whole.

When deciding whether we should use vertical or horizontal scaling, we should take into consideration **Amdahl's law**.[32] In a nutshell, this engineering law states that if a fraction α of a given calculation (measured by duration) cannot be parallelized, then the speedup for P nodes under ideal conditions* is given by the following formula:

$$\text{Speedup} = 1/(\alpha + (1 - \alpha)/P)$$

For example, let's take a case where 40% of an algorithm can be parallelized ($\alpha = 0.6$). In such a case, the speedup for using $P = 2$ computers would be $1/(0.6 + 0.4/2) = 1.25$. We have doubled the number of computers (from 1 to 2), but have only gained a 25% speedup, not 100% as we might have naïvely assumed. The reason is that 60% of the algorithm's duration cannot be parallelized, so the only speedup that can be gained is on the 40% that is parallelized. We have slashed the run time of these 40% in ideal conditions by half (the two halves being run in parallel by the $P = 2$ processors), so that it now takes only 20%. Our algorithm is now only 20% faster than before, in other words it takes 80% of the original time, which means a 1.25 speedup ($1/0.8 = 1.25$).

Adding additional computer nodes improves the speedup even less:

P (Number of Nodes)	Relative Run Time (%)	Speedup
1	100	1.00
2	80	1.25
4	70	1.43
8	65	1.54
∞	60	1.67

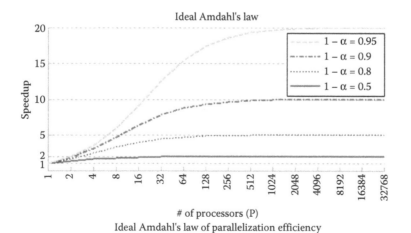

Ideal Amdahl's law of parallelization efficiency

---

* That is, no distribution/communication/assembly overheads; optimal work distribution, etc.

The higher the relative portion of parallelizable code $1 - \alpha$, the higher the potential speedup, but even extremely high $1 - \alpha$ values have a speedup limit that is based on the non-parallelizable portion α. For example, at the extremely high parallelization potential of $1 - \alpha = 0.95$ (i.e., only 5% of the code is non-parallelizable), the optimal speedup limit is only 20 (= 1/0.05). A 20× speedup is of course impressive, but considering the fact that it needs a massive investment in parallelization code and in computer nodes, the picture is not as bright as might appear at first glance.

Actual real-life speedups never even reach Amdahl's theoretical limits. The reason is that in practice there are communication delays, limited network bandwidth, load-balancing overheads, nonhomogeneous nodes, suboptimal jobs distribution and results assembly, resource contentions, and other effects that cause delays in the parallelized code.[33] In fact, at some point, adding additional computer nodes actually degrades performance (speedup <1, or *slowdown*), due to increasing management and communication overheads compared to negligible incremental processing power.[34]

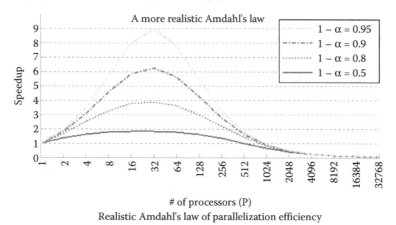

Realistic Amdahl's law of parallelization efficiency

The good news is that parallelism improves performance better for larger data sizes. The larger the problem size, the more processing can benefit from additional processors and the lesser the relative effects of the communication overheads.[35]

1.8 Perceived versus Actual Performance

Before starting to performance-tune any code, we should ask ourselves what is the ultimate goal of this tuning: Are we really interested in the program *running* faster, or do we only wish it to *appear* faster and more responsive? The distinction between these two motivations is not at all theoretical. In fact, it can have a dramatic effect on what and how we will tune our code.

Perceived performance[36] is the art of tailoring an application to appear faster, although in practice it would not in fact run faster. This may sound like an oxymoron, but is in fact based on deeply rooted psychological effects that have been extensively studied by usability experts over the past decades.

Whenever we performance-tune our code, we should consider whether it would be more cost effective to improve the perceived rather than the actual performance. In certain situations where the actual performance cannot really be improved, we can use perceived performance as a substitute, often with minimal or no degradation to the actual performance. Of course, we can always mix improved actual *and* perceived performance, in order to maximize our tuning effectiveness.

Actual performance improvements lower than 20% are not perceived, so don't bother. It is generally not worth the extra development, retesting, deployment and customer communication. We would probably be wiser to invest our time in additional features or in improving the perceived performance. On the other hand, if the 20% of actual overall improvement is concentrated in a single functionality that improved by 80%, or in very long-running applications where 20% makes a real difference, then this is significant and probably merits extra development and a separate release. Also remember that several separate 20% improvements might add up to a much larger, noticeable effect that will have a definite user impact.

Several techniques have been suggested over the years for achieving improved perceived performance. These include:

- Presenting continuous feedback on ongoing tasks
- Enabling user control of processing parameters
- Enabling user interaction while a task is processing
- Continuously displaying streaming data
- Ensuring that the program flow is intuitive and streamlined
- Reducing the program's run-time variability

We shall now proceed to discuss these aspects separately.

1.8.1 Presenting Continuous Feedback for Ongoing Tasks

Providing feedback to the user is arguably the most important advice for improved perceived performance. The basic idea is that when users receive sensual (typically visual) feedback about an ongoing task, they are much more at ease waiting for the task to complete, than without feedback. In this latter case, users do not know when the task is expected to end, or even that it is in fact handling the request properly. This heightens the users' psychological anxiety, which prolongs the perceived run time.

Tests have shown that users are willing to accept much longer-running tasks that provide continuous feedback, than shorter tasks without feedback. When asked to describe the duration of such tasks, users tend to describe feedbacked-tasks as taking a shorter time than the exact same task without feedback. The bottom line is that the simple modification of presenting feedback to the user for ongoing tasks will improve the program's *perceived* performance, even if the *actual* run-time remains unchanged.

MATLAB has many different ways by which we can present feedback:* We can send a message to the MATLAB console (Command Window); or modify the figure window's title or color; or display a pop-up message window; or update some graphics. The feedback can be as simple as temporarily disabling GUI controls, modifying the figure pointer (cursor), displaying animated icon, or displaying a progress bar with the expected run time.

Feedback should typically be presented immediately following the user action, and then continuously throughout the ongoing task. This is true in general, but particularly for long tasks when no other feedback is available. A typical example for this is MATLAB's long

---

* See §10.4.2.

start-up time for deployed (compiled) applications, where an immediate splash-screen would solve a well-known user concern.*

When presenting feedback, we might need to deliberately slow-down the feedback, otherwise it might change too quickly for the human eye.[37] We can do this by inserting deliberate *pause* calls,† or by only displaying feedback once every few steps.

1.8.2 Placing the User in Control

Enable the user to choose (possibly via some preference option, or maybe a GUI control) between Fast/Default/Comprehensive processing. This way, the user is in control of the performance, and can choose whether to sacrifice accuracy for performance, or vice versa. This *throttling* control reduces user frustration and aligns expectations with the results. In fact, users will be much more likely to accept a slow-running computation if they are placed in charge and consciously choose the "Comprehensive" option.

1.8.3 Enabling User Interaction during Background Processing

The idea here is to enable the user to continue interaction with the application, while it is still doing some processing. This gives the user an impression that the tasks complete faster, and that the program is more responsive.

For example, if a task needs to update some data tables and graphs, followed by updating an Excel file, it could make sense to return control to the user (i.e., re-enable the GUI controls) immediately after updating the tables and graphs, so that the user could continue interaction with the GUI while the program is updating the Excel file (something that could take many seconds). Users would get the feeling that the program has processed the request very quickly since they could continue their GUI interaction, and would not be bothered by the background update of the Excel file.

A variant of this idea is to delay long-running initializations or data-load until after the main program GUI is presented to the user, in cases where these initializations are not immediately required for the presentation. The actual long-running initialization or data load could be done in the background, after the GUI is presented. In some cases it could even be delayed until the user actually needs the data. This technique is a special case of the *lazy evaluation* technique.‡

Another incarnation of this technique is not to *delay* data loading, but rather to *advance* it. This is done after the main application has loaded, and is used to *pre-load* (or *pre-fetch*) data in the background, in anticipation of the user's future need for it. When the user eventually makes the request, the data is auto-magically available immediately. This technique is commonly used by websites (e.g., Google Maps) to enable extremely fast response to user actions, at the expense of network traffic and memory/local storage, which are very cheap.§

It is not just data that can be cached, but also complete GUIs, panels, images, and graphs. In fact, we can prepare figures and axes in the background, as invisible containers, and fill them with the anticipated data and graphics. When the user makes the relevant request, the corresponding figure or panel is made visible and presented. This would be much faster

---

* See §8.4.
† However, see §10.5.1.
‡ See §3.8.1.
§ Techniques for caching data in MATLAB are discussed in §3.2; pre-fetching is discussed in §3.8.2.

than generating the same container from scratch on-the-fly, so the application appears much faster and more responsive.

Yet another variant of this technique is to enable users to stop the processing in mid-stream, using some dedicated GUI control that is made available when a long-running task begins. This is especially important if GUI interaction is disabled during the processing.*

1.8.4 Streaming Data as It Becomes Available

Iteratively present data as it becomes available (rather than all at once at the end). This was one of the hard-earned lessons of early Internet Explorer versions that were considered extremely slow in rendering webpages. Much of this perception was due to the fact that HTML tables (on which the vast majority of webpages were based) were not displayed until the very last table element has been downloaded to the browser and processed. In contrast, other browsers were quick to display the top table rows as they were downloaded, even before the rest of the rows were received.[38]

A similar effect caused interlaced images to receive wide adoption in webpages.[39]

In MATLAB, we can update the data presented onscreen, while the information is still streaming or processing. For example, if we are searching for some optimized solution to a problem, we could present a quick initial rough estimate and periodically update the display with the best variant that the algorithm has found so far.[†]

When updating real time or streaming data, it is important to automatically process the updates wherever possible. Do NOT use dedicated action GUI controls (buttons, links, or menu items) that will only refresh the data when the user interacts with them. The main processing task (or some independent timer task) should be responsible for automatically updating the display as needed.

1.8.5 Streamlining the Application

In general, the more the user needs to think about what to do next, the slower the program is perceived. Therefore, spend time in making the program intuitive, so that there is no question at all during processing regarding what should be done next.

Streamlining the application means that we should reduce the requirements for user interaction to a minimum. The fewer user clicks needed, the better. Numerous usability studies have shown that users drop off with each additional click. This is the reason why professional websites reduce the purchase experience to as few clicks as possible. Amazon even filed a well-known patent for a one-click checkout process.[40]

We should attach callbacks to any GUI control used in your GUI, so that user interaction with any of them will automatically update the rest of the display. A typical design used by many nonprofessional GUIs is to add a custom <Refresh> button that will update the display based on the values of the rest of the controls. This should generally be avoided, because not only does it make the GUI harder to use, it also creates a situation where the display is inconsistent with the control values until the user clicks <Refresh>. Instead, the callbacks of all the controls should automatically update the display, making a <Refresh> button redundant. An exception to this rule is when the refresh action is relatively slow, in which case we might wish to postpone the action until the user has updated several controls.

§10.4.1.6 presents some additional advice for streamlining GUI applications.

---

* See §10.4.2.4.
† §10.2 presents several techniques for making such updates efficiently.

1.8.6 Reducing the Run-Time Variability

Usability studies have shown that users typically perceive an application's run-time duration not according to mean/median time, but somewhere close to the maximal wall-clock run time.* For this reason, dedicated analyses are often done to determine the program's worst-case execution time (*WCET*).[41]

For example, if a task usually runs for 5 s, but every now and then it takes 10 s, then the users would likely say that the application's run time is around 8 or 9 s. This is obviously "unfair", but is in fact how the human mind works.

One way around this problem is to display an actual timer displaying the elapsed time. This also answers the need for continuous feedback, explained in §1.8.1.

Another solution is to simply find a way to reduce the run-time *variability* by reducing WCET, even if it means to *increase* the average run time! WCET tuning is often more important (and surprisingly, also often easier) than improving the average performance.[42] In our example above, we could perhaps reduce the maximal run time from 10 to 7 s at the expense of increasing the average run time from 5 to 6 s. Most users would see this as an *improvement* to the overall program run time, reporting that the run time has drastically improved from 8–9 s to 6–7 s. It is funny how the human mind works (and how it can be manipulated …).

We can reduce run-time variability by other means: We could use a local disk rather than network I/O that is subject to unexpected latencies. We could also cache or pre-compute all possible data, rather than just the requested data. The precise mechanism is not very important, as long as we achieve the aim of reducing the maximal run time and achieving more-consistent run times (these are usually aspects of the same thing).

Conventional knowledge is that the *median* (rather than *mean*) time measurement should be reported during performance profiling, in order to eliminate spurious long-running measurements. However, in practice it is actually beneficial to *over*-weigh such measurements, or at least to report the mean (not median) time in order to not *under*-weigh them. When reporting responsivity, professionals sometimes use the second-worst or 95th percentile as benchmark.[43] This allows elimination of the really outlandish outlier measurements, while remaining close to the WCET perception. A more professional treatment of performance measurements should employ more rigorous statistical methodologies, including evaluation of confidence intervals.[44]

When monitoring, be careful to measure the actual end-to-end time duration, rather than just memory usage or CPU load, which are related but do not present the full picture.

In truth, selecting the benchmark is more a matter of estimate and applying good sense, than an exact science. After all, it's just a tool to help us get a feel for performance and improve it. The specific benchmark being used is less important than using it consistently during the iterative performance-tuning cycle.

1.8.7 Performance and Real Time

Application performance is often confused with the term *real time* (RT). These terms are related in the sense that they both refer to the application's responsivity and speed. However, they are completely independent of each other.

In fact, it may well be that an application that is considered highly performant and would typically react in sub-millisecond speed would still not be considered as a real-time

---

* See §2.1.5 for a discussion of wall-clock time versus CPU time.

application. Conversely, another application that reacts within a full second might still be considered real time.

The difference between the terms stems from their definition: A real-time system might mean (and the term is indeed often loosely used in this way) that the system reacts to events with a non-noticeable delay, "as-it-happens". But more precisely, a real-time system is defined as one that is deterministically guaranteed to react within a certain pre-defined time span (*deadline* or *constraint*).[45] This is different than *performance*, which is more interested in run-time average and variability (the user perception perspective), than in the worst-case execution time (deadline perspective).

There are several different cases of real-time systems (hard-, firm-, soft-, and near-RT). The difference between them depends on whether missing a deadline is acceptable and on the amount of time delay. The time span itself can be anything from microseconds to minutes, depending on the situation. It is customary to denote systems in which there is no "noticeable" delay as *true real time*, and other systems that meet the deadline constraint within a noticeable time span as *near real time*. Naturally, the exact definition is domain- and application specific.

When trying to determine whether our application is real time or not, we should consider its worst-case execution time (*WCET*), not its variability or average run time. In this respect, we are not interested in the user perception of the performance.

However, in many cases we are not interested in hard response-time constraints for 100% of the cases, but rather in "good" performance for 90% of the cases, as explained in the previous section. We should be aware of the fact that these two targets may have conflicting impacts on how we optimize our code, depending on the specific situation. We should never fool ourselves that by improving the performance for 90% of the cases, we have also automatically made our program real time.

MATLAB itself has several aspects that prevent it from being able to guarantee a response within a certain time span, thereby making MATLAB applications non-RT by definition. Moreover, MATLAB is often run on non-RT operating systems, so that even if MATLAB itself would be RT-enabled, the platform would not enable the running application to become RT. Still, we can make our MATLAB application fast and responsive enough that it becomes RT in a practical (if not scientific) sense.

2

Profiling MATLAB® Performance

Profiling is the act of monitoring an application's run in order to understand how much time is spent by the program and its internal components.[46] Profiling an application is extremely important when we wish to diagnose and improve a program in terms of performance. It is worth repeating the well-known adage that whatever cannot be monitored and measured, cannot really be improved.

Moreover, before profiling an application, we cannot be certain where its performance bottlenecks occur. Without profiling data, we might well spend time and energy optimizing code that has no significant effect on program performance.

Profiling an application is an important step in the development cycle, whereby we develop a first draft of the algorithm, which is complete yet not necessarily fast. We then profile it and tune its performance until it meets our needs. By profiling and tuning only after our first algorithm version works correctly, we remove the need for premature optimizations that increase code complexity, require extra development time, and potentially introduce new bugs. Profiling and tuning a working algorithm is, in contrast, a much more focused and productive effort.

Profiling application code can help the tuning process in a variety of manners:

- Identifying run-time bottlenecks that are candidates for focused tuning.
- Identifying code segments that are unnecessary or can be bypassed/cached.
- Identifying code segments that are called numerous times (e.g., within a loop) and can be optimized by caching, vectorization, or moving outside the loop.
- Help determine whether our application is CPU, memory, or I/O bound.

The last of these points is very important. It makes no sense to custom-tune a data-processing algorithm if our application is I/O bound: we should focus our efforts on the I/O hotspots. Similarly, if our application is memory-bound, then we should try to reduce memory usage and allocations, and perhaps the data, rather than tune the code flow.

This chapter explains the different ways in which we can profile MATLAB programs. MATLAB includes a couple of built-in tools and functions that support profiling:

- Profiler tool — This is the single most important tool for profiling and tuning MATLAB applications. A detailed description of its use will follow shortly.
- *tic/toc* — Easy-to-use functions that enable simple timing of code sections. Related built-in functions in this family are *cputime*, *etime*, and *clock*.

In addition to these MATLAB tools, we should not forget the age-old tested method of using timed log files and console printouts.

Last but not least, we can also use standard external (non-MATLAB) tools to profile the performance and behavior of our application, from various aspects: process behavior, system resources, and disk/network/memory usage.

Note: This chapter describes general profiling techniques, but many of the details focus on a standard non-parallel MATLAB installation. §6.1.5 provides information about profiling parallel MATLAB applications. §9.2 provides information about profiling MATLAB's memory usage.

2.1 The MATLAB Profiler

MATLAB's built-in Profiler tool[47] has existed for many years (at least since MATLAB 5, some 15 years ago). MathWorks has invested significant energy in making the Profiler accessible in video tutorials,[48] blog articles,[49] newsletters,[50] webinars,[51] and conference presentations. Still, I have a feeling that the Profiler remains underutilized by many MATLAB users for some unknown reason.

The MATLAB Profiler is the single most important tool in our arsenal when we need to improve MATLAB application performance. It is well worth spending the short time that it takes to learn how to use this tool effectively.

To run the Profiler, simply turn it on, run the MATLAB program in question, and when the program has ended, run the Profiler's report command. For example:

```
>> profile on        % or: profile('on')
>> surf(peaks);
>> profile off       % or: profile('off')
>> profile viewer    % or: profile('viewer')
```

In addition to these command-line operations, the Profiler can also be started and stopped interactively: select "Profiler" in the MATLAB Desktop's "Window" main menu, then click the "Start/Stop profiling" button in the Profiler window. In R2012b, click the "Run and Time" button in the Home or Editor tabs of the MATLAB toolstrip to display the Profiler GUI, then use the "Start/Stop" button as before.

Whichever way that the Profiler is started, whatever MATLAB code is run between start and stop (including GUI callbacks, timers, etc.) will be profiled and reported.

Regardless of the manner in which we stop the Profiler, interactively in the Profiler window or programmatically via *profile viewer*, a summary report will be displayed, showing the amount of time that MATLAB spent in each of the program's functions, as well as in external and built-in functions called by the program (see figure on page 27):

In this example, the profiled program (*surf*) took a total of 1.31 s to run, of which 0.249 s were *self-time*, time spent running *surf's* own code, excluding external function calls. The rest (1.061 s) was spent in other (so-called *child*) functions called by *surf's* code. Apparently, most of that time was spent in a single function, *newplot*, which took 873 ms to run, of which 57% was self-time. Clicking the "Self Time" header sorts the table by decreasing self-time, thereby highlighting those functions that are singularly responsible for most of the run time, and are therefore prime candidates for potential speedup.

Subfunctions are shown using main>sub notation. For example, one of *newplot's* subfunctions, *newplot>ObserveAxesNextPlot* is a top CPU user, at 359 ms total.

Private functions are shown using *package\private\func* notation. For example, *clo* (denoted as *graphics\private\clo*) is a private function of the graphics package.

MATLAB class member functions appear using *package.class.func* notation. For example, *surfaceplot* is a member method (function) of the graph3d.surfaceplot class.

```
Profile Summary
Generated 06-Sep-2012 18:51:40 using cpu time.
```

Function Name	Calls	Total Time	Self Time*	Total Time Plot (dark band = self time)
surf	1	1.310 s	0.249 s	
newplot	1	0.873 s	0.499 s	
newplot>ObserveAxesNextPlot	1	0.359 s	0.032 s	
cla	1	0.327 s	0.046 s	
graphics\private\clo	1	0.281 s	0.108 s	
graph3d.surfaceplot.surfaceplot	1	0.141 s	0.078 s	
setdiff>setdifflegacy	2	0.095 s	0.064 s	
setdiff	2	0.095 s	0.000 s	
findall	1	0.079 s	0.063 s	
plotdoneevent	1	0.047 s	0.016 s	
peaks	1	0.031 s	0.031 s	

MATLAB profiler summary report

External non-MATLAB (e.g., MEX or Java) functions are marked as such. For example, *"ismembc (MEX-file)"* or: *"java.lang.String (Java method)"*.

Built-in functions (e.g., *isempty, length, find, getPlotFigure*) are not monitored or displayed in the summary by default. However, we can ask the profile to monitor and display such functions using the **profile** function's optional *-detail* switch:[52]

```
profile on -detail builtin
profile('on','-detail','builtin') % an equivalent alternative
```

where the possible detail levels are 'mmex' (the default) and 'builtin'.*

The profiling summary report provides a lot of useful information. For instance, in the above example we might check why **newplot** takes so long to run, and whether it can be optimized, or even bypassed altogether. If we can find a way to bypass **newplot**, then **surf** will gain 873 ms and run at approximately 1.31 − 0.87 = 0.44 s, or 3 times faster! The profiling summary report helped us focus our attention on **newplot**, rather than trying to optimize **surf**'s own code, which in this case is much less important (only 249 ms altogether).

The Profiler's summary report also tells us how many times any of the functions was called. This can help us identify cases where a certain function can be optimized by running once, outside the loop, versus numerous times within it.

Each function in the Profiler's summary report is hyperlinked to a detailed report on that specific function, enabling drill-down all the way down to built-in functions.

---

* Old MATLAB releases also supported the 'operator' detail level, which provided profiling information about built-in operators such as +. However, this level of detail is no longer supported in recent MATLAB releases.

We can save profiling data for later use by using the ***profsave*** function, which saves a copy of the graphical profiling report on the local computer, including separate HTML pages for each of the functions that have a detailed profiling report. The saved profiling data can later be loaded into MATLAB or viewed in any web browser. When storing multiple profiling sessions, it may be wise to archive the session results in a zip file, rather than keeping ***profsave***'s folder with its multitude of HTML files:

```
profsave(profile('info'), 'profile-results');
zip('profile-results.zip', 'profile-results/*');
```

The Profiler can also be used to profile memory usage, in addition to CPU usage. This can be used to detect memory leaks and to tune memory-related performance issues.*

MATLAB's Profiler includes several other features that may be useful. For example, checking the code coverage, checking MLint (Code Analyzer) messages, and so on. It is worth the time to read the full documentation of the ***profile*** function[53] and of the Profiler tool[54] to learn how to use its features and programmatic interface effectively.

2.1.1 The Detailed Profiling Report

As noted above, each function in the profiling summary report is hyper-linked to a detailed profiling report. For example, let us click the ***newplot*** hyperlink:

```
newplot (1 call, 0.873 sec)
Generated 06-Sep-2012 19:45:45 using cpu time.
function in file C:\Program Files\Matlab\R2012a\toolbox\matlab\graphics\newplot.m
Copy to new window for comparing multiple runs
```

| Refresh |

☑ Show parent functions   ☑ Show busy lines   ☑ Show child functions

☑ Show Code Analyzer results ☑ Show file coverage   ☑ Show function listing

Parents (calling functions)

Function Name	Function Type	Calls
surf	function	1

MATLAB profiler detailed report (top)

The top section of the detailed report displays the number of invocations and overall run time of the function, its location on disk (the link opens the function in the MATLAB Editor), and the list of *parent functions* that have called this function. Each parent function is in turn hyperlinked to the detailed profiling report for that parent. In this particular case, we see that ***newplot*** was only called once, by the parent function ***surf***.

The top section includes a link to "*Copy to new window for comparing multiple runs*". Clicking this link stores the displayed report in a separate window, such that it will not be overwritten when the Profiler is re-run. This enables us to check whether changes that we make to a program improve or degrade its run-time performance.

The middle section of the detailed profiling report (see the screenshot below) displays a list of the specific lines and child functions that took the most time within this function's

---

* See §9.2.6.

run. In this case, we see that lines 61 and 74 account for practically the entire run-time. Looking at child functions, we see that the *ObserveAxesNextPlot* subfunction is responsible for 41% of the time (this corresponds to line #74), whereas the built-in *getPlotFigure* (corresponding to line #61) is responsible for almost all the rest. The child function names are hyperlinked to their own detailed profiling report.

This middle section helps us focus our attention on those lines and child functions that take most of the processing time, and are therefore most worthwhile to tune. There is no use tuning code that only accounts for a minor portion of the run time.

Lines where the most time was spent

Line Number	Code	Calls	Total Time	% Time	Time Plot
61	fig = getPlotFigure;	1	0.514 s	58.9%	▆▆▆▆▆
74	ax = ObserveAxesNextPlot(ax, h...	1	0.359 s	41.1%	▆▆▆
78	end	1	0 s	0%	
77	axReturn = ax;	1	0 s	0%	
76	if nargout	1	0 s	0%	
All other lines			0 s	0%	
Totals			0.873 s	100%	

Children (called functions)

Function Name	Function Type	Calls	Total Time	% Time	Time Plot
newplot>ObserveAxesNextPlot	subfunction	1	0.359 s	41.1%	▆▆▆
datamanager.schema	function	1	0.015 s	1.7%	▏
newplot>ObserveFigureNextPlot	subfunction	1	0 s	0%	
initprintexporttemplate	function	1	0 s	0%	
Self time (built-ins, overhead, etc.)			0.499 s	57.1%	▆▆▆▆▆
Totals			0.873 s	100%	

MATLAB Profiler detailed report (middle)

The report's bottom section (see screenshot below) is also illuminating. It displays the entire source code of the profiled function, with automatic hyperlinks to a detailed profiling report for each child function called within the code. The line numbers are hyperlinked to the MATLAB Editor: clicking a line number will open the file in the Editor and jump the cursor to the specified line position. Built-in functions (e.g., *getPlotFigure* in line #61) are neither profiled nor hyperlinked by default, unless the Profiler is started with the *-detail builtin* switch as explained above.

Code lines that have not run at all are colored gray and their line number is not hyper-linked. It is important to check such segments to determine whether the code should really have been skipped, or in fact should have run, indicating an algorithmic bug. The line numbers of comment lines, colored green, are also not hyperlinked.

A blue value next to the line number indicates the number of times that each particular line was ran. In deeply nested loops, this value can be huge. Seeing the values associated with each source-code line helps us identify cases where loops can be optimized to reduce the number of times that a particular line is called.

MATLAB Profiler detailed report

Each line that clocked 10 ms or more displays a red number specifying its run time. In our particular case, there were two such lines (61 and 74) that had run times of 514 and 359 ms, respectively.

Timed lines have a reddish background color, whose intensity is proportional to the relative CPU time for that line in the function. So, if 95% of the time is spent in a single line, only that line will have a visible red background, which will be dark red. On the other hand, if there are five lines that take 20% of the time each, then these lines will all have a light-red background. The background shade helps us focus our attention only to these lines that have the most impact in terms of performance.

By default, the background color is selected based on the time spent by the code lines. However, we can use the combo-box (drop-down) menu at the top of the function listing section to highlight according to the number of calls or other aspects.

The Profiler records information at a system-dependent time resolution, typically 1 ms. However, the profiling report only displays information at a granularity of 10 ms. I often find it useful to increase the reported resolution to 1 ms, and in fact the screenshot above reflects this change. This can easily be done by editing the ***profview*** function.* The following is from lines 1441 to 1446 of *profview.m* in R2014a:

```
if timePerLine > 0.01,
    s{end+1} = sprintf('<span style="color: #FF0000"> %5.2f </span>',...
        timePerLine);
elseif timePerLine > 0
    s{end+1} = '<span style="color: #FF0000">&lt; 0.01 </span>';
end
```

which could be replaced by the following (note the highlighted changes):

```
if timePerLine >= 0.001,
    s{end+1} = sprintf('<span style="color: #FF0000"> %6.3f</span>',...
        timePerLine);
elseif timePerLine > 0
    s{end+1} = '<span style="color: #FF0000">&lt; 0.001</span>';
end
```

When source-code is unavailable (e.g., p-files, built-in functions, or MEX/Java/DLL functions), the detailed profiling report only displays the information in the top section (including calling parent functions), not the middle and bottom sections.

On some systems, MATLAB has an internal bug that causes the code in the profiling reports to be rendered using a proportional font, rather than fixed width (monospace) font. This causes the code section in the detailed report to become nearly unusable. A small fix to the *%matlabroot%/toolbox/matlab/codetools/matlab-report-styles.css* file may solve this issue. Simply replace the following directive at the top of the file:[55]

```
PRE {
  font-size: 100%;
}
```

with this:

```
pre, tt {
  font-family: monospace
  font-size: 100%;
}
```

MATLAB Profiler detailed report before (left) and after (right) the font fix

---

* *%matlabroot%/toolbox/matlab/codetools/profview.m*. This function can be edited in the MATLAB editor via ***edit('profview')***. Depending on your MATLAB installation, you might need administrator privilege to edit the file.

2.1.2 A Sample Profiling Session

Let's use the Profiler to optimize the following simple MATLAB function:

```
function perfTest
    for iter = 1 : 100
        newData = subFunc(iter);
        result(iter) = max(max(newData)) + rand(1);
    end
    disp(max(result));
end % perfTest

function result = subFunc(iteration)
    fileData = load('data.mat');
    result = sin(fileData.data);
end   % subFunc
```

A simple run of the perfTest function takes too long. Let's profile the function:*

```
>> profile on; perfTest; profile off; profile report
```

Profile Summary
Generated 04-Dec-2012 02:29:49 using cpu time.

Function Name	Calls	Total Time	Self Time*	Total Time Plot (dark band = self time)
perfTest	1	34.942 s	0.797 s	
perfTest>subFunc	100	34.145 s	34.145 s	

Self time is the time spent in a function excluding the time spent in its child functions. Self time also includes overhead resulting from the process of profiling.

The profiling report shows that the parent function (perfTest) took 35 s, but most of this time was not spent as *self-time* but rather in other functions that perfTest has called, in this case the subFunc function of perfTest.† This is not surprising: we can easily understand that perfTest's loop overhead is minor compared to the time spent *within* the loop's 100 iterations.

This analysis is easily verified by drilling into perfTest's detailed profiling report:

```
 time     calls line
                  3 function perfTest
        1    4         for iter = 1 : 100
34.270  100    5             newData = subFunc(iter);
 0.672  100    6             result(iter) = max(max(newData)) + rand(1);
        100    7         end
        1    8         disp(max(result));
        1    9  end  % perfTest
```

---

* *profile report* and *profile viewer* are synonymous.
† Sub-functions are denoted as main>sub in MATLAB notation — see §2.1 above.

So for our first performance-tuning pass, we do not bother optimizing `perfTest`, only
`perfTest > subFunc`:

```
 time    calls line
                  11 function result = subFunc(iteration)
29.973    100   12     fileData = load('data.mat');
 3.937    100   13         result = sin(fileData.data);
 0.031    100   14 end  % subFunc
```

In `subFunc`'s detailed profiling report we see that close to 90% of the time, nearly 30 s,
were spent reading the data file. This happens 100 times. We immediately recognize that
we do not need to keep re-reading the same data file so many times, since the data remains
unchanged. We will therefore cache the data file's contents in the first iteration, and then
reuse the data on subsequent iterations:*

```
function result = subFunc(iteration)
    persistent fileData
    if isempty(fileData)
        fileData = load('data.mat');
    end
    result = sin(fileData.data);
end  % subFunc
```

This simple fix immediately reduced our total run time by nearly 30 s, down to 5.6 s (a 6×
speed-up). The detailed report shows that our caching was done correctly, and that the
data file is indeed being read only once:

```
 time    calls line
                  11 function result = subFunc(iteration)
          100   12     persistent fileData
          100   13     if isempty(fileData)
0.733       1   14         fileData = load('data.mat');
            1   15     end
4.060     100   16     result = sin(fileData.data);
          100   17 end  % subFunc
```

Now let's take a closer look at the second code line, which has now become the top
performance hotspot: 4.06 s (72% of the total run time) was spent in code line #16, which
recomputes a function of the same data over and over, 100 times.

Our natural next step is therefore to also cache these results, along with the caching of
the data-file read. Our total run is now reduced to 1.4 s, with `subFunc` taking only 0.9 s of
these (65%):

```
 time    calls line
                  11 function result = subFunc(iteration)
0.016     100   12     persistent fileData persistantResults
          100   13     if isempty(fileData)
0.850       1   14         fileData = load('data.mat');
0.032       1   15         persistantResults = sin(fileData.data);
            1   16     end
          100   17     result = persistantResults;
          100   18 end  % subFunc
```

---

* Read more about caching data in §3.2.

After two performance passes of subFunc, subsequent optimizations may not yield the same improvement factor. In fact, we see that 95% of subFunc's time is spent loading the data file just once, which can probably not be avoided without affecting the functionality. So let's move on to the parent (main) function, perfTest. After optimizing subFunc, perfTest now takes a larger relative chunk of self-time:

Function Name	Calls	Total Time	Self Time*	Total Time Plot (dark band = self time)
perfTest	1	1.386 s	0.488 s	
perfTest> subFunc	100	0.898 s	0.898 s	

This phenomenon is an expected by-product of the optimization process: by reducing the time spent in the slowest function(s), the other function(s) become more important in relative run time. So let's drill-down into the detailed perfTest report:

```
time      calls line
                  3 function perfTest
              1   4     for iter = 1 : 100
0.898       100   5         newData = subFunc(iter);
0.488       100   6         result(iter) = max(max(newData)) + rand(1);
            100   7     end
              1   8     disp(max(result));
              1   9 end  % perfTest
```

We see that virtually the entire run time is spent within the loop. But wait: did we not just see that subFunc always computes and returns the same (cached) values? We can easily optimize the loop by moving the constant terms outside the loop.* Our natural first step is to move the top-hotspot line (newData = subFunc(iter)) outside the loop. After all, it accounts for 65% of the total run time. We can do this by noting that subFunc does not really use its iter input parameter:

```
function perfTest
    newData = subFunc();
    for iter = 1 : 100
        result(iter) = max(max(newData)) + rand(1);
    end
    disp(max(result));
end  % perfTest
```

Unfortunately this does not have the desired effect: the total run time remains virtually unchanged (the slightly different timing values are well within the noise generated by repeated profiling runs):

---

* Loop optimizations are discussed in §3.1.

```
time     calls line
                3 function perfTest
0.866       1 ___4      newData = subFunc();
            1 ___5      for iter = 1 : 100
0.503     100 ___6          result(iter) = max(max(newData)) + rand(1);
          100 ___7      end
            1 ___8      disp(max(result));
            1 ___9 end  % perfTest
```

The reason for this is that while `subFunc` is now being run only once instead of 100 times, it is actually only the first iteration which takes any significant amount of time (at least now that we have implemented our data-load caching mechanism). We still spend nearly 0.9 s to read our data file, something that cannot be avoided.

Let's turn our attentions to the loop's remaining code line. Here we are recomputing the maximal value of a constant data matrix, and then adding a nonconstant (random) value to it. Instead, we can move the constant calculation term outside the loop:

```
function perfTest
    newData = subFunc();
    maxData = max(max(newData));
    for iter = 1 : 100
        result(iter) = maxData + rand(1);
    end
    disp(max(result));
end  % perfTest
```

This latest fix completely removed any measurable time from the entire loop, reducing the total run time to 0.9 s, a ~40-fold speed improvement over the original version.

Emboldened by our new performance, we modify the loop counter to run a million times. `subFunc` still takes only 0.9 s to run, but now the main loop — which previously took absolutely no measurable time — takes almost 3 full seconds:

Function Name	Calls	Total Time	Self Time*	Total Time Plot (dark band = self time)
perfTest	1	3.847 s	2.972 s	▰▰▰▰▰▰▭▭
perfTest>subFunc	1	0.876 s	0.876 s	▰▰

```
time     calls line
                3 function perfTest
0.876           1 ___4      newData = subFunc();
                1 ___5      maxData = max(max(newData));
                1 ___6      for iter = 1 : 1000000
1.8611000000    ___7          result(iter) = maxData + rand(1);
1.0951000000    ___8      end
                1 ___9      disp(max(result));
                1 __10 end  % perfTest
```

We can see that 1.095 s were reportedly spent on the end line. This is typical profiling behavior when memory is deallocated. A further hint that memory allocations/deallocations are involved can be seen by looking at the Code Analyzer (MLint) messages section of the detailed profiling report, where we see a related message for the loop contents (line #7):*

Code Analyzer results

Line number	Message
7	The variable 'result' appears to change size on every loop iteration. Consider preallocating for speed.

Let's follow the suggestion and preallocate the result array:†

```
function perfTest
    newData = subFunc();
    maxData = max(max(newData));
    result(1000000) = 0;
    for iter = 1 : 1000000
        result(iter) = maxData + rand(1);
    end
    disp(max(result));
end  % perfTest
```

This shaves a full second from the loop's run time.

```
time      calls line
                 3 function perfTest
0.875        1   4     newData = subFunc();
             1   5     maxData = max(max(newData));
             1   6     result(1000000) = 0;
             1   7     for iter = 1 : 1000000
1.0631000000 8         result(iter) = maxData + rand(1);
0.7491000000 9     end
             1  10     disp(max(result));
             1  11 end  % perfTest
```

We can still do better: Let's get rid of the loop altogether by vectorizing it.‡ The **rand** function enables us to specify the number of rows and columns of its output, and maxData is constant, so vectorization is quite easy and highly effective in this case:

```
function perfTest
    newData = subFunc();
    maxData = max(max(newData));
    result = maxData + rand(1,1000000);
```

---

* We could also use the Profiler's memory-monitoring feature here — see §9.2.6 for additional information. The complete list of MLint-reported performance-related warnings is presented in Appendix A.4.
† Data preallocation is discussed in §9.4.3.
‡ MATLAB vectorization is discussed in Chapter 5.

```
      disp(max(result));
 end  % perfTest
```

```
     time   calls line
                   3 function perfTest
    0.856       1    4      newData = subFunc();
                1    5      maxData = max(max(newData));
    0.079       1    6      result = maxData + rand(1,1000000);
                1    7      disp(max(result));
                1    8 end   % perfTest
```

Recapping, our new code of a million data elements is ~40 times faster than the original code that only had 100 elements. 15 min of work that were well-spent.

When performance-tuning applications, we should not generally expect to improve the run speed so easily and effectively. After all, the example presented here was purposely simple and contrived, for illustration purposes. However, the basic profiling techniques presented in this section can indeed be used in the general case.

A well-known MathWorks article from 2002 that is widely circulated online shows how to use the MATLAB Profiler to iteratively improve a simple Mandelbrot fractal display function.[56] Readers should be wary when reading this article, since many of its specific performance tips are no longer relevant today (remember that over a decade of MATLAB improvements have passed since the article was published!).* This is in fact an excellent example that we should not rely on outdated performance tips, but rather retest using our current platform, software, and environment. In the specific case of Mandelbrot fractals, there are in fact more recent resources.[57]

2.1.3 Programmatic Access to Profiling Data

The Profiler's preferences (sorting modes, which profiling data section to display, etc.) are stored in MATLAB's user preferences file ([*prefdir* '/matlab.prf']), in the preference group "profile". These preferences persist across MATLAB sessions on the same computer. They can be retrieved and modified programmatically, as follows:

```
>> prefsStruct = getpref('profiler')
prefsStruct =
                sortMode: 'totaltime'
         busyLineSortKey: 'time'
       parentDisplayMode: 1
     busylineDisplayMode: 1
     childrenDisplayMode: 1
        mlintDisplayMode: 1
     coverageDisplayMode: 1
      listingDisplayMode: 1
             hiliteOption: 'time'
             showJitLines: '1'
             profileIndex: '18'

>> setpref('profiler','sortMode','selftime');
>> setpref('profiler','hiliteOption','numcalls');
```

---

* A similar example (*Tony's trick*) is discussed in §9.4.3; also see §9.4.2.

The *profile* function also has a programmatic interface that enables us to develop custom tools and reports to analyze the profiling data. When the 'info' parameter is specified, *profile* returns a MATLAB structure that contains the full set of raw data used to generate the summary and detailed profiling reports:*

```
>> profData = profile('info')
profData =
        FunctionTable: [61x1 struct]
      FunctionHistory: [2x0 double]
       ClockPrecision: 0.001
           ClockSpeed: 2533333333
                 Name: 'MATLAB'
             Overhead: 0
```

The returned structure `profData` contains an array of sub-structures `FunctionTable`, which in turn contain the detailed profiling data for each of the called functions:

```
>> profData.FunctionTable(5)
ans =
              CompleteName: [1x65 char]
              FunctionName: 'surf'
                  FileName: 'C:\MATLAB\R2012a\toolbox\matlab\graph3d\surf.m'
                      Type: 'M-function'
                  Children: [15x1 struct]
                   Parents: [0x1 struct]
             ExecutedLines: [27x3 double]
               IsRecursive: 0
       TotalRecursiveTime: 0
               PartialData: 0
                  NumCalls: 1
                 TotalTime: 0.163841795214196
```

The `ExecutedLines` field is a numeric matrix that has three columns and a variable number of rows: each row represents one of the code lines that has executed; the column data represent the source-code line number, the number of times this line has run (= number of calls), and the execution time. Execution times smaller than the Profiler's resolution (`profData.ClockPrecision`, which is typically 1 ms) are specified as having a run-time of 0. In our example, for the *newplot* function (index 6 in the profiling report), we can easily spot our two main busy lines, 61 and 74:

```
>> profData.FunctionTable(6).ExecutedLines
ans =
            45           1                    0
            46           1                    0
            48           1                    0
            60           1                    0
            61           1            0.514342686
            62           1                    0
            64           1                    0
            66           1                    0
```

---

* The returned `profData` can be stored and loaded in a later MATLAB session using *save* and *load*; the profiler report can then be redisplayed via `profview(0,profData)`.

```
74              1        0.359399108
76              1                  0
77              1                  0
```

Note that the structure elements are not ordered in a way that we might expect. To access the data of a specific function (e.g., *peaks*), use the following code snippet:

```
>> funcNames = {profData.FunctionTable.FunctionName};
>> funcIdx = strcmp(funcNames, 'peaks');
>> profData.FunctionTable(funcIdx)
ans =
        CompleteName: [1x65 char]
        FunctionName: 'peaks'
            FileName: 'C:\Program Files\MATLAB\R2012a\toolbox\matlab\elmat\peaks.m'
                Type: 'M-function'
            Children: [1x1 struct]
                ...
```

The function's report's Children field is an array of sub-structures that contain the IDs of all the child functions called by this function:

```
>> profData.FunctionTable(5).Children(10)
ans =
         Index: 6     ← this is the index in profData.FunctionTable
      NumCalls: 1
     TotalTime: 0.163841795214196
```

2.1.4 Function-Call History Timeline

One of the interesting features that the programmatic profiling interface provides us, and which is not represented in the Profiler GUI, is a report of the exact order in which different functions were called during the profiling session (*call graph*[58]). This profiling history can be turned on using the *-history* switch.[59] There is also an optional switch of *-historysize*, which enables us to modify the history size from a default maximum of 1 million function entry and exit items. Here is a sample usage of this history feature:

```
>> profile on -history; surf(peaks); profile off
>> profData = profile('info');
>> history = profData.FunctionHistory
history =
  Columns 1 through 11
      0     0     0     1     0     1     0     1     1     1     0
     19     1    17    17    17    17    18    18     1    19     5
    ...
```

The history data is actually a numeric matrix, where the first row contains the values 0 (= function entry) or 1 (= function exit), and the second row is the corresponding index into profData.FunctionTable, indicating the called function. We can easily convert this matrix into human-readable form using the following code snippet:

```
offset = cumsum(1-2*history(1,:)) - 1;  % calling depth
entryIdx = history(1,:)==1;    % history items of function entries
funcIdx = history(2,entryIdx); % indexes of relevant functions
```

```
funcNames = {profData.FunctionTable(funcIdx).FunctionName};
for idx = 1: length(funcNames);
    disp([repmat(' ',1,offset(idx)) funcNames{idx}]);
end
```

which generates the following calling list in the MATLAB Command Window:

```
isempty
  isempty
    transpose
  meshgrid
    peaks
  nargchk
    error
  ishg2parent
...
```

Unfortunately, the history information does not by default contain specific timing of each function entry/exit. But we can still make good use of it, by looking at the sequence in which the functions were called by each other.

To retrieve actual history timing information, we can run *profile* with the undocumented/unsupported *-timestamp* switch, which stores the CPU clock next to the history information. The reported history matrix now has four rows rather than two, where the extra rows represent the timestamp of each function entry and exit:[*]

```
>> profile on -detail builtin -timestamp ; surf(peaks); profile off
>> profData = profile('info');
>> profData.FunctionHistory(:,1:3)
ans =
                   0                   0                   1
                   1                   2                   2
          1347473710          1347473710          1347473710
              453000              453000              468000
```

In this report, the 3rd row represents the timestamp in seconds, while the 4th row represents the fractional portion of the timestamp, in microseconds. In the example above, the first timestamp item corresponds to 1347473710.453 seconds. The seconds value appears to be related to the number of seconds since midnight of January 1, 1970 (the so-called *Epoch*), which is a standard time representation format in computing systems.[†]

The timeline of the profiling session can be visualized as follows:

```
histData = profData.FunctionHistory;
startTime    = histData(3,1) + histData(4,1)/1e6;
```

---

[*] You may wish to issue a *format long g* command in order to see the large integer values in the manner shown. Note that this only affects the display, not the actual values, so it is only relevant for debugging and does not affect the actual program.

[†] For an unknown reason (probably a bug), on Windows, this value appears to be off by over a day from the expected value, which is retrievable via getTime(java.util.Date)/1000. Since we are only interested in relative rather than absolute times when profiling, this difference is not really important. On Macs and Linux the actual values appear to match the expected.

```
relativeTimes = histData(3,:) + histData(4,:)/1e6 - startTime;
plot(relativeTimes);
```

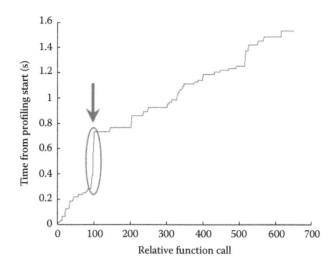

This report helps us see that a particular set of function calls, around the 100th call mark, is responsible for about 0.5 s, a prime candidate for tuning investigation. If we only relied on the standard profiling report we might have missed this because it might have been meshed into the same "bucket" as other invocations of the same function. As illustration, take the following simulated example:

Invocation #1 of func():	0.500 s
Invocation #2 of func():	0.013 s
Invocation #3 of func():	0.011 s
...	
<u>Invocation #10 of func():</u>	<u>0.012 s</u>
Total invocation time:	0.600 s

In this simulation, we would not have known that the 0.6 s invocation time of func() is not really evenly distributed across all 10 invocations. This could lead us to incorrect conclusions. For example, we might spend time unnecessarily on tuning the steady-state run-time performance, whereas we should really concentrate on just the first invocation. By looking at the actual timestamps we could see the large run time used by the first invocation and this information can possibly be used to tune it and significantly reduce the overall time taken by the function.

The *profile_history* utility on the MATLAB File Exchange[60] displays the function-call timeline graphically, in a highly interactive GUI that displays information about the invocations order and timing. Clicking a function label or graph segment displays the corresponding builtin detailed profiling report for that function:

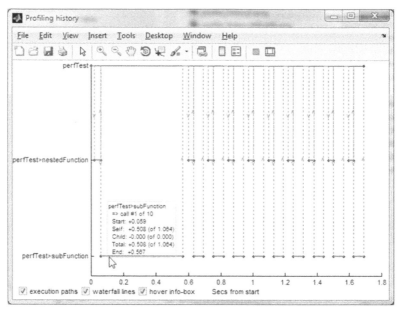

Function invocation timeline using the *profile_history* utility

2.1.5 CPU versus Wall-Clock Profiling

By default, the Profiler displays *CPU time*, rather than wall-clock time. This means that if a function is suspended for some time while external computer processes use the CPU, then that suspension time will not be reported. In most cases, it is the CPU time that is of interest to us when profiling code, since the external suspension times typically fluctuate across re-runs of the code, and are typically independent of the running MATLAB code. For this reason, CPU time is usually more important when trying to improve code speed, which is the major usage of the Profiler.

In §1.8.6, I explained that the perceived speed of a program depends on the actual run-time variability. To tune the perceived performance, we can use the Profiler's *wall-clock* timing mode, rather than its default *CPU time* mode:

```
>> profile on -timer real; myProg; profile report % wall-clock (real)
>> profile on -timer cpu; myProg; profile report % CPU time
```

Comparing the wall-clock time with CPU time could help us determine whether our application is CPU, memory, or I/O bound.

Note: On Windows platforms, in MATLAB releases at least as early as R2008a (possibly earlier), there is an internal MATLAB bug that causes the *cpu* option to have no effect.[61] In these Windows releases of MATLAB, both timer options actually use wall-clock timing. Since this has been officially recognized by MathWorks as a bug, it is expected to be fixed in an upcoming release.

2.1.6 Profiling Techniques

When profiling a program, ensure the results are reproducible, representative, and consistent. Otherwise, we would be tuning a moving target hopelessly. This section provides some tips to ensure that we maximize the profiling session's usefulness.

2.1.6.1 Relative versus Absolute Run Times

The Profiler instruments the running code and logs run-time information. The more run-time information is requested (e.g., function call history, or memory allocations and deallocations, which is an optional Profiler feature — see §9.2.6), the slower will the application become, as measured by the Profiler. Typically, we would not mind this extra overhead. We are generally more interested in comparing code versions, to decide whether a code change has improved performance compared to the previous version. For this reason, *relative* times in profiling sessions are typically more interesting than *absolute* times. Once we achieve the best possible code version in *relative* run time, we would expect its *absolute* run time to improve by a similar relative amount when the application is run without the Profiler.

Moreover, different platforms have different hardware, operating systems, MATLAB versions, run-time environments and externally running tasks. For this reason, it is meaningless to compare *absolute* run times across different platforms.* However, when you compare profiling sessions on the same platform, the *relative* timing behavior is likely to be similar on other platforms. If platform A exhibits a 2× speedup, then it is likely that platform B will exhibit a 1.5×–3× speedup. It is of course possible that due to some specific reason platform B might actually exhibit a slowdown, but such cases are indeed rare and usually happen when the original speedup was achieved by relying on a specific feature of platform A (e.g., SSD†).

In certain cases, we may be interested only in absolute run times during a profiling tuning pass, and none of the other information provided by the Profiler. In such cases we should use the low-overhead *tic/toc* functions (see below), rather than the Profiler.

2.1.6.2 Ensuring Profiling Consistency

For consistent results, it is advisable to run the application twice, discarding the results, and to only profile the third run.[62] Such a *warm-up phase* excludes MATLAB's compilation and caching overheads from the profiling results, ensuring repeatable and comparable results across different tuning runs. Even better would be to profile several runs of the same code, to average performance spikes.‡

A single run might provide misleading information about the actual application's hotspots and run-time behavior in real-world situations. Single runs are subject to momentary interrupts by the OS (operating system) or other processes, which may not occur in normal run time. In addition, the first run is frequently slower than normal, due to time spent to load code from disk into memory and then to compile and instrument it for profiling. The first run also does not benefit from multi-level caching (OS, disk, CPU, networking, database, application) as subsequent runs do. We should try to reduce the variability between profiling runs (more on this below). However, we can never avoid it altogether; profiling several runs smoothes-out this variability. Re-run the profiling to ensure consistency of the profiling conclusions.

To ensure repeatable consistent profiling results, we can restrict the OS (e.g., using Windows safe mode), so that as few processes and services as possible will interfere with the application. Significantly, turn off automatic OS updates and antivirus scans — such

---

* This is typically termed *YMMV** (*your mileage might vary*) in tech jargon.
† See §3.5.3.
‡ Both the warm-up phase and the averaging technique are performed by the *timeit* function — see §2.2.3.

processes often run at high priority, slowing down other concurrent processes (such as our MATLAB application).[63] Antivirus programs often update their definitions immediately following login or revival from sleep/hibernation, so be careful not to run the MATLAB application while this is still taking place.

We should close all nonessential active applications except MATLAB. For example, if we have an open word processor document, it might decide to perform a document auto-save at an unpredictable moment. Modern systems allow separate processes to run concurrently on separate CPU cores, so the effect of other processes running in parallel with MATLAB is not noticeable in general. But for the most accurate repeatable results, it is still better to isolate the MATLAB process.

On systems with Intel multi-core chips, MathWorks recommends restricting MATLAB to use only a single active CPU core for *"the most accurate and efficient profiling"*, by setting the MATLAB process's *CPU affinity*.[64] In practice, I believe this is rarely beneficial. In fact, we risk forgetting to unset the affinity after profiling, forcing MATLAB to run on a single CPU core, significantly degrading performance.*

Try to set up the environment as an actual typical usage and ensure that profiled runs use similar (if possible, identical) conditions (platform, workload, data, user actions, and configuration), so profiling results are predictable and statistically meaningful.[65] It is expected that consistent bottlenecks discovered this way, and their fix, would also be applicable to other similar environments.

Using a typical usage scenario, rather than an invented usage flow, is very important. After all, we are trying to profile what users would typically encounter, not some unrealistic scenario that would never happen in real life. Therefore, we need to ensure that we understand the typical usage patterns of our program.

As the program matures and its functionality changes, so too do usage patterns and in such cases we would need to re-profile our application. It is quite possible that new performance hotspots will emerge simply due to different usage patterns. It is also quite possible that we will need to revert some of the earlier performance fixes, in order for the performance to become optimal for the new typical workflow pattern, at the expense of the earlier workflow, which is now less typical.

It is important to ensure that either the data is not cached between re-runs, or cached in all runs, so that the runs' performance is comparable. If the application generates and uses random numbers, use the same random-number seed in all profiling runs, in order to ensure that exactly the same algorithmic processing path is done in all cases.

When developing code, it is natural to performance-tune it on the development machine, but it may behave differently than the target (deployment) machine or a typical user's machine. The application's performance profile may depend on computer memory, disk, CPU, GPU, and OS.[66] Complement your development tuning with at least one tuning cycle on a machine that mimics the target platform(s).

Whenever we profile, we should take care to profile MATLAB functions, rather than direct code lines in the Command Window. The reason is that the internal compilation and JIT behavior of MATLAB would mimic the application better, making profiling results more applicable. In this book I sometimes act contrary to this specific advice, showing command-window profiling. Readers should note that this is only done for the purpose of presentation within the limitations of a textbook. In real life, we should enclose your profiled code within a function and only then profile the function.

---

* CPU affinity can help overall performance if we are running multiple MATLAB instances (see §7.4), but degrades performance in the common case of running only a single MATLAB process.

That is, instead of doing this:

```
>> profile on; for idx = 1:100, a = a-b + c^d*log(e); end; profile report
```

do this:

```
>> profile on; myFunc(a,b,c,d,e); profile report
```

Where *myFunc.m* is a file that contains the profiled MATLAB code within a function:

```
function a = myFunc(a,b,c,d,e)
    for idx = 1 : 100
        a = a-b + c^d*log(e);
    end
end
```

2.1.6.3 Ensuring Compatibility with Real-World Conditions

A second type of profiling, no less useful, is to test a wide variety of scenarios, each with repeatable data/actions, rather than repeatedly profiling the same scenario. The reason for this is to ensure the validity of the discovered issues (bottlenecks) and fixes across a wide range of possible actual situations that the application would encounter in real life. It is sometimes surprising to discover that a performance tune that worked wonders for scenario A, causes unacceptable slowdowns in scenario B.

Yet a third type of profiling is testing our application's behavior under heavy load. "Load" can mean different things for different applications. It could mean a large amount of data, or frequent user actions, or a fast influx of data events that should be processed, or a combination of these. The common factor is that the application needs to churn more data at a faster rate than normal. A well-behaved application reacts well to increased load, displaying only *gradual degradation* of the functionality and/or performance. On the other hand, a non-professional application might freeze at heavy load, which is not an acceptable behavior. It is prudent to profile the application under heavy load, at least once during the performance-tuning cycle (toward the end, once the performance has more-or-less stabilized). Based on load/stress testing,* we can discover the application's limits and decide whether to take any corrective action.

Load testing is often considered a part of the quality assurance (QA) phase, along with functional testing, rather than part of the performance-tuning phase. However, I find that load testing can help improve the application's performance. Recall that the typical user measures performance based on maximal run time, not average run time.† I therefore advise to perform load tests as part of the performance-tuning cycle. To simulate heavy load we can artificially enlarge the data set; reduce timer periods so they run more frequently; and use `java.awt.Robot` to simulate user GUI actions.[67]

Performance tuning an application does not stop at improving the *average* run-time performance (latency). We also need to handle cases of performance spikes (*peak latency*) that can arise due to external factors and affect the *worst-case* performance. For example, network latencies, server load/downtime, deeply churning high-priority external processes such as drivers and antivirus services, garbage-collection cycles kicking in, and so on. To

---

* The semantic difference between *load testing* and *stress testing* is not in clear agreement (e.g., http://stackoverflow.com/questions/9750509/load-vs-stress-testing or: http://bit.ly/1lcRSJM). In the current context, we mean testing that discovers the program's behavior in the face of higher-than-normal load.

† See §1.8.6 for a detailed discussion of run-time variability.

diagnose such cases, after tuning our application for best average performance, we should then profile several single application runs, to identify cases and locations where instantaneous hotspots affect the overall run time. In many cases we can handle such hotspots without degrading average performance. The reason is that many of these hotspots are located in very specific locations within the code, where we can check for certain conditions and take preventive actions. Such pin-point checks typically have little or no visible effect on performance.

2.1.6.4 Profiling GUI and I/O

Try to profile the non-GUI parts separately from GUI parts, since GUI performance often varies widely based on user interactions and asynchronous callback interrupts. If we profile the GUI and non-GUI parts together, we might miss important hotspots in the non-GUI code. We might also miss some GUI code parts that vary *too* widely without apparent reason. Moreover, the techniques used to tackle GUI and non-GUI performance are often different, although there is indeed some overlap.*

While profiling GUI, interact with the application in a consistent manner, in order to achieve repeatable results. Even seemingly innocuous mouse movements, and mouse or keyboard clicks, trigger GUI events that take some MATLAB processing time and in extreme cases might affect the actual performance (e.g., if our application has callbacks that process mouse movements across a displayed image or graph axes).

It is often not useful to profile an entire GUI, from creation to end. Instead, we are usually interested in optimizing specific workflows of the GUI interaction. We could do this by starting the GUI, and setting it up in such a way that we are just ready for the workflow under analysis. Then start profiling, either from the MATLAB Command Prompt (*profile on*) or interactively (clicking the profile button in the Profiler tool), using dedicated GUI controls, or even programmatically.[68] Now start the relevant workflow, and stop profiling immediately when the workflow ends. In this manner, the profiling report will display only the workflow's profiling details, with minimal external overheads. Naturally, if our GUI contains periodic timers or MATLAB Workspace updates we would also see those in the report, but the main workflow should be easily identifiable and analyzable. As simple as the above procedure is, many MATLAB users are unaware that they can turn profiling on and off at will, interactively or programmatically, while a MATLAB program is running. Doing so significantly improves our profiling and performance-tuning effectiveness.

In fact, just like we can wrap analyzed code segments with *tic* and *toc*, we can also wrap them with *profile on* and *profile off; profile report*. This will ensure that we will profile only the time spent in the wrapped segment (and anything asynchronous that happens during this time), further minimizing external effects on the profiling results.

I/O is typically the slowest and least-repetitive part of a program. Therefore, as with GUI, try to profile and optimize CPU-processing (algorithmic) parts separately from I/O. Optimizing I/O typically presents large potential performance benefits, but is not always easy to do. Depending on the specific application, consider whether tuning I/O would be as cost effective as tuning the algorithm.†

---

* GUI performance tuning is discussed in Chapter 10.
† I/O performance tuning is discussed in Chapter 11.

2.1.6.5 Code Coverage

The Profiler can be used to discover different types of performance hotspots. These range from detecting code that runs more times than necessary (see §3.1), data that can be cached and reused (§3.2), checks that can safely be bypassed (§3.3), internal MATLAB helper functions that could be used (§4.3), to memory-related bottlenecks (Chapter 9). The Profiler helps us understand the application's data and code flow, helping us improve performance, and in some cases even identify functional bugs.

The Profiler provides, as part of its detailed report, a code-coverage report that tells us which code segments ran (and how many times), and which segments have not. This is important for both debugging and performance tuning: Looking at the report we easily see code segments that *should* have run according to our algorithm design, but in fact have not, due to some bug. It is then easy to review the source code up to the last line which *had* ran, and check why the lines following it have not. For performance tuning, the code-coverage report enables us to detect so-called *dead code* that can safely be eliminated.*

To verify that we are correctly profiling all the relevant code paths (use-case flows), we can look at the Profiler's code-coverage report. For example, in the following screenshot we see that only 56% of source-code lines have been profiled, the others have simply not run and so we have no profiling information on them:

Coverage results
Show coverage for parent directory

Total lines in function	109
Non-code lines (comments, blank lines)	61
Code lines (lines that can run)	48
Code lines that did run	27
Code lines that did not run	21
Coverage (did run/can run)	56.25 %

MATLAB Profiler's coverage report

Naturally, the closer the reported coverage is to 100%, the closer we are to full profiling coverage of all possible code paths. We can do this by running the profiled function with different input algorithms, or under varying environmental conditions. In practice, it is difficult to achieve more than 90% coverage in a profiling session. After all, many code lines are devoted to handling errors and exceptional cases that do not normally occur, and in fact are not very important to profile since they do not need to be optimized. But if we see a coverage value below 50%, we should really try harder, since tuning based on such partial coverage may be less than optimal.

---

* See §3.10.1.3 for a discussion of the performance aspects of dead code elimination.

The code listing section below the coverage report show us exactly which lines have not run (they have a light-gray color). This can help us understand which of the code paths we still need to run for a more-complete coverage:

```
              77       % use nextplot unless user specified an axes handle in pv pairs
              78       % required for backwards compatibility
           1  79       if isempty(cax) || ~hadParentAsPVPair
   0.032   1  80           if ~isempty(cax) && ~isa(handle(cax),'hg.axes')
              81               parax = cax;
              82               cax = ancestor(cax,'Axes');
              83               hold_state = true;
           1  84           else
   0.706   1  85               cax = newplot(cax);
           1  86               parax = cax;
           1  87               hold_state = ishold(cax);
           1  88           end
              89       else
              90           cax = newplot(cax);
              91           parax = cax;
              92           hold_state = ishold(cax);
              93       end
```

MATLAB Profiler's code-listing report showing which of the code lines were run/profiled

2.1.7 Profiling Limitations

The MATLAB Profiler has several inherent limitations that should be understood:[69]

- **Instrumentation noise** — Profilers function by modifying the run-time code such that function calls are timed and periodic checks are made of the stack trace to determine the number of times and duration that each code line has ran. The logging is pretty terse,* yet still takes some nonnegligible time (often called the *profiling overhead*). This overhead is naturally more dominant for short-duration functions than for longer ones, causing short-duration functions to be reported as taking a disproportionate time compared to longer functions.

- **Sampling rate effects** — Since the Profiler samples the stack trace at a constant rate (1 kHz, or once every 1 ms), the profiling report's accuracy is only as good as the sampling rate. Code lines that execute fast enough may well be missed by this sampling, and will appear with no profiling data in the detailed profiling report. Users should avoid the tendency to think that these lines take zero time. The more times that a code line is invoked, the higher the probability that the Profiler's sampling will happen to occur during the code line's execution, and in general the Profiler's report for each code line and function will more closely match the average real-life. In other words, the profiling report is basically just a statistical approximation, not 100% accurate.

- **I/O waits** — Whenever a program waits for I/O data, whether it is user input, or disk file, or network download, the wait time is logged on the MATLAB function that waits for the data. Such functions will have a disproportionate weight in the overall run-time report. When analyzing a profiling report, be aware of this over-weight before deciding to optimize these functions.

- **OS-induced latencies** — The operating system introduces latencies at random places throughout the profiling run. These latencies are due to context-switches to

---

* This can be seen in the raw-data of the function-call history, discussed in §2.1.4.

external (non-MATLAB) processes. They can be reduced,* but never entirely eliminated. Naturally, if a function takes a full minute to run, these latencies become
insignificant. But they can be significant if we measure sub-second profiling times
for a function.

- **Function call overheads** — If we look closely at the detailed profiling report,
  we will notice that the time reported for calling function B from function A is a
  bit longer than the total function time reported by function B. This is due to the
  function-call overhead, which is always attributed to the calling function (A).† We
  can use this difference to estimate the benefit of inlining functions.

- **JIT effects** — MATLAB's Profiler disables the JIT engine during profiling.[70] This
  usually causes noticeable slowdown compared to real life (*tic/toc*). But in those
  rare cases where JIT acceleration is actually detrimental to performance, it can also
  cause an unexpected speedup. §2.1.8 discusses this in some detail.

2.1.8 Profiling and MATLAB's JIT

MATLAB's JIT (*Just-in-time*) compiler‡ provides great speedup benefits in run time, but in
practice complicates our task of profiling performance. The reason is that JIT results can
vary widely with only minor changes to the code, which should not naïvely have caused
drastic performance differences.[71] Moreover, the data used by a loop's contents can in some
cases determine if and how JIT will be employed, causing subsequent runs to have widely
different performance results.

JIT optimizations could also cause some code lines to be skipped or rearranged in the
resulting machine code. In fact, the resulting machine code may be functionally equivalent to the original m-code or p-code, but may not look anything like the original source
code, once all the JIT optimizations have been applied.[72]

It is possibly for these reasons that JIT optimizations are turned off during profiling.§[73] In
some cases this could lead to speedups (see §3.1.15), although in most cases it will be slower
than unprofiled run-time executions. Whichever the case, one thing is certain: we cannot
rely on the *absolute* profiler timings. *Relative* timings (between different runs and code sections) are more likely to be representative of real-life (unprofiled) executions, although this
too should be taken with some reservations.[74] To get accurate timing results, we need to
use *tic/toc*, as explained in §2.2.

Some JIT support for statements in script files and the command prompt has been
added over the years, but JIT is still most effective for m-file function code.¶ For this reason, it is best to run performance checks using functions rather than the command line or
script files.**

Whereas the JIT optimizations are indeed disabled during profiling, MATLAB releases
R2010b (7.11) and earlier could display information within the profiler's detailed report

---

* See §2.1.6.2.
† Function call overhead can be estimated in R2013b and newer using ***matlab.internal.timeit***
  .functionHandleCallOverhead.
‡ See §3.1.15.
§ For similar reasons, JIT is also disabled when running with MATLAB's integrated debugger (i.e., up to a
  breakpoint).
¶ See §3.1.15.
** See §2.1.6.2.

about which code line is JIT-able.* This information was originally displayed by default in the Profiler, but was later made hidden in order not to encourage users to code against the ever-changing JIT,[75] and since it could mislead users into thinking that JIT-ted code lines are always better than unjitted ones.[76] This feature was finally removed from the Profiler in R2011a (7.12). For R2010b and earlier, the feature can be turned on by running the following before profiling:[77]

```
setpref('profiler','showJitLines',1);
```

The Profiler's detailed report will display a new "unjitted" column next to the code listing, from now on (the **setpref** operation need only be done once). As noted, this preference only affects MATLAB R2010b and earlier. On some even-older MATLAB releases, the "X" marks were linkable to a short explanation of why the specific code line was not JIT'ed.

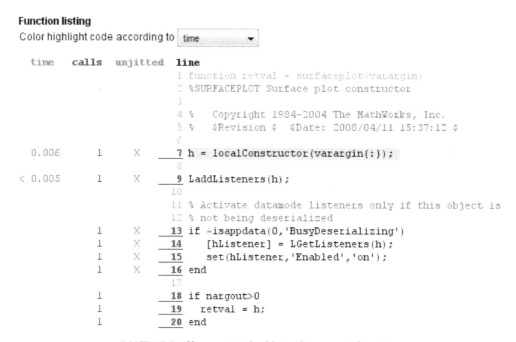

MATLAB Profiler report with additional memory information

2.2 *tic, toc* and **Relatives**

2.2.1 The Built-In *tic, toc* Functions

The built-in *tic* and *toc* functions are easy to use, and provide a very quick way to profile a particular code segment. The basic idea is to start a stopwatch timer at the beginning of the profiled code segment (*tic*) and report the timer value (elapsed time) at the end of the segment (*toc*). The elapsed time is reported to the MATLAB Command Prompt. For example:

```
>> tic, surf(peaks); toc
⇨ Elapsed time is 0.084146 seconds.
```

Or, within a function:

```
tic
for rowIdx = 1:size(data,1)
    for colIdx = 1:size(data,2)
        total = total + data(rowIdx,colIdx);
    end
end
toc
⇨ Elapsed time is 0.556565 seconds.
```

Such timing reports (*"Elapsed time is ..."*) will often be presented in this book. Where this will not cause confusion, the corresponding *tic/toc* calls that generated these timing reports will be elided, for the sake of brevity and clarity.

We should take care not to place the *toc* call within a loop, otherwise it will be executed in each loop iteration, reporting the latest timer value. The timer is only reset to 0 in *tic*, so successive *toc* calls will report successively higher elapsed time count.

If we are only interested in the timing for an internal portion of the loop, we can use *toc*'s programmatic output, which does not report an output to the Command Prompt:

```
elapsed = 0;
for rowIdx = 1:size(data,1)
    for colIdx = 1:size(data,2)
        tic
        total = total + data(rowIdx,colIdx);
        elapsed = elapsed + toc;
    end
end
fprintf('%f seconds passed in loop core\n', elapsed);
```

Be careful not to place multiple active *tic/toc* calls within our code in a way that one set would interfere with another. For instance, modifying the example above:

```
tic
elapsed = 0;
for rowIdx = 1:size(data,1)
    for colIdx = 1:size(data,2)
        tic
        total = total + data(rowIdx,colIdx);
        elapsed = elapsed + toc;
    end
```

```
end
fprintf('%f seconds passed in loop core\n', elapsed);
toc
⇨ Elapsed time is 0.000481 seconds.
```

What happened here? How come the elapsed time is so low? The answer is that the inner *tic*, within the loop core, resets the global timer whenever it is called. Therefore, the end-of-block *toc* effectively reports only the time of the latest core iteration, plus the external *fprintf* call. This is not only not useful, but actually misleading.

Instead, on MATLAB release R2008b (7.7) or newer, use *tic*'s output value to attach specific *toc*s to specific *tic*s. The basic format is *t = tic; toc*(t):[78]

```
tExternal = tic;
elapsed = 0;
for rowIdx = 1:size(data,1)
    for colIdx = 1:size(data,2)
        tInternal = tic;
        total = total + data(rowIdx,colIdx);
        elapsed = elapsed + toc(tInternal);
    end
end
fprintf('%f seconds passed in loop core\n', elapsed);
toc(tExternal);
⇨ Elapsed time is 0.557326 seconds.
```

On MATLAB R2006a (7.2) or older, *tic/toc* use the *clock* function with a resolution of 10 ms. Since R2006b (7.3), *tic/toc*'s reported resolution is about a microsecond:[79]

```
>> feature('timing','resolution_tictoc')
Resolution of Tic/Toc clock is 1.396826e-06 sec.
```

On R2014a (8.3), feature('timing','resolution _ tictoc') was replaced with the following command, which is part of the new *timeit* functionality (see §2.2.3). It is more dependable than the former *feature*, since it averages actual *tic/toc* calls:

```
>> matlab.internal.timeit.tictocCallTime
ans =
        9.85239066206834e-08
```

Note that this reported value is an averaged value, computed from a thousand consecutive calls to *tic/toc*.* This is a pretty good although inexact estimate of the *tic/toc* performance and resolution in actual programs. In any case, the value is so small that we can safely ignore its overhead in actual measurements.

Platform timing limitations may cause *tic/toc* to be unreliable when timing durations shorter than 1 ms. If we use *tic/toc* directly on faster functions, results might be inaccurate and nondependable; successive calls may report different durations.

When timing MATLAB code, we should take note of inherent timing inaccuracies. Sub-millisecond accuracy seems to be impractical, at least on Windows platforms.[80] Much larger inaccuracies exist for timers and intentional synchronous delays using the *pause* function.†[81] As a simple example, consider the following surprising result:

---

* Take a look at the function's m-code: *%matlabroot%\toolbox\matlab\timefun\+matlab\+internal\+timeit\ tictocCallTime.m.*

† The reported inaccuracy here is mostly due to the *pause* function (see §10.4.3.5) rather than loop overhead or *tic/toc* inaccuracy.

```
tic, for idx = 1:3, pause(2); end, toc %expected result: exactly 6 secs
⇨ Elapsed time is 6.040262 seconds.
```

Even with these limitations, *tic/toc* is more accurate at determining elapsed time than the Profiler. The following section compares these two measurement mechanisms.

2.2.2 Comparison between the Profiler and *tic, toc*

MATLAB's profiler is basically a *sampling profiler*, which provides timing information based on periodic sampling.[82] The Profiler also includes a call-graph history, but as we have seen in §2.1.4, it is unofficial and not easy to use. On the other hand, using *tic/toc* enables *structural* (or *instrumenting*) *profiling*, which can provide information on function entry/exit times and durations, as well as detailed call history.

Using the Profiler is complementary to using *tic/toc* (along with timed logs*). In some situations we may prefer *tic/toc*'s ease of use, whereas in others we may wish to use the Profiler's detailed information and drill-down ability.

Here is a qualitative comparison of these two profiling mechanisms, based on some subjective criteria that I find important. Readers should feel free to use other criteria or evaluations, in order to decide which of these alternatives (or both) to use:

- **Ease of use** — Running the Profiler is quite easy. However, some users find that the requirement to drill-down, sometimes numerous levels, hinders the ability to quickly diagnose memory hotspots in deeply nested code. A couple of *tic/toc* at the suspected location helps to quickly answer such questions, although the user is left in the dark regarding other parts of the running code.

- **Code modifications** — Using the Profiler requires absolutely no code modifications. On the other hand, using *tic/toc* requires manual modifications.

- **Realistic results** — The Profiler needs to sample the code at a high frequency (1 kHz, or once per 1 ms). Moreover, the Profiler turns off the JIT accelerator during profiling. The Profiler therefore affects the performance of the code, making it slower than the nonprofiled case. On the other hand, *tic/toc* calls have negligible effect on code run time, and present a much more realistic view of the run time behavior.

  However, if timed logging occurs at very high frequency (sub-millisecond), the associated overhead could become significant. While still providing more accurate *structural* (call-graph) results than the Profiler, these results might not match the actual *performance* of the uninstrumented code, and might even cause *heisenbugs* — bugs caused by the timing effects of observation.[83]

- **Access to results** — The Profiler enables easy GUI-based analysis of its profiling results (although not to its call history†). In addition, it enables output of the raw collected data, which can then be analyzed programmatically (although some users might find this task a bit difficult). On the other hand, *tic/toc*'s much simpler result is easily available programmatically, as the output of the *toc* function.

- **Automation** — it is easier to use *tic/toc* for automated performance tests.

- **Level of detail** — The Profiler provides information and detail unmatched by *tic/toc*, which only return a single numeric value (elapsed time). In contrast, the

---

* See §2.3.
† See §2.1.4.

Profiler provides this information for every line of code, in addition to providing information about the number of times it has been executed, the memory it has used,* the list of functions that it had called, and a distinction between the self-time and the total time of each function call.

- **Multiple code sections** — While it is possible to use *tic/toc* on multiple code blocks concurrently, using *tic*'s output value as an input to each of the blocks' *toc*, this is not trivial. In fact, it is quite easy to forget to use the *toc(param)* input argument syntax, thereby mixing the results of the concurrent blocks and providing misleading information. In contrast, the Profiler profiles multiple code blocks automatically and correctly.

- **Code coverage report** — The Profiler provides, as part of its report, a code-coverage report that tells us which code segments have run (and how many times), and which segments have not run at all. This is very important, for both debugging and performance tuning: Looking at the report we can easily detect segments that *should* have run according to our algorithm design, but in fact have not ran, due to some bug. It is then easy to go up the source code to the last line which *had* ran, and check why the following lines have not. For performance tuning, the code-coverage report enables us to detect the so-called dead code that can safely be skipped or eliminated to save processing time. All of this valuable information is not available if we use *tic/toc*.

- **CPU versus wall-clock timing** — The Profiler provides the ability to measure CPU timing, as opposed to the wall-clock timing that both the Profiler and *tic/toc* provide. If we wish to profile CPU-based timing information, we have no other option but to use the Profiler,† or the *cputime* function (see below).

2.2.3 Related Tools

MATLAB has several built-in functions that are related to *tic/toc*:

- *cputime* — This function returns the CPU time in seconds used by the MATLAB process since MATLAB started.[84] Note that the reported resolution is quite low (only 15.6 ms on Win7 64-bit running R2012b).[85]

```
>> cputime
ans =
     3061.890625

>> tOriginal = cputime;  surf(peaks(40));  elapsed = cputime-tOriginal
elapsed =
         0.0936006
```

MathWorks recommends measuring performance using *tic/toc* rather than *cputime*, due to some inaccuracies in CPU time reported on hyperthreaded CPUs.[86] This has been a bug that has in fact been fixed in R2006a (MATLAB 7.2),[87] so it is not clear whether the recommendation should still be followed on newer MATLAB releases. Proceed with caution …

---

* See §9.2.6.
† Note the Profiler's bug on Windows platforms that prevents this – see §2.1.5 for additional details.

- *clock* — This function returns the computer's current date/time, as a 6-element array of numbers, where the last (6th) element represents the seconds with millisecond resolution (e.g., 23.456).

```
>> clock % September 6, 2012 at 16:36:57.609
ans =
  Columns 1 through 3
            2012              9                   6
  Columns 4 through 6
              16             36             57.609
```

clock's actual resolution is provided on MATLAB R2013b and earlier by:[88]

```
>> 1/feature('timing','clocks_per_sec')
ans =
    0.001
```

- *now* — This function returns the computer's current date/time, as a floating-point number in units of days since the day before January 1, 0000. While the resolution of the value suggests micro-second resolution, the actual resolution is in fact lower — the same resolution as *clock*, on which *now* is based.

```
>> now
ans =
    735118.696029178

>> datestr(now)
ans =
06-Sep-2012 16:42:16
```

- *etime* — This function returns the time in seconds that has elapsed between two 6-element *clock* vectors T1 and T0.
- *bench* — A related MATLAB function that compares MATLAB's performance on the current platform compared to several other computer systems. While this gives us a rough idea of the *comparative* power of our platform, it does not really help us *improve* the speed of any specific MATLAB function.

A few MATLAB utilities were posted to the MATLAB File Exchange to facilitate profiling. One of the best of these, which uses *tic/toc* internally, is *timeit*, by MathWorker Steve Eddins.[89] *timeit* was written in such a way as to remove some common profiling artifacts. This was done by repeating the measurement multiple times, reporting the median elapsed time while discounting the *tic/toc* overhead, insignificantly short as it might be. *timeit* solves *tic/toc*'s resolution limitation cleverly by calling the timed function multiple times such that the timed duration is at least 1 ms. *timeit* was incorporated as a standard MATLAB function in the R2013b (8.2) release; on earlier releases it can be downloaded from the MATLAB File Exchange. One limitation of *timeit* is that it requires an input of a function handle that accepts no input parameters. This often means using slow anonymous functions:*

```
>> hFunc = @()datestr(now);
>> timeit(hFunc)
ans =
    0.000254059150743082
```

---

* See §4.6.1.2.

On MATLAB R2014a (8.3), the Desktop's Command History panel reports the elapsed (*tic/toc*) time for those commands that took more than 100 ms or so. The exact execution time (in ms) for all the commands entered at the Command Prompt are stored in the new *History.xml* file, located in the user's ***prefdir*** folder:

```
>> edit(fullfile(prefdir,'History.xml'))
...
<session>
<command time_stamp="142de19feb3">%-- 10/12/2013 22:01 --%</command>
<command execution_time="1224">hFig = figure;</command>
<command execution_time="2307">surf(peaks)</command>
...
```

Readers should avoid the natural tendency to rely too much on this feature for performance tuning. It is a very convenient feature for getting a general performance overview of different coding variants, but we should remember that Command Window timings are inaccurate and non-representative (for one thing, JIT is often not used for commands entered in the Command Window). As noted in §2.1.6.2, we should normally profile and performance-tune MATLAB functions (in m-files), rather than direct Command Window commands.

2.3 Timed Log Files and Printouts

Using log files with timestamps should not be discounted as a profiling tool. In fact, this age-old method is often used as a simple and effective debugging and performance memory tool. The basic idea is to attach a human-readable timestamp to every log message, and then analyze the log to determine applicative hotspots. For example, suppose we define the following logging function:

```
function log(severity,message,varargin)
    message = sprintf(message,varargin{:});
    timeStr = datestr(now,'yyyy-mm-dd HH:MM:SS.fff');
    logFid = fopen('application.log','at');
    fprintf(logFid, '%s %s\t %s\n', timeStr, severity, message);
    fclose(logFid);
end
```

In our application, it then becomes very easy to place log messages. For example:

```
log('Info', 'this is an informational message...');
log('ERROR', '...and this is error message #%d', id);
```

Analyzing the log file is quite straightforward:

```
...
2012-09-06 17:20:22.375 Info  this is an informational message...
2012-09-06 17:20:25.593 ERROR ...and this is error message #123
...
```

In this example, 3.218 s have passed between the log messages. Depending on our specific application, this may be normal or may indicate a problem.

In the implementation of the `log` function above, note the use of the `'yyyy-mm-dd HH:MM:SS.fff'` format. Depending on the specific situation, we may remove the date portion from this format string, leaving only the timestamp. We may also decide that millisecond resolution is not needed, in which case we can remove the `.fff` part.

A more realistic log function might check some application configuration to decide which (if any) messages to actually log in the file (possibly based on severity), as well as to use a dynamic (rather than hard-coded) log filename.

The MATLAB File Exchange includes two logging utilities that expand the above ideas and enable logging and filtering of message severity, and various other functionalities. These utilities are Gavin Paul's *log4matlab*[90] and Luke Winslow's *log4m*,[91] which is based on *log4matlab*. Both of these utilities are modeled after the well-known *log4j* utility and its variants in other programming environments, such as *log4cxx*, *log4net*, *log4perl*, and so on.[92]

2.4 Non-MATLAB Tools

In addition to the MATLAB tools mentioned above, we can also use standard external tools to profile the performance and behavior of our application.

Monitoring the MATLAB process during application run time provides invaluable insight about whether our application is CPU, memory, or I/O bound. MATLAB's built-in Profiler can provide information about memory (see §9.2.6), but the Profiler unfortunately does not provide information on the CPU usage, I/O traffic, and I/O waits caused by the monitored application. We need external monitoring tools to complete our picture. Based on this knowledge we can focus our tuning efforts on the most effective element.

It should be noted that all of these monitoring tools work at the Operating System process level. This means that there is no clear link between any specific MATLAB function or computation, and the tools' monitoring results. Despite this, external tools can still provide valuable information about environmental effects of our MATLAB application, such as disk and CPU usage, network traffic, GDI handles, and so on.

If we use these tools in conjunction with step-by-step debugging (using MATLAB's integrated debugger), we can often pinpoint specific places in our MATLAB code that cause some environmental issue reported by the external tools. We could use such tools to check the so-called "environmental footprint" of our application, and choose among alternative implementations based on this footprint (e.g., choose network I/O over disk I/O as the means of data import).

Here is a list of some popular external monitoring tools, grouped by type. The list is by no means comprehensive: there are numerous available tools, and for specific uses other tools could perhaps be more effective. The list below is of tools which are widely used and for which support can easily be found online:

- **Process monitoring** utilities, such as the *Task Manager* or *Process Explorer*[93] on Windows; *top* or *ps* on Linux. These utilities provide information on the overall CPU, memory, and swap/paging usage by the process.

  For command-line usage, the *JavaSysMon* utility[94] provides interesting statistics that can easily be queried from within MATLAB.[95] This utility is cross-platform and works on all platforms that run MATLAB.

On Windows, much greater profiling granularity can be achieved using the *Process Monitor* utility.[96] This utility enables monitoring file-system, registry, and network events, complementing the built-in Profiler's report, as well as its sibling *Process Explorer* utility (both are part of Microsoft's SysInternals suite of power utilities). Here is *Process Monitor*'s partial output when I issued a MATLAB *getpref* command:

SysInternals *Process Monitor*'s partial output for a MATLAB *getpref* command

An example of using these tools to monitor and debug memory-related performance in MATLAB is provided in §9.2.9.

Windows provides programmatic access to many performance counters via a dedicated API. MATLAB programs can easily access them via corresponding Dot-Net objects, for example System.Diagnostics.PerformanceCounter.

- **System monitoring** utilities, such as *Performance Monitor* on Windows; *sar* or *vmstat* on Linux.

- **Disk monitoring** utilities, such as *Process Monitor* again on Windows; *iostat* on Linux.

- **Network monitoring** utilities, such as hardware packet sniffers, the *netstat* utility (all platforms) or the *Wireshark* (previously called *Ethereal*; Windows and Macs)[97] and *Fiddler*[98] (Windows) applications.

- **Memory monitoring** utilities, such as the *Process Explorer* again on Windows, or its *VMMap* sibling.[99] For detailed information on profiling MATLAB's memory usage, refer to §9.2.

- **System-level performance monitoring** tools,[100] such as *VTune*,[101] *Valgrind/Callgrind*,[102] *Shark*[103]/*Instruments*,[104] *GlowCode*,[105] *PurifyPlus*,[106] *Insure++*,[107] and *Eclipse TPTP*[108] — all of them purportedly used by MathWorks themselves.[109] Note that some of these profiling tools may not be of any use without access to the MATLAB source code. However, they could be useful when tuning MEX or Coder-generated C-code (see Chapter 8).

3

Standard Performance-Tuning Techniques

Optimization techniques can very broadly be categorized into the following groups:[110]

- Optimization of iterative loops
- Caching and other uses of trading memory for performance
- Parallelization — running code in parallel on multiple cores/GPUs/computers
- Using compiled (binary) rather than interpreted code
- Reusing system resources and programmatic constructs
- Employing knowledge about the data's memory arrangement for optimized access using in-place manipulation, locality of reference, and preallocation
- Reducing code complexity
- Trading accuracy, code size, and latency for run-time throughput
- Removing unnecessary, redundant, or unused code
- Optimizing the most common program path
- Dynamic adaptation of program parameters based on run-time measurements

This grouping is very coarse: some techniques may belong to several categories, while others may perhaps not belong to any of the major groups above. Moreover, in some cases, conflicts may arise between techniques of various groups. For example, optimizing the most common code path (so-called *fast path*[111]) may come at the expense of less common paths, whose performance might degrade; memory optimization may come at the expense of parallelization.

The remainder of this book details specific optimization techniques.

This chapter explains standard performance-tuning techniques that are well-accepted practices in software programming in general, or in other words, are not MATLAB-specific. Performance tuning has been widely researched and practiced in computers and software long before MATLAB was invented. Many of the useful practices that were invented for non-MATLAB systems can often be applied in MATLAB programs.

The explanations in this chapter shall be MATLAB-oriented and in some cases may even rely on MATLAB-specific functions or constructs. However, in all cases, the underlying techniques can be traced back to existing non-MATLAB tuning practices.

In contrast, the following chapters describe techniques that are generally MATLAB-ish in origin, and even if they are sometimes used elsewhere, they are not in widespread use outside MATLAB.

3.1 Loop Optimization

Loop optimization is the generic name for multiple techniques that improve the performance of program loops.[112] Loops are important since they are used extensively in programs and typically take the lion's share of a program's run-time. Any specific loop iteration may take only a tiny amount of time, but running the loop numerous times results in a noticeable effect on the overall run-time. Any improvement to the iteration run-time, small as it may be, is multiplied by the number of loop iterations.

As of R2014b (MATLAB 8.4), these techniques need to be hand-coded, since they are not applied automatically by the MATLAB interpreter. However, the interpreter is constantly improving, so this may well change in future MATLAB releases.

The following loop optimizations techniques are described (among others) below:

- Move loop-invariant code out of the loop
- Minimize function calls within loop iterations by inlining
- Employ early bail-out conditions to skip unnecessary loop iterations
- Simplify loop contents
- Unroll simple loops by selectively repeating content
- Switch the order of nested loops and in general invest in tuning nested loops
- Minimize dereferenced access to object property and array element
- Postpone I/O and graphics updates until the loops end
- Merge independent loops, or split complex loops
- Loop over the shorter dimension
- Run loops backwards

 ### 3.1.1 Move Loop-Invariant Code Out of the Loop

Loop-invariant hoisting, or LIH (also called *loop-invariant code motion*),[113] is perhaps the most used optimization technique. The basic idea is that any expression or computation that is unnecessarily repeated in a loop should only be done once.

It often happens in loops that an expression is common to all loop iterations (i.e., invariant within the loop), and yet is evaluated separately in each iteration. The idea is to move this common expression outside the loop, so that it is only evaluated once.

3.1.1.1 A Simple Example

LIH was illustrated in the sample profiling session of §2.1.2. Recall that the original function looked as follows:

```
function perfTest
    for iter = 1 : 100
        newData = subFunc(iter);
        result(iter) = max(max(newData)) + rand(1);
    end
    disp(max(result));
end  % perfTest
```

In §2.1.2, we have seen that the first line within the loop (newData = subFunc(iter)) is an invariant expression, since it repeatedly loads the same data from file and runs the same *sin* function on the loaded data:

```
function result = subFunc(iteration)
    fileData = load('data.mat');
    result = sin(fileData.data);
end  % subFunc
```

We can safely infer that newData in the original loop retains a constant value and is therefore an invariant expression in all loop iterations.* Moving this expression out of the loop, we reduce the run-time from 35 s to 1.4 s (a 25× speedup):

```
function perfTest
    newData = subFunc(iter);   % LIH-moved from within the loop
    for iter = 1 : 100
        result(iter) = max(max(newData)) + rand(1);
    end
    disp(max(result));
end  % perfTest
```

3.1.1.2 I/O and Memory-Related Invariants

Loop invariants are not limited to numeric or algorithmic expressions. In fact, I/O (reading/writing data from/to disk, USB, or network) is often slower than pure-CPU computations. In such cases, as in the example above, LIH is very effective.

A typical example that seems innocent at first glance but in fact uses costly I/O is the use of the *getpref* / *ispref* / *setpref* set of built-in MATLAB functions, which under the hood access an uncached MAT file whenever they are used.† Instead of using these functions in a loop, we should read/update preferences just once, outside the loop.

LIH is also important with memory allocation. As shown in Chapter 9, memory allocation can have a significant impact on run-time performance. By moving common allocations outside the loop, we can prevent repeated reallocations:[114]

```
x = (1:3)'; for idx = 1:1e5, a = x(:,ones(1,10)); end % non-optimized code
⇨ Elapsed time is 0.182793 seconds.

x = (1:3)'; y = ones(1,10);for idx = 1:1e5, a = x(:,y); end % LIH-optimized
⇨ Elapsed time is 0.116191 seconds.
```

As explained in §9.3.3, even simple indexing has the effect of allocating (and deallocating) temporary arrays, which could be disastrous for loop performance:[115]

```
data = rand(1e6,10);
for idx = 1:1000, m = max(data(:,10)); end %1000 temp allocations
⇨ Elapsed time is 3.793339 seconds.

v = data(:,10); for idx = 1:1000, m = max(v); end %only 1 allocation
⇨ Elapsed time is 0.494257 seconds.
```

---

* This would of course be incorrect if the data file changes during the program, or if the processing function depended on time.
† See §11.2.

3.1.1.3 Subexpression Hoisting

It is not only subexpressions in <u>specific iterations</u> that can be hoisted out of the loop. LIH can also be used to hoist the result of a subexpression that spans the <u>entire loop</u>:

```
for idx = 1 : 100
    total = total + idx + data;
end
```

We see that `data` is repeatedly added to `total`. Applying LIH optimization, we get

```
for idx=1 : 100
    total=total+idx;
end
total=total+100*data;
```

After applying LIH optimization, it often becomes clear that the entire loop can be optimized away. In the example above, it is now clear that we can completely remove the loop using simple vectorization:

```
total = total + sum(1:100) + 100*data;
```

3.1.1.4 Loop Conditionals

A common source of loop invariants is function evaluation within the loop condition:

```
while result <= functionOf(someConstant)
    result=doSomething();
end
```

Counterintuitively, the loop condition is repeatedly re-evaluated in each loop iteration. The re-evaluated loop condition can often be replaced by a constant term:

```
loopLimit=functionOf(someConstant); % LIH-moved outside the loop
while result <= loopLimit
    result=doSomething();
end
```

Another common source of loop invariants is a loop that contains a simple internal conditional statement that does not depend on the loop iteration:

```
for idx = 1 : 1000
    process(idx);
    if someInvariantCondition
        doSomething(idx);
    else
        doSomethingElse(idx);
    end
end
```

In such cases, we can take the conditional statement out of the loop and duplicate the loop, as follows (a technique called *loop unswitching*[116]):

```
if someInvariantCondition
    for idx = 1 : 1000
        process(idx);
        doSomething(idx);
    end
else
```

```
    for idx = 1 : 1000
        process(idx);
        doSomethingElse(idx);
    end
end
```

The modified code, while somewhat larger and arguably a bit less maintainable, is better: it does not repeatedly re-evaluate the condition in the loop. The new loops are simpler, making it easier for the JIT to optimize and for us humans to vectorize.

3.1.1.5 Invoked Functions

Loop invariants may be difficult to detect when a loop calls functions internally. In the following example, it is easy to see that `array` is an invariant expression. However, it is more difficult to detect that in the specific context of this loop `arrayLen` is also an invariant that is unnecessarily re-evaluated 1000 times:

```
for loopIdx = 1 : 1000
    array = 1 : 100;
    process(array, loopIdx);
end

function process(array, loopIdx)
    arrayLen = length(array);
    for arrayIdx = 1 : arrayLen
        doSomething(array(arrayIdx), loopIdx);
    end
end  % process
```

In such cases, modifying the called function with extra parameters may obviate the need to re-evaluate the internal invariant (`arrayLen` in this case):

```
array = 1 : 100;
arrayLen = length(array);
for loopIdx = 1 : 1000
    process(array, arrayLen, loopIdx);
end

function process(array, arrayLen, loopIdx)
    % arrayLen is now an input parameter: no need to evaluate it here
    for arrayIdx = 1 : arrayLen
        doSomething(array(arrayIdx), loopIdx);
    end
end  % process
```

In a related matter, some functions can be called just once and do not need to be reinvoked in each loop iteration. For example, calling *hold* in the following example can be done just once following the loop, rather than multiple times within it:[117]

```
for idx = 1 : 100
    data(idx) = someFunctionOf(idx);
    hold all;
end
plot(data);
```

3.1.2 Minimize Function Call Overheads

Calling functions incurs a nonnegligible overhead, especially when done repeatedly within a loop.* The effect is especially important when calling functions (methods) of object-oriented classes or external objects such as COM or Java, when the loop iteration time is very small (several milliseconds or less), and when the loop runs numerous iterations. By minimizing function calls within the iterations, we can sometimes achieve significant speedups.

When developing programs, avoid creating numerous tiny functions that call each other, a design antipattern that is sometimes called the "nested-doll (or *matryoshka*) principle", after the Russian toy.[118] Instead, fold the code into the parent function. A traditional good programming practice is to divide the code into multiple separate subfunctions for maintainability and supportability. However, owing to the nonnegligible function call overhead in MATLAB, these benefits should be weighed against performance.[119]

When function calls cannot be avoided, we can minimize function call overheads by caching the function results, a technique that is called *memoization*.†

Minimizing function calls can also be done by *inlining* (copying) the function's contents directly within the loop, replacing the call to a function with its internal code. For example, instead of calling the **mean** function, we could inline its core algorithm, bypassing internal checks and removing the function call overhead. It is not only the function call overhead that is saved in this case: replacing a function call with a series of statements that only use built-in functions enables JIT to compile the code in run-time, thereby significantly improving loop performance:[120]

```
A = rand(1000,1);

for idx=1:100000, b=mean(A); end
⇨ Elapsed time is 1.638925 seconds.

for idx=1:100000, b=sum(A,1)/size(A,1); end
⇨ Elapsed time is 0.318459 seconds.    % 5x speedup
```

Inlining functions only makes sense for small functions that can easily be inlined and for which the invocation overhead is relatively significant compared to the overall iteration time. If we need to inline the same function in multiple loops, the reduced maintainability and modularity of the code should be weighed against the performance speedup.

Note that inlining is risky: the bypassed internal checks may be important in some use-cases, and JIT may evolve to better-tune the function rather than its inlined code. Also, the function may someday become built-in with better performance than inlined m-code, making our existing code suboptimal.

An example of this latter risk is the built-in **repmat** function, which (while it was still an m-file, before becoming built-in in R2013b‡) uses numerous internal checks for the number of elements, whether or not the data is sparse, and so on. We can use the Profiler code listing to determine the actual code coverage that is in use and just inline (execute) that coding core.§ For example, for regular numeric data, it turns out we can replace **repmat** calls with a simple vectorized index approach for dramatic speedup:[121]

---

* Function call overhead can be estimated in R2013b and newer using *matlab.internal.timeit. functionHandleCallOverhead*, part of the new *timeit* functionality (see §2.2.3). For example: *matlab. internal.timeit.functionHandleCallOverhead(@myFunc)*.
† See §3.2.4.2.
‡ See §5.4.2.
§ Smart checks bypass is discussed in §3.3.

```
x = (1:3)';

for idx=1:1e5, a=repmat(x,1,10); end          % non-inlined repmat
⇨ Elapsed time is 1.781095 seconds.

y=ones(1,10); for idx=1:1e5, a=x(:,y); end    % inlined: 15x speedup
⇨ Elapsed time is 0.116191 seconds.
```

This speedup significantly decreased (from 15x to 3–4x) when *repmat* became builtin in R2013b, changing the performance-maintainability tradeoff and highlighting the risk.

In many cases, built-in MATLAB functions perform internal sanity checks that may be irrelevant under a specific set of circumstances, before running an internal helper function for the core logic. Using such internal functions directly for improved performance is discussed in §4.3. MATLAB's built-in date/time processing includes a set of functions that can benefit from this technique (see §4.4).

3.1.3 Employ Early Bail-Outs

Early bail-out checks are conditional expressions that skip some of the loop processing and even entire loop iterations, when a certain condition occurs, using *break* and *continue*:

```
data = [];
newData = [];
outerIdx = 1;
while outerIdx <= 20
        outerIdx = outerIdx + 1;
        for innerIdx = -100 : 100
            if innerIdx == 0
                continue         % skips to next innerIdx (=1)
            elseif outerIdx > 15
                break            % skips to next outerIdx
            else
                data(end+1) = outerIdx/innerIdx;
                newData(end+1) = process(data);
            end
        end  % for innerIdx
end    % while outerIdx
```

MATLAB's *continue* command is used to skip directly to the next loop iteration (or end the loop if this is already the last iteration); *break* is used to skip out of all remaining loop iterations and jump directly to the code immediately following the containing loop. Such early bail-outs can save unnecessary computations of the loop iterations (in this case, evaluation of the process(data) function). Note that in this particular case, *break* effectively breaks out of both the inner and outer loops, since the bail-out condition (outerIdx > 15) is true for all subsequent outer loop iterations.

The speedup potential of the early bail-out mechanism is large since we can skip lengthy processing of irrelevant or nonsensical cases. Similar early bail-out is often employed at the beginning of lengthy functions (see §3.3).

Unfortunately, the use of such conditional expressions within a loop often makes it hard for the built-in JIT to properly optimize the code. Also, manual vectorization of the code for additional speedups becomes nearly impossible. Moreover, it may well turn out that computing the conditions in the 99% of the cases when they are not met is more expensive overall than the 1% of the cases when they are met and save some processing. For this reason, I suggest using early loop bail-outs with caution.

When we have multiple nested loops, we may sometimes wish to bail-out of an ancestor loop that is not the direct parent loop. Unfortunately, unlike some other programming languages, MATLAB does not enable **break** and **continue** to break out of anything other than the directly enclosing loop. We can use one of three alternatives (selection among these alternatives is mostly a matter of personal taste):

- We could use logical flags to indicate the condition to the containing loops. Within these loops, we would check the flags and skip iterations as needed:

```
bailOutFlag = false;
for outerIdx = 1 : 5
    middleIdx = 10
    while middleIdx <= 20 && ~bailOutFlag
        middleIdx = middleIdx + 1;
        for innerIdx = -100 : 100
            data = outerIdx/innerIdx + middleIdx;
            if data == SOME_VALUE
                bailOutFlag = true;
                break
            else
                process(data);
            end
        end  % for innerIdx
    end  % while middleIdx
    if bailOutFlag
        break
    end
end  % for outerIdx
```

  Although using such flags is the textbook approach to such bail-out needs, I personally dislike this method due to the extra coding complexity and the additional run-time checks within the outer loops, which affect performance.

- We could place the code segment that should be bailed-out within a dedicated function and return from the function when the bail-out condition occurs:

```
function bailableProcessing()
    for outerIdx = 1 : 5
        middleIdx = 10
        while middleIdx <= 20
            middleIdx = middleIdx + 1;
            for innerIdx = -100 : 100
                data = outerIdx/innerIdx + middleIdx;
                if data == SOME_VALUE
                    return
                else
                    process(data);
                end
            end  % for innerIdx
        end  % while middleIdx
    end  % for outerIdx
end  % bailableProcessing()
```

- We could place the segment of code that should be bailed-out within a *try-catch* block, and programmatically raise a trappable exception using *error*:[*]

---

[*] Read more on using exception handling as a performance-tuning technique in §3.4.

```
try
    for outerIdx = 1 : 5
        middleIdx = 10
        while middleIdx <= 20
            middleIdx = middleIdx + 1;
            for innerIdx = -100 : 100
                data = outerIdx/innerIdx + middleIdx;
                if data == SOME_VALUE
                    error('bail out!')
                else
                    process(data);
                end
            end  % for innerIdx
        end  % while middleIdx
    end  % for outerIdx
catch
    % ignore - simply continue normally...
end
```

As noted, I personally prefer using either the function-based or the exception-based approach, rather than the textbook flag-based approach.

When using early bail-outs, be careful to avoid a performance pitfall: sometimes the skipped code enables reusing existing calculations,* having a potential speedup that surpasses that of the bail-out. For example, an exhaustive search for best gene combination was sped up 6× by restoring code that was previously being skipped.[122]

3.1.4 Simplify Loop Contents

Conditional statements (if … end) prevent the JIT/accelerator from optimizing loops and prevent automatic instruction pipelining.[123] If possible, try to evaluate conditional expressions outside the loop, possibly skipping the loop entirely. Alternatively, modify the loop condition in a way that would remove the need for an inner conditional (helping both JIT and a manual vectorization):

```
outerIdx = 1;
while outerIdx <= 20
    outerIdx = outerIdx + 1;
    data = [];
    newData = [];
    if outerIdx <= 15
        for innerIdx = [-100:-1, 1:100]        % skip the 0 element
            data(end+1) = outerIdx/innerIdx;
            newData(end+1) = process(data);
        end  % for innerIdx
    end
end  % while outerIdx
```

Alternatively, consider evaluating the condition <u>after</u> the loop has ended, modifying the calculated data (this is not always possible and should be done with care):

```
outerIdx = 1;
while outerIdx <= 20
```

---

* Read more on data caching in §3.2.

```
        outerIdx = outerIdx + 1;
        data = [];
        newData = [];
        if outerIdx <= 15
            for innerIdx = -100 : 100        % don't skip anything here
                data(end+1) = outerIdx/innerIdx;
                newData(end+1) = process(data);
            end   % for innerIdx
            newData(isinf(data)) = [];   % remove the innerIdx==0 element
        end
end   % while outerIdx
```

If the conditional expression cannot be moved outside the loop, try replacing it with an equivalent conditional-less (branch-less) expression. For example, the expression

```
if y > SOME_VALUE
    data = sin(x);
else
    data = cos(y);
end
```

can be replaced by the equivalent statement (§4.6.3)

```
data = cos(y)*(y <= SOME_VALUE) + sin(x)*(y > SOME_VALUE);
```

If all else fails and a conditional expression is unavoidable, at least try to arrange the data and the conditional expression that depends on it such that the condition will be as simple as possible to evaluate, and will have only a single branch (i.e., without multiple *elseif* branches).[124] Moreover, the condition should either fail or succeed consistently, rather than randomly. This is very important for the CPU's internal *branch-prediction* and *speculative execution* logic.[125] The important thing here is not to make the condition always fail or always succeed, but rather to make the condition <u>consistent</u>, that is, fail a lot of times, then maybe succeed a lot of times, and so on. The longer the streak of identical condition outcomes, the more effective would be the CPU's branch prediction. On the other hand, random condition outcomes (i.e., fail and succeed with no apparent order) throw the CPU's branch prediction totally off-track, significantly degrading performance. For this reason, it may well be worthwhile to presort the data over which we are looping. Additional aspects of optimizing conditional expressions are discussed below (§3.1.4, §3.10.1).

Within the loop, try to simplify the expressions, splitting them into separate lines and using temporary variables if necessary. The benefits of simplifying loop contents, in addition to making it easier for the JIT to optimize the code, is that it makes it easier for us programmers to identify potential manual optimizations such as LIH, caching, or vectorization, or possibly a combination of these. By splitting a complex expression into simpler subexpressions, we can often apply different optimizations to each of the smaller expressions separately.

In one example, showcased by MathWorker Doug Hull in a short video tutorial,[126] an almost 2× speedup is achieved by simplifying one branch condition within a loop, at the expense of less generalized code (I simplified Doug's code even further here):

```
% Original code
value = zeros(1,1e6);
for idx = 1 : 1e6
    if round(rand)
        vec = rand([1,3]);
```

```
    elseif round(rand)
        vec = [0,0,0];
    else
        vec = [1,1,1];
    end
    value(idx) = sum(vec);
end
```
⇨ Elapsed time is **2.459548** seconds.

```
% Simplified branch code - 1.9x faster
value = zeros(1,1e6);
for idx = 1 : 1e6
    if rand >= 0.5
        value(idx) = sum(rand([1,3]));
    elseif rand < 0.5  % else value=0: unchanged from initialization
        value(idx) = 3;
    end
end
```
⇨ Elapsed time is **1.307274** seconds.

3.1.5 Unroll Simple Loops

A technique that is often mentioned in performance-tuning literature is *loop unrolling* (also called *loop unwinding*).[127] The idea is to replace a loop with repeated sets of the loop contents. This is sometimes done automatically by the JIT optimizer, but can also be done manually by the programmer. For example, the simple loop

```
for idx = 1 : 5
    a(idx) = sin(1/idx);
end
```

can be replaced with the following code:

```
a(1) = sin(1/1);
a(2) = sin(1/2);
a(3) = sin(1/3);
a(4) = sin(1/4);
a(5) = sin(1/5);
```

The potential benefit is the removal of the run-time loop checks and branches. Naturally, this technique is only feasible for very short loops and very simple contents. For such loops, the actual performance benefit is negligible. In fact, loop unrolling may even *hurt* performance on systems where instruction execution (CPU) speed is faster than instruction fetch (memory) speed.[128] On the other hand, the maintenance disadvantage of unrolling loops is clearly evident. For this reason, I suggest <u>not</u> to manually unroll loops in general. Leave this optimization to the JIT.

One case where loop unrolling should indeed be considered is when using the MATLAB Coder.* Another case is where the loop condition includes some nontrivial test whose evaluation takes nonnegligible time. In such cases, we might be tempted to repeat some loop iterations internally before reevaluating the loop condition:

```
% original code
idx = 1;
```

---

* See §8.2.4 for details. Even in this case, unrolling should be done with care, since the Coder might be able to employ loop optimizations that would not be done with unrolled code.

```
while someConditionThatTakesTimeToEvaluate()
    process(idx);
    idx = idx + 1;
end

% optimized code using 5x loop-unrolling
idx = 1;
while someConditionThatTakesTimeToEvaluate()
    process(idx);
    process(idx+1);
    process(idx+2);
    process(idx+3);
    process(idx+4);
    idx = idx + 5;
end
```

A variant of loop unrolling, called *loop peeling*,[129] takes one or more iterations that are special in some way out of the loop. This simplifies the loop's contents and prevents branching with its associated JIT and pipelining penalties. For example, let's compute Fibonacci's sequence in a naïve manner:*

```
for idx = 1 : 100
    if idx < 3
        fibonacci(idx) = 1
    else
        fibonacci(idx) = fibonacci(idx-1) + fibonacci(idx-2);
    end
end
```

and now with loop peeling of the initial two loop iterations:

```
fibonacci = [0, 1];
for idx = 3 : 100
    fibonacci(idx) = fibonacci(idx-1) + fibonacci(idx-2);
end
```

3.1.6 Optimize Nested Loops

Loop optimization is especially important in inner loops that execute numerous times within an outer loop. In fact, it would help performance even if we only move an invariant expression one loop upward, or improve performance by a small amount. The determining factor in the performance of the entire (outer) loop is the performance of its inner loop, so optimizing anything outside the innermost loop is usually a waste of time, unless the inner loop executes very few times.[130]

In the following example, we have two invariants: sin(x) is common to both loops, whereas the expression sin(x)/outerIter is common only to the inner loop:

```
function result = perfTest(x)
    result = 0;
    for outerIter = 1 : 1000
        for innerIter = 1 : 1000
            result = result + innerIter/outerIter * sin(x);
```

---

* Recall that in Fibonacci's sequence each value is the sum of its preceding two values: 1, 1, 2, 3, 5, 8, 13, 21, 34, and so on. Fibonacci sequence generation is also discussed in §3.2.4.2, §3.10.1.2, §5.7.7, and §9.4.1.

```
            end
        end
end    % perfTest
```

The improved version of this function calculates `sin(x)` only once (rather than a million times), since it does not rely on either of the loop iterations. Similarly, since `sin(x)/outer-Iter` does not rely on `innerIter`, it can safely be taken out of the inner loop and ran only 1000 times (rather than a million times):

```
function result = perfTest(x)
    result = 0;
    sin_x = sin(x);                 % LIH-moved outside both loops
    for outerIter = 1 : 1000
        temp = sin_x / outerIter;   % LIH-moved from within inner loop
        for innerIter = 1 : 1000
            result = result + innerIter*temp;
        end
    end
end   % perfTest
```

In this specific case, the function can be further optimized using vectorization (see Chapter 5). Optimizing loops often results in simpler loops that are easier to vectorize than the original loops. For this reason, LIH in MATLAB is often just a step toward vectorization, which results in removing loops altogether. But even when vectorization is not feasible, LIH often gives a significant performance boost.

Here is a real-life example. The original code was[131]

```
n1(1:lonc,1:latc)=0;
n2(1:lonc,1:latc)=0;
nn = length(lat2);
for j = 1:nn
   for k = 1:latc
      for l = 1:lonc
          latrad1 = k*step*pi/180;
          lonrad1 = l*step*pi/180;
          latrad2 = lat2(j)*pi/180;
          lonrad2 = lon2(j)*pi/180;
          londif  = abs(lonrad2-lonrad1);
          arclen1 = 6371*acos(sin(latrad2)*sin(latrad1) +
                          cos(latrad2)*cos(latrad1)*cos(londif));
          if (arclen1 < step*111)
             n(l,k)  = n(l,k) + 1;
             n1(l,k) = n1(l,k) + p(j);
             n2(l,k) = n2(l,k) + (sqrt(((s(j))*1000000)/3));
          end
      end
   end
end
```

This code can be optimized utilizing a combination of vectorization and LIH:

```
n1 = zeros(lonc,latc);
n2 = n1;
n  = n1;
nn = length(lat2);
```

```
d2rad = pi/180;
lat2 = lat2*d2rad;
lon2 = lon2*d2rad; % variables in calling routine so is safe
lat1 = step*d2rad.*[1:latc]; % compute the full vector once
lon1 = step*d2rad.*[1:lonc];
T1 = sin(lat2).*sin(lat1);
T2 = cos(lat2).*cos(lat1);
S = sqrt(s*1000000/3);
for j = 1:nn
    for k = 1:latc
        for l = 1:lonc
            londif = abs(lon2(j)-lon1(l)); % This makes it tougher...
            arclen1 = 6371*acos(T1(k)+T2(k)*cos(londif));
            if (arclen1 < step*111)
                n(l,k)  = n(l,k) + 1;
                n1(l,k) = n1(l,k) + p(j);
                n2(l,k) = n2(l,k) + S(j);
            end
        end
    end
end
```

Another usage example of this technique was recently discussed on the MATLAB Answers forum.[132] Also see §5.7.10.

3.1.7 Switch the Order of Nested Loops

Switching the order of nested loops (also known as *loop flipping, loop swapping*, or *loop interchange*)[133] is another corollary of the effect of memory on performance. In this case, the manner in which data is stored in memory means that it is more efficient to loop down columns as the external loop, rather than the reverse. The technical details are explained elsewhere (§9.3.2); here is a simple usage example:

```
data = rand(5000,5000);    % 25M elements, 200MB memory

% Row-first loop (natural order, bad for performance)
total = 0;
for rowIdx = 1:size(data,1)
    for colIdx = 1:size(data,2)
        total = total + data(rowIdx,colIdx);
    end
end
⇨ Elapsed time is 0.556565 seconds.

% Column-first loop (less natural, but much better performance)
total = 0;
for colIdx = 1:size(data,2)
    for rowIdx = 1:size(data,1)
        total = total + data(rowIdx,colIdx);
    end
end
⇨ Elapsed time is 0.320137 seconds.
```

The impact of loop order optimization is often surprising, and can sometimes be greater than other optimizations (e.g., branch prediction mentioned in §3.1.4 above[134]):

```
% Without loop inversion
function total = perfTest
    data = round(rand(1,1e6)*256);   % A million elements
    total = 0;
    for idx1 = 1 : 100
        for idx2 = 1 : length(data)
            if data(idx2) >= 128
                total = total + idx1 + data(idx2);
            end
        end
    end
end  % perfTest
⇨ Elapsed time is 6.710993 seconds.

% With loop inversion: 4x speedup
function total = perfTest
    data = round(rand(1,1e6)*256);   % A million elements
    total = 0;
    for idx2 = 1 : length(data)
        if data(idx2) >= 128
            for idx1 = 1 : 100
                total = total + idx1 + data(idx2);
            end
        end
    end
end  % perfTest
⇨ Elapsed time is 1.668971 seconds.
```

The importance of loop inversion is that quite often, after switching the loop order, it becomes much easier to apply other optimizations such as LIH or vectorization. In the specific case above:

```
% With loop inversion and LIH - 100x speedup
function total = perfTest
    data = round(rand(1,1e6)*256);   % A million elements
    total = 0;
    temp = sum(1:100);    % LIH optimization
    for idx2 = 1 : length(data)
        if data(idx2) >= 128
            % the entire inner loop is removed by LIH optimization
            total = total + temp + 100*data(idx2);
        end
    end
end  % perfTest
⇨ Elapsed time is 0.066588 seconds.
```

Switching loop order does not always improve performance — sometimes the reverse is true. Remember that other factors besides the memory allocation order affect performance. For example, switching loop order may mean that a conditional expression is now evaluated in the inner loop rather than the outer loop, which could have a devastating impact on performance that would far outweigh any benefit gained from the improved memory access.

3.1.8 Minimize Dereferencing

Accessing object properties/methods, array elements, and structure fields incurs a runtime processing overhead (including boundary and security checks). Depending on the frequency of the usage, the loop size and the average iteration run-time, this overhead can be a significant portion of the loop's total run-time.

The simple act of caching such indirect data accesses in a simple variable negates this performance penalty and can often bring about important speedups:*

```
% Simple loop accessing an array element
a = [0,0];
for idx = 1 : 1e8
    a(2) = a(2) + idx;
end
⇨ Elapsed time is 0.993558 seconds.

% The same loop using a temporary simple accumulator - 2x faster
a = [0,0];
b = 0;
for idx = 1 : 1e8
    b = b + idx;   % b instead of a(2)
end
a(2) = b;
⇨ Elapsed time is 0.466524 seconds.
```

Simplifying dereferencing applies to any array type, including cell arrays such as c{1} {2,3}(4). Similarly, when accessing structure, table or class object fields:[135]

```
% Example of updating a struct field
for idx = 1 : 1e6
    a(3).b(2,3).c(4).d = rand(1);
end
⇨ Elapsed time is 0.809273 seconds.

for idx = 1 : 1e6
    val = rand(1);   % val instead of a(3).b(2,3).c(4).d
end
a(3).b(2,3).c(4).d = val;
⇨ Elapsed time is 0.617945 seconds.        ← 25% faster

% Example of retrieving a struct field
for idx = 1 : 1e6
    val = a(3).b(2,3).c(4).d * rand(1);
end
⇨ Elapsed time is 12.035308 seconds.

v = a(3).b(2,3).c(4).d;
for idx = 1 : 1e6
    val = v*rand(1);   % v instead of a(3).b(2,3).c(4).d
end
⇨ Elapsed time is 0.700558 seconds.        ← 17x faster
```

---

* Caching is discussed in detail in §3.2.

Similarly, avoid complex/nested indexing such as data(dependencies(inner(2,3),idx)). It is often possible to hoist the inner index out, replacing it with a simple variable:[136]

```
innerIdx = dependencies(inner(2,3),:);
for idx = 1 : maxVal
    newData = data(innerIdx(idx));
    ...
end
```

In addition to the removed data access run-time checks, the simplified usage of a simple variable may help JIT automatically employ additional optimizations such as using registers or the stack, rather than the heap. However, I do not know whether the JIT actually uses this specific optimization.

3.1.9 Postpone I/O and Graphics until the Loop Ends*

One of the most disastrous performance pitfalls that I often see is code where I/O is performed within a loop. Many people are not aware that I/O is thousands of times slower than CPU. By forcing I/O within a loop, we actively force the loop to slow down. If the I/O is crucial to the computation within the loop, we may have little choice. But quite often the output is in the form of outputting computation results to some data file or log. Such outputs can usually be deferred until after the loop ends.

It might seem at first glance that since the same data is output to disk, no speedup should be expected. However, I/O works in such a way that a single large update is much more efficient than multiple small ones. This is especially important when updating an Excel worksheet using *xlswrite*, since every time that *xlswrite* is invoked it launches an Excel process, loads the relevant workbook, updates it, and then closes the workbook and Excel. Doing this in a loop is disastrous to performance:

```
data = magic(3);   % 3x3 magic square
% Performing I/O inside the loop - bad for performance
for col = 1 : size(data,2)
    for row = 1 : size(data,1)
        cellAddr = ['A'+col-1 '1'+row-1];
        range = [cellAddr ':' cellAddr];
        xlswrite('test.xls',data(row,col),range);
    end
end
⇨ Elapsed time is 40.801892 seconds.

% Performing the I/O outside the loop - 8x faster
lastCellAddr = ['A'+size(data,1)-1 '1'+size(data,2)-1];
range = ['A1:' lastCellAddr];
xlswrite('test.xls',data,range);
⇨ Elapsed time is 5.049598 seconds.
```

3.1.10 Merge or Split Loops

Loop merger, also called *loop fusion*,[137] identifies that two or more loops run over the same iteration range and yet are independent of each other. For example:

```
for idx = 1 : 100
    a(idx) = process1(idx);
```

---

* See Chapter 11 for additional details on I/O performance tuning in MATLAB.

```
end
for idx = 1 : 100
    b(idx) = process2(idx);
end
```

We can sometimes combine such independent loops into a single, faster, unified loop, thereby reducing loop overhead and enabling pipelining:[138]

```
for idx = 1 : 100
    a(idx) = process1(idx);
    b(idx) = process2(idx);
end
```

Unfortunately, merging loops sometimes *degrades* the performance, when the merged code prevents some JIT optimization, internal (automated) vectorization, or efficient memory-cache utilization. For this reason, we should carefully profile the code before and after the loop fusion to discover whether this specific technique is beneficial or detrimental in each specific case.

The opposite technique, *loop distribution* or *loop fission*,[139] can be attempted for the exact mirror image of the above. In this case, we try to split a single loop into two or more independent loops, in the hope that the new simpler loops will be more efficient in terms of JIT optimization, internal (automated) vectorization, and efficient memory-cache utilization. In some cases, we can apply separate (different) optimizations to the split loops, due to their different contents.

As above, using loop fission is a matter of trial and error based on the specific application, data set, and environmental setup. For example, certain data size or type may reduce CPU cache misses using one method or the other.[140]

3.1.11 Loop Over the Shorter Dimension

When looping over data having two or more dimensions (e.g., a matrix or tensor), it is often beneficial to loop over the shorter dimension when we have a choice. The idea is to reduce the overheads associated with the loop iteration and with any internal function calls, as well as improve CPU caching.[141] Within the loop, we would optimally vectorize the computation. The result would be fewer invocations of the vectorized computation, typically resulting in faster code.

The following example illustrates this idea. The example is contrived and unrealistic: after all, the entire operation could be vectorized using a single *sum()* function. Nevertheless, it clearly shows how running fewer loops can be significantly better:

```
data = rand(100,10000);

% Loop over the longer dimension (10K loops)
total=0; for col = 1:size(data,2); total = total + sum(a(:,col)); end
⇨ Elapsed time is 0.024386 seconds.

% Loop over the shorter dimension (100 loops) - 2x faster
total=0; for row = 1:size(data,1); total = total + sum(a(row,:)); end
⇨ Elapsed time is 0.011106 seconds.
```

Another usage example was provided by MathWorker Steve Lord, of looping over the smaller set of strings when comparing two sets of strings.[142] This idea of looping over the shorter dimension can be extended to **fun* functions (§5.2.4). For example:

```
results = cell2mat(cellfun(@(c)strcmp(c,longer),shorter,'uniform',0)');
```

Looping over the shorter dimension in this manner is not always practical in terms of the algorithm. Our first priority must always be given to correct and maintainable code, rather than to more performant code. Still, all other things being equal and if the algorithm allows it, then looping over the shorter dimension may be worthwhile.

A related (and sometimes conflicting) technique is to loop over a rightward dimension before a leftward dimension, when nested loops must be used. In the common case of 2D data matrices, this means looping down columns before looping down rows, regardless of which of them is the shorter dimension.* By careful preparation of our data and algorithm, we may be able to benefit from both of these related techniques.

3.1.12 Run Loops Backwards

As shall be explained in detail below (§9.4.3), running loops backwards (*loop reversal*) is the simplest and fastest way to preallocate data arrays and to prevent repeated dynamic reallocations that significantly degrade performance:†

```
for colIdx = 3000 : -1 : 1
    for rowIdx = 1000 : -1 : 1
        data3(rowIdx,colIdx) = someValue;
    end
end
```

3.1.13 Partially Optimize a Loop

In some cases, a loop repeats the exact same calculation in a large portion of the loop iterations. In such cases, it may be worthwhile to employ a special treatment for these blocks of calculations that may be more efficient than the regular loop calculation.[143]

As a very simplistic example, consider the following loop:

```
for idx = 1 : N
    doSomeCalculationWith(idx);
end
```

It may be possible to improve performance of this loop (at the possible cost of turning off JIT due to the additional conditional) by using a variant of the algorithm that is faster for even-numbered values:

```
for idx = 1 : N
    if mod(idx,2)==0
        doSomethingQuickWith(idx);    % even-numbered values
    else
        doSomeCalculationWith(idx);   % odd-numbered values
    end
end
```

In fact, we might even discover that we can vectorize the doSomethingQuickWith values, thereby reducing the loop size:

```
doSomethingQuickWith(2:2:N);    % even-numbered values, vectorized
for idx = 1 : 2 : N
    doSomeCalculationWith(idx);  % odd-numbered values
end
```

---

* See §3.1.7 and §9.3.2 for a detailed analysis of this technique.
† For a detailed discussion of memory allocations and preallocation variants, see §9.4.

3.1.14 Use the Loop Index Rather than Counters

During code development, code segments are often moved around. As a result, counters, remnants of a previous implementation, are often used within loops. The problem is that by not using the loop index, we unnecessarily make the code more complex and harder to maintain. Importantly, performance may also suffer:

```
data = rand(1,1e6);

% Original code: uses counter rather than the loop index
k = 0;  total = 0;
for idx = 1 : numel(data)
    k = k + 1;
    total = total + data(k);
end
⇨ Elapsed time is 0.015243 seconds.

% Optimized loop: uses only the loop index, faster and simpler
total = 0;
for idx = 1 : numel(data)
    total = total + data(idx);
end
⇨ Elapsed time is 0.012843 seconds.
```

3.1.15 MATLAB's JIT

MATLAB includes a built-in JIT (*just-in-time*) compiler.[144] The purpose of JIT is to analyze invoked MATLAB m-code in run-time, and cache the interpretation results (i.e., low-level binary machine code) in memory. Subsequent reruns of the same code will be significantly faster (up to 100× in some cases), since the machine code can be run directly, without requiring any interpretation. For this reason, JIT usually comes into play in repetitive loops, but not in code segments that only run once.[145]

When MATLAB's JIT was first introduced in MATLAB R13 (6.5), it only sped up some calculations in loops that obeyed certain characteristics, for example, scalar operations on double values, within m-file functions.[146] Subsequent MATLAB releases continuously improved JIT, adding support for numerous MATLAB constructs that are typically found within loops. Most MATLAB loops used in practice are now JIT-ted when using one of the latest MATLAB releases.*

Some JIT support for statements in script files and the command prompt has also been added over the years, although it is still far from being as effective as for m-file function code,[147] especially for console (command-line) statements.† The main reason for the discrepancy appears to be that MATLAB's functions provide a more controlled variable workspace environment, enabling JIT to use more assumptions on the nature of the code and data. In any case, since JIT is most effective in functions, it is advisable to run performance checks within functions rather than the command line.‡

JIT incurs a small overhead in the first few loop iterations, in order to assert the consistency of the compilation results (e.g., assumptions about variable type and dimensionality).

---

* See §9.4.2 for a discussion of how JIT's continuous improvements have affected performance; also see related §11.9.1.
† See §4.6.1.1 for a discussion of how recent JIT improvements have made script performance much faster than previously. Some statements run faster in the console than within a function, for some unknown reason. For example, for t = 1:1e6, x = 3; end
‡ See §2.1.6.2.

If the assumptions are ever broken, MATLAB falls back to using the interpreted version of code, but then the loop runs significantly slower (see §4.1.4). We can place a *tic-toc* pair (§2.2.1) within the loop to see JIT's effect on iteration run-time. We will see that iterations #10+ run much faster than iteration #1.

On a relatively modern MATLAB release, many simple loops now run at speeds comparable with vectorization (Chapter 5). Vectorization still has an edge, since it employs multithreading on multicore CPUs, as well as CPU-level SIMD accelerations. But when these are not a factor, JIT-ted loops can be as fast as the best vectorized code.[148]

In some cases, MATLAB's JIT compiler is unable to optimize looping constructs due to the presence of special functions inside a loop:[149]

```
a = rand(3,3000);
b = rand(3,3000);
sz_a = size(a);
sz_b = size(b);
C = zeros(sz_b(2), sz_a(2));
for i = 1:sz_b(2)
    for j = 1:sz_a(2)
        C(i,j) = norm(a(:,j) - b(:,i));
    end
end
```

This version takes **10.48** s to run. Let's inline *norm* using simple arithmetic:*

```
for i = 1:sz_b(2)
    for j = 1:sz_a(2)
        C(i,j) = sqrt((a(1,j) - b(1,i)).^2 + ...
                      (a(2,j) - b(2,i)).^2 + ...
                      (a(3,j) - b(3,i)).^2);
    end
end
```

This runs in **0.53** seconds! Apparently, JIT was unable to handle **norm** inside the loop. So another performance tip is: try to replace special MATLAB functions with straight arithmetic that gives more freedom to MATLAB's JIT to optimize the code.

Note that we can eliminate loops altogether here, using fully vectorized *bsxfun*:[†150]

```
c1 = bsxfun(), b(1,:)');
c2 = bsxfun(), b(2,:)');
c3 = bsxfun(), b(3,:)');
C = sqrt(c1.^2 + c2.^2 + c3.^2);
```

 In some cases, code that is un-JIT-able may cause useless extra work by JIT, since it would need to fall back into interpreted mode in any case. Such cases are indeed rare, but in those cases it is useful to turn off JIT.[151] In one specific case that I encountered, a MATLAB program consistently ran for 12 s in MATLAB's default mode, and 7.8 s when JIT was turned off. We can turn JIT off by running MATLAB with the -nojit startup parameter[152] or by issuing the following MATLAB command:

```
feature jit off
```

---

* Inlining is discussed in §3.1.2.
† *bsxfun* is discussed in §5.2.4.2 and §5.3. This vectorized version executed in 0.12 seconds on the same platform/ release.

Additional (non-JIT) acceleration optimizations, which are used by the MATLAB interpreter (e.g., automated use of CPU multicores), can be turned off using the following MATLAB command:

```
feature accel off
```

Although this is undocumented, a utility by then-MathWorker Bill McKeeman[153] implies that JIT optimizations are only a subset of the accelerations controlled by *feature('accel')*, so turning off accel implies turning off JIT (but not vice versa).*[154]

In addition, there are a series of specific JIT features that can be turned on or off using the jitallow feature. I strongly suggest not using this under any circumstance, except when specifically instructed to do so by MathWorks in order to work around some bug or to diagnose a problem that you have reported to them.

Note an open bug on SELinux systems[155] that prevents JIT from working unless we turn on the *allow_execheap* flag[156] for the entire operating system (OS):

```
setsebool -P allow_execheap = 1
```

JIT is extremely sensitive: minor changes to the code could disable some JIT optimizations and potentially even JIT itself.[157] Extra performance can sometimes be squeezed by hand-crafting the code to fully utilize JIT, but this technique (sometimes called *"coding against the JIT"*) could backfire if the program is used on another MATLAB version or platform, where the modification can actually hurt performance. Moreover, such code is often less maintainable and readable than "natural-flow" code, although in some cases it can actually help by becoming a step toward more efficient vectorization, as in the example above. Still, with all the drawbacks of hand-crafted JIT'ing, it is not surprising that MathWorks recommends against this practice.[158] I urge readers to use such techniques only when other optimization techniques are not readily available.

Still, there are certain classes of actions that seem to be widely applicable for improving loop performance via JIT. In general, the simpler the loop, the more likely we are to have it JIT-ted, and to have the JIT results employ the maximal number of internal optimizations. Try to minimize the following within loops:

- Conditional statements (*if, switch*)
- Statements that modify variable types (see §4.1.4)
- Dynamic evaluations (*eval* and its ilk)
- Time-dependent or randomized data that is evaluated on-the-fly[159]
- Using variable names that override built-in function names (e.g., *sum* or *min*)
- Invoking complex functions

MATLAB's JIT is constantly evolving. Optimizations that are perfect on one release might become less effective on the following release, and vice versa. Therefore, it is best to retune the application performance whenever we change MATLAB releases.

It should be noted that third-party JIT engines can also be used with MATLAB. One such implementation, *MaJIC*, was developed at the University of Illinois at Urbana-Champaign in 2000 and apparently predates MATLAB's JIT.[160] Another project, *McVM*, was developed in 2010 at McGill University and showed results comparable and in some cases superior, to MathWorks' JIT implementation.[161] MATLAB's JIT has no known connection with any of these projects and was presumably developed independently. Still, understanding the

---

* See the related footnote at the beginning of §9.4.2.

mechanisms that underlie *MaJIC* and *McVM* could provide insight into how MATLAB's JIT might also possibly work. Unfortunately, MATLAB's JIT is undocumented and so we have no better resource.

3.2 Data Caching

Caching[162] is the process of storing data that is expected to be reused multiple times, in a temporary repository (memory variable, disk file, and so on), in order to increase data access speed. The data is stored in the repository when it is first used; subsequent usages read the data from the repository rather than from the original location. A smart choice of cache repository is such that the multiple subsequent fetches are much faster than reloading or recomputing the data.

Whenever we have information that a naïve approach would recalculate or reload multiple times, it makes sense to utilize caching. There are many possible ways to implement caching, and all of them are preferable to the naïve noncached version.

Caching is arguably one of the simplest and yet probably the single most effective performance optimization technique. If there is one single optimization technique you can and should start using right away, it is caching. In MATLAB, caching benefits often surpass any other performance-tuning technique. Using both caching and the other techniques is of course even better.

Caching often increases code complexity, and makes the code slightly less maintainable. Another drawback is the increased memory usage as the program runs and consistently caches data. This can be alleviated by keeping track of the cached data size and erasing old (*stale*) cache elements when some limit is reached or after some predefined amount of time has passed. Caching is a typical example of the space–time trade-off in performance tuning (Chapters 9 and 11 provide additional examples).[163]

Yet another drawback is that caching is only effective when the *locality principle*[164] is observed by the data; in other words when a large segment of the data is repeatedly used often enough. There are basically two types of locality: *spatial locality* means that when we access data element X, we are also likely to access element $X \pm dX$, where dX is small; *temporal locality* means that when we access element X at time t, we are also likely to reaccess the same element X at time t + dt. The spatial locality principle is served by loading data in blocks (*pages*), even when only a tiny fraction of the block is requested. Temporal locality is served by the *MRU* (most recently used) principle: we purge the oldest accessed blocks from the cache, when we need to make room for new data blocks. When both locality types are not observed in a particular application, caching merely wastes memory and CPU resources.

We should be aware of the so-called *"fallacy of caching"*: when developing, debugging, and testing, we typically reuse the same data set, so caching brings enormous speedups. However, in the deployed system it is possible that none of the cached data elements is ever reused. In such a case, caching may actually *degrade* performance, due to the additional processing and memory overheads. When deciding whether or not to use caching, as well as during the performance-tuning cycle, we should attempt to mimic the target run-time environment as closely as possible in our system.

Another caching-related fallacy is the somewhat famous maxim[165] that

"Every problem in Computer Science can be solved by adding another level of indirection"

In fact, using caching as a brute-force performance-optimization technique, while often quite effective, is also quite often employed by engineers who lack the know-how, time, or will to refactor badly designed code.[166] The resulting code just becomes more complicated and less maintainable. Even when caching is effective, it may perhaps have been outperformed by optimizing the underlying code.

3.2.1 Read-Only Caches

LIH (§3.1.1) typically uses caching. In the following example, loadData() is repeatedly called to recalculate newData by loading data from file:

```
for iter = 1 : 100
    newData = loadData();
    result(iter) = max(max(newData)) + rand(1);
end
```

Instead, loadData() could be called just once into a cache variable (let's call it newData), which would then be reused in all the loop iterations. The same is true for all direct derivatives of the cached (hoisted) data, in this case, max(max(newData)):*

```
newData = loadData();
newData = max(max(newData));
for iter = 1 : 100
    result(iter) = newData + rand(1);
end
```

A variant of this idea: when we have an array that is repeatedly re-evaluated, do not recompute the entire array in each iteration; instead, only update the new data elements. In this sense, the array becomes a cached version of itself. For example, if we display a map and then pan eastward, we just compute the new map tiles in the east and add them to the existing map, rather than recompute the entire map.

As yet another variant, consider the common use-case of iteratively processing unique values of an array. A common expression of this idea is the following loop:

```
uniqueVals = unique(data);
for idx = 1 : length(uniqueVals)
    % Search for the current value within data
    dataIdx = (data == uniqueVals(idx));  % logical array

    % now do something useful with dataIdx
    b = someFunctionOf(dataIdx);
end
```

Instead, we could use *unique*'s second output argument to cache the unique values' index positions for later reuse within the loop, saving much processing time:[167]

```
[sortedData, sortedDataIdx] = sort(data);
[uniqueVals, sortedStartIdx] = unique(data);
sortedEndIdx = [sortedStartIdx(2:end)'-1, length(data)];
for idx = 1 : length(uniqueVals)
    % Direct access to cached data indexes - no need to search
    dataIdx = sortedDataIdx(sortedStartIdx:sortedEndIdx);
```

---

* In this simple example, the loop could easily be removed altogether by simple vectorization, but my aim here was merely to illustrate caching.

```
    % now do something useful with dataIdx
    b = someFunctionOf(dataIdx);
end
```

Another example of read-only cache is precalculated constants. These are values that remain constant throughout the program, although they might be different across different invocations of the program. A few simple examples of this are

```
time = 5/(24*60);  % =0.0034722 (meaning 5 minutes in datenum units)
folder = ['abc\', computer('arch'), '\def'];  % ='abc\win64\def'
```

3.2.2 Common Subexpression Elimination

Common subexpression elimination,[168] or CSE, is an optimization technique that caches identical (common) subexpression results in a variable. This is a variant of the read-only cache mechanism.

I often see common subexpressions in MATLAB code. This usually does not affect performance very much if the expression takes only a short time to be re-evaluated. But sometimes the common expression is costly, for example, reloading data from file, or recomputing a complex minimization problem. In such cases, it makes sense to cache the common expression result in a temporary variable that is later reused.

For example, in the following code, there is absolutely no need to call complicatedAlgo() multiple times:

```
% Non-optimized version
data1 = a*sin(b)/complicatedAlgo(c) + 2.3;
data2 = a*sin(b)/complicatedAlgo(c) + 5.8/c;
data3 = a*sin(b)/complicatedAlgo(c) - 8.1*b^2;

% Optimized version
temp = a*sin(b)/complicatedAlgo(c);
data1 = temp + 2.3;
data2 = temp + 5.8/c;
data3 = temp - 8.1*b^2;
```

Another case where CSE comes into play is in GUI callbacks (particularly mouse movement callbacks that fire very rapidly), where there is often a need to refer to common expressions such as get(gca,'XLimit'). In such cases, we simply store the property value in a variable (e.g., xlim) and then use this variable whenever we need to access the unmodified property value in the callback function.

3.2.3 Persistent Caches

3.2.3.1 In-Memory Persistence

Caching data in variables works well within the function. But what if the caching needs to work across multiple function invocations? It is not always possible, practical, or even wise, to inline a function's contents in the calling function, to enable regular variable caching. For example, we might want to cache information in a callback function of a timer or a graphics object.

In such cases, we could use one of several mechanisms to persist our cached data:*

---

* See §9.5.5 for a comparison of these mechanisms.

- Use the *persistent* command. *persistent* enables access to the data within the same function in which data is declared as *persistent*. *persistent* variables are initialized to [] (the empty array), making it easy to differentiate between pre-cached and post-cached data. Here is a simple usage example:

```
function myCallbackFunc(hObject,eventData)
    persistent data
    if isempty(data)
        data = loadData();  % cache the data for later reuse
    end
    % ...now do something useful with the data...
end  % myCallbackFunc
```

Clearing such caches can easily be done, even from outside the function:

```
clear myCallbackFunc % clear all persistent vars in this function
```

- Use the *global* command, which works exactly the same way as *persistent*, except that it enables access to the data anywhere in MATLAB, even across function boundaries. Like *persistent*, *global* variables are also initialized to the empty array []. Here is the equivalent example:

```
function myCallbackFunc(hObject,eventData)
    global data
    if isempty(data)
        data = loadData();  % cache the data for later reuse
    end
    % ...now do something useful with the data...
end  % myCallbackFunc
global data; data=[];  % clear the global data, anywhere in MATLAB
```

- Attach the data to a persistent object such as a figure handle, or the MATLAB desktop handle (0). We can do this by setting their **UserData** property, or using the built-in functions *getappdata/setappdata*,* or using a few other GUI-based alternatives.[169] For example:

```
% Using the UserData property
set(gcf,'UserData',cachedData);          % store cache data
cachedData = get(gcf,'UserData');        % load cache data

% Using getappdata/setappdata
setappdata(0,'CachedData',cachedData);   % store cache data
cachedData = getappdata(0,'CachedData'); % load cache data
```

3.2.3.2 Non-Memory Persistence

It is not always possible to cache the entire data set in memory. In some cases, the data is simply too big.† In other cases, we may wish to persist a memory cache for use by later MATLAB sessions or another invocation of a deployed (compiled) program.

---

* Under the hood, *getappdata* and *setappdata* simply update a structure of user-defined fields that is placed in the object's hidden **ApplicationData** property.

† One workaround for this is data packing, enabling more elements (keys) to fit in the cache. The extra processing required for packing/unpacking the data might well be worth the cost of not caching some important data elements. See Jon Bentley at http://www.new-npac.org/projects/cdroms/cewes-1999-06-vol1/nhse/hpccsurvey/orgs/sgi/bentley.html#timespace1 (or: http://bit.ly/1gyN5iO).

In such cases, we can cache the data in disk or database. This idea might sound strange at first. After all, isn't I/O so slow compared to CPU and memory that it should be avoided? But in fact, persisting cached data to disk or database is the only way to ensure that it could be reused in subsequent MATLAB sessions, or across different invocations of a deployed MATLAB application.

For performance reasons, it is better to use binary files on a local disk.* Loading a pre-parsed packed (possibly compressed) binary cache from file can be an order of magnitude faster than loading and parsing the master data. The master data might be on a remote server, span multiple data files in wasteful text format, and contain data that is irrelevant to our program.

The basic idea is to load the cached data from disk or database at the beginning of the program, and then save the loaded data in a memory-based variable (*persistent*, *global*, **UserData**, **ApplicationData**, or any other similar mechanism). Loading the initial cache from disk is still often faster than recomputing it from scratch. Of course, when this is not the case, it makes more sense to actually recreate the cache from scratch each time.

Unfortunately, using cached data files is especially tricky in terms of synchronization with the master data file(s).[170] Care must be taken to update the cache file(s) whenever the master data is updated, and vice versa. Doing this synchronization correctly is more difficult with I/O than with memory (see discussion in §3.2.4).

All MATLAB processing passes through MATLAB's single main thread, theoretically making cache synchronization easier. However, MATLAB callbacks (as from timers and GUI events) may interrupt one another, and so cache processing atomicity is not assured. The end result is that synchronization is especially tricky when different asynchronous processing threads may access the cache concurrently.

We should periodically clean cache files, otherwise they will keep getting larger and program execution speed will decrease over time. A related issue is a misperception by some users that programs start from scratch when MATLAB restarts; in fact, previously-saved data/output files may affect the execution speed of subsequent runs.

3.2.4 Writable Caches

In many real-life uses of caching, we need to update the cache in real-time, and cannot rely on a read-only cache described in preceding sections. Writable caches are naturally more difficult to set up than read-only caches. We need to check whether the requested key exists in the cache and retrieve the cached value if so (a *cache hit*), or update the cache with the uncached value if not (a *cache miss*). In some cases, we may wish to update an existing cached value with a more recent value (*cache update*). Implementing this logic correctly can be tricky, especially in cases where the cache can be accessed asynchronously and/or from multiple different program locations.

In a writable cache, when the cache becomes too large to fit into memory, it may be required to remove some data items from the cache. There are many algorithms that help to decide which data items to discard: oldest, newest, least frequently used, least recently used, most recently used, randomly, and so on.[171]

---

* See the discussion in §11.3.1 (binary vs. text files) and §11.9.2 (local vs. remote I/O).

3.2.4.1 Initializing Cache Data

When initializing a writable cache, we have two main options:

- Start with an empty cache and let the cache get populated during normal usage. This causes the so-called *cache warm-up* effect, whereby it takes some time for the cache to get filled up to such a degree that it becomes effective for performance.
- Prepopulate the cache with some initial values in order to speed up the cache's effectiveness, at the expense of somewhat longer set-up time. This was the option chosen for *datestr2* in the example below (§3.2.5). In that case, the *datestr2* cache is prepopulated with all the dates since January 1, 2000 at the expense of a 222 ms initial set-up time.

We have full control over cache initialization, so we can decide to scale down the prepopulation in order to achieve faster set-up time. The control is entirely ours. In the Fibonacci memoization example below (§3.2.4.2), a choice has been made to prepopulate only the first two sequence values, and have the rest be updated in run-time.

3.2.4.2 Memoization

A variation of writable caches is *memoization*,*[172] which caches the results of function calls based on their input arguments.[†] This enables reuse of results when the function would otherwise have been re-evaluated with the same set of input parameters. Memoization is preferable to straightforward caching within a function, since it avoids function call overheads. It is used internally by some MATLAB functions.[173]

Here is a simple example of memoization that calculates Fibonacci sequence values:[‡]

```
function value = fibonacci(n)
    persistent cachedData
    if isempty(cachedData)
        cachedData(1) = 1;
        cachedData(2) = 1;
    end
    try
        value = cachedData(n);
    catch
        %value = fibonacci(n-2) + fibonacci(n-1);
        if length(cachedData) < n-1
            % populate the cache for n-1,n-2; discard result
            fibonacci(n-1);
        end
        value = cachedData(n-2) + cachedData(n-1);
        cachedData(n) = value;  % update the cache
    end
end  % fibonacci
```

---

* The term *"memoization"* should not be confused with *"memorization"*.
† Lazy evaluation, discussed in §3.8.1 below, shares similar concepts with memoization.
‡ For didactic purposes, this example lacks the sanity checks that would be required in real-life applications: ensuring that n is specified, that n is a positive integer, and so on. Fibonacci sequence generation is also discussed in §3.1.5, §3.10.1.2, §5.7.7, and §9.4.1

Traditional recursion-based Fibonacci implementations usually call two functions in each recursion iteration.* In contrast, this memoized implementation calls either a single function or none, relying instead on cached data. Standard recursion would lead to a recomputation of many sequence elements, whereas efficient memoization never computes any element more than once.[174]

As an alternative to storing memoized data in persistent variables, we could use nested functions.†[175] As long as we keep the handle of the nested function in our workspace scope its inner variables will be kept like a cache. Modifying our example:

```
function fh = fibonacci_outer
    fib = [1, 1, 2];  % first few elements
    fh = @fibonacci_nested;  % handle for nested function

    function value = fibonacci_nested(n)
        try
            value = fib(n);
        catch
            if length(fib) < n-1
                % populate the cache for n-1,n-2; discard result
                fibonacci_nested(n-1);
            end
            value = fib(n-2) + fib(n-1);
            fib(n) = value;  % update the cache
        end
    end  % fibonacci_nested
end  % fibonacci_outer
```

The usage would then be as follows:

```
>> fh = fibonacci_outer
fh =
    @fibonacci_outer/fibonacci_nested

% Let's see the function's internal cache
>> ff = functions(fh); ff.workspace{1}
ans =
    fh: @fibonacci_outer/fibonacci_nested
    fib: [1 1 2]

% Get the Fibonacci value of 50
>> fh(50)
ans =
        12586269025

% Let's see the internal cache again: now includes 50 sub-results
>> ff = functions(fh); ff.workspace{1}
ans =
    fh: @fibonacci_outer/fibonacci_nested
    fib: [1x50 double]

% Get the Fibonacci value of 50 again: reuses the cache (super-quick)
>> fh(50)
ans =
        12586269025
```

---

* See §3.1.5 for example.
† See §9.5.6 for a discussion of nested functions.

A variation of this idea was provided John D'Errico in his tutorial on Fibonacci sequence computational performance.[176] John's code is less efficient for repeated calls (since it does not cache results), but is undoubtedly more elegant and succinct:*

```
function [value,valuePrev1] = fibonacci(n)
    if n <= 2
        value = 1;
        valuePrev1 = 1;
    else
        [valuePrev1, valuePrev2] = fibonacci(n-1);
        value = valuePrev1 + valuePrev2;
    end
end
```

More complex memoization is typically implemented using a *look-up table* (LUT), where the input arguments populate most of the table's columns, except the last column(s), which hold(s) the results.[177] Separate function calls having separate input arguments are cached in separate rows of the LUT. For values that do not exist in the precalculated LUT, interpolation can be used to extract the result. LUTs are probably the best-known subtype of programmatic caching.

A utility called *Micro-cache* by Christian Kothe on the MATLAB File Exchange[178] provides a well-documented framework for memoization in MATLAB code, which can easily be adapted for most needs.

3.2.4.3 Multilayered (Offline) Cache

When persisting cache data to disk in a multilayered caches (i.e., caches that exist in memory with a mirror image offline, on disk or database), it makes sense to minimize the I/O cost by storing the data in compressed binary format such as MATLAB's own MAT format. I suggest using the built-in load/save functions for simple serialization and deserialization:

```
save('cachedData.mat','cachedData');   % store cache data

data = load('cachedData.mat');         % load cache data
cachedData = data.cachedData;
```

Alternatively, rather than developing a dedicated serialization/deserialization function and file format, consider using one of the available utilities on the MATLAB File Exchange, such as *disk_cache*[179] or *cache_results*.[180]

Multilayered caches are usually classified as either *write-through* or *write-back* (or *write-behind*) backing-store policy. Write-through caches update the offline mirror immediately (synchronously) after the memory cache has been updated. Write-back caches delay updating the mirror to some later (asynchronous) time. They are typically more complex than write-through, but have less effect on run-time performance since the offline updates are asynchronous. Whichever policy we choose, we need to implement it ourselves, since MATLAB does not have a built-in cache persistency mechanism.

3.2.5 A Real-Life Example: Writable Cache

The built-in **datestr** function receives date values in MATLAB's numeric format and converts them to human-readable strings (e.g., datestr(735297.6459) returns '04-Mar-2013

---

* One could of course argue that this is an example of smart recursion, rather than memoization.

15:30:05'). The problem is that we may need to call *datestr* repeatedly, for multiple numeric dates, for example, in a callback that recomputes axis tick labels upon pan/zoom. Unfortunately, the *datestr* implementation is not vectorized, so computing *datestr* for a vector of 1000 dates takes almost 1000 as much time to calculate as a single date. Doing so repeatedly can be disastrous for performance.

If we are certain of the precondition that the output string format is the same, we can cache the results, and even use vectorization to further speed up processing. In my specific case,[181] I plotted historical daily stock quotes data and so I was assured that (1) all dates are integers (i.e., full and not partial dates) and that (2) I always use the same date-string format 'dd-mmm-yyyy'.

First, let us define the wrapper function *datestr2* with caching and vectorization:*

```
% Faster variant of datestr, for integer date values since 1/1/2000
function dateStrs = datestr2(dateVals,varargin)
    persistent dateValsCache
    persistent dateStrsCache
    persistent cachedVarargin

    if isempty(dateStrsCache)
        origin = datenum('1-Jan-2000');
        dateValsCache = origin:(now+100); %also cache 100 future dates
        dateStrsCache = datestr(dateValsCache,varargin{:});
        cachedVarargin = varargin;
    end

    [tf,loc] = ismember(dateVals, dateValsCache);
    if all(tf) && isequal(varargin, cachedVarargin)
        dateStrs = dateStrsCache(loc,:);
    else
        dateStrs = datestr(dateVals,varargin{:});
    end
end % datestr2
```

As can be seen, when *datestr2* is called for the first time, it computes and caches all *datestr* values for all the dates since January 1, 2000. Subsequent calls to *datestr2* simply retrieve the relevant cache values. Note that the input date entries need not be sorted.

When an input date number is not found in the cache, *datestr2* automatically falls back to using the built-in *datestr* for the entire input list. This could of course be improved to add the new entries to the cache — I leave this as an exercise to the reader.

The bottom line was a 150-times speedup improvement for a 1000-item date vector (50 ms → 0.3 ms):

```
% Prepare a 1000-vector of dates, starting 3 years ago until today
dateVals = fix(now) + (-1000:0);

% Run the standard datestr function: ~50 ms
s1 = datestr(dateVals);
⇨ Elapsed time is 0.049089 seconds.
s1 = datestr(dateVals);
⇨ Elapsed time is 0.048086 seconds.
```

---

* MATLAB's date/time functions are further discussed in §4.4; vectorization is discussed in Chapter 5.

```
% Now run our datestr2 function (caching already done before): 0.3 ms
s2 = datestr2(dateVals);
⇨ Elapsed time is 0.222031 seconds. %initial cache preparation: 222 ms
s2 = datestr2(dateVals);
⇨ Elapsed time is 0.000313 seconds. %subsequent datestr2 calls: 0.3 ms
s2 = datestr2(dateVals);
⇨ Elapsed time is 0.000296 seconds. %subsequent datestr2 calls: 0.3 ms

% Ensure that the two functions give exactly the same results
>> disp(isequal(s1,s2))
     1
```

3.2.6 Optimizing Cache Fetch Time

Employing a cache mechanism, any cache mechanism, is already a big step in performance tuning our application. But in certain cases, we can do even better by selecting a cache data structure that is adapted for fast data fetch (access) in all the subsequent uses of the cached data.

In the *datestr2* example above, we used the fact that our cache key was numeric (the *datenum* value), so searching for the key was as simple as using a standard *ismember* function call. Unfortunately, such simple lookups are not always the case.

For example, to cache a phone directory of thousands of names and phone numbers, we could indeed store the data in a regular cell array having N-by-2 cells (one column for names, another for numbers). But fetching the phone number associated with any specific name would mean that we would need to look for this name in the entire cell array of N names. Doing this repeatedly for a large value of N would be highly inefficient, unless we only cache a small number of the most frequently used names. As an alternative, we could use a more suitable data structure, namely, a hashtable.[182]

MATLAB releases R2008b (7.7) and newer contain the built-in *containers.Map* object that emulates a hashtable and can be used to efficiently store and fetch keyed data (i.e., data that is defined using a unique key, such as a person's name or ID). The benefit of *containers.Map* is that it can store any MATLAB construct associated with a key, including structs, class objects, and cell arrays.

```
persistent cachedData
if isempty(cachedData), cachedData = containers.Map; end  %initialize

% Examples of storing data in the cache
cachedData('key #1')    = 'my string data';
cachedData('2nd key')   = magic(5);
cachedData('third key') = {'cell array', magic(3)};

% Examples of querying the cache
flag = cachedData.isKey('key #1')    % true
size = cachedData.size               % =[3,1]

% Examples of retrieving data from the cache
str   = cachedData('key #1');        % string
data  = cachedData('2nd key');       % numeric matrix
cells = cachedData('third key');     % cell array
bad   = cachedData('no such key');   % raises an exception (error)
```

As an alternative to *containers.Map,* consider using Java Collections[183] that are much faster and available out-of-the-box in MATLAB releases as old as R12.1 (6.1):*[184]

```
persistent cachedData
if isempty(cachedData), cachedData = java.util.Hashtable; end  %init

% Examples of storing data in the cache
cachedData.put('key #1',    'my string data');
cachedData.put('2nd key',   magic(5));
cachedData.put('third key', {'cell array', magic(3)});

% Examples of querying the cache
flag = cachedData.containsKey('key #1')     % true
size = cachedData.size                      % =3

% Examples of retrieving data from the cache
str   = cachedData.get('key #1');           % string
data  = cachedData.get('2nd key');          % numeric matrix
cells = cell(cachedData.get('third key'));  % cell array
bad   = cachedData.get('no such key');      % returns empty array []
```

Note that cell arrays are stored in (and retrieved from) Java `Hashtable` as an array of Java objects. Therefore, to convert the data back into MATLAB cell array, we have to use the built-in *cell* function, as shown above. Also note that Java `Hashtable` objects return an empty array [] when a key is not found, unlike *containers.Map* that raises an exception (error). We need to handle these idiosyncrasies within our code.

Unlike *containers.Map,* Java Collection objects can natively store only basic MATLAB data types (numbers, numeric arrays, cell arrays, and strings), but not structs or class objects. To store structs and class objects in a Java `Hashtable` cache and then be able to load them back into MATLAB, we need to first convert them into Java objects, which the cache can then accept. To do this, we can use the *Lightspeed* toolbox (§8.5.6.1)'s *toJava()* function, which converts MATLAB's basic data types into corresponding Java classes (*double* into `java.lang.Double`); structs and class objects are converted to a `java.util.Hashtable`, where the field/property names serve as keys. The reverse is done via *from-Java()*. This is effective, generic, and very fast:

```
>> data.a = pi; data.b = 'sadf'; data.s = data; data.c = {1,2,'3a'}
data =
    a: 3.14159265358979
    b: 'sadf'
    s: [1x1 struct]
    c: {[1]   [2]   '3a'}
>> jData = toJava(data)   % convert MATLAB data=>Java Hashtable object
jData =
{b=sadf, a=3.141592653589793, _fields=[a, b, s, c],
 s={b=sadf, a=3.141592653589793, _fields=[a, b]},
 c=[[Ljava.lang.Object;@6bf2830}
>> cachedData.put('4th key', jData); % store the data in Java cache
>> data=fromJava(cachedData.get('4th key')); % get back MATLAB data
```

---

* For additional (non-Java) alternatives to *containers.Map,* see §4.9.6. Advanced users might consider using a different (faster) Java Collections class, `java.util.concurrent.ConcurrentHashMap` rather than `java.util.Hashtable`, as suggested by Joshua Bloch, *Effective Java,* 2nd Ed., Addison-Wesley, ISBN 0321356683, item #69, p. 274.

As an alternative, we can use ***getByteStreamFromArray, getArrayFromByteStream***
built-in functions, which convert any MATLAB data to/from a ***uint8*** (byte) array:[185]

```
wordStream = int16(getByteStreamFromArray(data)); %Java data is signed
cachedData.put('4th key', wordStream); % store the data in Java cache
wordStream = cachedData.get('4th key'); % retrieve the cached data
data = getArrayFromByteStream(uint8(wordStream)); %get the MATLAB data
```

3.3 Smart Checks Bypass

Checks of various types (bounds, sanity, validity, etc.) are an essential part of any real-life
program. We have to ensure that the data fits the expected parameters for the program to
function properly. Unfortunately, checks take time to evaluate in run-time, and in certain
cases (particularly for code that executes numerous times in a loop or rapidly fired callback) this could be significant.

In some cases, it is possible to know in advance that certain conditions would never happen and simply bypass the checks for these conditions. Since checks are so important to
program robustness and correctness, we need to double-check our assumptions carefully,
and only remove checks where the benefits of performance speedups outweigh the risk of
invalid or out-of-bounds run-time data.

A corollary of checks bypass is that we should not check unnecessarily, certainly not
within a loop. In the following example, both checks can safely be removed, the first
because it has no effect on the result, and the second because it always evaluates to true:

```
% Original version
for idx = -100 : 50
    if idx ~= 0                    % check #1: no effect on total1
        total1 = total1 + idx;
    end
    if idx < 200                   % check #2: always true
        total2 = total2 - 0.1;
    end
end

% Optimized version, without run-time checks
for idx = -100 : 50
    total1 = total1 + idx;         % this can now be vectorized
    total2 = total2 - 0.1;         % this can now be LIH-optimized
end

% Fully-optimized version
total1 = total1 + sum(-100:50);
total2 = total2 - 0.1*length(-100:50);
```

In some cases, conditional checks can be replaced by preprocessing, postprocessing, or
an equivalent arithmetic expression:

```
% Example of replacing a conditional check with pre-processing
for idx = -100 : 50
    if idx ~= 0
        total = total + 1/idx;
    end
```

```
end
                    →
for idx = [-100:-1, 1:50]  % pre-process the 0 index out of the loop
    total = total + 1/idx;
end

% Example of replacing a conditional check with post-processing
for idx = -100 : 50
    if idx ~= 0
        data(end+1) = 1/idx;
    end
end
                    →
for idx = -100 : 50
    data(end+1) = 1/idx;
end
data(isinf(data)) = [];  % post-process the 0 index out of the results

% Example of converting logical condition into arithmetic expression
if a > 5
    b = 3;
else
    b = 5;
end
                    →
b = 5-2*(a>5);
```

Such a process of *deconditionaling* is especially important in loops or in often-called utility or callback functions. Additional ideas are discussed in §3.1.4 and §3.10.1.

A closely related technique is *logic inversion*: It is sometimes easier to post-process the data for edge-cases than to pre-process it and employ bail-out conditions. An example of such a case was provided by Loren Shure in the calculation of the *sinc* function.[186] The original code tests for the special condition x==0 (where y should be 1). The modified (faster) code post-processes the data, enabling the calculation to be fully vectorized without any costly sub-indexing:*

```
% Original code - pre-process the data for x==0 edge case
y = ones(size(x));
i = find(x);
y(i) = sin(pi*x(i))./(pi*x(i));

% New (faster) code - inverted logic: post-process the data
y = sin(pi*x)./(pi*x);
y(x==0) = 1;
```

When using the Profiler to search for the internal core logic to possibly inline,† we sometimes encounter early bail-out checks‡ that we can utilize by passing an extra (optional) input parameter to the function, or by modifying the data in a certain manner (e.g., sorting it). Using the function's early bail-out in such a way can significantly improve the run-time performance.

---

* See §9.3.3.
† See §3.1.2.
‡ See §3.1.3.

For example, it turns out that `datenum('2013-02-15')` is an order of magnitude slower than `datenum('2013-02-15','yyyy-mm-dd')`, because the former needs to perform all sorts of checks and computations to determine the date format. By specifying the extra (optional) input parameter telling the function the data string format, we shave away these run-time computations.*

Similarly, if our date strings are properly formatted, we could use the internal ***dtstr2dtnummx*** directly, rather than the ***datenum*** function, bypassing all the latter's sanity and formatting checks and improving performance even further. Jan Simon's *DateConvert* used the reliance on a specific timestamp format to speed up ***datenum*** by a factor of over 100 in some cases.†

The idea of avoiding run-time checks and only using the core logic is also discussed in §4.3 below. It turns out that quite a lot of MATLAB functions are simply wrappers around internal helper functions, with additional sanity and early bail-out checks. In certain cases, we might determine that our program does not need these wrapper checks, and we could directly use the internal helper functions. A sample profiling/debugging session is presented in §4.3.1, explaining this idea.

As noted above, such changes can dramatically improve performance but come at a cost of reduced code robustness and generality. They should only be used after careful determination that the reduced checks do not affect the program's behavior.

Another similar example is using the ***interp1q*** function for quick 1D linear interpolation. Unlike the generic ***interp1*** function, ***interp1q*** avoids any checks of its inputs, which are expected to be a monotonically increasing column vector (x), a column vector or matrix with ***length***(x) rows (Y), and a column vector (xi). Although ***interp1q*** is fully supported and documented,[187] MathWorks recommends using the generic function ***interp1*** instead. An even faster (24×) implementation of the core logic was implemented by Bruno Luong using MEX, which he called *nakeinterp1*.[188] Jan Simon has also implemented a MEX variant, called *ScaleTime*,[189] which is dozens of times faster than ***interp1*** or ***interp1q***.

While we want to reduce unnecessary run-time checks, it is in fact advantageous to add early bail-out checks throughout our code, to reduce unnecessary computations. Early bail-out is especially important in long-running loops and was therefore discussed separately (§3.1.3). However, this approach can also be used to bypass entire loops, costly I/O and other long-running code sections.[190] For example, if an Excel file does not need to be read in some specific code-flow situation, then a check should be added that would only read the file in the actual cases where it is needed, rather than in all cases.

In practice, we should weigh the performance benefits of the reduced checks (in 100% of the function calls) versus the performance gains of early bail-out (in only a fraction of the function calls). Both alternatives should be profiled based on real-life data, to determine which is better.

We should place the bail-out checks as early as possible in the function, to prevent unnecessary computations in a bail-out situation. The more common bail-out cases should be checked before less frequent ones.[191] The decision of which case is more common should be based on empirical (profiled) evidence, not misleading intuition.

---

* Strictly speaking, in the specific case of ***datenum***, the function uses a different processing path when the format string is specified (in which case it uses the internal ***dtstr2dtnummx*** function), than when it is unknown (in which case it uses the built-in ***datevec*** and ***datenummx*** functions). The aforementioned format checks and computations are done in the ***datevec*** function.

† See §4.4 for a detailed description of ***dtstr2dtnummx***, *DateConvert* and other related alternatives.

3.4 Exception Handling

Exception handling is the programming mechanism that catches errors (exceptions) that are thrown in a code block. In MATLAB, following the convention of other programming languages, this mechanism uses the built-in keywords ***try-catch***:

```
try
    ... do some regular code that may throw an error ...
catch
    ... process a thrown exception (error)
end
```

Whereas the use of exception handling is common for code correctness and robustness, not many are aware of its potential benefits for performance. In fact, judicious use of exception handling can significantly improve performance by removing the need for programmatic checks of edge-case conditions:

```
for idx = 1 : 10000
    if exist('test.m','file')
        fid = fopen('test.m','rt');
        process(fid);
        fclose(fid);
    end
end
```
⇨ Elapsed time is **3.657024** seconds.

```
% Same code, now utilizing exception handling - 2x faster
for idx = 1 : 10000
    try
        fid = fopen('test.m','rt');
        process(fid);
        fclose(fid);
    catch
        % ignore
    end
end
```
⇨ Elapsed time is **1.822939** seconds.

Raising exceptions is expensive in terms of performance, so it makes sense to use exception handling only where the number of regular flow processing (in the ***try*** block) significantly outnumbers the number of exceptions (processed in the ***catch*** block), say by a ratio of 1000-to-1 or more. In other cases, we should replace ***try/catch*** with a conditional statement (*if*).[192]

Code that was written on relatively old MATLAB releases sometimes relies on the two-inputs functional form of the *eval* function to process some functionality in case of error. Using a ***try/catch*** block is preferable although generally not much faster.[193]

Exception handling incurs some processing overhead. For this reason, it makes sense to use exception handling only when the alternative programmatic check takes nonnegligible time. I/O checks (as above) are usually expensive enough to merit this.

For example, in the following code, using exception handling rather than programmatic checks actually degrades performance:

```
N = 1000000;    % 1 million
a = rand(1,N);
```

```
b = zeros(1,N);
n = 0;
tic
for idx = 1 : N
    ref = rand(1)*N - N/5000;
    if round(ref) > 0
        b(idx) = a(round(ref));
    else
        b(idx) = ref;
        n = n + 1;
    end
end
toc
⇨ Elapsed time is 0.118698 seconds.

% Same code, now utilizing exception handling - 17x slower
for idx = 1 : N
    ref = rand(1)*N - N/5000;
    try
        b(idx) = a(round(ref));
    catch
        b(idx) = ref;
        n = n + 1;
    end
end
⇨ Elapsed time is 2.072388 seconds.
```

As noted in §3.1.3, we can also use exception handling to implement a very efficient early bail-out mechanism out of several nested loops and functions. For example:

```
try
    for outerIdx = 1 : 5
        middleIdx = 10
        while middleIdx <= 20
            middleIdx = middleIdx + 1;
            for innerIdx = -100 : 100
                data = outerIdx/innerIdx + middleIdx;
                if data == SOME_VALUE
                    error('bail out!')
                else
                    process(data);
                end
            end  % for innerIdx
        end  % while middleIdx
    end  % for outerIdx
catch
    % ignore - simply continue normally...
end
```

3.5 Improving Externally Connected Systems

MATLAB program performance sometimes depends more on external components than on the MATLAB code itself. The first rule in this regard is to minimize the amount of integration to external components.* In cases where relying on external components is unavoidable, we can sometimes improve the overall performance by improving those components. This section provides some simple guidelines/ideas:

3.5.1 Database

3.5.1.1 Design

When designing a database, consider whether the target use is OLTP[194] or OLAP.[195] In a nutshell (seasoned database administrators (DBAs) will excuse my simplification), OLTP is oriented toward real-time interaction with the raw data, while OLAP is focused on aggregated statistics and trends. This decision should be taken already at the ER model.[196]

OLTP data should be arranged in normalized data tables.[197] The third normalized form (*3NF*)[198] is usually used to reduce data redundancy to a minimum and increase overall data integrity. However, for reasons of performance, it may be helpful to **denormalize**[199] data, sacrificing data redundancy (and database size) for improved retrieval performance, due to the reduced need for table joins.† When denormalizing manually, we must ensure that the redundant data is stored consistently in the data tables that we update. A good practice for this is to use a single transaction for the multiple updates, committing the changes only upon success of all updates.

A better way to denormalize is to use **materialized views** (or *snapshots*):[200] these are data tables that denormalize by joining the original (normalized) tables using a predefined SQL join query, storing the resulting data as a separate physical table. This effectively caches the result of expensive table joins, enabling faster data retrievals.‡ Further speedup can be gained by indexing the resulting table. The database engine ensures that whenever the original data is updated, so too is the materialized view, saving us the trouble of ensuring this consistency ourselves. To modify the denormalization, we need only modify the materialized view's originating SQL query. Materialized views are especially effective in read-intensive tables.

OLAP data is generally arranged differently, typically either star-shaped[201] or snowflake-shaped.[202] A star schema should be used when there is a single large fact table, whereas a snowflake schema should be used when there are multiple independent fact tables or large dimension tables that could be normalized. Since OLAP is mostly read-oriented while OLTP is generally update-oriented, materialized views are also highly effective in OLAP databases.

In both OLTP and OLAP databases, server-side **database jobs**[203] could be scheduled to run at off-peak hours in order to create or update *summary* (or *pivot*) *tables*[204] that could later be used to speed up the run-time user queries at peak usage times. An extension of this idea is to use a separate reporting server or data warehouse.

When defining data tables, it is important to use correct data types and length. For example, using a string to represent dates requires costly conversions whenever data needs to be fetched or compared. Similarly, using a VARCHAR2(2000) requires less I/O in SQL queries

---

* See §11.1 for a discussion on minimizing I/O.
† Note that in some cases denormalization can degrade performance due to increased I/O: more data pages need to be read from disk, since each page holds fewer records.
‡ See §3.2 for additional aspects of data caching.

than would be required if the data type was CHAR(2000), since VARCHAR2 only uses as much space as actually used by the data, whereas CHAR uses the specified length regardless of the data (blank-padding is used where needed).

In addition, using appropriate data types enables the database's *query optimizer* to run an SQL query more efficiently (see §3.5.1.5).

Finally, ensure that data columns that have an interrelationship (e.g., columns that are often combined or compared) have consistent data types, in order to reduce the amount of automated conversions that occur whenever these fields are used together.

3.5.1.2 Storage

Compress table data to accelerate queries. When the table is compressed, less I/O is required for reading the table. Reading a compressed table's data is much faster and outweighs the extra compressing and decompressing by the CPU (CPU is an order of magnitude faster than I/O). This makes a difference for tables that have nonnegligible size (e.g., >1MB); for small tables, the benefit is small and could actually be negative (i.e., decompressing data takes longer than the extra I/O time).

Different databases have different ways of displaying table size and for compressing the data. In SQL Server, for example, in the SQL Management Studio right-click the database name → Reports → Standard Reports → Disk Usage by Top Tables:

Table Name	# Records	Reserved (KB)	Data (KB)	Indexes (KB)	Unused (KB)
dbo.Measure	214,535	26,456	20,120	6,272	64
dbo.RunAttribute	221,302	13,584	11,088	2,472	24
dbo.BlobData	39	8,872	8,848	24	0
dbo.Topic	39,075	3,608	2,256	1,232	120
dbo.Run	36,887	2,640	2,024	584	32
dbo.ToolData	1	32	8	24	0
dbo.TopicPartition	0	0	0	0	0
dbo.sysdiagrams	0	0	0	0	0

SQL Server's tables usage report

Tables are not usually compressed by default. Compressing tables may require administrator privileges, and is done differently for different database types.[205] Moreover, different databases impose different limitations on compression. For example, MySQL's default table format (MyISAM) only allows compression via the *myisampack* utility, which makes the table read-only (i.e., nonupdatable).[206]

Use **partitions**[207] (where available) to place data in multiple data files that can be searched and accessed in parallel. This makes sense for large data tables (e.g., >10M records) that have a more or less uniform distribution of some key, thereby enabling efficient parallelization over the separate partitions. A DBA can place the partition files on separate disks, further improving speed by enabling parallel disk I/O at the hardware level.* Using partitions, administrative tasks for large data tables (maintaining tables and indexes, collecting tables and indexes statistics, purging old data, etc.) are made easier and faster. There are also fewer lock contentions and improved caching. We should choose a partitioning function that maximizes query parallelization, in order to improve the query performance in addition to administration tasks.

---

* Further speedup can be gained by placing partitions that are often updated on a write-optimized file system (e.g., RAID 1 or RAID 10), and partitions that should be optimized for data retrieval on read-optimized file systems (e.g., RAID 5).

Place the DB's *tempDB* on a separate disk from the data files, preferably SSD or RAM disk.[208] The OS's swap (page) files should also be placed on a separate disk. Failing to separate these from the data files will significantly degrade DB throughput.

3.5.1.3 Indexing

Create **indexes*** to speed up queries. Indexing works on one or more columns of a single table. Those columns are the columns by which we usually query (i.e., the columns usually involved in our query's WHERE clause). Indexes typically have special B-tree structures[209] that store the data in sorted manner to allow faster data lookup and access. The database's SQL engine looks for those index rows that match the SQL WHERE clause, and then use the index pointers to the actual data records to access just those records that match the requested criteria. This avoids scanning the entire table data: only index lists are searched. Index lists are smaller and optimized for fast access, so searching them is typically much faster.[210]

There are two main disadvantages of indexes that should be weighed against the performance benefits:

1. Indexes hold copies of the indexed data keys, thereby increasing the database size. In fact, the storage necessary for several indexes on a data table (a situation that is quite common) may well be larger than the storage needed by the data itself.

2. In addition, whenever the main data is updated, so too are the indexes, resulting in decreased run-time performance. This trade-off means that we often need to balance update performance (→ less indexes) against retrieval performance (→ more indexes) in our application.

Indexes are less effective when numerous indexes are used that store a large portion of the data, so searching them may actually involve more I/O than a full table scan. Another case is when the index table cannot be searched efficiently. In this case, a full index search may be required, which might be less effective than a simple table scan.

Creating an index on a small data table is both futile and bad for performance. In such cases, full table scans are often preferable to index scans or even index seeks. The reason is that the entire table can fit in a single I/O page, while an index scan would require loading a minimum of two I/O pages (one for the index, another for the data).

Creating a table index is again database dependent. For example, in SQL Server again right-click the table name → Design → right click on a column name → Indexes/Keys → Add. Then select the multiple columns on the right, as well as uniqueness feature and clustering.

Some databases enable specifying the file system used to store the index data files — we should consider placing these files on fast SSDs (solid-state disks): the main DB data is sometimes too large for SSD to be cost-effective, but the index data may well fit on a small SSD, resulting in considerable performance speedup. Even when we do not have an SSD, it is still beneficial to place the index data files on a separate file system, since this allows the database to parallelize the query's search (in the index file system) and data retrieval (in the main data file system).

Some databases enable customizing index characteristics. The index *fill-factor* is one such characteristic that can affect performance,[211] and so too could the index format (B-tree or bitmap; regular or reversed keys). We should normally let the database decide on the

---

* The correct English term is "indices", but practically all programmers use "indexes"...

optimal values, but to squeeze that extra drop of performance, we might consider fine-tuning these characteristics manually.

Unique indexes[212] accelerate data retrieval because once a matching record is found, the search ends. Indexes can be defined as unique provided they really are. For example, if we have a unique identifier for each record then it can be a unique index, but if some records have the same index value(s), then the index cannot be made unique. The uniqueness feature depends on the data — SQL will check it as we try to save the index file, and report an error if duplicate keys are detected, preventing uniqueness. Once a unique index is set, a nonunique value cannot be inserted or updated.

Primary keys are a specific type of unique index that have specific uses. A table can have multiple unique indexes, but only a single primary key. The smaller a unique (or primary) index, the more such index values can be stored in a single index I/O page, and therefore the faster the queries that use it. Counterintuitively, using primary keys can sometimes improve the performance of insert queries, not just retrievals.[213]

Surrogate keys[214] are a special type of unique index that use a compact data type (typically four bytes in size) for storing a unique ID that represents unique data rows. They are called differently in different databases (Sequence, Serial, Identity, Autoincrement, UUID, GUID, etc.), but the basic idea is the same in all cases: using a compact data column for faster index usage in query processing. The speedup results from the fact that index data files can store many more key values in a single I/O page than with *natural keys*[215] (especially with compound keys of several columns). In some cases, the database enables specifying an automatically generated surrogate key (e.g., an autoincrement column), but in other cases, we need to create the unique value in our program.

A **covering index**[216] is a regular index that contains all the information needed to answer a data query. For example, if a table index includes the ID, Name, and Value columns and our query only requests the number of records having Value = 5, such a query can be fully satisfied using the index, without having to dereference the original data table. This usually results in faster query retrieval, since the index data is much more compact than the original data table, reducing the required I/O. To ensure that our application makes optimal use of this feature, as many of the program's queries as possible should be fully covered by indexes. We can add data columns to the indexes as required, at the expense of increased storage and decreased update performance.

Indexing can also improve the performance of *computed* (or *virtual*) *columns,*[217] which are columns that hold computed values that rely on a function of some other values in the same data row.[218] By placing the computed value in the index, run-time queries are saved from the need to recalculate the value on-the-fly. On the other hand, once again the trade-off is increased storage and decreased update performance.

Clustering is an option within indexes. It allows SQL to use multiple processors to scour the database in parallel, which is much faster. Only a single index can have this feature turned on for any single table, since clustering rearranges the original table data in a way that would match the index (typically, storing data having different index ranges in separate files, sorting each data file separately). Once clustering is turned on for one index, it cannot be turned on for any other index in the same table. Clustered indexes in Oracle are called IOT (*Index-Organized Tables*)[219] and are somewhat different in implementation to clustered indexes on other DBs.[220]

While clustering can improve run-time *retrieval* and *update* performance, it has an adverse effect on *insert/delete* performance.* The reason for this is that the clustered data

---

* Update of the clustering key value would also be slower, for similar reasons.

files need to be rearranged and resorted. If our application continuously inserts or deletes data rows, then we should carefully check whether clustering is beneficial. In certain cases, when the entire result can be easily found in a nonclustered index, it could be faster than the clustered index. The reason is that the nonclustered index stores less data than a clustered index and so may require less I/O. Clustered indexing is typically better for range queries (e.g., date between X and Y), queries that use the primary key, and noncovered queries that retrieve multiple data columns.

If we do use clustering, we should ensure not to include the cluster index fields as part of the regular (nonclustered) indexes' keys: the cluster index is automatically used by all indexes, so adding it to other indexes is futile, redundant, wasteful in storage, and degrades run-time update performance (unnecessary index updates).

Indexed data tends to get *fragmented* over time, as the original data is updated, degrading indexing performance. From time to time (every few weeks or months, depending on the data update frequency), we need to *rebuild* (or *reorganize*) the index to restore optimal performance. This is done differently in each database, but can usually be done via both a DDL query and the database's administration GUI.

An important aspect in this regard is to perform regular measurements of database response performance, in order to detect deteriorating database conditions that require maintenance. For example, it may be discovered that as we add data to the main data tables, the application's queries become sluggish, requiring optimization, additional indexing, clustering, partitioning, or perhaps more drastic measures in order to bring the performance back in line. Similarly, changing application usage patterns may dictate removing, modifying or adding some indexes.

3.5.1.4 *Driver and Connection*

On Windows platforms, avoid using the standard ODBC connection bridge to connect MATLAB with the database. ODBC is very easy to set up and use, and users are often tempted to use this for their database connection. However, ODBC performance is horrible and it also has stability issues. Instead, use a dedicated driver, preferably JDBC since it interacts seamlessly with MATLAB. ODBC often uses JDBC under the hood, thereby adding another point of possible failure, a tie-in to a specific computer configuration, and worse performance than a direct JDBC connection.

MATLAB's Database Toolbox uses JDBC, but we can also connect MATLAB directly to the database using JDBC, without needing the DB Toolbox. The first step is to download the database's JDBC driver from the database's website. This typically includes a *.JAR file and possibly also a *.DLL file.

For JDBC to work properly, it must be added to the static Java classpath (the *classpath.txt* file) and possibly also to the *librarypath.txt* file. MathWorks has posted a technical description of how to do this for the DB Toolbox,[221] which also applies to stand-alone connections without the toolbox. Setting up a JDBC connection without the toolbox is not very difficult, but is beyond the scope of this book. Readers are referred to my MATLAB-Java programming book for details.[222]

The driver attempts to wait forever for the connection to succeed or fail. This can be limited by setting a timeout value in seconds prior to the connection attempt:

```
java.sql.DriverManager.setLoginTimeout(3);  % wait 3 seconds max
con = java.sql.DriverManager.getConnection(connStr,username,password);
```

A database connection is successfully established when a valid `java.sql.Connection` object[223] is returned by the call to `DriverManager.getConnection`.

Some DBs provide both a generic JDBC driver and a native driver, which is usually faster. However, unlike JDBC, which has cross-platform compatibility, native drivers do not. Our code would not be portable to platforms that do not support this native driver, but the performance speedup may justify this. I once consulted to a client who used JDBC to connect to an SQLite DB.[224] Switching to a native driver (Martin Kortman's *mksqlite*[225]) improved run-time from 7 s to 70 ms, a 100× speedup!

Some DBs enable encrypted TLS/SSL connections. Unless you have a good reason, avoid using this feature. Such connections can be up to an order of magnitude slower than regular nonencrypted ones.[226] If you need security, consider using a lower-layer networking encryption (e.g., VPN/SSH) rather than application-layer encryption.

For additional performance and resource considerations in applications that heavily utilize a database, consider using *Data Source* connections, *connection pooling,** and other advanced features available in the database and database driver.

3.5.1.5 SQL Queries

When processing just a subset of the data, it is better to apply the filtering as soon as possible. Therefore, update the SQL query's WHERE clause† such that it only returns the necessary data. This is preferable to the query returning the entire data to MATLAB and then filtering it in MATLAB code. The reason is reduced I/O cost, both within the database and between the database and MATLAB, not to mention the additional memory and CPU costs required by the MATLAB filtering. In short, offload as much as possible of the filtering and processing in general to the database, using a database stored procedure if necessary.‡

As an extension of this idea, it is preferable to employ data selection in the query's WHERE clause rather than its HAVING clause, when data aggregation is used on the result of the internal WHERE selection.

Likewise, if we have a data-conversion function (e.g., date formatting), it is better to return its results in the SELECT clause (on the small data subset returned by the WHERE selection), than in the WHERE clause (that runs on every input-data row).

Similarly, instead of paging results in MATLAB memory (displaying 50 results each time, e.g.), we should directly filter the result set in the database, using clauses such as LIMIT, TOP, OFFSET, ROWNUM, ROW_NUMBER(), FETCH FIRST, and RANK() OVER. Most databases have some sort of paging ability, although the specific syntax is sometimes different.

As another example, when combining data in result sets, it is faster to use the UNION ALL clause (which allows duplicate rows) than UNION DISTINCT (which removes them). However, this results in increased I/O and memory, and longer processing on the MATLAB side. Therefore, it is generally better to filter duplicates in database.§

When merging data, it is always faster to merge (JOIN) data tables in database than to merge them in MATLAB memory, after loading the separate table data.

Many programmers are familiar with basic SQL syntax (simple JOIN and GROUP BY queries). Unfortunately, my experience has been that few are aware of the power of aggregate *window* (or *analytic*) functions (OVER PARTITION BY). Using such functions can significantly speed up SQL processing, not to mention the reduced result-set I/O and MATLAB postprocessing that would otherwise be required.[227]

---

* Also see §4.7.1.
† Or, in a grouped (aggregated) query, also the HAVING clause. See §3.6.1 for some usage examples.
‡ Stored procedures are discussed later in this section; also see §3.6.
§ Applying smart table JOINs on all relevant data columns can often remove the need for filtering duplicate rows, which are often a symptom of imprecise joins that result in undesirable Cartesian products.

Limit the returned fields (columns) in the result set to only those that are actually needed by the application. Returning extra fields (e.g., by liberal use of SELECT * FROM...) results in unnecessary extra I/O and memory, in addition to forcing the database to use a less efficient *execution (explain) plan.*[228] For example, returning only fields that occur in an index enables the database to serve the entire query by only scanning the index, without needing to access the main table data.*

Additional speedups can be gained by fetching the result-set data in bulks, for example, of 2 K or 5 K records (rows).† This would reduce the memory required on both the Java and the MATLAB sides of the I/O. A utility that implements this idea is MathWorker Tucker McClure's *fetch_big* on the MATLAB File Exchange.‡[229]

Unfortunately, when using JDBC directly, each table value needs to be fetched into MATLAB separately, which could be slow when fetching large data. To fix this, we could create a Java class that acts as a connector to JDBC, processing all data items in Java and returning a single unified numeric or cell array to MATLAB.[230] Alternatively, use the MATLAB Database Toolbox which apparently does this internally.

On databases that process large volumes of both updates and queries, we may encounter slowdowns due to internal DB locks. These are automatically set by the DB to ensure data consistency (the well-known *ACID* principle[231]). Various DBs have different mechanism for setting update query locks at the table level, or record level, or even field level (Oracle). This would ensure that queries that access nonaffected records/fields will not need to wait for the update operations to complete.

Moreover, in many cases, we may not mind getting a 10-ms-old value in our query result, even for those records/fields that *are* updated (resulting in stale/dirty data), if this means that the query runs 10× faster. For this, we could use a database-specific directive in the SQL query, which tells the database to ignore any potential locks. For example, on SQL Server we can use the NOLOCK hint:[232]

```
SELECT field1, field2 FROM table WITH (NOLOCK) WHERE field3 > 10
```

Alternatively, we can disable locks at the DB connection level. Again for SQL Server:

```
SET TRANSACTION ISOLATION LEVEL READ UNCOMMITTED; -- turn it on
SET TRANSACTION ISOLATION LEVEL READ COMMITTED;   -- turn it off
```

Use the database's **SQL query profiler**[233] (where available) to identify SQL bottlenecks and workarounds in the query's *execution plan*. SQL query optimization can be achieved by examining alternative plans.[234] In general, when looking at execution plans, *index seeks* are faster than *index range scans* and *full index scans*, with *full table scans* normally being the slowest. However, in small tables or when the table is partitioned, table scans may actually be faster.

For example, we could simplify an SQL query to minimize table joins or convert between subqueries and joins. In different situations, subqueries may be preferable than joins, in other cases, the reverse.[235] A similar situation occurs with aggregation queries (queries that use GROUP BY) — such queries can sometimes be converted into subqueries or regular table joins; in different situations, one or the other of these variants is faster. Using the

---

* Refer to the topic of a covering index in §3.5.1.3.
† See related §3.6.1.
‡ *Fetch_big* also uses the additional optimization of preallocating the required MATLAB memory, as per §9.4. Tucker McClure states that the *fetch_big* optimizations have been incorporated into MATLAB Database Toolbox's **fetch** function in R2013a (8.1). Users who use an older MATLAB release, or who use a nontoolbox (direct) DB driver, would still benefit from this utility.

profiler helps determine the situation in specific cases. Note that table joins and subqueries are not inherently slow: they are often slow due to inefficient implementations more than due to any intrinsic slowness.[236]

As a special case, we often do not actually need to retrieve data from a table but only information about whether or not a certain condition exists in the data. In such cases, using the EXISTS or IN *semi-join* operators in a subquery can significantly improve the query performance compared to a regular table join.[237]

When deleting all data in a table, it is *much* faster to TRUNCATE it (a DDL action), than to DELETE * (a DML action) or to DROP TABLE and then re-CREATE it again (a DDL action, but much more expensive than TRUNCATE).

Use **SQL hints**[238] (where available*) to force the database to use a specific index or search strategy, utilizing *a priori* information about the data that is known to us but not to the database. We should be careful to use hints only when we are certain that the database's query optimizer decides to use suboptimal execution plans. In most cases, we should let the optimizer select the execution strategy automatically.

Prepared statements utilizing *bind variables* provide significant performance speedups for queries that are repeated multiple times. The idea is that the statement (SQL query or procedure call) is only compiled and placed in memory once — each subsequent invocation of the statement will merely update its parametric values, without necessitating the internal SQL interpreter and optimizer engines to reprocess the code, only to rerun it using the updated data. Bind variables significantly reduce CPU overhead and better utilize database memory. Repeated queries can reuse the same memory structures resulting in better performance and response time. When rerunning the same statement numerous times, the speedup can be amazing.†

Stored procedures (SPs)[239] have performance benefits over simple SQL queries. SPs are precompiled in the database, whereas SQL queries are parsed and compiled individually. In addition, stored procedures enable aggregation of data within the result set, something that might otherwise require multiple separate SQL queries, with their associated round-trip latency, parsing, I/O, and postprocessing overheads.

In stored procedures, some databases enable using a *data cursor*[240] to iterate over data records. Except when there is no other feasible way, cursors should not be used since they are generally much slower than regular set-based queries. An exception to this rule might be a case where a stored procedure could break down a complex SQL multitable join into separate smaller results that can be cached before being combined, thereby saving lots of I/O. However, most modern DBs can optimize SQL queries to such an extent that cases where hand-crafted cursors are faster are rare. In cases where we do use cursors, many of the performance tips in this chapter (including caching, LIH, etc.) are also applicable to the stored-procedure code.

3.5.1.6 Data Updates

Remove unnecessary data **triggers**.[241] For bulk inserts/updates/deletes, consider temporarily disabling the triggers before running the bulk operation. On the other hand, if multiple

---

* To the best of my knowledge, SQL hints are currently (2014) supported only on Oracle and MySQL databases, using different proprietary syntaxes.
† For usage examples and detailed explanations of using JDBC prepared statements (and JDBC code in general) in MATLAB, refer to Chapter 2 of my book *Undocumented Secrets of MATLAB-Java Programming*, CRC Press 2011, ISBN 9781439869031, http://undocumentedmatlab.com/matlab-java-book (or: http://bit.ly/Zqqojt).

database updates are needed to be done together, it makes sense to implement a trigger on the first operation (say, an insert into the main data table) that would then update the rest of the database without requiring the MATLAB program to send any additional SQL requests. Instead of a trigger, we could also elect to use a database stored procedure (see above), possibly placing all updates within a single database transaction,[242] for assured data consistency.

Remove unnecessary data **constraints**,[243] for a similar reason. Often-used constraints include NOT NULL, UNIQUE, and *foreign keys*, which are used to ensure the data integrity. Constraints carry a performance penalty, since they necessitate checking each data update to ensure compliance. If we can ensure the data integrity in our non-DB program, and are willing to reduce the safety level of our database data, then removing constraints can improve update performance.

However, using constraints can improve performance as well as data integrity: some databases use constraints metadata to optimize query execution plans,[244] so adding constraints could speed up retrievals, while slowing down updates/inserts. For this reason, it is often said that constraints are more important for OLAP databases such as data warehouses, where there are much more retrievals than updates, than for OLTP (transactional real-time) databases where updates outnumber retrievals. If updates do not significantly outnumber retrievals in our database, then using constraints will help improve both performance (retrievals) and data integrity (updates).

Disable transaction logging for long-running updates. Also, use **bulk updates** (especially imports; also called *batch processing* and *ETL*[245]) rather than separate SQL statements, where this is supported. Both of these suggestions can significantly improve performance when a large amount of data needs to be imported or updated. Different databases have different ways of employing bulk updates; refer to your database's manual for details. For example, the popular MySQL database enables to insert multiple data rows using a single INSERT operation:

```
INSERT INTO dataTable (colA,colB,colC) VALUES (1,2,3),(4,5,6),(7,8,9);
```

Defragment tables and indexes, remove old **transaction logs**, update **data statistics**, and in general **optimize** the database periodically (e.g., once a month).[246] These are normally part of a DBA's work but if we self-manage the database then it is up to us. In many cases, we can set the database on "auto-pilot" mode, so that most of the important maintenance tasks are done automatically by the database. In other cases, we can usually set up periodic maintenance *jobs* that do this.

There are numerous online and offline resources available for tuning databases. A partial list is provided in the Appendix (§A.1.2 and §A.2.3).

3.5.2 File System and Network

The physical file storage system used for both the MATLAB installation and for storing data can have a dramatic effect on performance.

MATLAB loads from disk and so the effect is first seen on the MATLAB startup times. There are many factors that affect MATLAB's startup times,* but the type of file system can be a major factor. I have one version of MATLAB that is identical on two laptops, which are similar in most respects except that one has a regular hard disk and the other uses an SSD (solid-state drive): The hard disk laptop takes about a minute to load MATLAB; on the SSD it only takes 10–15 s.

---

* See §4.8.

The speedup is not limited to MATLAB startup: MATLAB needs to load all data files and functions from disk. Functions are loaded from disk for the most part not when MATLAB starts but rather when they are first needed. The result is that MATLAB performance in general is much better on my SSD laptop than on my hard disk one.[247] Nowadays, the cost differential between SSDs and standard hard disks is not very large. In my opinion, the extra cost of SSD is justified by its superior performance and reliability (SSDs never crash, but over the years, three of my hard disks have). In most cases, a simple inexpensive SSD is sufficient;[248] enterprise-grade SSD is only needed in rare cases, if we store massive amounts of data and/or have huge throughput.[249]

I advise against using an externally connected disk (e.g., USB-connected flash disk) as an alternative to an internal hard disk. The reason is that the I/O would be limited by the external connection's throughput, which is typically much lower than for an internal connection. Modern connectors (e.g., USB 3.0) improve the connection bandwidth, but these are less common than the slower connectors, and in any case are usually still slower than internal connections. Technology advances at such a rapid pace that by the time you read this, the situation could be reversed. Still, it is safe to say that in general, an internally connected disk would be faster than the same disk connected externally, with an internal SSD being the fastest alternative.

If you use a non-SSD hard disk, consider using a RAM disk: the relevant data and function files would be copied from the hard disk to the RAM disk externally to MATLAB, and then MATLAB would run from the RAM disk (memory) rather than from the physical drive. In effect, this trades memory and start-up time in exchange for run-time performance.* Note that doing this for internal SSDs has little value.

To improve hard disk performance, defragment the disk periodically, especially data files and MATLAB installation folder. This would improve hard disk access time and overall performance, since I/O is often a limiting factor in performance.† As above, defragmentation only applies to hard disks, and has little or no benefit for SSDs.

I have seen several cases of systems where two or more antivirus/antispyware monitoring services were continuously running in the background. These can have a disastrous effect on performance, and quite often clash with each other, reducing performance even further. Consider using only a single such application rather than several. Also, check whether these applications enable exclusion of certain disk folders — if so, then consider excluding the MATLAB installation folder as well as the folder containing your application code and data files.

On the target platform, running a good antivirus/malware program and removing unnecessary services can dramatically improve overall computer performance. Appendix §A.1.2 includes a few resources containing other similar general tips.

Similarly, consider disabling code-versioning systems (CVS), at least when running in production (rather than development) mode. Such systems, especially when run off a network server, can also have a significant negative effect on MATLAB performance.

Running code or accessing data files off a network disk is almost always significantly slower than off a local disk. I once worked with a large industrial client, which installed MATLAB centrally, for maintenance and management reason. Users accessed MATLAB remotely over the company network. The slowdown compared to a local installation was enormous — speedups of 10×–20× were observed when time-critical code segments were moved to local disk.

---

* This is basically what the MATLAB Startup Accelerator (§4.8.1) tries to do.
† See Chapter 11.

To synchronize files between computer, a very efficient and cross-platform solution that I found was to use independent (non-MATLAB) synchronization services such as DropBox.[250] This is especially effective over a local network (LAN). However, if data files are constantly being updated, then such synchronization could significantly degrade overall performance due to the multiple concurrent I/O.

If you need enterprise-level storage, use an IT consultant to optimize the solution for your needs. Such an expert could advise on whether to use SAN/NAS/DAS, RAID, and striping, and compare various vendor models.[251] Many factors affect storage performance, including its required size, concurrency, and throughput.[252] The savings from good advice on cost/performance alternatives may well offset the expert's cost.

A specific advice that was referenced with regard to MATLAB on Linux is to periodically purge the file system cache using the following MATLAB command,[253] which can be run periodically via a MATLAB timer,* or directly in Linux via *cron*:

```
system('echo 1 >/proc/sys/vm/drop_caches'); % MATLAB variant
```

For optimal networking performance, ensure that you are using the latest TCP/IP stack (this is typically part of the operating system (OS)). Optimizing MTU (packet sizes)[254] may improve network throughput, but I do not believe it is worth the effort in most cases. A few other ideas for Linux networking were suggested by Maarten Van Horenbeeck.[255]

For additional suggestions for improving MATLAB I/O, see Chapter 11.

3.5.3 Computer Hardware

Computer hardware obviously affects performance, but some components affect performance to a larger degree. A major cost component of any system is its CPU, but this is often less important than a much less expensive component — the system memory (RAM). The importance of memory on performance is discussed in Chapter 9. It is sufficient to note here that an investment of only a few dozen dollars in RAM can have a dramatic effect on the overall system performance, MATLAB included.

I suggest having at least 8GB or 16GB of RAM. For the majority of use-cases, this should be enough.† On existing machines, we can either add memory cards (if we have empty memory slots‡), or replace existing memory cards with ones that contain more memory. Such upgrades are easy to do within minutes, unlike upgrading other memory components. So, we can purchase an 8GB machine, leaving empty slots for future upgrades. We do not need to purchase the full 16GB+ right from the start. The situation is different with other hardware components such as the main CPU or GPU, whose future upgrade is more difficult and much more expensive.[256]

When increasing the amount of memory, we should note the limitations imposed by the OS. 32-bit Windows only enables 2GB processes and total usage up to ~3.5GB;§ extra memory is simply ignored. 64-bit Windows running 32-bit MATLAB enables using more memory, but the MATLAB process itself would be limited to only 4GB. To gain full benefit of memory, we need to use 64-bit MATLAB on a 64-bit machine: this enables MATLAB to use up to 8TB (or less, depending on the amount of physical RAM and the page-file size

---

* See §7.3.5
† Naturally, the situation changes if we need to crunch massive amounts of data.
‡ Note that we need to ensure consistency between the memory sticks. Adding a memory module with lower frequency than the rest automatically degrades all memory modules to the lower frequency, thereby reducing their throughput (access speed).
§ Even this is only available by setting the nondefault /3GB switch — see §9.7.1 for details.

allocated for virtual memory[257]).* We should generally prefer a 64-bit CPU, running 64-bit OS and 64-bit MATLAB.[258]

Another important aspect of memory is its speed, typically measured in MHz. The higher this value, the faster will the memory send data to the CPU for processing. However, memory speed is limited by the bandwidth of the system bus and the support by the motherboards. Memory modules usually arrive in predefined speeds, and the price difference between speeds is usually very small. The specific memory speed values that we can use on any system is usually declared by the manufacturer. Given a choice between two supported memory modules, we should opt for the faster.

More and/or faster memory does not by itself guarantee faster processing. Numerous other factors affect performance, including OS, memory/bus speed, and CPU. Direct comparison between computers is meaningless if they differ by multiple factors. MATLAB includes the **bench** function that enables a generic comparison. We can run **bench** and say that for the *specific* **bench** application one computer may be faster than another. But for other programs, the other machine may well be faster than the first.

When it comes to CPUs, more cores are better than faster processing: the OS can delegate non-MATLAB processes to core 0 (for example) while MATLAB runs on core 1. We can force this using a procedure called *CPU affinity*, but it is normally best to let the OS figure out the best CPU core allocation per process. Multiple cores also enable MATLAB to parallelize some important functions such as linear algebra using implicit or explicit parallelization.† Explicit parallelization, at least the one used by the Parallel Computing Toolbox (PCT), uses heavyweight *worker* processes, which can run not only on separate cores, but also (using the Distributed Computing Toolbox) on other CPUs.‡ Implicit parallelization uses lightweight multithreading that runs on separate cores of the same CPU. However, only MATLAB R2007a (7.4) and newer releases employ this feature. Running an older MATLAB release on a multicore machine would only help in the context of the CPU process affinity, as explained above. For the most effective use of modern multicore CPU, use a modern MATLAB release.

Some Intel CPUs provide hyperthreading (HTT),[259] which uses microcoding and redundant CPU execution paths to expose two virtual CPU cores for each physical core. For HTT-supported OSes, this improves performance for multithreaded code portions.[260] However, reports on the efficacy of HTT have been mixed over the years and a consensus has not been reached. HTT is apparently ineffective for floating-point operations, which form the bulk of MATLAB processing.[261] HTT effectiveness highly depends on the nature of the running application and the other OS processes that run at the same time. For some applications, performance can actually improve by disabling the HTT feature. HTT is also relevant when setting the number of local workers in the PCT.

Some CPUs also enable *overclocking*, but this feature decreases CPU lifetime and is not recommended for extensive use.

All modern computers come with a GPU nowadays. When given a choice, we should opt for one of NVidia's GPUs, which is supported by PCT. PCT does not support all NVidia GPUs so we should be careful to choose one that is.§ We should do this even if we do not currently have PCT: we may possibly need it next year, and in this case it would be better to have a system that could make the most effective use of PCT.

---

* 64-bit Linux and Macs have different maximal process sizes, up to 128TB in RHEL 6 for example.
† See Chapter 5 for implicit parallelization, and Chapters 6 and 7 for explicit parallelization.
‡ As discussed in Chapter 6; see Chapter 7 for alternative implementations.
§ Any GPU with a *Compute-Capable* value of 1.3 or higher: http://developer.nvidia.com/cuda-gpus (or: http://bit.ly/1jBpNaU). Note that future MATLAB releases might only support GPUs having a higher minimal Compute Capable value (2.0 or more).

Regardless of whether we decide to use a supported NVidia GPU, we should at least ensure that the chosen GPU supports hardware-based OpenGL (almost all modern GPUs do). This is critically important for graphics performance, not just of MATLAB but many other applications, such as browsers, document-processing software, and so on.

MathWorks has posted a detailed advice on optimal hardware selection for MATLAB performance.[262] A general list of computer speedup tips is available here.[263]

Finally, ensure the computer's power management and CPU throttling is disabled.[264] If running on a laptop, ensure that it is connected to electric power and not running from battery, so that its power management will not slow down processing.

3.6 Processing Smaller Data Subsets

One of the typical pitfalls in performance tuning of any processing function, in any programming language or environment, is late filtering of data. Rather than loading and processing only the small subset of data that is of interest, many implementations load the entire data set and only filter the data late in the processing flow. This causes performance slowdowns in many places along the route: unnecessary I/O (both at the data source and in the network), unnecessary memory allocations and deallocations,* cache flushes,† and unnecessary CPU processing of irrelevant data.

Therefore, a fundamental advice in tuning any processing function is to filter the data as early as possible. Load, process, and save only the part of the data that is really needed.[265] If multiple filtering steps are needed, first apply the filter that causes the biggest reduction and only later the other filters, so that they work on less data.

3.6.1 Reading from a Database

When reading data from a database, apply the filtering at the source using SQL's WHERE and HAVING clauses, for example:

```
SELECT * FROM db_table WHERE price > 0

SELECT name, price, count(1)
FROM   db_table
WHERE  price < 100
HAVING count(1) > 1
```

SQL data query result sets can be further reduced by limiting the number of returned rows. The syntax is somewhat different in different databases:

```
SELECT TOP 5 * FROM db_table ORDER BY price DESC    %SQL Server
SELECT * FROM db_table ORDER BY price DESC LIMIT 5  %MySQL
SELECT * FROM (                                     %Oracle
   SELECT *, row_number()
            OVER (ORDER BY price DESC) r
   FROM db_table
)
WHERE r <= 5
```

---

* And potentially even memory paging, which would be disastrous for performance — see §9.1.
† The effects of cache flushes are discussed in §11.3.5.

When we need to process multiple database records in a loop, we can preload data in chunks by unifying SQL queries, rather than query the database separately for each record. Query unification improves performance by saving I/O communications and DB processing. However, if the chunk is too large, we may encounter memory problems.[266] In such cases, we should actually split the query into separate subqueries that return smaller result sets. If we use the Database Toolbox, we can use the *Fetch_Big* utility on the File Exchange[267] that automatically splits queries if needed.

3.6.2 Reading from a Data File*

When reading variables from a MATLAB *.mat file, only load the relevant variables:

```
load('filename.mat', 'varName');
```

If the saved variable is huge, consider loading only the relevant data subsection, using the new *matfile* function on release R2011b (MATLAB 7.13) and newer:[268]

```
matFileObj = matfile('filename.mat');
partialData = matFileObj.varName(11:13,21:23);
```

Similarly, we can update only a small part of a stored variable without needing to resave the entire MAT file:

```
matFileObj.Properties.Writable = true;
matFileObj.varName(11:13,21:23) = magic(3);
```

Note that in order to use *matfile*, the MAT file needs to be saved in the nondefault 7.3 format, which became available in MATLAB R2006b (7.3):†

```
save('filename.mat', 'varName', '-v7.3');
```

When reading a binary data file, we can use MATLAB's built-in I/O functions to read only the file segment of interest, rather than the entire file. This can be done by skipping the file index to the relevant file part, reading it, and then immediately bailing out and closing the file, without bothering to read the file's remainder:

```
fid = fopen('dataFile.dat','rb');
fseek(fid,20000000,'bof');            % skip 20MB from beginning of file
data = fread(fid,1000000,'int8');     % read 1MB
fclose(fid);
```

Similarly, for a text file, we could use *fseek* with either *fscanf* or *textscan*:[269]

```
fid = fopen('dataFile.txt,'rt');
fseek(fid,20000000,'bof');            % skip 20MB from beginning of file
data = fscanf(fid,'%5d',[6,8]);       % read up to 6x8 integers, width 5
fclose(fid);
```

Both *fscanf* and *textscan* support *skip-fields*, by using the %* field format specifier. Skipping fields save MATLAB from loading these fields from I/O and from allocating memory for them. The saving could be significant when reading only a single column in a large data file that contains many columns.

---

* Performance aspects of reading data files in MATLAB are discussed in detail in Chapter 11, and especially §11.3. In this section, we only mention tips that directly relate to processing smaller data sets.
† We can make v7.3 the default save format using the MATLAB Preferences (General → MAT files).

The *textscan* function further enables us to specify the 'HeaderLines' parameter to skip the specified number of lines, and we can use it instead of *fseek*.

If a binary data file has a known format, we can use the *memmapfile** to "map" the file's contents onto a variable, which does not consume any real memory. Accessing any specific element of the variable will directly access the relevant portion of the data file. In this manner, we can read and write only the elements of interest, rather than the entire data. This improves both performance and memory usage:[270]

```
>> m = memmapfile('dataFile.dat',    ...
                  'Offset', 1024,    ...
                  'Format', {'uint32' [4 10 18] 'x'});
>> A = m.Data(1).x;
>> whos A
   Name       Size                    Bytes  Class
   A          4x10x18                 2880   uint32 array
>> partialData = A(2:3,4:7,5);   % read a 2x4 data subsection
>> A(5:7,3:5,8) = magic(3);      % update a 3x3 data subsection
```

Sometimes, the specific data format is dynamic and not constant. In such cases, we could store the format at the top of the file when saving the data; when reading the file, we would first read the header with the format information, and then use this format to read the rest of the file data (using either *memmapfile, fscanf*, or *textscan*). A utility that implements this idea is available on the MATLAB File Exchange.[271]

Just as we can map a huge file onto a memory variable, so too can we do the reverse, namely, memory-map a huge variable onto a disk file.[†] The idea is to leverage *memmapfile*'s ability to directly access specific file positions in order to avoid the need to hold an entire huge variable in memory, something that would require extensive use of slow virtual memory and constant memory paging (thrashing) to and from the system disk. It turns out that using a memory-mapped file in this manner can be 100× or more faster than using standard MATLAB variables.

Many binary data files of specific formats (images, audio, video, or HDF) have dedicated MATLAB functions.[272] Some of these functions have optional input parameters that enable us to directly access only a small subsection of the entire file, based on the file's known format specification.

Finally, if we have to read an Excel file (heaven forbid, since using Excel files is a horrid performance bottleneck in general), then we should try to read only the relevant cell range, rather than the entire workbook or worksheet:

```
data = xlsread('dataFile.xls', 'Sheet2', 'D24:M37');
```

For additional I/O recommendations, refer to Chapter 11, especially §11.3.

3.6.3 Processing Data

Data processing often involves several consecutive steps. The sooner in the processing flow that we reduce the amount of data by filtering out uninteresting records or fields, the less processing would need to be done in later stages.

In one specific case, my client was retrieving an enormous amount of data from a database and then displaying it onscreen. Only a very small subset of the data was actually

---

* See §11.3.3.
[†] See the *vvap* utility, described in §9.4.3.

visible. The data-load part was quite fast (a few seconds), but the data processing and presentation took several minutes and during this entire time, the application appeared "hung". When I added a condition at the beginning of the processing phase that limited the amount of displayed data to the top 1000 records (down from the actual ~50 K), performance improved by two orders of magnitude, and the entire thing took seconds rather than minutes.* It was not just processing CPU time that was saved, but also memory allocation, I/O, and GUI updates.

When processing huge data, it sometimes makes sense to split the data into smaller, more manageable chunks that would be processed separately and then merged. While in general processing everything *en masse* in MATLAB improves performance due to vectorization and internal optimizations,† if the data is *too* large, then MATLAB will be forced to handle huge amounts of memory (again, possibly requiring memory paging) and the benefits of CPU cache and internal parallelization will no longer be as effective. The end result is that processing 10 chunks of 10 million elements can be faster than processing a single chunk of 100 million elements.‡

For example, to apply a specific filter to a large image, it may be faster to process independent image subsegments separately, rather than the entire image at once.[273]

Section 11.3.4 and §11.3.5 provide specific examples of using this mechanism for improving the performance of I/O, while §9.5.9 and §6.4.1 provide data processing examples.

The exact point at which the benefits of vectorized performance start to be offset by memory implications is different across applications and platforms. It is possible that an application on platform A will favor mass vectorization, while on platform B it would be faster to divide the data into four chunks, and on platform C into 1000 chunks.

To employ this technique effectively for streaming data, we can use *buffering* and *postponed processing*: data will not be processed immediately, but rather only after a certain amount of data has accumulated in the buffer, or a certain amount of time has passed. We then process the data in an efficient vectorized manner, thereby improving the data throughput. This concept is also often used in Graphics/GUI (see §10.2.2) and I/O (§11.3).

As another variant of the concept of data reduction, it makes sense to uniquify data before processing, to prevent duplicate processing.§ We can expand the processed data back to the original (nonunique) locations, but this extra step is often unneeded. Here is a simple example illustrating this concept, where we wish to get the string (human-readable) representation of numerous timestamps, having multiple recurrences:[274]

```
>> datenums = repmat(now:now+10,1,1000);   %1000 repetitions of 11 vals

>> tic; s1=datestr(datenums); toc
⇒ Elapsed time is 0.913491 seconds.

>> tic; [d2,m,n]=unique(datenums); s2=datestr(d2); s2=s2(n,:); toc
⇒ Elapsed time is 0.004601 seconds.          ← 200x faster!

>> disp(isequal(s1,s2))   % ensure that the results are equivalent
     1
```

---

* Additional speedup could be achieved by limiting the database query to the top 1 K records as explained in §3.6.1, but in this specific case it was difficult to do and would have provided only small additional benefit compared to that which was achieved.
† This fact is nonintuitive outside MATLAB. See, for example, http://java.dzone.com/articles/matlab-performance-testing (or: http://bit.ly/1beoh9e).
‡ Using explicit parallelization can modify this picture. See Chapter 6 for additional details.
§ See §3.2.1 for a discussion of using unique values to improve caching; see §3.9.1 for additional aspects of data preprocessing.

Processing smaller data sets is not only important for performance due to easier processing but also due to the effects of locality[275] and data caching at the hardware level. Modern computers employ a series of CPU[276] and disk caches[277] at several levels. These caches are optimized for speed but have limited storage capacity (typically only a few KB or MB, depending on the cache and platform). Processing smaller data chunks (*loop tiling*[278]) ensures that we maximize cache effectiveness by minimizing *cache misses*.[279] Such cache misses require loading the replacement data from main storage and have disastrous performance effect.[280] We may well discover that the simple act of reducing the processing chunk size significantly improves performance, an effect that amazes many people the first time they encounter it.

In some cases, the very act of looping over the last (right-most) dimension provides the necessary memory locality and chunking benefits for significant speedups.*[281]

In general, it is impractical to precompute the optimal chunk size. In my experience, the best practical solution is to test several different chunk sizes on the target platform. When testing, it is important to use an environment that behaves as close as possible to run-time.† The reason is that hardware caches (as opposed to application-specific caches) are not devoted only to our running program. They are shared among all processes running on the system, including the OS and other concurrent tasks. So, while we cannot rely on hardware caching when designing our application, we should also not neglect it by not optimizing the processed data.

If you have MATLAB's PCT, consider using ***spmd*** and ***drange / parfor*** loops to split the data processing among threads/cores/CPUs.‡

3.7 Interrupting Long-Running Tasks

It often makes sense to provide a timeout mechanism for long-duration tasks such as waiting for a certain condition to occur, or for a computation to reach a predefined accuracy. This is especially important when integrating with external systems,§ where the external system might be blocked or nonresponsive. The typical *synchronous* (fire-and-wait) flow may cause significant slowdowns in cases where the other side is busy or nonresponsive.

Rather than waiting a long time (potentially forever) for such tasks to complete, we could use a mechanism that interrupts this process after some time.

In MATLAB, this can be done using a single-shot timer object, which runs *asynchronously* of the main task.¶ This implements a fire-and-get-notified mechanism. The timer triggers its callback function after the specified timeout, if it has not been deleted earlier (in case the main task has finished in time). The timer's callback function simply sets some flag or property value that indicates to the main task that it should abort. Naturally, we need to have the main task monitor this flag internally for this mechanism to work.

---

* See §3.1.11 for a related technique, of looping over the shorter dimension of multidimensional data, and §9.3.2 for looping order. All these techniques originate from the way that MATLAB stores array data (§9.3), contrasting with looping overheads.
† See §2.1.6 for additional discussion.
‡ See Chapter 6.
§ See related §3.5 and Chapter 11.
¶ Also see §7.3.5 and §10.4.3.

Here is a sample implementation of this mechanism. First initialize the timeout flag:

```
% Initialize the global timeout flag
global timeoutFlag
timeoutFlag = false;
```

Next, set up a single-shot timer to trigger the timeout event:

```
% Create and start the separate single-shot timeout timer thread
hTimer = timer('TimerFcn',@timeoutEventFunc, 'StartDelay',timeout);
start(hTimer);
```

where `timeoutEventFunc` is defined to raise `timeoutFlag`, as follows:

```
% Function to raise the timeout flag
function timeoutEventFunc(hTimer,eventData)
    global timeoutFlag
    timeoutFlag = true;
end
```

Finally, in our main processing loop, we would monitor this `timeoutFlag`:

```
% Main processing loop that ends when timeout flag is raised,
% even if the mainCondition is not yet satisfied
global timeoutFlag
while mainCondition && ~timeoutFlag
    processData();
end
```

Using global variables is fast and effective. However, there are strong reasons to limit the usage of globals in applications.* As an alternative to globals, we can use the **UserData** or **ApplicationData** properties of the root object (handle 0).†

A variation of this mechanism is to wait for a MATLAB HG handle (or Java/ActiveX reference) to change an internal property value. This is normally done using the built-in *waitfor* function. *waitfor* is a blocking function, which means that it waits for the specified condition to occur before continuing any further MATLAB processing. The timer mechanism enables us to bypass this block, by asynchronously setting the condition on which *waitfor* blocks. Here is a skeleton of the mechanism:[282]

```
% Wait for data updates to complete
% (isDone = false if timeout, true if data ok)
function isDone = waitForDone(object, timeout, propName, propValue)

    % Initialize: timeout flag = false
    set(object,propName,propValue);

    % Create and start the separate single-shot timeout timer thread
    hTimer = timer('TimerFcn',@(h,e)set(object,propName,propValue),...
                   'StartDelay',timeout);
    start(hTimer);

    % Wait for the object property to change or for timeout,
    % whichever comes first
    waitfor(object, propName, propValue);
```

---

* See §9.5.5 below, which explains the reasons and provides several alternatives.
† **ApplicationData** should not normally be used directly, only via the *getappdata*, *setappdata*, and *isappdata* wrapper functions.

```
% waitfor is over because of timeout or because the data changed.
% To determine which, check whether timer callback was activated
isDone = (hTimer.TasksExecuted == 0);

% Delete the time object
try stop(hTimer);    catch, end
try delete(hTimer); catch, end

% Return the flag indicating whether or not timeout was reached
end  % waitForDone
```

We can then use this function as follows:

```
% 30 secs timeout on hFig's UserData property
waitForDone(hFig,30,'UserData',true);
```

A related performance-tuning technique, of preferring asynchronous callbacks to synchronous polling, is described in §10.4.3.5. As noted there, when using MATLAB timers, we should take their inherent timing inaccuracies into consideration.

3.8 Latency versus Throughput

When performance tuning any application, we need to ask ourselves which is more important: the program's initial response (latency) or its ongoing performance (throughput).[283] The answer may be different for different applications and even for the same application under different usage scenarios. Depending on the answer, we may wish to optimize different aspects of the program. In many cases, optimizing for low latency may impair the ongoing performance and vice versa.

Two standard tuning techniques that are directly affected by this trade-off are lazy evaluation and prefetching.

3.8.1 Lazy Evaluation

The idea behind the lazy evaluation technique is to delay long-running initializations or data-load until after the main program GUI is presented to the user, in cases where these initializations are not immediately required for the presentation. The actual initialization or data-load is done in the background, after the GUI is presented. In some cases, it can even be delayed until the program needs the data (possibly never).

This technique has many variations and names. It is often called *lazy initialization*[284] or *lazy loading*,[285] and is a special case of the *lazy* (or *delayed* or *deferred* or *demand-driven*) *evaluation* technique.[286] Note that there is a distinction between lazy initialization and lazy loading: lazy initialization is only one of several possible implementations of lazy loading. However, in practice, these terms are often used interchangeably. The opposite approach is often called *eager initialization*.

MATLAB itself uses lazy evaluation internally, within its JIT (by compiling m-files only as needed),* and for optimizing memory access.†

---

* See §11.9.1.
† Refer to the copy-on-write mechanism, discussed in §9.5.1.

A typical example in a real-world user application is loading a GUI that needs some data stored in an Excel file for one of the GUI's functionalities. Reading Excel data serves here as a generic example for any long-running initialization, since it typically requires launching a new Excel process. This could take quite some time,* much longer than required for processing the data or displaying the GUI.

My advice in this case is to display the GUI and enable user interaction before reading the Excel data. If our GUI was prepared using GUIDE, then we can postpone our Excel read to the last stages of the *_OutputFcn() function. As a reminder, GUIs prepared using GUIDE automatically have two initialization functions that can be customized with user code: *_OpeningFcn, which is called just before the GUI is made visible, and *_OutputFcn, which is called after the GUI is presented. By delaying slow initializations (such as reading Excel) to OutputFcn, we can ensure that the GUI is displayed sooner to the user, thereby improving the perceived performance.†

During the long Excel read, it would be wise to prevent user interaction with any functionality that relies on the data, until it is ready. This can be done by setting the relevant control's **Enable** property to 'off' when the GUI loads. Once the GUI is displayed, the Excel file is read and only then is the control's **Enable** property set to 'on':

```
% --- Executes just before doselab_v6 is made visible
function myGUI_OpeningFcn(hObject, eventdata, handles, varargin)
    % only simple initializations are performed here
    handles.version = 6.5;
    % Update handles structure
    guidata(hObject, handles);

% --- Outputs from this function are returned to the command line
function data = myGUI_OutputFcn(hObject, eventdata, handles)
    % any long-running initializations are performed here
    set(handles.ProcessDataButton, 'Enable','off');
    data = xlsread('dataFile.xls');
    set(handles.ProcessDataButton, 'Enable','on');
```

In this example, we could place the Excel read in a dedicated timer, in order to return control to the user immediately.‡ This may further improve the perceived performance, although the actual added value would be minor and might not justify the added code complexity. A template using this coding pattern might look as follows:

```
% --- Outputs from this function are returned to the command line
function data = myGUI_OutputFcn(hObject, eventdata, handles)
    % any long-running initializations are performed here
    set(handles.ProcessDataButton, 'Enable','off');
    % load data after a 3-sec delay (hObject is the figure handle)
    start(timer('TimerFcn',{@loadDataFcn,hObject}, 'StartDelay',3);

% Timer callback function to load the data asynchronously
function loadDataFcn(hTimer,eventData,hFig)
    handles = guidata(hFig);
    handles.data = xlsread('dataFile.xls');
    guidata(hFig, handles);  % update the figure's guidata
    set(handles.ProcessDataButton, 'Enable','on');
```

---

* See §11.5 for additional information and some workarounds.
† See §1.8.3.
‡ Also see §7.3.5 and §10.4.3.

In fact, we could even decide to delay reading the Excel file until after the user clicked on the relevant control that needs the data (`handles.ProcessDataButton` in our case). The trade-off here is that while most of the time the user would not need the functionality, thereby saving the time required for the Excel read and improving the ongoing performance, when the user actually *does* need the functionality, she would need to wait a relatively long time for the data to become available (latency). We have a choice whether to prefer the typical or the edge-case performance. We shall see in the next section that prefetching presents exactly the same dilemma.

As a variation, if we have a MATLAB class that needs the Excel data for some functionality (`MyClass.processData()`), we can load the data in `MyClass`'s constructor, thereby prolonging its initialization time. Alternatively, we can decide to delay reading the Excel data until `processData` is actually called by the user (if at all), saving time in the constructor, at the expense of run-time delay in `processData()`:[287]

```
% Slower constructor initialization, faster run-time processing
classdef MyClass < handle
    properties
        data = [];
    end

    methods
        % Class constructor - slow
        function classObj = MyClass()
            classObj.data = xlsread('dataFile.xls');
        end
        % Run-time processing - fast (uses existing data)
        function newData = processData(classObj)
            newData = sin(classObj.data);  % process the data
        end
    end
end

% Faster initialization, slower run-time processing
classdef MyClass < handle
    properties
        data = [];
    end
    methods
        % Run-time processing - slow first time (loads data)
        function newData = processData(classObj)
            if isempty(classObj.data)
                % Initial loading of the data, if not already done
                classObj.data = xlsread('dataFile.xls');
            end
            newData = sin(classObj.data);  % process the data
        end
    end
end
```

When using lazy initialization (and especially lazy class object instantiation), we should be careful to always check for the existence of the object before using it. In the code snippet above, the Excel data file is loaded if the class's `data` property is empty. Similar sanity checks should be made in all cases. We should be especially careful in cases where multiple timer threads might access the code independently, potentially causing errors when trying to access uninstantiated objects or data.[288]

An extension of lazy initialization is to delay update of nonvisible GUI components until all visible components are displayed. This includes preparing GUI controls and plot axes in nonvisible GUI tabs/panels/figures, loading images in invisible axes, and so on.*

Lazy evaluation does not apply only to GUI applications. The concept of delayed processing can be applied to non-GUI algorithms as well. For example, rather than precomputing and caching all the possible inputs to an algorithm, we could instead only calculate (and cache) them as they are encountered for the first time. The resulting code would be a bit more complex, but we gain speedup from not precomputing data that is in fact never needed. This technique also enables faster start-up of the algorithm (important for perceived performance†), by skipping the possibly lengthy precalculation phase.

One specific implementation of lazy initialization (or rather, lazy *loading*), is loading segments of a large binary data file only as needed, rather than all in advance. While this can be achieved programmatically (using the low-level I/O functions *fseek* and *fread*), it is faster and easier to use MATLAB's built-in support for memory-mapped files using the *memmapfile* function.‡ The memory-mapping mechanism ensures that I/O is only performed for those data pages that are actually referenced. We can design our program in such a manner that it only loads a small subset of the data at the beginning, with the rest being accessible upon demand.

MATLAB's Copy-on-Write mechanism is another example of non-GUI lazy evaluation. See §9.5.1 for its usage and additional details.

3.8.2 Prefetching

Prefetching data is the exact opposite of lazy evaluation. In this case, we actively load or calculate some data in anticipation of its possible future usage. Loading data can often be done in the background, in otherwise nonactive times, for example, while the user is looking at the presented information. When the user will eventually need the new data (e.g., by clicking a <Next> button), it would already be waiting, significantly improving the overall responsiveness and perceived performance of the application.

Prefetching is widely used in computer design, both for I/O and CPU (*instruction prefetch*[289] and *speculative execution*[290]). Prefetching typically involves using a cache (buffer) to store the prefetched data for later use by the main processing flow.

Prefetching hyperlinks has also been employed in web browsers, to increase their perceived performance at the expense of increased memory and network bandwidth. It is also used by OSes to increase performance in some situations.[291]

When Google Maps was first launched, it was not the first mapping applications. Other mapping apps were available with an existing user base. One feature that made Google Maps so effective was its innovative use of prefetching: When we view a map of any area, Google automatically and unobtrusively prefetches the mapping tiles that are adjacent to the presented tiles and just outside the visible map area. So, when we drag and move the map in any direction, the newly visible map tiles are already there for immediate display, and do not need to be downloaded. This presents a fluid uninterrupted natural map movement, an enormous usability improvement compared to the existing method of clicking an arrow button and waiting for the new map tiles to download and display.

---

* See Chapter 10 for additional GUI performance ideas; §10.4.1.5 discusses an idea similar to the one presented here.
† See §1.8.
‡ See §11.3.3.

Naturally, prefetching consumes a lot more I/O bandwidth, memory, and CPU compared to non-prefetched processing. A trade-off can be decided between these aspects and the potential usability improvement.

We might base our decision on the *likelihood* of the prefetched data actually being used. For example, a GUI that presents 100 possible data scenarios might not justify prefetching all of them into memory, but if we only have a single possible data source then we might be tempted to prefetch it.

Similarly, we might base our decision on the prefetching *cost*. For example, loading an Excel data file can be very expensive in terms of time/CPU, but precomputing some numeric data might be much cheaper in this respect.

When we need to process multiple database rows in a loop, we can preload data rows by unifying SQL queries, rather than query the database separately for each row. Refer to §3.6.1 for additional details.

3.9 Data Analysis

The book is meant as a generic reference for MATLAB performance tuning. As such, it does not include domain-specific topics such as numerical analysis, optimization, statistics, genetic algorithms, or image processing. All of these are topics are well worth discussing for performance aspects, but are outside the scope of this book. Readers are referred to dedicated works on these specific topics.

It is impossible to write a book on performance tuning without at least mentioning some important aspects of data analysis, which is the purpose of this section. However, readers should note that the following discussion is superficial, and only provides a subset of the possible techniques.

To improve the performance of data analysis algorithms, we often need to employ a combination of problem insight, general computing performance-tuning techniques (as shown elsewhere in this book), and manipulation of the problem formulation. Project Euler[292] is a splendid online repository of problems that teach methods in problem solving, enhancing our individual skills and ability to mesh the combination of techniques mentioned above. By spending time to solve Project Euler's problems, we train our mind to use a combination of mathematics, software engineering, and problem-domain insight to solve computing problems. Improved performance would be a natural by-product of this training. Project Euler reports[293] over 2000 users who use MATLAB to solve problems, a sizeable community.

One of the nice benefits from working on Euler problems is that we are forced to think about the complexity of the algorithm. Whereas a brute-force $O(n^m)$ approach may be good enough for small data sets, it quickly becomes unfeasible as the data set grows, requiring a different approach. The chosen approach may be different based on the data size: for example, for sets of up to 100 data elements, we might use an $O(n^2)$ algorithm, for 100–1000 elements, we might use an $O(n)$ algorithm, and for larger data sets, we might need to find an $O(\log(n))$ solution. A short video tutorial by MathWorker Doug Hull shows the process:[294] a problem having a naïve $O(n^2)$ solution was replaced by an $O(n)$ algorithm to accommodate larger data sets. Another well-known example is the various implementations of data sorting algorithms.[295]

When searching for an algorithmic solution to a problem, we should be careful to avoid *algorithm frenzy,** whereby we spend R&D time trying to find an O(n) algorithm, when an O(n^2) solution would be more than fast enough for any practical purposes.[296] A brute-force approach is quite often good enough, and may even be faster for small data sets.† As the well-known maxim says:

> *"Don't use a 20-pound hammer to drive a nail"*

3.9.1 Preprocessing the Data

In some cases, it makes sense to sort the data (O(n · log(n))) and then process the sorted data (O(log(n))) rather than multiple processings of the entire data (c · n). If the data is already sorted, this saves the preprocessing part, and in such cases we can utilize optimized binary search O(log(n)) algorithms rather than brute-force O(n) ones.[297]

In many cases, preprocessing the data can dramatically reduce the problem complexity, by reducing the input data set on which the algorithm works.‡ For example, we should generally preprocess the data to remove outlier data points that could cause the algorithm significant slow-downs. Such slow-downs can be due to a variety of reasons, such as the need for more complex and robust checks (to ensure that all the processed data items are valid), or slower convergence of an iterative algorithm (that might get stuck on some outlier data point at a far corner of the data domain). By preprocessing the input data, we ensure that the algorithm only works on valid data points that are of interest. This increases the algorithm's robustness, stability, and accuracy, as well as enabling faster processing due to fast iterative convergence and using simpler processing (without extraneous checks).

Reducing the input data set is not applicable merely to removal of outliers, in the sense of obviously invalid data points. In fact, we can also use this technique to discard valid data that may be of little use in the processing. For example, consider the problem of computing a rotated bounding rectangle around a set of data:[298]

```
xy = rand(1e5,2)  * [2 3;-1 4];
x = xy(:,1);
y = xy(:,2);
plot(x,y,'b.');

t = convhull(x,y);
hold on
plot(x(t),y(t),'ro-');
```

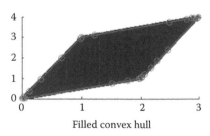

Filled convex hull

---

* See §1.5.4.

† O(n) means some constant factor c1 times n; O(n^2) means some other constant factor c2 times n^2; for small n and c1 ≫ c2, O(n) may well turn out to be slower than O(n^2). See http://en.wikipedia.org/wiki/Big_O_notation (or: http://bit.ly/1fvX5sP).

‡ See §3.6 for related aspects of data reduction for performance speedups

In fact, we can completely ignore all of the points in the convex hull's interior. To find a bounding rectangle, only the perimeter vertices matter. In this case, instead of working with 100000 points, we only care about the 30-odd vertices t that were identified as being on the convex hull polygon itself:

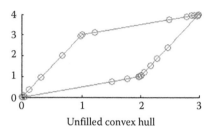

Unfilled convex hull

As a variant of this idea, it is often useful to reduce the data set size by removing duplicate input data entries, typically using *unique*. §3.6.3 showed an example where uniquifying the data produced a 200× speedup. In some cases, we may need to apply applicative filtering to identify duplicate entries, for example, case-insensitive string matching, or data items that are identical within a problem-specific tolerance.

We should remove data that does not contribute to the end result, even if it is valid. In a $1/\alpha$ filter, there is no use for 500 data points — 20 are usually quite enough; the extra 480 points take computation time but do not substantially improve the result accuracy.

A common performance pitfall when designing an algorithm is filtering its *results* rather than its *inputs*. There is usually no point running an algorithm over data points that can be filtered in advance. The only exception to this rule is when the postprocessing filtering is dynamic, based on user input that cannot be done before processing, for example, if the user needs to interact with the results in a GUI or plot.

A specific type of preprocessing that can often help improve performance is sorting the data. Sorted data is processed faster by a wide variety of analysis functions. So, if such analyses are performed multiple times (e.g., within a loop, or to generate various result vectors), then it may help the overall performance to presort the data:[299]

```
x = rand(1e3,1);
yp = rand(100,100,100);

% Find nearest neighbor of yp in x - slow iterative loop version
[K1,K2,K3] = size(yp);
xind = zeros(size(yp));  % pre-allocate the result data
for k1 = 1 : K1
    for k21 = 1 : K2
        for k22 = 1 : K2
            [~, xind(k1,k21,k22)] = min(abs(yp(k1,k21,k22)-x));
        end
    end
end
⇨ Elapsed time is 12.494134 seconds.

% Presort, then use binning (§5.3.5) on sorted midpoints: 100x faster
[x2,p] = sort(x);
[~,xind2] = histc(yp, [-inf; (x2(1:end-1)+x2(2:end))/2; inf]);
xind2 = reshape(p(xind2),size(yp));
⇨ Elapsed time is 0.122693 seconds.
```

For related aspects of the data preprocessing, see §3.6.

3.9.2 Controlling the Target Accuracy

In any iterative converging algorithm (optimization, numerical equation-set solution, etc.), a stop criteria needs to be set in order to define when the algorithm has converged to a solution that is "good enough". This is often formulated in terms of a tolerance on the variability. For example, if the algorithm finds that succeeding iterations modify a parameter so little, as to be effectively the same value.

The tricky part in many algorithms is formulating the problem in such a way that reduced parameter variability indeed indicates convergence to a real solution, avoiding numerical stability, accuracy, and singularity issues. This is not an easy task at all — numerous academic careers were devoted to this topic.

A tolerance that may be good for one problem (e.g., 1 m for GPS geo-location) may be too fine for another problem (spacecraft navigation), or too coarse for yet another problem (surgery robotics). Naturally, the higher the degree of required accuracy (i.e., the lower the tolerance), the more steps are required for the algorithm to converge to within the required tolerance. The accuracy versus performance trade-off is inherent in all iteratively converging algorithms. MATLAB's built-in optimization functions all contain parameters that enable user to control the required tolerance (see §4.5.5). MATLAB's default tolerance, where it exists, is typically 1e-6, which is often much too fine in real-life problems. Then again, there are problems in which it is too coarse. Any user code that implements an iterative algorithm needs to incorporate similar stop criteria.

When we develop our algorithm, we should be careful to set tolerances that meet the real-life accuracy criteria, and no more, since this would be a useless waste of processing time. On the other hand, be careful not to sacrifice too much accuracy, just for the sake of performance.

In multidimensional (multivariate) problems, it is often the case that not all tolerances have the same importance. It is often more important to set a tight tolerance on one dimension and relax the requirements on another, than the reverse. When we code our algorithm, we should take special care to decide on the exact tolerances for each variable separately, and not to use a global or default tolerance.

Some MATLAB functions enable specifying the target accuracy in either relative values (of some constant basis or of the previous iteration's value), or absolute values (or both). Depending on the specific situation, one or the other might be more appropriate.[300] We should carefully select a tolerance type and value that matches the situation at hand.

When we use solvers in our code, we should note that different solvers define their tolerances differently. For example, some solvers might treat the tolerance as a relative value, whereas other solvers might use an absolute value.[301] Therefore, if we switch solvers we need to re-evaluate the tolerances, even if the optimized function or the set of equations have not changed at all.

When we transform a problem formulation (e.g., by coordinate transformation), we need to ensure that we update the tolerances as well as the equations. This could help uncover design problems before we even run the algorithm once. Whenever we transform a problem, we should carefully re-evaluate its solution's numerical stability and accuracy.

An example of this was provided by Yi Cao in his *KNN (K-Nearest Neighbors) search* utility on the MATLAB File Exchange.[302] This utility includes different variants of the algorithm (see within the m-file): a vectorized approach suffered from numerical accuracy problems, finally being replaced by a simpler yet more stable loop.

As another example,[303] an old trick for speeding up an interpoint distance computation uses the idea that for two vectors A and B:

$$||A\text{-}B|| = \text{sqrt}(||A||^\wedge 2 + ||B||^\wedge 2 - 2^*\text{dot}(A,B))$$

This is a commonly used way to speed up Euclidean distance computations, but it suffers in accuracy when A and B are not near zero. Add any constant to both A and B, and the computation becomes unstable, although the distance does not change.

A related widely known variant for the fast evaluation of the *sqrt*(a^2 + b^2) function is max(A,B) + 0.5*min(A,B). This is inherently inaccurate, but for some use-cases, the performance speedup justifies the reduced accuracy. The inverse square root can also be calculated to a surprisingly high accuracy level, using a simple bit-shift and magic-number subtraction.[304] As another example illustrating the performance/accuracy trade-off, consider the pair of built-in numeric integration functions *trapz* and *quad*.[305]

An excellent example of trading accuracy for speed is an approximation of the *exp* function that provides a 2–8× speedup at the expense of up to 4% inaccuracy.[306]

Sometimes, a deliberately inaccurate or nondeterministic algorithm (e.g., neural nets or genetic algorithms) can drastically reduce the time to solve a problem compared to an accurate deterministic exhaustive-search algo. By randomly searching the domain space, such a nondeterministic algo can sometimes converge very quickly to the requested solution, in some problem domains. However, the speedup carries a cost of being nondeterministic: repeated invocations, or invocations with slightly different conditions will yield different search paths, possibly converging to different solutions, missing the overall optimal solution. To reduce the likelihood of such a case, we can let the nondeterministic algo run longer and scan more possibilities, effectively approaching the exhaustive search algos in concept and performance.

For related aspects of the accuracy versus performance trade-off, see §10.2.6.

3.9.3 Reducing Problem Complexity

We can often use a variant of the original problem that provides the same results but has a lower computational complexity (*strength reduction*[307]). *Computational complexity* (and its close relative: *algorithmic efficiency*) is a thriving academic research branch.[308] Generally speaking, the basic aim is to make the problem simpler for a computer program to solve, thereby saving processing time.

Strength reduction can be done at either the entire problem level, or lower levels such as functions or even specific assignments or operators. In general, the largest benefit is gained by reducing the complexity of the entire problem. Lower-level reductions only offer local speedups and single-operator reductions are seldom worth the effort.*

In general, we should try to minimize the use of expensive nonlinear functions (*sqrt*, *log*, *power*, etc.), possibly at the expense of some accuracy.

For example, it turns out that repeated multiplications are much faster than using the standard *power* function or operator for an integer exponent:[309]

```
a = rand(1,1e7)*10;

% Using standard builtin exponentiation
b = a.^4;
⇨ Elapsed time is 0.845275 seconds.
```

---

* Loop *induction variables* (http://en.wikipedia.org/wiki/Induction_variable or: http://bit.ly/1mfSCNN) are often provided as an example for possible strength reduction, but in fact this specific optimization has negligible or no effect in practice.

```
% Using repeated multiplication - 25x faster!
b = a.*a.*a.*a;
⇨ Elapsed time is 0.031922 seconds.
```

John D'Errico's *HPF* (Big Decimal Arithmetic) utility[310] includes a variety of similar optimizations for numeric computations, including optimizing Taylor expansions[311] and iterative Newton's method[312] for rapid convergence. A simple example of strength reduction is replacing a division by 5 by a decrement of the exponent (=/10) and a single shift of the mantissa (=*2).

The *HPF* utility is massive (~7000 lines of code) but extremely well documented, and is well worth reviewing. Quoting John from HPF's webpage:

> For example, multiplies are best done in MATLAB by **conv**. But divides take more work, so here I use a Newton scheme that employs only adds and multiplies, and is quadratically convergent. A similar trick is available for square roots. Or, look into how my exponential function works. Here I've used a few tricks to enhance speed of convergence of the exponential series. Of course, there are obvious range reduction tricks, but I've gone an extra step there. I also employ a different way of summing the series for exponentials (as well as the sine and cosine series) that minimizes divides. A lot of thought and research has gone into the methods of HPF.

A well-known simplification example is minimizing the distance between points: we do not really need to use the true distance (which includes a *sqrt* and possibly also a units factor). It is enough to minimize the sum of the basic squares — values that minimize this function will also be a solution (minimum) to the true distance function.

A variant of this example is finding all data points within a distance r of some center point. Rather than computing the actual distance *sqrt(v)* and checking $sqrt(v) < r$ for all points, it is faster to compute $v < r\textasciicircum2$.[313] We implicitly use a mathematical identity here ($sqrt(v) < r \Rightarrow v < r\textasciicircum2, \forall v, r \in \mathbb{R} > 0$). Mathematical identities can often be used in a similar manner to reduce the problem complexity, speeding up the computation time.*

In this respect, coordinate transformations can sometimes be very effective.[314] For example, problems having radial symmetry can often be described by much simpler functions in polar/spherical or cylindrical coordinates, than in Cartesian coordinates.[315] In such cases, converting the problem to a native coordinate system where the objective function is much simpler to optimize may well be worth the trouble.

In some cases, the problem formulation can be simplified by a different type of transformation. For example, transforming a time-series onto the frequency domain (FFT), or using transforms such as Laplace or Legendre, or using Lagrange multipliers (which are often used to solve constrained optimization problems[316]).

Another technique that is widely used to solve sets of math equations is separation of variables.[317] This technique enables reformulation of the problem in terms of simpler equations, which are also generally easier solve analytically or numerically.

As another simplification of the algorithm, it is better to use a function that is fully mathematical, that is, does not include any internal conditionals or other executable code that requires the MATLAB interpreter.† For example, if the problem includes a condition that the function value cannot be negative, then perhaps we could minimize the square of the values.[318] Also, if our original function has singularity points or discontinuities, perhaps the problem could be reformulated in terms of a smooth set of well-behaved differentiable functions.[319]

---

* See §3.10.1.2 for details and additional examples.
† See §4.6.3.

It is often useful to search the literature for alternative algorithms that may have better performance characteristics. For example, many are familiar with the following series:

$$\pi = 4 \cdot \left(1 - \frac{1}{3} + \frac{1}{5} - \frac{1}{7} + \frac{1}{9} - \ldots \right)$$

However, it turns out that there are literally dozens of much faster-converging series and algorithms for computing the value of *pi* to any level of requested accuracy.[320] Sometimes a problem cannot be simplified into simple vectorization or equation. However, a reduction in the algorithm's complexity, for example, by removing the need for some nested loops or iterative steps, could still yield significant speedups.

In the following code segment, the computation of the sum of factors required two nested loops, a *mod* operation (= costly modulus division), and a conditional (which prevented JIT optimizations). A modification of the algorithm computed the factors by looping over all possible factors and adding their repetitions to the lhs array (similarly to Eratosthenes' Sieve[321]), providing a 1150× (!) speedup:[322]

```
% Slow original version
for i = 1:1e5
    lhs = 0;
    for j = 1:i
        % find sum of factors
        if mod(i,j)==0
            lhs = lhs + j;
        end
    end
end
⇒ Elapsed time is 144.862746 seconds.

% Simplified algorithm - 1150x faster!
lhs = 1 + [1:1e5];
lhs(1) = 1;
for idx = 2:numel(lhs)/2
    lhs(2*idx:idx:end) = lhs(2*idx:idx:end) + idx;
end
⇒ Elapsed time is 0.125344 seconds.
```

When solving a set of equations, use known mathematical techniques to reduce their complexity before applying MATLAB's brute-force numerical approach. For example, if a system of 12 simultaneous equations only uses 3 independent unknowns and 9 dependent unknowns, then try to isolate the 3 equations for the three independent unknowns rather than trying to solve 12 equations with 12 unknowns.[323]

When solving several differential equations, merging them into a single ODE-solver call, rather than solving them separately, would be both faster and more accurate.[324]

Yi Cao has uploaded numerous utilities to the MATLAB File Exchange that solve data analysis problems very efficiently.[325] Cao's utilities are particularly optimized for performance, often outperforming alternative implementations by a wide margin,[326] and have received extremely high user rating for their usefulness, robustness, and readability. John D'Errico has also generated numerous excellent data analysis utilities,[327] in addition to being arguably the most prolific contributor on various MATLAB forums for the past two decades. Both Cao and D'Errico have for years starred at the top of the File Exchange

Author's list.[328] Readers are encouraged to review their submissions for implementation ideas. In addition to Cao and D'Errico, several other authors have also generated some highly rated utilities in this domain,[329] although Cao and D'Errico appear to be in a league of their own.

3.10 Other Techniques

Numerous additional performance-tuning techniques were suggested over the years. Here is a list of some that could be used in MATLAB, grouped by type.

3.10.1 Coding

3.10.1.1 Recursion

Recursion is the concept of a function calling itself until some condition is fulfilled.[330] A simple example of this is a recursive algorithm for searching a list: if the current list element matches, then return it; otherwise, return the search for the rest of the list:*

```
function foundIdx = searchList(list,elementIdx,value)
    if list(elementIdx) == value
        foundIdx = elementIdx;
    else
        foundIdx = searchList(list,elementIdx+1,value);
    end
end
```

Or, taking the example of the factorial (n!) function:

```
function result = factorial(n)
    result = 1;
    if n > 1, result = n * factorial(n-1);   end
end
```

Recursive code is highly maintainable. Unfortunately, it is also relatively slow. The reason is the numerous function call overheads, repeated recomputations, and allocated stacks that need to be preserved for the entire recursion length (limited by default to 500 in MATLAB). To improve performance, several variants are normally used. In different situations, each of these variants may perform better than the others:

- Use dynamic or static memoization (caching).†
- Have the recursion function return multiple values rather than just one.‡
- Convert the recursion into a format acceptable to *filter* or one of its siblings.§
- Convert the recursion into *tail-recursion*.[331] This typically saves memory allocations and enables JIT-based optimization of the function call overhead:

---

* This oversimplified example does not contain sanity checks and multiple possible optimizations. It is merely provided to illustrate the concept of recursion and should not be used as is.
† See §3.2.4.2.
‡ See §3.2.4.2 again.
§ See §5.7.7.

```
function result = factorial(n)
    result = factorial_tail(n,1);
end

function result = factorial_tail(n, result)
    if n > 1,  result = factorial_tail(n-1, result*n);   end
end
```

- Convert the recursion into a straightforward iteration, again saving memory and function call overheads:

```
function result = factorial(n)
    result = 1;
    while n > 1
        result = result * n;
        n = n - 1;
    end
end
```

3.10.1.2 Using Known Computational Identities

We can sometimes use known mathematical identities to prevent the need for costly computations. For example, using the identity $log(x*y) = log(x) + log(y)$, we can halve the number of costly *log* computations:

```
>> tic, for idx=1:1e6, a=log(idx)+log(2*idx); end, toc
⇨ Elapsed time is 0.100816 seconds.

>> tic, for idx=1:1e6, a=log(2*idx^2); end, toc
⇨ Elapsed time is 0.056854 seconds.
```

Similarly, simplify conditionals such as sqrt(x) > 3 with x > 9, and x^2 > 0 with x ~= 0.

Many such identities are known for a wide range of mathematical functions. It is well worth spending a few minutes to look them up before banging our heads in more complex performance optimizations.

A less trivial example is the well-known Fibonacci sequence,[332] which is examined in additional places in this book. The basic definition of this sequence is*

$$F_n = F_{n-1} + F_{n-2}; \quad F_0 = 0; \quad F_1 = 1;$$

This sequence has a mathematical identity known as *Binet's formula*[333] that enables to directly compute any sequence value, without having to compute its predecessors:

```
function values = fibonacci(n)
    persistent sqrt5 Psi
    if isempty(sqrt5)
        sqrt5 = sqrt(5);
        Psi = (1+sqrt5)/2;
    end
    values = round((Psi.^n - (-Psi).^-n) / sqrt5);
end
```

---

* Recall that in Fibonacci's sequence each value is the sum of its preceding two values: 1, 1, 2, 3, 5, 8, 13, 21, 34, ... Fibonacci sequence generation is also discussed in §3.1.5, §3.2.4.2, §5.7.7, and §9.4.1.

Implemented in MATLAB, this provides an extremely efficient computation of either a single value, or a vector of values:*

```
>> tic, values = fibonacci(1:10), toc
values =
     1     1     2     3     5     8    13    21    34    55
⇨ Elapsed time is 0.000383 seconds.

>> tic, value = fibonacci(7000); toc
⇨ Elapsed time is 0.000052 seconds.
```

An additional example of using an identity is the convolution theorem (§4.5.6).

3.10.1.3 Remove Unnecessary Computations ("Dead-Code" Elimination[334])

Remove computations from code-flow branches that do not need them. If necessary, duplicate some code for different conditional branches:

```
% Naïve code - compute data and then use it selectively
data = someComputationOrIO();
if someCondition
    doSomethingWith(data);
elseif anotherCondition
    doSomethingElseWith(data);
else
    doSomethingElseEntirely(); % data is not used here
end

% Improved code: only compute data where needed
if someCondition
    data = someComputationOrIO();
    doSomethingWith(data);
elseif anotherCondition
    data = someComputationOrIO();
    doSomethingElseWith(data);
else
    doSomethingElseEntirely(); % data is not used here
end
```

During program development, it often happens that we leave old code segments behind, in case we might need them later, or perhaps just out of forgetfulness. Such code segments may in some cases cause unnecessary delays in program execution and should be removed, or at the very least commented out. Modern compilers automatically eliminate such *dead-code* from the compiled output, but this is much harder for an interpreter such as MATLAB.

MATLAB's editor[†] displays a useful warning message (NASGU) about variables that are set without being used. We can use this to identify potential code lines to be removed:[‡]

```
data = someLongComputation();
⚠ The value assigned to variable 'data' might be unused.
```

---

* We use MATLAB vectorization in this case, using the element-wise .^ operator. Refer to Chapter 5 for additional details.
† Or rather, the Code Analyzer (MLint) component that is integrated in the MATLAB editor.
‡ The complete list of MLint-reported performance-related warnings is presented in Appendix A.4.

Note that *absence* of this warning does not imply that a code segment is actually used. For example, we are not warned about calling a function that does nothing useful.

MATLAB's editor (Code Analyzer) cannot detect all forms of unnecessary computations. For example, a very common MATLAB pitfall is multiple unnecessary data transpose. Quite often, the code will run exactly the same when the data is not transposed. Transposing data is very efficient for small data arrays, but can become a performance hog when repeated within a loop, or when applied to large data:

```
data = rand(10000);   % 10K x 10K = ~800MB
tic, data = data'; toc
⇨ Elapsed time is 0.514245 seconds.
```

A variation of this pitfall is applying multiple operations that have opposite effects on the data, canceling each other. Again, in MATLAB, the transpose operator is responsible for many such cases. In the following example, the extra (unnecessary) transpose operations take twice as long as the core **kron** function:*

```
N=100; a=rand(N); b=rand(N);
tic, c=kron(a',b')'; toc
⇨ Elapsed time is 0.771621 seconds.

>> tic, c=kron(a,b); toc
⇨ Elapsed time is 0.278749 seconds.
```

3.10.1.4 Optimize Conditional Constructs

Conditional branches (*if* and *switch* constructs) are processed from top to bottom. Placing the more frequent branch occurrences (*fast path*[335]) before less frequent ones (or for equally frequent conditions, placing the simpler conditionals before costlier ones) ensures that the CPU spends less time evaluating false conditions:[336]

```
if frequentOrSimpleCondition         % Occurs 70% of the time
    doSomething();
elseif lessFrequentOrCostlyCondition  % Occurs 25% of the time
    doSomethingElse();
else                                  % Occurs only 5% of the time
    doSomethingElseEntirely();
end
```

Try to convert simple *switch* constructs into a direct table lookup, which is much faster for the CPU then scanning through the different *switch* options. For example:†

```
% Old (slower) code
total = 0;
for idx = 1 : 1000000   % 1M iterations
    input = mod(idx,5);
    switch input
        case 1,     data = 12.3;
        case 2,     data = -12;
```

---

* Also see §5.7.6; *kron* itself is discussed in §5.2.4.2.
† If we place the following two code snippets in different functions, or if we use different variable names than data for the two code snippets, then the JIT optimizer would automatically and internally convert the first snippet from a switch-case into something faster (like the look-up-table mechanism in the second snippet), giving comparable run-times. But in a general switch-case when JIT does not optimize, we might help performance by doing this optimization manually, as we did here.

```
        case 3,     data = pi;
        case 4,     data = 4.567;
        otherwise,  data = -8.90;
    end
    total = total + data;
end
⇨ Elapsed time is 1.460352 seconds.

% New (optimized) code - 45x faster
total = 0;
data = [-8.90, 12.3, -12, pi, 4.567];   % pre-computed look-up table
for idx = 1 : 1000000  % 1M iterations
    input = mod(idx,5);
    total = total + data(input+1);
end
⇨ Elapsed time is 0.032284 seconds.
```

As a variant of this technique, some conditions can be unified, thereby simplifying the resulting conditional construct and improving performance. For example:[337]

```
a = find( (MT>=(X(n)-0.1)) & (MT<=(X(n)+0.1)) );   % original code
a = find( abs(MT-X(n))<=0.5 );                      % simplified code
```

See additional ideas for optimizing conditional expressions in §3.1.4, §4.2.4, and §3.3.

3.10.1.5 Use Short-Circuit Conditionals (Smartly!)

The logical operators || and && are *short-circuit operators* in MATLAB, as in many other programming languages.[338] This means that the right-hand portion of the logical operator will only be evaluated if the left-hand portion provides a value that does not uniquely determine the value of the entire expression. For example, in the expression

```
if pi < 0 && myFunction() > 3
```

the CPU can determine that the entire expression is false immediately when the left-hand portion (pi < 0) is evaluated, so there is no need to spend time evaluating the right-hand portion (myFunction() > 3). Similarly, in the expression

```
if pi > 0 || myFunction() > 3
```

the outcome is known to be true immediately after evaluating the left-hand portion (pi > 0), so myFunction() will not be evaluated.

In simple cases such as above, where the right-hand operator is a function that might take nonnegligible time to execute, using the || and && operators, rather than the corresponding *eager operators* | and &, improves performance.[339]

Short-circuiting can be used algorithmically, not just as a performance-tuning technique. For example, we can use it to prevent access to uninitialized data or trying to read from missing files:

```
if length(a) > 5 && a(idx-5) > 0
if ~exist(filename,'file') || readDataFile(filename) < 0
```

Note that it is not always advisable to use short-circuit evaluation. In some cases, we may depend on some side effect of the right-hand evaluation. Moreover, MATLAB's | and & operators are not always directly replaceable by || and &&, since | and & are element-wise array operators, whereas || and && only work on logical scalar values.

Even when considering performance, using || and && is only useful when the right-hand portion takes nonnegligible time to compute and when its evaluation is infrequent. In other cases, counterintuitively, using || and && may actually hurt some internal optimizations that can be done at the JIT and CPU levels.[340]

A different type of short circuiting arises when a costly conditional can be broken up into constituent parts, such that part of the costly computations is avoided if the first part of the conditional is enough to evaluate the entire expression. For example:[341]

```
% Standard, non-optimized code
if distanceX^2 + distanceY^2 < someValue

% Optimized code
dX = distanceX^2;
if dX < someValue
    % only calculate dX+distanceY^2 if it is relevant
    if dX + distanceY^2 < someValue
```

3.10.1.6 *Multiply Rather than Divide (or Not)*

Division is a much more computationally intensive operation than multiplication, and can take an order of magnitude longer at the CPU opcode level. The overheads of high-level languages such as MATLAB are such that we rarely see such a difference between regular multiplication and division operations. Where critical, consider precomputing the reciprocal value and transform a division into a multiplication. Modern compilers and interpreters often do this automatically, and I suspect that MATLAB's JIT does so as well. This is proved by running the following code once in the MATLAB Command Window (JIT disabled) and once in a function (JIT enabled):

```
% In the command Window (JIT disabled) - 6x speedup
>> a = 10; tic, for idx=1:1e6, b=a/pi; end, toc
⇨ Elapsed time is 0.018803 seconds.

>> tic, c=1/pi; for idx=1:1e6, b=a*c;  end, toc
⇨ Elapsed time is 0.002868 seconds.

% In a function (JIT enabled) - no speedup at all
function perfTest
    a = 10; tic, for idx=1:1e6, b=a/pi; end, toc
    tic, c=1/pi; for idx=1:1e6, b=a*c;  end, toc
end

>> perfTest
Elapsed time is 0.014184 seconds.
Elapsed time is 0.014064 seconds.
```

The effect is indeed visible when employing nonscalar data. In this case, presumably, JIT is not smart enough to employ this optimization and the speedup is noticeable:

```
% Non-scalar data - 14x speedup for multiplication over division
function perfTest
    a = rand(100);
    tic, for idx=1:1e4, b=a/pi; end, toc
    tic, c=1/pi; for idx=1:1e6, b=a*c; end, toc
end
```

```
>> perfTest
Elapsed time is 0.745946 seconds.
Elapsed time is 0.052900 seconds.
```

In most cases, the performance benefit of converting divisions into multiplications in MATLAB is negligible, and the computation might lead to floating-point inaccuracies (dividing by 3 is not always the same as multiplying by 0.333...). Therefore, while this optimization is well known in the C-world, it often has little impact in MATLAB.

3.10.2 Data

3.10.2.1 Optimize the Processed Data

A common performance-tuning pitfall is to try to optimize the code rather than the data on which the code operates. We have seen above (§3.6.3) how uniquifying data can produce a 200-times speedup at the expense of two additional code lines. We would never have been able to achieve a similar speedup by trying to optimize the code rather than the data. The lesson is that we should always check whether it is possible to simplify and/or reduce the data, before trying to optimize the code.

As another example, consider the case of a matrix that contains a huge number of empty (zero-value) cells and only a few valid (nonempty) ones. If full vectorization is not possible, then it makes sense to extract and process only the valid cells, rather than loop over all the data elements. Alternatively, consider using sparse data.*

3.10.2.2 Select Appropriate Data Structures

Some data structures are more suitable for insertion than for searching, extraction, deletion, and/or memory usage; the reverse may be true for other structures. No data structure is fully optimized for all usage aspects. For improved performance, we should select a data structure that suits the more frequent usage of the data. For example, storing strings in MATLAB cell arrays is natural, but a hashtable[†] or Huffman tree[342] may be more appropriate for specific needs.

There are many textbooks on data structures that can help select the optimal construct for specific needs. Adapting them for MATLAB is usually relatively easy, and working examples can often be found on the MATLAB File Exchange.

3.10.2.3 Utilize I/O Data Compression

Use I/O data compression[343] whenever possible. This will reduce the amount of I/O at the expense of some CPU time required to compress and decompress the data. Since I/O is many orders of magnitude slower than CPU, the performance speedup can be significant. A typical usage example is storing data in a regular binary (compressed) MAT format rather than in a textual (uncompressed) format. Similarly, use compressed GIF/JPG/PNG formats to store images rather than uncompressed BMP.[‡]

Note that using data compression is not generally beneficial for non-I/O computations. The additional development costs associated with debugging and maintenance, as well as the run-time overhead of compression/decompression is simply not worth the savings except in very rare cases (such as when memory usage is at a premium).

---

* See §4.1.2.
† See §3.2.6.
‡ See §11.4.

3.10.3 General

3.10.3.1 Reduce System Interferences

Avoid running CPU-crunching or memory-heavy non-MATLAB processes at the same time as MATLAB. This may sound trivial, but many people are not aware that

- Even a seemingly inactive browser is often using the CPU and memory extensively behind the scene.
- Our OS might be configured to automatically download and install updates and Murphy's law dictates that this will consistently happen when you run your app.
- Our OS may be demoting long-running MATLAB processes to low priority.[344]
- Our system's antivirus software is constantly downloading virus signature updates and scanning our file system and memory for infections.
- All sorts of installed services and daemons are silently working in the background, churning CPU, I/O, and memory for their work.

By being aware of these processes, we can decide whether to stop them or accept their performance penalty. For best performance, we should run our MATLAB application in as sterile an environment as possible (e.g., in Windows' Safe Mode).

3.10.3.2 Self-Tuning

A *self-tuning*[345] (or *adaptive*) application constantly monitors run-time parameters and automatically updates its internal operations to achieve a consistent objective, such as consistent performance.* Automated self-tuning relieves users from constantly tweaking parameters manually, and provides faster and more accurate feedback.

For example, an application could use *tic*/*toc* to constantly measure the timing of a critical code section. If the timing increases beyond some predefined threshold, the application's internal parameters would be tweaked to reduce the code section's accuracy, bringing performance back in line. Likewise, if things start getting faster for some reason, accuracy could be improved. The overall effect would be that regardless of the external situation, the application's performance is constant and predictable.

It is interesting to note that even core algorithms, such as FFTW, which is used by MATLAB's *fft* implementation, employ self-tuning mechanisms.[346] In the case of FFT, this self-tuning can be controlled via the built-in function *fftw* (also see §4.5.6).

3.10.3.3 Jon Bentley's Rules

Computer scientist Jon Bentley's seminal book *Writing Efficient Programs*[347] includes a series of rules for improving the performance of computer programs, independent of platform and programming language. Many of Jon's suggestions were discussed in this chapter; some others may be irrelevant on modern computers and/or MATLAB. Still, this book (along with his other two books) is a valuable resource well worth reading. A concise list of Bentley's rules is provided in several online locations.[348]

---

* The objective can also be a different constant measurable rate, such as CFAR (constant false alarm rate) in radars or constant BER (bit error rate) in communications.

4

MATLAB®-Specific Techniques

In addition to the general performance-tuning techniques listed in the previous chapter, we can employ several techniques that are, to a large extent, specific to MATLAB. These techniques shall be described in the following chapters:

- This chapter explains general MATLAB programming techniques.
- Chapter 5 explains how to use MATLAB's implicit parallelism (indexing and vectorization).
- Chapters 6 and 7 explain how to use explicit parallelism in a variety of manners.
- Chapter 8 discusses various ways of using compiled binary code in MATLAB.
- Chapter 9 details techniques that relate to MATLAB's memory management.
- Chapter 10 details techniques that are specific to MATLAB graphs and GUI.
- Chapter 11 details techniques that relate to I/O (especially file read and write).

4.1 Effects of Using Different Data Types

MATLAB's native data type is the 2D numeric multidimensional array, or matrix. In fact, MATLAB's name is an acronym of *MATrix LABoratory*, emphasizing MATLAB's origins in numeric matrices. MATLAB's built-in functions, and in particular its numeric analysis functions, are highly performance tuned.

Still, any half-serious MATLAB program needs to use some other data types at some point. These range from simple strings, cell arrays, and structs, through sparse and non-double numeric arrays, to class objects. Using these diverse data types can significantly improve the program's usability and maintainability. Unfortunately, not all data types are alike in terms of performance, and we would do well to consider performance implications when deciding which data type to use.

The performance difference between alternative implementations of an algorithm using different data types is often too small to justify the increased development and maintenance costs of using less appropriate data types. It usually makes sense to code initially using the more appropriate data type for the task, and to refactor the code using simpler types only in specific identified performance hotspots (particularly within code segments that execute numerous times, such as large loops).

4.1.1 Numeric versus Nonnumeric Data Types

Having said all this, a general rule of thumb is that numeric data is faster to use than non-numeric types (strings, cells, structs, or class objects). The more complex and nonnative the data type, the more internal processing is required to access its stored data, with class objects being the most expensive in this regard (see §4.7 for details).

There are several reasons why numeric data is generally faster than nonnumeric data: internal MATLAB and third-party library (for instance, BLAS which is used for numeric processing, or FFTW for FFT processing) functions are highly tuned for numeric* data and less-so for other types. Also, numeric data is stored compactly in a contiguous memory block, unlike other data types that require additional memory management and cannot use the benefits of CPU data caching and prefetching.

Here is a simple example illustrating the general advantage of using numeric data:

```
% First calculate using a numeric array
numericArray = rand(1e6,1); % 1M random numbers between 0-1

tic; total = sum(numericArray > 0.5); toc
⇨ Elapsed time is 0.001162 seconds.

% Now do the same with a cell array - 3000x slower, 15x more memory
cellArray = num2cell(numericArray); % prepare the data for our test
tic, total = sum(cellfun(@(element) element>0.5, cellArray)); toc
⇨ Elapsed time is 3.580406 seconds

% Now do the same with a struct array - 4500x slower, 15x more memory
for idx=1e6:-1:1, structArray(idx).a = numericArray(idx); end
tic, total = sum(arrayfun(@(element) element.a>0.5, structArray)); toc
⇨ Elapsed time is 5.200801 seconds.
```

In such cases, when the original data is stored in a cell array or struct array, we could temporarily convert them into a numeric array for the benefit of the computation:

```
% Cell array - now only 350x slower (9x speedup)
numericArray2=[cellArray{:}]; total=sum(numericArray2>0.5);
⇨ Elapsed time is 0.409988 seconds.

% Struct array - now only 335x slower (13x speedup)
numericArray3=[structArray.a]; total=sum(numericArray3>0.5);
⇨ Elapsed time is 0.389096 seconds.
```

The speedups in these examples (9× and 13×, respectively) may seem impressive, but in real-world applications I rarely see such large values. Moreover, these speedups are only gained at the expense of having to support data in different formats in the same program, and to code the program using a data type that may be less natural. This often leads to extra development and maintenance costs, which should be weighed against the performance benefits. Finally, in some specific cases, cell arrays may actually perform better than numeric arrays due to more optimized memory access.[349]

 ### 4.1.2 Nondouble and Multidimensional Arrays

When a cell array or struct array holds heterogeneous data types (e.g., strings and numbers), we could split them into separate arrays of consistent data types. It is often tempting to write code that uses a single two-dimensional cell array that contains some numeric key or index as well as some strings and/or structs. Instead, consider splitting the cell array

---

* Usually, noncomplex, nonsparse double-precision data. However, some functions are faster for sparse or non-double data.

into a single numeric array holding the key values, and corresponding arrays that hold the strings and structs data.

As a logical extension of this idea, a cell array holding a single numeric array in one of its cells (e.g., {[1,2,3,4]}), is both faster and uses less memory than a regular cell array that holds the same values ({1,2,3,4}). The reason is that each cell element is stored separately in MATLAB memory, and requires a separate ~100 bytes header.[*][350] Therefore, using {1,2,3,4} rather than {[1,2,3,4]} requires three additional headers for the extra cell elements.[351] The extra headers mean extra memory, as well as extra CPU time for dereferencing the data. It also means that the data elements reside in noncontiguous memory locations, preventing effective vectorization.

Similarly, a struct holding a single numeric array in one of its fields (s.n(1:5)) is both faster and uses less memory than a struct array with a field that holds only a single value (s(1:5).n).[352] In short,

A struct/cell of arrays is faster and smaller than an array of structs/cells

The additional memory required for storing the more complex data types has an immediate effect on performance when allocating and deallocating large arrays, in addition to the slower access and processing times as shown above.[†]

Using nondouble data types sometimes carries a performance penalty in unexpected places. For example, a caveat of using nested functions is that on MATLAB release R2011b (7.13) and earlier, accessing parent cell arrays is significantly slower than accessing numeric arrays (see §9.5.6). Another caveat is that pre-allocation of nondouble data types behaves somewhat differently than numeric doubles (see §9.4.5).

Multidimensional arrays are supported in MATLAB and are usually efficient, but for many if not most built-in functions a 2D array is even faster. If possible, try to see whether the data can be represented in a 2D matric rather than an N-D array. We can use the built-in functions *reshape, transpose*, and *permute* to transform our data, if necessary.[353] In many cases, smart application of these three functions can transform data without resorting to costly cell arrays or iterative element-wise loops.[‡]

4.1.3 Sparse Data

Sparse arrays are generally slower and use less memory than full arrays when the data density is low (few percent or less). When data density is higher than (say) 20%, the memory allocation and processing time for the sparse matrix is comparable or higher to a full matrix and so it would make sense to use a full matrix for both objective (performance and memory) and subjective (development time and maintainability) reasons. Cases of intermediate densities should be considered on a case-by-case basis.

Sparse arrays can be faster than full arrays, depending on the situation. If the data is indeed sparse (<1% of cells hold actual data), using sparse arrays can significantly improve memory-related performance, and possibly also processing performance.[354] Unfortunately, only a small subset of MATLAB functions support sparse arrays, and optimizations such as in-place data manipulations (§9.5.2) are often not performed on sparse data. But in

---

[*] Note that the *whos* command and the Bytes column in the MATLAB Workspace browser do not include the header size when reporting the memory used by MATLAB variables.

[†] See Chapter 9.

[‡] The idea of using *reshape* and *permute* is further discussed in Chapter 5, as a vectorization method. See §5.5 for additional performance aspects of multidimensional data.

cases where memory throughput or size is a limitation, using a sparse array can significantly improve performance:[355]

```
% Comparison of sparse vs. full matrices
S = sprandn(1000,1000,0.01); % 0.01 data density
F = full(S);
tic, B = S * S; toc % Sparse (10K non-zero elements, 163 KB)
⇨ Elapsed time is 0.004227 seconds.

tic, C = F * F; toc % Full (1M elements, 8 MB, 13x slower)
⇨ Elapsed time is 0.055718 seconds.
```

Running this for multiple values of data size and density yields the following graph:

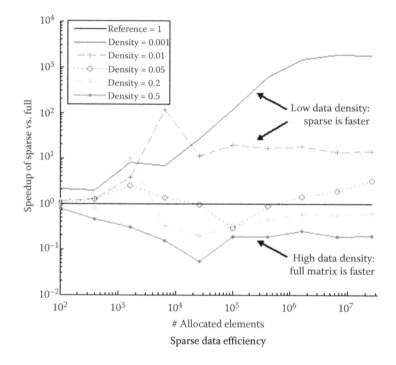

Sparse data efficiency

This graph illustrates that using sparse arrays is faster than full arrays, at low data density (<1%). In mid-range densities (5%–20%), the speedup depends on the specific case, and may be positive or negative (i.e., full data may be faster) depending on data size. In high data densities (50%+), a full matrix is always faster than a sparse matrix.

Note that sparse matrices have a larger memory footprint (16 bytes per element plus overhead) than regular (full) numeric arrays (8 bytes) for nonzero elements.* Also, sparse matrices currently (R2014b) only support double-precision numeric data, not singles, integers, or characters. Finally, not all internal functions support sparse data.

Sparse matrices have a significant overhead when updating/assigning values. Doing so in a loop should be avoided. The MATLAB Editor's Code Analyzer (MLint) includes a specific warning (SPRIX) when such a case is detected:†

---

* Zero-value elements are not stored in sparse format at all, yet require 8 bytes in full double-precision numeric arrays.
† The complete list of MLint-reported performance-related warnings is presented in Appendix A4.

```
data = sparse(1,1000);
for idx = 1 : 1000
    data(idx) = 0;    % #ok<SPRIX>
```

⚠ This sparse indexing expression is likely to be slow. (Details ▲)

Explanation   ↕   ▲

Code Analyzer detects an indexing pattern for a
sparse array that is likely to be slow. An
assignment that changes the nonzero pattern of
a sparse array can cause this error because
such assignments result in considerable
overhead.

Suggested Action

If possible, build sparse arrays using sparse
as follows, and do not use indexed assignments
(such as C(4) = B) to build them: ▼

A common pitfall when using sparse data is to allocate the sparse elements one by one in an iterative loop. Due to the way that sparse arrays are represented internally, each allocation of a new sparse element updates all other elements. For sparse data having n elements, the overall allocation performance would therefore be O(n^2), becoming a performance bottleneck that completely overshadows any potential speedup.[356]

Instead, we should be careful to create sparse arrays only using the *sparse* or *spdiags* functions, except in very rare cases. The general syntax for *sparse* is S = *sparse(i,j,v)* where i, j are vectors of the sparse data's index locations (row and column indices, respectively) and v is the vector of corresponding data values, such that S(i(k),j(k)) = v(k). Using this syntax, we can create the sparse data "virtually", by creating the three vectors i, j, and v; once they are ready we call S = *sparse(i,j,v)* to create the sparse matrix. In one simple example shown by Tim Davis, this modification caused a 200× speedup.[357]

When the data density is relatively high, or when testing indicates that the sparse overheads outweigh the performance benefits, then it might be worthwhile to spend our performance-tuning energies at finding a more efficient full-matrix algorithm, than trying to optimize the sparse matrix. A good example of this was discussed in StackOverflow a few years ago (it is worth reading the comments on that thread).[358]

There are several cases in which low-density sparse data can be used not just as a more efficient replacement for dense data, but as a dedicated data structure. These cases include representation of graph nodes and manipulation of large data sets.[359]

The internal data storage of sparse data stores triplets of values for each nonzero data element: its row index, column index, and data value.* For low data density this provides a large memory saving, in a similar manner that indexed images (e.g., GIF files) are smaller

---

* This is an inaccurate simplification: sparse data are actually stored in Compressed Sparse Column (CSC) format: http://mathworks.com/help/matlab/apiref/mxsetjc.html (or: http://bit.ly/1mlB0go). Readers interested in archeology might find interest in the following outdated (20+ years old) yet detailed technical report on MATLAB's initial sparse implementation: http://mathworks.com/help/pdf_doc/otherdocs/simax.pdf (or: http://bit.ly/1lAEqik); much has changed since then of course.

than true-color RGB images (e.g., PNG) for flat-color cartoons.* However, while no space is allocated for unused indexes, space is indeed allocated for all possible column indexes as a global overhead. For this reason, it is generally better to reshape the data such that the large dimension is stored as the sparse-data row rather than the column (i.e., ensure a smaller column dimension):[360]

```
s1 = sparse([],[],[],1,N); % row vector - slower, more memory
⇨ Elapsed time is 0.023915 seconds.

s2 = sparse([],[],[],N,1); % column vector - much faster, low memory
⇨ Elapsed time is 0.000031 seconds.

>> whos s1 s2
Name            Size              Bytes  Class     Attributes
s1              1x10000000     80000024  double    sparse
s2       10000000x1                  32  double    sparse
```

In addition to its use for binning, *accumarray* (§5.3.5) can also be used for efficient sparse-data multiplication,[361] although the reason may not be immediately obvious.

Two documented bugs in various MATLAB releases are related to sparse matrices and may be of interest to readers who use sparse data on the relevant releases:

- 306972 — Scalar times sparse matrix yields performance degradation (reported for R2006a; fixed in R2006b).[362]

- 535814 — Linear algebra operations involving sparse matrices or small full matrices may see a decrease in performance on some 8-core machines running on 32-bit Windows XP (reported for R2008b-R2011b; fixed in R2012a; workaround available for the earlier releases).[363]

As an alternative to using *sparse* matrices, consider using the *packed* format for symmetrical 2D matrices.[364] BLAS is slower in processing packed matrices than full ones, but the 2:1 memory saving may just be enough to avoid out-of-memory errors.

4.1.4 Modifying Data Type in Run Time

MATLAB is a weakly-typed language. This means that we do not need to pre-declare the type of any variable — MATLAB automatically assigns the correct type in run time, based on the assigned value. However, this flexibility has a performance impact: Run-time-type assignments reduce the ability to optimize code for performance.

The effect is even stronger if we happen to modify a variable's assigned type in run time. This can completely throw the JIT accelerator off, possibly to the point that JIT optimization will no longer be used and the entire program will be interpreted line-by-line. This in turn can have a significant impact on performance:[365]

```
for idx=1:1e6, if mod(idx,1), a=2; else a=3; end, end
⇨ Elapsed time is 0.209135 seconds.

for idx=1:1e6, if mod(idx,1), a=2; else a='b'; end, end
⇨ Elapsed time is 7.100596 seconds.
```

---

* See §10.1.12.

Setting an altogether different type, as above (double ⇔ char), can easily be detected and avoided. However, it is more difficult to detect/avoid cases such as int16 ⇔ int32 or double real ⇔ complex. If our program uses such data types anywhere, we should spend some time to carefully ensure that variable type is not modified in run time. If we detect that it is, then we should use separate variables for the separate data types.

Modifying the data type, especially between numeric types (e.g., int16 ⇔ int32) is sometimes unavoidable.* In some cases, we can directly specify the required data type.† In other cases we could use the built-in ***typecast*** function. The problem with ***typecast*** is that it creates a deep copy of the data before transforming it. If we do not mind that the original data will be modified, we can significantly speed up type casting using James Tursa's in-place *typecastx* utility.[366]

4.1.5 Concatenating Cell Arrays

When accessing cell arrays, it is better not to use the {:} format, since this generates extra memory allocations and is much slower than the alternative. For example,

```
c = num2cell(1:1e6);  % 1M elements

c2=[]; c2={c{:} 123};
⇨ Elapsed time is 0.332692 seconds.

c2=[]; c2=[c {123}]; toc
⇨ Elapsed time is 0.013609 seconds.  → 25x faster
```

Similarly,

```
c2=[]; c2=[{c{1:1000}} {123}];
⇨ Elapsed time is 0.000651 seconds.

c2=[]; c2={c{1:1000} 123};
⇨ Elapsed time is 0.000519 seconds.

c2=[]; c2=[c(1:1000) {123}];
⇨ Elapsed time is 0.000050 seconds.
```

Luckily, the MATLAB Editor's Code Analyzer (MLint) includes specific warnings (CCAT and CCAT1) about using the slower coding format:‡

```
c2={c{:} 123};
```
⚠ { A{:} B } can often be replaced by [A {B}], which can be much faster. ⟨ Details ▾ ⟩

Sample MLint (Code Analyzer) message

```
c2={{c{:}} 123};
```
⚠ { A{I} } can usually be replaced by A(I) or A(I)', which can be much faster.

Sample MLint (Code Analyzer) message

---

* See §5.5, for example.
† See §9.4.3 and §11.3.4.
‡ The complete list of MLint-reported performance-related warnings is presented in Appendix A.4.

4.1.6 Datasets, Tables, and Categorical Arrays

MATLAB's Statistics Toolbox includes a data type called *dataset*,[367] which is similar to cell array in its ability to store data of heterogeneous types. Cell arrays can hold any data type in any cell element. Like struct arrays, datasets are more structured in that the data type of any column is homogeneous; heterogeneous types are only allowed between columns. Datasets are often used to hold multiple observations (rows) of several different variables (columns), each having its own unique data type.

Dataset alternatives include cell arrays or struct arrays (each struct is an observation, where each struct field is a variable).* Datasets and struct arrays are quite similar, and I see no significant advantage to datasets over struct arrays from a development and maintenance standpoint (cell arrays are less "user-friendly" in this respect, since the programmer needs to keep track of the meaning of each separate column).

Performance-wise, datasets incur a huge performance penalty (100x or more!) compared to regular numeric arrays, struct arrays or even cell arrays.[368]

Like datasets, the new *table* and *categorical*[369] types are basically MCOS wrappers for primitive arrays, and are similarly slow.[370] The *categorical* data type stores values using integer indices to the unique values, thereby ostensibly saving storage space for duplicate values, especially when these values are large (e.g., long strings) and repeated often. However, the memory and performance overheads associated with the *categorical* class often outweigh any savings. MathWorks webinars sometimes use a data file with thousands of records of flights into and out of Logan airport, Boston. Naturally, these records will have numerous repetitions of the origin and destination airport names — prime candidates for storing as compact categorical arrays.[371]

Having said all this, performance is not the only important aspect in development. Ease of development, debugging, and maintenance should also be considered, and the new data types do provide important benefits in those regards.

4.1.7 Additional Aspects

Among numeric data, there are often performance differences between using double precision, single precision, integer or Boolean (logical) data. In this case, it cannot be generalized that one type is generally faster than the others. The savings due to a lesser memory footprint are sometimes offset by slowdowns due to extra CPU processing. In one application that I wrote, changing the data type from *double* to *single* sped up the program enough to make a difference between real-time and non-real-time performance. But in many other cases it may not be worth the effort.[372]

Among the functions that support multiple numeric data types (not all functions do), some functions are faster for doubles while others may be faster for singles or integers. This may depend on the internal implementation, the size of the data, whether or not JIT is being used,[373] the Operating System,[374] and the CPU. There are so many affecting factors in this case that there is no general advice other than to test the different variants on your target platform using a representative data set.

MATLAB uses Intel's Integrated Performance Primitives Library (IPPL)[375] to optimize some image processing functions. IPPL uses SSE and AVX CPU instructions, whose processing paths are shorter for 32-bit singles than 64-bit doubles, making double-precision

---

* Recall (§4.1.2) that a struct of arrays is faster, yet perhaps less intuitive, than an array of structs; both are alternatives to *dataset*.

calculations 50%–100% slower than single precision.* Unsigned integers (*uint8* and *uint16*) are even faster, while signed integers (which do not use IPPL) are much slower.†

As a counter-example, the standard *fft* function is faster for doubles than singles, since the underlying FFTW library is better optimized for doubles.

For maintainability reasons I advise not to customize the numeric data types to the functions being used. However, this can be done for very specific point cases (e.g., within long loops) where we have identified that a performance hotspot can be solved using a different data type. Such cases need to be checked individually, since as noted there is no general rule about which data type is faster — we need to decide based on hard empirical data, by testing the different possible combinations.

When assigning logical values, it is better to directly use the *true* and *false* functions rather than the equivalent but slower *logical(1)* and *logical(0)*.

MATLAB strings are examined in detail in §4.2. For the purposes of this section it is sufficient to say that we should not use strings except for text processing. Most other uses (dates, timestamps, enumerations, input selectors, and so on) can be accomplished much faster using numeric equivalents.

4.2 Characters and Strings

A general rule of thumb in performance tuning is not to use strings except for text processing. Most other uses are performed faster using numeric equivalents. We have seen in §4.4 how using numeric *datenum* values is faster than using date strings:

```
% Using string timestamps
flag = strcmp(datestr(dateNum1),datestr(dateNum2));
⇨ Elapsed time is 0.001005 seconds.

% Using numeric date values - hundreds of times faster
flag = (dateNum1 == dateNum2);
⇨ Elapsed time is 0.000002 seconds.
```

A similar thing can be said of enumerations and input selectors. That is, instead of using input parameters that specify a selector as a string (e.g., 'stage3' or 'detailed'), we can use a numeric representation such as 3 or –1. Note that while this would be faster for performance, the obvious drawback would be degraded maintainability.

MATLAB's command syntax[376] is often used to send input parameters to functions. This causes the input parameters to be considered as strings, although the user's intent may possibly have been to use them as numeric values. This might affect both the correctness and the performance of the invoked function. For this reason, it is preferable to use the equivalent functional form, which makes the usage explicit:

```
% Command syntax
func Yair 123
```

---

* Until R2012a, doubles were much slower than singles (even 20–30× slower). The reason was that IPPL was not used for doubles, only singles and integers. This was fixed in R2012b to also include doubles; R2013a saw even higher speedups due to added GPU support. See http://mathworks.com/matlabcentral/answers/68620-performance-issue-of-imrotate-in-double-precision-mode (or: http://bit.ly/13GNv0Q) and also the internal documentation (help section) of the built-in *ippl* function.

† Although, again, both signed and unsigned integers are supported in general by IPPL.

```
% Functional syntax (equivalent)
func('Yair','123');

% Functional syntax (faster-numeric)
func('Yair',123);
```

Unfortunately, we do not always have a choice of processing the data in any numeric variation, forcing us to use regular string processing. In this case, the nature of strings implies that LIH (§3.1.1) and caching (§3.2) can often be used to reduce the amount of expensive string processing. We should always try these techniques before modifying the algorithm to use faster string processing — in general caching and LIH are both more effective and less risky (in terms of the possibility of creating bugs).

4.2.1 MATLAB's Character/Number Duality

Strings are implemented in MATLAB as regular arrays of characters that are in essence simple integer values representing the corresponding Unicode (UTF-8) value.* Such characters can be used and manipulated just like any other numeric array, and the same functions that can be used with numeric arrays can be used with strings with few exceptions.

We can use this fact for very efficient operations such as the following, for converting a string into a number (assuming the input is a valid string representing an integer):[†]

```
>> str = '54321';
>> str + '0'
ans =
    5    4    3    2    1

>> 'abcde' - 'a'
ans =
    0    1    2    3    4
```

Such "string arithmetic" may not be intuitive. So whenever using this technique, we should include a comment explaining it, to avoid code maintenance problems.

Jan Simon's *DateConvert* utility (see §4.4) uses this technique to efficiently parse date/time strings of a known format.

4.2.2 Search and Replace

We can extend the character/number duality principle to searching for a specific value or sub-set in a numeric array. We can treat the numeric array as a character array and use the *strfind* function to find a numeric "sub-string" in that array.[377] This turns out to be faster than using loops or the *find* function, presumably since *strfind* skips the intermediate step of computing and allocating a logical array.

For example, searching for newline characters in a million-element array:

```
data = randi(1000,1,1e6); % 1M elements between 1-1000

% Using find
n = find(data==10);
⇨ Elapsed time is 0.002824 seconds.
```

---

* For standard English (ANSI) characters this corresponds to their ASCII code.
† See §5.7.14 for an additional example that is a bit more complex.

```
% Using strfind - 2.5x faster
n = strfind(data,10);
⇨ Elapsed time is 0.001177 seconds.
```

We can even use this technique to easily search for subsets — something that would be harder to do otherwise:

```
n = strfind(data,[5,10]);
⇨ Elapsed time is 0.001224 seconds.
```

Note *strfind*'s limitation that its inputs must be single-row arrays (not matrices or column vectors), and the matching needs to be exact (not isprime(data) or data < 10).

Urs (us) Schwartz, who was the first to report the *strfind* tip back in 2002,[378] created a long list of excellent utilities on the MATLAB File Exchange that use this and other string-processing performance tricks. Specifically, users may be interested to try his *strpat*,[379] *fpat*,[380] *assort*,[381] *rude*[382] (which inspired several other utilities), and last but not least — *grep*,[383] which is one of my favorite utilities in the entire File Exchange.

Regular expressions[384] are a very powerful tool, and MATLAB's set of regex functions (esp. *regexp*, *regexpi*, and *regexprep*) enable very versatile pattern matching of a sub-string (pattern) within a search string. However, this versatility comes at a performance cost. For static pattern strings that do not need any regex feature, it is better to use the static and more limited alternatives *strfind* and *strrep*:*

```
% Using regexp
str = 'This is sentence #1. This is sentence #2.';
idx = regexp(str,'sentence'); % = [9,30]
⇨ Elapsed time is 0.000198 seconds.

% Using strfind - 13x faster
idx = strfind(str,'sentence');
⇨ Elapsed time is 0.000015 seconds.
```

Similarly, when replacing singular strings:

```
% Using regexprep
str = regexprep(str,'sentence','question');
⇨ Elapsed time is 0.000206 seconds.

% Using strrep - 13x faster
str = strrep(str,'sentence','question');
⇨ Elapsed time is 0.000016 seconds.
```

When replacing strings, we should be aware of *strrep*'s limitation of being able to process only a single pattern at a time.† If we wish to make multiple replacements in the same input strings, we need to either use *strrep* repeatedly, or to use *regexprep* (which accepts a cell-array of patterns). We should test both cases separately, but in

---

* Eric Sampson, a former MathWorker, commented that *"folks who work on MATLAB's regexp engine have made performance improvements over time, and continue to do so. Not something that would make it into the release notes, but more 'behind the scenes' improvements".* (http://mathworks.com/matlabcentral/newsreader/view_ thread/330388#907584 or: http://bit.ly/1672nRn). Unfortunately, these improvements have still not made regex performance be on par with the simpler string functions.

† *strrep* actually accepts a cell array of patterns, but the result is a cell array of strings corresponding to each of the separate patterns, rather than a single string modified according to all the patterns. In the vast majority of cases this is not what we need.

general it can be said that if more than ~10 patterns are needed for the same input string, then it is probably better to use *regexprep*, if only for the improved readability and maintainability:

```
oldPatterns = {'This','is','sentence','#'};
replacements = {'That','was','question','number'};
str = regexprep(str, oldPatterns, replacements);
```

Despite being able to process only a single pattern at a time, *strrep* is able to process multiple input strings in a cell array (*regexprep* can also do this).[385] Using this cell-array input format enables vectorized processing that is faster than using a loop:

```
% Standard loop over all 1000 input strings in cell-array strArr
for idx = 1 : length(strArr)
    strrep(strArr{idx},'is','was');
end
⇨ Elapsed time is 0.001296 seconds.

% Using a vectorized cell-array input - 2x faster
strrep(strArr,'is','was');
⇨ Elapsed time is 0.000597 seconds.
```

Other possible alternatives to regex functions are the *sscanf* and *textscan* functions, when *regexp* or *regexprep* are used to extract sub-strings from a larger input string based on some nonstatic pattern (hence the inability to use the static *strfind/strrep*).

For example, we sometimes need to employ a different processing logic based on the MATLAB release currently in use. MATLAB releases have version names that are strings that look like "6.5", "7.6", "7.12", or "8.2". In such cases, we cannot simply convert the version string into a numeric value and compare it versus some value, since it would turn out that 7.6 > 7.12 (numerically), although "7.6" < "7.12" (version-wise). For this reason, we need to process the major version number ("6", "7", or "8") separately from the minor version number. In R2007a onward we can use the built-in function *verLessThan*, but if we need to support R2006b and earlier, we can use the following snippet (*verLessThan* uses similar, although not identical, logic):

```
% Using regexp
% Get the version number corresponding to '8.1.0.604 (R2013a)'
[a,b,c,d] = regexp(version,'(\d+)[.]');
versionVector = str2double(d)';
multipliers = 10.^(0:-3:-3*(length(versionVector)-1));
versionNumber = multipliers * versionVector; % =8.001000604
⇨ Elapsed time is 0.000475 seconds.

% Using sscanf - 6x faster
versionVector = sscanf(version,'%d.'); % [8; 1; 0; 604]
multipliers = 10.^(0:-3:-3*(length(versionVector)-1));
  % multipliers = [1, 1e-3, 1e-6,...]
versionNumber = multipliers * versionVector; % =8.001000604
⇨ Elapsed time is 0.000070 seconds.
```

This example may not be particularly useful as-is. After all, the MATLAB version number does not change during a program's execution and so the processing result (version-Number) can be cached with a maximum one-time performance penalty of only 0.4 ms. However, the point I am trying to make here is more general: using *sscanf* or *textscan* can often be more efficient than using a regex function. We shall consider this phenomenon again in the next section.

When using regex functions, it is best not to use complex regular expressions (look-ahead/look-behind operators,[386] conditional evaluations, dynamic expressions, etc.) if this is not strictly necessary. Such constructs heavily tax the regex engine. In the vast majority of cases we can make do perfectly well with simpler regex patterns.

An example of a bad design choice is ***strjoin*** and ***strsplit***'s reliance on the private ***strescape*** function, which in turn uses regex dynamic command expression[387] to convert an escape sequence into the corresponding control character (e.g., '\t' to char(9)). This turns out to be ~100× slower than a much simpler equivalent call to ***sprintf***, causing ***strjoin*** and ***strsplit*** to be much slower than necessary.*

As an alternative to ***strjoin***, use Jan Simon's *CStr2String*, referenced in §4.2.5.2 below.

In 2002, Peter Boettcher wrote a simple string splitter in MEX C-code (*splitstr*) that to this day (R2014b) outperforms the built-in ***strsplit*** by a wide margin:[388]

```
% strsplit - MATLAB builtin
strs = strsplit('this is a test',' ');
⇨ Elapsed time is 0.008728 seconds.

% splitstr - MEX variant, 170x faster (note the transposed result)
strs = splitstr('this is a test',' ')';
⇨ Elapsed time is 0.000051 seconds.
```

Having said that, regex functions enable us to perform in a single function call what would otherwise necessitate multiple string-processing functions, so that using a smart pattern and using more of the regex functions' many output parameters can actually save us a fair bit of processing run time:

```
% Find a file's direct parent-folder name ('iofun')
str = which('fileread'); % = ...\toolbox\matlab\iofun\fileread.m

% Using non-regexp processing
[dummy,folderName] = fileparts(fileparts(str));
⇨ Elapsed time is 0.000251 seconds.

% Using regexp look-ahead processing - 3x faster
folderName = regexp(str,'\w+(?=\\\w+\.[mp])','match');
⇨ Elapsed time is 0.000076 seconds.
```

In some cases, regexp functions are both faster and more readable and maintainable:[†]

```
% Find all words that contain i..s => {'This','is','first'}
str = 'This is the first of many an idiom';

% Using non-regexp processing
strs = {};
startIdx = strfind(str,'i');   % =[3,6,14,33,35]
endIdx   = strfind(str,'s');   % =[4,7,16]
spaceIdx = [0, strfind(str,' '), length(str)+1];
```

---

* For all practical purposes ***strescape*** is functionally equivalent to ***sprintf***, although some edge-case handling may be different.

† This can be further vectorized for slightly improved performance at the expense of degraded legibility; the end-result will still be slower than the simple ***regexp*** expression. Note that the code does not take care of punctuation marks, while ***regex*** does.

```
wordStartIdx = spaceIdx + 1;
wordEndIdx   = spaceIdx - 1;
while ~isempty(startIdx)
    thisWordEndIdx = min(endIdx(endIdx>startIdx(1)));
    if isempty(thisWordEndIdx), break; end
    thisWordStartIdx = max(wordStartIdx(wordStartIdx<=startIdx(1)));
    thisWordEndIdx   = min(wordEndIdx(wordEndIdx>=thisWordEndIdx));
    strs{end+1} = str(thisWordStartIdx: thisWordEndIdx);
    startIdx(1) = []; % continue to the next potential match
end
⇨ Elapsed time is 0.000250 seconds.

% Using regexp - simpler, more readable/maintainable, and 3x faster
strs = regexp(str,'\<\w*?i.*?s\w*?\>','match');
⇨ Elapsed time is 0.000072 seconds.
```

When using *regexp*, consider using the optional 'once' parameter, so that the regex engine will stop processing once it found the first match. This tip is only actually relevant for very large input strings (thousands of characters) or within loops, or when memory is at a premium — the difference for shorter strings is small and has no real meaning in a stand-alone case (outside of a loop).*

For example, on my computer, a 1000-character string was processed in 0.2 ms by a regular *regexp* versus 0.03 ms for *regexp(…,'once')*; a 1-million character string was processed by a regular *regexp* in 114 ms, while *regexp(…,'once')* only took 1.5 ms:

```
str = char('0'+randi(10,1,1e6));

idx = regexp(str,'(\d+):');
⇨ Elapsed time is 0.114033 seconds.

idx = regexp(str,'(\d+):','once'); % 75x faster
⇨ Elapsed time is 0.001487 seconds.
```

If you find yourself using *strtok* repeatedly, this could be a prime candidate for using either *textscan* or *regexp*. Even a single *strtok* is twice as slow as a simple *regexp*:

```
str = strtok(str);
⇨ Elapsed time is 0.000084 seconds.

str = regexp(str,'\w+','match','once'); % 2x faster
⇨ Elapsed time is 0.000047 seconds.
```

If MATLAB's powerful regex engine is insufficient for a specific need, or on an old MATLAB release that lacks regex features, consider using Java's regex instead.[389] It would be slower than MATLAB's, but it might provide the missing functionality:

```
flag = java.util.regex.Pattern.matches('(\d+):',java.lang.String(str));
```

Finally, note Jan Simon's MEX-based utility *CStrAinBP*,[390] which significantly outperforms *intersect, ismember, setdiff*, and *union* for cell-arrays of strings.

---

* A similar advice also applies for using the optional input argument of *find*, specifying the maximal number of results to return.

4.2.3 Converting Numbers to Strings (and Back)

When converting numbers to strings, we basically have four alternatives: use *num2str,* *sprintf,* some external conversion function (in Java, C/C++, or C#), or manually construct the string from the numerical digits by adding them to '0' and applying the *char* function.* The latter alternative is very efficient, but can only be used in very rare cases, and needs careful programming to prevent run-time errors. The external conversion function alternative is rarely more efficient to justify the added development and maintenance cost.

Among the mainstream alternatives (*num2str* and *sprintf*), *sprintf* is usually faster than *num2str*. This is not surprising, considering the fact that *num2str* uses *sprintf* internally. Note the reversed order of input parameters between *num2str* and *sprintf*:

```
% Using num2str
str = num2str(3.141593,'%f');
⇨ Elapsed time is 0.000313 seconds.

% Using sprintf - 6x faster
str = sprintf('%f', 3.141593);
⇨ Elapsed time is 0.000048 seconds.
```

Since *sprintf* is much more versatile than *num2str* in its ability to process a complex string format with multiple data inputs, it really makes sense to use only *sprintf* in any number-to-string conversion, not *num2str*. If you prefer using *num2str* over *sprintf* for some reason, then at least consider using the faster *int2str* function for conversion of integer values (*int2str* has exactly the same interface as *num2str*). Note that *sprintf* is still faster than *int2str*.

When converting numbers to hexadecimal strings and vice versa, it is usually faster to use the *sprintf* and *sscanf* functions than the corresponding *dec2hex* and *hex2dec*:[391]

```
value = sscanf('3fe5b000','%X'); % faster than hex2dec('3fe5b000')
str = sprintf('%X',1072017408);  % faster than dec2hex(1072017408)
```

When considering the usage of the low-level *sprintf* and *sscanf* functions, we should keep in mind that these internal library functions employ very limited input-data checks. The higher-level MATLAB wrapper functions (*num2str* and *str2double,* respectively) check their inputs providing a higher degree of program robustness, although at the possible expense of performance.[†]

 When converting a numeric array into a cell array of strings, we have several alternatives, but the fastest appears to be using the internal helper function *sprintfc*.[392]

When converting in the reverse direction, from string to number, we have a choice of many alternatives: *str2num, str2double, sscanf, textscan,* using external conversion functions or handling individual digit characters.[‡] Among the mainstream alternatives, *sscanf* is the fastest and the most versatile in its ability to process the input string into various numeric types; *textscan* is only slightly slower and quite versatile, yet awkward to use

---

* There are other esoteric conversion functions (e.g., *mat2str* or *num2hex*) but these are only used in rare cases.
† See the discussion about using internal functions in §4.3, and the discussion on bypassing sanity checks in §3.3.
‡ Plus a few other esoteric conversion functions (e.g., *str2mat* or *hex2dec*) but as above, these are only used in rare cases. There is also the *eval* function, which is in fact used internally by *str2num*, but directly using *eval* is strongly discouraged as bad programming, even if it does provide a performance improvement compared to using *str2num*.

since it places its outputs in a cell array;* *str2double* is even slower and the least versatile (only accepts scalars);† *str2num* is the slowest and yet the most forgiving, accepting any expression that evaluates to a number, including variables and functions:

```
value = str2num('3.141593');
⇨ Elapsed time is 0.000115 seconds.

value = str2double('3.141593');
⇨ Elapsed time is 0.000074 seconds.

cells = textscan('3.141593','%f'); value = cells{1};
⇨ Elapsed time is 0.000037 seconds.

value = sscanf('3.141593','%f');
⇨ Elapsed time is 0.000016 seconds.
```

Of all these alternatives, *str2num* and *str2double* are by far the most widely used in practice. Fortunately, the MATLAB Editor's Code Analyzer (MLint) includes a specific warning (ST2NM) about using the slower *str2num*:‡

```
v=version('-Release');
if str2num(v(1:end-1)) >= 2011
```
⚠ STR2DOUBLE is faster than STR2NUM; however, STR2DOUBLE operates only on scalars. Use the function that best suits your needs.

ST2NM MLint (Code Analyzer) message

4.2.4 String Comparison

Performing a case-insensitive string comparison or regex is more expensive than performing exact–case matching, so it should not be done when the string is known to keep a consistent case (upper or lower or mixed in a consistent manner):§

```
% Using case-insensitive strcmpi (~12 µsec per comparison)
% Note: str1 and str2 are both 1000-character strings
for idx=1:1e6; flag = strcmpi(str1,str2); end
⇨ Elapsed time is 12.493068 seconds.

% Using case-sensitive isequal (~1.3 µsec per comparison) - 9x faster
for idx=1:1e6; flag = isequal(str1,str2); end
⇨ Elapsed time is 1.353475 seconds.

% Using case-sensitive strcmp (~1 µsec per comparison) - 12x faster
for idx=1:1e6; flag = strcmp(str1,str2); end
⇨ Elapsed time is 1.014530 seconds.
```

---

* I assume that *textscan* uses *sscanf* internally. Both *textscan* and *sscanf* are both internal library functions (on Windows platform the library file is *libmwbuiltins.dll*), whose source is not provided.
† *str2double* uses *sscanf* internally, so it is no surprise that it is slower than *sscanf*.
‡ The complete list of MLint-reported performance-related warnings is presented in Appendix A.4. I do not know why MLint does not suggest using the even-faster *textscan* or *sscanf*; perhaps because they are a bit more difficult to use than the simpler *str2num/str2double*.
§ Whereas *strcmp* is a bit faster than *isequal*, *strcmp* expects string inputs, otherwise it croaks. In this respect, using *isequal* enables safer programming, transparently handling the case of non-string (numeric etc.) data.

I often see code segments where the input string's casing is known and yet either the case-insensitive versions of the string functions (*strcmpi, strncmpi, regexp(...,'-ignorecase')* etc.) are used,[2] or the *upper / lower* function is applied, "just in case". Such operations are wasteful and should be avoided in performance-critical sections.

In functions where we repeatedly need to use case-insensitive string functions, we should consider transforming our inputs into a consistent casing using *upper / lower*, and then proceed using the case-sensitive version of the string functions:*

```
newStr = lower(originalStr);
flag1 = strcmp(newStr(1:20), lower(str1)); % case-sensitive
flag2 = strcmp(newStr(10:15),lower(str2)); % case-sensitive
```

On the other hand, do not do this unless the case-insensitive operations are actually used repeatedly, since a single *strcmpi* is faster than any single *strcmp(upper(...))*:

```
% Using strcmp(upper(...))
for idx=1:1e6, flag = strcmp(upper(str1),upper(str2)); end
⇨ Elapsed time is 17.949020 seconds.

% Using strcmpi - ~50% faster
for idx=1:1e6, flag = strcmpi(str1,str2); end
⇨ Elapsed time is 12.523642 seconds.
```

Readers might be tempted to think that the *strncmp* and *strncmpi* functions will be faster than the corresponding *strcmp, strcmpi*. Perhaps counter-intuitively, this is incorrect. For some unknown reason, at least on my system, using the standard *strcmp, strcmpi* functions is actually ~20% faster. We should therefore use *strncmp* and *strncmpi* only when the logic dictates their usage, not for performance.

For comparing strings from their *end* (toward the left), rather than from their start (toward the right), consider using Jan Simon's efficient MEX-based *strcmpr* utility.[393] This can be useful, for example, for comparing filename extensions.

Acute readers may have noticed that the preceding discussion has left out the *strmatch* function, which is an alternative for the *strcmp* family of functions. This omission was intentional, since *strmatch* was deprecated and may be discontinued in a near-future MATLAB release. Using *strmatch* is discouraged, and issues a warning (MATCH2) in MATLAB Editor's Code Analyzer (MLint).† *strmatch* is much slower than *strcmp*, so performance is certainly not a factor in favor of using *strmatch*.

As a final consideration, rather than comparing multiple strings in an *if-elseif-else* construct, it is often faster to use a *switch-case-otherwise* construct:‡

```
if strcmpi(str,'abc')           switch lower(str)
    doSomething();                  case 'abc', doSomething();
elseif strcmpi(str,'def')   =>      case 'def', doSomethingElse();
    doSomethingElse();              otherwise,  doSomethingOther();
else                            end
    doSomethingOther();
end
```

---

* One might consider newStr to simply be a cached version of lower(originalStr) in this case.
† The complete list of MLint-reported performance-related warnings is presented in Appendix A.4.
‡ Also see §3.10.1.4.

4.2.5 Additional Aspects

4.2.5.1 Deblanking

When we need to deblank (remove leading and trailing spaces from) a string, a cell-array of strings or a multi-line matrix of characters, it is better to use *strtrim* rather than a combination of *deblank* and *fliplr* or *strjust*. These latter forms are often found in legacy code and should be replaced with the faster and newer *strtrim* function.

4.2.5.2 Concatenating Strings

The standard [...] construct is faster than the *strcat* function to concatenate strings:*

```
str = char ('0' + randi (10, 1, 1e3));

s3 = strcat(str, str);
⇨ Elapsed time is 0.000247 seconds.

s2 = [str, str];      % 20x faster
⇨ Elapsed time is 0.000012 seconds.
```

Similarly, when concatenating strings vertically, it is better to use cell arrays than *strvcat*. These two alternatives do not provide the same result, but in many cases they are quite interchangeable and in such cases using the cell-array variant is better:

```
% Using strvcat
>> tic, str = strvcat('leave','me','alone'); toc, str
⇨ Elapsed time is 0.000060 seconds.
str =
leave
me
alone

% Using cell arrays - 15x faster
>> tic, str = {'leave','me','alone'}; toc, str
⇨ Elapsed time is 0.000004 seconds.
str =
    'leave' 'me' 'alone'
```

Readers should also note Jan Simon's MEX utilities *CStrCatStr*[394] and *CStr2String*,[395] which are 3–10× faster than their corresponding built-in MATLAB functions.

4.2.5.3 Converting Java Strings into MATLAB

When encountered with a Java `String` that needs to be turned into a MATLAB string (*char* array), there are two alternatives. The simple method is to use the built-in *char* function. Surprisingly, this turns out to be slower than using the built-in *cell* function:

```
jString = java.lang.String(mString); % some Java string

mString = char(jString);
⇨ Elapsed time is 0.000066 seconds.
```

---

* Java-savvy developers might be tempted to use `StringBuffer` to concatenate strings (this is a standard Java-performance tip). However, using `StringBuffer` in MATLAB turns out to be even slower than *strcat*.

```
mString = cell(jString){1}; % 2x faster
⇨ Elapsed time is 0.000037 seconds.
```

4.2.5.4 Internationalization

When considering internationalization (i18n) support in a MATLAB program, it is usually faster to use Java's built-in support[396] than a manually programmed variant. There is really no point in reinventing the wheel, here and elsewhere.

4.3 Using Internal Helper Functions

MATLAB uses numerous internal functions to help its documented built-in functions perform their tasks. Many of these internal functions reside in stand-alone files that can be accessed directly. In general, the internal functions are not very useful for users developing MATLAB programs.

It turns out that in specific cases, directly using the internal functions, rather than their documented wrapper functions, can significantly improve performance. The reason is that the wrapper function often performs numerous checks and data conversions for the sake of generality and robustness that may be unnecessary in our specific cases.

Using internal functions has drawbacks: Internal functions do not normally perform any sanity checks or data conversions (after all, that's what we are trying to avoid in their wrapper function in the first place), so if our input data is not *exactly* as expected by the internal function we might get an error, a crash, or even worse — incorrect results.

In addition, note that internal functions are not officially supported by MathWorks. MathWorks may decide to modify these functions, or to remove them altogether, in a future release without any prior notice. MathWorks' only commitment is to support the wrapper functions, not their internal implementation. So while our program may work well today using internal functions available on the current MATLAB release, if we ever upgrade to a newer MATLAB release, we stand the risk that something will break in our program. This could be difficult to debug without official support.

Having said that, internal functions do provide the promise of significant performance speedups in certain cases, which may well be worth the extra trouble.

The easiest way to identify potential candidates for directly using internal functions is to profile the code, preferably with the *-detail builtin* option (see §2.1). Naturally, we should not search for such functions in noncritical sections of our code, but only in those areas identified by the MATLAB Profiler as performance hotspots. If we see a built-in MATLAB function that takes up a large time portion, this would be a prime candidate. Step into that function's profile report and try to detect a pattern of multiple checks followed by a core algorithm that is done by some helper function.

We can place a breakpoint in our code and step through the MATLAB wrapper function until we get to that internal helper function. We then need to carefully check which inputs this function expects (they might very well be different than the wrapper function's*).

---

* See the example of ***dtstr2dtnummx*** below, where for some inexplicable reason the timestamp uses a different format string than ***datestr, datenum***.

Finally, check where the target helper function is located and make it accessible to your code by either adding its folder to the MATLAB path (this is not always possible), or copying the function to your code folder.

4.3.1 A Sample Debugging Session

To illustrate the technique of finding a candidate internal help function, let's analyze the following example:[397]

```matlab
function perfTest
    % Initial setup
    n = 1e6;
    a = randi(n,1,n);    % 1M random integers
    b = randi(n,1,n);    % 1M random integers
    c = ismember(a,b);   % 1M logical values
end
```

When running this in the MATLAB Profiler,* we see that most of the time is spent in MATLAB's built-in *ismember* function:

Function Name	Calls	Total Time	Self Time*	Total Time Plot (dark band = self time)
perfTest	1	0.384 s	0.053 s	
ismember	1	0.331 s	0.051 s	
ismembc (MEX-file)	1	0.280 s	0.280 s	

Let's drill into the *perfTest* function:

```
time    calls  unjitted  line
                        1 function perfTest
                        2      % Initial setup
         1           3        n = 1e6;
0.03     1      X    4        a = randi(n,1,n);    % 1M random integers
0.03     1      X    5        b = randi(n,1,n);    % 1M random integers
0.33     1           6        c = ismember(a,b);   % 1M logical values
         1           7 end   % perfTest
```

Space here is limited to show the full screenshots, but if we drill into *ismember* we can see that it does numerous checks for sparse data, NaNs, unsorted data, number of input and output parameters, and so on.† The core logic, which uses 280 ms of *ismember*'s 331 ms, takes only a single line out of *ismember*'s nearly 400:

```matlab
tf = ismembc(a,s);
```

---

* As noted above, it is advised to run the profiler with the *-detail builtin* option, in order for the internal built-in functions to appear in the profiling report (see §2.1).
† In R2013a, *ismember* passes through a new internal function *ismemberR2012a*, which does the heavy lifting. Then again, in R2013a, MathWorks has replaced the direct call to *ismembc* within an unprofilable *builtin* ('_ismemberoneoutput',a,b). For didactic reasons, I decided to use a profiling session using R2010b to illustrate the technique.

```
        123    % Two C-Helper Functions are used in the code below:
        124
        125    % ISMEMBC  - S must be sorted - Returns logical vector indicating which
        126    % elements of A occur in S
        127    % ISMEMBC2 - S must be sorted - Returns a vector of the locations of
        128    % the elements of A occurring in S.  If multiple instances occur,
        129    % the last occurrence is returned
        130
        131    % Check for NaN values - NaN values will be at the end of S,
        132    % but may be anywhere in A.
        133
      1 134    nana = isnan(a(:));
        135
< 0.01  1 136    if (any(nana) || isnan(s(numelS)))
        137        % If NaNs detected, remove NaNs from the data before calling ISMEMBC.
        138        ida = (nana == 0);
        139        ids = (isnan(s(:)) == 0);
        140        if nOut <= 1
        141            ainfn = ismembc(a(ida),s(ids));
        142            tf(ida) = ainfn;
        143        else
        144            locl = ismembc2(a(ida),s(ids));
        145            tf(ida) = (locl > 0);
        146            loc(ida) = locl;
        147            loc(~ida) = 0;
        148        end
      1 149    else
        150        % No NaN values, call ISMEMBC directly.
      1 151        if nOut <= 1
0.28  1 152            tf = ismembc(a,s);
        153        else
```

At the top of the screenshot we can see a detailed explanation about the usage of *ismembc* and its sibling function *ismembc2*. Unfortunately, internal helper functions are seldom documented: we often need to use educated guesses and trial-and-errors.

The next step is figuring out the inputs to the internal function. In this case, either via simple static code inspection or by step-by-step debugging, we easily conclude that these are the same two inputs sent to the wrapper function *ismember*.

We next check whether the internal function is on the path. Luckily, it is in this case:*

```
>> which ismembc
C:\Program Files\MATLAB\R2010b\toolbox\matlab\ops\ismembc.mexw64
```

(in case it was not on the MATLAB path, we would need to add its folder to the path, or to copy the file to somewhere on the path).

All that now remains is to test our code using the newly found internal function. Note that we need to check both the performance *and* the correctness — checking correctness is especially important when using internal functions that may require special treatment of their inputs (in this case, the second input has to be sorted):

```
n = 100;
a = randi(n,n,1);
b = randi(n,n,1);

% Using the standard ismember
for idx = 1 : 100000, c = ismember(a,b); end
⇨ Elapsed time is 3.525726 seconds.

% Using the internal ismembc (10x faster)
b = sort(b); % b must be sorted for ismembc to work properly!
```

---

* Until R2009b the folder also included the source code files *ismembc.cpp* and *ismembc2.cpp*.

```
for idx = 1 : 100000, c2 = ismembc(a,b); end
⇨ Elapsed time is 0.348144 seconds.

sameResults = isequal(c,c2) % ensure that we get the same results
⇨ 1
```

As noted, we should only use internal helper functions after carefully checking their expected parameters and limitations. In the specific case of *ismembc*, the second input must be sorted nonsparse non-NaN values. If we cannot ensure that our data is so, then we should programmatically correct it (as done above, using *sort*), or use the slower wrapper function, which does it for us.

Just in case you were wondering: The difference between the two MEX-file siblings *ismembc* and *ismembc2* is that whereas *ismembc* returns an array of logical values, *ismembc2* returns the index locations of the found members.

Another example of using internal functions for performance tuning was provided by Jérôme Lecoq, who explained how he profiled MATLAB's image-reading functions, improving load time of a multi-image TIFF file from 200 s down to 1.5 s.[398]

Some additional examples of using the core logic of built-in MATLAB functions (*inlining*) was provided in §3.1.2. For example,

```
A = rand(1000,1);
tic, for idx=1:100000, b=mean(A); end, toc
⇨ Elapsed time is 1.638925 seconds.

tic, for idx=1:100000, b=sum(A,1)/size(A,1); end, toc
⇨ Elapsed time is 0.318459 seconds.    % 5x speedup
```

As noted there, inlining functions only makes sense for small functions that can easily be inlined and for which the invocation overhead is relatively significant compared to the overall iteration time. If we need to inline the same function in multiple loops, the reduced maintainability should seriously be weighed against the performance speedup. Refer to §3.1.2 for additional details and examples; §3.3 provides additional discussion on the related technique of smart checks bypass.

Another example of a fast internal MEX function having an m-code wrapper is the *sortrowsc* function, used by *sortrows* function when 4+ full real rows are sorted.[399]

Until R2009b (MATLAB 7.9), the internal MEX functions' C source code was often provided alongside the compiled binary. However, since R2010a (7.10), the source code is no longer provided,* and in many cases (such as *sortrowsc*) the binary code was moved from a standalone MEX binary file into one of the built-in DLLs (in *sortrowsc*'s case, *libmwmathcore.dll*). We can always access an earlier MATLAB release for the MEX source code, noting the potential for bugs in earlier versions.

---

* Perhaps I may be blamed for this, as I started blogging about these internal helper functions in 2009: http://bit.ly/XsYaEv ...

4.4 Date and Time Functions

MATLAB has built-in function to convert date and time strings to numeric format (*datenum* and *datevec*) and vice versa (*datestr*). These functions need to accommodate a wide range of possible string representations. For this reason, the internal checks that are done by the wrapper functions take a significant part of the overall processing time. Multiple calls to these functions may occur frequently in MATLAB programs, for example, when we need to update dates in a list or when updating a plot's X-axis tick labels.

The best advice for using MATLAB's date/time functions is to try to avoid them wherever possible. In many cases it is possible to use the numeric values (such as the results of the *now* and *clock* functions). Numeric data processing is much faster than string processing or the date/time conversion functions.

For example, if all we need to do is to check whether less than 5 min have passed, we can use simple numeric arithmetic code such as the following, which is much faster than using *datenum*, *datestr*, or even *etime*:

```
persistent lastTime
if isempty(lastTime)
    lastTime = now;
end

thisTime = now;
if thisTime - lastTime > 5/(24*60) % 5 minutes
    % do something useful here...
    lastTime = thisTime;
end
```

When we cannot avoid converting date/time values, we can use caching (see §3.2.5) and uniquification (see §3.6.3), both of which can improve performance by a factor of up to several hundred (!).

In any case, it is faster and more efficient to store timestamps as arrays of numeric values, rather than their string representations (which then need to be parsed), and certainly not financial time series,[400] which are even less efficient.[401] Numeric arrays are naturally less maintainable and harder to debug than timestamp strings or time-series objects, so the tradeoff against performance should be carefully considered.

As noted in §3.3, datenum('2013-02-15') is an order of magnitude slower than datenum('2013-02-15','yyyy-mm-dd'), because the former needs to perform all sorts of checks and computations to determine the date format. By specifying the extra (optional) data-format input parameter, we shave away these run-time computations.*

```
a = datenum('2013-02-17');
⇨ Elapsed time is 0.003163 seconds.

a = datenum('2013-02-17', 'yyyy-mm-dd');
⇨ Elapsed time is 0.000357 seconds.
```

If the timestamp happens to be is in the common 'dd-mmm-yyyy HH:MM:SS' format, consider using Jan Simon's *DateConvert* utility on the MATLAB File Exchange.[402]

---

* Strictly speaking, in the specific case of *datenum*, the function uses a different processing path when the format string is specified (in which case it uses the internal *dtstr2dtnummx* function), and when it is unknown (in which case it uses the built-in *datevec* and *datenummx* functions). The aforementioned format checks and computations are done in the *datevec* function.

DateConvert does not check its inputs and relies on the assumption that the format is preset. In return, it converts a string to a date number or vector (and backwards) 100 times faster than *datenum, datevec,* and *datestr,* using pure MATLAB code.[403]

Even with a different timestamp format, we can still improve performance by using internal functions, as explained in §4.3. It turns out that *datenum* uses the built-in function *dtstr2dtnummx* for the actual processing — converting a date from text to floating-point number.[404] We can use this function directly for improved performance: On my particular computer, *dtstr2dtnummx* is 3× faster than the standard *datenum*:

```
% Using the standard datenum
for idx = 1 : 1000
    dateNum = datenum({'2010-12-12 12:21:12.123'},...
                       'yyyy-mm-dd HH:MM:SS');
end
⇨ Elapsed time is 0.658352 seconds.

% Using the internal dtstr2dtnummx (3x faster)
for idx = 1 : 1000
    dateNum = dtstr2dtnummx({'2010-12-12 12:21:12.123'},...
                            'yyyy-MM-dd HH:mm:ss');
end
⇨ Elapsed time is 0.218423 seconds.
```

While the difference in timing may appear small in absolute terms, if we use *datenum* to parse a text file with thousands of lines, each with its own timestamp, then these seemingly small time differences quickly add up. Of course, it only makes sense to do the replacement if the Profiler reports that this date parsing is a performance hotspot in our particular application. It was indeed such a hotspot in some of my applications, and was also reported to be so by some users on StackOverflow.[405]

The speedup is even more pronounced when using cell-array input rather than a loop. For example, for a 1000-cell array the speedup is almost 40× on my system:

```
% Prepare 1000 date strings in a 1000x1 cell-array
timestamps = datestr(now+(1:1000),'yyyy-mm-dd HH:MM:SS');
timestamps = num2cell(timestamps,2);

% Using the standard datenum
dateNums = datenum(timestamps);
⇨ Elapsed time is 0.324127 seconds.

% Using the internal dtstr2dtnummx (38x faster)
dateNums = dtstr2dtnummx(timestamps,'yyyy-MM-dd HH:mm:ss');
⇨ Elapsed time is 0.008505 seconds.
```

The reason for the large speedup is that *datenum* cannot know in advance that all the input timestamp strings use the same format, so for each of the 1000 timestamp string it needlessly and separately checks the timestamp, recomputes the format string, and calls the internal function. In fact, the majority of the processing time is spent on these unnecessary checks and format-string computations, and not in the internal function calls. Since we know that we can reuse the same format string for all timestamps, we can use it in a single vectorized call to *dtstr2dtnummx* that is far more efficient. The speedup only grows as the number of timestamps increases.

Like *ismembc, dtstr2dtnummx* is an internal mex function. On my Windows system it is located in *C:\Program Files\MATLAB\R2013a\toolbox\matlab\timefun\private*

dtstr2dtnummx.mexw32. It has a different extension on non-Windows systems, but you will easily find it in its containing folder beneath the MATLAB installation root. You can access this folder using the following MATLAB command:*

```
winopen([matlabroot '\toolbox\matlab\timefun\private'])
```

To gain access to **dtstr2dtnummx**, copy the *dtstr2dtnummx.mexw32* file to another folder that is already on your MATLAB path (as reported by the **path** command). In this case, we cannot simply add the folder to the MATLAB path using the **addpath** function, since private folders are not allowed on the path.

Note that the string format is inconsistent between **dtstr2dtnummx** and **datenum**: In the test case above, **dtstr2dtnummx** used 'yyyy-MM-dd HH:mm:ss.SSS', while **datenum** required 'yyyy-mm-dd HH:MM:SS.FFF'.† Also note that the first input parameter to **dtstr2dtnummx** has to be a cell array, otherwise, MATLAB will crash when the mex function tries to access an invalid memory location. Because of these, we need to be extra careful when using **dtstr2dtnummx**.[406]

Jan Simon wrote an even faster date/time converter called *DateStr2Num*, available on the MATLAB File Exchange.[407] It is dozens of times faster than **dtstr2dtnummx** and hundreds of times faster than **datenum**. The speedup is based on two methods:[408] (1) the format has to be specified by a very limited set of six common timestamp formats, and (2) the value is not checked for validity: While **datenum** and **dtstr2dtnummx** recognize '2011-04-180' more or less correctly as the 179th day after 2011-04-01, *DateStr2Num* fails and does not even catch '2011-04-AB' as an error. So, using *DateStr2Num* makes sense only if the date string is known to be valid. It is a MEX file that can either be compiled from source,[409] or downloaded in binary format.[410]

Just as **dtstr2dtnummx** provides the core algorithm for processing date/time strings into date numbers, so too does its sibling function **dtstr2dtvecmx** (located in the same folder) provide the core algorithm for converting date/time strings into date-component vector, as the **datevec** function does:

```
% Using the standard datevec
datevec(datestr(now));
⇨ Elapsed time is 0.001569 seconds.

% Using the internal dtstr2dtvecmx (9x faster)
dtstr2dtvecmx({'2013-03-22 17:25:13'},'yyyy-MM-dd HH:mm:ss')
ans =
     2013        3       22       17       25       13
⇨ Elapsed time is 0.000168 seconds.
```

As before, the speedup is much higher when using an input of multiple timestamps:

```
% Prepare 1000 date strings in a 1000x1 cell-array
timestamps = datestr(now+(1:1000),'yyyy-mm-dd HH:MM:SS');
timestamps = num2cell(timestamps,2);

% Using the standard datevec
dateVecs = datevec(timestamps);
⇨ Elapsed time is 0.221857 seconds.
```

---

* **winopen** is only available on Windows, but you can easily access the same folder on non-Windows systems.
† The format string conversion is done in the helper function *cnv2icudf.m*, located in the same folder as **dtstr2dtnummx**.

```
% Using the internal dtstr2dtvecmx (25x faster)
dateVecs = dtstr2dtvecmx(timestamps,'yyyy-MM-dd HH:mm:ss');
⇨ Elapsed time is 0.008838 seconds.
```

The *datenum* function uses another internal function that we can utilize in a certain situation. Specifically, if our input uses a 6-element numeric date-vector format,* then using *datenummx* will be more efficient:[411]

```
dateVecs = clock;
for idx=1:1000, dateNums = datenum(dateVecs); end
⇨ Elapsed time is 0.015782 seconds.

for idx=1:1000, dateNums = datenummx(dateVecs); end
⇨ Elapsed time is 0.001152 seconds.
```

datenummx used to be a separate mex file whose source code was provided, but in recent MATLAB releases it is an internally compiled function that is not available as a separate file but rather part of an internal MATLAB library (*libmwbuiltins*). As such, it is directly available anywhere in MATLAB and there is no need to update the MATLAB path or copy any file to use it.

Like *datenummx*, there is also a corresponding *datevecmx* in the same shared library. However, the equivalent *dtstr2dtvecmx* is faster and so I see no reason to ever use *datevecmx* directly, in contrast to *datenummx*.

The folder that contains *dtstr2dtnummx* and *dtstr2dtvecmx* also contains the internal function *addtodatemx*, which is the actual computation core for *addtodate*, which does simple date arithmetic:

```
% Using the standard addtodate
t = datenum('2013-03-22 17:25:13');
t2 = addtodate(t,5,'day'); % add 5 days to specified date
⇨ Elapsed time is 0.000223 seconds.

% Using the internal addtodatemx (2x faster)
% 3rd input arg = addition type index: year,month,day,hour,min,sec,msec
t2 = addtodatemx(t,5,3);
⇨ Elapsed time is 0.000106 seconds.
```

The same folder also contains the *getmonthnamesmx*, *getweekdaynamesmx*, and *getampmtokensmx* internal functions. These functions accept no input arguments (or the 'short'/'long' input argument[†]), and return a cell array of the requested strings (12 month names, 7 weekday names, and the AM/PM strings), based on the computer's locale. Users can save a bit of execution time by caching their output or using a hard-coded cell array of corresponding strings.

Another variant suggested by Jan Simon as a faster replacement for *datestr* in cases where the input is the 6-element numeric date-vector format (the output of *clock, datevec,* or *dtstr2dtvecmx*) is to use *sprintf*:[412]

```
% Using the standard datestr()
for idx = 1 : 1000, s2 = datestr(now); end
```

---

* The output of *clock, datevec,* or *dtstr2dtvecmx* — [yyyy mm dd hh mm ss.###].

† 'short' and 'long' return the English terms (e.g., 'Sun' and 'Jan'); 'shortloc' and 'longloc' return the corresponding computer locale strings (the regional settings when MATLAB was started; changing regional settings while MATLAB is running does not affect the running session). The default input value is 'short'. Read more on MATLAB's usage of the locale here: http://mathworks.com/help/releases/R2012a/techdoc/matlab_env/brj_w4w-2.html (or: http://bit.ly/15H5X4X).

```
⇨ Elapsed time is 0.253474 seconds.

% Using sprintf() - 10x faster
for idx = 1 : 1000
    tnow = clock;
    s1 = sprintf('%.2d-%s-%.4d %.2d:%.2d:%.2d',...
                  tnow(3), months(tnow(2),:), fix(tnow([1,4,5,6])));
end
⇨ Elapsed time is 0.024396 seconds.
```

This example shows that we should sometimes "think outside the box": even if the underlying data represents a timestamp, we do not necessarily need to use costly date/time functions to process it.

As another example for this, consider the common use case of transforming one date/time format to another. For example, suppose that we have data in the format 'm/d/yyyy' (e.g., '3/15/1987') and wish to convert it into the numeric yyyymmdd format (i.e., 19870315). The standard (slow) solution would be to use a pair of *datestr* and *datenum / str2num* calls to convert the data; instead, we could use a faster call to *datevec* with some simple numeric arithmetic:[413]

```
% Some sample data
dateStrs = {'12/31/1989';'1/31/1990';'2/28/1990';'3/31/1990'};

% Standard solution using datestr() & str2num()
dateNums = str2num(datestr(dateStrs,'yyyymmdd'));
⇨ Elapsed time is 0.002250 seconds.

% Faster solution using datevec() and simple arithmetic (65% faster)
dateVecs = datevec(dateStrs);
dateNums = dateVecs(:,1)*10000 + dateVecs(:,2)*100 + dateVecs(:,3);
⇨ Elapsed time is 0.001354 seconds.
```

If we are assured that the input data has consistent format ('03/15/1987', for example), then we can avoid even *datevec* and use much faster and simpler string functions:

```
% Faster solution using datevec() and simple arithmetic (12x faster)
strsMat = cell2mat(dateStrs);
strsMat = [strsMat(:,7:10) strsMat(:,1:2) strsMat(:,4:5)];
dateNums = str2num(strsMat);
⇨ Elapsed time is 0.000192 seconds.
```

As a final tip, for those applications that interact with Java object that use `java.util.Date`, my advice is to cache the `java.text.DateFormat` object reference that is used for the Java date/time manipulations.[414] In my tests I have not seen any significant speedup by using `java.util.Date` rather than MATLAB functions, particularly with the tips provided earlier in this section.* Therefore, `java.util.Date` should really be used only when interfacing to Java, or when needing to use a specific feature that Java provides (such as the ability to easily switch time zones and locales within a running program†). Here is a sample usage using a cached `DateFormat` object:

```
persistent dateFormat
if isempty(dateFormat)
```

---

* Specifically, caching, uniquifying, and using internal functions.
† Recall that MATLAB uses the same locale throughout its session.

```
        dateFormat = java.text.SimpleDateFormat('yyyy-MM-dd HH:mm:ss.SSS');
    end

    currentDateTime = java.util.Date;
    dateStr = dateFormat.format(currentDateTime).char;    % Date to string
    dateObj = dateFormat.parse(dateStr);                  % string to Date
```

Note that the Java date format corresponds to the format used by ***dtstr2dtnummx*** and ***dtstr2dtvecmx***, not the format used by ***datenum, datestr,*** and ***datevec.***

Jan Simon provided a very nice m-code wrapper for the internal MEX functions (****mx***) in his *DateConvert* utility on the MATLAB File Exchange.[415] This utility provides an all-in-one faster replacement for the built-in ***datenum, datevec,*** and ***datestr*** functions. Users should note that *DateConvert* does not include data checks nor support for multiple date/timestamp formats as the builtins. With these limitations in mind, *DateConvert* is one or two orders of magnitude faster than the built-in functions.

4.5 Numeric Processing

4.5.1 Using inf and NaN

NaN and inf are special numeric values that have a unique behavior when used in the context of MATLAB calculations,[416] following the IEEE standard for these values. The infinity value (+inf and –inf) is a real number that can be used in regular numeric and logical expressions (a/inf, b == inf, c < inf). It turns out that while we can indeed test for an infinity value using the regular equality (==) operator, it is somewhat faster to use the built-in ***isinf*** function:

```
% data has 1M double-precision elements
n = sum(data==inf);
⇨ Elapsed time is 0.002019 seconds.

n = sum(isinf(data));
⇨ Elapsed time is 0.001704 seconds. % isinf is 15% faster than ==
```

There is a corresponding built-in function ***isfinite***, which returns the complementary value for ***isinf*** (namely, those elements that are non-inf). The main drawback of ***isinf*** is that it does not differentiate between +inf and –inf. If the algorithm needs this information, then we should use the == operator instead (if data==-inf...).

Contrary to infinity values, NaN (*Not a Number*) values cannot be used directly in numeric/logical expressions:* data==nan will always return false, even when data is in fact NaN. We can use this to test instead for data==data (this returns true only in the non-NaN locations, odd as this may sound). However, as with inf, we should test for NaN values using the built-in ***isnan*** function rather than the equality operator.†

---

* Strictly speaking, NaN *can* be used in such expressions, but the results may be unexpected and so it *should* not be used.

† In old MATLAB releases, and especially with high NaN density, using the == or ~= operator used to be faster and more memory-efficient than ***isnan*** (see http://mathworks.com/matlabcentral/newsreader/view_thread/81509#207718 or: http://bit.ly/1fjzvvV — read the full discussion in that thread). However, this is no longer the case today.

NaNs (and sometimes inf) are often used to indicate missing or invalid values. In plots, NaN data elements are plotted as empty elements (line breaks — §10.1.1.14). The Statistics toolbox contains a set of functions for analyzing data having NaNs.[417]

In general, avoid using inf and NaN values in algorithms. Their presence makes the code more complex and harder to debug/maintain than regular finite values. Moreover, nonfinite values may slow down code by up to 10×,[418] especially NaNs on older processors or 32-bit Windows/Intel.[419] On modern CPUs running a 64-bit OS, the effect is still visible, but is normally more limited; this may or may not be significant, depending on the specific situation. If we find in our specific case significant performance degradation due to non-finite values, we can replace NaN/inf data with marker values (e.g., 0 or −1). Subsequent operations will run at the speed of regular numeric processing, and the special data values can be processed separately. If you do need to use either inf or NaN, then use inf, which can be noticeably faster.[420]

Note that NaNs can also *speed up* algorithms, by enabling vectorized processing with reduced memory impact, for example, on large finite-element lattices.[421]

4.5.2 Matrix Operations

A well-entrenched notion in the MATLAB community is that the **transpose** function (`transpose(data)`) is faster than the ' operator (`data'`), due to the fact that the ' operator does not only transpose the data, but also computes the complex conjugate. While this is indeed true for complex data values, the performance impact (at least in modern MATLAB releases) is not very significant:

```
a = sqrt(randn(1e7,1) * 2i); % 10M complex data values
tic, b = a'; toc
⇨ Elapsed time is 0.075430 seconds. % baseline perf., conjugated

tic, b = transpose(a); toc % or: b = a.'
⇨ Elapsed time is 0.062589 seconds. % 20% faster, not conjugated

tic, b = conj(a'); toc
⇨ Elapsed time is 0.132219 seconds. % 75% slower, de-conjugated
```

Note that the **transpose** function is equivalent to the element-wise transposition operator (.'). Rather than using the much slower form *conj*(a'), which is sadly often encountered in MATLAB code, we should use the equivalent a.' or **transpose**(a).

For real (noncomplex) data, **transpose**, .' and ' all have the same performance and functionality, since conjugation is not applicable.

When operating on complex data, alternative approaches that yield the same result (up to roundoff errors) sometimes have significantly different performance:[422]

```
tic, b = sum(abs(a).^2); toc
⇨ Elapsed time is 0.249919 seconds. % baseline performance

tic, b = sum(a.*conj(a)); toc
⇨ Elapsed time is 0.176502 seconds. % 1.4x faster

tic, b = sum(abs(a.^2)); toc
⇨ Elapsed time is 0.162168 seconds. % 1.5x faster

tic, b = sum(real(a).^2+imag(a).^2); toc
⇨ Elapsed time is 0.141809 seconds. % 1.8x faster

tic, b = a'*a; toc
⇨ Elapsed time is 0.011326 seconds. % 22x faster
```

Additional performance aspects of complex computations are presented in §4.5.3.

As seen in the last example, the symmetric multiplication a*a' has by far the fastest performance. A little-known aspect of this, discovered by Mike Croucher, is that symmetric data multiplications are faster than nonsymmetric multiplications (a*b):[423]

```
a = randn(3000); % 9M elements
tic, b = pi*a*a'; toc
⇨ Elapsed time is 1.809753 seconds. % baseline performance

tic, b = pi*(a*a'); toc
⇨ Elapsed time is 0.906156 seconds. % 2x faster
```

The symmetric product variant is faster since the underlying matrix multiplication routine (BLAS *GEMM[424], or perhaps using BLAS DSYRK[425] in this specific case as other systems do[426]) is able to employ an optimization, presumably based on the fact that a symmetric product yields a symmetric result, thereby halving the number of elements that need to be computed.[427] Note that a*a (without the transpose) does NOT yield a symmetric result and does not enjoy this optimization.

In the example above, adding parentheses ( ) forces MATLAB to compute the symmetric product a*a' before multiplying by pi. The default operator precedence used by the top variant (pi*a*a'), first creates a temporary matrix for pi*a, which is then multiplied by a' nonsymmetrically, preventing the symmetry optimization.

A related optimization that we can employ is based on the fact that the order of matrix multiplication does not alter the result (a.k.a. *associativity*): A*(B*C) = (A*B)*C. On the other hand, performance, which is based on the number of numeric operations, heavily depends on the multiplication order and the relative sizes of the matrices, as noted by Martin Cohen.[428] If the relative matrix sizes are

- A: n-by-m
- B: m-by-p
- C: p-by-q

Then the result of D = C*A*B will be a matrix of size n-by-q. There are basically two ways to compute D:

- D = A*(B*C) — this takes $mq(n + p)$ numeric operations
- D = (A*B)*C — this takes $np(m + q)$ numeric operations

Let's take a numeric example to see this in action: A is 300×200, B is 200×10000, C is 10000×400. The resulting D = A*B*C is 300×400. The expected speedup of A*(B*C) versus (A*B)*C is therefore $3 \cdot 100 \cdot (2 + 4)/2 \cdot 4 \cdot (3 + 100) \approx 2.2$. And indeed:

```
A = rand(300,200); B = rand(200,10000); C = rand(10000,400);
D = A*B*C;
⇨ Elapsed time is 0.120633 seconds.
D = A*(B*C);
⇨ Elapsed time is 0.051404 seconds. % ~x2.3 faster
```

This example shows that depending on the relative matrix sizes, it may well be worthwhile to group the computation's components in a nondefault manner, to reduce the numerical complexity, thereby improving performance.

It should be noted that while the numerical complexity may decrease, in some cases this may come at the expense of increased memory usage. This is not always the case. In fact, in our example above, the temporary matrix created by B*C is 200×400 (or 625 KB), which is significantly smaller than the temporary matrix for A*B (300×10000, or 23 MB). This may be the reason that the speedup we received (~2.3) was slightly faster than the theoretical expected value (2.2). As the matrices size increases, so too does the impact of the memory effect (also see Chapter 9).

The effect of multiplication order increases when the multiplicands have significantly different dimension sizes. This is the case when matrices are multiplied by a vector on the right, which is very common. For example, if A and B are N×N matrices and v is a N×1 vector, then D = A*(B*v) will be O(N) faster than D = A*B*v. In general, the rule is to group the elements such that the data size is reduced as soon as possible.

Owing to MATLAB's efficient implementation of matrix operations compared to iterative loops, it makes sense to use such operations even when the data is not properly aligned. In many cases it is less expensive to adapt the data for vectorized matrix operation than it is to use loops. For example, if data vectors do not have a constant length, we could still place these vectors within a zero-padded bounding matrix, then act on this matrix in a single vectorized operation.

Linear algebra performance is undergoing rapid improvements in recent years and is a hot topic in academic research.[429] The corresponding MATLAB routines are expected to benefit from these improvements. Readers may find interest in an introduction to matrix multiplication performance by the late Allan Snavely of UCSD/SDSC.[430]

As MATLAB uses highly optimized BLAS, LAPACK, and MKL library functions for its core linear algebra functions, I would be wary of trying to out-smart it using manual modifications of the computations. In simple changes such as forcing a specific computation order for a 2-times speedup, as shown above, this may be worthwhile. However, I suggest not to extend this idea to more complex modifications, except in rare cases. As one study has found,[431]

> *"Libraries are far better optimized, their outstanding performance against mere mortals implementations are unbelievable."*

As MATLAB's basic low-level linear algebra uses highly optimized library code, there is little point in trying to create manually optimized MATLAB or even C++ alternatives for the simple cases. However, in certain cases we can benefit from inside information. For example, we can improve performance by reusing existing matrix data, or by performing operations in-place (see §9.5.2).

MATLAB's documentation explains how to link MEX code against MATLAB's pre-bundled BLAS/LAPACK libraries, to call their routines from within MEX.[432] A File Exchange utility by Tim Toolan provides the MEX code and m-code wrapper to easily call BLAS/LAPACK from our MATLAB m-code, without having to mess around MEX C-code.[433] There are several FEX utilities that interface with BLAS/LAPACK,[434] such as James (Jim) Tursa's *mtimesx* (for matrix multiplications):[435]

```
% Usage examples for mtimesx
a = sqrt(-rand(1000)); % 1000x1000 complex
v = rand(1,1000);      % 1x1000 real

% Run standard MATLAB arithmetic using conj, transpose and mtimes
b = conj(a)*v.';
⇨ Elapsed time is 0.008618 seconds.
```

```
% Now use mtimesx for the same result - 5x faster!
c = mtimesx(a,'g',v,'t','speed');
⇨ Elapsed time is 0.001722 seconds.

% Ensure that we got the same results
>> disp(isequal(b,c))
    1                              ← true
```

In addition, *mtimesx* enables vectorized multiplication of N-dimensional numeric data (*tensors*). MATLAB's standard multiplication (***mtimes*** function or the * operator) only supports scalar, 1D, and 2D data, so multiplying N-D data requires expensive loops. *mtimesx* can be orders of magnitude faster in such cases:[436]

```
a = rand(2,3,1000,2000);
b = rand(3,5,1000,2000);

% Using standard MATLAB N-D matrix multiplication
c = zeros(2,5,1000,2000); % pre-allocate - see §9.4
for m = 1 : size(a,3)
    for n = 1 : size(a,4)
        c(:,:,m,n) = a(:,:,m,n) * b(:,:,m,n);
    end
end
⇨ Elapsed time is 9.496173 seconds.

% Using mtimesx - 18x faster
c = mtimesx(a,b);
⇨ Elapsed time is 0.517339 seconds.
```

Tursa's *mtimesx* includes a detailed document that explains the underlying algorithms, lists the BLAS functions being used, and details the cases where *mtimesx* is expected to provide speedups or not. *mtimesx* auto-compiles itself when first run and includes a variety of test scripts having extensive internal documentation.[437]

Unlike *mtimesx*, which lets BLAS handle the multi-threading, Yuval Tassa's *mmx* utility[438] employs threads* to explicitly parallelize data dimensions, using BLAS in single-threaded mode. This ensures *embarrassingly parallel* problems† that are independent of each other and are therefore simpler and faster.[439] It turns out that *mmx* outperforms *mtimesx* for a wide range of use cases. In our last example above:

```
% Using mmx - 31x faster
c = mmx('mult',a,b);
⇨ Elapsed time is 0.305624 seconds.
```

With sparse data arrays, ***accumarray*** (§5.3.5) is faster than simple multiplication.[440]

Peter Boettcher developed *ndfun*,[441] which is also a MEX-based BLAS interface for N-D MATLAB arrays, not just for multiplication. *ndfun* was not updated since 2007. So while the speedup potential is significant, using such an old MEX function could crash MATLAB due to changes in its internal memory representation since then.[442] Use *ndfun* only after careful checks, and possibly adaptations to its MEX C code.

---

* See §7.3.1 for an example of using threads in MEX code.
† The term *"embarrassingly parallel"* is discussed in §5.1.1.

Marcel Leutenegger's *MATLAB Toolbox* (§8.5.4) contains a few functions for vector addition, cross/dot product, scaling and norm on Intel-based CPUs.[443] Being assembler based, the resulting MEX functions are highly efficient.[444] However, they were not updated since 2005 and do not use Intel's latest vectorization instructions. Users may well attempt to update the provided source code for extra performance.

A highly rated *Arraylab* (a.k.a. *MultiProd*) toolbox for manipulating N-D tensor data was developed by Paolo de Leva, and posted on File Exchange.[445] The toolbox includes an extensive performance-testing framework, as well as an article describing the performance and memory aspects with comparative results.

A more extensive *MATLAB Tensor Toolbox* was developed at Sandia National Labs by Tamara (Tammy) Kolda et al.[446] Much effort was invested in performance tuning this toolbox,[447] for example, employing special workarounds for sparse data and using dedicated C++ versions of kernels to access strides in BLAS to avoid run-time permutations.[448] A highly cited paper from 2005 (published 2006) describes the toolbox;[449] many additional related papers are available on Kolda's homepage.[450]

A MEX utility that can help with memory-intensive matrix calculations is Bruno Luong's *InplaceArray*,[451] which enables in-place data computation, thereby avoiding the need for expensive memory allocations or interim computation results.* Used judiciously with Tursa's, Tassa's and Kolda's tricks, very efficient MEX implementations can be achieved.

Instead of trying to out-smart the low-level linear algebra functions, we could intelligently apply linear-algebra rules to compute matrix operations more efficiently. An example of this is to use the Kronecker product for matrix multiplications.†[452] Another example: repeated multiplication is much faster than exponentiation (§3.9.3). Yet another example is to pre-factor a matrix into L,U components before looping.[453]

Many additional aspects of matrix performance are vectorization- and memory related. Refer to Chapters 5 and 9 for additional details.

When dividing matrix data, the reader is reminded that division of a matrix by a scalar is much more expensive than multiplication by the scalar's reciprocal value.‡ Be careful when extending this idea to division by nonscalar values: Matrix division can be tricky, error prone, and subject to numerical inaccuracies and instabilities.

The MATLAB documentation has the following passage in the doc-page for *inv*:§[454]

> *It is seldom necessary to form the explicit inverse of a matrix. A frequent misuse of **inv** arises when solving the system of linear equations Ax = b. One way to solve this is with x = inv(A)*b. A better way, from both an execution time and numerical accuracy standpoint, is to use the matrix division operator x = A\b. This produces the solution using Gaussian elimination, without forming the inverse. See **mldivide** (\) for further information.*

The same *inv* doc-page provides an example, which shows that using the backslash operator (A\b) is faster and yet just as accurate as *inv*(A)*b.[455] The corresponding 2–3× speedup occurs across a wide range of platforms, MATLAB releases, and data.[456] Yuval Tassa's *mmx* utility (see above) includes an even faster multithreaded backslash.

---

* See §9.5.2.

† See §5.2.4.2 for additional information on Kroneker product implementation; using linear algebra is also discussed in §5.7.16.

‡ See §3.10.1.6.

§ Note that in MATLAB, running *inv(A)* is the same as running A^(-1); both share the same drawbacks compared to *mldivide*.

4.5.3 Real versus Complex Math

Complex data is inherently supported by a large portion (if not the majority) of MATLAB's numeric functions and operators. However, it should be noted that "complex" is not by itself a primitive type, but rather an attribute of the data. Under the hood, MATLAB stores the real and imaginary parts of the data separately, and handles the associated arithmetic internally:[*][457]

```
>> format debug

>> a = 1:10
a =
Structure address = ac81670
m = 1
n = 10
pr = 22c4e3e0
pi = 0
     1     2     3     4     5     6     7     8     9    10
>> b = sqrt([-1,-2])
b =
Structure address = acce240
m = 1
n = 2
pr = 22c7d8c0
pi = 22c7a200
   0.0000 + 1.0000i      0.0000 + 1.4142i

>> whos a b
  Name        Size              Bytes  Class     Attributes
    a         1x10                 80  double
    b         1x2                  32  double    complex
```

As can be seen in the snippet above, the real data part (referenced by the `pr` pointer) is always set, even when the data is purely imaginary. On the other hand, the imaginary part (referenced by the `pi` pointer) is only set if the data has any imaginary portion.

It stands to reason, then, that we should keep calculations in the purely real domain for as long as possible, since adding any imaginary portion will multiply our data and the associated computations. In other cases, it might help to separate (*unroll*[†]) the computations manually, into real and complex parts.[458] However, recent MATLAB releases have improved complex math performance,[‡] so this is no longer as important today as it was in MATLAB releases of the early 2000s. You may well discover, as I have on several occasions, that unrolling complex math or modifying the computation order has actually degraded performance… As a specific example, one user reported that unrolling an *fftshift* operation resulted in an 8× speedup,[459] whereas on my system the same code had no noticeable effect.[§] So you should test this using your specific program and MATLAB release, on your target platform.

A different advice, which is still very much relevant today as it was a decade ago, is to use the built-in *i* constant, rather than the ubiquitous *sqrt(–1)*. I suggest using the more

---

[*] *format debug* is discussed in §9.2.10.
[†] Compare: loop unrolling (§3.1.5).
[‡] Owing to a combination of improvements in the underlying math libraries, the CPU and MATLAB's internal processing.
[§] Unfortunately, the stated report does not specify the running environment (platform, OS, MATLAB release, etc.), so it is difficult to perform a direct comparison to isolate the cause of the performance difference.

verbose *1i* rather than just *i*, because i is often used as a local variable (loop index, etc.) in MATLAB code:[*]

```
for idx=1:1e6, a=sqrt(-1)*pi; end
⇨ Elapsed time is 0.714479 seconds.

% directly use 1i: 14x faster than sqrt(-1)
for idx=1:1e6, a=1i*pi; end
⇨ Elapsed time is 0.051491 seconds.
```

It is interesting to note that the computation order, which does not affect performance in the case of real scalars,[†] does indeed affect performance for complex data:

```
% 21x faster than sqrt(-1); 50% faster than 1i*pi
for idx=1:1e6, a=pi*1i; end
⇨ Elapsed time is 0.033789 seconds.
```

These differences become less distinct when running within a function, where JIT optimizations come into play, rather than in a script or the Command Window as above. In this case, using *1i* is only 4–5× faster than *sqrt(–1)*, and the computation order is of no discernible importance.

MATLAB's Code Analyzer (MLint) advises using *1i* rather than *i* for performance reasons. In my tests I have not in fact seen that this has a performance impact, at least on the latest MATLAB releases. However, *1i* is still preferable for code robustness.

When we are assured that our data is real (noncomplex), we can use a few built-in MATLAB functions to compute some low-level mathematical functions faster than their generic (not-necessarily-real) counterparts. These functions include *reallog*, *realpow*, and *realsqrt* (corresponding to the generic *log*, *power*, and *sqrt*). Unfortunately, I have found that surprisingly, at least on my specific system running R2013b, these dedicated functions are often *slower* than their generic counterparts. A possible explanation is that the generic functions directly call CPU counterparts, while the dedicated functions rely on software algorithms, which will typically be slower than a CPU function. However, this is worth checking on your target platform — it is quite possible that due to a different MATLAB release, BLAS version and/or CPU, your application might actually be faster by employing these dedicated functions. Since the dedicated functions are direct replacements of the generic functions, this should be fairly easy to test.

4.5.4 Gradient

In many engineering algorithms, we need to numerically differentiate data, that is, determine its derivative/slope. The simplest, fastest, and least accurate way to do this is using the *diff* function. A more accurate yet much slower result can be achieved using the *gradient* function:[460]

```
dt = 0.0001;
t = 0 : dt : 2*pi;
y = sin(t);
realGradient = cos(t);

% Calculate gradient - relatively slow but accurate
y2 = gradient(y,dt);
⇨ Elapsed time is 0.002454 seconds.
```

---

[*] Using i as a variable name is a practice that is highly discouraged, but is widely entrenched in practice.
[†] As opposed to the case of nonscalar computations — see §4.5.2.

```
disp(max(abs(realGradient - y2)))
    5.932e-09

% Use diff - 3x faster but much less accurate than gradient()
y3 = [1,diff(y)/dt];
⇨ Elapsed time is 0.000709 seconds.

disp(max(abs(realGradient - y3)))
        5e-05
```

The *gradient* function computes at each data point the central difference between the data points on both sides of the point. This provides a more accurate result than *diff*, which simply computes the difference of each adjacent pair of data points. Some alternative algorithms have been suggested for varying data sizes.[461]

As an alternative, consider using Jan Simon's *DGradient* utility on the File Exchange.[462] Unlike *gradient* which is a pure m-file implementation, *DGradient* uses speed-optimized C (MEX) code. In addition, *gradient* always processes all the data dimensions, while *DGradient* only computes the requested dimension. The result is faster than *diff* (up to 10–20× faster than *gradient*), yet just as accurate as *gradient*!

To use *DGradient*, first compile the mex file using the mex -O DGradient.c command, then place the resulting *.mex** file in MATLAB's path. We can then use *DGradient*:

```
% Use DGradient - 15x faster than gradient() but just as accurate
y4 = DGradient(y,dt);
⇨ Elapsed time is 0.000163 seconds.

disp(max(abs(realGradient - y4)))
    5.9344e-09
```

I expect the speedup to be even higher in multidimensional cases, due to *DGradient*'s smart avoidance of calculating unnecessary dimensions (unlike *gradient*). The future addition of multi-threading and AVX/SSE2 support, which are listed in the utility's TODO list, promise even greater speedups.

4.5.5 Optimization

Optimization functions are typically function-functions, in other words, they are functions that accept other functions (constraints and objective) as input parameters. As noted in §4.6.1.2, we should avoid using anonymous function handles. Instead, we should place our input functions in actual m-code functions, using their direct handle.

The Optimization Toolbox's *fmincon* function[463] is often used in a variety of optimization problems to find the minimum value of a constrained nonlinear multivariate objective function. Finding this minimum value can be time consuming if no a priori information about the nature of the optimized function is supplied: *fmincon* would then need to scan numerous possibilities that it could otherwise have skipped as irrelevant. In fact, most optimization algorithms used by MATLAB's functions are variants of Newton's optimization method,[464] which requires a computationally intensive numerical estimation of the partial derivatives using finite-difference approximations. By specifying the optional Hessian and Jacobian/gradient partial-derivative matrices to *fmincon*, we can significantly speed up its convergence. This is because it enables fast computation of the second-order Taylor expansion:

$$f(x + \Delta x) \approx f(x) + \nabla f(x)'\Delta x + \frac{1}{2}\Delta x'H\Delta x$$

where Δx is the iteration delta, ∇f is the gradient, and H is the Hessian

$$H(f) = \begin{bmatrix} \dfrac{\partial^2 f}{\partial x_1^2} & \dfrac{\partial^2 f}{\partial x_1 \partial x_2} & \cdots & \dfrac{\partial^2 f}{\partial x_1 \partial x_n} \\ \dfrac{\partial^2 f}{\partial x_2 \partial x_1} & \dfrac{\partial^2 f}{\partial x_2^2} & \cdots & \dfrac{\partial^2 f}{\partial x_2 \partial x_n} \\ \vdots & \vdots & \ddots & \vdots \\ \dfrac{\partial^2 f}{\partial x_n \partial x_1} & \dfrac{\partial^2 f}{\partial x_n \partial x_2} & \cdots & \dfrac{\partial^2 f}{\partial x_n^2} \end{bmatrix}$$

In one example, provided in a MathWorks newsletter,[465] specifying the gradient and Hessian functions to *fmincon* provided a speedup of two orders of magnitude (85×–300×, depending on the specific configuration). That example used symbolic arithmetic,* but the advice is also valid for regular nonsymbolic computations. Another example using the Symbolic Toolbox for optimizing a financial model optimization problem, provided a 400× speedup.[466]

The drawback of this solution is that generating a high-fidelity gradient and Hessian can take a relatively long time. In some rare cases, the slowdown in the preparation stage might be longer than the resulting speedup of the optimization.

Here is a much simpler example, illustrating the usefulness of specifying gradient and Hessian. In this case, the problem is simple enough to enable a fully analytic calculation. We wish to minimize the material cost of a soda can by minimizing its surface area subject to a constraint that the contained volume be 330 mL (~11 oz),[467] and both height and radius <100 cm. First, we define the cost (objective) function:

```
% Simple solution via fmincon(), no gradient/Hessian
function S = can_surface(x)
    r = x(1);
    h = x(2);
    S = 2*pi*r.*(r+h);
end
```

And now the corresponding nonlinear constraints function. There are no nonlinear inequality constraints so we leave those empty ([]), just returning the 330 mL equality:

```
function [Vcne, Vceq] = can_volume(x)
    r = x(1);
    h = x(2);
    Vcne = []; % no inequality
    Vceq = pi*r.*r.*h - 330; % equality, [cm]
end
```

We now search for an optimal set of can radius (x(1)) and height (x(2)) that satisfy the constraints to within 1e–10 (default = 1e–6), starting with initial guess [1,1]:

```
options = optimset('Display','off', 'TolFun',1e-10, 'TolCon',1e-10);
x = fmincon(@can_surface,[1;1],[],[],[],[],[0;0],[100;100],...
            @can_volume, options)
x =
            3.74493850391498
            7.48987700857537
⇨ Elapsed time is 0.061368 seconds.
```

---

* Also see §4.9.11.

We get a result (r ≈ 3.7 cm, h ≈ 7.5 cm) satisfying the constraints in ~61 ms. We can get the same result 1.7× faster by specifying the analytic gradient, which is easy here:

```
% Solution with gradient but no Hessian
function [s,g] = can_surface(x)
    r = x(1);
    h = x(2);
    s = 2*pi*r.*(r + h);
    if nargout > 1
        g = [4*pi*r + 2*pi*h; 2*pi*r]; %gradient: 1st derivatives
    end
end

function [Vcne, Vceq, gcne, gceq] = can_volume(x)
    r = x(1);
    h = x(2);
    Vcne = []; % no inequality
    Vceq = pi*r.*r.*h - 330; % equality, [cm]
    if nargout > 1
        gcne = []; % no inequality gradient
        gceq = [2*pi*r.*h; pi*r.*r]; %gradient: 1st derivatives
    end
end

options = optimset('Display','off', 'TolFun',1e-10, 'TolCon',1e-10,...
                   'GradObj','on', 'GradConstr','on');
x = fmincon(@can_surface,[1;1],[],[],[],[],[0;0],[100;100],...
            @can_volume, options);
⇨ Elapsed time is 0.036325 seconds.
```

And even faster (for a total speedup of 2.7×) by specifying the Hessian:*

```
function hess = can_hessian(x, lambda)
    r = x(1);
    h = x(2);
    hessf = [4*pi,    2*pi;    2*pi,    0]; % Hessian of can_surface
    hessc = [2*pi*h, 2*pi*r; 2*pi*r, 0]; % Hessian of can_volume
    hess = hessf + lambda.eqnonlin*hessc;
end

options = optimset('Display','off', 'TolFun',1e-10, 'TolCon',1e-10,...
                   'GradObj','on', 'GradConstr','on',...
                   'Hessian','user-supplied', 'HessFcn',@can_hessian);
x = fmincon(@can_surface,[1;1],[],[],[],[],[0;0],[100;100],...
            @can_volume, options);
⇨ Elapsed time is 0.022737 seconds.
```

In the above example, we chose to reduce the default tolerances of 1e–6 to 1e–10. We can improve convergence speed by increasing the tolerances, thereby requiring fewer steps to

---

* Different optimization strategies (which can be set in the options struct) have different manners of specifying the Hessian; refer to the documentation for details. The method for the default strategy (interior-point) is explained here: http://mathworks.com/help/optim/ug/fmincon-interior-point-algorithm-with-analytic-hessian.html (or: http://bit.ly/1bf0yus).

converge to an acceptable solution. Naturally, the result will not be as accurate, and might converge to an invalid/suboptimal solution.*

Similarly, use known information about the problem domain to intelligently set the initial search points and constraints, both of which improve the convergence time. In some cases, the effect is nonintuitive. For example, it turns out that setting a starting point at the optimal point, may actually severely degrade optimization speed![468]

Since *fmincon* repeatedly evaluates both the objective function and the constraints, it would help to optimize the performance of both. The objective function is typically more problematic in this respect, constraints are often simpler and faster to evaluate. We can employ vectorization, caching, LIH, or any other technique mentioned elsewhere in this book, in order to improve the evaluation speed and thereby speed up the optimization convergence in general. One aspect that is often overlooked by *fmincon* users in this respect, is that the objective and constraint functions can share their workspace variables and use commonly cached data.[469]

fmincon is a generic function that can handle linear and nonlinear constraints of various forms. It is rather easy to pass such constraints to *fmincon* as input parameters (as shown above), but this is not always the fastest approach. Instead, we might try to see whether we can adapt our problem to a simpler constrained problem that requires a less computationally-intensive optimization.[470] In general, bounds are the easiest to satisfy, followed by linear equalities, linear inequalities, nonlinear equalities and finally nonlinear inequalities, which are the most expensive.[471]

It should be noted that many problems that would appear to require slow iterative optimization at first glance, could in fact be reformulated using a simpler optimization function (e.g., *lsqlin* rather than *fmincon*), or much-faster vectorized linear algebra.[472]

fmincon can handle functions of unknown type. This generality is great when the optimized function is complex or perhaps not even well-known, but it carries a performance penalty. If the problem is known in advance to be adequately described via up to second-order derivative (i.e., second-order Taylor expansion, or in other words if the problem is linear or quadratic within good accuracy), then consider using *quadprog* rather than the more-generic *fmincon*. In the newsletter optimization problem mentioned above, this provided another 4–10× speedup compared to the already-optimized *fmincon* implementation. A run time of 1:40 hours using unoptimized *fmincon* was reduced to 5 s with *quadprog*![473]

fmincon, *quadprog*, and other MATLAB optimization functions can use a variety of internal solvers to converge to the requested optimization. Depending on the specific problem, different solvers will exhibit different accuracies and speed. Refer to the documentation for a description of the various solvers and recommendations on when each of them might be better than others.[474]

In some cases, even the fast *quadprog* may be too slow to converge. In those cases, consider using other solver implementations such as

- Swiss Federal Institute of Technology, Zurich (ETHZ)'s MATLAB interface (CPLEXINT) to IBM's CPLEX solver,[475] or another CPLEX interface by Claude Tadonki of Mines ParisTech/University of Orsay.[476]
- MPT toolbox by the Automatic Control Lab at ETHZ.[477] MPT provides an interface to several external solvers, including the open-source CDD, GLPK, CLP, qpOASES, qpSpline, YALMIP, and SeDuMi, as well as the commercial GUROBI and CPLEX.

---

* See §3.9.2 for a discussion of the performance versus accuracy tradeoff, with specific treatment of optimization tolerances.

- QPC toolbox by the SPM group at the University of Newcastle, Australia.[478]
- Mark Mikofski's Newton–Raphson solver on the MATLAB File Exchange.[479]
- IPOPT package, part of the COIN-OR open-source project.[480]
- YALMIP toolbox by Johan Löfberg of Linköping University, Sweden.[481]
- Cornell's Murillo–Sánchez's MATLAB interface (BPMD_MEX) to the BPMD solver by Csaba Mészáros at the Hungarian Academy of Sciences, Budapest.[482]

Note that these solvers target different types of problems. For example, some solvers target multi-parametric problems; others target nonlinear problems, and so on. In some cases there is an overlap between them, in others not. This is by no means an exhaustive list. Optimization is an area of ongoing research and solvers are constantly being developed to target different sets of problem types. A nice utility that enables to easily install multiple different solvers is *tbxManager*.[483]

In many cases, a local optimization is sufficient, and there is no need to search for a global optimization, which takes longer to evaluate since it needs to verify that we are not converging onto a suboptimal local minimum. The smoother a function is, the easier it is to search for a global optimum, but even in cases of relatively smooth functions, searching for a local minimum within a predefined range of parameter values would be faster than finding a global optimum over the entire range. In a related matter, the importance of avoiding over-optimization is never over-stated.

When searching for an optimum target (i.e., global minimum), we should note that if we rely on a single run, then the algorithm might well converge onto a local (or even trivial) minimum. To increase the probability of a successful optimization, we should rerun the search using several starting points. These are typically either random or arranged as equi-distance lattice points. The convergence algorithm's parameters as well as the algorithm itself could be modified in a random or regular manner.[484] The more starting points and convergence modifications, the more likely we are to arrive at a true global optimum, but this naturally has a run-time expense.

If we have the Parallel Computing Toolbox (PCT), we can use the UseParallel flag to enable parallel optimization search (see §6.4.2 for additional details):

```
parpool(4);
opts = optimset('fmincon');
opts = optimset('UseParallel','always');
[x, fval] = fmincon(objfun,x0,[],[],[],[],[],[],confun);
```

A related tip is to use the execution time (as reported by the *toc* command) as part of the optimization criteria, to enable optimization timeout.[485] A few optimization functions also honor the associated MaxTime parameter.[486]

A series of hour-long MathWorks webinars explain MATLAB optimization,[487] from the basics,[488] through global optimization,[489] and parallelized optimization.[490]

A great resource that is well worth reading is John D'Errico's *Optimization Tips and Tricks* guide,[491] a very detailed HTML document complete with code snippets and graphics, as well as separate runnable m-flies. Reading the entire thing can take some time, but you will definitely be much more informed at the end. John D'Errico is/was very active in the MATLAB CSSM newsgroup on optimization issues, and contributed several other notable submission to the File Exchange that are worth perusing, esp. *fminspleas*[492] and *fminsearchbnd*.[493] In general, D'Errico's submissions are some of the most downloaded and most highly rated on the File Exchange.[494] John is also a great educator, taking the time to explain elusive fine points clearly and thoroughly.[495] Correspondingly, John's code is very well documented. We can learn a lot from the way he coded his submissions, not less than benefit from directly using them. The File Exchange contains several additional optimizers for specific uses.[496]

Another interesting resource for MATLAB optimization is the relevant page on the Mathtools.net website.[497] Also, consider using the TOMLAB (§8.5.6.3) or NAG (§8.5.2) toolboxes, which overlap and complement many MATLAB functions.

4.5.6 Fast Fourier Transform

Fast Fourier transform (FFT)[498] and its offshoots (IFFT, etc.) are extensively used with sampled natural signals such as audio, video, and sensorial inputs. An enormous amount of academic research was done over the years in optimizing FFT algorithms for speed and memory usage. MATLAB has licensed the FFTW (*"Fastest Fourier Transform in the West"*)[499] implementation starting in MATLAB 6.0 (R12).[500]

Over the years, MATLAB has incorporated progressively advanced versions of FFTW. MATLAB releases have generally kept pace with the latest FFTW versions available at that time. But if we currently use an out-of-date MATLAB release then it might be worthwhile to download the latest FFTW version and use it directly,* rather than MATLAB's built-in *fft** functions that rely on the older FFTW. The version of FFTW used by MATLAB can be seen using the *version* function:[501]

```
>> version -fftw % R2013b
ans =
FFTW-3.3.3-sse2-avx
```

We can also use the following round-about way:

```
>> fftw('wisdom')
ans =
(FFTW-3.3.3 fftw_wisdom #x3c273403 #x192df114 #x4d08727c #xe98e9b9d)
```

As MATLAB's documentation says,[502] the execution time for the FFT set of functions (*fft** and *ifft**) depends on the length of the transform. It is fastest for powers of 2 (e.g., 256, 512, 1024); it is often almost as fast for lengths that have only small prime factors (e.g., 3×5×7); it is typically several times slower for lengths that are prime or which have large prime factors (e.g., 541×547). In many cases, it turns out that lengths of repeated small prime factors are faster than the next higher power-of-2 (e.g., $81 = 3^4$ would be faster than $128 = 2^7$).[503] The specific comparison between potential window sizes should be done on the target computer, since it is highly affected by hardware issues such as cache size and memory boundaries/alignment.

In the general case, the smart thing to do, performance-wise, is to pad the input data to the nearest power of 2. We can use the *nextpow2* built-in function for this:

```
x = rand(1,1e6); % 1M elements test data

y = fft(x,541*547);
⇨ Elapsed time is 0.022332 seconds.

y = fft(x, 2^nextpow2(541*547)); % 2x faster
⇨ Elapsed time is 0.011956 seconds.
```

---

* FFTW is defined as Free Software (under GPL license) for non-commercial use. See http://fftw.org/faq/ section1.html.

We might be able to increase the speed of the FFT functions using the built-in utility function *fftw*, which controls how MATLAB (or more specifically, its FFTW library) optimizes the algorithm used to compute an FFT of particular size and dimension.[504] Depending on our data and on how willing we are to sacrifice initial performance in return for subsequent speedups, we can choose one of several self-tuning strategies.*

When computing multi-dimensional inverse-FFTs, note that the built-in *ifftn* function is generally equivalent and faster to an iterative loop over all data dimensions via *ifft*:

```
% ifftn variant
y = ifftn(x);

% ifft loop variant
y = x;
for p = 1 : length(size(x))
    y = ifft(y,[],p);
end
```

FFT can be used to significantly speed up some calculations, by transforming the data from the natural time domain into the frequency (spectral) domain (with *fft, fft2,* or *fftn*), applying the required processing, finally returning results to the time domain using *ifft, ifft2,* or *ifftn* (respectively). By converting the problem into the frequency domain, we can often employ vectorized processing, rather than a much slower iterative (sliding-window) approach that would be necessary in the time domain.

An example of using this technique for improved performance is image noise filtering using a high-pass filter.[505] Another example, perhaps less obvious, is Bruno Luong's implementation of efficient convolution[506] using the Convolution Theorem: $conv(a,b) = ifft(fft(a,N)*fft(b,N))$,[507] rather than a sliding-window approach used by the built-in *conv** functions.† When using this identity, remember to zero pad the data:[508]

```
% Prepare the input vectors (1M elements each)
x = rand(1e6,1);
y = rand(1e6,1);

% Compute the convolution using the builtin conv()
z1 = conv(x,y);
⇒ Elapsed time is 360.521187 seconds.

% Now compute the convolution using fft/ifft: 780x faster!
n = length(x) + length(y) - 1; % we need to zero-pad
z2 = ifft(fft(x,n).*fft(y,n));
⇒ Elapsed time is 0.463169 seconds.

% Compare the relative accuracy (the results are nearly identical)
>> max(abs(z1-z2)./abs(z1))
ans =
      2.75200348450538e-10
```

Another mathematical identity can sometimes be used to speed up FFT calculations: $fft(x) = conj(ifft(conj(x)))*N$, where N is the data length. Whereas in the majority of cases the simpler (left-side) format is faster, in some cases the reverse may be true.[509] This also applies for the reverse calculation: $ifft(x) = conj(fft(conj(x)))/N$.

---

* Also see §3.10.3.2.
† Bruno's utility was made even more efficient by using MEX in-place data manipulations. Also see §8.1, §5.7.7, and §9.5.2.

FFT functions are generally faster for double-precision data than for singles, since the underlying FFTW library is better-optimized for doubles, which is also MATLAB's default data type.*

An interesting theoretical breakthrough in FFT performance was announced by MIT researchers in 2012.[510] The resulting library, called *sFFT* (sparse FFT), is available for download in source-code format from the sFFT portal, under an MIT open-source license.[511] A version of sFFT that is hardware optimized (and 2×–5× faster than the generic reference sFFT) was developed in ETH and released under GPL license,[512] as part of the *Spiral* open-source project.[513] The basic idea of sFFT is that in any naturally occurring signal, there are only a limited number of principle frequencies, which together account for 90%+ of the information. By analyzing only those key frequencies, we can significantly speed up data processing while sacrificing just a minimal amount of data.† The result is that sFFT is significantly faster than FFTW for a wide range of natural signal sources: sound, images/video, and so on.[514] Users who find that a large portion of their program time is spent on FFT analysis of natural signals, may benefit from directly utilizing sFFT rather than MATLAB's built-in functions. sFFT is most effective for repeated FFTs having the same window length, since in this case the up-front overheads are amortized over multiple calls.

As a related suggestion, we could pass our data through a frequency filter (low-pass, high-pass, or band-pass filter), to remove noise due to irrelevant frequencies, before processing the data. Once we remove irrelevant frequencies, processing becomes both faster and more accurate.[515]

In a related matter, one user has reported slowdowns when using the *fftshift* function, possibly due to the use of suboptimal internal *repmat* calls.[516] The user reported that unrolling the *fftshift* operation into separate real and complex components resulted in an 8× speedup. However, such tricks are apparently not quite required in the latest MATLAB release, possibly due to *repmat*'s improvements in R2013b.‡

4.5.7 Updating the Math Libraries

MATLAB bundles several well-known third-party numeric libraries that help it process data. Depending on the specific MATLAB installation, these may include (among others) FFTW for FFT-related functions,§ MAGMA[517] for GPU processing, and Intel's IPP[518] and MKL[519] for math/algebra functions.¶

Each release of MATLAB bundles its own version of these libraries. Progressively advanced MATLAB releases bundle progressively advanced library versions. The version of these libraries in any MATLAB release can be found using the following commands (the example below is for MATLAB 8.1 R2013a):**,[520]

```
>> version -fftw
ans =
FFTW-3.3.1-sse2-avx
```

---

* Also see §4.1.

† In many respects this is similar to JPG's lossy image compression. Compare the performance versus accuracy tradeoff in §3.9.2.

‡ See §5.4.2.

§ See §4.5.6.

¶ See §5.1.1 for some additional detail. Old MATLAB releases used other libraries (e.g., ATLAS for BLAS, or SPL for Solaris).

** We can also look at the file properties of the individual library files.

```
>> version -blas
ans =
Intel(R) Math Kernel Library Version 10.3.11 Product Build 20120606 for
Intel(R) 64 architecture applications

>> version -lapack
ans =
Intel(R) Math Kernel Library Version 10.3.11 Product Build 20120606 for
Intel(R) 64 architecture applications
Linear Algebra PACKage Version 3.4.1

>> [isIpplEnabled, ipplVersion] = ippl
isIpplEnabled =
                    1% == true
ipplVersion =
    'ippie9_t.lib 7.0 build 205.58 7.0.205.1054 e9  Apr 18 2011'
```

Both the third-party libraries and MATLAB constantly issue new versions. If you have a relatively old MATLAB release and for any reason cannot upgrade to a newer release, you may find it beneficial to use newer versions of the external libraries.

For example, in the example above we see that we use IPP version 7.0, whereas version 7.1 offers important performance improvements.[521] One user has reported[522] achieving a 30% code speedup, without any code modification, using a newer and more optimized version of BLAS (BLAS and LAPACK are normally included in the MKL library that is bundled with MATLAB).

However, note that replacing the MATLAB-bundled libraries is entirely unsupported by MathWorks and carries a real risk of causing severe computational problems and inaccuracies in the results. Also, while some of the libraries (e.g., FFTW) are open-source, others (e.g., IPP or MKL) are commercial and require having a valid license. MathWorks' licenses for these libraries may possibly not extend to MATLAB users.* Even if they do, if we need new versions of IPP or MKL, we would definitely need to pay for them ourselves.

There are basically two ways to use the new library versions:†

- The safer way is to add the external libs and call them directly from within our MATLAB code. In this case, the built-in MATLAB functions will continue to use the default libraries, but our application code will use the new ones. This is relatively safe since it does not impact the existing MATLAB functionality.

    Section 4.5.2 has shown that we can link our MEX code with MATLAB's built-in BLAS/LAPACK libraries. We can relatively easily extend this to statically link our MEX code with newer versions of these libs, rather than MATLAB's standard libs. Intel even has official support pages explaining how to do this for MKL.[523] One user has reported[524] a 150% run-time discrepancy (i.e., 2.5× speedup) between different BLAS variants.‡ Users may find this compelling enough to try MKL alternatives, such as locally compiled versions of the open-source ATLAS,[525] BLIS/FLAME,[526] or GotoBLAS/OpenBLAS[527] libraries.

---

* Note: I am an engineer, not a lawyer, and this is not legal advice. Contact the relevant companies directly for licensing info.
† See §8.1.7.3 for the related aspect of using functions in the bundled libraries, for example, to use MKL's *fft* rather than FFTW's.
‡ Note that this user's report did not mention MATLAB, nor MKL, which is quite highly optimized.

Another possibility is to use an entirely separate external (third-party) library, such as Marcel Leutenegger's MATLAB Toolbox of math functions,* Tom Minka's Lightspeed toolbox,† NAG,‡ GSL,§ or sFFT.¶ Such toolboxes or libraries can be used in a selective way, only for certain specific functions.

- A faster, easier but much riskier alternative is to tell MATLAB to automatically use a nondefault library by its internal functions. This can be done in several ways (the first two only apply to BLAS/LAPACK):[528]

  o Place the new dynamic library file (*.dll, *.so, or *.dylib, depending on platform) in MATLAB's main lib folder *%matlabroot%/bin/%arch%/* (e.g., *C:\Program Files\ MATLAB\R2013b\bin\win64*). Then modify the *blas.spec* and *lapack.spec* files in the same folder, using any text editor. A typical **.spec* file may look as follows, where we would replace the term "mkl.dll" with the name of the new library file:

  ```
  GenuineIntel Family * Model * mkl.dll # Intel processors
  AuthenticAMD Family * Model * mkl.dll # AMD
  ```

  Multiple libraries can be loaded in this manner. For example, to load *mkl.so* followed by *mklcompat.so*, use "mkl.so:mklcompat.so".[529]

  o Set the global environment variable BLAS_VERSION to the new library's full file path, or simply to the file name if you placed it in the main libraries folder as above. This alternative has the benefit of being simpler as well as enabling placement of the new lib file outside the MATLAB installation folder, and without affecting any files in the installation folder. In this manner, we can use a new library even if we cannot modify MATLAB installation folder. This may also enable easier deployment. As above, multiple libraries can be loaded:[530]

  ```
  setenv BLAS_VERSION mkl.so:mklcompat.so
  ```

  o The simplest and riskiest alternative is to directly replace MATLAB's library file with a new version. For example, on Windows, the MKL library file (*mkl. dll***) can be replaced with a newer DLL version.

  Note that directly replacing library files will work for any library, not just for BLAS/LAPACK. The same folder also holds IPP and other libraries that can be directly replaced in a similar fashion.††

Using nondefault external libs needs to be done <u>EXTREMELY CAREFULLY</u>, since the MATLAB functions are tied to the external libraries quite tightly, and so any interface or functionality change in the new libraries might cause MATLAB to crash, misbehave, or even worse — to return incorrect results without any sign of error! I therefore urge extreme caution when doing this, and only do this for library versions that are supposed to be compatible (e.g., Intel's MKL version 10 is partially incompatible with earlier versions[531]). Use this technique only as a last resort, after trying all the other techniques in this book. I am not just saying this as mere formality due to the unsupported nature of such library replacements: I strongly discourage it.*

If you do decide to use a new library, test MATLAB carefully to ensure that the replacement did not break anything important. The risk is rarely worth the benefits. So be aware of this option, but only use it if a specific feature or improvement in the replacement outweighs the risk.

To ensure that the correct libraries are used by MATLAB, use the following:†

```
>> version -modules
C:\Program Files\MATLAB\R2013b\bin\win64\libut.dll Version <unknown>
C:\Program Files\MATLAB\R2013b\bin\win64\libmwfl.dll Version <unknown>
C:\Program Files\MATLAB\R2013b\bin\win64\libmx.dll Version 8.2.0.627
C:\Program Files\MATLAB\R2013b\bin\win64\zlib1.dll Version <unknown>
...
```

4.5.8 Random Numbers

When using a pseudo-random-number function (**rand, randn, randi, sprand** etc.) in MATLAB, many people are not aware of the fact that in MATLAB 7.4 (R2007a) the default random-number generation (RNG) algorithm changed from SWB (*Subtract With Borrow*, used for uniform distribution **rand**) and SHR3 (Marsaglia's *Shift Register* generator, used for normal distribution **randn**), to Mersenne Twister (MT).[532] In recent years, MT has become the default RNG in a variety of programming languages; it is currently considered a de-facto programming standard.

SWB and SHR3 are faster than MT, but produce a stream of pseudo-random numbers that are less random than MT.‡ In many applications, the functional distinction between these RNGs is of no practical consequence, and the stream produced by SHR3 is random enough for all practical purposes. In such cases, switching MATLAB's RNG to SHR3 could produce a very nice speedup:[533]

```
new_RNG_seed = sum(100*clock);
rand('state', new_RNG_seed); % R2007a to R2010b
rng(new_RNG_seed, 'v5normal'); % R2011a onward

% Using the default RNG (Mersenne Twister)
rng(0,'twister'); for i=1:100000, x=randn(1000,1); end
⇨ Elapsed time is 1.410599 seconds.

% Using the legacy RNG (SHR3) - 2x faster
rng(0,'v5normal'); for i=1:100000, x=randn(1000,1); end
⇨ Elapsed time is 0.623054 seconds.
```

---

* Readers who are aware of my reputation for using undocumented aspects know that I do not normally use such strong language.

† The -modules feature was removed in R2014a. (8.3). There is no known way to retrieve library version information since then, except by looking at the file properties of the individual library files.

‡ SWB's randomness period is 2^{1492} (~1E150), MT's period is 2^{19937} (~1E2000) and has much better equi-distribution. SWB produces a flawed (nonequi-distribution) stream (see http://www2.cs.cas.cz/~savicky/papers/rand2006.pdf).

Note that MATLAB 7.12 (R2011a) added the *rng* function that allows setting the RNG algorithm and seed.* The older *rand* syntax is still supported, but is deprecated and in fact sets RNG to a slower variant of SHR3:

```
rand('state',0); for i=1:100000, x=randn(1000,1); end
⇨ Elapsed time is 0.808543 seconds.
```

MATLAB offers several RNG alternatives (*combRecursive, multFibonacci, v5uniform* (= SWB) and *v4*). They are all slower than *v5normal* and should be used in special cases such as in parallel/GPU code (*combRecursive* or *multFibonacci*).[534]

In most cases, the random-number generation performance is not very important. Such numbers are often used to drive simulations of data, and the associated computations typically take much longer to compute than the RNG. In fact, for some applications it is more important to consider the type of randomness distribution (pseudo or quasi[535]). If you are considering changing the RNG algorithm, consider whether it is worth the dev time and the decreased randomness. Consider using Peter Li's *RNG* utility that uses MT, and yet is faster than MATLAB's internal implementation.[536]

4.6 Functional Programming

4.6.1 Invoking Functions

4.6.1.1 Scripts versus Functions

One of the first performance-improvement suggestions that users hear when starting to use MATLAB is that functions are faster than scripts.† The often-stated explanation is that MATLAB's JIT caches JIT-compiled functions, but not scripts,[537] and that JIT can compile functions but must interpret scripts.[538] This is all very common knowledge, and the suggestion to use functions rather than scripts still appears on MATLAB's official performance-improvements webpage.[539]

While all this was true until the late 2000s, recent advances in MATLAB's JIT have made the performance difference between scripts and functions effectively indiscernible in recent releases:

```
% scriptA.m
a=1; b=2; c=3; d=4;
for k = 1:1e6
    a = b*c*sqrt(d);
end

% funcA.m
function funcA
```

---

* MATLAB 7.7 (R2008b) added the *RandStream* class that enables detailed RNG customization but in most cases this is not needed and adds unnecessary complexity. See http://blogs.mathworks.com/ loren/2011/07/07/simpler-control-of-random-number-generation-in-matlab (or: http://bit.ly/1fZhFCw). RandStream.getGlobalStream returns the current RNG.

† The reader may recall that scripts are simply a set of MATLAB commands placed directly in an m-file, which affect the base workspace; functions are very similar except that they begin with the keyword "function" and have a dedicated workspace. In both cases, the MATLAB code is placed within m-files. Directly entering commands in the MATLAB console (Command Window) is usually (but not always) significantly slower than either of these options (see §2.2.3) and is not covered in this section.

```
    a=1; b=2; c=3; d=4;
    for k = 1:1e6
      a = b*c*sqrt(d);
    end
end

% Now the test run: no discernible performance difference
clear; for idx=1:100, scriptA; end
⇨ Elapsed time is 1.325592 seconds.

clear; for idx=1:100, funcA; end
⇨ Elapsed time is 1.325472 seconds.
```

There are two notable exceptions:*

- Scripts that call other scripts are somewhat slower than direct scripts (i.e., using inlined code) or the equivalent functions:

```
% scriptAs.m
scriptInit % call another script
for k = 1:1e6
    a = b*c*sqrt(d);
end

% Now the test run: script calling another script is a bit slower
clear; for idx=1:100, scriptAs; end
⇨ Elapsed time is 1.429214 seconds.
```

- Functions that call scripts that modify a variable used by the function are much slower than the corresponding script or function. This is an open bug in MATLAB since at least R2006a (MATLAB 7.2):[540]

```
% scriptInit.m
a=1; b=2; c=3; d=4;

% funcAs.m
function funcAs
    scriptInit % call a script
    for k = 1:1e6
        a = b*c*sqrt(d);
    end
end

% Now the test run: functions calling scripts are *MUCH* slower
clear; for idx=1:100, funcA2; end
⇨ Elapsed time is 75.009764 seconds.
```

Once again, we see an example where common knowledge and intuition fail in real life, highlighting the need to test our assumptions. In this case, we see that functions either provide no significant performance gain over scripts, or, in one edge case, actually degrade performance significantly.

---

* Aside from these notable exceptions, differences between the way that the JIT works in scripts and functions may affect some use cases. For example, http://mathworks.com/matlabcentral/newsreader/view_thread/325638 (or: http://bit.ly/1mqRTJE).

However, even without the performance benefits, I still highly recommend converting every MATLAB script into a function. The reasons for this are the other benefits provided by functions, including their ability to define inputs and outputs, and their use of self-contained "undirtied" workspace.

4.6.1.2 Function Types

Try to avoid function handles in general and anonymous functions in particular, if possible, since they are significantly slower than direct function invocation.* Sometimes we cannot avoid using function handles, for example, when defining callbacks or using function-functions such as *fminbnd* (§4.5.5) or *arrayfun* (§5.2.4). In such cases, using a handle of a direct function is faster than using an anonymous function:

```
% Direct function invocation: fastest approach
for idx=1:1e5, a=sin(idx); end
⇨ Elapsed time is 0.003811 seconds.

% Invoking using a function handle: 55x slower
fh = @sin; for idx=1:1e5, a=fh(idx); end
⇨ Elapsed time is 0.210944 seconds.

% Invoking using an anonymous function: 187x slower
fh = @(x)sin(x); for idx=1:1e5, a=fh(idx); end
⇨ Elapsed time is 0.711505 seconds.
```

Some old-school MATLAB programmers sometimes use *eval* to evaluate functions or expressions. However, *eval* is always the slowest approach, by a long margin:

```
% Using eval(): ~1860x slower!
for idx=1:1e5, eval(sprintf('a=sin(%f);',idx)); end
⇨ Elapsed time is 7.086479 seconds.
```

We should also consider using the *feval* function rather than *eval*.

```
% feval() with function name: 80x slower, ~24x faster than eval()
for idx=1:1e5, a=feval('sin',idx); end
⇨ Elapsed time is 0.306193 seconds.

% feval() with function handle: 127x slower, ~15x faster than eval()
for idx=1:1e5, a=feval(@sin,idx); end
⇨ Elapsed time is 0.482897 seconds.

% feval() with anonymous func: 836x slower, ~2.2x faster than eval()
for idx=1:1e5, a=feval(@(x)sin(x),idx); end
⇨ Elapsed time is 3.186314 seconds.
```

As a general rule, we should always attempt to replace *eval* calls in our code with other types of function invocations, if only for the sake of performance (there are many additional reasons for avoiding *eval*, which are unrelated to performance).

---

* Pre-MATLAB 7 code used the *inline* function, which was much slower than even anonymous functions. If you have any remnant *inline* call in your code, replacing it with a function handle (even anonymous) will significantly improve performance. See http://www.csc.kth.se/utbildning/kth/kurser/DN1240/numi12/matlab_perf.pdf#page=3 (or: http://bit.ly/1lADwlH).

If the result of the calculation needs to be assigned to a variable, then we should either assign it directly, or modify our code to use dynamic struct field names, or cell arrays.[541] One might think that using the **assignin** function would be better than **eval**, but this is not in fact the case:

```
% Direct assignment: fastest approach
for idx=1:1e3, a = 3.1416; end
⇨ Elapsed time is 0.000015 seconds.

% Using eval(): 170x slower
for idx=1:1e3, eval('a=3.1416;'); end
⇨ Elapsed time is 0.002606 seconds.

% Using assignin(): 190x slower
for idx=1:1e3, assignin('caller','a',3.1416); end
⇨ Elapsed time is 0.002851 seconds.

% Using cell-arrays: 56x slower
for idx=1:1e3, c{1} = 3.1416; end
⇨ Elapsed time is 0.000841 seconds.

% Using dynamic struct fields: 59x slower
for idx=1:1e3, s.('a') = 3.1416; end
⇨ Elapsed time is 0.000885 seconds.

% Using containers.Map: ~1300x slower!
c = containers.Map; for idx=1:1e3, c('a')=3.1416; end
⇨ Elapsed time is 0.019309 seconds.
```

MathWork's Loren Shure commented as follows:[542]

> *If you specify the function of interest as a string, MATLAB must search the path to find the function. If you are performing a calculation in a loop, perhaps to find roots near multiple locations, MATLAB must search for the target function each time it passes through the loop in case the path has changed and a different version of the target is now relevant. A function handle locks down the target version, enabling MATLAB to call the target function without searching for it each time it passes through the loop.*

However, careful readers might notice the odd fact that for **feval** in the example above, using function name inputs actually appears to be faster than function handles. This is indeed surprising, but since **feval** is a built-in (not m-file) function, we do not have insight as to the reason for this.

Another exception to the rule that using function handles is faster than function names, is in the special case where these strings are used as special input parameters to built-in functions. An example of this (for **cellfun**) is discussed in §5.2.4.1:

```
c{1e6} = 1; % cell-array 1x1M
a = cellfun('isempty',c);        % string input: fastest
⇨ Elapsed time is 0.003908 seconds.

a = cellfun(@isempty, c);        % function handle: 130x slower
⇨ Elapsed time is 0.506266 seconds.

a = cellfun(@(c)isempty(c), c); % anonymous function: ~1100x slower
⇨ Elapsed time is 4.350575 seconds.
```

Among the "standard" (nonobject-oriented) types of functions, namely regular functions, subfunctions, and nested functions, I have found no discernible performance difference on recent MATLAB releases. For a discussion of object-oriented functions (methods), see §4.7.3.

When specifying callbacks for asynchronous events (e.g., timer invocations or interaction with GUI controls), we normally have a choice between using a string callback and specifying a callback function handle. In general, using function handles is preferable to strings that evaluate at run time. String used to be the only way to specify callbacks in the old days (pre-MATLAB 7), before function handles became available. But in modern MATLAB releases we should replace the string syntax with function handles. In practice, the performance implications of using a string function name is small compared to a function handle: ***str2func*** (which converts a string function name into a runnable function handle) can be faster than 0.1 ms, which is usually negligible compared to the callback execution time. However, using function handles enables passing parameters to the callback function more easily and in general results in a more readable and maintainable code.

It is very unfortunate that to this day (MATLAB 8.4 R2014b), the built-in Graphical User Interface Design Environment (GUIDE) still uses runtime-evaluated strings as the default callback mechanism.* In fact, GUIDE uses an indirect manner of calling callback functions, by assigning a default callback which is an anonymous function that runs the GUI's main function that then converts the specified function name input into a callable function handle using ***str2func*** which is then invoked. This is a very indirect and inefficient way of calling callbacks:

```
@(hObject,eventdata)
    guiName('hButton_Callback', hObject, eventdata, guidata(hObject))
```

So, in the case of GUIDE-generated callbacks, users are advised to modify the standard GUIDE-generated callback properties. We could, for example, assign a direct function handle to some external function. Unfortunately, in GUIDE we cannot directly assign a function handle to a sub-function in the main m-file. Therefore, we need to move that callback function to an m-file of its own, so that we could reference it directly in GUIDE.

With external interfaces (C/C++/C#/Java/COM/…), it may be possible to call a single merged function rather than separate simpler functions, to reduce the interface's function invocation overheads. Note that the expected speedup here is small.

4.6.1.3 Input and Output Parameters

In some cases, we can call built-in functions with nondefault optional input arguments that enable the function to bypass some internal checks and processing, thereby speeding up its processing. A typical example of this is to specify the date/time format for the ***datestr*** and ***datenum*** functions, as shown in §4.4:

```
a = datenum('2013-02-17');
⇨ Elapsed time is 0.005505 seconds.

a = datenum('2013-02-17', 'yyyy-mm-dd');
⇨ Elapsed time is 0.000430 seconds.
```

---

* Possibly to preserve backward compatibility with very old (pre-2000) MATLAB releases that did not support function handles.

Accelerating MATLAB® Performance

As another example, the built-in *optimget* function has an undocumented usage for fast access with no error checking:*

```
d = optimget(o,'TolX',1e-6);
⇨ Elapsed time is 0.000640 seconds.

d = optimget(o,'TolX',defaultOpts,'fast');
⇨ Elapsed time is 0.000298 seconds.
```

A variant of this idea is to use optional input arguments to limit the internal processing, thereby preventing the need for additional explicit external checks. For example,[543]

```
% Standard code, using explicit check for a surface handle
flag = ishghandle(h) && strcmp(get(h,'type'),'surface');
⇨ Elapsed time is 0.000088 seconds.

% Alternative, not any faster
flag = ishghandle(h) && isa(handle(h),'surface');
⇨ Elapsed time is 0.000117 seconds.

% Faster, simpler, more compact code
flag = ishghandle(h,'surface');
⇨ Elapsed time is 0.000018 seconds.
```

This example demonstrates that checking if *ishghandle(h,'surface')* is faster than checking if *ishghandle(h) && strcmp(...)* or *&& isa(...)* (also see §4.9.9).

Whenever possible we should attempt to use functions that do not modify their input data, since this enables in-place data manipulation and prevents the need for MATLAB to allocate space and copy the data. This is particularly true if the modified input data is very large. Read-only functions will be faster than functions that do even the simplest of data updates. See §9.5.1 for additional details.

When calling functions, we should not request/assign output arguments that are not needed. In many cases, this can save time if the function is smart enough to check for this and not compute these output arguments unnecessarily:

```
str = char('0' + randi(10,1,1e6)); % 1M characters

[idx,a,b,c,d,e] = regexp(str,'(\d+):');
⇨ Elapsed time is 0.207075 seconds.

idx = regexp(str,'(\d+):'); % 2x faster
⇨ Elapsed time is 0.095514 seconds.
```

This idea applies to two separate but related aspects: (1) we should not request unnecessary output parameters when calling functions; (2) in our functions, we should use *nargout* to compute only the requested output parameters, and no others:

```
function [a,b] = myFunc
    ...
    if nargout > 0
        a =...; % only compute a if it is requested by the caller
    end
    if nargout > 1
```

---

* *optimget* is used in optimization problems — see §4.5.5.

```
        b =...; % only compute b if it is requested by the caller
    end
end
```

An alternative to using predefined (named) output parameters is to use *varargout* with a variable number of output parameters. This can help us compute only those output args that are required. Naturally, we still need to ensure this programmatically (with *nargout*) — it is not an automatic by-product of using *varargout*.[544]

```
function varargout = magicfill()
    nOutputs = nargout;
    varargout = cell(1,nOutputs);
    for k = 1 : nOutputs;
        varargout{k} = magic(k);
    end
end
```

MATLAB R2009b (7.9) introduced the syntax of tilde (~) parameter.[545] When we use the ~ sign as parameter placeholder, rather than using a parameter name, the parameter is ignored. We can use this syntax to ignore unneeded output parameters:

```
[filePath, ~, fileExt] = fileparts(filename);
[~, ~, rawData] = xlsread(filename);
```

Unfortunately, while this syntax prevents the need to assign the relevant output(s) to a local variable, it does *not* really help performance since the output parameters are still being computed by the function. It is too bad that MATLAB does not have a function corresponding to *inputname*, which would enable us to check for such unassigned outputs within our functions, to prevent the needless computations of ~ outputs.

Given the above, it should now be understood why

```
[param1, ~, ~] = someFunction();
```

is slower in general than the equivalent

```
param1 = someFunction();
```

In fact, trailing ~ signs should be removed from the code wherever they appear.

As a final note, MathWork's Fundamentals course says that calling a function with a single output argument enclosed in square brackets ([outVar] = func) is slightly less efficient than without the square brackets (outVar = func). In my experiments, I saw no difference between these alternatives, so take this advice with a grain of salt...

4.6.1.4 Switchyard Functions Dispatch

MATLAB code often employs so-called *switchyard functions*, to dispatch one of several alternative functions based on one of the inputs (which is typically a string). This is essentially a small state machine, where the string input represents the state and the switchyard function is the state-machine interpreter. In many cases, the switchyard function dispatches a sub-function in the same file. The general format is

```
function mainFunc(inputState)
    switch inputState
        case 'stateA', doSomethingA();
        case 'stateB', doSomethingB();
```

```
        case 'stateC', doSomethingC();
    end
end
```

In such cases, it is often tempting to call `mainFunc('stateA')` within the actual processing functions, when we wish to activate some other state. However, this involves extra function-call overheads and input-data parsing compared to the much more direct call to `doSomethingA()`.

We should be especially wary of this in GUIDE-generated GUI functions: GUIDE creates generic m-files that accept a sub-function name as input parameter and parse it internally (via *str2func*), dispatching the function as necessary. This is a very useful mechanism when we wish to invoke the internal sub-function from outside the m-file. However, when we wish to call it from within the m-file (as in the vast majority of cases), it is much faster to directly invoke the other sub-function. In other words, within our GUIDE-generated m-file, rather than calling another sub-function indirectly:

```
guiName('hButton_Callback', gcf, eventdata, guidata(gcf));
```

call it directly:

```
hButton_Callback(gcf, eventdata, guidata(gcf));
```

4.6.2 *onCleanup*

MATLAB 7.6 (R2008a) introduced the *onCleanup* function (or more precisely, class object) that enables us to define a code segment that should be run whenever the function exits for any reason — normally, due to error, or a user's Ctrl-C. This is a very convenient feature that enables more robust code using a very simple so-called *finalization* mechanism. Typical usages: close any open I/O files/channels, GUI reset.[546]

Unfortunately, the underlying mechanism uses MCOS (*MATLAB Class Object System*) and specifically a very simple class object that automatically runs its internal *delete* method when its instance is being destroyed.[*] The implementation is simple,[†] but it has nonnegligible overhead. On my specific platform, using *onCleanup* costs an extra 0.2 ms for creating the object instance and then running its finalizer method.[‡]

It is possible that at some point in the future, MathWorks will improve *onCleanup*'s performance by implementing it as an internal function rather than using the MCOS destruction workaround. But until then, if we run a function multiple times in a loop, and *onCleanup*'s performance overhead is significant compared to the containing function's overall run duration, then we could use the following alternative:

```
% using onCleanup - slower
function myAlgorithm()
    oco = onCleanup(@onCleanupFcn);
    % now do the main processing
    ...
end
```

---

[*] Note that this could also happen by using *clear* or by overriding the variable holding the object handle with any new value.

[†] Readers are encouraged to look at the source code: *%matlabroot%/toolbox/matlab/general/onCleanup.m*

[‡] See related §4.7 for additional aspects of MCOS performance.

```
% direct alternative - faster
function myAlgorithm()
    try
        % do the main processing
    catch
        % process any possible error
    end
    onCleanupFcn(); % we're done, so run finalizer function directly
end
```

Using the direct alternative has the additional benefit of being runnable on MATLAB releases 7.5 (R2007b) and earlier, where *onCleanup* was not available.* We could also use it to pass input parameters to the cleanup function, which cannot be done with *onCleanup*. However, we must ensure that all possible program flows result in running the finalizer function (in the example above, `onCleanupFcn`). Overlooking function early bail-out branches is quite common. Also note that the faster direct alternative does not run the finalizer in case the user stopped execution in mid-processing via Ctrl-C or the debugger, while *onCleanup* does. I therefore advise using the slower *onCleanup* mechanism in all cases, except where profiling indicates that *onCleanup* is a performance hotspot.

4.6.3 Conditional Constructs

As noted in §3.1.4, MATLAB's JIT has a harder time optimizing loops and functions that have conditional branches. We have seen that one way around this is to simplify loops so that they do not contain such conditionals, either directly (in the loop's code) or indirectly (in functions that are called within the loop). In §3.1.4, we have seen that using the functional (branch-less) form is simpler, faster, and often more maintainable than straightforward conditional branches. For example, the expression

```
if y > SOME_VALUE
    data = sin(x);
else
    data = cos(y);
end
```

can be replaced by the equivalent statement:

```
data = cos(y) * (y <= SOME_VALUE) + sin(x) * (y > SOME_VALUE);
```

Such statements are also useful for creating function handles, which many built-in MATLAB functions expect as input. For example, Jiro Doke on the File Exchange blog presented a problem that required using *ode45* for solving a problem of a damped harmonic oscillator with a force applied at a certain time: $m\ddot{y} + b\dot{y} + ky - F(t) = 0$.[547] The *ode45* function expects a function handle as input. We could indeed create a separate function whose handle could be passed to *ode45*,

```
function dY = odeFcn(t, y, m, k, b, t0, F)
    if t > t0 % after time t0, apply force F
        dY = [y(2); -1/m * (k * y(1) + b * y(2) - F)];
    else % standard damped oscillator equation: ym"+by'+ky=0
        dY = [y(2); -1/m * (k * y(1) + b * y(2))];
    end
end
```

---

* Such earlier MATLAB releases use another type of MATLAB object-oriented system (*schema*-based); interested users can re-implement *onCleanup* using that system (which still works to this day, side-by-side with MCOS).

However, it would be more convenient to use a one-liner anonymous function. Unfortunately, MATLAB does not enable defining anonymous functions that use conditionals. Instead, we could use the equivalent functional form:[548]

```
odeFcn = @(t,y)  [y(2);  -1/m * (k*y(1) + b*y(2)  -  F*(t > t0))];
```

Not only does this form enable JIT'ization and avoids the function-call overhead, it is also arguably simpler and easier to maintain and debug, as well as lending itself rather easily for vectorization.*

However, as Steve Lord correctly pointed out,[549] multiplication by logical values, as done here for masking,† may cause inconsistent results if the input data happens to include unexpected values (integers, complex, nonnumeric, inf/NaN, etc.). Therefore before we employ this technique, care should be taken to validate the inputs.‡

4.6.4 Smaller Functions and M-files

A surprising aspect of MATLAB is that very large m-files (thousands of lines) can cause the MATLAB Editor to slow down significantly and impair the overall MATLAB performance. The problem becomes worse with increasing file size, typically for files larger than 200 KB (3-5 K lines). This is apparently due to the *mtree* function,[550] which constantly scans the file after each small modification (as small as typing a single character). This is used by the MLint/Code Analyzer, cell mode, identifier highlighting, and other Editor features. The larger the file, the longer it takes to parse, the more memory is needed by the parser (potentially requiring memory paging), and the longer the resulting slow-down. The problem affects MATLAB releases as old as R2009b (7.9) and possibly earlier. Aside from a detailed technical solution article,[551] this was never listed in MATLAB's official list of bugs.

A good rule of thumb is to limit function length to between 20 and 80 lines, of which about 2/3 are lines of code and the rest comments and empty spacer lines. Smaller functions are good candidates for inlining, to reduce the function-call overhead.§ Larger functions should be split into smaller blocks of code in separate smaller functions, in either the same m-file or in separate m-files. In cases where splitting a function appears to require complex code surgery, consider using a nested function: this prevents the need to pass input and output variables, since the nested function can directly access and manipulate the variables in its parent function.

A related rule of thumb is to limit m-file length to no more than several hundred lines, up to 2000 lines in rare cases. When the m-files start getting bloated (which is typical for GUIDE-generated GUI applications in MATLAB), split off the algorithmic code blocks into separate m-files to keep the m-file size compact.[552]

Both of these rules improve Editor and general MATLAB performance. But no less importantly, they also help improve code readability, maintainability, and reusability.

However, if for some reason you cannot avoid editing large m-files, consider editing it in an external editor (e.g., Notepad++ has an excellent MATLAB syntax-highlighter). This will not provide the benefits of the MATLAB Editor's integrated debugger and MLint/Code Analyzer, but you could still use the MATLAB Editor for these features and just make the editing changes in the external editor.

---

* Chapter 5 discusses MATLAB vectorization.
† See §5.1.3 for a related discussion of using logical values for masking.
‡ A similar warning was discussed in §4.3.1.
§ In some cases, inlining functions could also help JIT to efficiently vectorize the code — see §3.1.15.

As another alternative, the above-mentioned technical solution article lists detailed workarounds and patches that could help alleviate the problem. These workarounds need to be manually applied whenever you are editing a large file, and requires a MATLAB restart between each update, so they should only be used as a last resort.

As a final suggestion, the article mentions that some users have found that switching MATLAB's Java engine[553] to Oracle's JRE has improved Editor performance.

4.6.5 Effective Use of the MATLAB Path

In some cases, a MATLAB application might contain functions having the same name in different paths. To let MATLAB know which of these functions to use, we can either update the MATLAB path (using ***addpath*** and ***rmpath***), or temporarily *cd* into the relevant folder. Updating the path would seem to be a more "professional" approach, but *cd*-ing may prevent some inconsistencies in deployed (compiled) applications.[554]

As far as performance is concerned, contrary to intuition and widespread perception, I cannot find a discernible difference between running a function in the current folder or somewhere on the MATLAB path, at least on any recent MATLAB release. For the same reason, it placing a function higher on the MATLAB path, or in one of MATLAB's core library folders on the MATLAB path, has little impact. All of these suggestions used to help performance in old MATLAB releases, but no longer have any discernible effect in recent releases, at least for non-bloated paths.[555]

The situation may indeed be different when the MATLAB path contains numerous folders, or folders that contain numerous m-files that need to be searched in order for a particular function. In such cases, it may be beneficial to place a frequently accessed function higher up in the search path. The best alternative is using imported package functions, although admittedly this is slightly awkward. Using nested functions or sub-functions within the same file is both simpler and one of the fastest alternatives, followed by placing the function in a *private*/sub-folder, or within the current folder. The full list of MATLAB function precedence order is quite long, but is well documented.[556]

One suggestion that still works today, is saving the current path for reuse in all future MATLAB sessions, via the Desktop's <Set path...> functionality (in the main menu/toolstrip). This can save some time during MATLAB start-up, since the saved path loads faster than programmatically using ***addpath*** to modify the dynamic path.*[557]

In some cases, we cannot avoid the need for dynamically modifying the path. In such cases, I suggest using the vectorized form of ***addpath***, rather than multiple separate addpath calls:

```
addpath(genpath(baseFolderName));
```

The final path-related suggestion comes directly from MathWorks' own page dedicated to performance-improvement ideas:[558] MathWorks recommends to avoid overloading MATLAB built-in functions on any standard MATLAB data classes because this can negatively affect performance. However, as we shall see in the following section, this technique can indeed be a useful speedup technique.

4.6.6 Overloaded Built-In MATLAB Functions

MATLAB's built-in functions, as shown in various places in this book, can often be optimized for improved performance. They might benefit from a compiled-binary

---

* Also see §4.8.3.

implementation (as *repmat* was improved in R2013b), or reduced sanity checks in certain use-cases (as *ismember* could*), or a faster algorithm (as *conv* could†), or one or more of the numerous other techniques presented in this book. A list of several such built-in MATLAB functions was recently suggested by Jan Simon.[559]

It is possible to edit those MATLAB functions that are provided in m-file form. However, this has several drawbacks. First, this requires administrator OS privileges, something that is not always available. Second, the fix is not scalable: it needs to be redone on each MATLAB installation separately, including when installing another MATLAB release (recall that MathWorks release MATLAB updates twice a year). Finally, not all built-in functions are provided in m-file format. In short, modifying MATLAB files is not maintainable and not recommended.‡

We could naturally save the modified MATLAB functions under a different name (e.g., *sin* → *mySin*) and place the newly modified file in the MATLAB path. The problem with this approach is that all existing MATLAB functions will continue to refer to the existing function name (*sin*) and our improved function will not be invoked in all possible places within the application's code path.

A solution to these problems can be to overload MATLAB's implementation with our own variant of each overloaded function. This is very easy to do. For example, assuming that we wish to overload the *sin* function for the *double* data type. In this case, we create a sub-folder named *@double* under one of the user folders in our MATLAB path, ensuring that this folder is higher up the path hierarchy than the built-in MATLAB folders (run *pathtool* if you're not certain). We then simply place our variant of *sin.m* in the *@double* sub-folder and that's it.§ Whenever any function on the MATLAB path invokes *sin()* with a *double* input, our overloaded variant will be called. Of course, we need to ensure that our variant has exactly the same interface signature as the builtin (in this case, $y = sin(x)$), otherwise ugly run-time errors will occur. If we wish to invoke the built-in function at any time, we can simply call *builtin('sin',value)*.

MathWorks suggests to avoid overloading built-in MATLAB functions on any standard MATLAB data class,[560] for performance reasons.[561] For example, overloading *plus* to add integer values differently than the built-in implementation might degrade performance whenever we use this function or operator in our code.

However, in certain cases, overloading MATLAB's built-in functions can provide a real performance boost, so this technique is worth keeping in our speedup tool-chest.

---

* See §4.3.1.
† See §4.5.6.
‡ See the related discussion in §4.5.7.
§ We could also place *sin.m* directly in the user path but this is more problematic and is not recommended.

4.7 Object-Oriented MATLAB

MATLAB's *MCOS* (MATLAB Class Object System) enables maintainable, modular, reusable code using a modern Object Oriented Programming (OOP) paradigm.[562] MATLAB code that used to be encapsulated in script files and functions can now be encapsulated in class *methods*.* While disciplined developers can achieve modularity also with the older procedural programming paradigm, OOP forces programmers to think in an object-oriented manner, often resulting in more maintainable code. OOP objects encapsulate data with its relevant actions, again improving maintainability.

Unfortunately, using MATLAB MCOS carries some performance penalties that need to be considered when deciding whether to code in the new paradigm or keep using the older, simpler procedural paradigm. A major resource in this regard is a detailed post from 2012 by Dave Foti, who heads MCOS development at MathWorks.[563] The discussion below will parallel Dave's, with some additional new observations.

As Dave pointed out, the importance of MCOS's overheads only comes into play when our program uses many class objects and calls many short-duration methods. In the majority of cases, this is actually not the case: Our performance bottlenecks, as discussed elsewhere in this text, are normally I/O, GUI, memory, and processing algorithm — not object manipulation and function-call overheads.

MCOS performance used to be an important issue in some MATLAB releases following its official debut in R2008a. However, significant performance improvements have been incorporated into MCOS since then, to a point where it is no longer a critical factor for most use-cases using a recent MATLAB release. If your code is heavily MCOS-oriented and suffers from performance issues, upgrading to a newer MATLAB release might well alleviate the performance pains.

4.7.1 Object Creation[564]

When creating class objects, each creation needs to chain the default values and constructors of its ancestor superclasses. The more superclasses we have, the more modular our code, but object creation becomes slightly slower. The effect is typically measured in a few milliseconds, so unless we have a very long super-class chain or are creating numerous objects, this has little impact on the overall program performance.

Objects are typically created at the beginning of a program and used throughout it. In such cases, we only pay the small performance penalty once. In cases where objects are constantly being destroyed and created throughout the program's duration, consider using *object pooling* to reuse existing objects.[565] The basic idea is to create a set of ready-to-use objects at the beginning of the program. A static (*singleton*[566]) dispatcher (*factory*[567]) object would create this pool of objects in its constructor, possibly with a few new objects ready for use. The factory's only public interface would be public *pull*/*recycle* methods to retrieve and return objects from/to the pool. The program would have no direct access to the objects pool (except via these public methods), since the pool is stored in the factory object's private data. The *pull* method would create new objects only when asked to retrieve objects from an empty pool; otherwise, it would simply return a reference to one of the unused pool objects (which need to be *handle class objects*[568] to be reference-able).

---

* Class functions are called *methods*. See http://en.wikipedia.org/wiki/Method_(computer_programming) (or: http://bit.ly/1dJeDkW).

Object pooling entails programming overheads that only make sense when a large number of short-lived objects are constantly being created and destroyed, or when object creation is especially expensive. It is often used in databases (*connection pooling*), since programs often connect to a database numerous times for short-lived SQL queries.* Similar ideas can be found in GUI† and I/O‡ programming (also see §9.5.3).

Here is a simple implementation of such a system.[569] The singleton factory class is Widgets and it holds a pool of reusable Widget objects. We can only access the pool via the widget = Widgets.pull() and Widgets.recycle(widget) methods:

```matlab
% Manage a persistent, global, singleton list of Widget objects
classdef Widgets < handle
    properties (Access=private)
        UnusedWidgets
    end

    methods (Access=private)
        % Guard the constructor against external invocation.
        % We only want to allow a single instance of this class
        % This is ensured by calling the constructor from the
        % static (non-class) getInstance() function
        function obj = Widgets()
            % Initialize an initial set of Widget objects
            for idx = 1 : 5
                try
                    obj.UnusedWidgets(idx) = Widget;
                catch
                    obj.UnusedWidgets = Widget; % case of idx==1
                end
            end
        end
    end

    methods (Static) % Access=public
        % Get a reference to an unused or new widget
        function widget = pull()
            obj = getInstance();
            try
                % Try to return widget from the list of UnusedWidgets
                widget = obj.UnusedWidgets(end);
                obj.UnusedWidgets(end) = []; % remove from list
            catch
                widget = Widget; % create a new Widget object
            end
        end

        % Return a widget to the unused pool, once we are done with it
        function recycle(widget)
            obj = getInstance();
            obj.UnusedWidgets(end+1) = widget;
        end
```

```
      end
   end

% Concrete singleton implementation
% Note: this is deliberately placed *outside* the class,
% so that it is not accessible to the user
function obj = getInstance()
   persistent uniqueInstance
   if isempty(uniqueInstance)
      obj = Widgets();
      uniqueInstance = obj;
   else
      obj = uniqueInstance;
   end
end
```

Another consideration when designing classes is that while handle classes are slightly slower to create (due to multiple super-class overheads), they are typically much faster to use. The reason is that handle classes are passed to functions *by reference*, whereas value classes are passes *by value*. Whenever we modify a handle class property within a function, we directly manipulate the relevant property memory. On the other hand, when we manipulate a value class property, a copy of the class needs to be created and then the modified class needs to be copied back to the original object's memory.[*][570] Since we cannot normally anticipate all usage patterns of a class when we create it, I suggest creating any new user class as handle class, unless there is a specific good reason to make it a value class. All it takes is to add the ***handle***[571] (or ***hgsetget***[572]) inheritance to the class definition:

```
classdef MyClass < handle
```

When implicit expansion of class-object arrays takes place, an abbreviated version of object instance creation takes place, which bypasses the constructor calls and just copies the instance properties.[573] For example, `array(9) = Widget` creates an array of nine separate `Widget` objects, but the `Widget` constructor is only called for `array(1)` and `array(9)`; `array(1)` is then expanded (copied-over) to the remaining `array(2:8)`.[†]

A general concern with MCOS objects is that standard built-in MATLAB operators (indexing, sub-referencing, concatenations, etc.) are much less efficient than for the primitive data types. Users can override such methods for improved performance.[574]

4.7.2 Accessing Properties

When accessing a class object's properties, it is apparently faster to use dot-notation (*object. property*) than the built-in ***get*** and ***set*** methods, for those classes that support it (i.e., inherit *hgsetget*).[575] For example, let's define a simple handle class:

```
classdef TestClass < hgsetget
   properties
         name
   end
end
```

---

[*] MATLAB's *copy-on-write* mechanism is described in §9.5.1.
[†] For some additional aspects of class object creation, see §9.4.5.

Now let's set the property value:

```
obj = TestClass;

% Using the standard set() function
for idx=1:10000, set(obj,'name','testing'); end
⇒ Elapsed time is 0.138276 seconds.

% Using class.propName notation - 72x faster!
for idx=1:10000, obj.name = 'testing'; end
⇒ Elapsed time is 0.001906 seconds.
```

And similarly for retrieving a property value:

```
% Using the standard get() function
for idx=1:10000, a=get(obj,'name'); end
⇒ Elapsed time is 0.105168 seconds.

% Using class.propName notation - 6.5x faster
for idx=1:10000, a=obj.name; end
⇒ Elapsed time is 0.016179 seconds.
```

The general conclusion is that we should always strive to use the dot notation, rather than the familiar *get/set*. In MATLAB's new graphics system (R2014b), all GUI and graphic objects use MATLAB classes, so this will become more important than ever.*

Accessing individual properties is very fast (sub-millisecond). Therefore, using dot-notation is usually only important for performance in time-critical sections that access numerous properties in a loop. Personally, I prefer using dot-notation not just for performance but also for readability/maintainability, but this is a subjective opinion.

When accessing properties in MEX, we could use the built-in *mxGetProperty* function. Unfortunately, *mxGetProperty* returns a new copy of the original data rather than its pointer. This could be problematic when the property is very large (multi-MB). James Tursa has created MEX variant *mxGetPropertyPtr*, which returns a pointer that directly accesses the property. This enables faster data retrieval, as well as in-place updates even for value classes.†[576] On my system, computing sum((z.x-z.y).^2) where x,y are 100 MB each, took 44 ms with pure m-code, 98 ms using a MEX file that uses *mxGetProperty*, and only 16 ms using Tursa's *mxGetPropertyPtr*.

A very common programming paradigm is to use a property's setter method (*set.propName*) to validate the requested new value and reject it if the value has a wrong type or outside an allowed value range.‡[577] At least the type validating part can be accomplished much simpler and faster, using the propName@type syntax:[578]

```
% Regular syntax, validation using setter function
classdef TestClass
    properties
        name
    end
    methods
        function obj = set.name(obj,value)
```

---

* Also see §10.2.4.
† Also see §9.5.2.
‡ Note that we can use setter methods in all MCOS class types — they do not need to inherit *hgsetget*.

```
            if ~ischar(value)
                error('name property must be a string!');
            end
        end
    end
end

% Simpler & faster syntax, no need to test type in setter function
classdef TestClass
    properties
        name@char
    end
end
```

The dimensionality of the property can also be specified, as follows:

```
properties
    value1@double scalar % 0D value only
    value2@struct vector % 1D array only
    value3@uint32 matrix % 2D array or higher dimensions

    innerObject1@UserClass
    innerObject2@packageName.UserClass
end
```

Note that at the moment (MATLAB 8.4, R2014b), this `propName@type` syntax is undocumented and may possibly change in some future MATLAB release. It has existed in this form at least as far back as R2012a (MATLAB 7.14).

Also note that when the property type is defined as some `UserClass`, then it may need to use *.empty()* to initialize, rather than simply setting to [] using propName = []:[579]

```
properties
    innerObject@UserClass
end
methods
    function initialize()
        % innerObject = []; % bad - avoid
        innerObject.empty();
    end
end
```

If we use a setter function for whatever reason, we should not automatically add a corresponding getter method (and vice versa). Getters and setters are independent; if we pay a performance penalty in one, we are not required to also pay it in the other...

In general, the more checks that a property access needs to do in its getter and setter methods, the slower it will be. For this reason, consider removing checks from class code for performance reasons, with a drawback of reduced robustness and potentially incorrect results in edge cases. Dave Foti's article gave a simple example where run-time checks resulted in a 50% slowdown. When Dave's same code is reimplemented using the propName@type syntax, it runs even <u>faster</u> than the simple code that had no error validations, perhaps because specifying data type enables JIT optimizations.

In some cases, we cannot avoid using setter methods. While type checks can and should indeed be performed at the property-definition level (propName@type syntax), data values cannot apparently be specified in this manner, only in a setter method.* The problem here is that these setter methods will be used by both *external* code (that attempts to update the object properties with some new values that must be validated), as well as *internal* class code (methods) where we do not need to make such validations since we can rely on the existing property values.

One solution to this dilemma, as suggested by Dave Foti, is to use a duplicate set of properties: the actual values will be stored in private properties and a duplicate set of *dependent properties*[580] will be made public. The publicly accessible setter methods will only update the public-dependent properties, whereas all the internal class methods will directly update the private copies without setter overheads:

```
classdef TestClass
    properties (Dependent)
        name
    end
    properties (Access=private)
        name_
    end
    methods
        function initialize(obj)
            % Example internal update, no need to pass via setter func
            name_ = 'unknown';
        end
        function obj = set.name(obj,value)
            if isempty(value)
                error('name cannot be empty!');
            end
            name_ = value;
        end
        function name = get.name(obj)
            name = obj.name_;
        end
    end
end
```

When designing a class, we may be tempted to widely use dependent properties, especially when some computation is required, unlike above where we merely access the internal private data mirror. Dependent properties increase code robustness by ensuring data consistency. For example, a settable Meters property might have an associated dependent Feet property that is automatically recomputed whenever Meters is modified. The drawback is that the dependent property is recomputed whenever it is referenced, even if the components on which it depends have not changed. So when such a property is read numerous times, consider using a nondependent property (which is only set when updated), or caching.† An alternative is to update the "dependent" value in setter methods for all of the property's dependency components, thereby ensuring the property is always in-sync with its components.

---

* As noted in http://UndocumentedMatlab.com/blog/class-object-tab-completion-and-improper-field-names#related (or: http://bit.ly/1tjb7Cd), there is a documented mechanism for limiting the specific values that a class property is allowed to accept. However, this mechanism has severe limitations (e.g., it cannot be used with *hgsetget* classes) and is not very flexible.
† See §3.2.

If you decide to use a regular (nondependent) property, consider adding property-change listeners on the dependency component, which will automatically update the "dependent" property.[581] This will likely out-perform a dependent-property implementation, but result in increased code complexity and reduced maintainability, so its performance benefits should be weighed against these drawbacks. Also, note that using listeners on property-change events disables many JIT optimizations on these properties, so it is not self-evident that we will always improve performance.[582]

If you are not using class-object property listeners, avoid using the SetObservable property attribute, since this can have a disastrous effect on performance.[583]

Within a class method, we often need to access multiple object properties, multiple times. A trick that can help performance is to load all relevant object property values into local variables at the top of the method, process these variables in the method's body, and finally update the updated object properties at the end. Doing this decreases the property access times, since local variable access is much faster than accessing class object properties. One user has reported that this resulted in a 40× speedup.[584]

```
function myMethod(obj)
    data = obj.Data;
    result = processData(data); % use a local variable here & below
    result = processSomeMore(result,data);
    ...
    obj.Result = result;
end
```

Finally, when deciding whether to use a class instance or a regular standalone function, remember that there are multiple ways to encapsulate data even in nonclass functions. For example, we could use a *persistent* cache (e.g., hash-table)* to store information that will be preserved across function invocations, but will only be accessible within the function, not outside it.

For additional performance aspects of accessing object properties, see §4.9.3 and §9.5.7.

4.7.3 Invoking Methods

Unlike property access, it seems that method invocation using the *f(obj,data)* syntax is generally faster than the equivalent *obj.f(data)* syntax, because of extra overhead involved within MATLAB in interpreting the dot-notation.[585]

Even when using the *f(obj,data)* syntax, the resulting performance is still significantly slower compared to calling a direct non-MCOS function.†[586] For this reason, it may be useful to place time-critical functionality in regular functions, rather than object methods. Consider the following simple example:

```
classdef TestClass
    properties
        Width = 12;
        Height = 5;
    end
    methods
```

---

* See §3.2.3.

† Function call overhead can be estimated in R2013b and newer using *matlab.internal.timeit.functionHandleCallOverhead*, part of the new *timeit* functionality (see §2.2.3). For example: *matlab.internal.timeit.functionHandleCallOverhead(@myFunc)*.

```
        function value = getSize(obj)
            value = obj.Width * obj.Height;
        end
    end
end
function value = getSizeExt(width, height) % regular m-function
    value = width * height;
end

C = TestClass2;
tic, for idx=1:1000; v = C.getSize(); end, toc
⇨ Elapsed time is 0.024017 seconds.

% getSize(C) is 1.3x faster than C.getSize()
tic, for idx=1:1000; v = getSize(C); end, toc
⇨ Elapsed time is 0.018949 seconds.

% getSizeExt() is even faster (7x)
tic, for idx=1:1000; v = getSizeExt(C.Width,C.Height); end, toc
⇨ Elapsed time is 0.003361 seconds.

% Avoiding the function call is the fastest alternative (1100x faster)
tic, for idx=1:1000; v = C.Width*C.Height; end, toc
⇨ Elapsed time is 0.000021 seconds.
```

If encapsulation is important, consider using sub-functions or nested functions in the class's m-file (such as the *getInstance()* function in §4.7.1). Calling these functions from within class methods provides a speedup compared to calling class methods. Dave Doti himself used this technique in his article. However, note that such mixing of OOP code (*classdef* contents) and procedural code (sub-functions) is frowned upon from a design perspective. The textbook approach is to make such helper functions private or protected class methods, and have the m-file contain nothing but OOP class-code. Then again, in the fight versus performance, code beauty often suffers…

Owing to the OOP method-invocation overhead, it is best to design class methods that loop over the data internally (using a single method call), rather than looping externally over multiple method invocations that process a single data item at a time.

A very interesting comparison of the overheads of various types of MATLAB functions and class methods was reported by Andrew Janke on StackOverflow.[587] It is one of the (if not *the*) highest-rated MATLAB-related responses on StackOverflow, highlighting the significance that readers have placed on his comparison. The results of Janke's benchmark utility[588] on my system are as follows (other platforms and MATLAB releases may well behave differently, although the general trends hold):

```
Matlab R2014a on PCWIN64
Matlab 8.3.0.532 (R2014a)/Java 1.7.0_11 on PCWIN64 Windows 7 6.1
Machine: Core i5-3210M CPU @ 2.50 GHz, 8 GB RAM (62724GU)
```

Operation	Time (µsec)
nop() function:	1.07
nop() subfunction:	1.07
@()[] anonymous function:	5.94
nop(obj) method:	4.84

```
nop() private fcn on @class             1.54
classdef nop(obj):                      5.74
classdef obj.nop():                    10.56
classdef pivate_nop(obj):              10.73
classdef class.static_nop():           21.71
classdef constant:                      7.08
classdef property:                      3.45
classdef property with getter:         21.04
+pkg.nop() function:                    7.12
+pkg.nop() from inside +pkg:            4.43
feval('nop'):                           3.57
feval(@nop)                             1.57
eval('nop'):                           59.05
Java obj.nop():                       226.50
Java nop(obj):                          8.47
Java feval('nop',obj):                 14.23
Java Klass.staticNop():                11.93
Java obj.nop() from Java:               0.06
MEX mexnop():                           1.05
builtin j():                            0.03
struct s.foo field access:              0.18
isempty(persistent):                    0.00
```

For details about the specific meaning of each row, refer to Janke's report. Some tentative conclusions from the report are that using package namespaces (+pkg.classname) degrades performance,[589] and also that the method-call overhead may be higher on 64-bit MATLABs than on 32-bit releases (this is uncorroborated).

In a related report, Dimitrios Korkinof has demonstrated how we can implement an m-file that consists of a parent function and nested functions, mimicking a class while running significantly faster than even a handle class, updating property values.[590]

4.7.4 Using System Objects

MATLAB System Objects (SO) are several hundred MCOS classes that encapsulate related functionality used in some MATLAB toolboxes, such as the Communications System Toolbox and Computer Vision System Toolbox.[591] SO-enabled toolboxes have been renamed "XYZ System Toolbox" by MathWorks. The term "System Object" is reserved for those specific class objects that process streaming signals data, using a common interface, defined by the superclass `matlab.System`. This enables users of one SO to easily learn and use other SOs. Over the past several releases, MathWorks has enlarged the number and enhanced the functionality of these SOs, and it appears to be an area of interest at MathWorks, expected to continue growing in the near future. In other words, it's not going away anytime soon...

In addition to their functional benefits, SOs have performance benefits. Most SOs are pre-compiled into MEX files for speed;[592] 300+ SOs are supported by the MATLAB Coder* for potential extra speedup; some SOs also support GPU parallelization.†

In addition, SOs optimize data buffering and other internal algorithms for improved performance and reduced memory footprint. Many streaming data functions can

---

* See §8.2.
† See §6.2.

improve performance by depending on the data, state, and processing results of previously streamed data, by avoiding costly initializations, and by processing the data in small blocks (chunks).*

For example, in video-image processing it may be possible to process only the deltas between successive frames, rather than the entire image. By looping over the images using regular image processing functions, we would basically be recomputing each frame independently from scratch; in contrast, the SO remembers its prior data, state, and processing results and therefore has much less processing to do.

Consider an edge-detection algorithm: Successive video frames do not need to scan the entire frame image for edges, only the pixels that are adjacent to the locations of the previous frame's edges. This would enable the SO's *step()* method to become very efficient at processing real-time streaming data.†

MathWorks demonstrated a 4× speedup using Communications System Toolbox's SOs compared to the standalone function implementation.[593] A similar speedup was demonstrated in a MathWorks webinar using Computer Vision Toolbox's `vision.DCT` method (formerly called `video.DCT2D`), compared to the standalone *dct2d* function.[594]

The bottom line is that if an algorithm can be implemented using either built-in toolbox functions or SOs, then using SOs would usually be faster to both develop and run.

4.8 MATLAB Start-Up

4.8.1 The MATLAB Startup Accelerator

The MATLAB Startup Accelerator is a process introduced in R2011b (MATLAB 7.13) whose aim is to decrease the MATLAB application startup time, on Windows platforms (only). The process works by setting up scheduled tasks that load the MATLAB libraries into memory immediately after any login, as well as twice a day. Doing so forces Windows to have the MATLAB libraries resident in the physical memory, so that when a user actually starts the MATLAB application, these libraries would not need to be loaded, saving some time in the startup process.

The Startup Accelerator is a fully documented and supported functionality.[595] It appeared in R2011b's pre-release notes and What's New page,[596] although for some reason (probably simple oversight) not in R2011b's official release notes.[597]

MathWorker Philip Borghesani provided the rationale behind the mechanism:[598]

> *The MATLAB startup accelerator is quite simple, it reads in the files that will be loaded/run when MATLAB is started then the process exits. Testing showed that this can dramatically improve the startup times of MATLAB. There is no effect on system performance other than the file io while it is running and if you had other data in system file cache it might get flushed and need to be reloaded. Now my opinion's, not Mathwork's, on when this is likely to be most helpful and when it may not be as helpful.*
>
> *I believe you will see the biggest improvements on a machine with large quantities of memory that is not frequently used for other memory hungry applications. Running a memory or disk intensive program between the time MATLAB starter has run and when you start MATLAB may nullify any improvements.*

---

* See §3.6.
† Computer Vision System Toolbox's `vision.EdgeDetector` SO does not necessarily implement this specific algorithm.

I expect little benefit or harm from the accelerator if MATLAB is installed on a local SSD.

Windows 7 superfetch dynamically performs a similar operation if you frequently use MATLAB then superfetch may be doing the same thing for you already.

In the case of multiple installed MATLAB versions I suggest disabling the accelerator on all but the version you use most often unless you have >8GB memory and frequently use multiple versions.

In conclusion this is a trade-off that may or may not be to your liking or helpful in any specific usage scenario: MATLAB Startup time is reduced at the expense of system overhead when the scheduled process runs and possibly when it runs again at user login.

Within Control Panel's Scheduled Tasks utility, we can see the Startup Accelerator as separate tasks for each of the MATLAB R2011b or newer releases that is installed:

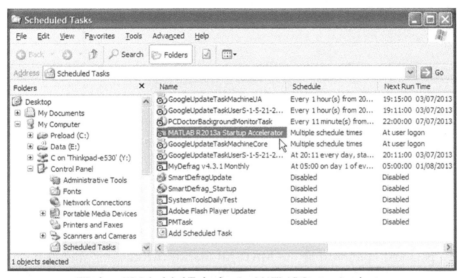

Windows XP Scheduled Tasks showing MATLAB Startup Accelerator

On Windows7, the functionality is basically the same; just the utility interface is somewhat different:

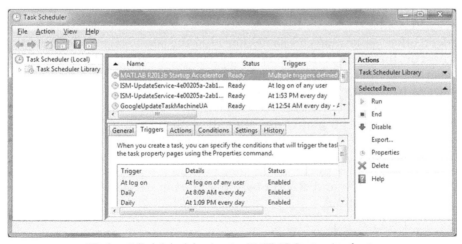

Windows7 Task Scheduler showing MATLAB Startup Accelerator

We can delete the task right there, if we wish to do so (click <Delete> or select the Delete option from the right-click context menu). Alternatively, we can modify the scheduling by double clicking the task or selecting Properties from the context menu:

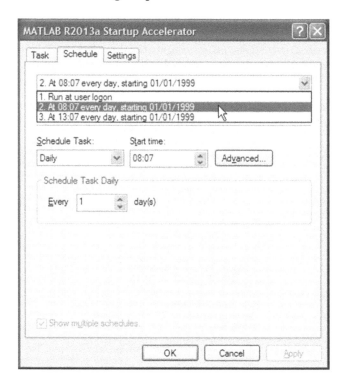

It is my personal opinion that the MATLAB Startup Accelerator should never have been released, certainly not as part of the default installation. It is in fact one of the first things that I disable when installing a new MATLAB release. The reason is that when we have multiple MATLAB releases installed (as I do, for different clients who use different releases), then all these accelerators start at once, forcing my poor machine to its knees.*

When I did have the Startup Accelerator installed, it shaved a few measly seconds off my MATLAB startup time, at the expense of adding a significant number of seconds to the Windows startup time.[599]

If I work on other applications for some time after the latest Startup Accelerator scheduled task has run, then the MATLAB libraries get paged-out to disk. If I then start MATLAB, all the potential gains of pre-fetching the libraries, are meaningless since the libraries need to be re-fetched from disk.

Moreover, if I start Windows and do not plan to work on MATLAB for some time (MATLAB is not the only program I use, after all…), then I suffer reduced memory (due to the startup tool's JVM caching), increased startup time, and disk cache purging — all for something that is in fact not used at all. The end result is that my overall system becomes **slower** than before![600] In fact, one user even reported[601] that on his system, MATLAB releases R2010b and R2011a (without startup acceleration) load *faster* than R2011b (with the acceleration).

---

* I have seen unconfirmed traces that this may indeed have been fixed in R2013b, such that only the latest release's startup accelerator task is active. Please check this on your specific system to ensure.

Since the default install is to make the startup tool schedule several times a day even after login, it has happened several times that I was doing some work when suddenly the computer became sluggish for a noticeable time, which I later traced to these scheduled tasks. I find the official statement that *"the scheduled task is not a running process and will not consume system resources"*[602] quite ironic in this light.

Finally, nowadays I often work on a computer that has an SSD, where the Startup Accelerator brings none of the benefits it might otherwise have, leaving only the drawbacks. As SSDs become cheaper, larger, and more ubiquitous, MATLAB users will be increasingly harmed by the Startup Accelerator.*

For all of the reasons above, my suggestion for future MATLAB releases is to make the Startup Accelerator an elective preference, with the default being off.

In the meantime, we need to be able to prevent or disable the Startup Accelerator feature. Luckily, this is quite easy to do. There are several ways to disable the MATLAB Startup Accelerator.

The easiest method is to delete or disable the Startup Accelerator task directly in the Scheduled Tasks utility, as shown in the screenshots above, after we install MATLAB.

We can prevent the task from being installed in the first place, by installing MATLAB using a dedicated setup parameter:[603]

> When you install MathWorks products, the installer configures the MATLAB Startup Accelerator as a scheduled task on your computer. If you would prefer to install MathWorks products without configuring the startup accelerator, use the createAccelTask option on the installer command line or in an installer input file. A sample installer input file, namedinstaller_input.txt, is included in the top-level folder on the MathWorks product DVD. You can customize this file to suit your needs.
>
> For example, to install MathWorks products without configuring the startup accelerator, run the installer and specify the createAccelTask option on the command line, setting the value to false.
>
> setup.exe -createAccelTask false

As an alternative to starting *startup.exe* with this parameter, we can modify the *installer_input.txt* configuration file located in the top-level installation folder:

```
## The MATLAB Startup Accelerator installer creates a
## system task to preload MATLAB into the system's cache
## for faster startup.
##
## NOTE: By default, a MATLAB Startup Accelerator task will
## automatically be created.
##
## If you want a MATLAB Startup Accelerator task to be created,
## do not edit this section.
##
## Set createAccelTask value to false if you do not want to
## create an Accelerator task and uncomment the line.

# createAccelTask =        # default line
createAccelTask = false
```

Naturally, both of these latter alternatives are not relevant for users who install MATLAB directly from its DVD disk. For such users I suggest deleting the task in the Scheduled Tasks utility following MATLAB's installation, as explained above.

---

* Read more on SSD-related performance aspects in Chapter 11.

4.8.2 Starting MATLAB in Batch Mode

We can run MATLAB scripts in batch (noninteractive) mode by using MATLAB's *-r* startup option. This enables us to automatically run MATLAB scripts at specified times using the Operating System's scheduler (e.g., Task Scheduler on Windows; *cron* on Linux; *launchd* or *iCal Automator* on Macs):

```
matlab -r scriptName.m
```

In many cases, we can (and often also wish to) skip the display of MATLAB's splash screen, which is normally shown during MATLAB start-up. This can be done by adding the *-nosplash* startup option. The effect is mostly functional (preventing an annoying out-of-the-blue splash popup) — the performance impact is rather small.

In many of these cases, the scripts run in unattended noninteractive mode, and therefore do not require the MATLAB Desktop GUI. Unlike the *-nosplash* option, running MATLAB with the *-nodesktop* startup option will dramatically reduce its resource consumption (memory and graphic handles) and improve startup time.[604]

Finally, if the script does not use any Java-related features (such as GUI, plots, or network I/O), then additional resource and performance improvements can be achieved by adding the *-nojvm* startup option:

```
matlab -nosplash -nodesktop -nojvm -r scriptName.m
```

If Java is required for non-GUI features (e.g., network I/O or Container classes), then instead of *-nojvm* we should use the *-noawt* startup option. This loads the Java virtual machine (JVM) in MATLAB, but not its GUI-related stuff.

Note that running MATLAB without Java (using either *-nojvm* or *-noawt*) disables all GUI in MATLAB, including plotting and Desktop features such as the Editor or Profiler. In essence, we get a bare-bones (*headless*) MATLAB computation engine. This is excellent for unattended batch processing, but is really unusable in interactive mode.

In a related note, starting MATLAB with the *-norootrecord* option fixes a bug in R2011a (MATLAB 7.12) that slows down the performance of certain indexed assignments.[605] This bug only exists on R2011a, so this startup option should not be used with any other MATLAB release.

4.8.3 Slow MATLAB Start-Up

Multiple possible causes for slow MATLAB startups have been reported over the years, and are listed below. Note that some of these may not be applicable to all MATLAB releases or platforms. Most of the listed causes have workarounds in the specified end-note references:

- Licensing-related issues:
  - Badly configured LM_LICENSE_FILE environment variable[606]
  - Unreachable license manager server[607]
  - Network disconnection on a platform that uses a network license manager[608]
  - Network connection on a platform that uses a non-network license manager[609]
  - Clashes with other software that uses the same FLEXnet license manager as MATLAB[610]
  - License manager lockouts (various sorts)

- User-specific start-up:
  - Errors in user startup files[611]
  - User startup files or MATLAB path or startup folder that reference hung drive mounts[612]
  - User startup files or MATLAB path or startup folder that reference folders with numerous files[613]
  - Some lengthy process or calculation being done in a user startup file
  - User start-up causing numerous variables to be displayed in a visible Workspace Browser (see §4.9.1)[614]
  - Multiple path settings in the user start-up[615]
- MATLAB launch:
  - Deployed (compiled) MATLAB application takes up to a minute to launch due to MCR start-up (see §8.4)
  - MATLAB application shortcut's target folder using UNC pathname[616]
  - Failure to identify a My Documents\MATLAB\ folder on Windows (MATLAB 7.4 R2007a)[617]
  - Running MATLAB via an X-server[618] or SSH[619]
  - Running MATLAB from a network disk (see §11.9.2)
  - Repeated MATLAB start-ups in batch mode (*-r* ...) for processing multiple scripts, rather than a single invocation that runs all scripts.[620]
  - MATLAB start-up using a too-high value of memory shielding[621]
  - MATLAB start-up could be faster by using the startup parameters *-automation, -nojvm, -nodesktop* and/or *-noFigureWindows*.[622]
- MATLAB preferences:
  - Badly configured MATLAB preferences[623]
  - Disabled toolbox caching in MATLAB's general preferences[624]
- MATLAB Desktop:
  - Help Browser visible when MATLAB was last closed[625]
  - MATLAB Editor having numerous loaded files visible when MATLAB was last closed[626]
  - Some users have reported that switching MATLAB to use the Oracle JRE instead of the one with OpenJDK helped speed up the Editor[627]
- Operating environment:
  - JIT unavailable on SELinux unless we set the *allow_execheap* flag[628]
  - Overly active antivirus scans[629]
  - Antivirus or OS updates, which often occur at high priority*
  - Unnecessary web-proxy autodiscovery[630]
  - Using a vast number of fonts on Windows[631]
  - Misbehaving CVS (code versioning system) causing slow file access
  - Using network folders in MATLAB's path or current folder

---

* See §2.1.6.2.

- Hardware-related issues:
    - Incompatibility of MATLAB 7.0 (R14) and some graphic cards[632]
    - Windows platforms having 32 or more logical CPUs (R2008b-11a)[633]
    - Insufficient available physical RAM when running MATLAB[634]
- Miscellaneous:
    - Initialization of the MuPAD kernel when running *syms* (R2008b-9a)[635]
    - Installation of MATLAB in the same folder as an earlier release[636]

Some of the root causes listed above are related to the MATLAB license manager. By using a local license manager, or a compiled (deployed) MATLAB program (which does not require a license), we remove these potential slowdown causes. However, compiled (deployed) programs are notoriously slow, due to MCR launch.*

Another large group of slowdowns is due to inefficient user start-ups. By minimizing startup code to the bare minimum we can eliminate this slowdown cause.

Java-related performance customizations can be controlled via the *java.opts* file.† This file is loaded by MATLAB during its startup process, unless the *-nojvm* runtime parameter is specified.

4.8.4 Profiling MATLAB Start-Up

Unfortunately, MATLAB does not include a detailed trace-log file that enables us to pin-point the direct cause if and when MATLAB happens to start-up too slowly. In fact, such slow start-ups may be due to several entirely unrelated factors.

We can start MATLAB with the *-timing* startup option to see a list of trace points during the MATLAB startup process. While this does not pin-point the exact cause of a slow start-up, it may provide hints. For example, if the slow start-up is due to licensing issues, then we would expect the LM Start-up row to take a long time. Likewise, if the problem is related to MATLAB path, we would see a definite time gap in the relevant path rows. Here is the command-window output when I start my MATLAB using *-timing*:[637]

```
Opening timing log C:\Users\Yair\AppData\Local\Temp\timing_log.9104..
 MATLAB Startup Performance Metrics (In Seconds)

 total   item    gap       description
 == = == = == = == = == = == = == = == = == = == = == = ==
  0.00   0.00    0.00    MATLAB script
  0.00   0.00    0.00    main
  0.07   0.06    0.01    LM Startup
  0.08   0.00    0.01    splash
  1.30   0.81    0.40      InitSunVM
  2.50   1.03    0.17      PostVMInit
  2.50   2.01    0.40      mljInit
  3.83   1.33    0.00      StartDesktop
  3.83   3.34    0.40    Java initialization
  3.85   0.03    0.00    hgInitialize
  3.89   0.01    0.03    psParser
  4.07   0.08    0.10    cachepath
```

---

* See §8.4 for a discussion and alternatives for improving compiled programs' startup time.
† See §4.8.5.

```
    4.60    0.19    0.33        matlabpath
    5.47    0.00    0.87        matlabpath
    5.72    0.03    0.21        matlabpath
    5.78    0.01    0.05        matlabpath
   21.18   17.35    3.83        Init Desktop
   22.46   18.38    0.00        matlabrc
   == = == = == = == = == = == = == = == = == = == = == = ==
   Items shown account for 174.8% of total startup time [TIMER: 2 MHz]
```

Note that the *-timing* startup option was documented and supported up to R2009a (MATLAB 7.8).[638] It was removed from the documentation and became unsupported in the following release (7.9, R2009b) although in practice it continues to work to this day (MATLAB 8.4, R2014b).* This information is contained in the log file whose path name appears at the top of the list. It can also help a MathWorks technical support engineer help diagnose the root cause of slow MATLAB start-ups.

4.8.5 Java Start-Up

MATLAB's GUI is Java-based, at least as of release R2014b. Therefore, the more GUI-oriented our program, the more memory that we should allow the Java engine (JVM) to use. This is controlled by the *Java heap size*. MATLAB sets a relatively low value by default: 64–256 MB, depending on release and platform. It is a good idea to increase this for MATLAB programs that use GUI. In MATLAB R2009b (7.9) and earlier, add the line -Xmx512m to a *java.opts* file in our MATLAB's startup folder (*512m* indicates 512 MB; this can be changed to any other value).†[639] On MATLAB R2010a (7.10) and later, use MATLAB's main Preferences window (Preferences → General → Java Heap Memory).[640] MATLAB starts Java and allocates its heap on start-up, so a MATLAB restart is needed for the new heap size value to take effect.

Increasing the Java heap space can improve performance for any MATLAB program that uses Java heavily. This includes GUI applications, as well as applications that interact with databases[641] and network I/O,[642] which typically use Java libraries.

Whereas in general it is better to increase the Java heap size, if the value is increased too much, this could cause MATLAB's start-up to become slower or even to fail.[643] Also note that Java's memory is only part of MATLAB's process memory, so increasing Java memory reduces the memory available to MATLAB variables. A larger Java heap size may also imply longer garbage collections, causing noticeable short "hangs" of our application (one way around this is forcing a synchronous *java.lang.System.gc()* in places that the application knows that a relatively long pause is acceptable). To avoid virtual memory thrashing,‡ set a heap size below 50% of the physical available RAM.

There are several additional performance-related Java startup parameters that can be specified in *java.opts*.[644] These include -client,§ various -XX parameters (such as

---

* And is even referenced in an official technical solution (1-186EP) that was last updated in June 2013: http://mathworks.com/matlabcentral/answers/91507 (or: http://bit.ly/1b0GOEq).

† Be careful to name the file *java.opts*, not *Java.opts* nor *java.opts.txt* (as would be the default in Windows Notepad if we are not careful — see http://mathworks.com/matlabcentral/newsreader/view_thread/154954 or: http://bit.ly/197wgDC).

‡ See §9.1.

§ For faster startup and smaller memory footprint, but a somewhat slower runtime performance. The corresponding server parameter starts JVM in server mode (slower startup, more memory, faster run time), which is used in MATLAB by default.

-XX: + AggressiveOpts and -XX: + UseLargePages), -Dapple.awt.graphics. UseQuartz (MacOS) and -Dsun.java2d.pmoffscreen = false (MacOS and Linux).* The list of Java startup parameters used by MATLAB can be determined in run time as follows:

```
>> mxBean = java.lang.management.ManagementFactory.getRuntimeMXBean;
>> params = cell(mxBean.getInputArguments.toArray)
params =
    '-XX:PermSize=32m'
    '-Xms64m'
    '-XX:NewRatio=3'
    '-XX:MaxPermSize=128m'
    '-Xmx256m'
    '-Dsun.java2d.noddraw=true'
    ...
```

4.9 Additional Techniques

4.9.1 Reduce the Number of Workspace Variables

MATLAB automatically updates the Desktop Workspace's variables when we run MATLAB commands in the Command Window or a script file. This update is done automatically and normally takes only a negligible amount of time. The length of this update is dependent on the number of Workspace variables.

Unfortunately, the update is also done when we run MATLAB functions that do not directly affect any Workspace variable.[†] The end result is that if we have a large number of Workspace variables then profiling might show that the Workspace updates took a considerable amount of time of the total. Even when the number of variables is minimal, Workspace updates can still take a hundred msecs:

workspacefunc	4	0.132 s
workspacefunc>getShortValueObjectsJ	2	0.101 s
workspacefunc>getShortValueObjectJ	34	0.085 s
workspacefunc>num2complex	32	0.056 s
workspacefunc>getAbstractValueSummaryJ	20	0.038 s
opaque.double	12	0.016 s
workspacefunc>getclass	20	0.013 s
workspacefunc>getWhosInformation	2	0.013 s
workspacefunc>createComplexVector	6	0.007 s

Profiler results showing the non-negligible effect of Workspace updates

---

* See §4.9.13 for details on these MacOS/Linux-specific parameters.

† Perhaps, the reason is that MATLAB cannot really be certain when a function might update the base Workspace, although theoretically there are only certain access points where this could happen in our code and workflow.

We can reduce the update time: closing the Workspace and Variable Editor panels in run time is the best option, since this eliminates the updates completely. This can be done from the panels' title bar or from the Desktop's toolbar/toolstrip. Oddly enough, this was the official MathWorks workaround for this unresolved problem, since R14SP2 (MATLAB 7.0.4), until it was finally fixed in R2014a (MATLAB 8.3).[645]

Alternatively, we can hide/minimize the Workspace panel during the application run — this will significantly reduce the number and length of the Workspace updates, although not eliminate them altogether. Another alternative is to drastically reduce the number of variables in the base Workspace — the lower the number (preferably zero), the shorter the Workspace updates.[646] We can often replace multiple separate variables with a single cell array or a single struct having multiple data fields.

4.9.2 Loop Over the Smaller Data Set

It often happens that we need to set data elements in a large array to one of several values. We have two basic ways of doing this: We can loop over all array elements and set the relevant value in each position, or we could initialize the entire array to some value and then only modify some of them. Since initialization can be done in a vectorized manner in MATLAB,* the latter method turns out to be more efficient, especially if we initialize using the more-common value and then only change a small portion of the data.

Let's consider a simple case, where we need to set a large array to the value 2 in 80% of the positions, and to the value 3 in the rest:

```
% Slowest method - update all data elements in a loop
data = ones(1,1e6); % initialize all data (preallocation: see §9.4.3)
for idx = 1 : numel(data)
    if rand > 0.2 % update 80% of the time
        data(idx) = 2;
    else          % update the remaining 20%
        data(idx) = 3;
    end
end
⇨ Elapsed time is 0.036537 seconds.

% Directly update 80% of the data - faster
data = ones(1,1e6) + 2; % initialize all data to 3
for idx = 1 : numel(data)
    if rand > 0.2 % update 80% of the data
        data(idx) = 2;
    end % no need to update 20% of data elements
end
⇨ Elapsed time is 0.032659 seconds.

% Directly update only 20% of the data - fastest
data = ones(1,1e6) + 1; % initialize all data to 2
for idx = 1 : numel(data)
    if rand < 0.2 % update only 20% of the data
        data(idx) = 3;
    end
end
⇨ Elapsed time is 0.030188 seconds.
```

---

* Refer to Chapter 5 for a discussion on MATLAB vectorization.

Note: this example could easily be made fully vectorized for even better performance. Unfortunately, in the general case full vectorization is not always possible. So while this specific example may seem contrived, its underlying conclusions are valid in the general case.

4.9.3 Referencing Dynamic Struct Fields and Object Properties

Dynamic field names can be used in MATLAB 7 onward as a direct replacement for the built-in functions *setfield* and *getfield*, to access properties of structs and objects. While dynamic field names are still much slower than using direct field referencing, they are an order of magnitude faster than the *setfield* and *getfield* functions:

```
% Set a field value
for idx=1:1e6, a.b=pi; end              %direct referencing: fastest
⇨ Elapsed time is 0.017099 seconds.

for idx=1:1e6, a.('b')=pi; end          %dynamic fields: 100x slower
⇨ Elapsed time is 1.118628 seconds.

for idx=1:1e6, a=setfield(a,'b',pi); end %setfield: 1000x slower
⇨ Elapsed time is 17.506818 seconds.

% Get a field value
for idx=1:1e6, c=a.b; end               %direct referencing
⇨ Elapsed time is 0.013722 seconds.

for idx=1:1e6, c=a.('b'); end           %dynamic fields: 60x slower
⇨ Elapsed time is 0.862631 seconds.

for idx=1:1e6, c=getfield(a,'b'); end   %getfield: 1000x slower
⇨ Elapsed time is 13.473074 seconds.
```

Since dynamic field referencing was unavailable before MATLAB 7, we still find these functions being used in legacy code that runs to this day. For improved performance, and unless we need to preserve backward-compatibility with MATLAB 6 or older, it is therefore advisable to replace all instances of these functions with the corresponding direct referencing or dynamic field referencing.

Fortunately, MATLAB's Editor (or rather, MLint — the MATLAB Code Analyzer) displays a warning message (GFLD/SFLD) when we edit code with *getfield/setfield*:

```
tic, for idx=1:1e6, a=setfield(a,'b',pi); end, toc
⚠ Use dynamic fieldnames with structures instead of SETFIELD.
```

Note that despite the 100× slowdown of dynamic fields (a.('b')) compared to direct referencing (a.b), MLint does not report those cases where we could replace the former by the latter. We should take care in our code never to use dynamic field names when the field name is static and constant (as was 'b' in the above example). Dynamic field names should only be used when the field name is really dynamic and should be evaluated in run time (as from a variable, or by concatenating dynamic run-time values into a field-name).

When a field name is dynamic but can only be one of several values, we might be tempted to use a conditional construct to directly reference one of the several possible fields, rather than dynamic field names. However, in the majority of cases this would not produce any speedup, since evaluating the conditional normally takes at least as long as evaluating the dynamic field name.

4.9.4 Use Warning with a Specific Message ID

Another potential slowdown that is detected and reported by MLint/Code Analyzer (WNON/WNOFF) is when we use *warning('on')* or *('off')* in our code without specifying a specific message ID as the second optional argument to the *warning* function:

```
70 -    warning off                          % supress warnings
71 -    ⚠ WARNING('OFF',msgID) is faster than WARNING('OFF').  [Details ▲]
72 -
73 -    Explanation
74 -
75 -    The warning('off') command turns off all warnings.
76 -    This action can be slow and might turn off
77 -    warnings that you want to continue receiving.
78 -
79 -    Suggested Action
80 -
81 -    MathWorks recommends that you turn off specific
82 -    messages by adding the appropriate warning
83 -    message IDs to the warning call. For more
84 -    information, see "Warnings".
85 -
```

4.9.5 Prefer *num2cell* Rather than *mat2cell*

I often see the usage of the built-in *mat2cell* function for conversion of a numeric array into a cell array. It always amazes me that so many people use this function rather than the much simpler and faster equivalent *num2cell*:

```
data = rand(1000,1000); % 1 K x 1 K elements
cells = mat2cell(data, ones(1,size(data,1)), ones(1,size(data,2)));
⇨ Elapsed time is 2.750814 seconds.

% Using num2cell - 9x faster (and much simpler to use)
cells = num2cell(data);
⇨ Elapsed time is 0.313791 seconds.
```

Luckily again, MLint/Code Analyzer comes to our rescue, detecting and reporting such cases in our code within the MATLAB Editor (MMTC):

```
tic, c2 = mat2cell(data,ones(1,size(data,1)),ones(1,size(data,2))); toc
⚠ This use of MAT2CELL should probably be replaced by a simpler, faster call to NUM2CELL.
```

4.9.6 Avoid Using *containers.Map*

As shown in §4.6.1.2, *containers.Map* is much slower than the alternatives for storing data. On the other hand, *containers.Map* is a very useful class that increases code maintainability and reduces development time. Therefore, in code sections that are not performance critical, it would be wise to use *containers.Map*. But if your profiling session shows that *containers.Map* is indeed a bottleneck in code-critical sections, there are several possible replacements that could be much faster.

Two such pure-MATLAB alternatives are cell arrays and struct arrays. With cell arrays, we would store records in cell rows, with the record key in column 1 and the value in column 2. For example,*

```
data = {'John',  12.1;
        'Mary',   4.7;
        'Bill',  -8.4};
```

Accessing a particular record could then be done as follows:

```
idx = strcmp(data(:,1), userName); % logical indexing, see §5.1.3
if any(idx)
    oldValue = data{idx,2};  % get
    data{idx,2} = -oldValue; % set
else
    oldValue = []; % indicates that the value was not found
    data(end + 1, :) = {'Ann', 3.1416};
end
```

In a struct array, records are stored in separate structs, having key and value fields:

```
>> data=struct('key',{'John','Mary','Bill'}, 'value',{12.1,4.7,-8.4})
data =
1x3 struct array with fields:
    key
    value

>> data(end)
ans =
      key: 'Bill'
    value: -8.4
```

Accessing the records could then be done as follows:

```
idx = strcmp({data.key}, userName); % logical indexing, see §5.1.3
if any(idx)
    oldValue = data(idx).value;  % get
    data(idx).value = -oldValue; % set
else
    oldValue = []; % indicates that the value was not found
    data(end+1).key = 'Ann';
    data(end).value = 3.1416;
end
```

We should also consider using one of the super-fast Java Collection classes.[647] One of the most useful of these is `java.util.Hashtable`, as shown in §3.2.6:

```
>> hash = java.util.Hashtable;
>> hash.put('key #1','myStr');
>> hash.put('2nd key',magic(3));
>> disp(hash) % same as: hash.toString
{2nd key=[[D@59da0f, key #1=myStr}
```

---

* Both keys and values could of course have other data types, this example is merely used to illustrate the concept.

```
>> disp(hash.containsKey('2nd key'))
     1                          % = true

>> disp(hash.size)
     2

>> disp(hash.get('key #2')) % key not found
     []

>> disp(hash.get('key #1')) % key found and value retrieved
myStr

>> values = hash.values.toArray
values =
java.lang.Object[]:
     [3x3 double]
     'myStr'

>> values(1)
ans =
     8     1     6
     3     5     7
     4     9     2

>> values(2)
ans =
myStr
```

See §3.2.6 for a comparative implementation of a simple cache mechanism using either ***containers.Map*** or `java.util.Hashtable`. There are many different kinds of Java Collections that we can use out-of-the-box. While `Hashtable` may seem appropriate for most uses, other classes may be even faster for specific uses.[648]

While Java classes do not accept nonsimple MATLAB data (e.g., class objects or structs) as parameters, we can easily convert these into Java-acceptable forms using either the *Lightspeed* toolbox or the builtin ***getByteStreamFromArray*** and ***getArrayFromByte-Stream*** functions, as shown in §3.2.6.

Over the years, users have contributed multiple utilities that implement a hashtable-like mechanism to the MATLAB File Exchange.[649] Most of these utilities use one of the alternatives discussed above. The benefit of using such a utility is that we get a wrapper utility for accessing the data without having to program the details ourselves. Note, however, that these submissions are of varying levels of quality, robustness, and performance. Therefore, test carefully before deciding to use any of them.

4.9.7 Use the Latest MATLAB Release and Patches

MATLAB includes many bug fixes and small improvements in each new release. I therefore suggest using the latest MATLAB release if possible. A non-comprehensive list of performance-related bugs and fixes is listed in Appendix A.3. Note that this list is not comprehensive — many MATLAB bugs are not listed anywhere.

Even if we do not use the latest MATLAB release, we might still find a workaround in MATLAB's official online bugs list, which is updated on a constant basis.[650] A significant portion of the issues have patches or workarounds that can be used even on non-latest MATLAB releases (see, e.g., §11.6).

4.9.8 Use *is** Functions Where Available

MATLAB has several built-in functions to test for specific value types, dimensionality and some special conditions: *iscell, ischar, islogical, isstruct, isnumeric, isnan, isinf, isfinite, isscalar, ismatrix, isrow, iscolumn, isempty*. These built-in functions are highly efficient pre-compiled functions that outperform any equivalent user code. So, whenever we need to check for any of these conditions, it is always better to use the built-in functions rather than user code. Fortunately, MLint/Code Analyzer again helps, detecting and reporting some of the common equivalent m-code constructs:

```
if length(data)==0
```
⚠ Using ISEMPTY is usually faster than comparing LENGTH to 0.   [Fix]

Of course, MLint cannot be expected to detect and report *all* the potential equivalents. For this reason, we should not solely rely on it. Appendix A.3 contains a list of the MLint messages that are in fact detected and reported.

One might think that using the *isa* function would also be highly efficient, but this is in fact <u>not</u> the case. This function needs to make numerous checks for possible super-classes, and so on when it compares a value's type to a specified type. So, for instance, *islogical(flag)* would generally be faster than the equivalent *isa(flag,'logical')*.

4.9.9 Specify the Item Type When Using *ishghandle* or *exist*

As explained in §4.6.1.3, when testing Handle Graphics (HG) object types, it is faster to use *ishghandle(h,'line')* than the equivalent *ishghandle(h) && strcmp(get(h,'type'),'line')*, or *ishghandle(h) && isa(handle(h),'line')*, and similarly for any other HG object type: figure, axes, surface, patch, text, and so forth. Note that while *ishghandle* is a fully documented and supported built-in function, the useful fact that it accepts an optional-type argument is undocumented and unsupported.

Similarly, *exist(name,'type')* is faster and more robust/maintainable than using *exist(name)* without specifying the second optional-type argument. In this case, the second (type) argument and its use for performance are both documented. MLint/ Code Analyzer even detects and reports cases where this second argument is omitted:

```
if exist(filename)
```
⚠ EXIST with two input arguments is generally faster and clearer than with one input argument.   [Details ▲]

Explanation

Using exist with one argument means if MATLAB does not find that argument immediately, it must search all categories to which that argument might belong. This can be time consuming.

Suggested Action

To speed up the search and make your code more readable, specify a second argument to the exist call to indicate the category to which the first argument belongs. Valid values are: 'builtin', 'class' (for Java classes), 'dir', 'file', and 'var'.

4.9.10 Use Problem-Specific Tools

In some cases, tools may be available for the specific problem at hand, which may be faster and more powerful than a generic general-purpose tool. The MATLAB PDE (partial differential equation) Toolbox provides an example for this,[651] where the fast Poisson solver function *poisolv* is twice as fast as the standard general-purpose *assempde* in numerically solving Poisson's equation[652] for some boundary conditions.

$$\Delta\phi(x) = \nabla^2\phi(x) = f(x)$$

4.9.11 Symbolic Arithmetic

When computing long-running operations with the Symbolic Toolbox, consider using *vpa* with a specified number of digits in order to reduce the problem complexity. This is usually much faster than searching for symbolic equivalences and will likely require far fewer terms.[653] Alternatives to *vpa* are discussed in §8.5.5.

When computing operations on symbolic expressions, use numeric computations wherever it is possible for them to replace symbolic ones. For example, in one example a user computed the values of a *limit* of some functions at some points. A much faster approach is to directly evaluate the function at the required points, assuming no discontinuities.[654] This would then enable additional speedups using standard numeric vectorization (see Chapter 5).

A related recommendation is to explicitly convert symbolic results to regular numeric format (using the *double, single, int** or *matlabFunction* functions), once the analytic code portion finished. This enables MATLAB to process the rest of the program with regular numeric values, rather than unnecessarily applying the slower symbolic math.[655]

There is a bug on MATLAB R2008b (7.7) and R2009a (7.8) that causes initialization of MuPAD kernel on Windows to take long minutes.[656] The bug occurs when running *syms* or any other Symbolic Math Toolbox function that requires MuPAD. This was fixed in MATLAB R2009b (7.9) and a patch for the buggy releases is available.[657]

Symbolic Math Toolbox's includes dedicated documentation on performance tips.[658]

In a related matter, one user has reported (unconfirmed) that the MATLAB R2010a (7.10) Symbolic Toolbox may be faster in some cases than in R2012b (8.0).[659]

4.9.12 Simulink[660]

This book attempts to cover mainly the MATLAB core product and closely related toolboxes (such as the PCT, MDCS, and Coder). However, MathWorks has created a dedicated documentation section,[661] newsletter articles,[662] webinars,[663] and multiple blog posts[664] on improving simulation performance using the Simulink product. It therefore makes sense to briefly mention them here. They cover the following ideas:

- Select Accelerator or Rapid Accelerator simulation mode.[665]
- Enable the options for Compiler optimization, automatic Block Reduction and others in the Optimization pane of the Configuration Parameters window.[666]
- Switch the Compiler optimization from *faster build* to *faster run*.[667]
- Disable the options for debugging/animation support, overflow detection* and echoing expressions without semicolons in the Simulation Target pane of the Configuration Parameters window.[668]

---

* Except for fixed-point data, where overflow detection is important; if your model includes FP data, keep this option enabled.

- Disable similar configuration options in Stateflow's Configuration Parameters window (which is similar but separate from the Simulink window).[669]

- Keep Stateflow charts and Simulink blocks/models closed during simulation to prevent run-time updates (this relates to the animation feature, if enabled).

- Keep output scopes closed during simulation run time; open the scopes only after the simulation ends. If a scope must remain open in run time, reduce the number of data points (via decimation, reduced time range, and reduced number of plotted signals), increase the plotting refresh period, disable scope scrolling, remove data markers and legends, limit the history size, and use reduced fidelity in the viewer parameters (cf. §10.2).[670]

- Vectorize processing by combining multiple signals into a vector and applying the processing on the vector rather than on the separate signals.

- Sample data in batches (rather than one-by-one) using frame-based processing (MATLAB R2011b and newer; cf.: §3.6 and §9.5.9).[671] This can speed up simulations by 10× or more, at the expense of just a little extra memory.

- Aggregate small blocks into larger ones.

- Load the model in memory using **load_system** instead of **open_system** and simulate it using the **sim** command, then post process/display the outputs.*[672]

- Avoid MATLAB S-Function and Interpreted MATLAB Function blocks. Instead, use MATLAB Function, System and C-MEX blocks.†

- Avoid blocks that do not support code generation, since they would have to be run in slower interpreted mode.

- Avoid algebraic loops where possible.

- If your simulation does not use linear algebra (matrix arithmetic), disable BLAS library support in the Simulation Target pane. But if it does use it, then ensure that BLAS is enabled for optimal performance.

- Reduce, simplify, or eliminate initialization and termination phases.

- Use the Mask Editor to reduce block images resolution and size.[673]

- Store configurations in loadable MAT files rather than programmatically.[674]

- Consolidate multiple **set_param** calls into a single call with name/value pairs.

- Use a stronger platform (esp. increase the amount of memory — see §9.1).

- Limit or disable log output, and disk I/O in general (cf. §11.1).[675]

- Use the **sldiagnostics** and **slprofreport** functions and the Simulink Profiler,[676] to identify and eliminate simulation bottlenecks.

- Use the **performanceadvisor** function and the Simulink Model Performance Advisor[677] to automatically adjust model configurations for best performance (MATLAB R2012b and newer).

---

* This and the following three tips (along with others) are described in the referenced blog article by MathWorker Guy Rouleau.
† The naming of these blocks is indeed confusing: *MATLAB Function* refers to a subset of the MATLAB language previously called *Embedded MATLAB* that can be directly converted into C-code without requiring the MATLAB run-time (MCR); this is basically the language subset used by the MATLAB Coder Toolbox (see §8.2). *Interpreted MATLAB Function* refers to the full MATLAB functionality and requires the MCR — it was previously called *MATLAB Function*.

- Manually adjust model solver parameters for optimal performance, based on a priori knowledge about the model behavior.
- Use a faster solver that can still process the model. Different models may require different solvers to run optimally.[678]
- Tune solver parameters (decrease solver order, increase step size,[679] and error tolerance) to improve solution convergence speed at the expense of some accuracy. One user reported a 20× speedup by simply increasing step size.[680]
- Prevent excessive zero-crossing or disable zero-crossing detection.[681]
- Use simpler models or models with reduced fidelity.
- Store the simulation state at specific points, then load these states for repeated later simulations, thus saving the simulation time of the loaded portion.[682]
- Run simulations in parallel using Parallel Computing Toolbox (Chapter 6).[683] An easy way to do this is to run the *sim* command within a *parfor* loop.[684]
- Move some calculations onto FPGA (§8.3) or GPU (§6.2).[685]
- Use a real-time system and set simulation speed to real-time (high priority).[686]
- Ensure that MEX S-functions contain exception-free code and set the SS_OPTION_EXCEPTION_FREE_CODE S-function option.[687]

Some additional ideas can be found in the blog posts and webinars mentioned in the end notes at the top of this section. One of the newsletter articles[688] concludes with an example of how applying some of these techniques to a specific model resulted in reduced simulation time, from 453 s (7.5 min) down to 5 s.

4.9.13 Mac OS

Over the years, several suggestions for improving MATLAB performance that are specific to Macintosh computers have been reported. Note that some of these suggestions refer to Mac OS/MATLAB versions that may perhaps not be applicable to your specific system.

An aggregation of these Mac-specific suggestions for Mac OS X 10.5 (Leopard) was compiled by Fred Nugen at the University of Texas:[689]

- Apparently, MATLAB Editor and GUI on OS X 10.5 is especially slow, so try using OSX 10.4 (Tiger) or 10.6 (Snow Leopard)
- Run MATLAB directly from an X11 terminal, rather than from an app icon:

```
$/Applications/MATLAB_R2007b/bin/matlab &
```

- Apply the latest available Java update. Scrolling performance is improved in Java for Mac OS X 10.5 Update 4.
- Use the official MathWorks workaround for bug #412219.[690] The bug* reportedly affects MATLAB releases R2007a (7.4) through R2008a (7.6) and was fixed in R2008b (7.7).† The workaround for affected releases is to add the following

---

* MathWorks claims that it is actually an Apple OSX bug, not a MATLAB bug, due to 10.5's the use of the Sun2D renderer rather than 10.4's Quartz2D. Apparently, Quartz2D is slower than Sun2D for some GUI operations such as scrolling.

† Well, sort of fixed: MathWorks simply started adding the UseQuartz line to the *java.opts* files shipped with MATLAB…

line to MATLAB's *java.opts* file, which instructs Java to use the Quartz version of Graphics2D when MATLAB starts:

```
-Dapple.awt.graphics.UseQuartz = true
```

Notes:

○ MATLAB needs to be restarted whenever *java.opts* is updated

○ Figure drawing may become slower using Quartz. To disable the workaround during extensive figure drawing, temporarily rename *java.opt* and restart MATLAB (restart again when you rename back).

○ Some users find that Quartz rendering causes odd shifting behavior with variable-width fonts. A workaround for this is to use a monospaced font such as Monospaced, Andale Mono, or Courier New.

Some additional suggestions that do not appear in Nugen's list include:

• Running MATLAB directly on the X-server, rather than on a client.[691] In general, it is better to run MATLAB directly on the server (e.g., using VNC) rather than via a SSH or X client on a remote terminal. If working on a remote system cannot be avoided, try adding the following line to *java.opts*:[692]

```
-Dsun.java2d.pmoffscreen = false
```

Note: a Ubuntu Linux has reported that a solution to the remote-running slow-down is to downgrade the libx11 library.[693] This may or may not be applicable to Mac users.

• Running MATLAB on relatively new MacBooks — one user has reported[694] that MacBooks produced mid-2009–2010 (and possibly even 2012[695]) seem noticeably slower when running MATLAB.

• Running MATLAB via BOOTCAMP or even Parallels (as a Windows application) seems to improve the performance significantly.[696]

• Mac OS X 10.5 (Leopard) seems to have numerous Java compatibility issues that affect MATLAB, some of which are performance related. For example, some MATLAB releases will not start, and MATLAB may hang, if *libtiff*.dylib* is not removed from *$matlabroot/bin/mac*, so that the system version of LibTIFF is loaded.[697]

• On Mac OS X 10.6 (Snow Leopard), a bug in MATLAB's license manager (#581959) prevents MATLAB from launching, causing a long timeout.[698] This bug affects MATLAB R2009b (7.9) through R2010b SP2 (7.11.2), and was fixed in R2011a (7.12). The suggested workaround is to start the license manager manually by opening Terminal from Application/Utilities, then running the *lmgrd* command manually. For example,*

```
$ $MATLAB/etc/maci64/lmgrd -c $MATLAB/etc/license.dat -l
/var/tmp/lm_TMW.log
```

---

* Use *maci* rather than *maci64* on a 32-bit Mac.

4.9.14 Additional Ideas

Some ideas suggested by MLint/Code Analyzer (along with their message ID):[*]

- MIPC1 — On Windows platforms, calling *computer* with an argument returns 'win32' or 'win64', but never 'PCWIN', so checking for this condition is futile.
- TLEV — A *global/persistent* statement could be very inefficient unless it is a top-level statement in its function
- FLUDLR — Replace *flipud(fliplr(x))* or *fliplr(flipud(x))* by a faster *rot90(x,2)*
- FLPST — For better performance in some cases use *sort* with 'descend' option[†]
- PSIZE — *numel(x)* is usually faster than *prod(size(x))*
- MRPBW — To use less memory, replace *bwlabel(bw)* by *logical(bw)* in a call of *regionprops*
- SPEIG — *eig* function is called in an invalid manner with a sparse argument
- GRIDD — Consider replacing *griddata* with *TriScatteredInterp* for better performance (also see *griddedinterpolant*). Note: *TriScatteredInterp* will be removed in a future release; use *scatteredInterpolant* instead.[699] While you are at it, consider replacing *griddata* with John D'Ericco's *gridfit* utility.[700]

When creating shared libraries using the MATLAB Builder NE toolbox, be sure to specify the -S parameter in order to create a singleton MCR, rather than separate MCRs for each DLL.[701] Unfortunately, this switch is not accepted by the other compiler/builder toolboxes.

On MATLAB releases prior to R2013b (MATLAB 8.2), using *docsearch* is faster than *doc* (and provides better results), when a direct function match is not found.

Some of the items that affect MATLAB startup times (§4.8.3) also affect the MATLAB run-time performance. A partial list of examples includes:

- User files or MATLAB path that reference hung drive mounts[702]
- User files or MATLAB path that reference folders with numerous files[703]
- Unreachable license manager server[704]
- Running MATLAB via an X-server[705]

---

[*] The complete list of MLint-reported performance-related warnings is presented in Appendix A.4.
[†] Also see §4.6.1.3.

5

Implicit Parallelization (Vectorization and Indexing)

5.1 Introduction to MATLAB® Vectorization

5.1.1 So What Exactly Is MATLAB Vectorization?

MATLAB contains very powerful internal mechanisms that enable code to run much faster by automatically parallelizing arithmetic and logical operations on data. This is called *implicit (automatic) parallelization*,[706] since we do not need to explicitly tell MATLAB to parallelize the operations; MATLAB does it auto-magically when we pass vectorized (nonscalar) data to most built-in functions, using multithreading.*

Implicit parallelization relies on the fact that many loops have iterations that are independent of each other, and can therefore be processed in parallel with little effort (this is called *pleasingly* or *embarrassingly parallel* tasks).[707] As a design goal, MATLAB implicitly parallelized whatever fits this model, and enables users to easily annotate loops for explicit parallelization using minimal effort (see Chapter 6).[708]

However, in order to benefit from this feature, we need to use a specific syntax format to generate and use nonscalar data. The act of transforming code to use nonscalar data is called "vectorization"[709] and is the subject of this chapter. Because its performance impact can be enormous (10× speedup is common; 1000× possible[710]), vectorization is often considered to be the "holy grail" of MATLAB programming.

The technical details of MATLAB's implementation of implicit parallelization are unknown. We do know that MATLAB started to use multithreading in R2007a.[711] Multithreading enables parallel processing of data segments on multiple logical CPUs (physical CPUs, CPU cores, and hyperthreading).[712] When running external profiling applications (e.g., *Process Explorer* — see §2.4), we see indeed that some MATLAB functions make good use of the multiple cores in modern CPUs.† For this reason, MATLAB releases R2006b and earlier used only a small portion of the available CPU (e.g., 25% of a dual-core hyperthreaded processor[713]) whereas on newer releases this portion increases when multithreaded functions are used by MATLAB (remember that many functions remain

---

* In contrast, *explicit parallelization*, as well as parallelization techniques that require additional software or hardware, is discussed in detail in Chapter 6. These typically involve running parallel MATLAB processes, rather than lightweight threads.

† The command *feature('numcores')* displays information on the number of cores that are detected and usable by MATLAB. The *-singleCompThread* startup option or *maxNumCompThreads* command control the number of cores used by MATLAB — see http://blogs.mathworks.com/loren/2007/09/12/controlling-multithreading (or: http://bit.ly/1coqnqi). It is not advisable to limit threading under normal (single process) conditions in modern MATLAB releases (see technical solution 1-BYXN3A: http://mathworks.com/matlabcentral/answers/103133 or: http://bit.ly/1dLdhzX), but it may help multi-process runs (see §7.4).

single-threaded to this day). In R2007a and R2007b, multithreading was an optional configuration,[714] which became enabled by default starting in R2008a.[715]

MATLAB's release notes[716] only mention multithreading additions to some built-in functions,* although we know that other functions are parallelized as well. A detailed list of parallelized internal functions is maintained in MathWorks' technical solution 1-4PG4AN.[717] A more extensive unofficial list is maintained by Mike Croucher.[718]

1-4PG4AN tells us that internal parallelization only kicks in when the processing is CPU-bound, rather than I/O or memory-bound, which makes perfect sense. Another criterion is for the data to be large enough for parallelization benefits to outweigh the overheads.[719] This threshold is normally thousands of data elements; Mike provides some concrete numbers, which are often 20 K, 40 K, or 200 K data elements. 1-4PG4AN provides[720] the following speedup graph for *acos*, from which we learn that the specific threshold for *acos* is 20 K:[†]

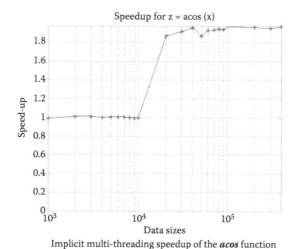

Implicit multi-threading speedup of the *acos* function

Vectorization results in extremely fast MATLAB code for a variety of reasons. Implicit parallelization (multithreading) is indeed an important mechanism. But vectorization also uses other mechanisms to achieve its performance goals:

From its very early beginnings in the late 1970s, MATLAB was highly optimized for vectorized data manipulation. MATLAB was in fact started as wrapper functions around standard linear algebra (matrix) library functions: initially, EISPACK and LINPACK, which were at the time state-of-the-art linear algebra libraries,[721] and in later years BLAS, LAPACK, and MAGMA, which are today's state-of-the-art libraries.[‡] These libraries employ a wide variety of professional optimization techniques that are beyond the capabilities of the vast majority of programmers.[722] Using MATLAB's matrix operations via built-in operators and functions that use these libraries can easily outperform any hand-crafted code in another

---

* *airy, psi*, and *bessel** functions in R2012b; *fft, conv2*, and some arithmetic in R2010a; *sort, bsxfun, mldivide, filter, gamma**, *qr*, and *erf** in R2009b; FFT functions, *prod, sum, max*, and *min* in R2009a; *rdivide, ldivide, log, log2*, and *rem* in R2008a.

† See §9.4.3 for a similar performance discontinuity (break-out) graph. This mechanism is not unique to MATLAB: Intel also uses it in MKL (http://software.intel.com/en-us/articles/parallelism-in-the-intel-math-kernel-library or: http://intel.ly/1lhzIDu).

‡ PBLAS (http://netlib.org/scalapack/html/pblas_qref.html) and ScaLAPACK (http://netlib.org/scalapack) are used for dense linear algebra distributed computations (http://link.springer.com/article/10.1007/s10766-008-0082-5 or: http://bit.ly/1ghexMt).

language.[723] Of course, to make efficient use of these libraries, we need to use them in the vectorized manner for which they were optimized. This is the purpose of this chapter.

MATLAB presumably uses optimized C code that loops efficiently over adjacent memory locations of the input data, without having to worry about bounds-checking, type-checking, function-call overheads, and other similar issues that plague interpreted coding loops. In some cases, MATLAB's C code may be smart enough to use bulk operations on adjacent memory blocks, rather than having to iteratively process each memory location separately. Optimized C code was the reason that vectorization was so fast in MATLAB even before multithreading was introduced in R2007a. For those MATLAB functions that are still not multithreaded, optimized C code underlying vectorized code still provides significant benefit over iterative non-JIT'ed loops.

On platforms having modern Intel® CPUs, MATLAB uses Intel's Integrated Performance Primitives Library (IPPL)[724] to optimize some image processing functions, and Intel's Math Kernel Library (MKL)[725] for linear algebra functions on which many vectorized functions rely (MKL includes both BLAS and LAPACK).[726] Both IPPL and MKL use SSE[727] and AVX[728] CPU instructions that are SIMD (*Single Instruction, Multiple Data*)[729]-vectorized by the CPU without requiring explicit multithreading code. SIMD optimization can bring enormous speedup,[730] but in addition to that the libraries employ multithreading internally to improve performance.[731]

MATLAB uses MKL on both Intel and AMD® CPUs, despite AMD's promotion of its AMD Core Math Library (ACML).[732] In a related note, MATLAB 8.2 (R2013b)'s release notes state that "the performance of the linear algebra functions improves on computers with AMD processors supporting the Intel AVX instruction set",[733] due to significant improvement in the latest MKL release that R2013b incorporated.

MATLAB possibly uses SSE/AVX for other core computational functions: this would only require using a C compiler (e.g., Intel Studio XE) that can do this automatically.*

Over the years, the number and performance of MATLAB functions that support vectorization has consistently increased. Today, most MATLAB functions are vectorized to some degree. What started out as highly tuned sequential Fortran code, is now implemented using multithreaded Fortran and C-code,† compiled with the very latest optimizing compilers.

In summary, MATLAB's excellent performance with vectorized code is actually due to four entirely separate aspects:

- Implicit (automated, internal) parallelization (multithreading)
- Highly tuned math libraries employing state-of-the-art algorithms and performance optimizations techniques
- Highly tuned core MATLAB functions
- CPU-level SIMD vectorization

So when we say that vectorized MATLAB code is fast, it could be due to any or all of these aspects used on our data. We have no way of knowing which exact mechanism is employed, and we should not really care. In fact, it could be counterproductive to rely on

---

* MathWorks uses Intel IPPL, MKL, as well as the VTune profiler (see http://mathworks.com/company/jobs/opportunities/10433 or: http://bit.ly/VSmlyq, a job listing for a performance-tuning engineer), so it would make sense for MathWorks to use Intel's Studio XE compiler package (http://software.intel.com/en-us/intel-xe-product-comparison or: http://intel.ly/1bpewan).

† BLAS and LAPACK remains Fortran-based to this day, although there are ports to other languages such as C. Intel's MKL apparently uses Fortran, and provides interfaces for C/C++ (http://software.intel.com/en-us/articles/intel-math-kernel-library-intel-mkl-blas-cblas-and-lapack-compilinglinking-functions-fortran-and-cc-calls or: http://intel.ly/17FNTL4).

such knowledge even if we had it, since the internal implementation changes between releases and across platforms. In any case, we have little control over this.*

In recent years, advancements in the JIT engine have made simple loops almost as fast as their vectorized equivalents.[734] In some cases, vectorization actually *decreases* performance compared to the JIT'ed code, due to some inherent optimizations that the JIT can make. Moreover, vectorization can sometimes become tricky to implement correctly,[†] as well as to debug and maintain. This is not always so: in many cases vectorization makes the code simpler and easier to maintain, but the reverse situation of undecipherable vectorized code is not uncommon. Another potential drawback of vectorized code is increased memory in some cases.

It should be noted that improvements to the internal JIT interpreter have occurred over the years,[‡] in parallel to the increased use of multithreading, and implicit parallelization in general. While the JIT improvements favor iterative loops, parallelization favors vectorized code. The performance balance sometimes shifts from one to another, depending on multiple factors: data type and data size, function(s) being performed, platform, release, and other environmental aspects.

Still, taking these potential drawbacks into note and carefully testing different code variants for both correctness and performance, the ability to vectorize MATLAB code can have a significant positive impact on program performance.

The important message here is that we should become familiar with vectorization, and then carefully test vectorized versus nonvectorized variants for correctness and performance on our target platform(s), using the anticipated target data. This chapter aims to make readers familiar and comfortable enough with vectorization, for it to become a readily available tool in their toolbox of performance-tuning alternatives.

5.1.2 Indexing Techniques[§]

Indexing into MATLAB arrays is a powerful technique for avoiding iterative loops and improving code performance by vectorization. MATLAB has plenty of documentation on indexing techniques and variations.[735] It is assumed that the reader is familiar with the basics of data indexing in MATLAB. If you are not, it is highly recommended that you review the referenced material. Here, we shall mainly focus on those aspects that pertain to performance.

We begin by considering several different ways to list all elements in a row array:

MATLAB Command	Effect
`x = [1:5] + 10 % create the row array`	11 12 13 14 15
`% several ways to access all array elements`	
`x`	
`x(:)`	
`x(1:5)`	11 12 13 14 15
`x(1:end)`	
`x([1 2 3 4 5])`	

---

* We could force MATLAB to use only a single computational thread using the -singleCompThread command-line parameter (see http://mathworks.com/help/matlab/ref/matlabwindows.html or: http://bit.ly/17FftYL). This will degrade performance, so its use is limited to checking whether speedup is due to multithreading or not, and when submitting cluster jobs (see §6.0).
† Recall John Ousterhout's adage (§1.6) that a wrong answer is always infinitely slower than a correct one…
‡ See §9.4.2 for example.
§ This section is based in part on Vincent Cheung's Quick Reference webpage (http://vincentcheung.ca/research/matlabindexrepmat.html or: http://bit.ly/1nGTixb).

Indexing can also be done on "singleton" dimensions (dimensions with size 1). For example, x is a row vector and its dimensions are [1 5], which means that its first dimension is a singleton. Hence the following two statements are equivalent:

MATLAB Command	Effect
x(:)	
x(1,:) % equivalent	11 12 13 14 15

Interesting results are obtained when indexing a singleton dimension multiple times. This is called "Tony's Trick".* The following code repeats all columns of x in the first row 3 times.

MATLAB Command	Effect
x([1 1 1],:)	1 2 3 4 5
	1 2 3 4 5
x(ones(3,1),:) % equivalent	1 2 3 4 5

Instead of indexing all array elements in order, we can index into the array in arbitrary ways. Here are a few examples:

MATLAB Command	Effect
% reverse the array	
x(end:-1:1)	15 14 13 12 11
% every second element	
x(1:2:end)	11 13 15
% specific elements in the array	
% (including elements duplication)	
x([1 1 4 2 5 3 5])	11 11 14 12 15 13 15
% random elements from the array	
x(floor(rand(4,1)*5) + 1)	13 11 15 13

When using indexing, we must ensure that we use numeric index values that are either real positive integers, or logicals (read more on logical indexing below). For example, to extract and swap the two halves of an array, as *fftshift* does, we might be tempted to write the following code, but MATLAB complains:

```
% extract and swap the halves of x
n = numel(x);
x([n/2+1:end 1:n/2])
⇨ Subscript indices must either be real positive integers or logicals
```

The reason for the error is that in our example, n/2 translates into 2.5, which is not an acceptable index value. Instead, we need to convert our indexing into integers:

```
>> x([ceil(n/2)+1:end 1:ceil(n/2)])
ans =
    14   15   11   12   13
```

---

* Named after a MATLAB user who showed this trick to Loren Shure in 1990; Loren then promoted its usage in her blog: http://blogs.mathworks.com/loren/2006/02/22/scalar-expansion-and-more-take-2#6 (or: http://bit.ly/17GvcGZ). See §9.4.3.

Note the use of the reserved keyword end in these examples. MATLAB automatically converts end to the size of the corresponding dimension (in our example, the value 5). We can even use this value to compute index values relative to the dynamic run-time end of the array. For example, we can shift the array elements to the right using the following code (note the corresponding built-in MATLAB function *circshift*):

```
>> x([end 1:end-1])
ans =
    15 11 12 13 14
>> circshift(x,[0,1]); % equivalent to above
```

In this example, end-1 translated into 4, so the indexing variant was essentially x([5 1:4]), that is, the last (5th) element followed by the 1st through 4th.

A common pitfall with direct array indexing is forgetting the operator precedence rules, which are basically (see the reference[736] for details):

1. Grouping via [], (), or {}
2. Arithmetic operators such as transpose ('), +, -, *, /
3. The colon operator (:)
4. Logical operators such as <, >, ~, ==, &, |

In the example above, 1:end-1 translated into 1:(end-1), so there is no need to group end-1 using () or []. But if we actually need (1:end)-1, then we would need to group.

Similarly, to create a simple column vector of values between 1 and 5, we cannot simply use x = 1:5', since that would translate into x = 1:(5'), which is the row vector [1,2,3,4,5]. Instead, we need to group as follows: x = (1:5)' or: x = [1:5]'.

Indexing techniques also work for multidimensional and cell arrays:*

MATLAB Command	Effect
A = zeros(3,4); A(:) = 1:12	1 4 7 10 2 5 8 11 3 6 9 12
% reverse the rows and columns % Note: end==3 for dim 1; end==4 for dim 2 A(end:-1:1, end:-1:1)	12 9 6 3 11 8 5 2 10 7 4 1
% get only rows #1,3 and columns #2,4 A([1 3], [2 4])	4 10 6 12
% get the diagonal elements* A(1 : size(A,1)+1 : end)	1 5 9
% shift matrix rows by k positions right k = 2; n = size(A, 2); A(:,[n-k+1:n 1:n-k])	7 10 1 4 8 11 2 5 9 12 3 6
% tile the data over a 7x6 area with % wrap-around A([1 2 3 1 2 3 1], [1 2 3 4 1 2])	1 4 7 10 1 4 2 5 8 11 2 5 3 6 9 12 3 6 1 4 7 10 1 4
% or alternatively: A(mod([0:6], size(A,1))+1,... mod([0:5], size(A,2))+1)	2 5 8 11 2 5 3 6 9 12 3 6 1 4 7 10 1 4

(continued)

---

* This technique relies on the order in which numeric matrix values are stored in memory. See §9.3 for additional details.

MATLAB Command	Effect
`A(3, :)         % extract only the 3rd row`	3 6 9 12
`A(:,[2 4]) = [] % remove columns #2,4`	1 7
`A(2,:) = []     % remove row #2`	3 9

`% cell-array indexing example`
`c = {'red',[0,1,0],'blue','cyan',[0,0,0]};`
`d = c([3,1,5])`
	{'blue', 'red',
	[0,0,0]}

Another indexing variation is *logical indexing* (see below). In logical indexing, we use a single, logical array for the matrix subscript. MATLAB extracts matrix elements corresponding to true values of the logical array.

MATLAB Command	Effect
`% all matrix elements greater than 9` `A(A > 9)`	10 11 12
`% more complicated logical indexing` `A(and(A > 1, A.*A < 10))`	2 3

The benefits of indexing are multiple: Indexing enables more compact and maintainable code; it also improves performance compared to simple loops that access the indexed elements one at a time, by enabling MATLAB to parallelize the operation internally and prevent looping and function-call overheads.

MATLAB code that accesses array elements using vectors of indices, as shown above, is called *vectorized code*. The process of transforming MATLAB code from simple iterative loops to vectorized code is called *vectorization*. Except in rare cases, vectorizing MATLAB code will usually result in much faster performance.

5.1.3 Logical Indexing

Logical indexing,[737] also called *array masking,* is the act of using *logical* (true/false) values to access selected elements of a data array.* The usage is quite intuitive:

```
% Convert all negative values to 0
data(data<0) = 0;
```

In this example, data < 0 returns an array of *logical* values corresponding to each of the elements in data (true if negative, false otherwise). This *logical* array is used to index (*mask into*) the numeric data array, returning only those elements of data where the corresponding *logical* value is true. The concept is easily extendable:

```
% Convert all out-of-bounds values to 0
data(data<0 | data>100) = 0;

% Convert all negative odd values to 0
data(data<0 & mod(data,2)) = 0;

% Remove the smallest value from data
data(data==min(data)) = [];

% Subtract 3 from any element larger than 5 738
data(data>5) = data(data>5) - 3;
```

---

* The data array can be of any type: numeric, cell, struct, handle reference, and so on.

A well-known MATLAB maxim is that *"logical indexing is faster than numeric indexing"*. Unfortunately, this maxim is inaccurate. A more correct form would be that *"logical indexing is faster than **find**-generated numeric indexing, but **slower** than direct numeric indexing"*:

```
data = 1 : 1e6; % 1M elements
logicalIdx = (mod(data,2)==0);

% Numeric indexing using find
c1 = data(find(logicalIdx));
⇨ Elapsed time is 0.009874 seconds.

% Logical indexing (2x faster than find)
c2 = data(logicalIdx);
⇨ Elapsed time is 0.004781 seconds.

% Direct numeric indexing (6x faster than find)
c3 = data(2:2:end);
⇨ Elapsed time is 0.001721 seconds.
```

Direct numeric indexing is faster, perhaps because the internal interpreter is able to convert the index values into direct memory access reference using *pointer arithmetic*.[739] The reason that *find* is so much less efficient is the time that it takes the *find* function to scan the logical values in logicalIdx and produce a numeric array.

The odd thing here is that even if we factor out *find*, the overall time is still slow:

```
numericIdx = find(logicalIdx);
⇨ Elapsed time is 0.003236 seconds.

c4 = data(numericIdx);
⇨ Elapsed time is 0.006581 seconds.
```

The reason for this oddity is that using a variable to store array indices requires the interpreter to separately access each array element in order to access the target data element. On the other hand, when we use direct (static) numeric indexing, the interpreter can optimize the operation such that no memory access other than the target data is needed. In other words, the term 2:2:end does <u>not</u> create a temporary numeric array in memory — the interpreter is able to optimize this away by directly using the numeric values. Note that this optimization appears to be highly dependent on the platform, MATLAB release, and the particular situation.[740] When it is in fact used, it reduces memory access overheads and improves CPU cache efficiency.

Even when the numeric index is placed in an interim variable, calculating the direct indexing can sometimes provide huge performance gains. In one example, an 11-line double-loop that calculated a nearly diagonal matrix was replaced by a 2-line vectorized version that reportedly ran 850× faster.[741] The trick was to find the pattern responsible for the nonzero data elements, calculate their associated indexes, and then assign the nonzero values directly to those indices, in a single operation.

Another optimization, which in some cases may generate a small speedup, is to use integer indexes, that is, *uint8, uint16,* or *uint32* rather than the standard *double*.[742]

To summarize, in cases where array indexes can be specified using direct numeric values, then do so. Otherwise, prefer the logical indexing approach. In any case, avoid using *find* unless you actually need the numeric position values of the indices (e.g., to implement your own index arithmetic).

Luckily, the MATLAB Editor's Code Analyzer (MLint) includes a specific warning (FNDSB) about using *find* when it detects that our code is not actually making use of the numeric index position values other than to index:*

```
idx = find(data<0.2);
c2 = data(idx);
```

⚠ To improve performance, use logical indexing instead of FIND. [Details ▼] [Fix]

Sometimes, the potential for replacing *find* with direct logical indexing is not immediately apparent. Such was the case highlighted by Doug Hull in a short video about optimizing a simple binning algorithm using logical indexing, which provided an order-of-magnitude (9×) speedup:[743]

```
% Original non-optimized code
isToRight = find(x > binEdges(binNum));
isToLeft  = find(x > binEdges(binNum+1));
isInBoth  = intersect(isToRight, isToLeft);
count(binNum) = numel(isInBoth);

% Optimized code using direct logical indexing (9x faster)
isToRight = (x > binEdges(binNum));
isToLeft  = (x > binEdges(binNum+1));
isInBoth  = (isToRight & isToLeft);
count(binNum) = sum(isInBoth); % or: nnz(isInBoth)
```

Logical indexing uses less memory than numeric indexing, since logical values use only a single byte of memory compared to the 8 bytes that double values use. On the other hand, debugging logical-indexed arrays is much more difficult in practice than debugging numerically indexed arrays. Both of these factors should be considered when deciding which indexing method to use — performance is just one criterion.

A widely used MATLAB programming idiom is using *length(find(condition))* to determine the number of items that match some criteria (condition). As explained above, the *find* function is inefficient compared to direct logical indexing. Therefore, we should replace *length(find(condition))* with *sum(condition)* whenever possible:

```
len = length(find(data==10));         % data has 10M elements
⇨ Elapsed time is 0.033219 seconds.

len = sum(data==10);
⇨ Elapsed time is 0.022408 seconds. % logical indexing: 50% faster
```

As a side benefit, Coder generated C-code of such code will also be significantly simpler and faster (see §8.2).

When using *find* in general, we are often only interested in the first result. Once this is found, there is often no need to continue searching the rest of the data for another occurrence. We can improve *find*'s performance by limiting the number of indexes k returned, by specifying k as the second optional parameter. In fact, this could even be faster than vectorized logical indexing, which needs to process the entire data:

```
len = length(find(data==10,1));
⇨ Elapsed time is 0.014895 seconds. % 50% faster than logical indexing.
```

---

* The complete list of MLint-reported performance-related warnings is presented in Appendix A.4.

And similarly,

```
firstRelevantData = find(data==10,1);
if ~isempty(data),...
```

This is especially important in cases where we are not really interested in the data matching the condition, but only in the question "is there *ANY* element that matches some condition in the data?" This can be answered using only the first element; we often do not care exactly how many matches there are.

If we need to retrieve the last k indexes, rather than the first ones, then use *find*'s third optional input parameter, with the string 'last' — this would still be faster than retrieving the entire list of results and then selecting the last k items:

```
last3Indexes = find(data==10, 3, 'last');
```

A common pitfall when using logical functions (i.e., functions that test for some condition and return a logical true/false value) is to compare their result to 1/0/true/false. This is entirely unnecessary (degrading performance), since the function already returns a logical value, and yet I often see this pitfall in MATLAB code:[744]

```
len = sum(isnan(data)==1);
⇨ Elapsed time is 0.031567 seconds.

len = sum(isnan(data));
⇨ Elapsed time is 0.018667 seconds. % no unnecessary ==: 70% faster
```

Examples that portend to demonstrate that logical indexing is faster than nonlogical equivalents are often actually showing the benefits of vectorization and memory preal-location* rather than those of indexing. For example:[745]

```
data = rand(1000);    % 1000x1000 elements

% Non-vectorized
B1 = [];              % note: no preallocation here
counter = 1;
for col = 1 : size(data,2)
    for row = 1 : size(data,1)
        if data(row,col) < 0.2
            B1(counter,1) = data(row,col);
            counter = counter + 1;
        end
    end
end
⇨ Elapsed time is 0.897209 seconds.

% Vectorized (80x faster)
B2 = data(data < 0.2);
⇨ Elapsed time is 0.011357 seconds.
```

A common logical indexing pitfall is to repeat a search operation for complementary conditions, rather than simply negating the resulting logical array. For example:

```
badIdx = data>9; goodIdx = data<=9; % alternative #1 (slow)
badIdx = data>9; goodIdx = ~badIdx; % alternative #2 (fast)
```

---

* See §9.4 below.

Logical indexing is indeed very useful for vectorization and they are commonly used in tandem,[746] but these are two entirely separate concepts that should not be confused. Vectorization can be done in many manners other than using logical indexing.

5.2 Built-In Vectorization Functions

5.2.1 Functions for Common Indexing Usage Patterns

Some common indexing techniques have been assigned dedicated built-in MATLAB functions for convenience and even performance. These include, for example:

- *fliplr* and *flipud* — for reversing array data
- *rot90* — for rotating matrices, exchanging dimensions
- *circshift* — for shifting array contents
- *diag* — for retrieving the diagonal elements
- *repmat* — for replicating array data

For example, a very efficient way to initialize MATLAB arrays is to use *repmat* to replicate (tile) input data into a larger array:

MATLAB Command	Effect
`% replicate a column vector 2 times vertically`	1
`x = [1:3]';`	2
`repmat(x, [2 1])`	3
	1
	2
	3
`% replicate a matrix using 2-by-2 tiles`	1 3 1 3
`A = zeros(2, 2); A(:) = 1:4; % initialize 2x2`	2 4 2 4
`A = [1,3; 2,4];    % equivalent initialization`	1 3 1 3
`repmat(A, [2 2]); % extend to 3 adjacent tiles`	2 4 2 4

In the past, *repmat* function's implementation made efficient use of smart indexing (Tony's Trick) to replicate the data.* For example:

```
>> x = 1:3; y = x([1,1],:)
y =
     1     2     3
     1     2     3
```

In R2013b (MATLAB 8.2), *repmat* was reimplemented from being an m-file to a built-in (compiled c-code) function, gaining 10%–20% speedup in some cases.[747]

Similarly, in R2013b, *fliplr* and *flipud* have changed their implementation from using indexed reversals as shown in the text above, to using the new built-in *flip* function, also providing ~10% performance improvement. In contrast, the *flipdim* m-function was not

---

* Data replication is discussed in detail in §5.4 below.

changed; it now issues a Code Analyzer/MLint warning (DFLIPDIM) that *flipdim* will be removed in the future and that *flip* should be used instead.[748]

Since these core functions are used by numerous other MATLAB functions, improvements in them directly translate into performance improvements across the board in MATLAB applications. In future MATLAB releases, I expect more of the core MATLAB functions to be reimplemented as built-in compiled code for improved performance.

In past MATLAB releases, when these functions were implemented in m-code, it was beneficial to use direct indexing techniques to bypass the need for calling these functions, thereby saving function-call overheads (see §3.1.2) and possibly unnecessary input-data checks (see §3.3). However, now it is faster to use the optimized compiled built-in functions rather than direct indexing. But this is only true for the compiled built-ins — internal m-functions are still slower than direct indexing.

The lesson here, as elsewhere, is that we should not assume which method is faster, but rather test on the target system. The results are quite often counterintuitive.

5.2.2 Functions That Create Arrays

MATLAB provides several built-in functions that enable creating arrays of varying sizes and dimensions. These arrays can then be used by those functions that accept and process vectorized data. In addition to the **fun* functions that shall be discussed in §5.2.4, functions in this group include:

Function	Description
zeros	Creates an array of numeric elements having the value zero (0)
ones	Creates an array of numeric elements having the value one (1)
inf	Creates an array of numeric elements having the value Inf
nan	Creates an array of numeric elements having the value NaN
cell	Creates a cell array of the specified dimensions having empty elements
eye	Creates an identity (I) matrix (1 in diagonal elements, 0 elsewhere)
linspace	Creates a numeric vector of linearly spaced values
logspace	Creates a numeric vector of logarithmically spaced values
magic	Creates a numeric magic-square matrix
hadamard	Creates a numeric Hadamard matrix
hankel	Creates a numeric Hankel matrix
toeplitz	Creates a numeric Toeplitz matrix
repmat	Replicates data in multiple dimensions

In the case of an array of structures, arrays of data from the constituent fields can be created using the [] and {} operators:[749]

```
data(1) = struct('name','Bill', 'male',true, 'age',34};
data(2) = struct('name','Ann', 'male',false, 'age',26};
names = {data.name}; % cell array:    {'Bill', 'Ann', ...}
flags = [data.male]; % logical array: [true, false, ...]
ages  = [data.age];  % numeric array: [34, 26, ...]
```

5.2.3 Functions That Accept Vectorized Data

In addition to indexing methods, MATLAB provides a number of utility functions to help implicit parallelization of the code. These functions operate on vector data and can be used as a replacement for many looping constructs.

Some of the most commonly used functions to work with vector data are listed below:

Function	Description
all	Test whether all elements in input data are nonzero
any	Test whether any element in input data is nonzero
arrayfun	Run a specified function on all elements of input numeric array
bsxfun	Run a binary function on corresponding elements of two input arrays
cellfun	Run a specified function on all elements of input cell array
cumsum	Return cumulative sum
diag	Return diagonal elements of an input numeric matrix
diff	Return element differences and approximate derivatives of numeric data
find	Find indices and values of nonzero elements
fliplr	Flip elements of input array or matrix left-to-right
flipud	Flip elements of input array or matrix up-down
histc	Group elements into bins based on value (also see: *accumarray*)
ind2sub	Convert from linear index to subscripts (also see: *sub2ind*)
ipermute	Inverse permute dimensions of a multidim array (also see: *permute*)
logical	Convert numeric values to logical
meshgrid	Generate X and Y arrays for 3-D plots
ndgrid	Generate arrays for multidimensional functions and interpolation
permute	Rearrange dimensions of a multidimensional array (also see: *ipermute*)
prod	Return product of input numeric elements
repmat	Replicate and tile an array
reshape	Change the shape of an array without modifying its data
rot90	Rotate matrix by an integer number of 90 degrees (also see: *permute*)
shiftdim	Shift array dimensions
sort	Sort array elements in ascending or descending order
squeeze	Remove singleton dimensions from an array (also see: *reshape*)
structfun	Run a specified function on all elements of input struct array
sub2ind	Convert from subscripts to linear index (also see: *ind2sub*)
sum	Return the sum of input numeric elements
transpose	Nonconjugate transpose of a data matrix (also see: *ctranspose*)

This list is by no means exhaustive. MATLAB has hundreds of functions that act on vectorized data, including math (*sin*, *log*, etc.), I/O (*fprintf*, *textscan*, etc.), string processing (*strcmp*, *regexp*, etc.), and graphics (*plot*, *delete*, etc.). In all such cases, vectorization can significantly improve code performance compared to simple loops, especially for data having many elements.[750]

In fact, a very large portion of MATLAB's functions are vectorized in the sense that they both accept vectorized data (arrays, matrices, or multidimensional arrays in general), and also act on this data in a vectorized manner (i.e., without looping). User applications are well advised to take advantage of vectorized functions when possible.

Many of the functions that act on vectorized data have optional input parameters that facilitate their use in a vectorized manner. One such common optional parameter is the dimension parameter, which tells the function over which of the input data's dimensions it should work in a vectorized manner. For example:

```
>> data = randi(9,3,4)
data =
     4     6     4     5
     2     3     7     8
     1     7     4     5
```

```
>> sum(data) % act on columns (default)
ans =
      7    16    15    18
>> sum(data,2) % act on rows
ans =
     19
     20
     17
>> mean(data) % act on columns (default)
ans =
       2.3333       5.3333          5          6
>> mean(data,2) % act on rows
ans =
       4.75
          5
       4.25
```

5.2.3.1 reshape

reshape is a very useful built-in function that enables us to reshape a data block by modifying its dimension sizes and order such that the new data block has exactly the same number of elements, but is arranged differently. This enables effective use of linear algebra rules to process the data.*

Reshaping data is extremely efficient since it does not change the data storage in memory, only the internal metadata that describes how this data should be used (size of each dimension).[751] As a corollary, *reshape* requires its data to exactly fit the new dimension sizes. For example, a 3×4 data matrix can be reshaped as 2×6 or 12×1 but not as 5×2 — trying to do so will generate a run-time error:

```
>> reshape(rand(3,4),5,2)
⇨ Error using reshape
To RESHAPE the number of elements must not change.
```

reshape has a useful feature of enabling the user to *not* specify the exact size of one of the dimensions. If the size of any of the data dimensions is set to [], then *reshape* calculates the missing size based on the data size and the size of all the specified dimensions. This calculation naturally takes some time. In fact, since *reshape* is so efficient, this calculation could take much longer than the actual reshaping! The moral: if we know the data's dimensions, we should specify them:

```
data(1,:,:) = magic(3);
data(2,:,:) = magic(3)*10;

% Using reshape(data,[],3): fast
tic, d = reshape(data,[],3); toc
⇨ Elapsed time is 0.000218 seconds.

% Using reshape(data,N,3): much faster
tic, d = reshape(data,6,3); toc
⇨ Elapsed time is 0.000014 seconds.
```

---

* See §5.5 and §5.7.16.

In cases where we might consider using *squeeze* to remove singleton dimensions from data, it is in fact faster to directly use *reshape*, if the data size is known.[752] In fact, *squeeze* uses *reshape* in its internal m-file implementation:

```
function b = squeeze(a)   % simplified (core) logic from squeeze.m
    siz = size(a);
    siz(siz==1) = [];  % Remove singleton dimensions.
    siz = [siz ones(1,2-length(siz))];  % Make sure siz is at least 2-D
    b = reshape(a,siz);
end
```

When calling reshape, it is faster to use the multiparameters syntax than the syntax that accepts an array of dimension sizes:

```
data = rand(1000,1000);
```

```
for idx=1:1e6,  d = reshape(data, 5000,200); end
⇨ Elapsed time is 0.590540 seconds.
```

```
for idx=1:1e6,  d = reshape(data, [5000 200]); end
⇨ Elapsed time is 0.673690 seconds.
```

MATLAB's native data type is a 2D matrix. It is therefore understandable that reshaping to a higher-dimensional array is slower than a 2D array. The effect is more pronounced using the array-of-dimensions syntax, than the multiparams syntax:

```
for idx=1:1e6,  d = reshape(data, 50,20,40,25); end
```

```
⇨ Elapsed time is 0.789814 seconds.
```

```
for idx=1:1e6,  d = reshape(data, [50 20 40 25]); end
⇨ Elapsed time is 2.164455 seconds.
```

The conclusion is that it is generally faster to use 2D arrays in MATLAB, rather than multidimensional (N – D) arrays,* and to *reshape* using the multiparams syntax rather than the array-of-dimensions syntax.

The *reshape* function is often used together with *transpose* and *permute*. An example of this is the *ntimes* File Exchange utility[753] for multidimensional matrix product:[†]

```
% C=NTIMES(A,B) computes the matrix product of A and B. When A, B are
% both two dimensional matrices, the result C will be identical to the
% output of Matlab's MTIMES function. If A and B have higher dimension
% than two, then this function treats A and B as matrices of 2-D
% matrices, and performs matrix-wise multiplication element-by-element
% Written by Ampere Kui, 2009.
```

```
function c = ntimes(a, b)
    % Calculate the size of both matrices.
    s_a = size(a);
    s_b = size(b);

    % Using the size of the matrices, calculate the permutation
    % indices that we will need to perform matrix multiplication:
    % C(i,j) = dot(A(i,r), B(r,j))
    i = 1:s_a(1);  % Row indices of A.
    r = 1:s_a(2);  % Col indices of A.
    j = 1:s_b(2);  % Col indices of B.
```

---

* For additional performance aspects of multidimensional arrays, see §5.5.
† Additional aspects of matrix multiplication were discussed in §4.5.2.

```
    i = repmat(i, [1 s_b(2)]);
    j = repmat(j, [s_a(1) 1]);

    % Perform a transpose on B by switching dimensions 1 and 2.
    b = permute(b, [2 1 3:ndims(b)]);

    % Set up P, Q matrices such that when we compute the dot product,
    % we obtain the matrix-wise product of A, B across all dimensions
    p = a(i(:), r, :);
    q = b(j(:), r, :);
    c = reshape(dot(p, q, 2), [s_a(1) s_b(2) s_a(3:end)']);
end
```

5.2.4 Functions That Apply Another Function in a Vectorized Manner

5.2.4.1 arrayfun, cellfun, spfun, and structfun

As extensive as MATLAB's list of built-in vectorized functions is, it is sometimes difficult to find a combination of vectorized functions that will achieve our needs. For example, we may wish to apply a complicated user function to each element of an input matrix or cell array.

In such cases, rather than reverting to an iterative loop, we can use MATLAB's set of built-in explicit vectorization functions. All these functions look alike and act in the same manner — applying a specified target function to each element of the input data. The functions differ only in the type of their inputs:*

Function	Description
arrayfun	Run a specified function on all elements of input numeric array
bsxfun	Run a binary function on corresponding elements of two input arrays
cellfun	Run a specified function on all elements of input cell array
spfun	Run a specified function on all elements of a sparse matrix
structfun	Run a specified function on all elements of input struct array

For example, if we have the following target function, which acts on scalar data:

```
function scalarResult = someFunctionOf(scalarData)
```

we can apply it in a vectorized manner to all elements of an input matrix like this:

```
resultMatrix = arrayfun(@someFunctionOf, inputMatrix);
```

The `resultMatrix` output will have exactly the same dimensions as the `inputMatrix`. Each `resultMatrix` element will be the result of `someFunctionOf` on the corresponding input element: `resultMatrix(2,3) = someFunctionOf(inputMatrix(2,3))`.

Importantly, *fun* functions are often *slower* than simple *for*-loops.[754] They appear to be little more than convenience functions to make the otherwise-bulky loops simpler to code and maintain. Despite their appearance, they are not truly vectorized:‡

---

* *bsxfun* is different — see §5.2.4.2 for details.

† MATLAB vectorization guru Bruno Luong took this to an extreme, stating that *"ARRAYFUN and CELLFUN … are faked vectorized code. Nothing is vectorized inside, and they are usually slower than the for-loop. Avoid using them when speed is critical"* (http://mathworks.com/matlabcentral/newsreader/view_thread/328618#903281 or: http://bit.ly/1hgOzvr). Perhaps these functions are vectorized, but have some unknown high vectorization threshold value, as explained in §5.1.1.

‡ When *gpuArray* data is used, *arrayfun* uses the GPU for processing (see §6.2.2), which could definitely provide speedup.

```
data = rand(1,1e6);
% Using pseudo-vectorized arrayfun()
d = arrayfun(@sqrt,data);
⇨ Elapsed time is 1.368263 seconds.

% Using standard iterative loop - 28x faster than arrayfun()
for idx = numel(data) : -1 : 1 %loop backwards to preallocate: §9.4.3
    d(idx) = sqrt(data(idx));
end
⇨ Elapsed time is 0.049434 seconds.
```

Whenever the applied function can accept vectorized data directly, using that truly vectorized syntax is generally the fastest alternative:[755]

```
% Using truly vectorized sqrt() function - 370x faster than arrayfun()
d = sqrt(data);
⇨ Elapsed time is 0.003709 seconds.
```

Still, when performance is not as important as coding clarity and maintenance, the set of **fun* functions are very handy pseudo vectorization functions. In the future, it is possible that MathWorks will modify their implementation to provide real multithreaded vectorization. The remainder of this section shall detail their usage.

Note that the order by which MATLAB evaluates the input elements is not guaranteed.* It is possible that it will be done from top to bottom, or vice versa, or in some random order. Therefore, if our function relies on the processing order, we should either test carefully or better yet opt for the (potentially slower) iterative loop whose processing order is indeed guaranteed.

Using **fun* in the manner above works well for results that are numeric scalars, but what if scalarResult is a string or some other object that cannot be concatenated in a simple [] matrix? In such a case, the results need to be grouped in a cell array. This is done using the 'UniformOutput' parameter (this parameter name can be abbreviated to 'Uniform' or even 'Un' for brevity):

```
>> resultMatrix = arrayfun(@num2str, magic(4))
⇨ Error using ==> arrayfun
Non-scalar in Uniform output, at index 1, output 1.
Set 'UniformOutput' to false.

>> cellArray = arrayfun(@num2str, magic(4), 'UniformOutput',false)
cellArray =
    '16'    '2'     '3'     '13'
    '5'     '11'    '10'    '8'
    '9'     '7'     '6'     '12'
    '4'     '14'    '15'    '1'
```

We can use the 'UniformOutput' parameter to output a cell array, even when a regular (uniform) matrix output is possible:

```
>> cellArray = arrayfun(@sqrt, magic(4), 'UniformOutput',false)
cellArray =
    [      4]    [1.4142]    [1.7321]    [3.6056]
```

---

* This implies that the internal implementation is parallelized, conflicting with the findings above. Perhaps the parallelization kicks in only at some high data-size threshold (as explained in §5.1.1), or perhaps it is merely a warning for future releases where these functions may become fully multithreaded, so that existing code will not depend on the processing order.

```
    [2.2361]      [3.3166]      [3.1623]      [2.8284]
    [     3]      [2.6458]      [2.4495]      [3.4641]
    [     2]      [3.7417]      [ 3.873]      [     1]
```

If our target function accepts multiple inputs, we can pass these from several input arrays. Vectorization will apply the target function on all corresponding elements:

```
function scalarResult = someFunctionOf(scalar1, scalar2, scalar3)

data1 = magic(4);
data2 = -3*magic(4);
data3 = zeros(4,4);
resultMatrix = arrayfun(@someFunctionOf, data1, data2, data3);
```

Here, resultMatrix(2,3)=someFunctionOf(data1(2,3),data2(2,3), data3(2,3)).
There are several alternatives for specifying the target function: we can use a regular function handle as shown above, or an anonymous function:

```
>> arrayfun(@(x) x^2, magic(4))
ans =
   256      4      9    169
    25    121    100     64
    81     49     36    144
    16    196    225      1
```

Within the anonymous function definition, we can use any variable that has a known value at the time of the definition. For example:

```
>> v=10; arrayfun(@(x) x^2+v, magic(4))
ans =
   266     14     19    179
    35    131    110     74
    91     59     46    154
    26    206    235     11
```

In general, using anonymous functions is slower than using handles of "real" (m-file, sub- or nested) functions, so try to use real rather than anonymous function handles:

```
data = rand(1000); % 1000x1000 numeric matrix

% Anonymous function handle
data2 = arrayfun(@(x) sqrt(x), data);
⇨ Elapsed time is 3.169824 seconds.

% "Real" function handle - 2x faster
data2 = arrayfun(@sqrt, data);
⇨ Elapsed time is 1.609221 seconds.
```

Similarly to *arrayfun*, we have the *cellfun*, *structfun*, and *spfun* vectorization functions that act on corresponding elements of a cell array, struct array, and sparse matrix, respectively. These functions are similar to *arrayfun* except that

- *spfun* does not accept the 'UniformOutput' parameter.
- *cellfun* accepts some string values as target functions, in addition to the regular function handles (anonymous/"real"): 'isempty', 'islogical', 'isreal', 'length', 'ndims', 'prodofsize', 'size', and 'isclass'. Using the string

values provides significant performance benefits over the corresponding function handles, since MATLAB implemented them internally and does not need to pass through the generic function-handle mechanism:[756]

```
cellData = num2cell(rand(1000));  % 1000x1000 cell array

logicalMatrix = cellfun(@isempty, cellData);
⇨ Elapsed time is 0.657060 seconds.

% Using 'isempty' rather than @isempty - 70x faster!
logicalMatrix = cellfun('isempty', cellData);
⇨ Elapsed time is 0.009129 seconds.
```

arrayfun, cellfun, and *structfun* (but not *spfun*) have an optional 'ErrorHandler' input parameter that accepts a function handle. If specified, then this function will be invoked whenever an error (unhandled exception) occurs in the target function for any of the elements. If the ErrorHandler function is not specified and an error occurs, the entire vectorized operation fails. It would be a pity to waste the ability to vectorize an operation on a million elements just because of a few bad elements. ErrorHandler enables treating rare cases separately, without having to resort to costly loops.

Still, it should again be stressed that using **fun* functions are not generally faster than the corresponding loops.

Where possible, try to merge multiple calls to **fun* functions into a single call, in order to reduce the internal overheads. Try to avoid using costly anonymous function handles, so that the merge benefits will not be offset by the latter's bad performance.

The *accumarray* function is similar to the **fun* functions, in the sense that it accepts vectors of data, an optional function handle, and acts in a vectorized manner to distribute the data values into the specified index elements, combining multiple element values using the function handle. See §5.3.5 for a usage example.

5.2.4.2 bsxfun

The *bsxfun* (Binary Singleton eXpansion) function, introduced in MATLAB 7.4 (R2007a), is similar to the other **fun* vectorization functions in that it accepts a target function handle, which is then applied in a vectorized manner on corresponding elements of the input arrays. The importance of *bsxfun* for performance increased in MATLAB 7.9 (R2009b), when *bsxfun*'s implementation was made multithreaded (read: implicit parallelization). *bsxfun* is incredibly useful and its use pattern occurs frequently in computing problems. It is arguably the most important vectorization function. It should be considered wherever *repmat* is currently (ubiquitously) used.

bsxfun is unique in that it only accepts handles to binary functions (i.e., functions that accept two inputs and return a single value, such as @plus or @eq), and exactly two data inputs. *bsxfun* is also special since it automatically expands scalar dimensions for any of the two inputs, so that they effectively become the same size.

The example used in the official documentation is illustrative:[757] The idea is that we have a numeric 2D matrix from which we wish to subtract the mean values of each of the columns respectively, that is, subtract mean(data:,1)) from data(:,1), and similarly for all columns in data. We can not do a simple subtraction, since the mean values are stored in a 1D vector while data is a 2D matrix:

```
data = rand(1e6,5);      % 1M rows, 5 cols
dataMeans = mean(data);  % 1 row, 5 cols
```

```
data - dataMeans
⇨ Error using ==> minus
Matrix dimensions must agree.
```

In order to apply vectorization, we can use *repmat* or direct indexing to expand (tile) dataMeans to the same size of data, after which we can do the simple subtraction:

```
% dataMeans expansion using repmat (R2013a)
dataMeansBig = repmat(dataMeans, size(data,1), 1);
result = data - dataMeansBig;
⇨ Elapsed time is 0.045655 seconds.
```

```
% dataMeans expansion using direct indexing (R2013a)
dataMeansBig = dataMeans(ones(1,size(data,1)), :);
result = data - dataMeansBig;
⇨ Elapsed time is 0.046013 seconds.
```

The alternative is to use *bsxfun* directly, removing the need to create the memory-consuming temporary dataMeansBig array:

```
% Direct singleton expansion via bsxfun (3x faster than repmat, R2013a)
result = bsxfun(@minus, data, dataMeans);
⇨ Elapsed time is 0.016337 seconds.
```

Since data expansion (and the resulting memory allocations) seem to be a large factor contributing to *bsxfun*'s superiority over indexed expansion or *repmat*, it stands to reason that when the data size is small enough, indexed expansion can indeed be faster than *bsxfun* in some cases.[758] The only way to know for sure is to test on your target platform and using representative data.

In R2013b (MATLAB 8.2), *repmat* was changed from being an m-file function to a built-in (compiled binary) function, which significantly improved its run speed. As a result, *bsxfun* is no longer faster than the *repmat* variant (the above example produces equivalent run-times in R2014a). Sometimes *bsxfun* is much *slower* than *repmat*:[759]

```
A = uint64(2^64 * rand(10, 1, 1000));
B = uint64(2^64 * rand(10, 1000, 1));

bsxfun(@bitxor, A, B);
⇨ Elapsed time is 1.532619 seconds.
```

```
% Using repmat: 26x faster! (R2014a)
bitxor(repmat(A, [1 1000 1]), repmat(B, [1 1 1000]));
⇨ Elapsed time is 0.058308 seconds.
```

Still, *bsxfun* provides a very clean intuitive programming paradigm that is simpler than the *repmat* variant and therefore improves code readability and maintainability.

One might think that when we do not need any data expansion (i.e., both inputs are of the same size), then it is faster to use the built-in element-wise arithmetic operators .*, .+, and so on than it is to use *bsxfun*, right? Think again:

```
result2 = data - result; % both are 1e6 x 5 matrices
⇨ Elapsed time is 0.017466 seconds.
```

```
% Using bsxfun - 50% faster than arithmetic minus!
result2 = bsxfun(@minus, data, result);
⇨ Elapsed time is 0.011815 seconds.
```

This surprising result depends on platform, release, and data-size. The same code on a different platform produces timing results that are the same for both variants. This highlights once again the need to test performance on the target platform.

bsxfun's automatic singleton expansion enables easy generation of an *outer product* (also called cross-product or tensor product)[760] of two vectors, creating a matrix having dimensions that are the respective lengths of the two vectors.[761] For example, to create an outer product from the difference between two vectors (i.e., a matrix whose elements represent the difference between corresponding elements in the input vectors), we would do the following (note the transpose of y, to enable singleton expansion of x in the rows dimension and y in the columns dimension):

```
>> x = 1:5;   y = [3,5,7];
>> z = bsxfun(@minus, x, y')
z =
      -2    -1     0     1     2
      -4    -3    -2    -1     0
      -6    -5    -4    -3    -2
```

We can use *bsxfun*'s automatic singleton expansion to create an outer product of higher-dimensional data (e.g., 2D matrices, not vectors). The trick here is to permute the data's dimensions such that a higher dimension gets swapped with another. For example, a 2D matrix has a size of N×M. But this is actually the same as a 3D tensor having a size of N×M×1. Permuting dimensions is the higher-dimensionality equivalent of the simple transpose that we did with the 1D vectors. So we could permute the data matrix dimensions to N×1×M and 1×N×M to enable *bsxfun*'s automatic singleton expansion to kick in:[762]

```
z = bsxfun(@minus, permute(x, [1 3 2]), permute(y, [3 1 2]));
```

This idea was used in Bruno Luong's implementation of Kronecker's (tensor) product[763] (A ⊗ B) for full matrices.[764] It was so efficient that it was included in the official MATLAB implementation (*kron* function) in R2013b (MATLAB 8.2).* Here is a pared-down snippet of the relevant code for full (nonsparse) data:[†]

```
function K = kron(A,B)
    [ma,na] = size(A);
    [mb,nb] = size(B);
    A = reshape(A, [1 ma 1 na]);
    B = reshape(B, [mb 1 nb 1]);
    K = bsxfun(@times,A,B);
    K = reshape(K, [ma*mb na*nb]);
end
```

Note: The *kron* function attracts a lot of attention in the MATLAB community and several alternative implementations have been contributed, in addition to Bruno's: One implementation extends Bruno's solution for sparse data;[765] others use MEX,[766] object-oriented,[767] multidimensional,[768] and other techniques.[769] Comparing these separate implementations could be an interesting reader exercise. Not less interesting is an exercise in trying to find

---

* *%matlabroot%/toolbox/matlab/ops/kron.m*; the earlier approach used *meshgrid* rather than *bsxfun* — this was slower and also allocated a lot of temporary memory. See §9.1 for a discussion on memory's effect on performance.
† Sparse data was also *bsxfun*'ed for performance in R2013b.

interesting ways by which *kron* can solve mathematical problems faster than alternative solutions: there are quite a lot of these, it seems.[770]

As another *bsxfun* example, I was once asked by a client to speed up a stock trading strategy. The existing strategy was slow to compute and ineffective for real-time trading. I was able to drastically reduce the strategy run-time, enabling real-time usage:

```
% Some test data
signals = rand(300,200,400);  % 24M elements, ~180 MB
scales = rand(200,400);
siz = size(signals);
for i = 1 : siz(1)            % num_histogram_bins
    for j = 1 : siz(2)        % num_days
        for k = 1 : siz(3)    % num_stocks
            signals(i,j,k) = buy_signals(i,j,k) / scales(j,k);
        end
    end
end
⇨ Elapsed time is 30.972008 seconds.

% Using bsxfun - 165x speedup!
signals = permute(signals,[2,3,1]);   % permute to enable bsxfun
signals = bsxfun(@rdivide, signals, scales);
signals = permute(signals,[3,1,2]);   % permute back to original
⇨ Elapsed time is 0.187522 seconds.

% Even simpler: reshape scales, not signals - 400x speedup!
signals = bsxfun(@rdivide, signals, reshape(scales,1,siz(2),siz(3)));
⇨ Elapsed time is 0.076220 seconds.
```

bsxfun has the potential for significant code speedup and improved readability/maintainability, as these and other examples show (e.g., §5.3.8, §5.3.10). On the other hand, *bsxfun* is indeed sometimes more difficult to set up correctly (typically with some *reshapes* and *permutes**) than a vectorized solution, which is based on *repmat*. For this reason, it is sometimes better to try the *repmat* approach first, and then try a *bsxfun* variant only if we have still not met our performance goal.

If reshaping/permuting input dimensions turns you off, you are in good company! The concept takes some time getting used to. Consider using Paulo de Leva's *baxfun* utility on the File Exchange,[771] which uses and expands *bsxfun* to easily operate on shifted-dimensionality inputs. This is basically a convenience wrapper for using *shiftdim/reshape/ permute*.

In order to feel more comfortable with *bsxfun*, something that is highly recommended for its performance speedup potential, I suggest to review and try to understand some of the *bsxfun* examples in this book[†] and online.[772]

When working with *bsxfun*, we should note that it normally works on regular arrays, not cell arrays. With cell array inputs, when *bsxfun* would evaluate the result of c(i,j) = func(a(i,j),b(i,j)), func would receive as inputs two cell arrays of a single element (i.e., {5} rather than simply 5). In most cases, binary functions do not handle cell inputs and therefore the operation (and *bsxfun*) will croak.[773] It will only work if we use a custom-made binary function that handles cell array inputs, or use *cellfun*.

---

* The *shiftdim* function is also often used for this. This function is basically a simple wrapper for *reshape* and *permute*, depending on whether the dimensional shift is to the right or the left (respectively). See, for example, http://bit.ly/18IrzA0.

† See §5.3.

Another possible drawback of *bsxfun* is the fact that its singleton expansions may create temporary data of the size of the end result, thereby potentially hogging the system down due to memory considerations,* and possibly even throwing an out-of-memory error. Naturally, *repmat* has a similar effect, but in some cases, where memory is an issue, a vectorized solution that does not rely on either *bsxfun* or *repmat* can be found.[774]

bsxfun was added to MATLAB in R2007a (7.4). On older MATLAB releases, we can use the utility functions written by MATLAB veterans Doug Schwarz[775] or James Tursa[776] (both utilities rely on MEX for performance but use different algorithms, so each of them might be faster in different situations). Tursa also contributed the MEX-based *bsxarg* utility, which returns the singleton-expanded arrays of the *bsxfun* inputs (i.e., the input arrays expanded to the same dimensional size).[777]

Bruno Luong has contributed a set of highly efficient functions to the File Exchange, which complement *bsxfun*. These include *bsxcat* (multidimensional array concatenations with singleton expansion)[778] and *bsxops* (forcing built-in arithmetic operators to automatically expand singletons à la *bsxfun*).[779]

5.2.5 Set-Based Functions

MATLAB has several built-in functions that provide important basic functionality for sets of data. In this case, a "set" means any array of numeric ([1,2,3]) or nonnumeric (typically string — {'a','b','c'}) data. The data is not assumed to be sorted and may well contain duplicate values. Functions in this group include:

Function	Description
intersect	Intersection of two arrays (i.e., unique elements that exist in both arrays)
setdiff	Difference between two arrays (i.e., elements in the first but not second)
setxor	Exclusive OR of two arrays (i.e., element existing in only one of them)
sort	Sorted elements of an array
union	Union of two arrays (i.e., full merge of the elements in both arrays)
unique	Unique elements of an input array (sorted by default)
isempty	Test whether an input array has any elements
isfinite	Test which elements of a numeric array are finite (non-Inf, non-Nan)
isinf	Test which elements of a numeric array are infinite (+Inf or -Inf)
isnan	Test which elements of a numeric array are NaN
isprime	Test which elements of a finite numeric array are prime values
issorted	Test whether an input array's elements are sorted
ismember	Test which element in array A also exists in array B (also see: *ismembc*)
ismembc	Faster *ismember* alternative for sorted "regular" numeric data (see §4.3.1)

These functions accept both 1D arrays and multidimensional data. Multidimensional arrays are treated as 1D column vectors (similar to a `reshape(data,[],1)` operation†), before the set-based operation is done. For example:

```
>> setdiff(magic(3), magic(2))
ans =
    5
```

---

* See §9.1.

† See §9.3.1 for details on MATLAB's internal memory storage of multidimensional data.

```
        6
        7
        8
        9
```

The set-based functions normally accept both numeric and nonnumeric (typically string) data, but not mixed data (i.e., numeric and nonnumeric together):

```
>> setxor(1:7, [2,3,5,8,11])
ans =
        1      4      6      7      8     11

>> setxor({'abc','def','ghi'}, {'ghi','DEF','123','zxc'})
ans =
      '123'     'DEF'     'abc'     'def'     'zxc'

>> setxor({'abc','def','ghi'}, {'ghi','DEF',123,-pi})
⇨ Error using cell/setxor > cellsetxorR2012a (line 292)
Input A of class cell and input B of class cell must be cell arrays of
strings, unless one is a string.
```

The relevant set-based functions can be overridden by custom user classes, thereby extending the basic built-in support to new data types.

Most set functions return optional output arguments that indicate the relevant index positions of the returned result element, within the original data arrays. For example:

```
>> [newData, idxInA, idxInB] = setxor(1:7, [2,3,5,8,11])
newData =
        1      4      6      7      8     11

idxInA =
        1
        4
        6
        7
idxInB =
        4
        5
```

Many of the set functions have optional input parameters that facilitate their use in a vectorized manner. Most of them have the optional 'rows' input that tell them to act on rows rather than columns (also see *sortrows*); some of the functions have additional optional input arguments for specific uses.

An interesting speedup trick for locating rows in a data matrix was posted by Johan Löfberg in 2005. The basic idea is to convert data rows into a single scalar hash-code value, and then to simply compare hash-codes (this idea can naturally be extended to any data structure):[780]

```
A = randn(1e6,3);
B = A([11 12 13 14 15], :); % rows #11-15

% Standard implementation #1
[~, ~,idx] = intersect(B,A,'rows');
⇨ Elapsed time is 0.795593 seconds.
```

```
% Standard implementation #2
[~,idx] = ismember(B,A,'rows');
⇨ Elapsed time is 0.780576 seconds.

% Using matrix-multiplication hash-coding trick - 45x faster!
hash = randn(size(A,2),1); % random hash seed
b = B * hash; % convert all rows into unique scalar hash-codes
a = A * hash; % convert all rows into unique scalar hash-codes
[~,idx] = ismember(b,a);
⇨ Elapsed time is 0.017041 seconds.
```

Note Jan Simon's very efficient MEX-based utility *CStrAinBP*,[781] which significantly outperforms *intersect, ismember, setdiff,* and *union* for cell arrays of strings.

5.3 Simple Vectorization Examples

The following examples, of increasing complexity, illustrate the power of vectorization. The University of Cambridge's Department of Engineering has a very nice webpage with lots of additional simple vectorization examples.[782] Acklam's vectorization guide (see §5.7.15) contains many examples. Many more can be seen in MATLAB's implementation of stock functions, which readers can read and reuse.*

When reviewing vectorization examples, note that there are often multiple ways to achieve vectorization, which can vary widely in their performance and memory usage. Different variants may be faster for any particular data and running platform. Faster variants are not necessarily more elegant or memory-efficient.

5.3.1 Trivial Transformations

The easiest and most direct way to vectorize loops, which works in a surprisingly large number of cases, is to simply remove the *for* keyword and its corresponding *end*. For example:†

```
% Loop variant
t = 0 : 0.01 : 10;
for idx = 1 : length(t)
    y(idx) = sin(t(idx));
end

% Vectorized variant
t = 0 : 0.01 : 10;
idx = 1 : length(t); % add semi-colon to prevent console output
y(idx) = sin(t(idx));
```

which can then be rewritten in a simpler (although not faster) manner:

```
t = 0 : 0.01 : 10;
y = sin(t);
```

---

* For example: ***rot90*** (*%matlabroot%/toolbox/matlab/elmat/rot90.m*), ***toeplitz*** (*%matlabroot%/toolbox/matlab/ elmat/toeplitz.m*), ***nthroot*** (*%matlabroot%/toolbox/matlab/elfun/nthroot.m*), and many other functions in the *%matlabroot%/toolbox/matlab/* folder.

† This specific variant is not faster in modern MATLAB releases, since JIT can completely vectorize the looped version in run-time. See §3.1.15 and http://www.matlabtips.com/matlab-is-no-longer-slow-at-for-loops (or: http://bit.ly/1dTm7MD).

As another example:

```
% Loop variant
data = rand(10,20,30);  % 10x20x30
for idx = 1 : size(data,3)
    a(idx) = sum(sum(data(:,:,idx)));
end

% Vectorized variant
idx = 1 : size(data,3)
a(idx) = sum(sum(data(:,:,idx)));

% Simpler variant
a = reshape(sum(sum(data)), 1, []);  % reshape 1x1x30 → 1x30
```

Sometimes the transformation is not so simple, since there are dependencies between the loop iterators:

```
% Loop variant
for i = 1 : num_total_days
    for j = 1 : num_days_later-1
        if (i+j) <= num_total_days+1
            for k = 1 : num_stocks
                if data(i,j+1,k) >0
                    data2(i,j,k) = (data(i,j+1,k) / data(i,1,k)) - 1;
                end
            end
        end
    end
end

% Vectorized variant, in 2 phases:
% Phase 1 - divide all elements by the relevant divisor
data2 = zeros(size(data));   % [num_total_days num_days_later num_stocks]
for j = 1 : size(data,2)-1
    data2(:,j,:) = data(:,j+1,:) ./ data(:,1,:);
end
data2 = data2 - 1;

% Phase 2 - zero all cells where data <= 0
for i = 1 : num_total_days
    maxJ = min(num_days_later-1, num_total_days-i+1);
    j = 1 : maxJ;
    data2(i,j,:) = data2(i,j,:) .* (data(i,j+1,:)>0);
end
```

Note that the loop in Phase 1 can be removed altogether using *bsxfun*:

```
data2 = bsxfun(@rdivide, data(:,2:end,:), data(:,1,:)) - 1;
data2(:,end+1,:) = 0;
```

5.3.2 Partial Data Summation

In the following example, the aim is to sum all even-indexed elements of an array:

```
% Slow, standard version
sumValue = 0;
for idx = 2 : 2 : length(data)
    sumValue = sumValue + data(idx);
end
```

```
% Vectorized version
sumValue = sum(data(2:2:end));
```

In the following example, we remove matrix rows having some duplicate values:[783]

```
>> A = [2 4  6 8; 3 9  7 9; 4 8  7 6; 8 5 4 6; 2 10 11 2];
>> A = A(all(diff(sort(A'))), :) % rows 2,5 are stripped away
A =
     2      4      6      8
     4      8      7      6
     8      5      4      6
```

In this example, `sort(A')` returns a matrix of sorted data on a row-by-row basis; `all(diff())` then returns a logical array of those rows that have `diff()==0` (i.e., one or more duplicate value). This is then used as a logical index into the original data rows.

We transposed A before sorting, because by default *sort, diff,* and *all* act on columns rather than rows. This is standard for most internal MATLAB functions that act on vectorized data.* Luckily, most functions have optional input parameters that enable us to change this default behavior, such that they act on a different dimension. So, instead of transposing, we can tell these functions to act on the rows dimension:

```
A = A(all(diff(sort(A,2),[],2),2), :)
```

When removing the rows with duplicate data, we chose above to *copy the valid* rows back into A. When the data is large and the number of removed rows is expected to be small, then it might be more efficient to *remove the invalid* rows instead:

```
A(~all(diff(sort(A,2),[],2),2), :) = [];
```

5.3.3 Thresholding

In this example, we wish to limit the data values to a specified range of allowed values, clamping extreme values to the specified limits:

```
% Slow, standard version
for row = 1 : size(matrix,1)
    for col = 1 : size(matrix,2)
        dataValue = matrix(row,col);
        if dataValue > maxValue
            matrix(row,col) = maxValue;
        elseif dataValue < minValue
            matrix(row,col) = minValue;
        end
    end
end
```

In the vectorized version, we use logical indexing to set all extreme matrix elements to the corresponding limit, leaving all other data elements unchanged:

```
% Fast vectorized version #1
matrix(matrix>maxValue) = maxValue;
matrix(matrix<minValue) = minValue;
```

In this specific case, we can make the code even more compact, by combining the two MATLAB statements above into a single statement:[784]

---

* The underlying reason is that MATLAB's data is stored column-wise in memory — see §9.3.2 and §5.2.3.

```
% Fast vectorized version #2
matrix = min(max(matrix,minValue), maxValue);
```

Some programmers consider this more cumbersome and less maintainable; others actually like more compact code. In terms of performance, there is no discernible difference between these variants.

As a variant of this example, we can test whether any data value exceeds a limit:

```
flag = any(matrix>maxValue | matrix<minValue);
flag = ~all(matrix<=maxValue & matrix>=minValue); % equivalent
```

5.3.4 Cumulative Sum

The next example (modified cumulative sum) shows that quite often we can vectorize our code by the simple act of removing the *for* keyword (and its corresponding *end*):

```
% Slow, standard version
N = 1e6;
x(1) = 1;
for idx = 1 : N-1
    x(idx+1) = x(idx) + N - idx;
end

% Fast version (80x faster)
N = 1e6;
x(1) = 1;
idx = 1 : N-1; % simply remove the 'for' keyword
x(idx+1) = N - idx;
x = cumsum(x);
```

Here is another example of using *cumsum*, this time to attach unique values to index ranges, for example, value 1 to array indexes 1–25, value 2 for indexes 26–50, and so on:[785]

```
ind = [1,25,50,70,100,110]; % index boundaries
value = [1 2 4 6 10];        % index groups values
data = [ind(1) zeros(1,ind(end)-1)];
data(ind(2:end-1)+1) = diff(value);
data = cumsum(data);
```

The following code segment provided a 18800x (!) speedup when vectorized:[786]

```
% Slow original version: right hand side (rhs) of Riemann Hypothesis
rhs = zeros(1,1e5); % preallocate - see §9.4.3
for i = 1:1e5
    harmonicsum = 0;
    for j = 1:i
        harmonicsum = harmonicsum + 1/j; % compute harmonic sum
    end
    rhs(i) = harmonicsum + log(harmonicsum) * exp(harmonicsum);
end
⇨ Elapsed time is 72.890128 seconds.

% Vectorized version using cumsum (18800x faster!)
harmonicsum = cumsum(1./(1:1e5));
rhs = harmonicsum + log(harmonicsum) .* exp(harmonicsum);
⇨ Elapsed time is 0.003882 seconds.
```

5.3.5 Data Binning

Given two vectors x,y of integer values in the range 1..N, let's determine the number of occurrences of each combination of values and assign it to the correlation matrix L. For example, L(12,34) will contain the number of times that x==12 and y==34 occurred in corresponding vector positions:[787]

```
% generate some test data
N=100; x=randi(N,1,1e5); y=randi(N,1,1e5);   % 100K data values 1-100

% Slow, brute-force version
L = zeros(N,N);
for i = 1 : N
    for j = 1 : N
        %value = find(x == i & y == j);
        %L(i,j) = length(value);
        % This is a slightly faster alternative to the 2 lines above
        L(i,j) = sum(x == i & y == j);
    end
end
⇨ Elapsed time is 6.451564 seconds.

% Vectorized version using histc (550x faster!)
X = sub2ind([N,N],x,y);
H = histc(X,min(X)-0.5:max(X)+0.5);
L = zeros(N,N);
L(min(X):max(X)) = H(1:end-1);
⇨ Elapsed time is 0.011752 seconds.

% Vectorized version using accumarray (2000x faster!!!)
X = sub2ind([N,N],x,y);
L = reshape(accumarray(X',1), N, N);
⇨ Elapsed time is 0.003239 seconds.

% Vectorized version using arrayfun (slower than the JIT'ed loops!)
[i,j] = meshgrid(1:N,1:N);
L = arrayfun(@(ii,jj) sum((x==ii).*(y==jj)), i, j)';
⇨ Elapsed time is 7.493849 seconds.
```

This example shows that *histc* and *accumarray* can be enormously effective at vectorized data binning (550× and 2000× speedups, respectively). Yet, I have a personal feeling that for some reason they are relatively underutilized by the MATLAB community.[788] Perhaps the somewhat opaque documentation has something to do with it.[789] I hope the speedups illustrated here will drive readers to start using them. For readers interested in exploring *accumarray* and *histc*, I suggest reviewing Bruno Luong's snippets here:[790] and Vincent Odongo's snippet here:[791]

Another interesting aspect of this example is that the **fun* functions (in this case, *arrayfun*, §5.2.4.1) do not always provide any speedup compared to regular JIT'ed loops!

In addition to its use for binning, *accumarray* can also be used for very efficient sparse-data multiplication,[792] although the reason may not be immediately obvious.

5.3.6 Using *meshgrid* and *bsxfun*

Consider the case of generating the Euclidean (2-) norm of an input numeric matrix, based on each element's row and column index:

```
% Slow, standard version
rows = 10; cols = 20;
for row = 1 : rows
    for col = 1 : cols
        A(row,col) = sqrt(row^2 + col^2);
    end
end
```

The vectorized version uses *meshgrid* to generate two temporary arrays, colsGrid and rowsGrid. These matrices include values corresponding to the column/row index, respectively, of each element. We can square and combine them to get the requested vectorized result:

```
% Vectorized version #1
[colsGrid,rowsGrid] = meshgrid(1:cols, 1:rows);
A = sqrt(colsGrid.^2 + rowsGrid.^2);
```

Instead of using vectorized squaring, addition, and *sqrt*'ing, we could use *bsxfun* by applying the *hypot* (2-norm) binary function on all corresponding elements of our temporary arrays, colsGrid and rowsGrid:

```
% Vectorized version #2
[colsGrid,rowsGrid] = meshgrid(1:cols, 1:rows);
A = bsxfun(@hypot, colsGrid, rowsGrid);
```

This second vectorized variant is slower than the simpler version #1, and in fact is comparable to the original (nonvectorized) version in terms of performance. This example shows that different ways of vectorizing the algorithm can achieve significantly different performance results, and in some cases even slowdowns, compared to nonvectorized code.

The topic of Euclidean distances is further discussed in §5.3.8 below. An additional example of vectorization using *meshgrid* and *bsxfun* can be found here:[793]

5.3.7 A *meshgrid* Variant

As a variation of the previous example, let us plot the function $f(x,y) = x^* \exp(-x^2 - y^2)$:

```
% Slow, standard version for f(x,y)=x*exp(-x2-y2)
x = -2 : 0.1 : 2;
y = -2.5 : 0.1 : 2.5;
for row = 1 : length(x)
    for col = 1 : length(y)
        f(row,col) = x(row) * exp(-x(row)^2 - y(col)^2);
    end
end

% Vectorized version #1
[colsGrid,rowsGrid] = meshgrid(x,y);
f = colsGrid.*exp(-colsGrid.^2 -rowsGrid.^2);

% Vectorized version #2 (fastest): f(x,y)=x*exp(-x2)*exp(-y2)
[colsGrid,rowsGrid] = meshgrid(x,y);
f = ((x'.*exp(-x'.^2)) * exp(-y.^2))';

% Plot the data on the mesh-grid
surf(colsGrid,rowsGrid,f);
```

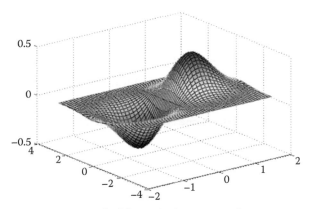

meshgrid vectorized usage example

Additional examples of using *meshgrid* can be found in the references.[794] *meshgrid* can create 2D and 3D matrices; for higher dimensions, we can use *ndgrid*.

5.3.8 Euclidean Distances

Given two sets A,B of coordinates (x,y,z), let's find the Euclidean distance r from each point in A to each point in B:[795]

```
% Common problem definition/initialization
A = rand(1000,3);  % 1000 points
B = rand(3000,3);  % 3000 points
sizeA = size(A,1); % # of points (rows) in A = 1000
sizeB = size(B,1); % # of points (rows) in B = 3000

% Slow, standard version
r = zeros(sizeA,sizeB); % preallocate (see §9.4)
for idxA = 1 : sizeA
    for idxB = 1 : sizeB
        coordDiff = A(idxA,:) - B(idxB,:);       % 3-element array
        r(idxA,idxB) = sqrt(sum(coordDiff.^2));
    end
end
⇨ Elapsed time is 4.622756 seconds.

% Vectorized version using repmat (20x faster)
tempA = repmat(A,sizeB,1);
tempB = repmat(B,1,sizeA);
tempB = reshape(tempB',3,sizeA*sizeB)';
r = reshape(sqrt(sum((tempA-tempB).^2,2)),sizeA,sizeB);
⇨ Elapsed time is 0.238358 seconds.
```

The vectorized version using *bsxfun* is even faster than this (this is often the case when *bsxfun* replaces *repmat* in vectorized code):*

```
% Vectorized version using bsxfun (100x faster)
c1 = bsxfun(@minus, A(:,1), B(:,1)');
```

---

* http://blog.accelereyes.com/blog/2010/04/05/converting-matlab-loops-to-gpu-code   (or: http://bit.ly/UfDAsF). The purpose of this webpage was to show that the *bsxfun* implementation on GPU using Jacket is the fastest alternative. But we can see that even using regular MATLAB *bsxfun* we gain significant speedup (×4 vs. *repmat,* ×170 vs. loops). Jacket is described in §7.1.

```
c2 = bsxfun(@minus, A(:,2), B(:,2)');
c3 = bsxfun(@minus, A(:,3), B(:,3)');
r = sqrt(c1.^2 + c2.^2 + c3.^2);
⇨ Elapsed time is 0.045484 seconds.
```

This can be further reduced into a one-liner but it turns out to be less effective, possibly due to the extra matrix reshaping that is required for full vectorization:[796]

```
r = squeeze(sqrt(sum(bsxfun(@minus,A,reshape(B.',1,3,[])).^2,2)));
⇨ Elapsed time is 0.080987 seconds.
```

This is a good example for the fact that sometimes it is better NOT to fully vectorize a problem, in order to eliminate some unnecessary calculations and/or to take advantage of some JIT optimizations.*

5.3.9 Range Search

Given a set X of coordinates (x,y,z,…), let's find all points within a certain Euclidean distance r from position c:[797]

```
% Common problem definition/initialization
nDims = 2;                    % number of dimensions
nPoints = 10000;             % number of points
X = rand(nPoints,nDims);     % random data points
c = [0.5 0.5];               % center of search
r = 0.2;                     % Radius of search

% Slow, brute-force version
[nPoints,nVariables] = size(X);
fidx = false(1,nPoints);
dist = zeros(1,nPoints);
r2 = r*r;
for k = 1 : nPoints
    x = X(k,:) - c;
    s = 0;
    for d = 1 : nVariables
        s = s+x(d)^2;
    end
    if s < r2
        dist(k) = sqrt(s);
        fidx(k) = true;
    end
end
idx = find(fidx);
dist = dist(fidx);
⇨ Elapsed time is 0.012760 seconds.

% Partially vectorized version by Yi Cao (60x faster)
nVariables = numel(c);
s = 0;
for d = 1 : nVariables
    s = s + (X(:,d)-c(d)).^2;
end
```

---

* For a simple variant of the Euclidean distance problem using *bsxfun*, see http://stackoverflow.com/questions/20280157/in-matlab-find-the-distance-for-every-matrix-element (or: http://bit.ly/1bvLJBt).

```
fidx = s < r*r;
idx = find(fidx);
dist = s(fidx);
⇨ Elapsed time is 0.000211 seconds.

% Fully vectorized, less efficient than partially-vectorized version
%s = sum((X-repmat(c,nPoints,1)).^2, 2);  % less efficient than bsxfun
s = sum(bsxfun(@minus,X,c).^2, 2);
fidx = s < r*r;
idx = find(fidx);
dist = s(fidx);
⇨ Elapsed time is 0.000358 seconds.

% Visualization
t = 0 : 0.02 : 2*pi;
x = c(1) + r*cos(t);
y = c(2) + r*sin(t);
plot(X(:,1),X(:,2),'b.', X(idx,1),X(idx,2),'go', ...
     c(1),c(2),'m+',       x,y,'r-','linewidth',2);
```

Range search vectorized example

5.3.10 Matrix Computations

Given an N×K matrix A and N×P matrix B, compute the N×P matrix $B - A \cdot f(A,B)$:[798]

```
% Sample data
N=300; K=400; P=500;
A = rand(N, K);
B = rand(N, P);

% Slow, brute-force version (can be improved by LIH etc.)
X(N,P) = 0; % preallocate space
for p = 1:P
    for n = 1:N
        for k = 1:K
            gamma(p,k) = dot(A(:,k),B(:,p))/sum(A(:,k).^2);
        end
        X(n,p) = B(n,p) - dot(gamma(p,:),A(n,:));
    end
```

```
end
⇨ Elapsed time is 0.913815 seconds.

% Vectorized version: 220x faster and much easier to read/maintain
X = B - A * bsxfun(@rdivide, B.'*A, sum(A.^2)).';
⇨ Elapsed time is 0.004160 seconds.
```

5.4 Repetitive Data*

Replicating data is an important component of vectorization. We can very often trade memory for performance, by expanding (tiling) data in certain ways, then acting on it in some vectorized manner (e.g., linear algebra computations and transformations), in order to get our desired output. In fact, the notion of data replication for vectorization is used so often that it deserves special treatment, which is the purpose of this section.

As explained briefly in §5.2.1, *repmat* can be used to replicate (tile) data along any dimension. The implementation of *repmat* until R2013a (MATLAB 8.1) basically consisted† of using direct indexing techniques to extend the data, as shown in §5.1.2. Until R2013a, using direct indexing was generally faster than using *repmat*, since it avoided the function-call overhead and all its internal checks. In R2013b (MATLAB 8.2), when *repmat* was reimplemented as a built-in (compiled C-code) function, this may no longer be true, and needs to be checked on a case-by-case basis.

Here are a few sample usages of *repmat*, and their equivalent direct indexing techniques. In this example, we assume that x is a column vector:

```
>> x = (1:5)' % 5x1 array
x =
     1
     2
     3
     4
     5
```

repmat Command	Equivalent Direct Indexing Command	Resulting Data Size
%tile horizontally repmat(x, [1 3]) repmat(x, 1, 3)	x(:, [1,1,1]) x(:, ones(1,3))	5 x 3
%tile vertically repmat(x, [3 1]) repmat(x, 3, 1)	idx = [1 : size(x,1)]'; x(idx(:, [1,1,1]), :)	15 x 1
	idx = [1:size(x,1)]' * [1,1,1]; x(idx, :)	

Note that when we specify a multidimensional array as the index into an array (as in the last line of the table above), MATLAB will usually convert it to a column vector (i.e., automatically convert idx into idx(:)). The exception to this rule is when no other index is provided, in which case the result size will be the same as the idx size:

---

* This section is based in part on Vincent Cheung's Quick Reference webpage (http://vincentcheung.ca/research/matlabindexrepmat.html or: http://bit.ly/1nGTixb).

† Disregarding the sanity checks, inputs verifications, edge cases, and so on.

```
>> x(idx) % 5x3 data, same size as idx
ans =
      1       1       1
      2       2       2
      3       3       3
      4       4       4
      5       5       5
>> x(idx,:)' % 15x1 column array, since idx is only one of two dims
ans =
      1
      2
      3
      4
      5
      1
      2
    ...
```

Similarly, for multidimensional data:

```
>> A = reshape(1:12, 4, 3)' % 3x4 array
A =
      1       2       3       4
      5       6       7       8
      9      10      11      12
```

repmat Command	Equivalent Direct Indexing Command	Resulting Data Size
	`A([1 2 3 1 2 3],` `[1 2 3 4 1 2 3 4 1 2 3 4])`	
	`A([1:end 1:end],` `[1:end 1:end 1:end])`	
`repmat(A, [2 3])`	`rowIdx = [1 : size(A,1)]';` `colIdx = [1 : size(A,2)]';` `A(rowIdx(:, ones(2,1)),` `colIdx(:, ones(3,1)))`	6 x 12
	`A([1:size(A,1)]' * ones(1,2),` `[1:size(A,2)]' * ones(1,3))`	
`repmat(A, [1 1 3])`	`A(:, :, ones(3, 1))`	3 x 4 x 3

In this table, the direct indexing equivalents use successively more intelligent constructs. In the third alternative:

```
rowIdx = [1:size(A,1)]';
colIdx = [1:size(A,2)]';

repmatRowIdx = rowIdx(:, ones(2,1));
repmatColIdx = colIdx(:, ones(3,1));

A(repmatRowIdx(:), repmatColIdx(:))
```

However, since we are indexing into multiple dimensions, MATLAB will automatically change `repmatRowIdx` into `repmatRowIdx(:)` and `repmatColIdx` into `repmatColIdx(:)`, so we can simplify the above:

```
A(repmatRowIdx, repmatColIdx)
```

In the fourth alternative, we use matrix multiplication to obtain the array indices in order to do everything in a single line of code:

```
A([1:size(A,1)]' * ones(1,2),
  [1:size(A,2)]' * ones(1,3))
```

5.4.1 A Simple Example

Consider the following problem: we have a numeric matrix A and wish to normalize each row independently, so that each row sums to 1. We can loop over all data rows, sum the values in each row, and then divide each element in that row with its total:

```
A = rand(1e6,3); % 1Mx3 test data

% loop over all rows
for row = 1 : size(A, 1)
    % sum all elements in the "i"th row
    total = 0;
    for col = 1 : size(A, 2)
        total = total + A(row, col);
    end

    % divide all elements in the "i"th row by the row's total
    for j = 1 : size(A, 2)
        A(row, col) = A(row, col)/total;
    end
end
⇨ Elapsed time is 0.128452 seconds.
```

The vectorized (loop-less) version would look something like this:

```
totals = sum(A, 2);
A = A ./ repmat(totals, [1 size(A,2)]);
⇨ Elapsed time is 0.022395 seconds.
```

This requires fewer lines of code, is more concise, is arguably easier to read, and most importantly, runs faster than the looped version. The basic idea is that we compute the column vector `totals` in a vectorized manner, then tile it sideways (horizontally, in the columns dimension) so that it becomes a matrix of the same size as A, and then we divide all elements in A by their respective elements in the expanded totals matrix.

On R2013a (MATLAB 8.1) and earlier, when *repmat* was implemented as a simple m-file, it was sometimes faster to use direct indexing instead:*

```
A = A ./ totals(:, ones(size(A,2), 1));
```

And similarly:

```
% normalize the columns
totals = sum(A, 1);
A = A ./ totals(ones(size(A,1), 1), :)

% normalize a multi-dimensional array over one of the dimensions
totals = sum(C, 3);
C = C ./ totals(:, :, ones(3,1));

% normalize over multiple dimensions
```

---

* The difference is inconsequential or nonexistent in this specific case.

```
totals = sum(sum(D, 3), 4);
D = D ./ totals(:, :, ones(3,1), ones(5,1));
```

5.4.2 Using *repmat* Replacements

The specific example of rows-normalization could be made even faster and simpler with *bsxfun*, which expands the singleton dimension (=columns) of totals without actually allocating the memory for it:

```
A = bsxfun(@rdivide,A,sum(A,2));
⇨ Elapsed time is 0.017764 seconds.
```

bsxfun provides faster and simpler equivalents to *repmat*, although admittedly it is sometimes more difficult to find the correct formatting of parameters.*

In the examples above, we have used the *ones* command. It is faster to use the dedicated built-in functions *ones, zeros, inf, nan, true*, and *false* to create repetitions of 1, 0, inf, NaN, true, and false (respectively), than to use *repmat* for the same effect:

```
data = repmat(0,3,5); % 3x5 matrix of zeros
data = zeros(3,5);     % equivalent, faster
```

In fact, if we attempt to use the *repmat* variant in our code, the MATLAB Editor's Code Analyzer (MLint) will issue a corresponding warning (RPMT0, RPMT1, RPMTI, RPMTN, RPMTT, and RPMTF, respectively):[†]

```
data = repmat(0,3,5);
```
⚠ Consider replacing repmat(0,x,y) with zeros(x,y) for better performance. [Fix]

Direct indexing is faster than *repmat* in R2013a and earlier (before *repmat* became built-in), but may be faster even in R2013b for small data, due to function-call overheads:[799]

```
x=(1:3)'; for idx=1:1e5, a=repmat(x,1,10); end
⇨ Elapsed time is 0.239108 seconds.
```

```
x=(1:3)'; y=ones(1,10); for idx=1:1e5, a=x(:,y); end
⇨ Elapsed time is 0.073770 seconds.
```

This is a contrived example that highlights the overheads in *repmat* compared to core indexing. But it illustrates that for those cases when profiling reports that a *repmat* call is a performance hotspot, direct indexing or *bsxfun* may be the solution.

Another alternative is to use the outer product, but this is often slower:[‡800]

```
x=(1:3)'; y=ones(1,10); s=size(x).*size(y);
for idx=1:1e5, a=reshape(x(:)*y,s); end
⇨ Elapsed time is 0.215145 seconds.
```

Finally, consider using the Lightspeed toolbox (§8.5.6.1), which includes a MEX-based direct replacement for *repmat*. In my tests, Lightspeed's *repmat* was 3× faster than MATLAB's version on R2013a and earlier, although a bit slower on R2013b.

---

* See §5.2.4.2. Note that since *repmat* became built-in in R2013b, *bsxfun* is often *slower* than *repmat*.

† The complete list of MLint-reported performance-related warnings is presented in Appendix A.4.

‡ The speedup effect compared to *repmat* is more evident in R2013a and earlier. In some cases, the matrix multiplication here could be made faster using the *mtimesx* utility (see §4.5.2). See §5.2.4.2 for additional aspects of the outer product.

5.4.3 Repetitions of Internal Elements

repmat (or internal indexing) can be used to replicate an entire data block in any (or several) dimension(s). Consider the following situation:

```
>> data = [0,1,-2; 3,-4,-5]
data =
       0      1     -2
       3     -4     -5

>> Rows=2;  Cols=3;  repmat(data,Rows,Cols)
ans =
       0      1     -2      0      1     -2      0      1     -2
       3     -4     -5      3     -4     -5      3     -4     -5
       0      1     -2      0      1     -2      0      1     -2
       3     -4     -5      3     -4     -5      3     -4     -5
```

Now what if we need the internal elements to be replicated in sequence, rather than the entire data block as is? A replication of the data can achieve the desired effect by reshaping the data appropriately:[801]

```
[m,n] = size(data);
data = reshape(data,1,m,1,n);
data = repmat (data,R,1,C,1);
data = reshape(data,m*R,n*C)
data =
       0      0      0      1      1      1     -2     -2     -2
       0      0      0      1      1      1     -2     -2     -2
       3      3      3     -4     -4     -4     -5     -5     -5
       3      3      3     -4     -4     -4     -5     -5     -5
```

We can use *repmat* to replicate *indexes* rather than *data*, to achieve the same effect:[802]

```
rowIdx = repmat(1:m,R,1);
colIdx = repmat(1:n,C,1);
data = data(rowIdx,colIdx);
```

The idea here is that the data is indexed by [1,1, 2,2] in the rows dimension, thereby duplicating row #1, then row #2, and so on.* Similarly, the data is indexed in [1,1,1, 2,2,2, 3,3,3] in the columns dimension. We can achieve the same result with direct indexing, as follows:

```
data = data([1,1, 2,2], [1,1,1, 2,2,2, 3,3,3]);
```

Or, more generally:[803]

```
rowIdx = ceil((1:R*m)/R);
colIdx = ceil((1:C*n)/C);
data = data(rowIdx,colIdx);
```

A more sophisticated solution relies on matrix multiplication, or more precisely, the Kronecker (tensor, or outer) product:[804]

```
>> data = kron(data,ones(2,3));
```

---

* repmat(1:m,R,1) generates a 2D (m x R) matrix, but when used as an index into data, it is converted into a 1D vector (see §5.4.0).

As noted in §5.2.4.2, *kron* uses *bsxfun* for efficient computation of the tensor product for full matrices. In this case, we could slightly improve performance by saving some function-call overheads:*

```
A = reshape(data,[1,2,1,3]);
B = ones(2,1,3);
data = reshape(bsxfun(@times,A,B),4,9);
```

The use of *kron* for replicating matrix elements is discussed in detail in Acklam's vectorization guide[805] (also see §5.7.15 below in this book).

Another (N – D) approach is used by MATLAB veteran Matt Fig's *expand* utility:[806]

```
S = [R,C];
SA = size(data);           % Get input size +number of dimensions
for ii = length(SA):-1:1
    H = zeros(SA(ii)*S(ii),1);% One index vector into A for each dim
    H(1:S(ii):SA(ii)*S(ii)) = 1; % Put ones in correct places
    T{ii} = cumsum(H);     % Cumsumming creates the correct order
end
B = A(T{:});% Feed the indices into A.
```

When comparing *expand*'s direct-indexing approach to *kron*, direct indexing is faster and uses less memory than *kron* on R2013a and earlier.[807] The results on R2013b and later are less conclusive,† and need to be tested on a case-by-case basis.

Yet another File Exchange utility that we can use is Mark Ruzon's *replicate* utility,[808] which uses *ndgrid* to generate the necessary direct indexing for 3D expansions. I do not believe that it provides much benefit over *kron, bsxfun,* or *expand* in terms of performance or memory usage, but I have not tested this thoroughly.

5.5 Multidimensional Data

Multidimensional (N – D) arrays are supported in MATLAB and are usually efficient. However, for many if not most built-in MATLAB functions, using MATLAB's native data type of a 2D array is even faster.‡ If possible, check whether the data can be represented by a 2D matrix rather than an N – D array. We can use the built-in functions *reshape, transpose,* and *permute* to transform our data, if necessary.§ In many cases, smart application of these three functions can transform data without resorting to costly cell arrays or iterative element-wise loops.[809] Even when converting the data into 2D is impractical, reducing the dimensionality could still improve performance, due to simplified indexing and improved *spatial locality.*¶

---

* This would improve performance significantly on R2013a and earlier, where the *bsxfun* trick was still not incorporated into *kron*'s implementation (see §5.2.4.2).
† Owing to *kron*'s improved implementation on R2013b.
‡ See §5.2.3.1.
§ The performance of *permute* for 3D+ arrays has been improved in R2013a (MATLAB 8.1), making this technique even more cost-effective. See http://mathworks.com/help/matlab/release-notes.html#btsiwqu-2 (or: http://bit.ly/12hoEK4)
¶ See §9.3.2 and http://stackoverflow.com/questions/24956776/performance-penalty-for-4d-arrays (or: http://bit.ly/Y8ExIt)

Here is a relatively simple example, taken from my *ScreenCapture* utility on the File Exchange:[810] The basic idea of this utility is to enable taking automated (programmatic) or interactive screenshots of any GUI control, entire figure window, or specified screen area from within MATLAB. The underlying mechanism uses the Java Robot class[811] to take the screenshot, and then converts the resulting int32 integer data into an M×N×3 3D matrix (width, height, RGB index). During the course of its development, this data conversion has taken three distinct forms, with ever-increasing performance. These variants were incorporated in an official MathWorks technical article explaining how to convert Java images into MATLAB images:[812]

```
% Use java.awt.Robot to take screen-capture the specified screen area
rect = java.awt.Rectangle(pos(1), pos(2), pos(3), pos(4));
robot = java.awt.Robot;
jImage = robot.createScreenCapture(rect);
h = jImage.getHeight;
w = jImage.getWidth;

% Convert the Java image into a MATLAB image - variant 1 (slowest)
imgData = zeros([h,w,3],'uint8');
pixelsData = uint8(jImage.getData.getPixels(0,0,w,h,[]));
for row = 1 : h
    base = (row-1)*w*3+1;
    data = pixelsData(base:(base+3*w-1));
    imgData(row,1:w,:) = deal(reshape(data,3,w)');
end
⇨ Elapsed time is 0.034534 seconds.

% Variant 2: 3x speedup based on feedback from Urs Schwartz
jImageData = jImage.getData.getDataStorage;
pixelsData = reshape(typecast(jImageData,'uint32'),w,h).';
imgData(:,:,3) = bitshift(bitand(pixelsData,256^1-1),-8*0);
imgData(:,:,2) = bitshift(bitand(pixelsData,256^2-1),-8*1);
imgData(:,:,1) = bitshift(bitand(pixelsData,256^3-1),-8*2);
⇨ Elapsed time is 0.012237 seconds.

% Variant 3: Another 2x speedup based on feedback from Jan Simon
pixelsData = reshape(typecast(jImageData, 'uint8'), 4, w, h);
imgData = cat(3, transpose(reshape(pixelsData(3, :, :), w, h)),...
                 transpose(reshape(pixelsData(2, :, :), w, h)),...
                 transpose(reshape(pixelsData(1, :, :), w, h)));
⇨ Elapsed time is 0.006068 seconds.
```

Note: This could be made even faster using James Tursa's in-place *typecastx* utility.[813]

One expert in multidimensional transformations is Bruno Luong, who has posted numerous code snippets utilizing them.[814] Here is a typical example to compute an autocorrelation matrix of 3D input data:[815]

```
% Generate data
noise = 3*randn(64,64,100);
cc = bsxfun(@plus, peaks(64), noise); % 64 x 64 x 100
cellsize = 16;
[m n p] = size(cc);
for r = 1 : p
    bb(:,:,r) = mat2cell(cc(:,:,r),...
```

```
                                    cellsize*ones(1,64/cellsize),...
                                    cellsize*ones(1,64/cellsize));
end
r = p;

% Non-vectorized method
C1 = zeros(4,4,r,4,4,r);
for ir=1:4
    for ic=1:4
        for is=1:r
            for jr=1:4
                for jc=1:4
                    for js=1:r
                        temp = cellfun(@(x,y) x.*y,  ...
                                       bb(ir,ic,is), ...
                                       bb(jr,jc,js), ...
                                       'UniformOutput', false);
                        temp1 = cell2mat(temp);
                        temp2 = mean(temp1(:));
                        C1(ir,ic,is,jr,jc,js) = temp2;   % 6-D
                    end
                end
            end
        end
    end
end
⇨ Elapsed time is 181.4119 seconds.

% Vectorized method (~3000x faster!)
mm = m/cellsize; % =4
nn = n/cellsize; % =4
b = reshape(cc, [cellsize mm cellsize nn p]);
b = permute(b, [1 3 2 4 5]); % 16 x 16 x 4 x 4 x 100
b = reshape(b, cellsize^2, [])/cellsize; % 256 x 1600
corr = b'*b; % 1600 x 1600
C2 = reshape(corr, [mm nn p mm nn p]); % 6-D
⇨ Elapsed time is 0.062182 seconds.
```

In this example, the naïve approach is to process the 3D input cell array bb (where each element is actually a 16×16 2D matrix) one cell element at a time, correlating it with every cell element in the same 3D input array bb to get a resulting 6D data array.

The vectorized approach is almost 3000× faster for an input size of 64×64×100, and linearly faster for larger data sizes. The idea is to reshape the data into a 256×1600 2D matrix, which is then multiplied by itself using vectorized matrix multiplication to get a 2D correlation matrix. The result is then reshaped into the expected 6D format.

The tricky part is to *permute* the input data's dimensions in such a way that the reshaping operation does not affect the computation results. In our case, two separate *reshapes* and a switching of the second and third dimensions are done before the correlation matrix is actually computed.

While this vectorization example may indeed seem difficult to understand at first, the performance benefits are self-evident. When dealing with multidimensional data, learning the tricks of *reshape* and *permute* is often very cost-effective.

A related idea that might be worth pointing out is that while MATLAB usually includes the capability to operate over specific dimensions of an array, it can be faster to *transpose* or *permute* the data first, then perform the operation on the contiguous dimension, then *permute* back.*

A detailed analysis of multidimensional data manipulation performance was provided in the 2002 paper *"Fast manipulation of multidimensional arrays in Matlab"* by Kevin Murphy of MIT's AI Lab.[816] Keep in mind that this paper is relatively old (MATLAB 7 was only released in 2005); MATLAB underwent many changes since then. Still, some of Murphy's insights are still applicable today. Anyone who extensively uses multidimensional arrays would do wisely to study this paper.

5.6 Real-Life Example: Synthetic Aperture Radar Matched Filter[†]

In this section, we shall start our real-life "running" example to illustrate how the proper usage of vectorization techniques can boost the performance. Our criteria for selecting a running example were based on the observation that the target algorithm should be computationally intensive and relatively complex to enable both task-level and data-level parallelism (and, thus, not giving preference to any particular parallel architecture). At the same time, the algorithm should be quite compact if written in MATLAB. We shall revisit this example several times in Chapters 6 and 7.[‡]

5.6.1 Naïve Approach

Matched filter is a technique to process raw synthetic aperture radar (SAR)[817] phase history data to obtain images useful for target location. SAR systems consist of a moving platform (such as an aircraft or satellite) with a mounted antenna, which repeatedly illuminates pulses of radio waves. These waves are echoed from a target scene and received by another antenna in a superimposed form. The purpose of SAR image formation algorithms is to compress this target information in range (frequency) and along-track (azimuth) directions to obtain interpretable 3D images. This enables images of the target area even under cloud cover or in darkness.

For details on the matched filter algorithm, refer to the paper "SAR image formation toolbox for MATLAB" by L.A. Gorham and L.J. Moore from the US Air Force Research Lab (AFRL).[818] Description of the relevant data fields is provided in the table below:

Field Name	Size	Description
Np	scalar	The number of azimuth points
K	scalar	The number of frequency samples per point
deltaF	scalar	Step size of frequency data (Hz)
minF	Np×1	Vector containing start frequency for each pulse (Hz)
x_max,y_max,z_max	Sx×Sy	The x, y, z positions of each pixel (m)
AntX,AntY,AntZ	Sx×Sy	The x, y, z positions of the sensor at each pulse (m)
R0	Np×1	The range to scene center (m)
phdata	K×Np	Phase history data (frequency domain)
im_final	Sx×Sy	The complex image value at each pixel

---

* See §9.3.2.

[†] This section was authored by Pavel Emeliyanenko of Max-Planck Institute for Informatics (http://mpi-inf.mpg.de/~emeliyan).

[‡] The source files for all variants of this running example throughout the book can be downloaded from the book's webpage: http://UndocumentedMatlab.com/books/matlab-performance (or: http://bit.ly/1pKuUdM).

The initial (naïve) implementation is the direct translation of a "C-style" algorithm to MATLAB. Certainly, we do not believe that somebody will really write such a code in MATLAB. Instead, our intention is to demonstrate that just by rewriting the code with looping constructs to utilize MATLAB's inherent ability to work with arrays, one can already achieve very good performance improvement.[819]

```
function data = matched_filter(data)
    % define speed of light
    c = 299792458;

    % determine the size of the phase history data
    K  = size(data.phdata,1);   % number of frequency bins per pulse
    Np = size(data.phdata,2);   % number of pulses

    % initialize the image with all zero values
    data.im_final = zeros(size(data.x_mat));

    % loop through every pulse
    for ii = 1 : Np

        % get antenna location/distance to scene center for this pulse
        AntX = data.AntX(ii); AntY = data.AntY(ii);
        AntZ = data.AntZ(ii); R0 = data.R0(ii);

        % loop through all pixels of an image
        for j1 = 1 : size(data.x_mat,1)
            for j2 = 1 : size(data.x_mat,2)

                % calculate differential range for each image pixel
                dR = sqrt((AntX-data.x_mat(j1,j2))^2 + ...
                          (AntY-data.y_mat(j1,j2))^2 + ...
                          (AntZ-data.z_mat(j1,j2))^2) - R0;

                % accumulate frequency samples for each pixel
                pix = 0;
                % start from the minimum frequency for this pulse
                freq = data.minF(ii);

                % perform the Matched filter operation
                for jj = 1 : K
                    phdata = data.phdata(jj,ii);
                    pix = pix + phdata * exp(1i*4*pi*freq/c*dR);
                    freq = freq + data.deltaF;
                end

                % update the pixel value
                data.im_final(j1,j2) = data.im_final(j1,j2) + pix;
            end % for j2
        end % for j1
    end % for ii = 1:Np
end
```

For experiments, we have taken a sample SAR data available for download from AFRL's Sensor Data Management System (SDSM) website.* The phase history data is stored in raw

---

* To obtain the sample data, complete a free registration at the website https://www.sdms.afrl.af.mil. Then, from the *"SDMS Public Data Products"* section, download *MSTAR Public Targets Part 1* (about 100 MB). Unpack the archive and go to a subdirectory "TARGETS\TEST\15_DEG\ T72\SN132". In this directory, each file is a part of X-band imagery data (T-72 target) that can be used for testing.

binary format consisting of 128 azimuth points with 128 frequency samples each of which results in a 128×128 complex floating-point image.

Throughout this section, we shall use the following machine for benchmarks:

- 4x Intel® Xeon® CPU E7-4860 (40 cores/80 hyperthreads)
- 256 GB RAM (32×8 GB)
- 4x NVidia Tesla S2050 (GF100)
- MATLAB R2012a running on a 64-bit Linux platform

On this platform, the matched filter algorithm requires more than **20 minutes (!)** to reconstruct the following 500×500 image:

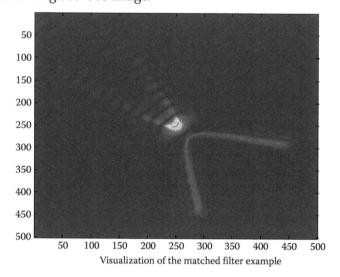

Visualization of the matched filter example

Such a long running time is not surprising since our implementation, consisting of *four* nested loops, is embarrassingly inefficient. The reason for this is because MATLAB's core is designed for high-performance vector and matrix operations while the looping constructs are not particularly efficient. Indeed, from the task manager, we can see that our algorithm utilizes only a *single* CPU core out of 40.

5.6.2 Using Vectorization

Using the techniques of implicit parallelization discussed in the preceding subsections, we can rewrite our matched filter algorithm to eliminate most looping constructs, thus taking advantage of code vectorization:

```
%%%%%%%%%%%%%%%%%%%%%%%%%%%%%%%%%%%%%%%%%%%%%%%%%%%%%%%%%%%%%%%%%%%%%%%%%%%
% Written by LeRoy Gorham, Air Force Research Laboratory, WPAFB, OH
% Email: leroy.gorham@wpafb.af.mil
% Date Released: 8 Apr 2010
% Gorham, L.A. and Moore, L.J., "SAR image formation toolbox for
% MATLAB", Algorithms for Synthetic Aperture Radar Imagery XVII
% 7669, SPIE (2010).
%%%%%%%%%%%%%%%%%%%%%%%%%%%%%%%%%%%%%%%%%%%%%%%%%%%%%%%%%%%%%%%%%%%%%%%%%%%
function data = matched_filter_vectorized(data)
    % define speed of light
    c = 299792458;
```

```
% determine the size of the phase history data
K = size(data.phdata,1); % number of frequency bins per pulse
Np = size(data.phdata,2); % number of pulses

% initialize the image with all zero values
data.im_final = zeros(size(data.x_mat), 'double');

% loop through every pulse
for ii = 1 : Np

    % Calculate differential range for each image pixel (m)
    dR = sqrt((data.AntX(ii)-data.x_mat).^2 + ...
              (data.AntY(ii)-data.y_mat).^2 + ...
              (data.AntZ(ii)-data.z_mat).^2) - data.R0(ii);

    % Calculate the frequency of each sample in the pulse (Hz)
    freq = data.minF(ii) + (0:(K-1)) * data.deltaF;

    % Perform the Matched Filter operation
    for jj = 1 : K
        data.im_final = data.im_final + ...
                data.phdata(jj,ii) * exp(1i*4*pi*freq(jj)/c*dR);
    end
end % for ii = 1:Np
end
```

Note how compact and easy to understand the algorithm looks when written in MATLAB's vector syntax. There are some loop constructs that cannot be readily replaced by matrix/vector operations. But even with this, the modified algorithm only needs about **57 s** to render the image. From the task manager, we can see that MATLAB utilizes up to **12 CPU cores** on our testing platform which is already a good performance improvement.

Experienced MATLAB users, however, may object that the above code is not yet fully optimized for performance. Indeed, we can go further by eliminating the inner loop using *repmat* command and vectorization techniques. In other words, the inner loop may be replaced by

```
% Perform the Matched Filter operation
dR = reshape(dR, [1 size(dR)]);
dR = repmat(dR, K, 1); % replicate data for bsxfun
t = exp(1i*4*pi/c*bsxfun(@times, freq', dR));
im_slices = im_slices + bsxfun(@times, phdata (:,ii), t);
```

where im_slices is a temporary array of size K×Sx×Sy (i.e., "image slice"). Surprisingly, such aggressive optimizations do not necessarily imply better performance. First, they come at the expense of substantially higher memory consumption, which can become just another bottleneck (replicating large matrices in memory is not for free!).* Second, owing to steady improvements in MATLAB's JIT compiler, loops are no longer as computationally expensive as they used to be in earlier MATLAB versions. As a matter of fact, our "fully vectorized" matched filter algorithm actually runs *slower* than the initial version. However, we shall find this algorithm useful later in our discussion: namely, in the context of GPU processing in MATLAB.†

---

* For information about the performance implications of MATLAB memory aspects, see Chapter 9.
† Refer to Chapter 6.

Sometimes we can improve performance by telling MATLAB explicitly how to distribute work across the multiple processing units (*explicit parallelism*) — CPU cores, CPUs, and GPUs. This topic will be covered in Chapters 6 and 7.

5.7 Effective Use of MATLAB Vectorization

5.7.1 Vectorization Is Not Always Faster

Despite a commonly entrenched belief, vectorization is not always faster than the simpler looped equivalent. We have already seen an example of this in §5.3.6 and §5.3.9, where the fully vectorized version did not outperform a simple JIT-ted loop. This effect is sometimes surprising. Consider the following example:[820]

```
% Slow version
N = 100;
A = rand(N,N,N);
L = zeros(N^3,4);
ind = -250 : 250;
for di = 1 : N
    for dj = 1 : N
        for dk = 1 : N
            L((N^2)*(di-1)+N*(dj-1)+dk,:) =...
                [ind(di),ind(dj),ind(dk),A(di,dj,dk)];
        end
    end
end
⇨ Elapsed time is 1.196058 seconds.
```

A simple vectorization attempt results in code that is 7× faster:[821]

```
% Fast (vectorized) version #1 (7x faster)
X = repmat({1:N},3,1);
[X Y Z] = ndgrid(X{:});
L2 = [ind([Z(:) Y(:) X(:)]) reshape(permute(A,[3 2 1]),[],1)];
⇨ Elapsed time is 0.169503 seconds.
```

It turns out that if we make loops simpler for the JIT to process, then the performance speedup becomes 13× , despite the fact that we have not used any vectorization![822]

```
% Faster (non-vectorized) version #2 (13x faster)
N = 100;
A = rand(N,N,N);
L = zeros(N^3,4);
ind = -250:250;
cnt = 0;
for di = 1 : N
    for dj = 1 : N
        for dk = 1 : N
            cnt = cnt + 1;
            L(cnt,1) = ind(di);
            L(cnt,2) = ind(dj);
            L(cnt,3) = ind(dk);
```

```
            L(cnt,4) = A(di,dj,dk);
        end
    end
end
⇨ Elapsed time is 0.092661 seconds.
```

Yi Cao, who authored the §5.3.9 example, has contributed several utilities[823] to the MATLAB File Exchange, which illustrate JIT's dominance over vectorization with regard to performance.

JIT often outperforms vectorization for small data sizes and a small number of loop iterations. While loops may be slightly faster than vectorization for small problem sizes, the same loops are often significantly slower when the problem size increases. The exact point at which vectorization outperforms JIT'ed loops obviously depends on the specific task, MATLAB-release, and platform. As Dennis Jaheruddin notes:[824]

> *"In practice using a loop that beats the vectorized equivalent will rarely bring down your total code runtime by a second. On the other hand a good vectorization can often save seconds or more. Thus unless you are optimizing something that is called a huge amount of times don't bother testing loops to gain performance."*

5.7.2 Applying Smart Indexing

Array indexing can sometimes be vectorized in nonintuitive ways. For example, consider the problem where we have a list of indices (row and column index) to requested elements in a 2D data matrix. For example, in the following 2D data matrix:

```
>> B = magic(5)
B =
    17    24     1     8    15
    23     5     7    14    16
     4     6    13    20    22
    10    12    19    21     3
    11    18    25     2     9
```

If we have the following list of indices:

```
>> A = [1,2;  3,4;  4,5;  5,3]
A =
     1     2
     3     4
     4     5
     5     3
```

Then we would expect to get the data elements 24, 20, 3, and 25. The naïve loop-based approach is relatively simple but far from optimal in terms of performance. Let's run our data with 10 K index pairs (i.e., A is 10 K×2 in size):

```
results = zeros(1,size(A,1));   % preallocate (see §9.4.3)
for row = 1 : size(A,1)
    rowIdx = A(row,1);
    colIdx = A(row,2);
    results = B(rowIdx,colIdx);
end
⇨ Elapsed time is 0.736695 seconds.
```

To vectorize this code, we can convert the 2D indices into corresponding 1D indices.* Oddly, this results in significant slowdown rather than speedup:

```
N = size(B,1);
indices = arrayfun(@(i) A(i,1) + N*(A(i,2)-1), 1:size(A,1));
results = B(indices);
⇨ Elapsed time is 3.956874 seconds.
```

The slowdown is due to the anonymous function, which cause severe slowdowns in MATLAB.† This explanation is supported by the fact that two alternatives for converting the 2D indices into 1D ones without using anonymous functions are much faster:

The first alternative, which is both much faster and simpler (if perhaps less intuitive), is to use vectorized matrix multiplication, obviating the need for *arrayfun*:[825]

```
N = size(B,1);
indices = [1 N] * (A'-1) + 1; % vectorized matrix multiplication
results = B(indices);
⇨ Elapsed time is 0.035077 seconds.
```

Alternatively, we can use *sub2ind*'s vectorized inputs:

```
indices = sub2ind(size(B),A(:,1),A(:,2)); % vectorized sub2ind inputs
results = B(indices);
⇨ Elapsed time is 0.031848 seconds.
```

Here is another example by Bruno Luong, of extracting a 2D slice from 3D data:[826]

```
% Original (unvectorized)
for idx=1:N,  x(idx,:)=V(idx, :, z(idx));  end

% Vectorized version using sub2ind (see reference for another variant)
[m n p] = size(V);
x = reshape(permute(V, [1 3 2]), [], n);
x = x(sub2ind([m p],1:m,z),:)  % transpose here if z is a column-vector
```

5.7.3 Breaking a Problem into Simpler Vectorizable Subproblems

A newsgroup user posted the problem of identifying consecutive elements in a numeric data array.[827] So, for example, the array [1 2 3 7 8 9 10 13 14 20 21] should report the result [1 1 1 2 2 2 2 3 3 4 4]. This problem could be solved relatively easily via a *for* loop, but for better performance we can vectorize the problem by breaking it into three separate, simpler subproblems:

- Getting an array of numeric differences between subsequent data elements
  `diff(data)` → [1 1 4 1 1 1 3 1 6 1]

- Getting an array of 0/1s based on whether the corresponding difference is 0
  `diff(data)~=1` → [0 0 1 0 0 0 1 0 1 0]

- Summing up the array of 0/1s in a cumulative manner

The end result by James Bejon is elegant, simple, and highly performant:[828]

```
results = cumsum([1, diff(data)~=1]) → [1 1 1 2 2 2 2 3 3 4 4]
```

---

* This technique relies on the order in which numeric matrix values are stored in memory. See §9.3 for additional details.
† See §4.6.1.2.

5.7.4 Using Vectorization as Replacement for Iterative Data Updates

Whenever MATLAB data is modified in structure or size, the new data is allocated a new space in memory, and the old data's memory is deallocated.* This may take nonnegligible time, especially when the data is large and when done in a loop.

For example, consider the need to delete rows in a large 2D matrix, based on some condition, such as the sum of the column values being <355 and range (max–min)[†] of the values being >20.[829] The iterative loop solution is

```
for i = length(data) : -1 : 1
    if (sum(data(i,:))<355) & range(data(i,:))>20
        data(i,:) = [];
    end
end
```

The corresponding vectorized solution is much faster, simpler, and more maintainable:

```
data(sum(data,2)<355 & range(data,2)>20, :) = [];
```

In other cases the transformation may not be as obvious. But it often helps to prepare a shadow array of logicals the size of the original data, initialized to false values, which are then updated. As James Tursa (whom we shall meet again later) notes:[830]

> "The statement dati(i,:) = [] causes the entire data of dati (minus one row) to be copied to another brand new part of memory. Doing so in a loop can easily dominate the running time. For small sizes you don't notice it of course, but for large cases it can easily consume 99.99% of the running time. I would always opt for logical indexing to mark the data you want deleted, then delete it all in one fell swoop. That way the data is only copied once, not a gazillion times. Even if you know the dataset is small, there is the chance you would reuse the code later on for a large dataset and *forget* you had that inefficient loop in there. Bottom line is I would always avoid the loop construct. It is the same as building up an array size in a loop. Use a loop to mark the rows if you have to (or use a one-liner if you can), but don't do the data deletion itself in the loop."

5.7.5 Minimizing Temporary Data Allocations

Using vectorization often implies creation of multiple temporary data arrays, which can be very large, requiring time to allocate and deallocate.[‡] All other things being equal, the fewer such intermediate data arrays are created, the faster. For example:[831]

```
data = rand(1e6,1);

% 2 intermediate arrays (diff, >)
idx = find(diff(data)>0.999, 1);
⇨ Elapsed time is 0.005801 seconds.

% 3 intermediate arrays (data, data, >) - slower
idx = find(data(2:end)-data(1:end-1)>0.999, 1);
⇨ Elapsed time is 0.010815 seconds.
```

---

* This is true for numeric arrays but not necessarily for cell arrays or other data types. See §9.4 for additional details.

† *range* is a function in the Statistics Toolbox. Users who do not have this toolbox can use the *max* and *min* functions.

‡ Chapter 9 discusses the relationship between memory and performance.

5.7.6 Preprocessing Inputs, Rather than Postprocessing the Output

In many cases, the input data arrays are smaller than the result of an operation. If you need to postprocess a large result, it is often more efficient to preprocess the smaller input arrays, since this would involve less microoperations. For example, if we need to transpose the result of a Kronecker (tensor) product:[832]

```
N=100; a=rand(N); b=rand(N); % test data

% Transpose the output
c = kron(a,b)';
⇨ Elapsed time is 0.816603 seconds.

% Pre-process the inputs (~x2.5 faster)
c = kron(a',b');
⇨ Elapsed time is 0.338405 seconds.
```

This idea can be generalized: it is generally faster to modify small data inputs in any manner (transpose or any other operation) than to modify a much larger result.

There is always a sweet spot for optimization: if the resulting data is not much larger than the inputs, then it could actually be faster to process a single result array en-bulk than to preprocess the inputs. The larger the result data is compared to the inputs, the more likely it is for preprocessing the inputs to be faster. See §3.3 and §3.9.1 for related ideas.

5.7.7 Interdependent Loop Iterations

Converting a code loop into vectorized code is easiest when the loop iterations are independent of each other. When iterations depend on one another, finding a vectorized equivalent is more difficult. Users should be aware that such dependencies might well lead to incorrect code, and should therefore carefully test their vectorized version for correctness and equivalence with the original version. However, the benefits of correctly vectorizing such loops or recursions can well be worth the effort.

Using internal MATLAB functions can help in some cases. For example, Aurélien Queffurust recently recovered an old MATLAB tip for using the *filter* function for vectorized computation of interdependent data sequences:[833]

```
% Slow version
L = 1e6;
A = 1;
for i = 1 : L
    A(i+1) = 2*A(i) + 1;
end
⇨ Elapsed time is 0.205670 seconds.

% Vectorized version: 12x faster
A2 = filter([1], [1 -2], ones(1,L+1));
⇨ Elapsed time is 0.017608 seconds.
```

Note: *filter* can only be used when there is only a single dependent variable;[834] *filter2* and *conv2* can possibly be used for sequences having two interdependent variables.

Loren Shure has shown[835] how the *filter* function can be used to significantly speed up the computation of the Fibonacci sequence:*

---

* Fibonacci sequence generation is also discussed in §3.1.5, §3.2.4.2, §3.10.1.2, and §9.4.1; *filter* is also discussed in §10.4.2.6.

```
% Slow recursive version
fibonacci = [0, 1];
for idx = 3 : 1000
    fibonacci(idx) = fibonacci(idx-1) + fibonacci(idx-2);
end
⇨ Elapsed time is 0.000562 seconds.

% Vectorized version: 10x faster
x = [0,1,zeros(1,98)];
fibonacci = filter(1, [1,-1,-1], x);
⇨ Elapsed time is 0.000059 seconds.
```

In another case, using *filter* improved a loop performance by 180× (!):[836]

```
nobs    = 1e5;              % number of observations
p       = 2;               % number of lags
ro      = [0.1 0.5];       % autoregressive coefficients
noise   = rand(nobs,1)-p;  % white noise
pseudoX = zeros(nobs,1);   % vector to fill in
pseudoX(1:p) = randn(p,1); % seed the initial vector positions

% Slow recursive version
for ii = p+1 : nobs
    pseudoX(ii) = ro * pseudoX(ii-p:ii-1) + noise(ii-p);
end
⇨ Elapsed time is 0.266240 seconds.

% Vectorized version: 180x faster
P2 = filter(1, [1-ro(end:-1:1)], [pseudoX(1);
                                  pseudoX(2)-ro(2)*pseudoX(1);
                                  noise(1:end-2)]);
⇨ Elapsed time is 0.001477 seconds.
```

In R2014a (MATLAB 8.3), *filter*'s performance improved by ~10%–15%, compared to earlier releases. The improvement may seem small, but it is pretty consistent and only highlights the benefit of using *filter* over nonvectorized loops.

Jan Simon provided a MEX implementation of the *filter* function, called *FilterM*, which is 20%–30% faster than R2014a's version, and accepts exactly the same inputs.[837]

Similarly, Jan Simon suggested[838] using the Signal Processing Toolbox's *buffer*[839] or *filtfilt*[840] functions to efficiently solve additional difference problems. Simon's *FilterM* submission also includes a *FiltFiltM* function, which can be used as a direct and faster replacement for *filtfilt*, even without the Signal Processing Toolbox.

MATLAB's built-in *conv** functions enable easy and efficient vectorized smoothing/filtering of time-series data, saving lengthy and complex loop constructs:[841]

```
window = 5;
mask = ones(1,window)/window;
movAvgY = conv(noisy, mask, 'same');
plot(x,y, x,noisyY, x,movAvgY);
```

Bruno Luong found[842] that it is faster to use FFT/IFFT pairs than sliding windows in the implementation of *conv**, especially when GPU parallelization is available.* Bruno's

---

* FFT functions, as well as their application to convolution, are discussed in §4.5.6. MATLAB's *conv** code cannot be analyzed since it is purely built-in (compiled c-code).

convnfft utility is based on the convolution theorem (Conv(a,b) = *ifft*(*fft*(a)*fft(b))).[843] It is usually 3–5× faster than MATLAB's built-in functions, but one user has reported a 350× speedup (!) for a specific case.[844] A simple example:

```
a = rand(1e5,1);
b = rand(1e5,1);

c1 = conv(a,b);
⇨ Elapsed time is 2.433974 seconds.

c2 = convnfft(a,b);
⇨ Elapsed time is 0.019030 seconds. % 130x faster

disp(max(abs(c2-c1))) % ensure identical results within numerical accuracy
⇨ 7.89441401138902e-10
```

convnfft apparently does not cover the case of *conv2(vector1, vector2, matrix)*,* but does cover other variants of *conv*, *conv2*, and *convn* quite nicely in a single function. Also notable is Ilias Konsoulas' utility:[845] it uses three different transforms in pure m-code, but works only on real data and is slower than *convnfft* (yet faster than *conv**).

In one case, Bruno used *conv2* as a replacement for a double loop:[846]

```
% Slow version using double loop
for i = 3 : dimmX-2
    for j = 3 : dimmY-2
        sur = aod(i-2:i+2, j-2:j+2);
        surNan(i,j) = length(find(isnan(sur)==1));
    end
end

% Vectorized version using conv2
surNan = zeros(size(aod));
surNan(3:end-2,3:end-2) = conv2(double(isnan(aod)),ones(5),'valid');
```

Similarly, newsgroup veteran Matt J and Bruno (again…) have suggested[847] using *conv2* as a faster replacement for both *imtransform* and *interp2*(…'linear').[†] Bruno's interesting conclusion was that[848]

> *"one should put the large array as first argument of conv/conv2, which is probably the majority of the cases in practice".*

The *filter* function has been suggested as a faster alternative for *conv* in some use-cases.[849] Similarly, Image Processing Toolbox's *imfilter* function was suggested as a substitute for *conv2*.[850] Also note Dirk-Jan Kroon's GPU-based solution *gpuconv2*.[851]

5.7.8 Reducing Loop Complexity

Simpler loops are much easier and safer to vectorize. Therefore, we should first employ any LIH optimization (see §3.1.1) and loop simplification (§3.1.4) techniques, before attempting to vectorize. Conditionals (*if/while*) in particular hinder vectorization. Try to move conditionals outside loops (switch loop/condition order):

```
% Original (non-vectorized) form
for idx = 1 : 100
```

---

* This specific usage of *conv2* was made faster in R2014a (MATLAB 8.13).
† *interp2* is faster than *imtransform* (http://bit.ly/1ea0omI), but still slower than *conv2*.

```
        if someCondition
            doSomething();
        else
            doSomethingElse();
        end
    end

% Option #1: move conditional outside the loop
if someCondition
    for idx = 1 : 100
        doSomething();
    end
else
    for idx = 1 : 100
        doSomethingElse();
    end
end
```

Resulting in the following vectorized format:

```
if someCondition
    doSomethingVectorized();
else
    doSomethingElseVectorized();
end
```

Alternatively, we can often convert the conditional into vectorized form:[852]

```
% Option #2: vectorize the conditional
for idx = 1 : 100
    result = someCondition*doSomething() +...
            ~someCondition*doSomethingElse();
end
```

For example, consider the following loop:

```
% Original (non-vectorized) loop
result = 0;
for x = 0 : 0.01 : 1
    if x < 0.5
        result = result + sin(x);
    else
        result = result + cos(x);
    end
end

% Vectorized version
x = 0 : 0.01 : 1;
result = sum(sin(x)*(x<0.5) + cos(x)*(x>=0.5));
```

As yet another example, consider the following loop, in which we essentially break up the internal conditional (if) into two separate double loops, which are easily vectorized independently:[853]

```
% Original (non-vectorized) loop
h = zeros(dimV);
for a = 1 : dimV
    for b = 1 : dimV
        if a ~= b && C(a,b) == 1
```

```
            % First part - all matrix cells except the diagonal (a==b)
            h(a,b) = C(a,b)*exp(-1i*A(a,b)*L(a,b))*(sin(k*L(a,b)))^-1;
        else if a == b
            % Second part - add the diagonal elements
            for m = 1 : dimV
                if m ~= a && C(a,m) == 1
                    h(a,b) = h(a,b) - C(a,m)*cot(k*L(a,m));
                end
            end
        end
    end
end

% Vectorized version:
% First part - all matrix cells except the diagonal (a==b)
diag_ind = 1 : dimV+1 : numel(C);
C_neq1 = C~=1;
parte1_2 = (sin(k.*L)).^-1;
parte1_2(C_neq1 & (L==0)) = 0;
h = exp(-1i.*A.*L) .* parte1_2;
h(C_neq1) = 0;

% Second part - add the diagonal elements
parte2 = cot(k*L);      % no need to multiply by C, since only if C==1
parte2(C_neq1) = 0;     % logical indexing
parte2(diag_ind) = 0;   % integer indexing
h(diag_ind) = -sum(parte2,2);
```

5.7.9 Reducing Processing Complexity

When performing any kind of mathematical processing, even highly vectorized code can sometimes be sped up by using simpler mathematical equivalents. In the order of complexity levels, + and − are naturally simpler than * and /, which are in turn simpler than exponentiation. A simple example was provided by Mike Croucher:[854]

```
% Naïve approach using a.^n
a = 0.9999; n = 10000;
tic, for idx=1:1000, y=a.^(1:n); end, toc
⇨ Elapsed time is 0.851664 seconds.

% Using exp(log(a)*n) - 8x faster
tic, for idx=1:1000, y=exp(log(a)*(1:n)); end, toc
⇨ Elapsed time is 0.111581 seconds.

% Using cumprod(a) - 20x faster
tic, for idx=1:1000, y=cumprod(zeros(1,n)+a); end, toc
⇨ Elapsed time is 0.042769 seconds.
```

5.7.10 Nested Loops

Loops are often nested within other loops, the inner loops executing numerous times. So, even when we cannot vectorize the entire loops construct, or when JIT makes it less efficient to fully vectorize, we can still benefit by vectorizing the inner loops.

I was once asked by a client to optimize a program that had the following piece of nested loops, which profiling reported as a performance hotspot:

```
% This is the original version: a bit too slow for our needs...
for i = 1 : num_total_days
    for j = 1 : num_days_later
        if (i+j) <= num_total_days+1
            evaluation_tensor_all(i,j,:) = actuals(i+j-1,:);
        end
    end
end

% ... and this is the equivalent vectorized version:
for i = 1 : num_total_days
    maxJ = min(num_days_later, num_total_days-i+1);
    evaluation_tensor_all(i,1:maxJ,:) = actuals(i-1+(1:maxJ),:);
end
```

Unfortunately, this loop is much more difficult to correctly vectorize further, due to the interdependencies between loop iterations. But having vectorized the inner loop saved 97% of the overall loop run-time, achieving our performance goal. Vectorizing the external loop (for the other 3%) did not seem to be cost-effective.

After vectorizing an inner loop, we sometimes discover that it is easier to find a vectorized solution to containing loops as well, as an added bonus. Naturally, this is only worth pursuing in case we have not yet reached our performance goal. Vectorizing in steps, one loop at a time, is easier to achieve and safer to implement.*

5.7.11 Analyzing Loop Pattern to Extract a Vectorization Rule

Sometimes vectorization is not trivial and requires understanding a loop's underlying aim. In such cases, successful vectorization depends more on using a vectorization mechanism that implements the core idea, and not on a direct translation of loop code.

For example, the following seemingly complex loop:

```
for j=1:size(index_vals,1)
    idx = index_vals(j,1);
    [R,C] = find(openPoints==idx);
    [R1,C1] = find(regionSet==idx);
    if isempty(R) && isempty(R1)
        openPoints = [openPoints;idx];
    end
end
```

is actually just selecting those points that are neither in openPoints nor regionSet:[855]

```
allSelectedPoints = union(openPoints(:), regionSet(:));
openPoints = setdiff(idx(:), allSelectedPoints, 'stable');
```

A more complex example was provided for the following loop:

```
N = 2; T = 3;
Trials = rand(N,T);
```

---

* Also see §3.1.6.

```
for i = 1 : N
    for j = 1 : T
        Trials(i,j) = Trials(i,j) - mean2(Trials);
    end
end
```

As user gevang explained,[856] by observing the pattern created by the recursive subtraction of the mean, we can infer the following recursive relationship:

```
m(k) = m(k-1) - m(k-1)/n % n=numel(Trials)=N*T
     = m(k-1) * ((n-1)/n)
     = m(1)   * ((n-1)/n)^(k-1)
```

where m(1) is the average of the original matrix. The subtraction is applied iteratively, for each element of the input matrix: the first element subtracts m(1), the second subtracts m(2), and so on. This leads to the following optimized (vectorized) form:

```
n = numel(Trials);
Trials = Trials - mean2(Trials)*reshape(((n-1)/n).^(0:n-1), T, N)';
```

5.7.12 Vectorizing Structure Elements

As we shall see in §9.7, it is better to use a *struct* of arrays rather than an array of *struct*s. However, if we do have the less efficient form (array of *struct*s), we can still use vectorization of constituent fields using the square brackets [] notation:

```
>> s(1).a=pi; s(2).a=-pi; s(3).a=0
s =
1x3 struct array with fields:
    a

>> vector1 = [s.a]
vector1 =
        3.1416        -3.1416         0
>> vector2 = [s(2:end).a]
vector2 =
    -3.1416                0
```

If our *struct* array is large, using the vectorized [s.a] notation can be an order of magnitude faster (and simpler) than an iterative loop that collects the field elements.

The separate values can be assigned to a cell array in one of two ways: we could use the [] notation and convert to a cell array, or we could use *deal* to assign the multiple outputs to the corresponding cell array:[857]

```
% Alternative 1: use [] notation
cellArray = num2cell([s.a]);

% Alternative 2: use deal()
cellArray = cell(1,length(s));
[cellArray{:}] = deal(s.a);       % Note: no [] around s.a
```

A similar construct can be used to assign values to a *struct* array:

```
[s.a] = deal(1,2,3);              % assign 3 separate element values
[s(2:end).a] = deal(2*pi);        % assign same value to 2 separate elements
```

Note that we must use the ***deal*** function rather than a simple assignment statement, because only functions can assign to multiple left-hand side values. In this case, ***deal*** copies its input into each element of the output. As a general rule, whenever we need to assign to or from a comma-separated list, we should use ***deal*** in conjunction with the [] concatenation operator.

The [s.a] vectorization technique was used to achieve an order of magnitude speedup in a problem of iterating over structure elements to find busy airports, by summing the number of flights to or from each airport in a lengthy list of flights:[858]

```
% Original code
function list = Problem10(flights, airports)
    a = zeros(1,(length(airports))); % pre-allocate
    for id = 1 : length(airports)
        a(id) = FlightsToFrom(flights, id);
    end
    a(a==0) = [];
    list = sort(a);
end
function tofrom = FlightsToFrom(flights, ID)
    sum = 0;
    for ii = 1 : length(flights)
        if (isequal(flights(ii).from_id, ID)) || ...
           (isequal(flights(ii).to_id,   ID))
            sum = sum + 1;
            if sum > 1000
                break;
            end
        end
    end
    if sum <= 1000
        tofrom = 0;
    else
        tofrom = ID;
    end
end
```

```
% Vectorized code: much faster and simpler
function list = Problem10( flights, airports )
    numAirports = length(airports);
    a(numAirports) = 0;   % preallocate: faster than zeros()
    from_ids = [flights.from_id];
    to_ids   = [flights.to_id];
    for id = 1 : numAirports
        a(id) = sum(from_ids==id | to_ids==id) > 1000;
    end
    list = find(a);
end
```

5.7.13 Limitations of Internal Parallelization

An important difference should be noted between implicit (automatic, internal) and explicit parallelization: Implicit parallelization (i.e., automatic multithreading) only works for some specific built-in functions. On the other hand, explicit parallelization can be used to distribute generic code across CPU cores or external CPUs/GPUs in a much more efficient manner, even if the distributed code is single-threaded. Consider the following example, provided in a recent MathWorks newsletter:[859]

```
y = zeros(1000,1);
for n = 1:1000
    y(n) = max(svd(randn(n)));
end
```

In this code snippet, only the *svd* function is multithreaded, whereas **max** and **randn** are not. Therefore, while multithreading does somewhat improve the loop's performance, the speedup is only by a few percent. If we change this loop into a parallel loop (by simply modifying the keyword *for* to *parfor*), we can achieve much higher speedups (3.2× using four parallelization labs).

The moral of this example is that we should not place too much emphasis on vectorization and using built-in functions that are implicitly multithreaded, to the exclusion of other parallelization techniques (or other speedup alternatives in general).

5.7.14 Using MATLAB's Character/Number Duality

MATLAB's inherent character/number duality* can be used to apply operations en-bulk on strings, converting the results to numbers (or vice versa). An example of this was provided by a CSSM newsgroup user who wanted to efficiently create a random permutation of ones and zeros in a 2D matrix.[860] Her naïve implementation involved no less than four nested loops in 17 lines of code, and was inefficient for large matrix sizes. A responder on that thread provided a single-line vectorized solution that was as elegant as it was efficient:[861]

```
>> N=3; randmat = dec2bin(randperm(2^N)-1, N)' - '0'
randmat =
    1    0    0    1    1    1    0    0
    1    1    1    1    0    0    0    0
    1    1    0    0    0    1    0    1
```

In this solution, `randperm(2^N)-1` returns a random permutation of all values between 0 and $2^N - 1$ (i.e., [7,3,2,6,4,5,0,1]). This is then turned into binary representation by dec2bin, resulting in a 2^N-by-N character matrix:

```
>> dec2bin(randperm(2^N)-1, N)
ans =
111
011
010
110
100
101
000
001
```

Transposing this result and subtracting the character '0' from each character results in the requested numeric matrix above (subtracting strings always results in numeric results: characters are simply converted into their ASCII or Unicode values).

---

* See §4.2.1.

5.7.15 Acklam's Vectorization Guide and Toolbox

Peter John Acklam[862] wrote an extensive MATLAB vectorization guide called *MATLAB array manipulation tips and tricks*, which has been updated several times between the years 2000 and 2003. The latest version[863] is 63 pages long and is packed with tips and examples for effective vectorization. Despite being originally developed for MATLAB 6 and not updated since 2003, the guide has withstood the test of time. It is still being referenced today as the go-to guide for MATLAB vectorization. I have found Acklam's guide print-out on several of my clients' work shelf.

Acklam's work in his guide and MATLAB CSSM newsgroup posts have gained him widespread recognition as the leading authority on MATLAB vectorization. Some of his work even entered stock MATLAB functions.* The term *"acklamization"* (*to acklamize*), originally suggested humorously in 2001 as a vectorization synonym,[864] has since then entered mainstream usage within the MATLAB community, and was even included in the semi-official MATLAB FAQ until late 2012.[865]

In addition to his guide, Acklam also posted a toolbox of utility m-functions, which illustrate his vectorization techniques.[866] Since 2003, he has removed many utilities from his website for some reason, but a cached version from 2003 is available.[867]

Unfortunately, some of Acklam's tips have become outdated over the years, as MATLAB has improved its built-in functions (e.g., **kron** in R2013b), added some new functions (e.g., **bsxfun** in R2007a) and added features to existing functions. Keeping this in mind, Acklam's work could still be an excellent starting point for MATLAB vectorization. Anyone who is serious about perfecting their indexing and vectorization skills should carefully study Acklam's guide and toolbox functions.

5.7.16 Using Linear Algebra to Avoid Looping Over Matrix Indexes

Matrix operations can often be vectorized by employing linear algebra matrix-processing rules. In such cases, the speedups are often significant compared to interactive loops, even when JIT is being used (also see §4.5.2).

For example, to compute the asymmetry of a square matrix:[868]

```
% Slow, standard version
N = size(a,1);
asymmetry = zeros(N,N);
for row = 1 : N-1;
    for col = rowIdx+1 : N
        asymmetry(row,col) = a(row,col) - a(col,row);
        asymmetry(col,row) = -asymmetry(row,col);
    end
end

% Vectorized version
asymmetry = abs(a-a');
```

As another example, the problem of rotating N×M 3D coordinates about the X axis:[869]

```
% Slow, standard version
R = [1,0,0; 0,cos(theta),sin(theta); 0,-sin(theta),cos(theta)];
```

---

* For example, **nthroot, erfinv, erfcinv** functions and several image I/O functions under *%toolbox%/matlab/imagesci/private/*.

```
for i = 1 : height
    for j = 1 : width
        tmp3(i,j,:) = reshape(tmp(i,j,:),1,3)*R;
    end
end

% Vectorized version
tmp3 = reshape(reshape(tmp,[],3)*R, [size(tmp,1), size(tmp,2), 3]);
```

Additional examples include computing the sum of squared matrix elements,[870] a grid mesh,[871] and complex product.[872] Linear algebra functions that can help vectorization are listed in the following table:

Function	Description
blkdiag	Create a matrix with specified diagonal
diag	Return the diagonal elements of a matrix
flipdim	Flip elements of input N – D array in any dimension (in R2013b+ use *flip*)
fliplr	Flip elements of input array or matrix left-to-right
flipud	Flip elements of input array or matrix up-down
permute	Rearrange dimensions of a multidimensional array (also see: *ipermute*)
reshape	Reshape an array into any size
rot90	Rotate matrix by an integer number of 90 degrees (also see: *permute*)
shiftdim	Shift array dimensions (also see: *permute, squeeze, reshape*)
squeeze	Remove singleton dimensions from an array (also see: *reshape*)
transpose	Return the transposed matrix
tril	Return the lower triangular part of a matrix
triu	Return the upper triangular part of a matrix

5.7.17 Intersection of Curves: Reader Exercise

As a concluding reader exercise, review the vectorized curves-intersection algorithm from File Exchange utility *InterX*.[873] Try to understand and explain the algorithm:*

```
%...Preliminary stuff
x1 = L1(1,:)';   x2 = L2(1,:);
y1 = L1(2,:)';   y2 = L2(2,:);
dx1 = diff(x1);  dy1 = diff(y1);
dx2 = diff(x2);  dy2 = diff(y2);

%...Determine 'signed distances'
S1 = dx1.*y1(1:end-1) - dy1.*x1(1:end-1);
S2 = dx2.*y2(1:end-1) - dy2.*x2(1:end-1);

C1 = feval(hF,D((bsxfun(@times,dx1,y2)-bsxfun(@times,dy1,x2)), S1), 0);
C2 = feval(hF,D((bsxfun(@times,y1,dx2)-bsxfun(@times,x1,dy2))',S2'),0)';

%...Obtain the segments where an intersection is expected
[i,j] = find(C1 & C2);
if isempty(i), P = zeros(2,0); return; end;
```

---

* The *InterX.m* file contains an explanation of the algorithm's core, but try to see if you can understand it without this assistance.

```
%...Transpose and prepare for output
i=i'; dx2=dx2'; dy2=dy2'; S2 = S2';
L = dy2(j).*dx1(i) - dy1(i).*dx2(j);
i=i(L~=0); j=j(L~=0); L=L(L~=0); %...Avoid divisions by 0

%...Solve system of eqs to get the common points
P = unique([dx2(j).*S1(i) - dx1(i).*S2(j),...
            dy2(j).*S1(i) - dy1(i).*S2(j)]./[L L],'rows')';
function u = D(x,y)
    u = bsxfun(,1:end-1),y).*bsxfun(,2:end),y);
end
```

The usage, as shown in the utility's help section, is as follows:

```
t = linspace(0,2*pi);
r1 = sin(4*t)+2;
x1 = r1.*cos(t);
y1 = r1.*sin(t);

r2 = sin(8*t)+2;
x2 = r2.*cos(t);
y2 = r2.*sin(t);

P = InterX([x1;y1],[x2;y2]);
plot(x1,y1,x2,y2,P(1,:),P(2,:),'ro')
```

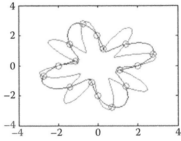

Visualization of the curves intersection example

For those interested in the topic of curves intersection, there are a few additional curve-intersection utilities on the File Exchange that could be compared, including semivectorized *intersections* by veteran MATLAB user Doug Schwartz.[874]

6

Explicit Parallelization Using MathWorks Toolboxes[*]

Several years ago, when multicore multi-CPU machines became mainstream, parallel processing was recognized as an indispensable tool for solving many scientific and engineering problems.[875] Furthermore, modern graphics processing units (GPUs) have gained popularity for high-performance computing (HPC) since, with the release of the CUDA (Nvidia's parallel computing architecture)[876] and OpenCL[877] frameworks, it is no longer necessary to program the GPUs through a graphics API. The ability to take advantage of multiple CPUs, cores and GPUs is called parallel processing, or *parallelization*.[878]

MATLAB has traditionally been a single-threaded uniprocessor engine. In fact, a 1995 article by MathWorks cofounder Cleve Moler explains *"Why there isn't a parallel MATLAB"*,[879] detailing early parallelization attempts that failed to meet expectations. However, the situation has drastically changed since then,[880] with multicore and GPU processors now standard on all computers. Accordingly, MathWorks has invested considerable resources in recent years to improve its parallelization offering. In addition to builtin support in the core MATLAB product, MathWorks now offers two MATLAB toolboxes dedicated to parallelization: the Parallel Computing Toolbox (PCT), and the MATLAB Distributed Computing Server (MDCS).[†]

MATLAB parallelism can very broadly be classified into five broad categories:[881]

- *Implicit multithreading:* One instance of MATLAB automatically generates multiple simultaneous multithreaded instruction streams, working in parallel on different parts of a vectorized data array. Multiple processors or cores, and sharing the memory of a single computer, execute these streams, spreading the computational load. Implicit multithreading is automatically employed by numerous built-in MATLAB functions. See Chapter 5 for details.

- *Explicit multithreading*: This can be achieved in MATLAB using a variety of means, but has no official support in MATLAB (see §7.3).[‡]

- *Explicit multiprocessing:* Multiple instances of MATLAB run on several processors, often with shared memory, and execute a single MATLAB command or M-function concurrently. Supported by dedicated PCT programming constructs (see §6.1) as well as other solutions (see §7.2, §7.4).

- *GPU-delegated processing*: Massively-parallel tasks are delegated to a GPU for processing within its multitude of numeric computation cores. Supported by several PCT constructs (see §6.2) as well as other solutions (see §7.1, §7.2).

- *Distributed computing:* Multiple instances of MATLAB run multiple independent computations on separate computers, each with its own memory, typically in a

---

[*] This chapter was authored by Pavel Emeliyanenko of Max-Planck Institute for Informatics (http://mpi-inf.mpg.de/~emeliyan).

[†] The history of MATLAB parallelism is discussed at http://walkingrandomly.com/?p=1214 (or: http://bit.ly/1it34hI)

[‡] Explicit multithreading was not mentioned in Cleve Moler's above-referenced article, for obvious reasons.

cluster, grid or cloud. In most cases, a single program is run many times with different parameters or random number seeds. Supported by dedicated MDCS constructs (see §6.3).

These categories coexist:* a distributed job might invoke multithreaded functions on each machine,[†] as well as local GPUs, and then use a distributed array to collect the results. For implicit multithreaded parallelism, threads number can be explicitly set in MATLAB's Preferences.

In this chapter, we shall discuss various ways to explicitly leverage MATLAB parallelism. To illustrate how different parallelization techniques compare to each other, we shall continue the matched filter example introduced in §5.6. While reading this chapter, we recommend perusing the numerous webinars and examples available on the MathWorks website.[882]

6.1 The Parallel Computing Toolbox — CPUs

PCT is a collection of high-level, easy-to-use MATLAB functions and language constructs that enable solving computationally intensive and data-intensive problems using multi-core CPU processors and GPUs. The key PCT features are summarized below:

- *Parallel for*-loops (*parfor*) for running task-parallel algorithms on multiple processors.

- Ability to run multiple workers[‡] on a multicore system. Using more workers than available cores generally *decreases* performance due to increased task-switching and communication overheads.[§] However, extra workers can be useful when workers are blocked on I/O for a major part of their active time. Also, additional workers can take advantage of CPU hyperthreading in certain use cases.[¶]

---

* This overlap is not unique to MATLAB. The form of interaction between processing entities is often used as a differentiator: the use of shared memory defines *parallel computing* (http://en.wikipedia.org/wiki/Parallel_computing or: http://bit.ly/Nog3VV), whereas the exchange of messages defines *distributed computing* (http://en.wikipedia.org/wiki/Distributed_computing or: http://bit.ly/NocKxT). However, this differentiation is very loose and may not be applicable to all cases.

† If permitted: some clusters limit the number of CPU cores used by submitted jobs/tasks. Use MATLAB's-singleCompThread start-up switch (http://mathworks.com/help/matlab/ref/matlabwindows.html or: http://bit.ly/17FftYL) to force MATLAB not to use multithreading on remote cluster nodes, to avoid a possible clash with such a limit.

‡ *Workers* are MATLAB processes that run in parallel without GUI or Desktop, also called "*headless MATLABs*". The number of supported local workers was increased from 8 to 12 in R2011b (extended to 256 by MDCS). The maximal number of local PCT workers increased to 512 in R2014a, without requiring MDCS.

§ See §1.7 for a discussion of Amdahl's law and the limitations of parallelization speedups.

¶ By default, MATLAB allocates a single worker per *physical* core. Modern Intel CPUs include hyperthreading technology that exposes several *logical* cores. Assigning workers to all logical cores (i.e., doubling MATLAB's default number of workers) could in some cases improve performance (see, e.g., http://mathworks.com/matlabcentral/newsreader/view_thread/334485#919894 or: http://bit.ly/1nybyZF). However, hyperthreading is a highly contentious issue, and Intel themselves advise using only as many numeric-processing threads as there are physical cores (http://software.intel.com/en-us/articles/parallelism-in-the-intel-math-kernel-library or: http://intel.ly/1lhzIDu). Also see http://stackoverflow.com/a/14469415/233829 (or: http://bit.ly/1sIPNIC). See §3.5.3 for additional details.

- Computer cluster and grid support (together with the MDCS toolbox).
- Interactive and batch execution of parallel applications.
- Distributed arrays and single-program multiple-data (*spmd*) construct for large data set handling and data-parallel algorithms.
- Parallel computing on GPUs using *gpuArray* and *CUDAKernel* objects.
- Automatic parallelization support for many functions in MATLAB toolboxes.[883] In fact, the number of MATLAB functions and toolboxes supported by PCT increases with each MATLAB release.[884]

Our discussion of PCT is split into two parts: This section covers multicore parallelization using *parfor*-loops, *spmd* and *pmode* functions, as well as profiling the parallel code with *mpiprofile*. The second part, §6.2, is devoted to parallel computing on the GPU. Distributed (multicomputer) computing using MDCS is discussed separately, in §6.3.

Those interested in the technical details underlying PCT may find interest in a very detailed paper by MathWorkers Gaurav Sharma and Jos Martin from 2009.[885] Some technical aspects are by now a bit outdated, but the core description remains valid.

6.1.1 Using *parfor*-Loops

The syntax and semantics of a *parfor*-loop in MATLAB software are the same as of the standard *for*-loop: MATLAB executes a series of statements of the loop body over a range of values. The necessary data on which *parfor* operates are sent from the client to a group of *workers* (computing processes), where the actual computation happens, and the results are sent back to the client and assembled together. The workers should be identified and reserved with the *parpool* command.*

Unlike threads in multithreading, workers are entirely separate MATLAB processes[886] that typically run in *headless* (non-GUI) mode, in their own memory space. Spawning such additional processes, and synchronizing their execution and data, is a nontrivial task that introduces more overhead than with in-process threads. However, to parallelize non-vectorized loops, branches, and other programming constructs that require the usage of MATLAB's single-threaded interpreter/JIT, there is no alternative to spawning additional processes to run on separate CPUs or cores.†

Each execution of the body of a *parfor*-loop is an *iteration*. MATLAB workers evaluate iterations in *unspecified* order, and independently of each other. Since each iteration is independent, there is no guarantee that the iterations are interleaved in any way, nor is there any need for this. If the number of workers is equal to the number of loop iterations, each worker performs a single loop iteration. If there are more iterations than workers, some workers perform several loop iterations. This is handled in an entirely automated manner, and we have no control over the allocation.

Moreover, the workers act as independent MATLAB processes, and we need to ensure that they can access all the functions and I/O resources that they require.[887]

---

* *parpool* replaced *matlabpool* in R2013b: http://www.mathworks.com/help/distcomp/parpool.html (or: http://bit.ly/1btgU32).
† A possible future improvement to the MATLAB m-code interpreter would be to make it thread safe. This would enable automatic conversion of *for* loops into multiple threads running on multiple local CPUs/cores, significantly improving MATLAB's standard performance and essentially eliminating the need for a separate *parfor*.

A *parfor*-loop is useful in situations where we need to run many loop iterations of a simple calculation, which cannot be vectorized, such as a Monte Carlo simulation. Even when looping over a fully vectorized function, *parfor* still helps, since multithreading only kicks in when the data size is large enough,* and also since the loop overheads are all single-threaded. In the following example, although both *max* and *svd* are vectorized and theoretically multithreaded, using multiple threads does not improve the overall performance by more than a few percent. On the other hand, *parfor* achieves a 3.2x speedup with four workers,[888] which is rather efficient.[†]

```
y = zeros(1000,1); % pre-allocate (see §9.4.3)
for n = 1 : 1000    % replace for with parfor for parallelization
    y(n) = max(svd(randn(n)));
end
```

parfor-loops are also recommended when we have loop iterations that take a long time to execute, since workers execute iterations simultaneously. The headless workers run independently of each other, typically on separate local CPUs/cores, processing whatever loop iterations are assigned to them. However, the major requirement is that there must be *no data dependencies* between loop iterations. Such dependencies would complicate the interprocess communication enormously; so, MATLAB simply refuses to execute such loops in parallel, to ensure consistent results. After all, *correct* and *consistent* is always better than *faster*.

We should also consider the *communication overhead*, since data need to be transferred between the main MATLAB process and the workers. There might be no advantage to using *parfor*-loops when we have only a small number of simple calculations.[889] On the other hand, there is also a maximal limit to the data size per transfer: 600 MB for 32-bit machines and 2 GB for 64-bit machines.[890]

There is one limitation, which differentiates *parfor* from conventional *for*-loops: the iteration values of a *parfor* statement must be *increasing consecutive integers*. The examples in the following table illustrate this:

parfor Statement	Validity
`parfor i = 1 : 100`	Valid: this range is ok.
`parfor i = -20 : 20`	Valid: this range is ok.
`parfor i = 1 : 2 : 25`	Invalid: 1, 3, 5,… are not consecutive.
`parfor i = -7.5 : 7.5`	Invalid: −7.5, −6.5,… are not integers.
`A = [3 7 -2 6 4 -4 9 3 7];` `parfor i = find(A > 0)`	Invalid: the resulting range, 1, 2, 4,… has nonconsecutive integers.

To use *parfor*-loops, we first need to start a MATLAB worker pool with the *parpool* command:

```
>> parpool local
Starting parallel pool (parpool) using the 'local' profile ... connected
to 4 workers.
```

---

* See §5.1.1.
[†] See §1.7 for a discussion of Amdahl's law and the limitations of parallelization speedups.

This command creates a pool of 4 workers on the client machine using the default *local* profile.* Alternatively, one can use a custom profile to start workers remotely on a computing cluster and specify the number of workers explicitly:

```
parpool('CustomProfile', 10)
```

which starts a worker pool using a cluster profile CustomProfile with 10 workers. If you encounter problems while opening *parpool*, validate your cluster profile using the Desktop menu item **Parallel > Manage Cluster Profiles**. Also, ensure that MATLAB has enough permissions to access the directory where it keeps temporary data.[891]

Now, assume that A, B, and C are variables and that f, g, and h are functions:

```
parfor idx = 1 : n
    t = f(A(idx));
    u = g(B(idx));
    C(idx) = h(t, u);
end
```

If evaluating the functions f, g, and h is expensive, *parfor* will be significantly faster than the corresponding *for* statement, even if n is relatively small. Also note that the assignments to the variables idx, t, and u do *not* affect variables with the same name in the context of the *parfor* statement. This is because the body of the *parfor*-loop is executed in parallel and hence there is no deterministic way to guess the "final" values of these variables. In contrast, the values of vector C are set deterministically.

As mentioned above, *parfor* does not support loops with data dependencies. There is, however, one exception to this rule, when the dependent variable appears in a simple *reduction* assignment.[892] A reduction is an accumulation across iterations of a loop. The following examples demonstrate valid *parfor* constructs:

```
% computing factorial using parfor
x = 1;
parfor idx = 2:100
    x = x * idx;            % reduction assignment
end

% reduction assignment with insert operator
d = [];
parfor idx = -10 : 10
    d = [d, idx*idx*idx];   % reduction assignment
end

% reduction assignment of a row vector
v = zeros(1,100);
parfor idx = 1:n
    v = v + (1:100)*idx;    % reduction assignment
end
```

With a bit of thinking, we can make several different MATLAB functions run simultaneously within the body of a *parfor*-loop, for example, as follows:[893]

```
funcList = {@fun1,@fun2,@fun3};    % list of function handles
dataList = {data1,data2,data3};    % and their parameters
```

---

* The default workers pool size is the number of physical cores. See the detailed footnote at the top of §6.1. The term *cluster profile* should not be confused with the term *code profiling*, which relates to the process of analyzing code runtime behavior and speed (see Chapter 2).

```
parfor idx = 1 : length(funList)
    funcList{idx}(dataList{idx}); % this is run in parallel
end
```

When the computations are done, the workers pool should be closed or released:

```
delete(gcp)
```

Variables that appear in the body of a *parfor*-loop can be attributed to one of the classes listed in the following table:

Data Class	Description
Loop	A loop index variable for arrays
Sliced	An array whose segments are manipulated on different loop iterations
Broadcast	A variable defined before the loop whose value is used inside the loop but never modified
Reduction	Accumulates a value across loop iterations, regardless of iteration order
Temporary	Variable created inside the loop, but unlike sliced or reduction variables, not used outside the loop

These variable classes are exemplified in the following code fragment:

```
c = pi;  s = 0;
X = rand(1,100);
parfor k = 1 : 100
    a = k;              % a - temporary variable; k - loop variable
    s = s + k;          % s - reduction variable
    if i <= c           % c - broadcast variable
        a = 3*a - 1;
    end
    Y(k) = X(k) + a;    % Y - output sliced var; X - input sliced var
end
```

Probably, the most important data class in this classification are *sliced* variables. They reduce communication between the client and workers, since only those slices needed by a particular worker are sent when it starts working on a particular range of indices. Variables are sliced if all the following conditions are met (loop variable is denoted by i):

- The first level of indexing is either parentheses '()' or braces '{}': For example, A.q{i} or A.s(i,12) are not sliced, while the variables A{i}.q or A(i,12).s are.

- Within the first-level parentheses or braces, the list of indices is the same for all occurrences of a given variable: for example, h(A(i),A(i+1)) is not sliced, while f(A(i),A{i}) is sliced (to slice the former, we could do: a=A(i:i+1); h(a(1),a(2))).

- Within the list of indices for a variable, *exactly* one index involves a loop variable in a simple expression, and every other index is constant, a broadcast variable, colon, or end: for example, A(i,i + 1) or A(i + 1,20:30,end) are not sliced, while A(i,:,end) or A(i + k,j,:,3) are sliced.

- The shape of an output-sliced variable is not changed (no deletion or insertion operators allowed): for example, in the operators A(i,:) = [] or A(end + 1) = i, the variable A is not sliced.

Variables that do not fulfill these conditions, and are not affected by an assignment inside a loop, will be declared as *broadcast* variables and sent to all workers. In some cases of misuse, an error will be evoked saying that *parfor* cannot be run.[894] In other cases, the full data will simply be broadcast to all workers, degrading performance.[895]

Special rules also apply for reduction variables that appear in a reduction assignment within a *parfor*-loop. The reduction assignment can take one of the following forms:

```
X = X op expr
X = expr op X
X = f(X, expr)
X = f(expr, X)
```

where X is a reduction variable, `expr` is a MATLAB expression, `op` is an associative binary operation (arithmetic, logic, or set operations (such as set insertion, intersection, or union), and `f` is a general function (defining some associative binary operation) that is not changed within the *parfor* body. The rules for reduction are:

- For any reduction variable, the same reduction function or operation `op` must be used in *all* reduction assignments for this variable: for example, it is not allowed to have both A = A + i and A = [A,4 + i] in the same loop.

- If the reduction assignment uses * or the insertion operator, then in every reduction assignment, the assignment variable must be consistently specified as either the first or second argument: for example, it is not allowed to have both A = [A,i] and A = [r(i),A] in the same loop.

- Operators && and || are not allowed in reduction assignments.

Other variables that appear in an assignment within a *parfor*-loop but do not obey the above rules will be classified as *temporary* variables. Temporary variables exist only in the worker's workspace and are not transferred back to the client. Having a clear understanding of your *parfor*-variables will prevent many surprises when the parallel performance suddenly degrades.[896] In case of any doubt, try to restructure your code or rethink whether using a *parfor* construct is in general appropriate.

Note that MATLAB commands that update the GUI (graphical user interface) are not allowed inside the body of a *parfor*-loop and will report a runtime error.[897] There are two reasons for that. First, the workers that run the body of the *parfor*-loop are MATLAB sessions that are independent of the MATLAB session in which the GUI runs. Second, since the iterations of the *parfor*-loop are executed in parallel, the simultaneous graphical output from different workers would not be of much use.

Monitoring the progress of long-running *parfor*-loops is not easy, since workers run independently of each other and cannot communicate. Luckily, there are several free monitoring tools available, both for MATLAB GUI and command-line modes.[898]

To debug *parfor* loops, move the loop body into a separate function, use *parfor(…, 0)* (or force the pool not to open automatically), then set a breakpoint in that function. There is no way to run the MATLAB debugger on a worker, but we can attach native debuggers (such as gdb) to MEX code on workers. To debug worker data transfer, use *parfor(…,'debug')*: this forces *parfor* to run locally in the host instance with the same serialization behavior as for remote *parfor*. This may help if the transferred data has custom *saveobj/loadobj* behavior, or transient class properties.[899]

Finally, note that starting with release R2012b (8.0), the MATLAB Coder toolbox can generate parallel MEX code from *parfor*-statements.*

---

* Using MEX programs and MATLAB Coder are discussed in §8.1 and §8.2, respectively.

We next discuss *spmd* constructs, which are somewhat similar to *parfor*-loops, but provide more direct control over parallelism and data distribution between workers. Readers interested in some technical background into the inner working of *parfor* versus *spmd* are invited to read CSSM thread #306658.[900]

6.1.2 Using *spmd*

Along with *parfor*-loops, PCT also provides the *spmd* *(single program, multiple data)* language construct, which allows seamless interleaving of serial and parallel programming. The *spmd* statement defines a block of code to be run simultaneously on multiple workers that should be reserved using *parpool*. The "multiple data" aspect of *spmd* means that although the *spmd* statement runs identical code on all workers, each worker can have different, unique data for that code. Therefore, multiple data sets can be accommodated by multiple workers.

The general form of the *spmd* statement is

```
spmd
    statements
end
```

Single-line definition is also allowed:

```
spmd, statements, end
```

Optionally, one can also specify the minimal and maximal number of workers to be used for executing the parallel code, for example, *spmd (3)* or *spmd (2,4)*. Typically, *spmd* is used for simultaneous execution of a program on multiple data sets when communication or synchronization between the workers is required. Some common use- cases are:

- Long-running programs: Several workers compute solutions simultaneously
- Programs operating on large data sets: Data is distributed to multiple workers

In contrast to *parfor*-loops, each worker executing an *spmd* statement is assigned a unique value of *labindex* that lets us address the worker explicitly, to specify the code to be run on certain workers only, or for the purpose of accessing unique data.* The total number of workers executing the block in parallel can be obtained using the *numlabs* value. This level of control, which is not available in *parfor*, enables allocating data among workers intelligently, for example to achieve *load balancing*.[901] The model makes it possible to exploit *data-level parallelism* if one can be extracted from a problem at hand. For instance, the following code initializes random arrays of different size depending on the value of *labindex*:[902]

```
spmd (3) % use exactly 3 workers
    if labindex == 1
        R = rand(9,9); % on worker #1
    else
        R = rand(4,4); % on workers #2,#3
    end
end
```

---

* Within *parfor* loops, *labindex* is always 1, but we can use the *getCurrentTask* function to get a unique task object. See http://mathworks.com/help/distcomp/getcurrenttask.html (or: http://bit.ly/1bQlqDw); http://mathworks.com/matlabcentral/newsreader/view_thread/306207#831053 (or: http://bit.ly/1bQlsLE).

The following example uses *labindex* to build several magic squares in parallel:

```
spmd
    q = magic(labindex + 2);
end
```

Values computed inside the *spmd* statement's body are returned in the form of *Composite* objects on the MATLAB client. A composite object contains references to the values stored on the remote MATLAB workers, and these values can be retrieved using *cell-array* indexing, that is, for the above example, we have:

```
>> q{1} % q is a Composite with one element per worker
     8     1     6
     3     5     7
     4     9     2
>> q{2}
    16     2     3    13
     5    11    10     8
     9     7     6    12
     4    14    15     1
```

The computed values are retained on the workers until the corresponding composites are cleared on the client or until the MATLAB pool is closed. Multiple *spmd* statements can continue to use the same variables defined in previous *spmd* blocks:

```
>> spmd (3), AA = labindex; end % Initial setting
>> disp(AA(:)) % Composite
    [1]
    [2]
    [3]
>> spmd (3), AA = AA * 2; end % Multiply existing value
>> disp(AA(:)) % Composite
    [2]
    [4]
    [6]
```

Note that the workers of an *spmd* statement are aware of each other. In other words, one can directly control data transfer between them, and use *codistributed* arrays among them. For example, consider the following code snippet:

```
spmd (3)
    RR = rand(30, codistributor());
end
```

This creates a *codistributed* array of size 30×30, where each worker gets a 30×10 segment of it. Codistributed array is a mechanism by which MATLAB partitions and distributes data across multiple workers. For instance, with four workers, an 80×1000 array can be partitioned by columns into 80×250 segments, or by rows into 20×1000 segments. Distributed and codistributed arrays are discussed in §6.1.3 in detail.

In addition to distributed arrays, MATLAB provides a number of communication functions to enable fine-grained control over parallel job execution within *spmd* blocks. The most important functions are listed in the following table:*

---

* The function names display an unmistakable affinity to PCT's origins in MPI and the MatlabMPI research project (see §7.2.4).

Communication Function	Description
labBarrier	Block execution until all workers reach this call
labBroadcast	Send data to all workers or receive data sent to all workers
labProbe	Test to see if messages are ready to be received from other lab
labReceive	Receive data from another lab
labSend	Send data to another lab
labSendReceive	Simultaneously send data to and receive data from another lab
gcat	Global concatenation of an array across all workers. The result array is duplicated on all workers, unless *targetlab* is specified.
gop	Global reduction using binary associative operation across all workers. The result is duplicated on all workers (unless *targetlab* is specified). For example, *gop(@max,data)*.
gplus	Global addition performed across all workers. The result is duplicated on all workers (unless *targetlab* is specified).

For example, consider the following function that reorders the elements of an array:

```
function out = shift_array(N)
    spmd
        if labindex == 1
            % generate & send the sequence 1..N to all workers
            A = labBroadcast(1,1:N);
        else
            % receive data on other workers
            A = labBroadcast(1);
        end

        % get an array chunk on each worker
        I = find(A > N*(labindex-1)/numlabs & ...
                A <= N*labindex/numlabs);

        % shift the data to the right among all workers
        to   = mod(labindex,   numlabs) + 1; % one to the right
        from = mod(labindex-2, numlabs) + 1; % one to the left
        I = labSendReceive(labTo, labFrom, I);

        % reconstruct the shifted array on the first worker
        out = gcat(I, 2, 1);
    end
end
```

Executing this function on eight workers produces the following results:

```
>> out = shift_array(20);  disp(out{1})
18  19  20  1  2  3  4  5  6  7  8  9  10  11  12  13  14  15  16  17
```

Note that **labSend** is asynchronous up to a certain data size. This means that it does not wait for the receiving lab to **labReceive** before returning, if the data size is small enough. The limit changes across MATLAB releases and possibly platforms: it is 64 KB on R2014a running on 64-bits Win7, but was higher on earlier releases.[903] Therefore, to increase lab performance, it may be better to chunk the data being sent.

Communication functions might also be useful if, for example, we have a parallel algorithm comprising two stages with an explicit synchronization point in between:[904]

```
step1Func = {@fun1,@fun2}; % define the function pointers for
step2Func = {@fun3,@fun4}; % our algorithm steps
dataList1 = {data1,data2}; % the corresponding parameter lists
dataList2 = {data3,data4};
spmd 2
    % run the first step of the algorithm in parallel
    step1Func{labindex}(dataList1{labindex});
    labBarrier; % barrier synchronization

    % run the second step of the algorithm in parallel
    step2Func{labindex}(dataList2{labindex});
end
```

As in *parfor*-loops, *spmd* workers are headless (no-display) MATLAB sessions. As a result, they cannot create plots or other graphic outputs on our desktop. Therefore, MATLAB commands that update the GUI are not allowed inside *spmd* statements.

R2013b (MATLAB 8.2) introduced the *parfeval* function, which runs MATLAB functions on parallel pool workers, without blocking the parent process (Desktop).[905] This enables asynchronous (nonblocking) parallelized programs, unlike those using *parfor* or *spmd*, which block processing. In addition, *parfeval* can use MATLAB GUI/graphics (plots, *waitbar*, etc.), which is impossible in *parfor/spmd* loops. Its syntax is:

```
F = parfeval(p,fcn,numout,in1,in2,...)
```

This requests asynchronous execution of function *fcn* on a worker contained in the optional parallel pool p (default = current pool), expecting numout output arguments and supplying as input arguments in1, in2, and so on. The asynchronous evaluation of *fcn* does not block MATLAB. F is a *parallel.FevalFuture* object, from which the results are obtained via *fetchNext(F)*, when any worker has completed evaluating *fcn*. Evaluation of *fcn* proceeds until it ends or is explicitly canceled by calling *cancel(F)*.

In the following example, we create a vector of multiple evaluation requests in a loop, and then retrieve, process, and report the individual results as they become available:

```
parpool('local', 4);
% create a vector of multiple evaluations in a loop:
for idx = 1 : 10
    % the size of the magic square is determined by idx
    f(idx) = parfeval(p,@magic,1,idx);
end

% collect the results as they become available:
magicResults = cell(1,10);
for idx = 1 : 10
    % fetchNext blocks until next results are available.
    [completedIdx,value] = fetchNext(f);
    magicResults{completedIdx} = value;
    fprintf('Got result #%d\n', completedIdx);
end
```

6.1.3 Distributed and Codistributed Arrays

A general idea behind distributed arrays is to provide a level of abstraction over a data distributed across server workers such that the data can be accessed as a single array from MATLAB workspace. A distributed array resembles a normal MATLAB array in the way we index and manipulate its elements but none of the elements exists on the client.

To remove any confusions, the difference between *codistributed* arrays, already encountered in the previous section, and *distributed* arrays, is just a matter of perspective: a codistributed array that exists on the workers is accessible on the client (the main MATLAB session) as a distributed array and vice versa.

Distributed arrays are created using the ***distributed*** function to distribute an existing array from the client workspace to the workers of an open MATLAB pool. We can then access the data as a codistributed array within an ***spmd*** block:

```
parpool('local',2)     % Create a parallel pool
W = ones(6,6);
W = distributed(W);    % Distribute to the workers
spmd
    T = W*2;        % Calculation performed on workers, in parallel
                    % T and W are both codistributed arrays here
end
T                   % View results in client.
whos                % T and W are both distributed arrays here
delete(gcp)         % Stop pool
```

Additionally, we can use overloaded methods of the ***distributed*** object to directly construct a distributed array on the workers without a preexisting array on the client side. These functions operate in the same way as their nondistributed counterparts in the MATLAB language. Therefore, we describe them very briefly here:

Distributed Method	Description
`distributed.cell(m,n,...)`	Create a distributed cell array
`eye(m,...,class,'distributed')`	Create a distributed identity matrix of given type
`distributed.spalloc(m,n,nzmax)`	Allocate space for a sparse distributed matrix
`distributed.speye(m,n)`	Create a distributed sparse identity matrix
`ones(m,n,...,'distributed')`	Create a distributed array of ones
`zeros(m,n,...,'distributed')`	Create a distributed array of zeros
`rand(m,n,...,'distributed')`	Generate a distributed array of uniformly distributed pseudo-random numbers
`randn(m,n,...,'distributed')`	Generate a distributed array of normally distributed pseudo-random numbers
`randi(m,n,...,'distributed')`	Generate a distributed array of distributed pseudo-random integer numbers
`true(m,n,...,class,'distributed')`	Create a distributed array of logical ones
`false(m,n,...,class,'distributed')`	Create a distributed array of logical zeros

When we distribute an array to a number of workers, MATLAB partitions the array into segments and assigns one segment of the array to each worker. A two-dimensional (2D) array can be partitioned horizontally, by assigning columns of the original array to the different workers, or vertically, by assigning rows, respectively. Generally, an array with

N dimensions can be partitioned along any of its N dimensions that is specified in the constructor. Consider the example:

```
>> spmd(4)
        A = zeros(80, 1000);
        D = codistributed(A)
    end
Lab 1: This lab stores D(:,1:250).
Lab 2: This lab stores D(:,251:500).
Lab 3: This lab stores D(:,501:750).
Lab 4: This lab stores D(:,751:1000).
```

Each worker has access to *all* segments of a codistributed array. However, access to the local segment is faster than to the remote segments, since the latter requires sending and receiving data between workers (especially when running on a parallel cluster). A local portion of the codistributed array can be explicitly accessed using *getLocalPart* function. The following example creates a codistributed array among four workers and prints local parts for each of them:

```
>> spmd(4)
        A = magic(4); % replicated on all workers
        D = codistributed(A, codistributor('1d', 1));
        L = getLocalPart(D)
    end
Lab 1: L = [16    2    3 13]
Lab 2: L = [ 5 11 10    8]
Lab 3: L = [ 9    7    6 12]
Lab 4: L = [ 4 14 15    1]
```

In the code, `codistributor('1d',1)` tells MATLAB to distribute the array A along its first dimension (rows). We shall consider different distribution schemes below.

The nondistributed form of a codistributed array can be restored using *gather* function. For instance, it can be used to consolidate the results of parallel computations on the client side: *gather* takes the segments of an array that reside on different workers and combines them into a replicated array on all workers, or into a single array on one worker.

Similar to distributed arrays, MATLAB provides several array constructor functions to build codistributed arrays of specific values, sizes, and classes "on-the-fly". These functions distribute the resultant array across the workers using a desired distribution scheme:

Distributed Method	Description
`codistributed.cell(m,n,...,codist)`	Create a codistributed cell array
`codistributed.colon(a,d,b)`	Generate a codistributed array from the vector a:d:b
`eye(m,...,class,'codistributed')` `eye(m,...,class,codist)`	Create a codistributed identity matrix of given type
`sparse(m,n,codist)`	Create a sparse codistributed sparse matrix
`codistributed.speye(m,...,codist)`	Create a sparse codistributed sparse identity matrix
`codistributed.linspace(m,n,...,codist)`	Generate a codistributed linearly spaced vector
`codistributed.logspace(m,n,...,codist)`	Generate a codistributed logarithmically spaced vector
`ones(m,n,...,'codistributed')` `ones(m,n,...,codist)`	Create a codistributed array of ones
`zeros(m,n,...,'codistributed')` `zeros(m,n,...,codist)`	Create a codistributed array of zeros

(continued)

Distributed Method	Description
`rand(m,n,...,'codistributed')` `rand(m,n,...,codist)`	Generate a codistributed array of uniformly distributed pseudo-random numbers
`randn(m,n,...,'codistributed')` `randn(m,n,...,codist)`	Generate a codistributed array of normally distributed pseudo-random numbers
`randi(m,n,...,'codistributed')` `randi(m,n,...,codist)`	Generate a codistributed array of distributed pseudo-random integer numbers
`true(m,n,...,class,'codistributed')` `true(m,n,...,class,codist)`	Create a codistributed array of logical ones
`false(m,n,...,class,'codistributed')` `false (m,n,...,class,codist)`	Create a codistributed array of logical zeros

In the table, `codist` parameter denotes a *codistributor* object specifying the desired distribution scheme (MATLAB supports 1D [one-dimensional] and 2D distributions with additional parameters). The versions with a 'codistributed' parameter create a codistributed array using the default distribution scheme.

Codistributor object can be created using one of the following constructors:

```
codist = codistributor()
codist = codistributor('1d', dim, part)
codist = codistributor('2dbc', lbgrid, blksize)
```

The first constructor, without parameters, is the same as `codistributor('1d')`. The scheme denoted by the string `'1d'` creates a codistributor object to distribute an array along a single specified subscript, the distribution dimension, in a noncyclic, partitioned manner. In the following example, we change the direction of distribution of an already-existing codistributed array using the ***redistribute*** function and a specific codistributor object:

```
spmd(4)
    % create a codistributed with column-wise distribution
    D = rand(8, 16, codistributor());

    % get the size of the local part on each worker
    size(getLocalPart(D))

    % redistribute the array along the rows
    X = redistribute(D, codistributor('1d',1));

    % get the size of the local part on each worker
    size(getLocalPart(D))
end
```

returns on each worker:

```
ans =
    8     4
ans =
    2    16
```

The last codistribution scheme, denoted by the string `'2dbc'`, creates a 2D block-cyclic codistributor object. In contrast to 1D distribution, where the segments of the codistributed array comprise a number of complete rows or columns of a matrix, here, the segments of the codistributed array are 2D square blocks. For example, let us consider a 6-by-6 matrix with ascending element values:

```
>> A = reshape(1:36, 6, 6)
ans =
    1     7    13    19    25    31
    2     8    14    20    26    32
    3     9    15    21    27    33
    4    10    16    22    28    34
    5    11    17    23    29    35
    6    12    18    24    30    36
```

We can distribute this array among four workers in 3-by-3 square blocks as follows:

```
spmd(4)
    codist = codistributor('2dbc', [2 2], 3);
    D = codistributed(A, codist)
end
```

This distributes the array among the workers in the following manner:

```
              Lab 1                    Lab 2
      1     7    13       19    25    31
      2     8    14       20    26    32
      3     9    15       21    27    33
      4    10    16       22    28    34
      5    11    17       23    29    35
      6    12    18       30    30    36
              Lab 3                    Lab 4
```

If the grid of workers (2 by 2 in our case) does not perfectly "overlay" the dimensions of the codistributed array, the grid of workers is repeatedly overlaid in both dimensions until all the matrix elements are included. To demonstrate this, we partition the same 6-by-6 matrix using 2-by-2 square blocks:

```
spmd(4)
    codist = codistributor('2dbc', [2 2], 2);
    D = codistributed(A, codist)
end
```

The first "row" of the matrix is distributed among workers 1 and 2. The second row among workers 3 and 4, and so on. The process continues until all the elements of the matrix are distributed:

```
              Lab 1         Lab 2         Lab 1
              1  7         13  19        25  31
              2  8         14  20        26  32
                          15  21
      Lab 3   3  9        16  22        27  33    Lab 3
              4  10       (Lab 4)       28  34
              5  11       17  23        29  35
              6  12       18  30        30  36
              Lab 1         Lab 2         Lab 1
```

Indexing into a nondistributed array is straightforward: each dimension is indexed within the range of 1 to the final subscript given by MATLAB *end* keyword. Accordingly, the length of any dimension is determined using either *size* or *length* function. In contrast, for codistributed arrays, these values are not so easily obtainable since the index range depends on the actual distribution scheme. For this purpose, MATLAB provides the *globalIndices* function, which tells us the relationship between indices on a local part and the corresponding index range in a given dimension on the distributed array.[906]

6.1.4 Interactive Parallel Development with *pmode*

The *pmode* functionality is very similar to *spmd*, and there should be little noticeable difference between running the code in *pmode* or *spmd*.[907] The *pmode* function lets us work interactively with a parallel job running simultaneously on several workers. Commands that we type at the *pmode* prompt in the *Parallel Command Window* are executed on all workers at the same time. Each worker executes the commands in its own workspace on its own variables. The variables can be transferred between the MATLAB client and the workers. All communication functions supported in *spmd* blocks (*labindex*, *labSend*, *labReceive*, etc.) can also be used in *pmode* sessions.

However, the main difference to *spmd* is that *pmode* does not allow us to freely interleave serial and parallel work as *spmd* does, and thus, we have less control over parallel execution. When we exit the *pmode* session, its job is effectively destroyed, the worker tasks exit, and all information and data on the workers are lost. Starting another *pmode* session always begins from a clean state.

Keep in mind that workers that run parallel tasks of a *pmode* command are MATLAB sessions without a display. As a result, the workers *cannot* interact with the GUI elements of the main (client) MATLAB session; so, plotting data is not possible within *pmode*. To achieve this, we need to transfer the data back to the client session using a special *pmode* command. The list of *pmode* commands is

- pmode start [prof] [numworkers]

  Starts *pmode* using (optionally) the PCT profile prof to define the cluster and number of workers (otherwise the default local profile is used). The number of active workers can be optionally specified by numworkers.

- pmode quit/pmode exit

  Stops the *pmode* job, deletes it, and closes the Parallel Command Window.

- pmode client2lab clientvar workers [workervar]

  Copies the variable *clientvar* from the MATLAB client to the variable *workervar* on the workers identified by *workers*, which can be either a single index or a vector of worker indices (lab numbers).

- pmode lab2client workervar worker [clientvar]

  Copies the variable *workervar* from the worker identified by *worker*, to the variable *clientvar* on the MATLAB client.

- pmode cleanup prof

  Deletes all parallel jobs created by *pmode* for the current user running on the cluster specified in the profile *prof*, including jobs that are currently running.

We can invoke *pmode* as either command or function. The following are equivalent :

```
>> pmode start myProfile 4
>> pmode('start', 'myProfile', 4) % alternative
```

A sample *pmode* usage follows. First, we start *pmode* using the local profile with 4 local workers:

```
>> pmode start local 4
```

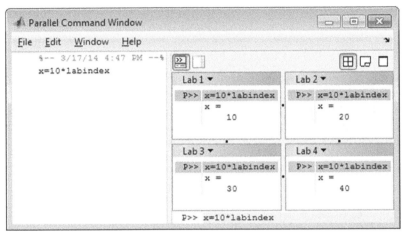

The *pmode's* interactive Parallel Command Window

In *pmode's* Parallel Command Window's prompt, we enter the following command:

```
P>> A = rand(1024, 1024, codistributor());
```

which creates a codistributed 2D array of random data on four workers. Each worker stores its 1024×256 part of the array locally. To collate the results to be transferred back to the client, we use the *gather* command:

```
P>> wholeA = gather(A);
```

Finally, the result array `wholeA` is copied back from worker #1 to the MATLAB client and we then close the *pmode* session:

```
>> pmode lab2client wholeA 1
>> pmode quit
```

We cannot transfer a codistributed array `A` to the client using *pmode lab2client* as is, since only a local portion of the array on the specified worker will be transferred. To transfer the entire array, we need to first use *gather* command to assemble the entire array into the workspaces of all workers, as the `wholeA` variable.

If we wish to plot distributed data obtained with *pmode*, we should first use the *gather* function to collect the entire array, and then transfer this array from the worker to the client using *pmode lab2client*, where the graphics functions can be used.

The benefit of using the interactive PMode window over *matlabpool/parpool* + *spmd* is its interactive graphic interface and the ability for simple interactive debugging. Once we are done, we can select the relevant commands in PMode's commands history panel and save them to an m-file script using the right-click context menu.

6.1.5 Profiling Parallel Blocks

Profiling helps identify and optimize various performance and memory usage bottlenecks in our program by tracking the execution time of each MATLAB function. The default MATLAB Profiler is activated with the *profile* command that was discussed in detail in §2.1. To profile parallel blocks, MATLAB provides an extension to the default Profiler and the Profiler viewer, which allows us to see how much time each worker spends evaluating each function versus time spent on communicating or waiting for communications with the other workers.[908] The MATLAB parallel profiler is enabled using the *mpiprofile* function:[909]

```
mpiprofile on <options>
```

This command can only be used within a communicating job task or entered at the *pmode* command prompt. It starts the parallel profiler and clears previously recorded profile statistics. Note that the parallel profiler does *not* work on *parfor*-loops. When the profiler is activated, it starts collecting statistics about the code execution on each worker and communications between workers. The collected information includes:

- Execution time of each function on each worker
- Execution time of each line of code in each function
- Amount of data transferred between each worker
- Amount of time each worker spends waiting for communications

mpiprofile collects data in a way similar to the standard *profile* command: meaning that for each MATLAB function, local function, or MEX function, *mpiprofile* records information about execution time, number of calls, parent functions, child functions, code line hit count, and code line execution time. There are several options to control *mpiprofile* behavior that are listed in the table below:

mpiprofile Option	Description
`-detail mmex` `-detail builtin`	Specifies the set of functions for which profiling statistics are gathered. The *mmex* option (default) records, records information about functions, local functions, and MEX functions. The *builtin* option also records information about built-in functions.
`-messagedetail default` `-messagedetail simplified`	Controls the level of detail at which communication data is stored. The *default* option collects information on a per-lab instance. *simplified* turns off collection of per-lab data fields which reduces the profiling overhead. This option might be useful while profiling on a very large cluster.
`-history`	Records the exact sequence of function calls (up to a million function entry and exit events by default).
`-nohistory`	Disables further recording of the history (exact sequence of function calls). All other profiling statistics continue to be collected.
`-history -historysize int`	Specifies the number of function entry and exit events to record. See §2.1.4 for details.

The parallel profiler will collect information about the running jobs until one of the following commands is issued:

```
mpiprofile off
mpiprofile reset
```

The `mpiprofile off` command stops the profiler without clearing previously recorded statistics. The `mpiprofile reset` command stops the profiler, clears up all gathered profile information and resets the profiler to the default state. To view the profiling results, use the command

```
mpiprofile viewer
```

which should be called in *pmode* after running a parallel program with *mpiprofile on*. Calling the viewer stops the profiler and opens the graphical profile browser with the function summary report. For each function, the following information is displayed in sortable columns:

Profiling Statistic	Description
Calls	How many times the function was called on this worker
Total Time	The total amount of time this worker spent executing this function
Self-Time	The time this worker spent inside this function, not within children or local functions
Total Comm Time	The total time this worker spent transferring data with other workers, including waiting time to receive data
Self-Comm Waiting Time	The time this worker spent during this function waiting to receive data from other workers
Total Interlab Data	The amount of data transferred to and from this worker for this function
Computation Time Ratio	The ratio of time spent in computation for this function versus total time (which includes communication time) for this function
Total Time Plot	Bar graph showing relative size of Self-Time, Self-Comm Waiting Time, and Total Time for this function on this worker

To display more details on a particular function, click on its name in the profile browser. We can display information for each worker, or use the comparison controls to display information for several workers simultaneously. Two buttons provide *Automatic Comparison Selection*, allowing us to compare the data from the workers that took the most versus the least amount of time to execute the code, or data from the workers that spent the most versus the least amount of time in performing interworker communication. *Manual Comparison Selection* enables us to compare data from specific workers or workers that meet certain criteria.

Let us consider a simple profiling session with *pmode*. First, we open a *pmode* session with four local workers:

```
pmode start local 4
```

Now, run the profiler by typing the following code in the Parallel Command Window:

```
P>> R1 = rand(1000, codistributor())
P>> R2 = rand(1000, codistributor())
P>> mpiprofile on
P>> P = R1*R2
P>> mpiprofile off
P>> mpiprofile viewer
```

The graphical profile view (for lab 3) is shown below (see top figure on page 304).

By clicking on `codistributor1d.hMtimesImpl` in the profile browser, we can see more detailed information on the selected function (see bottom figure on page 304).

Showing **all functions** called in lab 3

Automatic Comparison Selection	Manual Comparison Selection		No Plot
Compare (max vs. min TotalTime)	Go to lab: 3 ▼	Show Figures (all labs):	Plot Time Histograms
Compare (max vs. min CommTime)	Compare with: None ▼		Plot All PerLab Communication
			Plot CommTimePerLab

** Communication statistics are not available for ScaLAPACK functions, so data marked with ** might be inaccurate.

Function Name	Calls	Total Time	Self Time*	Total Comm Time	Self Comm Waiting Time	Total Interlab Data	Computation Time Ratio	Total Time Plot (dark band is self time and orange band is self waiting time)
codistributed.mtimes	1	0.256 s	0.040 s	0.021 s	0 s	11.44 Mb	91.8%	
codistributor1d.hMtimesImpl	1	0.191 s	0.149 s	0.021 s	0.009 s	11.44 Mb	89.0%	
distcomp.pInterPPromptFcn	2	0.034 s	0.034 s	0 s	0 s	0 b	100.0%	
codistributor1d.globalIndices	5	0.024 s	0.012 s	0 s	0 s	0 b	100.0%	
codistributed.display	1	0.021 s	0.002 s	0 s	0 s	0 b	100.0%	
codistributor1d.hDispImpl	1	0.018 s	0.003 s	0 s	0 s	0 b	100.0%	
codistributed.mtimes>iValidateInputArgs	1	0.015 s	0.001 s	0 s	0 s	0 b	100.0%	
...tor1d>codistributor1d.codistributor1d	1	0.015 s	0.003 s	0 s	0 s	0 b	100.0%	
codistributor1d.hDispImpl>getIndexExpr	1	0.014 s	0.002 s	0 s	0 s	0 b	100.0%	
codistributed.size	6	0.009 s	0.008 s	0 s	0 s	0 b	100.0%	
...codistributor1d>iParseConstructorArgs	1	0.009 s	0.003 s	0 s	0 s	0 b	100.0%	
CodistParser>CodistParser.isa	2	0.008 s	0.002 s	0 s	0 s	0 b	100.0%	
codistributor1d.hGlobalIndicesImpl	5	0.008 s	0.005 s	0 s	0 s	0 b	100.0%	
...or1d.hDispImpl>getIndexExprInDistrDim	1	0.008 s	0.002 s	0 s	0 s	0 b	100.0%	
codistributed.isaUnderlying	2	0.006 s	0.004 s	0 s	0 s	0 b	100.0%	
isPositiveIntegerValuedNumeric	15	0.005 s	0.005 s	0 s	0 s	0 b	100.0%	
Allocator>Allocator.create	1	0.005 s	0.001 s	0 s	0 s	0 b	100.0%	
codistributed.issparse	2	0.004 s	0.004 s	0 s	0 s	0 b	100.0%	
...odistributor.pVerifyGlobalIndicesArgs	5	0.004 s	0 s	0 s	0 s	0 b	100.0%	
...stParser>CodistParser.isValidLabindex	5	0.004 s	0.003 s	0 s	0 s	0 b	100.0%	

Function Name	Function Type	Calls
codistributed.mtimes	M-function	1

Lines where the most time was spent.

Line Number	Code	Calls	Total Time	Data Sent	Data Rec	Comm Waiting Time	Active Comm Time	% Time	Time Plot
121	LPC = LPC + LPA*LPB(k, :);	3	0.059 s	0 b	0 b	0 s	0 s	30.9%	
112	LPC = distributedutil.Allocato...	1	0.030 s	0 b	0 b	0 s	0 s	15.7%	
119	LPA = labSendReceive(to, from,...	3	0.024 s	5.72 Mb	5.72 Mb	0.009 s	0.012 s	12.6%	
114	LPC(:, :) = LPA*LPB(k, :);	1	0.024 s	0 b	0 b	0 s	0 s	12.6%	
111	codistrC = codistributor1d(2, ...	1	0.016 s	0 b	0 b	0 s	0 s	8.4%	
All other lines			0.038 s	0 b	0 b	0 s	0 s	19.9%	
Totals			0.191 s	5.72 Mb	5.72 Mb	0.009 s	0.012 s	100%	

** Communication statistics are not available for ScaLAPACK functions, so data marked with ** might be inaccurate.

Children (called functions)

Function Name	Function Type	Calls	Total Time	Data Sent	Data Rec	Comm Waiting Time	Active Comm Time	% Time	Time Plot
codistributor1d.globalIndices	M-function	4	0.018 s	0 b	0 b	0 s	0 s	9.4%	
...tor1d>codistributor1d.codistributor1d	M-subfunction	1	0.015 s	0 b	0 b	0 s	0 s	7.9%	
Allocator>Allocator.create	M-subfunction	1	0.005 s	0 b	0 b	0 s	0 s	2.6%	
codistributor1d.hLocalSize	M-function	1	0.004 s	0 b	0 b	0 s	0 s	2.1%	
...te/pRedistribute2DMatrixToDimOneOrTwo	M-function	2	0 s	0 b	0 b	0 s	0 s	0%	
Self time (built-ins, overhead, etc.)			0.149 s	5.72 Mb	5.72 Mb	0.009 s	0.012 s	78.0%	
Totals			0.191 s	5.72 Mb	5.72 Mb	0.009 s	0.012 s	100%	

To analyze profiling information from a parallel job outside *pmode* (in the MATLAB client), we first need to get the profiling results from each MATLAB worker:

```
info = mpiprofile('info'); % save profiling data for each worker
```

and then transfer it back to the client session. On the client side, the profiling data can be accessed as a cell array and displayed in the profile viewer as follows:

```
mpiprofile('viewer',[info{:}]); % convert cell array to vector
```

This form of *mpiprofile* is useful when we need to analyze the performance of an *spmd* job, since individual MATLAB workers cannot produce any graphical output.

The reader is referred to the documentation of the MPI profiler for additional features and usage information.[910]

The MPI profiler is designed to be used with *pmode*. If we are not using *pmode*, we could use the standard or MPI profiler commands within *spmd* blocks:[911]

```
spmd, profile('on', '-timer', 'real'); end  % or mpiprofile('on')
parfor, ..., end  % do some parallel computations on the workers
spmd, p = profile('info'); profile('off'); end  % or mpiprofile(...)
profview(0, p{1}); % or p{2} etc., or mpiprofile('viewer', p{1})
```

In a related matter, MathWorker Sarah Wait-Zaranek has contributed a *parfor* performance-visualization utility called *parTicToc* that displays the workers' iteration times and utilization. It complements the MPI profiler, although it does not provide the same detail level.[912]

Let us now return to the matched filter algorithm to see if there is a good fit for *parfor*-loops or *spmd* constructs. We shall not discuss the running example for *pmode* command separately since it is very similar to the one for *spmd*.

6.1.6 Running Example: Using *parfor* Loops

We can improve our serial implementation of the matched filter algorithm by noticing the fact that the variable data.im_final appears in a reduction assignment in the inner loop: see a "vectorized" version of the algorithm in §5.6.2. Although we can use *parfor* on the inner loop, it is better to parallelize the outer loop for reasons explained in §6.1.1. The new version of the algorithm is given below:*

```
function data = matched_filter_parfor(data)
    c = 299792458;        % speed of light
    R0     = data.R0;     % copy data fields to local variables
    AntX   = data.AntX;
    AntX   = data.AntX;
    AntX   = data.AntX;
    minF   = data.minF;
    x_mat  = data.x_mat;
    y_mat  = data.y_mat;
    z_mat  = data.z_mat;
    deltaF = data.deltaF;
    phdata = data.phdata;
```

---

* This file can be downloaded from: http://UndocumentedMatlab.com/books/matlab-performance (or: http://bit.ly/1pKuUdM).

```
% determine the size of the phase history data
K  = size(phdata,1);    % the number of frequency bins per pulse
Np = size(phdata,2);    % the number of pulses

% initialize the image with all zero values
im_final = zeros(size(x_mat), 'double');

% parallel loop through every pulse
parfor ii = 1 : Np
    % Calculate differential range for each image pixel (m)
    dR = sqrt((AntX(ii)-x_mat).^2 +...
              (AntY(ii)-y_mat).^2 +...
              (AntZ(ii)-z_mat).^2) - R0(ii);

    % Calculate the frequency of each sample in the pulse (Hz)
    freq = minF(ii) + (0:(K-1)) * deltaF;

    % Perform the Matched Filter operation
    for jj = 1 : K
        % Perform "reduction assignment" of im_final
        im_final = im_final + phdata(jj,ii)*exp(4i*pi*freq(jj)/c*dR);
    end
end

% return the computed image
data.im_final = im_final;
end
```

Note that we first copy relevant data to local variables, to ensure that MATLAB correctly *slices* the variables and does not *broadcast* them to all workers. Similarly, we use the im_final local variable for the output to ensure MATLAB recognizes it as a *reduction* variable.

Using a local MATLAB pool with **12** workers, the matched filter algorithm now runs in **25.2 s** to compute the 500×500 image, a 2× speedup.

6.1.7 Running Example: Using *spmd*

As an alternative to **parfor**-loops, we can use **spmd** blocks to execute loop iterations simultaneously. This requires only slightly more programming work since we need to consolidate the results computed by different workers manually. Yet, **spmd** provides us a greater control over the parallel execution:*

```
function data = matched_filter_spmd(data)
    % define speed of light
    c = 299792458;

    % determine the size of the phase history data
    K  = size(data.phdata,1);  % number of frequency bins per pulse
    Np = size(data.phdata,2);  % number of pulses

    % create distributed arrays from local data
    minF   = distributed(data.minF);
```

---

* This file can be downloaded from: http://UndocumentedMatlab.com/books/matlab-performance (or: http://bit.ly/1pKuUdM).

```
AntX    = distributed(data.AntX);
AntY    = distributed(data.AntY);
AntZ    = distributed(data.AntZ);
R0      = distributed(data.R0);
phdata = distributed(data.phdata);

% start parallel execution using spmd block
spmd
    im_slices = zeros(size(data.x_mat), 'double');

    % Loop through every pulse
    for ii = drange(1:Np)   % distributed range !
        % Calculate differential range for each image pixel (m)
        dR = sqrt((AntX(ii)-data.x_mat).^2 +...
                  (AntY(ii)-data.y_mat).^2 +...
                  (AntZ(ii)-data.z_mat).^2) - R0(ii);

        % Calculate the frequency of each sample in the pulse (Hz)
        freq = minF(ii) + (0:(K-1)) * data.deltaF;

        % Perform the Matched Filter operation
        for jj = 1:K
            im_slices = im_slices + phdata(jj,ii) *...
                                    exp(1i*4*pi*freq(jj)/c*dR);
        end
    end
end % spmd
% assemble the final image from "slices"
data.im_final = zeros(size(data.x_mat), 'double');
for ii = 1:numel(im_slices)
    data.im_final = data.im_final + im_slices{ii};
end
end
```

There are only several places where this algorithm differs from the *parfor*-loop version from §6.1.6. Namely, we have used **distributed** constructor to create distributed arrays from local data. Also, notice the usage of **drange** construct in the outer *for*-loop that instructs MATLAB that the loop is to be executed over a *distributed* range. Finally, at the end of the algorithm, we need to sum up the individual image "slices", available as a cell array, to yield the final result.

The *spmd* algorithm runs in **23.8** s using **12** workers on a local machine. This is comparable in speed to the *parfor*-loop version above.

6.2 The Parallel Computing Toolbox — GPUs

Besides multicore machines and computing clusters, PCT also lets us perform computations on CUDA-enabled Nvidia GPUs directly from MATLAB, using a special array type *gpuArray* and associated functions. It also enables us to program CUDA kernels* in C, using *CUDAKernel* objects.[913]

GPU acceleration was first introduced in MATLAB release R2010b (7.11). GPU array interface was provided for several built-in math functions, as well as support for calling CUDA kernels. R2011a (7.12) extended GPU arrays support to a larger set of built-in functions, supported GPU array indexing (via *subasgn* and *subsref* functions), and provided basic *arrayfun* support. R2011b (7.13) added support for GPU random number generation and MCR deployments. Subsequent releases further extended GPU support in terms of functionality, performance, and function coverage (in MATLAB and its toolboxes).[914] GPUs enable massive speedups of data-intensive algorithms: 100× and even 1000× speedups are not uncommon for some use cases.

I recommend watching MATLAB's GPU-computing tutorial[915] and webinars,[916] as well as reviewing PCT's code examples,[917] resources,[918] and numerous GPU programs on the File Exchange,[919] and especially the Mandelbrot Set example.[920]

6.2.1 Introduction to General-Purpose GPU Computing

We begin with a short introduction to the GPU-programming background.[921] Those readers who are familiar with the GPU architecture and CUDA framework can safely skip this section and go directly to §6.2.2 where we discuss *gpuArray*.

GPU is basically a stand-alone computation chip, similar to the computer's main CPU(s). GPUs are sometimes placed on the computer's motherboard, but are typically located on stand-alone cards (*devices*) having a parallel array of processors with dedicated memory. Originally used for accelerating computer graphics, GPUs were quickly found to have enormous potential for general/scientific programming.

The main difference between CPUs and GPUs lies in the fact that the major part of a GPU die area is dedicated for actual data processing (ALUs) rather than sophisticated flow control or data caching. GPUs are designed for *massive* numeric data processing where the same operation is applied to a large number of data elements. Large memory access latencies can be effectively hidden as long as the GPU can overlap ALU operations with memory access. Therefore, *arithmetic intensity* is one of the main prerequisites for high-performance GPU computing.

General-purpose GPU computing (GPGPU)[922] has been recognized as an important part of research and engineering with the release of Nvidia's *Tesla unified architecture* where vertex and fragment processors are unified in the so-called *Streaming Multiprocessors* (SMs). Tesla was later superseded by *Fermi* and *Kepler* GPU architectures but the underlying principles remained the same. Each SM consists of a number of CUDA cores (scalar in-order processors), special function units, load/store units, texture units, instruction fetch units, register file, and a block of shared memory. Lightweight GPU threads are executed on CUDA cores synchronously in groups of 32 threads called *warps*. Thousands of threads can simultaneously be scheduled for execution on the GPU.

---

* *Kernel* is a code that is compiled for, and runs on, the GPU rather than the CPU, enabling the same processing to run in parallel on multiple data (SPMD) using the numerous GPU cores.

CUDA is a heterogeneous serial–parallel programming model. This means that serial execution on the *host* (CPU) is interleaved with parallel execution on the device (GPU). A program running on the GPU across a large number of parallel threads is referred to as *kernel*. Kernels are written in the C language extended with additional keywords to express parallelism.

At the highest level, multiple GPU processing threads are grouped in a *grid* of *thread blocks*, which is launched for a single CUDA kernel. Block and grid configurations for each kernel call are set by the programmer. Threads within a block can communicate via SM's shared memory and synchronize using barriers. Threads from different blocks *cannot* communicate with each other within one kernel call. CUDA thread and memory hierarchies are schematically depicted in the following figure:

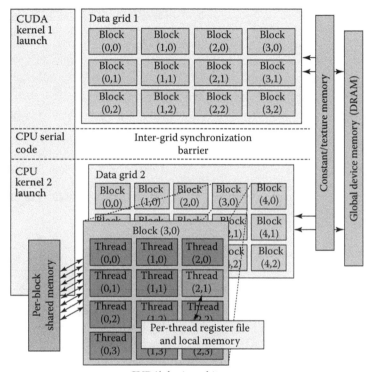

CUDA's basic architecture

CUDA programming model introduces six memory spaces shown in the figure above. The most important of them are *register file* which, together with *local memory*, forms each thread's private memory storage; on-chip *shared memory* used for interthread communications within a block; and *global memory* (external) that is visible to the entire grid of thread blocks and has a lifetime of an application.

Using a GPU to solve a problem is most effective when our problem is:[923]

- *Massively parallel:* Meaning that the computations can be broken down into many thousands of independent units of work. These units of work are created by distributing large sets of data across multiple GPU cores so that each core performs the *same* task on different pieces of data (*data-level parallelism*). Many linear algebra and simulation problems naturally fit into this category.

- *Computationally intensive:* The time spent on computations is significantly larger than the time spent for GPU–host memory transfers. Since the GPU is attached to the host CPU via the PCI Express bus, the memory access is slower than with a traditional CPU, implying that our overall computational speedup is limited by the amount of data transfer that occurs in our algorithm.

GPU-based computing in MATLAB can indeed be a very powerful alternative to multi-core processing, when the above conditions are satisfied.[924] One of MathWorks' presentations includes a representative example computing an N-body simulation on CPU versus GPU, with results showing GPU's effectiveness only for very large data:[925]

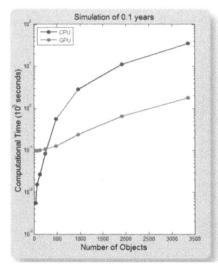

Objects	CPU (10³s)	GPU (10³ s)	Speed up
60	0.015	0.099	0.15
120	0.027	0.099	0.27
240	0.083	0.108	0.76
480	0.559	0.126	4.42
960	2.83	0.241	11.77
1920	11.3	0.655	17.17
3360	35.3	1.822	19.38

Let's now return to PCT to see its GPU-computing capabilities.*

6.2.2 Parallel Computing with GPU Arrays[926]

The first thing to do before starting with the GPU computations is to check whether our graphics card supports CUDA and has a sufficiently high *compute capability*. Note that not all GPUs are supported in MATLAB. To use GPU-accelerated computations in MATLAB, we need a CUDA-enabled Nvidia graphics card that has *compute capability* 1.3 (SM 1.3) or higher.† This limitation comes from the fact that, starting from SM 1.3, Nvidia GPUs support IEEE 754-compliant double-precision arithmetic; so, we are guaranteed to get the same computation results on the CPU and on the GPU. Most modern Nvidia GPUs have SM 1.3 or higher.[927] As a clarification, non-Nvidia GPUs are *not* currently supported by PCT.

To verify our GPU's compute capability, use the ***gpuDevice*** command:

```
>> gpuDevice
ans =
   parallel.gpu.CUDADevice handle
   Package: parallel.gpu
   Properties:
```

---

* Note that MATLAB is not the only mathematical software having integrated GPU support, see http://walkingrandomly.com/?p=3436 (or: http://bit.ly/18UESj1).

† While PCT requires Nvidia GPUs having compute capability 1.3 or higher, Jacket (§7.1) does not have these requirements. As new GPUs emerge that may also be IEEE compliant, it is expected that they too will become supported by PCT. MathWorks has announced that it will stop supporting SM 1.3 in some future version after R2014a, and will only support SM 2.0 or higher.

```
                   Name: 'Tesla S2050'
                  Index: 1
      ComputeCapability: '2.0'
         SupportsDouble: 1
          DriverVersion: 4.2000
      MaxThreadsPerBlock: 1024
       MaxShmemPerBlock: 49152
      MaxThreadBlockSize: [1024 1024 64]
            MaxGridSize: [65535 65535]
              SIMDWidth: 32
            TotalMemory: 2.8180e+09
             FreeMemory: 2.7349e+09
     MultiprocessorCount: 14
           ClockRateKHz: 1147000
            ComputeMode: 'Default'
    GPUOverlapsTransfers: 1
  KernelExecutionTimeout: 0
       CanMapHostMemory: 1
        DeviceSupported: 1
         DeviceSelected: 1
```

This command reports the GPU's **ComputeCapability** and other relevant properties (in particular, see the **DeviceSupported** flag) of the currently used GPU device.

Data stored on GPU is represented as a *gpuArray* object in MATLAB. A *gpuArray* object can be initialized from any of MATLAB's built-in data types: *single, double, int8, int16, int32, int64, uint8, uint16, uint32, uint64,* or *logical*. For instance:

```
X = rand(1000);   % 1000x1000 double created on the CPU
G = gpuArray(X);  % data converted into a gpuArray, transferred to GPU
```

This code snippet transfers a 1000-by-1000 data matrix X to the GPU and returns a *gpuArray* object G. Then, the GPU data can be manipulated using one of the methods defined for *gpuArray* objects, or it can be passed to the *feval* method of a CUDA kernel object. Note that cell arrays are *not* supported on the GPU.

Additionally, a *gpuArray* object provides a number of static methods to construct arrays directly on the GPU without data transfer from the MATLAB workspace. This is more efficient since it does not require CPU memory allocation, nor data transfer between the main memory and GPU. Some of these methods are listed in the following table:

gpuArray Method	Description
colon(a, d, b)	Generate a vector a:d:b
eye(m,..., class, codist)	Create an identity matrix of given type
linspace(m, n,..., codist)	Generate a linearly spaced vector
logspace(m, n,..., codist)	Generate a logarithmically spaced vector
ones(m, n,..., codist)	Create an array of ones of given type
zeros(m, n,..., codist)	Create an array of zeros of given type
rand(m, n,..., codist)	Generate an array of uniformly distributed pseudo-random numbers
randn(m, n,..., codist)	Generate an array of normally distributed pseudo-random numbers
true(m, n,..., class, codist)	Create an array of logical ones
false(m, n,..., class, codist)	Create an array of logical zeros

where codist is one of: 'distributed', 'codistributed', or 'gpuArray'.

Once the data is initialized and transferred to the GPU, there are three options for performing GPU computations in MATLAB:*

1. Use one of numerous built-in MATLAB functions that support *gpuArray* data.
2. Implement a custom MATLAB function of element-wise operations to be performed on the GPU, via *arrayfun* or *bsxfun*.
3. Create a custom GPU kernel that will process MATLAB's *gpuArray* data.†

In general, option #1 is simplest, while #3 is the most difficult to implement. Correspondingly, writing a dedicated kernel has the potential of delivering the best performance. MATLAB includes a working example comparing an implementation of these alternatives.[928] In this example, #1 provided a speedup of 18×, #2 provided 205×, and #3 provided a whopping 1279×. The decision of which method to use depends on whether all MATLAB functions in our code support GPUs and whether we wish to invest time rewriting the code to meet the requirements of element-wise processing.

The first approach, using MATLAB's built-in functions, is straightforward and does not require modification of the original program, except converting the input data into *gpuArray* type. It also requires no CUDA knowledge. A large subset of built-in functions with *gpuArray* support will be executed right away on the GPU (and return *gpuArray* as the result) if at least one of the input parameters is *gpuArray*. We can mix input from *gpuArray* and MATLAB workspace data in the same function call.

The basic method for transforming a CPU-based MATLAB program into one that runs on the GPU using the supported built-in MATLAB functions is almost trivial:

1. Move the input data from MATLAB to GPU using the *gpuArray* function, or directly create the input data on the GPU using static *gpuArray* methods.[929]
2. Use exactly the same processing code as for the CPU implementation.
3. Move the results back into MATLAB memory using the *gather* function.

Here is a simple usage example:

```
d = magic(1000);        % create data in MATLAB (CPU)
a = gpuArray(d);        % push the data to the GPU
f = fft(a);             % run FFT on GPU: fft() supports gpuArray natively
f = gather(f);          % bring the results back into MATLAB
R = norm(fft(d) - f);   % calculate difference between the CPU and GPU results

b = gpuArray.rand(1000,1); % create data directly on GPU
x = a \ b;      % backslash operator also supports gpuArray, runs on GPU
x = gather(x);      % bring the results back into MATLAB
```

As of MATLAB R2014b (8.4), 396 built-in MATLAB functions support *gpuArray* inputs, including math functions, FFT, element-wise operations, and linear-algebra operations. The list of supported functions (in core MATLAB and toolboxes) grows with each MATLAB release.[930] The complete list of functions supporting *gpuArray* in our current MATLAB release can be retrieved using the following command:

```
methods('gpuArray')
```

---

* Some toolboxes make GPU computing even simpler, by simply specifying the 'UseParallel' parameter (see §4.5.5, §9.5.9). See also a study of using GPU with PCT and Jacket libraries: http://walkingrandomly. com/?p=4062 (or: http://bit.ly/1dK3OvQ).
† See §6.2.5

Whereas the core MATLAB uses the BLAS and LAPACK linear-algebra packages for matrix computations, similar *gpuArray* operations use the MAGMA library.[931]

Aside from the built-in functions with *gpuArray* support, there are also MATLAB functions that are not methods of the *gpuArray* class, but work with *gpuArray* data. Some of these functions are listed in the table below:[932]

Small Subset of the Functions That Work with *gpuArray* Data (There Are Many More)				
angle	*flipud*	*iscolumn*	*kron*	*squeeze*
blkdiag	*flipdim*	*ismatrix*	*mean*	*std*
cross	*fftshift*	*isrow*	*perms*	*rot90*
fliplr	*ifftshift*	*isvector*	*rank*	*trace*

In most cases, if any of the input arguments to these functions is a *gpuArray*, their output arrays are also a *gpuArray*. If the output is always scalar, it is MATLAB data in the workspace. If the result is a *gpuArray* of complex data and all the imaginary parts are zero, these parts are retained and the data remains complex.

The second approach is based on invocation of **arrayfun** and **bsxfun** methods on user functions that accept *gpuArray* data. The semantics of **arrayfun** and **bsxfun** here is much the same as that of the corresponding MATLAB built-in functions,* with the exception that the evaluation runs on the GPU. Any required data not yet present on the GPU is moved to GPU memory. The *arrayfun* method has the following syntax:

```
[A, B,...] = arrayfun(FUN, C,...)
```

where FUN is a function handle to a user MATLAB function, which takes several arrays (at least one of them should be *gpuArray*) as input parameters, and returns *gpuArray*s A, B, and so on. For example, assume that we have the following user function:

```
function [out1, out2] = proc_fun(a, b, c)
    out1 = a.* b + c;
    out2 = out1.* a + 123;
end
```

The function gpu _ fun performs element-wise arithmetic on three input arrays, and returns two *gpuArray*s as a result. Now, use *arrayfun* to GPU-process data in parallel:

```
a = gpuArray(rand(1000));
b = gpuArray(rand(1000));
c = gpuArray(rand(1000));
[out1, out2] = arrayfun(@proc_fun, a, b, c);
```

Finally, use the *gather* function to retrieve the data from the GPU.

In this case, proc_fun is easily vectorized: (a.*b+c).*a+123. But the resulting multi-threaded CPU operation would still be outperformed by running proc_fun on the GPU as shown above, due to the numerous GPU cores compared to the few CPU cores.

---

* See §5.2.4. Note that not all MATLAB functions are supported for GPU executions, nor some of the *arrayfun* functionality. For example, indexing and singleton expansion is not supported on R2012b and earlier, and even then only in nested functions (http://mathworks.com/matlabcentral/newsreader/view_thread/333670 or: http://bit.ly/1ppM7qv).

MATLAB's documentation includes a complete working example, which shows how using a GPU with only *gpuArray* functions achieves a 12× speedup, while adding *array-fun* improves the speedup to 73×.[933] Mike Croucher has a similar example.[934]

The *bsxfun* method has the following syntax:

```
C = bsxfun(FUN, A, B)
```

This applies a binary element-wise *built-in* function[935] with the handle FUN to arrays A and B (at least one of them should be a *gpuArray*) and returns the output *gpuArray* C, which can be retrieved with the *gather* function. Either of the input arrays can be expanded along a singleton dimension (if there is one) to match the corresponding dimension of the other array. For example:

```
>> R1 = gpuArray.rand(2,5,4);
>> R2 = gpuArray.rand(2,1,4,3);
>> R = bsxfun(@plus,R1,R2);
>> disp(size(R))
   2    5    4    3
```

In this case, array R1 was expanded along its fourth (virtual) dimension to match array R2, and R2 was expanded along its second dimension to match R1.

When rewriting the code to be evaluated on the GPU using *arrayfun* or *bsxfun*, keep in mind the following limitations:[936]

- Nested and anonymous functions cannot access their parent function workspace.
- Overloading the supported functions is not allowed.
- The code cannot call scripts, nor functions in command syntax (e.g., clear all).
- Only simple data constructs are supported in the invoked code.
- Indexing (*subsasgn, subsref*) is not supported.
- The following language features are not supported: persistent or global variables; *parfor, spmd, switch*, and *try/catch*.
- All double-precision calculations are IEEE compliant, but because of hardware limitations, single-precision calculations (for which GPUs are optimized) are not.
- The supported data-type conversions are *single, double, int8, uint8, int16, uint16, int32, uint32*, and *logical*.
- Like *arrayfun* in MATLAB, matrix exponential power, multiplication, and division (^, *,/, \) perform only element-wise calculations.
- There is no *ans* variable to hold unassigned computation results. Explicitly assign the results of all calculations that we need to access to variables.
- When generating random matrices with *rand, randi*, or *randn*, we do not need to specify the matrix size, and each element of the matrix has its own random stream. Michael Weidman explained how this can be expanded to *correlated* random number generators on GPU, with interesting performance insights.[937]

Along with *bsxfun*, MATLAB release R2013b (8.2) also offered a new function *pagefun*,[938] which iterates over the pages of a *gpuArray* and applies the same function to each page.* The function has the following syntax:

---

* By page here, we mean a two-dimensional slice of a multidimensional array. For example, rand(2,3,4,5) has 4×5 = 20 pages of size 2×3.

```
A = pagefun(FUN,B,C,...)
[A,B,...] = pagefun(FUN,C,D,...)
```

The first variant invokes a function with a handle FUN using pages of the arrays B, C, and so on (with scalar expansion enabled). Any scalar dimensions of the input page are virtually replicated to match the size of the other arrays in that dimension:

```
A(:,:,I,J,...) = FUN(B(:,:,I,J,...),C(:,:,I,J,...))
```

The second *pagefun* variant applies a function FUN that returns multiple outputs, that is, *gpuArray*s A, B, and so on, each corresponding to one output argument of FUN. The function is applied using pages of input arrays C, D, and so on, with as many outputs as there are in the call to *pagefun*. The following example multiplies an M × K matrix (replicated P times) by a set of P matrices of size K × N yielding an array of size M × N × P:

```
A = gpuArray.rand(M,K);      % a matrix of size M x K
B = gpuArray.rand(K,N,P);    % P matrices of size K x N

% invoke matrix multiply for P pages with scalar extension for A
C = pagefun(@mtimes,A,B);
s = size(c); % return M x N x P
```

PCT provides multi-GPU support that can best be utilized through integration with other parallel constructs such as *spmd* or *pmode*. The information about the currently used GPU device can be obtained as follows:*

```
D = gpuDevice(deviceIndex);
```

This makes the GPU device specified by deviceIndex (>0) active, and returns its parameters as a *GPUDevice* object. The total number of GPU devices found in the system is given by the *gpuDeviceCount* function. Using *gpuDevice*, we can run different jobs on multiple GPUs if present and detectable in the system.

Note that the time for GPU–host memory transfers initiated by *gpuArray* calls can be noticeably large, especially if the latter calls appear many times in the loop. Therefore, a usual practice is to preload the data on the GPU "all at once", and then manipulate it using the methods of *gpuArray*. To check if some data already exist on the GPU, the function *existsOnGPU* can be used:

```
is_valid = existsOnGPU(G);
```

This returns *true* if the *gpuArray* or *CUDAKernel* object, passed as a parameter, points to valid data (data can be invalidated if the current device has been reset). We shall discuss *CUDAKernel* objects later in this section.

Starting with release R2013b (8.2), MATLAB also provides the *gputimeit* function to measure the running time of a single GPU routine.[939] This function ensures that all GPU operations have finished before recording the time, and compensates for the overhead. Like *timeit*,[†] it takes a function handle as input and returns the elapsed time:

```
A = gpuArray.rand(12000,400);
B = gpuArray.rand(400,12000);
f = @() sum(A.' .* B, 1);
t = gputimeit(f)
```

---

* This works well for a limited number of GPUs; with multiple GPU boards and dozens of GPUs, consider using dedicated solutions such as www.acceleware.com

† See §2.2.3.

Users of previous MATLAB releases can use the *wait* function for a selected GPU device in combination with *tic/toc* as an alternative to *gputimeit*:

```
tic
C = sum(A.'.* B, 1);
wait(gpuDevice);
toc
```

6.2.3 Running Example: Using GPU Arrays

It is now time to evaluate the performance of *gpuArray* on a real problem. The code for the matched filter (§5.6.2) adapted for GPU processing is as follows:*

```
function data = matched_filter_gpuarrays(data)
    % define speed of light
    c = 299792458;

    % determine the size of the phase history data
    K  = size(data.phdata,1); % number of frequency bins per pulse
    Np = size(data.phdata,2); % number of pulses

    % transfer data to the GPU
    minF   = gpuArray(data.minF);
    AntX   = gpuArray(data.AntX);
    AntY   = gpuArray(data.AntY);
    AntZ   = gpuArray(data.AntZ);
    R0     = gpuArray(data.R0);
    phdata = gpuArray(data.phdata);
    deltaF = gpuArray(data.deltaF);

    % initialize the image slices with all zero values
    im_slices = gpuArray.zeros([K size(data.x_mat)], 'double');

    % loop through every pulse
    for ii = 1:Np
        % Calculate differential range for each image pixel (m)
        ax = bsxfun(@minus,AntX(ii),data.x_mat);
        ay = bsxfun(@minus,AntY(ii),data.y_mat);
        az = bsxfun(@minus,AntZ(ii),data.z_mat);

        % computes sqrt(ax^2 + ay^2 + az^2)
        dR = bsxfun(@hypot, ax + 1i*ay, az) - R0(ii);

        % Calculate the frequency of each sample in the pulse (Hz)
        freq = minF(ii) + 0 : deltaF : deltaF*(K-1);

        % Perform the Matched Filter operation
        dR = reshape(dR, [1 size(dR)]);
        dR = repmat(dR, K, 1);  % replicate data for bsxfun
        t = exp(1i*4*pi/c*bsxfun(@times, freq', dR));
```

---

* This file can be downloaded from: http://UndocumentedMatlab.com/books/matlab-performance (or: http://bit.ly/1pKuUdM).

```
            im_slices = im_slices + bsxfun((@times, phdata(:,ii), t);
        end % for ii = 1:Np

        % sum up image slices and copy the results back to CPU
        data.im_final = gather(squeeze(sum(im_slices,1)));
    end
```

The algorithm features a completely vectorized inner loop that we have already seen while discussing implicit parallelization techniques in MATLAB. Here, in contrast to a CPU version, this approach appears to be just the right choice for computing with *gpuArray*, which is thoroughly optimized for element-wise operations. Also remark the usage of *bsxfun* to perform the most expensive computations in the algorithm: namely, computing differential range and the matched filter operation. The code runs in **19.3** s on our platform.

Going further with optimizations, we next try to distribute computations across multiple GPU devices using the features of PCT.

6.2.4 Running Example: Using Multiple GPUs with *spmd* Construct

To enable multi-GPU support, we need to launch several MATLAB CPU workers, such that each of them will utilize a different GPU. Since our testing platform has four GPUs, we use a MATLAB pool with four workers:*

```
function data = matched_filter_gpuarrays_spmd(data)
    % define speed of light
    c = 299792458;
    % determine the size of the phase history data
    K  = size(data.phdata,1);  % number of frequency bins per pulse
    Np = size(data.phdata,2);  % number of pulses
    chunks = 8;  % partition data in several chunks
    Kc = K/chunks;
    spmd(4)   % launch 4 MATLAB workers
        gpuDevice(labindex);  % set current GPU device to worker's ID
        % transfer data to the GPU
        minF    = gpuArray(data.minF);
        AntX    = gpuArray(data.AntX);
        AntY    = gpuArray(data.AntY);
        AntZ    = gpuArray(data.AntZ);
        R0      = gpuArray(data.R0);
        phdata  = gpuArray(data.phdata);
        deltaF  = gpuArray(data.deltaF);
        % initialize the image slices with all zero values
        im_slices = gpuArray.zeros([Kc size(data.x_mat)], 'double');
        % loop through every pulse
        for ii = drange(1:Np)
            % Calculate differential range for each image pixel (m)
            ax = bsxfun(@minus,AntX(ii),data.x_mat);
            ay = bsxfun(@minus,AntY(ii),data.y_mat);
            az = bsxfun(@minus,AntZ(ii),data.z_mat);
            % computes sqrt(ax^2 + ay^2 + az^2)
```

---

* This file can be downloaded from: http://UndocumentedMatlab.com/books/matlab-performance (or: http://bit.ly/1pKuUdM).

```
        dR = bsxfun(@hypot, ax + 1i*ay, az) - R0(ii);
        dR = reshape(dR, [1 size(dR)]);
        dR = repmat(dR, Kc, 1); % replicate data for bsxfun
        % loop for each chunk of size Kc
        for jj = 0 : Kc : K-Kc
            % Calculate the frequency of each pulse sample (Hz)
            freq = minF(ii) + jj*deltaF + 0:deltaF:deltaF*(Kc-1);

            % Perform the Matched Filter operation
            t = exp(1i*4*pi/c*bsxfun(@times, freq', dR));
            im_slices = im_slices + bsxfun(@times,...
                        phdata(jj+1:jj+Kc,ii), t);
        end
    end  % for ii = 1:Np
    % sum up image slices and copy the results back to CPU
    im_slices = gather(squeeze(sum(im_slices,1)));
end % spmd
% assemble final image from pieces computed by MATLAB workers
data.im_final = zeros(size(data.x_mat));
for ii = 1 : numel(im_slices)
    data.im_final = data.im_final + im_slices{ii};
end
end
```

A few comments are due here. The first command in the *spmd* body sets the current GPU device for each worker according to its *labindex*. *gpuArray* objects cannot be initialized from distributed arrays on the host. Therefore, we need to copy the complete arrays to each GPU device independently.

Additionally, we have found that performing the matched filter operation over several groups (chunks) of size Kc turns out to be more efficient than processing the entire data in a single step, possibly due to reduced memory consumption on the GPU. The algorithm now runs in **7.4** s to reconstruct the image of 500×500. This is the best result we have been able to achieve so far, 8× faster than our CPU-vectorized version (§5.6.2).

Therefore, as a rule of thumb:

> *Once you have finished vectorizing your code, put the loop back in order to process data in several (large) blocks. This often results in additional speedup.*

In addition to using *spmd* for assigning different GPUs to different PCT workers, we could also use *parfor* loops:

```
spmd   % launch MATLAB workers
    gpuDevice(labindex);  % set current GPU device to worker's ID

    % transfer data to the GPUs
    data = gpuArray(data);
end

parfor idx = 1 : N
    % Do some calculations on the GPU
    results = process(data);

    % Gather the data back to Matlab host
    allResults(idx,:) = results;
end
```

When PCT workers share a single machine having multiple GPUs, by default MATLAB automatically assigns a different GPU to each worker. You can customize the way that MATLAB assigns workers to GPUs by overriding MATLAB's *selectGPU* function. Note: *selectGPU* is an internal function, whose documentation is provided within the m-file (*%matlabroot%/toolbox/distcomp/gpu/selectGPU.m*), and not as part of the official PCT documentation.

Also note that PCT does not currently support accessing multiple GPUs on the same device (card), as opposed to accessing a single GPU on each of several devices (which is supported).

6.2.5 Executing CUDA Kernels from MATLAB

PCT enables us to execute CUDA kernels directly from the MATLAB using *CUDAKernel* objects. *CUDAKernel* object is initialized from CU (CUDA source) and PTX* files.

Once we have a CU file containing our CUDA kernel to be executed on the GPU, we need to compile it to create a PTX file. One way to do this is with the nvcc compiler in the standard Nvidia CUDA Toolkit.[940] If we have a CU file called *myfun.cu*, the PTX file can be created with the following shell command:

```
nvcc -ptx myfun.cu
```

This will generate the file named myfun.ptx (or *myfun.ptxw64* on 64-bit Windows). The next step is to create a *CUDAKernel* object as follows:

```
kernel = parallel.gpu.CUDAKernel(PTXFILE, CUFILE, FUNC)
```

where PTXFILE is the name of the file that contains the PTX code, or the contents of a PTX file as a string; CUFILE is the CUDA source file that contains a kernel definition starting with __global__ (that's a "global" with two underscores on either side). MATLAB uses CUFILE to search for a function prototype for the CUDA kernel that is defined in PTXFILE. The optional parameter FUNC must be a string that unambiguously defines the appropriate kernel entry name in the PTX file (if the latter one contains several entry points). For instance:

```
kernel = parallel.gpu.CUDAKernel('myfun.ptx', 'myfun.cu');
```

We can set various properties of the resulting *CUDAKernel* object:

```
kernel.ThreadBlockSize = [512,1,3];
kernel.SharedMemorySize = 8*prod(kernel.ThreadBlockSize);
kernel.GridSize = [ceil(size(data,1)/512), size(data,2)]; %[rows,cols]
```

Having a kernel object, we evaluate the kernel on the GPU using the *feval* command:

```
out = feval(kernel, gpuArray(data));
```

This command evaluates the CUDA kernel on a vector of random data. Here, we assume that our kernel takes one input parameter (in our case, a pointer to global memory array) and returns one output (a pointer to another global memory array). The out variable is a *gpuArray* that can be used in other GPU computations using CUDA kernels or the

---

* Parallel Thread Execution (PTX) is an assembly language used by CUDA nvcc compiler. The nvcc compiler translates a C-like CUDA program to an intermediate PTX format. Finally, the graphics driver compiler translates PTX file to native GPU code.

methods of the ***gpuArray*** object. When all GPU computations are done, the result is transferred to MATLAB workspace via the ***gather*** command:

```
out = gather(out);
```

It is important to understand the relation between the C-prototype of a CUDA kernel and the input/output parameters supplied to the *feval* command. That is, when calling

```
[out1, out2] = feval(kernel, in1, in2, in3)
```

the inputs in1, in2, and in3 correspond to each of the input argument to the C function within our CU file. The out1, out2 outputs store the values of the *non-const* pointer input arguments to the C function after the CUDA kernel has been executed. For example, if the CUDA kernel within a CU file has the following signature:

```
void mykernel(const float *pIn, float *pInOut1, float *pInOut2, float c)
```

then the corresponding kernel object (kernel) in MATLAB will have the following properties:

```
MaxNumLHSArguments: 2
NumRHSArguments:    4
ArgumentTypes:      {'in single vector' 'inout single vector' 'inout
single vector', 'in single scalar'}
```

Thus, we can use *feval* on the ***CUDAKernel*** object with the following syntax to execute a CUDA kernel:

```
[y1, y2] = feval(k, x1, x2, x3, c)
```

where x1, x2, and x3 are vectors and c is a scalar.

Before launching the kernel on the GPU, we should set the kernel object's properties that control its execution behavior. These are listed in the following table:

Property Name	Description	Read-only ?
ThreadBlockSize	1×3 vector defining the size of a thread block along 3 dimensions, respectively	No
MaxThreadsPerBlock	Maximal number of threads per block on selected GPU device	Yes
GridSize	1×2 vector specifying the dimensions of a grid of thread blocks	No
SharedMemorySize	Amount of dynamically allocated shared memory (in bytes) to used by a CUDA kernel	No
EntryPoint	Name of the entry point of a CUDA kernel	Yes
MaxNumLHSArguments	Maximal number of input parameters	Yes
NumRHSArguments	Number of output parameters	Yes
ArgumentType	A vector describing the formal parameters of a CUDA kernel	Yes

The most important properties are `ThreadBlockSize` and `GridSize`, which need to be set explicitly for each CUDA kernel. `ThreadBlockSize` defines the geometry of a single-thread block. Threads from the same block can communicate with each other using shared memory and synchronize with barriers. The total number of threads must not exceed the value of `MaxThreadsPerBlock`. `GridSize` specifies the dimensions of a grid of thread blocks on which a CUDA kernel is launched.* Additionally, we can specify some

---

* Consult *CUDA C Programming Guide* (http://docs.nvidia.com/cuda/index.html or: http://bit.ly/1nzA1Yv) for additional information on kernel execution parameters.

amount of *constant memory* to be used by a CUDA kernel using the following method of
CUDAKernel object:

```
setConstantMemory(kern, sym1,val1, sym2,val2,...)
```

which sets the constant memory in the CUDA kernel kern with symbol name sym1 to
val1, with symbol name sym2 to val2, and so on. val1, val2, and so on can be any numeric
arrays, including *gpuArray*.

Let us now step through a full CUDA kernel workflow. We shall use the following example, which performs an element-wise linear interpolation between two vectors:

```
__global__ void vector_interp(double *out, const double *a,
                              const double *b, const double c) {
    //evaluate global index for each thread
    int idx = blockIdx.x * blockDim.x + threadIdx.x;

    //perform element-wise interpolation
    out[idx] = a[idx] * c + b[idx] * (1.0 - c);
}
```

We begin by compiling this source file vector _ interp.cu to PTX using nvcc:

```
nvcc -ptx vector_interp.cu
```

Next, a CUDA kernel object needs to be created:

```
kernel = parallel.gpu.CUDAKernel('vector_interp.ptx',...
                                 'vector_interp.cu');
```

The kernel object has the following properties:

```
MaxNumLHSArguments: 1
   NumRHSArguments: 4
```

meaning that our kernel takes four input parameters and returns one output. As the next
step, we initialize the input parameters and set the kernel object's required properties:

```
n_elems = 10000;
a = gpuArray(rand(n_elems,1));
b = gpuArray(rand(n_elems,1));
out = gpuArray.zeros([n_elems, 1]);
% set the kernel parameters
kernel.ThreadBlockSize = [256, 1, 1];
kernel.GridSize = [ceil(n_elems/kernel.ThreadBlockSize(1)), 1];
```

For the kernel, we set 256 threads per block since, in our case, thread cooperation is not
required, and therefore, it is reasonable to use moderately small blocks. The grid size is
chosen according to the number of elements in the input vectors. Finally, we launch the
CUDA kernel on the GPU and return the results to MATLAB workspace:

```
out = feval(kernel, out, a, b, 0.5);
out = gather(out);
```

Keep in mind that all CUDA kernel invocations are *asynchronous* meaning that after the
call to *feval*, the control is returned to the CPU *immediately*, before the actual processing
on the GPU is complete. On the contrary, the call to *gather* function is *blocking* since the

CPU must wait until all computations on the GPU side are done before the results can be retrieved. Sometimes, it may result in very confusing errors if there is a flaw in the logic of your CUDA kernel.[942] Therefore, to obtain reasonable time estimates for the execution of a CUDA kernel, it is recommended to use the *wait* function to block program execution until all operations on the GPU device are finished:

```
tic
out = feval(kernel, out, a, b, 0.5);
wait(gpuDevice); % wait until the kernel launch is completed
toc
```

When using the MATLAB Compiler to deploy MATLAB applications, note that PTX and CU files need to be added as shared resources to the build project, since MATLAB does not automatically understand that they need to be included.

A typical programming paradigm when coding CUDA kernels is to use the kernel for the most compute-intensive algorithmic core, and use MATLAB m-code for the so-called "glue code": data setup and handling, I/O, GUI, and visualization. We would often design the core algorithm also in MATLAB m-code, for its development, ease of use, and RAD benefits, then optimize performance using a CUDA variant, and finally validate the CUDA implementation by comparing the results to our original MATLAB results.

CUDA kernels promise much greater speedups than possible using pure m-code, even when employing *arrayfun*. However, although programming a CUDA kernel relieves us from the effort of coding any GPU glue code, it is still admittedly non trivial. This is seldom worth the extra development, debugging, and validation effort, except for highly compute-intensive tasks that take a large portion of the overall profiling "cake" and takes an unacceptable amount of time to complete. In most other cases, we should stick to standard vectorization and the use of *gpuArray* with built-in MATLAB functions.

Recent MathWorks newsletter article[942] and webinar[943] show an example of using a CUDA kernel for white balancing an image. A complete working example of calling CUDA code from MATLAB was recently posted on Nvidia's development blog.[944]

To debug and profile CUDA code, consider using Nvidia's Visual Profiler tool[945] and Nvidia's Nsight plugin for Visual Studio or Eclipse.[946]

CUDA programming is a very broad topic on its own right, which goes beyond the scope of this book. There are many parameters governing GPU performance that need to be accounted for. For those readers not familiar with the CUDA framework, start by visiting Nvidia's CUDA Zone.[947]

A solid introduction to CUDA programming with plenty of examples can be found in the book *CUDA by Example* by J. Sanders and E. Kandrot.[948]

6.2.6 Running Example: Using CUDA Kernels

Perhaps, one of the strongest arguments toward using CUDA kernels in MATLAB is that we are no longer restricted to process data in an *element-wise* manner on the GPU but, instead, can develop any specific data flow suitable for our problem. Often enough, it implies that many vectorization constructs (such as *repmat, reshape, arrayfun*, etc.) can be eliminated in favor of more direct data processing.

We shall demonstrate this on our matched filter algorithm. Recall that in §6.2.3 and §6.2.4, we have used *repmat* function to replicate a differential range matrix dR. This was required since, the inner loop of the algorithm, each "copy" of dR must be multiplied by a different frequency value `freq`. Now, we can easily model this data flow in the CUDA kernel:*

```
const double c = 299792458.0;//speed of light
//the core of the Matched filter algorithm
__global__ void mf_core(double2 *out, const double *dR,
          const double *freq, const double2 *phdata, const int numel)
{
    //calculate a global thread index within a matrix dR
    int idx = blockIdx.x*blockDim.x + threadIdx.x;
    if (idx >= numel) //early bail-out for unused threads
        return;
    double f = freq[blockIdx.y];
    double2 ph = phdata[blockIdx.y], r;
    double t = 4.0 * M_PI * f * dR[idx]/c, re, im;
    sincos(t, &im, &re);
    r.x = ph.x*re + ph.y*im; //multiply two complex numbers
    r.y = ph.x*im - ph.y*re;
    idx = idx + blockIdx.y * numel;
    out[idx] = r;
}
```

Note that in CUDA, complex numbers are represented by the `double2` number type that is composed of two `double` values. We also use a CUDA built-in function *sincos* to compute sine and cosine in a single operation. The new version of the matched filter algorithm that uses the above CUDA kernel looks as follows:

```
function data = mf_gpuarrays_cuda(data)
    % Determine the size of the phase history data
    K  = size(data.phdata,1);  % number of frequency bins per pulse
    Np = size(data.phdata,2);  % number of pulses
    minF   = gpuArray(data.minF);
    AntX   = gpuArray(data.AntX);
    AntY   = gpuArray(data.AntY);
    AntZ   = gpuArray(data.AntZ);
    R0     = gpuArray(data.R0);
    phdata = gpuArray(data.phdata);
    deltaF = gpuArray(data.deltaF);
    mf_core = parallel.gpu.CUDAKernel('mf_core.ptx', 'mf_core.cu');
    mf_core.ThreadBlockSize = [256, 1, 1];
    n_elems = numel(data.x_mat);
    mf_core.GridSize = [ceil(n_elems/mf_core.ThreadBlockSize(1)), K];
    data.im_final = gpuArray.zeros(size(data.x_mat), 'double');
    im_slices = complex(gpuArray.zeros([size(data.x_mat) K], 'double'));
    for ii = 1 : Np % Loop through every pulse
        % Calculate differential range for each image pixel (m)
        ax = bsxfun(@minus,AntX(ii),data.x_mat);
        ay = bsxfun(@minus,AntY(ii),data.y_mat);
        az = bsxfun(@minus,AntZ(ii),data.z_mat);
        dR = bsxfun(@hypot, ax + 1i*ay, az) - R0(ii);
        freq = minF(ii) + 0 : deltaF : deltaF*(K-1);
        % Perform the Matched Filter operation
```

---

* The files in this section can be downloaded from: http://UndocumentedMatlab.com/books/ matlab-performance (or: http://bit.ly/1pKuUdM).

```
    im_slices = feval(mf_core, im_slices, dR, freq,...
                      phdata(:,ii), n_elems);
    data.im_final = data.im_final + squeeze(sum(im_slices,3));
  end
  data.im_final = gather(data.im_final);
end
```

The `mf_core` CUDA kernel constitutes the core of the matched filter algorithm. We launch the kernel on a grid of size `n_elems`×K, so that the first grid dimension indexes the elements of the matrix `dR` while the second dimension indexes different frequency values `freq`. In this way, we avoid the need to replicate the matrix `dR` multiple times.

The algorithm executes in **3.4** s using just a single GPU, which is a new best running time, 17× faster than our CPU-vectorized version (§5.6.2) and 2× faster than our *gpuArray*-based version (§6.2.4). Note that on the GPU devices with a compute capability below 2.0, it would make sense to access the arrays `freq` and `phdata` in the CUDA kernel through GPU's constant memory, or to preload them in shared memory first. However, on Fermi and higher GPU architectures, this optimization is unnecessary since constant data can be accessed via uniform cache. We leave it as an exercise to the reader to extend the algorithm to multiple GPUs.

6.2.7 Programming GPU Using MATLAB MEX

In the early days of GPU programming, MEX files that called CUDA were the only way to program the GPU in MATLAB. These files had to transfer data explicitly from MATLAB to the GPU. MEX did not support NVCC and CUDA, and so, third-party solutions provided the ability to compile and link CUDA code with MEX.

With the introduction of *CUDAKernel* and *gpuArray* objects in MATLAB, MEX files are needed less often, as *CUDAKernel* objects relieve MATLAB developers from having to deal with GPU memory and kernel execution, while performance is usually on par with that attained using the MEX interface. Therefore, we shall keep our discussion on using MEX interface for GPU programming concise.*

However, there are several occasions where, due to *gpuArray* and *CUDAKernel* limitations, using MEX to program the GPU is useful. As of MATLAB R2014a (8.3), *gpuArray* cannot access peer GPU memory, if the system has several GPU devices. As a result, a *gpuArray* allocated on one device cannot be used on other devices; so we need to copy data from the host to each GPU device separately. Another possible situation happens when a CUDA application needs to access GPU *texture* memory, something not currently supported by *CUDAKernel*. MEX may also be used on old MATLAB versions without *CUDAKernel* support, or when PCT is unavailable.

An important additional use case: MEX provides access to the large set of highly tuned C++ libraries developed by the GPU community[949] including CUFFT, CUBLAS, CURAND, CUSPARSE, NPP,[950] Thrust,[951] ArrayFire,[952] OpenACC,[953] CULA,[954] and so on. We can always call C++ library functions directly from within the *.cu* kernel file, but some users might find it easier to code a standard C++ MEX file that calls these functions, rather than deal with kernels, even if we do have PCT.[†] Also, note that MEX can call host-callable libraries such as CUFFT, while *CUDAKernel* files cannot. Finally, professional GPU libraries often outperform hand-crafted kernel code. MATLAB documentation includes a dedicated page[955] and usage example[956] devoted to implementing MEX–CUDA code.

---

* We assume that the reader is already familiar with MEX programming — see §8.1.
† The whitebalance webinar (referenced for the NPP library above) shows examples for both of these uses.

MEX did not support NVCC compilation directly until R2013a (8.1). On earlier MATLAB releases, we need to provide additional compile options in the *mexopts.sh* or *mexopts.bat* file.[957]

These steps can also be automated using Janaka Liyanage's[958] *nvmex* tool.[959] The *nvmex_tool.zip* file contains five files:

- *nvmex.pl* — copy this file to MATLAB's "bin" folder
- *nvmex_helper.m*
- *nvmex.m*
- *nvmexopts.bat* — for MATLAB releases 2007 and up*
- *nvmexopts_old.bat* — for pre-2007 MATLAB releases

With this tool, we can compile CU files from within MATLAB as follows:

```
nvmex -f nvmexopts.bat my_cuda_kernel.cu -IF:\path\to\cuda\includes
    -LF:\path\to\cuda\libraries -lcudart -lcutil
```

The flags `-I`, `-L`, and `-l` specify CUDA include and library folders, and link libraries used in the compilation process. `-lcutil` may be unnecessary in some cases.[960]

The file *my_cuda_kernel.cu* will then be compiled into *my_cuda_kernel.mexw32* (the MEX file extension may be different, depending on the compilation platform).

Alternatively, we can compile CU files in two stages as shown below. First, a CU file is compiled to a CPP file with embedded GPU codes (fatbin):

```
nvcc -I/path/to/mex/includes --cuda my_cuda_kernel.cu
    --output-file my_cuda_kernel.cpp
```

In the second step, the produced CPP file is compiled to a binary MEX file:

```
mex my_cuda_kernel.cpp -L/path/to/cuda/libraries -lcudart
```

A list of some common problems that occur when programming CUDA via MEX, and their workarounds, is provided here.[961]

Let us see how to sort an array of doubles on the GPU using the Thrust library.[962] Thrust is provided as part of the standard CUDA toolkit;[963] so, we do not need to install any additional libraries. The C code of `mex_gpusort` routine is given below:

```
#include "mex.h"
#include "cuda_runtime.h"
#include <thrust/host_vector.h>
#include <thrust/device_vector.h>
#include <thrust/sequence.h>
#include <thrust/sort.h>
/* The gateway function */
void mexFunction(int nlhs,        mxArray *plhs[],
                 int nrhs, const mxArray *prhs[]) {
    if (nrhs != 1)
        mexErrMsgIdAndTxt("MyMEX:gpusort:nrhs", "One input required");
    if (!mxIsDouble(prhs[0]))
        mexErrMsgIdAndTxt("MyMEX:gpusort:nrhs", "Array must be double");
    // obtain pointer to the input array and number of elements
    double *g_in = mxGetPr(prhs[0]);
    size_t numel = mxGetNumberOfElements(prhs[0]);
```

---

* Janaka's website does not clarify whether R2007a or R2007b, but I presume R2007b.

```
    // allocate page-locked host memory for zero-memory copy
    double *cpu_ptr, *dev_pinned_ptr;
    cudaHostAlloc(&cpu_ptr, numel*sizeof(double), cudaHostAllocMapped);
    cudaHostGetDevicePointer(&dev_pinned_ptr, cpu_ptr, 0);

    // copy data from the input array
    memcpy(cpu_ptr, g_in, numel*sizeof(double));

    // init thrust pointer to device memory
    thrust::device_ptr<double> d_vec =
        thrust::device_pointer_cast(dev_pinned_ptr);

    //sort the vector on the GPU and wait for completion
    thrust::sort(d_vec, d_vec + numel);
    cudaThreadSynchronize();

    //create the output matrix and obtain the pointer
    plhs[0] = mxCreateDoubleMatrix((mwSize)numel, 1, mxREAL);
    double *g_out = mxGetPr(plhs[0]);

    //copy the results from the GPU
    memcpy(g_out, cpu_ptr, numel*sizeof(double));
    cudaFreeHost(cpu_ptr);
}
```

Note how we use "zero memory copy" (pinned memory) to avoid explicitly copying data between the host and GPU. Instead, the hardware will automatically initiate data transfers as needed, possibly overlapping memory operations with kernel execution.

This method of accessing data on the GPU is not necessarily faster than direct GPU–host memory copy but it might give an additional speedup if a program being executed does not have sufficiently high arithmetic intensity (and it will definitely be faster on integrated GPU devices).

To evaluate the performance of mex_gpusort, we have compared it with the analogous algorithm that uses *gpuArray*:

```
X = rand(10000000);
tic, Y = mex_gpusort(X); toc
⇨ Elapsed time is 0.219109 seconds.

tic, Z = gather(sort(gpuArray(X))); toc
⇨ Elapsed time is 0.234807 seconds.
```

As we can see, the running times are nearly the same. One of the main drawbacks of the GPU–MEX interface is the fact that each call to a MEX routine incurs at least two GPU–host data transfers. As a result, each MEX routine needs to perform large volume of computations on the GPU to tolerate memory access latencies.

Fortunately, starting with MATLAB release R2013a (8.1), we can manipulate data contained in *gpuArray* directly, using MEX functions.[964] In other words, user-defined CUDA kernels can access *gpuArray* without the need for memory transfer. This is, however, only supported on 64-bit platforms.

An example of using CUDA kernels to speedup 2D convolution was provided by Dirk-Jan Kroon, in his File Exchange utility *gpuconv2*.[965] Readers are encouraged to review the code and adapt it for their own needs.*

MATLAB's PCT–GPU interface was introduced in MATLAB release R2010b (7.11). For older releases, as far back as R2007b (7.5), Nvidia used to offer platform-specific MEX files that

---

* Additional performance aspects of convolution are discussed in §5.7.7.

directly interfaced with its GPUs (CUDA 1.0 and 1.1 — naturally, not compute-capable 1.3 at that time).[966] The MEX files were direct replacements for the corresponding MATLAB functions (e.g., *fft*). Unfortunately, these plugins are no longer officially offered on the Nvidia website, but if you contact Nvidia, they might be willing to provide you with the necessary plugin.

Readers are encouraged to review MathWorks' guide *"Using GPUs to Accelerate Complex Applications"*[967] and Nvidia's whitepaper *"Accelerating MATLAB with CUDA™ Using MEX Files"*[968] for additional CUDA MEX examples. Note that Nvidia's whitepaper dates from 2007; much has changed since then, so be careful!

6.2.8 Accessing GPUs from within Parallel Blocks

MATLAB PCT supports GPU access by workers, subject to the limitation that each worker can only access a single GPU.[969] In fact, **gpuArray** can (at least as of R2014a) only use a single GPU. It is possible for all PCT workers to access the same GPU, but this can lead to hard-to-diagnose conflicts, performance bottlenecks (memory-transfer latencies, GPU serialization across workers, etc.), and GPU out-of-memory errors.[970]

If we have the luxury of several available GPUs that can be used by the workers, we can attach specific GPUs to specific workers, using the workers' *labindex*.

```
spmd
    gpuDevice(labindex);
    % GPU-based code goes here
end
```

In this case, we are assured that each GPU will only be used by a single worker, thereby maximizing its effectiveness, enabling true resource parallelization, and removing the potential for conflicts or errors. Naturally, this technique requires having as many GPUs available as workers. If this is not the case, we can slightly modify the previous command, with reduced (but not eliminated) conflict likelihood:

```
spmd
    gpuDevice(mod(labindex-1,gpuDeviceCount())+1);
    % GPU-based code goes here
end
```

This technique relies on the *labindex* function, which is only available in *spmd* blocks. How could this be adapted to *parfor*-loops? The trick, as shown on Cornell's webpage for MATLAB GPU-coding best practices,[971] is to combine the two blocks, using the fact that persistent variables are stored locally by each worker:

```
function use_GPU_in_parfor()
    spmd
        selectGPUDeviceForLab(); % see function definition below
    end
    parfor i = 1 : 10
        % Each iteration will generate some data A
        A = rand(5);
        if selectGPUDeviceForLab()
            A = gpuArray(A);
            disp('Do it on the GPU')
        else
            disp('Do it on the CPU host')
        end
```

```
        % Replace the following with whatever task you need to do
        S = sum(A,1);
        % Collect data back from GPU (gather is a no-op if not on GPU)
        S = gather(S);
    end
end
function ok = selectGPUDeviceForLab()
    persistent hasGPU

    if isempty(hasGPU)
        devIdx = mod(labindex-1,gpuDeviceCount()) + 1;
        try
            dev = gpuDevice(devIdx);
            hasGPU = dev.DeviceSupported;
        catch
            hasGPU = false;
        end
    end
    ok = hasGPU;
end
```

Attaching dedicated GPUs to PCT workers in this manner enables taking full advantage of the available computing resources, parallelizing both across CPU cores, as well as across GPUs, simultaneously.

6.3 The MATLAB Distributed Computing Server

As useful as PCT is, it only allows us to access computing resources (CPU cores or GPUs) on the same machine (*node*). It does not enable work parallelization across separate computing nodes (cluster, grid, or cloud).

MATLAB Distributed Computing Server (MDCS) is a separate MathWorks product (toolbox). MDCS enables execution of MATLAB code (*jobs*) on multiple computing nodes in clusters, clouds, and grids, thus enabling *distributed computing*.*[972] The server supports batch jobs, parallel computations, and distributed large data. The scheduling of parallel jobs on computing clusters is fully customizable via a generic *scheduler* interface. The server includes a default built-in job scheduler (MJS) and provides support for commonly used third-party schedulers.[973] The key features of MDCS are:

- Provides licenses for all our licensed toolboxes and blocksets, so that we can run MATLAB programs on a cluster without the need for additional product-specific licenses for each computer in the cluster.

- Execution of GPU-enabled functions on distributed computing resources (subject to the limitation that each worker can only use a single GPU.†)

---

* Until R2013b (MATLAB 8.2), MDCS provided the additional benefit of increasing PCT's maximal local workers pool size. In R2014a (8.3), PCT's pool size is no longer effectively limited; so, this is no longer an MDCS benefit over PCT.

† See §6.2.8.

- Execution of parallel code from applications and software components generated using MATLAB Compiler on distributed computing resources.
- Support for all hardware platforms and operating systems supported by MATLAB and Simulink.
- Task scheduling using either a built-in job scheduler or third-party schedulers.

6.3.1 Using MDCS

MDCS runs on actual or virtual machines in a cluster or cloud service. The server dispatches and controls multiple headless MATLAB workers, as with PCT.* These workers receive and execute MATLAB code independently, and return results to the clients via standard MAT files (except in the case of the MJS scheduler).[974] Multiple users can run their applications on the server simultaneously.

Users interact with the server via PCT. Programs are executed using interactive sessions or by submitting jobs for batch execution. Additionally, we can build stand-alone executables or shared libraries, using the MATLAB Compiler, which can distribute computations to MATLAB workers.

It should be noted that the communication overhead and latency with local workers (workers running on local machine cores) is lower than that for remote workers.

The central object in a distributed computing system is a *scheduler*. Schedulers manage, monitor, and distribute workload and administer resources across distributed computing systems comprising disparate hardware and software resources. The scheduler interface provided by MathWorks parallel-computing products is a high-level abstraction that lets us submit jobs to our computation resources without having to deal with differences in operating systems and environments.

A MATLAB scheduler can serve several MATLAB client sessions at the same time. The following figure shows typical interactions between the components of a distributed computing system:

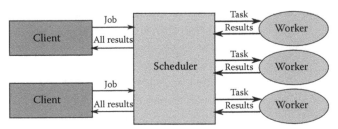

Each MATLAB worker is given a task from the running job by a scheduler. A worker executes the task, returns the result to the scheduler, and is then given another task. When all tasks for a running job have been assigned to workers, the scheduler starts running the next job on the next available worker.

The functionality of a scheduler is encapsulated in the so-called *cluster* object (*parallel. cluster*). Technically speaking, the cluster object provides an interface from a MATLAB client session to a cluster, which controls the job queue, and distributes tasks to workers for execution. MATLAB supports two categories of clusters: the MATLAB Job Scheduler (MJS) and the Common Job Scheduler (CJS). MJS is available in the MDCS; CJS clusters encompass

---

* The MDCS server only coordinates between MATLAB clients and workers, by scheduling/dispatching tasks and then synchronizing/overseeing their execution. The server does not directly run MATLAB code.

all other scheduler types, including the local, generic, and third-party schedulers. The table below lists different types of cluster objects (schedulers) supported in MATLAB:*

Cluster Type	Description
`parallel.cluster.MJS`	Interact with MATLAB job scheduler (MJS) cluster on-premises
`parallel.cluster.Local`	Interact with CJS cluster running locally on client machine
`parallel.cluster.LSF`	Interact with CJS cluster running Platform LSF
`parallel.cluster.HPCServer`	Interact with CJS cluster running Windows Microsoft HPC Server
`parallel.cluster.PBSPro`	Interact with CJS cluster running Altair PBS Pro
`parallel.cluster.Torque`	Interact with CJS cluster running TORQUE
`parallel.cluster.Generic`	Interact with CJS cluster using the generic interface
`parallel.cluster.Mpiexec`	Interact with CJS cluster using mpiexec from local host

The cluster objects can be constructed directly using the specified cluster type (as listed in the table above) or created from a *cluster profile*† as follows:

```
c = parcluster
c = parcluster(profile)
```

The first constructor creates a cluster object with the default cluster profile while the second one uses a cluster profile to create the cluster object of the specified type. A cluster profile identifies the type of cluster to use and its specific properties such as how many workers a job can access, where the job data is stored, where MATLAB is accessed, and so on. Once a cluster object is initialized, we can create a job to be executed on a parallel cluster and assign tasks within this job to individual workers.

Each cluster object has a number of methods and properties that differ depending on the cluster type. The important properties shared by all cluster types are

Property	Description
ClusterMatlabRoot	Specifies path to MATLAB for workers to use
Host	Host name of the cluster head node
JobStorageLocation	Location where cluster stores job and task information
Jobs	List of jobs contained in this cluster
Modified	True if any properties in this cluster have been modified
NumWorkers	Number of workers available for this cluster
OperatingSystem	Operating system of nodes used by cluster
Profile	Profile used to build this cluster
Type	Type of this cluster
UserData	Data associated with cluster object within client session

We shall use some of these cluster properties later in this section. Programming jobs for different types of parallel clusters is very similar, provided that the cluster is configured correctly. Therefore, we shall only discuss a generic scheduler interface in detail. The information about the other cluster types can be found in the MATLAB documentation,[975]

---

* Beginning with release R2012a (7.14), MDCS enables using Amazon EC2 and other cloud-computing services; see http://mathworks.com/discovery/matlab-ec2.html (or: http://bit.ly/1jecU8D) for details.
† The term *cluster profile* should not be confused with the term *code profiling*, which relates to the process of analyzing code runtime behavior and speed (see Chapter 2).

and readers are also referred to the PBS Pro scheduler example.[976] However, before this, we will make a short introduction to parallel jobs.

6.3.2 Parallel Jobs Overview

A *job* is some large operation that we need to perform in our MATLAB session. It is further divided into smaller segments called *tasks*. The tasks into which we divide a job do *not* need to be identical. Depending on a problem at hand, we can divide a job into tasks of different sizes.

Jobs managed by a scheduler can be of two types: *independent* and *communicating*. An independent job is one whose tasks do not directly communicate with each other. These tasks do not need to run simultaneously, and a worker might run several tasks of the same job in succession. Typically, all tasks perform the same or similar functions on different data sets in an *embarrassingly parallel* configuration.* An example of an independent job is evaluating a function for different input parameters (in parallel, if enough workers are available) when the results of the function evaluations are independent.

In contrast, communicating jobs are those in which the workers communicate with each other during the evaluation of their tasks. A communicating job consists of only a single task that runs simultaneously on several workers, usually with different data. More specifically, the task is duplicated on each worker, so that each worker can perform the task on a different set of data, or on a particular segment of a large data set. The workers can communicate with each other as each worker executes its task. Examples of communicating jobs are the ones created with **parfor** or **spmd** constructs where the results computed in different iterations are accumulated in a reduction assignment. There are two distinct types of communicating jobs: Pool (default) and SPMD. In an SPMD job, each worker runs a copy of the task function, and the task function itself calls **labSend** and so on. Basically the task function acts as the body of an **spmd** block. In contrast, in a Pool job, one worker takes on the role of the MATLAB client talking to a pool, and the task function contains **parfor** loops and **spmd** blocks.

The key differences between independent and communicating jobs are given in the following table:

Independent Job	Communicating Job
MATLAB workers perform the tasks but do not communicate with each other	MATLAB workers can communicate with each other during the running of their tasks
We define any number of tasks in a job	We define only one task in a job. Duplicates of that task run on all workers that run the communicating job
Tasks need not run simultaneously: they are distributed to workers as the workers become available. Hence, a worker can perform several of the tasks in a job	Tasks run simultaneously; so, we can run the job only on as many workers as are available at run time. The start of the job might be delayed until the required number of workers is available

From the client session, we can access a newly created job through a *job object* (**parallel. Job**). Similarly, tasks that we define for a job in the client session can be accessed through *task objects* (**parallel.Task**). Once we have created and launched a job on a cluster, it goes through a number of stages: *pending*, *queued*, *running*, and *finished* (or *failed*). The current stage is reflected in the job object's **State** property. The job's life cycle is schematically depicted in the following figure:

---

* The term *"embarrassingly parallel"* is discussed in §5.1.1.

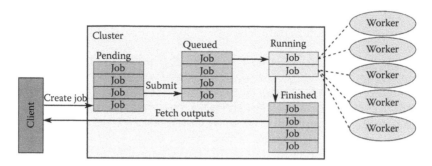

The job stages are

- **Pending:** Initially, when we create a job using the *createJob* function and assign tasks to it, its state is *pending*.

- **Queued:** Queued jobs are those that were submitted for execution using the *submit* function. A scheduler executes jobs in a queued manner in the sequence they were submitted.

- **Running:** Whenever some worker becomes available, the scheduler takes a job from the top of the queue and executes it. Then, the job's state changes to *running*. The scheduler can run several jobs at the same time.

- **Finished:** Once all job's tasks are evaluated, the job is moved to the *finished* state. The results of the job execution can be retrieved using the function *fetchOutputs*.

- **Failed:** A job might fail if, for example, a scheduler encounters an error while executing its commands or is unable to access necessary files.

Once a job is finished, its data remains in a folder specified by the scheduler's **JobStorageLocation** property, where information about the job can be retrieved later.

A very useful tool for displaying and monitoring parallel jobs is the *Job Monitor* GUI, which is included as part of MDCS.[977] We can launch it from the MATLAB Desktop via the Parallel drop-down control on the toolstrip's Home tab, Environment section.

A TCP/IP distributed progress bar can also be used to monitor the progress of distributed parallel jobs.[978] Also of note are the *Visual Timing Report for Distributed Tasks*,[979] and the *taskMap* utility that it inspired:[980]

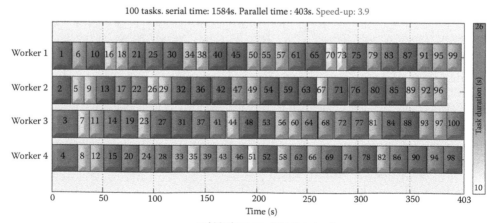

taskMap's report of distributed tasks

6.3.3 Setting Up a Scheduler Interface

MDCS provides a generic interface that lets us interact with third-party schedulers, or use our own scripts for distributing tasks to other nodes on the cluster for evaluation.

In this section, we describe how to set up a generic scheduler interface. Those readers not interested in such low-level details, can safely skip this section to the next one, where we discuss programming independent and communicating jobs.

When determining the type of scheduler to use with MDCS, consider the following rules of thumb:

- If your cluster is small, your number of users is small, your users are fine with FIFO type of job processing, and you do not expect to run anything else other than MATLAB/Simulink jobs, then you should probably use the MJS bundled with MDCS.

- If any of the assumptions above breaks, then we are better off with a more sophisticated scheduler. There is a wide variety available: IBM Platform LSF, Altair PBS Pro, Microsoft HPC Server, TORQUE, Moab, UNIVA Grid Engine (formerly SGE), or its Open-Grid Scheduler variant. These schedulers can provide extensive control on how resources are allocated for running computations (e.g., queue management, allocation of hardware, security, etc.).

- The Mpiexec scheduler, which has been shipped with MDCS in previous releases, has started generating warning messages in R2013b to the effect that using it is no longer recommended; MDCS will stop supporting it in some future release.

Since each job consists of several tasks, the purpose of the scheduler is to allocate a cluster node for the evaluation of each task, or to *distribute* each task to a cluster node. The scheduler starts remote MATLAB worker sessions on the cluster nodes to evaluate individual tasks of the job. To evaluate its task, a MATLAB worker requires access to certain information, such as where to find the job and task data. The generic scheduler interface controls the way tasks are managed between the cluster nodes.

To initialize the scheduler, first, we create a generic parallel cluster object as follows:

```
c = parcluster('generic_profile');
```

The cluster object has several properties that need to be set before jobs can be submitted. The first is **IndependentSubmitFcn** that identifies the submit function and has three main purposes:

- Identify the decode function that MATLAB workers run when they start
- Make information about job and task data locations available to the workers via their decode function
- Instruct the scheduler how to start a MATLAB worker on the cluster for each task of our job

The submit function takes three input parameters by default:

```
function my_submit_func(cluster, job, props)
```

Then, the `IndependentSubmitFcn` property can be set as follows:

```
c.IndependentSubmitFcn = @my_submit_func
```

We can extend the submit function prototype with additional parameters, for example:

```
function my_submit_func(cluster, job, props, time_limit, location)
```

The additional parameters need to be supplied in the property assignment:

```
cl.IndependentSubmitFcn = {@my_submit_func, 300, 'Plant3000'}
```

The first and the second arguments to the submit function specify the cluster profile and a job being executed. The third argument, props, is the object with the properties listed in the following table:

Property Name	Description
StorageConstructor	String. Used internally to indicate that a file system is used to contain job and task data.
StorageLocation	String. Derived from the cluster JobStorageLocation property.
JobLocation	String. Indicates where this job's data are stored.
TaskLocations	Cell array. Indicates where each task's data are stored. Each element of this array is passed to a separate worker.
NumberOfTasks	Double. Indicates the number of tasks in the job. We do not need to pass this value to the worker, but we can use it within our submit function.

These properties are automatically set by the toolbox, so that we can program our submit function to forward them to the worker nodes via environment variables.

The client's submit function also needs to identify the corresponding *decode function* (both functions work together as a pair), which is done by setting the environment variable MDCE_DECODE_FUNCTION. This variable should contain the name of the decode function *on the path of the MATLAB worker*. MATLAB provides standard decode functions for independent and communicating jobs. For instance, to use the standard decode function for independent jobs, set MDCE_DECODE_FUNCTION to 'parallel. cluster.generic.independentDecodeFcn'. An example of the submit function is shown below:

```
function mysubmitfunc(cluster, job, props)
    % values to be sent to the MATLAB workers
    decodeFcn = 'mydecodefunc';
    jobLocation        = get(props, 'JobLocation');
    taskLocations      = get(props, 'TaskLocations'); % a cell array
    storageLocation    = get(props, 'StorageLocation');
    storageConstructor = get(props, 'StorageConstructor');

    % set the corresponding environment variables
    setenv('MDCE_DECODE_FUNCTION',     decodeFcn);
    setenv('MDCE_JOB_LOCATION',        jobLocation);
    setenv('MDCE_STORAGE_LOCATION',    storageLocation);
    setenv('MDCE_STORAGE_CONSTRUCTOR', storageConstructor);

    % set the task-specific variables and scheduler commands
    for i = 1 : props.NumberOfTasks
        setenv('MDCE_TASK_LOCATION', taskLocations{i});
        % the code to execute scheduler's submit command
        constructSchedulerCommand;
    end
end
```

The *constructSchedulerCommand* call represents the code to execute our scheduler's submit command. This is typically a string that combines the scheduler command with flags, arguments, and values derived from our object properties. The command is looped; so, our scheduler starts a MATLAB worker on the cluster for each task.

Let us now consider the decode function. The aim is to read job and task information into the MATLAB worker session. The decode function reads this information from the environment variables transferred from the client to the worker nodes. Then, it sets the property values of the props object that is passed to the decode function as an argument. The property values are essentially the same as those in the corresponding submit function (see above). An example of the decode function is provided below:

```
function props = workerDecodeFunc(props)
    % Read the environment variables:
    storageConstructor = getenv('MDCE_STORAGE_CONSTRUCTOR');
    storageLocation    = getenv('MDCE_STORAGE_LOCATION');
    jobLocation        = getenv('MDCE_JOB_LOCATION');
    taskLocation       = getenv('MDCE_TASK_LOCATION');

    % Set props object properties from the local variables:
    set(props, 'StorageConstructor', storageConstructor);
    set(props, 'StorageLocation',    storageLocation);
    set(props, 'JobLocation',        jobLocation);
    set(props, 'TaskLocation',       taskLocation);
end
```

When the props object is returned from the decode function to the MATLAB worker session, its values are used internally for managing job and task data. As noted above, the standard decode function usually provides all required functionality.

Let's summarize the steps needed to run jobs on a parallel cluster using a generic scheduler interface.* First, create a scheduler object using *parcluster*:

```
c = parcluster('generic_profile');
```

If our cluster uses a shared file system for workers to access job and task data, set the **JobStorageLocation** and **HasSharedFilesystem** properties to specify where the job data is stored, and that workers should access job data directly in a shared file system:

```
c.JobStorageLocation = '\\share\scratch\jobdata';
c.HasSharedFilesystem = true;
```

If MATLAB is not on the worker's system path, to set the **ClusterMatlabRoot** property to specify where the workers are to find the MATLAB installation:

```
c.ClusterMatlabRoot = '\\apps\matlab\';
```

We now specify the submitted function, using the **IndependentSubmitFcn** property:

```
c.IndependentSubmitFcn = @mysubmitfunc;
```

Some schedulers enable setting remote copy (rcp) and shell (rsh) commands.[981] I suggest not to specify encrypted commands (*scp* and *ssh*, respectively) if security is not an issue, since the encryption increases interworker communication overheads and latencies.†

---

* See also an introductory video at http://blogs.mathworks.com/pick/2006/11/20/speed-your-code-with-distributed-computing (or: http://bit.ly/18UGUjp) about scheduling independent jobs.

† Also see §11.9.2.

That is all! Our parallel cluster is now fully set up. In the following section, we discuss how to create jobs and submit them to the parallel cluster for processing.

6.3.4 Programming Independent Jobs

An *independent* job is created in the client session using the cluster's *createJob* method:

```
j = createJob(c);
```

When a new job is created, its **State** property is set to pending, meaning that the job has not been queued for running yet. The job has no tasks; so, its **Tasks** property is an empty array at first.[982]

By default, since MATLAB R2013a (8.1), code dependencies are automatically scanned when the job is later submitted, to automatically send any dependent code file(s) to the workers.[983] This can hurt run-time performance; so, if we are certain that the workers already have all the required files, we can bypass this phase:[984]

```
j.AutoAttachFiles = false;
```

The next step is to create tasks for the job. Tasks define the functions to be evaluated by the workers during the job execution. Often, the tasks of a job are identical except for different arguments or data. In this example, we create five tasks where each of them generates a 3-by-3 matrix of random numbers:

```
T = createTask(j, @rand, 1, {{3,3} {3,3} {3,3} {3,3} {3,3}});
```

Note that the 5x1 cell array specifies the input parameters for each worker. In this case, T is a 5x1 matrix of task objects. Now, we can submit the job to the scheduler's job queue and force the MATLAB client session to wait until the job is complete:

```
submit(j);
wait(j, 'finished')
```

The results of each task's evaluation are stored in the **OutputArguments** property of a task object as a cell array. Finally, we use *fetchOutputs* method of a job object to retrieve the results from all the tasks in the job:

```
results = fetchOutputs(j);
```

The results variable is a 5-by-1 cell array containing the results of the task computations.

Individual tasks and entire jobs can be canceled or deleted using the *cancel* and *delete* methods, respectively. By default, these methods only affect the job data stored on the disk. To cancel or delete a job or task that is currently running or queued, we must provide the instructions to the scheduler what it should do when a particular task or job is to be canceled or deleted. This is accomplished by setting the corresponding properties of the cluster object: **CancelJobFcn, DeleteJobFcn, CancelTaskFcn,** and **DeleteTaskFcn.**

Note that unless all the workers have access to shared network resources, we need to explicitly define for each worker the search paths where to look for job files. This can be done in two ways: either by setting the **AdditionalPaths** property of a job to the search paths that should be added to the command search path for this job, or by adding the *path* command in any of the appropriate startup files for the worker, that is:

matlabroot\toolbox\local\startup.m

matlabroot\toolbox\distcomp\user\jobStartup.m

matlabroot\toolbox\distcomp\user\taskStartup.m

Additionally, there are several properties of task and job objects used to pass code and data between client and workers. These properties are listed in the following table:

Property Name	Description
InputArguments	A task property containing the input data provided to the task constructor. This data is passed into the function when the worker performs its evaluation.
OutputArguments	A task property containing the function output results.
JobData	A job property that specifies data to be sent to every worker that evaluates tasks for that job. This data is passed to a worker only once per job, saving time if that worker is evaluating more than one task for the job.
AttachedFiles	A job property that lists all the folders and files (as a cell array) to be zipped and sent to the workers. On the worker, the data is unzipped, and the entries specified in the property are added to the search path of the worker session.

Finally, we can provide the files *jobStartup.m, taskStartup.m, poolStartup.m,* and *taskFinish.m* at the worker's search path that will be executed by a worker at the respective times to initialize and cleanup worker session.

6.3.5 Programming Communicating Jobs

The principles of using the generic scheduler interface for communicating jobs are the same as those for independent jobs. Coding a communicating job for a generic scheduler involves the following steps:

1. Create an object representing our cluster with the *parcluster* function.
2. Set the appropriate properties (esp. **CommunicatingSubmitFcn**) on the cluster object, if they are not defined in the cluster profile. We can write our own communicating submit and decode function, or use the ones provided by MATLAB for various schedulers and platforms.
3. Use *createCommunicatingJob* to create a communicating job object.
4. Create a task, run the job, and retrieve the results as usual.

The only major difference is that communicating jobs are created using *createCommunicatingJob* (rather than *createJob*), with one of the following formats:

```
j = createCommunicatingJob(cluster,'Type','pool',...)
j = createCommunicatingJob(cluster,'Type','spmd',...)
```

The first parameter for both calls is the cluster object of the specified type. The first variant creates a communicating job of type "pool", meaning that the job runs the specified task function with a MATLAB pool available to run the body of *parfor*-loops or *spmd* blocks. The second variant creates a communicating job of type "spmd", where the specified task function runs simultaneously on all workers, and "lab" functions (*labSend, labReceive,* etc.) can be used for workers communication.

As an example, let us consider the following function computing the sum of maximal eigenvalues of random matrices:

```
function result = maxSumEigenvals(N)
    result = 0;
    parfor ii = 1:N
```

```
        result = result + max(eig(rand(ii)));
    end
end
```

We start by creating a communicating job with the specified cluster profile:

```
c = parcluster('generic_profile');
j = createCommunicatingJob(c,'Type','pool');
```

Next, we add the task to the job, supplying the number of matrices as input:

```
createTask(j, @maxSumEigenvals, 1, {100});
```

Remember that only *one* task is created for a communicating job, so we will always specify a cell array of exactly one element (which could itself be a cell array of several input values). We now set the number of workers required for parallel execution and submit the job:

```
j.NumWorkersRange = [5 10];
submit(j);
```

Finally, once the job is complete, we retrieve the results:

```
wait(j, 'finished');
out = fetchOutputs(j);
```

An *spmd*-type job is created in a similar way. First, we define a function to be executed by all MATLAB workers and place it within a *colsum.m* file:

```
function total_sum = colsum
    if labindex == 1
        A=labBroadcast(1,magic(numlabs)); %send data to coworkers
    else
        A=labBroadcast(1);  % receive data from coworker #1
    end

    % Compute sum of a column identified by labindex
    column_sum = sum(A(:,labindex));

    % Compute total sum by combining column sum from all workers
    total_sum = gplus(column_sum);
end
```

Next, we create a communicating job, attach the required files, and assign a task for it:

```
c = parcluster('generic_profile');
j = createCommunicatingJob(c,'Type','spmd');
cjob.AttachedFiles = {'colsum.m'};
cjob.NumWorkersRange = 4;
createTask(cjob, @colsum, 1, {});
```

Finally, we submit the job and retrieve the results upon completion:

```
submit(cjob);
wait(cjob);
results = fetchOutputs(cjob);
```

PCT provides several submit and decode functions for use with the generic scheduler interface. The *%MATLABROOT%/toolbox/distcomp/examples/integration* folder contains several subfolders with usage examples for different types of schedulers.

MATLAB uses the MPICH2 shared library as the underlying implementation for its inter-process MPI communication. This is a standard generic cross-platform library, hence its benefit for heterogeneous clusters. For homogeneous clusters using dedicated networking hardware, using a targeted MPI implementation may reduce the communication overheads. Replacing the MPI library can be done using the ***mpiLibConf*** function. The new library must be byte-compatible with MPICH2.

Additional tips for debugging and running parallel jobs can be found online.[985]

6.3.6 Using Batch Processing on a Cluster

Sometimes, it is useful to off-load work from the MATLAB session to another session, so that the job can run in the background without blocking the input. Or, alternatively, we might want to submit a job to a remote cluster, close down the MATLAB session, and collect the results later. This can be achieved using the ***batch*** command that has the following syntax:[986]

```
j = batch(cluster, 'aScript')
j = batch(cluster, fcn, N, {x1,...,xn})
```

The first command starts the script *aScript.m* on a specified cluster identified by the cluster object. The second variant runs the function with a function handle or function name *fcn* on a specified cluster. The function is evaluated with the given arguments, *x1,...*, *xn*, returning *N* output arguments. The function file for *fcn* is added to the **AttachedFiles** property of the job and copied to the worker.

The batch command can also take additional parameter-value pairs that modify the behavior of the job such as **Profile**, **AttachedFiles**, **CurrentFolder**, **Pool**, and so on.[987]

For example, the following command starts a batch job to generate random matrices using workers assigned on a specified cluster (five workers will be assigned in addition to the worker running the batch job itself, for a total of six workers on the cluster):

```
j = batch(c, @rand, 1, {100,100}, 'Pool',5)
```

Upon completion, the results of computation can be obtained in the usual way:

```
wait(j)                % wait for a job to be finished
r = fetchOutputs(j)    % returns a cell array of the results
celldisp(r)            % display the returned results
```

We can also capture the job's Command Window output with the ***diary*** command:[988]

```
diary(j)               % display the command window output
```

Boston University's research-computing portal[989] contains a tutorial[990] for working with the SGE (Sun Grid Engine) scheduler.* Converting to the PBS scheduler should be relatively

---

* Boston University actually uses Open-Grid Scheduler (http://gridscheduler.sourceforge.net), an open-source SGE variant. SGE itself was renamed Oracle Grid Engine (http://bit.ly/1jIS284) when Oracle acquired Sun, but the term SGE is still widely used. SGE/OGE was subsequently acquired by Univa in late 2013 (http://www.univa.com/products/grid-engine).

straightforward. Together, SGE and PBS probably cover most batch users who use a third-party scheduler, rather than MJS.

We now return to the matched filter running example to demonstrate how to launch parallel jobs on a computing cluster.

6.3.7 Running Example: Using Communicating Jobs on a Cluster

To run our matched filter algorithm on a cluster, we will use the *Mpiexec* third-party scheduler (additional schedulers are also supported.[991]) To use the scheduler, we need to ensure that the **smpd** daemon (not to be confused with *spmd*) is running on our host machine and is configured for *passwordless* communication. The daemon is started using the command:

```
matlabroot/bin/mw_smpd -s
```

which, by default looks for a passphrase in a file *.smpd* created in our home directory.

The **smpd** process is a part of the parallel infrastructure used by *Mpiexec* (and other schedulers) to launch MATLAB workers. The following screenshots of the *Process Explorer* utility (§2.4) show two MATLAB workers launched by **spmd**:

Process	PID	CPU	CPU Time	Start Time	Handles	GDI...	Private...
⊟ MATLAB.exe	5340	2.31	0:02:23.484	16:12:06 25/09/2012	1,535	888	576,840 K
⊟ smpd.exe	6904		0:00:00.296	16:14:07 25/09/2012	145	4	3,508 K
⊟ smpd.exe	2644		0:00:00.078	10:01:17 27/09/2012	346	4	3,964 K
MATLAB.exe	6620	5.38	0:00:08.937	10:01:17 27/09/2012	889	32	176,768 K
MATLAB.exe	3612	3.85	0:00:07.625	10:01:17 27/09/2012	931	32	177,740 K
mpiexec.exe	7044		0:00:00.093	10:01:16 27/09/2012	125	4	2,480 K

Command Line
"C:\Program Files\Matlab\R2010b\bin\win32\MATLAB.exe" "-nosplash"
"C:\Program Files\Matlab\R2010b\bin\win32\smpd" -anyport -phrase MATLAB_5340 -smpd
"C:\Program Files\Matlab\R2010b\bin\win32\smpd.exe" -p 4140 -mgr -read 000015C4 -writ
"C:\Program Files\Matlab\R2010b\bin\win32\matlab.exe" -dmlworker -noFigureWindows -c
"C:\Program Files\Matlab\R2010b\bin\win32\matlab.exe" -dmlworker -noFigureWindows -c
"C:\Program Files\Matlab\R2010b\bin\win32\mpiexec" -configfile "C:\Documents and Settir

Once the worker daemon is running, we can create and validate an *Mpiexec* scheduler profile in MATLAB to ensure that everything is set up correctly. We may encounter problems with *Mpiexec* (and other schedulers) if the machine hostname is not properly set. To work around this problem, use the *pctconfig* command to change the default parameters. For example, the following command:

```
pctconfig('hostname', 'ganymede')
```

changes the machine hostname, used in PCT, to 'ganymede'. We should also check if all MATLAB workers can access a shared directory where the files required for a job reside. Otherwise, the files must be made available for each worker using **AttachedFiles** or **AdditionalPaths** properties.

The pseudocode of the algorithm now consists of two parts. The first part is a script used to create and launch parallel jobs on a cluster:*

```
function data = matched_filter_mpiexec(data)
    % create the cluster object
    cl = parallel.cluster.Mpiexec;

    % create a communicating job of type "spmd"
    j = createCommunicatingJob(cl,'Type','spmd');

    % set the number of workers for this job
    j.NumWorkersRange = [1 20];

    % create a new task
    createTask(j, @mf_spmd_core, 1, {data});

    % submit the job and wait for completion
    submit(j);
    wait(j);

    % fetch the results and delete the job
    out = fetchOutputs(j);
    delete(j);

    % assemble the final image from parts computed by the workers
    data.im_final = zeros(size(data.x_mat));
    for ii = 1 : numel(out)
        data.im_final = data.im_final + out{ii};
    end
end
```

Note that we have set the number of workers in the range from 1 to 20. MATLAB can now utilize up to 20 workers, depending on job complexity. Also, we have used a communicating job of type "spmd". This is because the script mf_spmd_core.m we are going to run on a cluster represents the body of an *spmd*-statement:

```
function data = mf_spmd_core(data)
    % Define speed of light
    c = 299792458;

    % Determine the size of the phase history data
    K  = size(data.phdata,1);   % number of frequency bins per pulse
    Np = size(data.phdata,2);   % number of pulses

    % Create an output "image slice" for each worker
    im_slices = zeros(size(data.x_mat), 'double');

    % Loop through every pulse
    for ii = drange(1:Np)
        % Calculate differential range for each image pixel (m)
        dR = sqrt((AntX(ii)-data.x_mat).^2 +...
```

---

* The files in this section can be downloaded from: http://UndocumentedMatlab.com/books/ matlab-performance (or: http://bit.ly/1pKuUdM).

```
              (AntY(ii)-data.y_mat).^2 +...
              (AntZ(ii)-data.z_mat).^2) - R0(ii);

        % Calculate the frequency of each sample in the pulse (Hz)
        freq = minF(ii) + (0:(K-1)) * data.deltaF;

        % Perform the Matched Filter operation
        for jj = 1 : K
            im_slices = im_slices + phdata(jj,ii) *...
                        exp(1i*4*pi*freq(jj)/c*dR);
        end
    end
    im_final = im_slices; % save the computed "image slice"
end
```

Here, we have used the pseudocode for the *spmd* running example from §6.1.7 with minor modifications. Hence, we can omit further comments. The only thing that requires attention is the fact that we transfer the results computed on each worker back to the client and reconstruct the final image in the MATLAB client session.

Another possibility would be to use the *gplus*(*im_slices*, *1*) command to consolidate the results on a single worker and send it back to the client. However, this latter solution is actually *slower*, probably due to large communication overhead.* The algorithm runs in **21 s** on our machine.

Comparing the execution time with those obtained in the previous sections, we see that our distributed parallel algorithm is not faster than the original *parfor* and *spmd* versions and clearly loses to the GPU-based solutions. This, however, should not be taken as a general rule. It turns out that our problem is still relatively small to be executed on a large parallel cluster, and thus, communication overhead between workers brings a noticeable slowdown.

In this sense, MDCS is more appropriate for parallelizing problems on a "large scale". For instance, by combining a GPU-based approach with MDCS, that is, by letting each worker run its part of the algorithm on the GPU, we could achieve substantially better performance as was already demonstrated in §6.2.4, where we have used a local MATLAB scheduler.

6.4 Techniques for Effective Parallelization in MATLAB

In this section, we summarize a number of suggestions and "best practices" for writing efficient parallel code in MATLAB. Some of these have already been discussed in previous sections. Here, we have tried to collect all these techniques into a single location.

We start with general tips for effective code parallelization in MATLAB that apply equally well to CPU- and GPU-based algorithms. Then, we outline techniques that are relevant while using PCT to program for a specific parallel architecture.[992]

6.4.1 General Performance Tips

Consider the following "rules of thumb" when developing a parallel application:[993]

---

* Note that *gplus* and other global reduction commands are based on "tree-like" parallel reduction that requires intensive communication between workers.

- *Importance of benchmarks.* Benchmarks are simple programs that capture the "essence" of your parallel algorithm (or parts of it). They can be very useful to explore the effects of adding new processors or monitor performance regression during the algorithm development.

- *Minimize overhead.* Sometimes, it is useful to compare the single-processor performance of a program using parallel constructs (and distributed arrays) to a sequential program. If the parallel program is slower by even 10%, this can have a large performance impact when the program is run on many processors.

- *Embarrassingly parallel implies linear speedup.* If an application is embarrassingly parallel (requires no communication), then it should speed up *linearly* with the number of processors. If this is not the case, then there is something wrong with the application (e.g., load imbalance) or the underlying distributed array library (if it uses distributed arrays).

- *Algorithm and mapping are orthogonal.* Specifying the algorithm and specifying the parallelism should be orthogonal. This means that the way the data is partitioned among the computing units should only change the performance and *not* the correctness of the computed result.

It is also always advisable to adhere to the following coding practices that can save us much effort during our performance optimization:

- Preallocate arrays if possible; avoid using dynamically growing variables.*
- Respect the column-wise storage order of array data elements.† This is especially important for GPU computations due to memory coalescing.994
- Use *profile* and *mpiprofile* to find algorithmic and communication hotspots.‡
- Use vectorized indices/subscripts to deal with conditional code.§

Using good vectorized code, we can automatically profit from parallel computations in MATLAB, both on the CPU and GPU platforms, without needing explicit parallelism. By "good vectorized code" we mean MATLAB code that uses implicit multithreading using commands such as *bsxfun, arrayfun, repmat, find, prod, sum, cumsum,* and so on as shown in Chapter 5. Specifically, we should distinguish the *bsxfun* function (available both for CPU and GPU) that should often be preferred to the *repmat* command, since *repmat* can significantly increase memory usage.¶

In fact, on a single machine that does not employ a GPU, implicit multithreading will generally be faster than explicit PCT parallelization, since the former uses light-weight in-process threads while the latter uses full-blown headless MATLAB processes that require hefty communication and setup overheads.995 Since both the implicit multithreading and the workers would use the same platform memory and CPU resources, it is obvious why implicit multithreading is both simpler and faster.

On the other hand, vector optimizations should be applied generously without too much overdoing. For instance, some MATLAB developers have a "pathological" fear of

---

* See §9.4.
† See §9.3.
‡ See §2.1 and §6.1.5, respectively.
§ See §5.1.
¶ See §5.2.4.2 and §5.4.

loops and will try to optimize out any occurrence of them, sometimes to the point of absurdity. Instead, it should be understood that many loop constructs can be optimized by MATLAB's JIT compiler and our code might essentially run *faster* with them.* As an example, let us consider the following rather efficient piece of code:

```
K = 700;
N = 800;
X = rand(N);
S = rand(K,N);
R = zeros(N);
for kk = 1 : K
    R = R + acos(bsxfun(@times, S(kk,:)', X));
end
```

This code runs in **6.9** s on a laptop with Intel Core i5 3210M (dual core), 4 GB RAM, and GT 650M graphics card (GK107) under 64-bit Linux platform. We could go further by removing the loop completely:

```
D = reshape(X, [1 size(X)]);
D = repmat(D, K, 1);
R = acos(bsxfun(@times, S, D));
R = squeeze(sum(R, 1));
```

Unfortunately, this code fails even to start on our machine with 4 GB RAM due to excessive memory requirements. Instead, quite often, it is wise to use a loop to run several iterations over smaller chunks of data:†

```
n = 20;
Kc = K/n;
R = zeros([Kc size(X)]);
D = reshape(X, [1 size(X)]);
D = repmat(D, Kc, 1);
for jj = 0 : Kc : K-Kc
    R = R + acos(bsxfun(@times, S(jj+1:jj+Kc,:), D));
end
```

The code now executes in **6.0** s proving that loops (or rather, processing smaller data chunks) might be useful. In fact, the amount of speedup derived from GPU parallelization is closely tied to the amount of data being processed. In one example discussed on Stack Overflow,[996] a *convn* of data arrays produced a speedup on the GPU varying between 2.8× and 11.6×, based on the data size.

Certainly, clever use of MATLAB vectorization functions requires some practice and is not that easy in the beginning. Loop vectorization is even more important on the GPU since only MATLAB functions performing element-wise operations on a large set of data are GPU accelerated. Readers are referred to Chapter 5 for more details.

Finally, when considering whether to invest time in GPU (*gpuArray*) or CPU (*parfor*) parallelization, consider the following rules of thumb:[997]

1. GPUs are simpler, but run massive numbers of concurrent numeric threads.

2. GPUs are better for numeric processing but cannot handle strings, cell arrays, structs, class objects, and other nonprimitive MATLAB types.

---

* See §5.7.1.
† See §3.6.3 and §9.5.9.

3. GPUs can handle only simple arithmetic/indexing, and few toolbox functions.

4. GPUs can run code branching (if/else, logical indexing) but slower than CPUs — GPUs work much better when there is no code branching.

5. GPUs work better with 100K-1M threads, not less (due to overhead costs).

6. GPUs do not work well with memory-intensive operations; so, they are better for simple independent element-wise operations, but not to others.

7. GPUs are more memory limited than CPUs, although modern GPUs constantly increase the amount of onboard RAM.

8. CPU parallelization of implicitly parallelized code (vectorized linear algebra, etc.) is pointless; only GPU parallelization might help to improve performance.

9. For both CPUs and GPUs, parallelization is only effective if the amount of data transfer is low, such that data transfer does not become a performance bottleneck.

10. Be skeptical of GPU-versus-CPU benchmarks, which are easily manipulated.[998]

6.4.2 Performance Tips for Parallel CPU Programming

In this section, we discuss some best practices specific to parallel programming on the CPU with PCT. If you feel the need for explicit parallelism, the following steps should help to build a parallel application from the original algorithm:

- *Run the code normally on a local machine.* Ensure that your code is already vectorized and delivers optimal performance on one MATLAB worker. Use techniques and programming suggestions from §6.4.1 to optimize your code.

- *Decide which type of parallel job to use.* If your algorithm operates on large data sets for which you need to perform simultaneous computations, you might benefit from a communicating job. If, instead, your algorithm uses repetitive calculations (e.g., in a loop) that can be performed independently from each other, an independent job might be an option.

- *Modify your algorithm for division.* For an independent job, determine how best to divide it into separate tasks: for instance, each loop iteration can define a task. For a communicating job, decide how to take advantage of parallel processing: for example, a large array might be distributed across workers. This may be useful in some cases, but in others it might cause unnecessary extra communication overhead compared to using local worker variables.[999]

- *Use* **pmode** *to develop parallel functionality.* *pmode* running with a local scheduler can be particularly useful to understand data distribution between workers and to debug your application, especially with the MPI profiler.

- *Run the independent or communicating job with a local scheduler.* This verifies that your code is correctly set up for batch processing, and for an independent job, that the computations are properly divided into tasks.

- *Run the independent job on a single cluster node.* Run an independent job with one task to ensure that remote distribution is working correctly between the client and the cluster, and to verify proper transfer of additional files/paths.

- *Run the independent or communicating job on multiple cluster nodes.* Scale-up your job to include as many tasks as you need for an independent job, or as many workers as you need for a communicating job.

- *Ensure worker independence* as much as possible of any shared resource (such as memory, disk, or database) that could become a bottleneck.*

For optimization techniques specific to *parfor*-loops, revisit §6.1.1 to ensure that variables, which are indexed within a loop, obey the rules for *sliced* variables and are not *broadcast* to all workers. Also ensure that variables used in reduction assignments are not "technically" classified by MATLAB as *temporary* variables since the latter ones are not returned back to the client (they exist only inside a *parfor*-loop).

Always use *parfor* on the external loop in a multiloop (nested loops) construct. Since *parfor*-loops cannot be nested if the *parfor*'ed loop contains fewer elements than the available workers (e.g., a four-element loop on an eight-core system), then combine loops in order to enable more workers to work in parallel:[1000]

```
% Original code: two outer loops, parfor only on outer-most loop
parfor a = 1 : 4              % only 4x parallelization
    for b = 1 : N
        % do something useful here...
    end
end


% Combine two loops for improved parallelization
parfor ab = 0 : (4*N-1)       % better parallelization
    a = floor(ab/N) + 1;      % =1:4
    b = rem(ab,N) + 1;        % =1:N
    % do something useful here...
end
```

To improve the performance of a *parfor*-loop, it is sometimes better for MATLAB workers to create their own set of data in parallel, rather than to create a large array in the client before the loop and then dispatch it out to all workers. Having each worker create its own copy inside the loop saves the data-transfer time from the client to workers, since all the workers can create the data at the same time.

The decision whether to create arrays before or inside a *parfor*-loop depends on the size of the arrays, the time needed to create them, whether the workers need all or part of the arrays, the number of loop iterations that each worker performs, the amount of available physical memory, and other factors. Also note that when we run some algorithm with *local* workers (running on the same machine), we might not see any performance improvement from using a *parfor*-loop. This can depend on many factors, including how many processors and cores our machine has. For example, consider the following two code fragments:

```
n = 200; % arrays are created on the client side
M = magic(n);
R = rand(n);
parfor i = 1:n
    A(i) = sum(M(i,:).*R(n+1-i,:));
end

n = 200;  % each worker creates arrays locally
parfor i = 1:n
    M = magic(n);
```

---

* In MDCS, we can specify FileDependencies for a job, but this should be reserved for unavoidable or low-overhead cases.

```
    R = rand(n);
    A(i) = sum(M(i,:).*R(n+1-i,:));
end
```

When running on a local cluster, the execution times of these two parfor-loops will not differ much. However, on a remote cluster, you can experience different behavior since, for the second code fragment, workers can create arrays simultaneously and hence save on the transfer time. Therefore, code that is optimized for local workers might not be optimized for cluster workers, and vice versa. The same considerations also hold when using *spmd* statements.

Note that we can benefit from the PCT not only through direct use of *spmd, parfor,* or *pmode* commands, but also by using functions from other MATLAB toolboxes that have PCT support.[1001] However, sometimes, parallelization can make things run even *slower*. Thus, for many toolboxes, it is *not* enabled by default. For instance, suppose we wish to use *fmincon* function from the Optimization Toolbox:*[1002]

```
x0 = [1 1];
objfun = @(x)exp(x(1))*(4*x(1)^2+2*x(2)^2+4*x(1)*x(2)+2*x(2)+1);
confun = @(x)deal([1.5+x(1)*x(2)-x(1)-x(2);-x(1)*x(2)-10],[]);
[x, fval] = fmincon(objfun,x0,[],[],[],[],[],[],confun);
```

This code fragment runs in **0.049 s** on our laptop. We modify it to enable parallelization as follows:

```
parpool(4);
opts = optimset('fmincon');
opts = optimset('UseParallel','always');
[x, fval] = fmincon(objfun,x0,[],[],[],[],[],[],confun);
```

Now, the running time increases to **0.51 s** (10x slower), even without considering the nonnegligible worker setup time in *parpool*. The reason for this is that when a problem is trivial, the communication overheads between workers dominate and eliminate any benefits of parallel processing. Therefore, before using explicit parallelization, ensure that the problem is indeed worth the parallelization effort.

Still, in many cases parallelization can indeed improve the performance of many built-in functions in external toolboxes, such as *fmincon* from the Optimization Toolbox (§4.5.5), or the *blockproc* function in the Image Processing Toolbox (§9.5.9). In cases were a toolbox function does not accept the UseParallel flag, we can sometimes convert the problem into an equivalent representation using a function that does support UseParallel (for example, converting *fminunc* to *fmincon*).

When using highly vectorized code on a relatively modern (R2007a+) MATLAB release, the vectorization uses implicit parallelization (multithreading) to a large degree (see Chapter 5). In such cases, explicit parallelization on the computer's CPU is meaningless, since the code is already multithreaded.[1003] In fact, such cases might actually exhibit a slowdown when using *parfor*-loops, compared to the vectorized counterparts, due to the explicit parallelization overheads. The situation changes when our problem (or data) is large enough that GPU or multiplatform (distributed) parallelization benefits outweigh the communication overheads.

When communicating large data variables between PCT workers, it may be useful to use Joshua Dillon's *sharedmatrix* utility or one of its variants; so, workers access the same data in common memory without needing lengthy communication (see §9.5.2).

---

* Also see §4.5.5.

6.4.3 Performance Tips for Parallel GPU Programming

General best practices for using GPU in MATLAB can be summarized as follows:

- Minimize data transfers between the main memory and the GPU; create data directly on the GPU using the methods of *gpuArray*.[1004]
- Work on large-enough data sets, to amortize data-transfer overhead. GPUs are only effective for massively parallel data processing (as in 1M threads). Do not test algorithms using small data sets, to prevent misleading results.[1005]
- GPU computations only pay off when a problem has a large degree of data-level parallelism;[1006] otherwise, communication overhead will offset benefits.[1007]
- Remember: *for*-loops are not parallelized on GPU, resulting in slow sequential data transfers and processing. Instead, try to vectorize, possibly using *arrayfun*.
- Some GPU calculations work significantly faster when the data is sliced differently; simply transposing the data may yield a 10× speedup![1008]
- Code vectorization is critical for GPU effectiveness;[1009] loops, branching, and sometimes even indexing run on the CPU, not the GPU.
- Employ MEX and/or kernel implementations only if the standard *gpuArray* and *arrayfun* functions and vectorization fail to deliver the required speedup.
- Use *existsOnGPU* to query the existence of data on GPU, to avoid data transfers.
- Combine multiple element-wise calculations into a single *arrayfun* call.
- GPU support is not available for strings, cell arrays, structs, and class objects.
- Use GPU-optimized data types: *double* on Tesla; *single* on GeForce; and so on.[1010]
- Use the latest GPU drivers: This could significantly improve performance.[1011]

MATLAB's documentation includes a detailed webpage with related resources, dedicated to measuring and improving GPU performance using PCT.[1012] A dedicated webinar discusses when PCT/GPUs are and are not expected to provide speedup.[1013]

Using *arrayfun*, we can perform operations across a large number of scalar values and to a certain extent handle bulk processing of small arrays. However, sometimes, it is required to do repetitive computations (on small data sets) that are not yet implemented for the GPU. Still, with a bit of creative thinking, we can get a nice speedup on the GPU even for such operations. Suppose we have a sequence of points in 2D (represented by two arrays X and Y), and for each pair of consecutive points we wish to compute the coefficients (a and b) of a straight line passing through these points. First, we define a function calculating a line equation:

```
% computes coefficients for an equation: y = a*x + b
function [a,b] = eqline_fun(x1, y1, x2, y2)
    a = (y2 - y1)/(x2 - x1);
    b = (x2*y1 - x1*y2)/(x2 - x1);
end
```

The next step is to prepare the data for GPU processing with *arrayfun*. Here, the idea is to group each pair of consecutive points together in the following way suitable for bulk processing: [x1 y1 x2 y2], [x2 y2 x3 y3], and so on. The resulting code is given below:

```
Y = sin(X);  % generate a sequence of points in 2D
X = gpuArray(X);
```

```
Y = gpuArray(Y); % cast data to the GPU
[a,b] = arrayfun(@eqline_fun,X(1:N-1),Y(1:N-1),X(2:N),Y(2:N));
```

The algorithm runs in **0.55** s on our machine with GT 650M graphics card.

Sometimes, it makes sense to break up CPU vectorization (*devectorization*) to be able to benefit from using ***arrayfun*** on the GPU.

Using a GPU, it is likely that sooner or later we may get a confusing error such as

```
Error using ==> gpuArray at 28
Out of memory on device. You requested: 2.67 Gb, device has 1.73 Gb free.
```

This error indicates that we cannot allocate the GPU device's memory, due to memory fragmentation and the need to create temporary work arrays. In some cases, the GPU (or CUDA) only reports a generic error, without even bothering to tell us that it was memory related:[1014]

```
Error using gpuArray
An unexpected error occurred during CUDA execution. The CUDA error was:
unknown error
```

A possible workaround for all such memory-related issues is to free GPU memory using ***clear, reset,*** or ***gpuDevice*** functions. An alternative would be to use ***arrayfun***,[1015] or to break our problem into small subblocks, as shown in §6.4.1.*

Another quite confusing GPU error message while using ***arrayfun*** could be:[1016]

```
Warning: An unexpected error occurred during CUDA execution. The CUDA
error was: CUDA_ERROR_LAUNCH_TIMEOUT.
Error using arrayfun
The kernel execution failed because the CUDA driver timeout was encountered.
```

This error is triggered by the operating system when ***gpuArray*** or a CUDA kernel code runs for a long time on a GPU (possibly, due to an algorithm bug), to limit the time that the GPU dedicates to computations and graphics-rendering tasks.

Let us now discuss performance considerations when using double-precision arithmetic on the GPU. If there is no particular need for computations with double precision, it is often better to specify *explicitly* which data types to use, instead of relying on default settings. The reason for this is because double-precision values consume twice as much GPU memory bandwidth than 32-bit integers or single-precision data. Furthermore, on modern graphics cards, integer and single-precision computations are at least 2x faster than double precision. For example, the following function allocates an integer array of zeros on the GPU:

```
Z = gpuArray.zeros(8192, 'uint32');
```

This could also be used to obtain a rough estimate of the computation results of some lengthy algorithm, by running it using the faster integer or single precision. We could then fine tune the parameters based on the result and rerun with full double precision.

Sometimes, it is useful to check if the performance of our algorithm is memory bound (i.e., performance is limited by the speed GPU–host memory transfer). The following code snippet measures memory access speed:[1017]

```
sizes = power(2, 12:26);
repeats = 10;
sendTimes    = inf(size(sizes));
```

---

* Also see §3.6.3 and §9.5.9.

```
gatherTimes = inf(size(sizes));
gpu = gpuDevice; % get the current GPU device
for ii = 1 : numel(sizes)
    data = randi([0 255], sizes(ii), 1, 'uint8');
    for rr = 1 : repeats
        timer = tic();
        gdata = gpuArray(data);
        wait(gpu); % wait until memory operations are finished
        sendTimes(ii) = min(sendTimes(ii), toc(timer));
        timer = tic();
        data2 = gather(gdata);
        gatherTimes(ii) = min(gatherTimes(ii), toc(timer));
    end
end

sendBandwidth = (sizes./sendTimes)/2^30;
[maxSendBandwidth,maxSendIdx] = max(sendBandwidth);
fprintf('Peak send speed is%g GB/s\n', maxSendBandwidth)

gatherBandwidth = (sizes./gatherTimes)/2^30;
[maxGatherBandwidth,maxGatherIdx] = max(gatherBandwidth);
fprintf('Peak gather speed is%g GB/s\n', max(gatherBandwidth))
```

This produces the following results on a laptop with GT 650M graphics card:

```
Peak send speed is 3.03509 GB/s
Peak gather speed is 2.36025 GB/s
```

Given that the average device memory bandwidth for GT 650M is about 50 GB/s, one can see that frequent GPU–host memory copies can become a serious bottleneck.

Another useful test is to estimate floating-point performance of our graphics card in GFlop/s (floating-point operations per second) to see if our algorithm can utilize the full potential of the GPU hardware. One way to do this is by using the *GPUBench* utility.[1018] The utility runs several tests including backlash operator \, *mtimes*, and *fft*, both on the CPU and GPU, and then creates a detailed report as shown below:

	Results for data–type 'double' (In GFLOPS)			Results for data–type 'single' (In GFLOPS)		
	MTimes	Backslash	FFT	MTimes	Backslash	FFT
Quadro 6000	313.85	149.76	65.26	632.16	404.14	149.67
Tesla C2075	314.06	151.91	60.43	653.35	406.09	155.72
GeForce GTX 580	195.23	170.32	66.57	974.22	593.33	200.66
GeForce GTX 680	115.35	100.86	45.96	966.24	572.52	100.64
GeForce GTX 670	102.02	90.10	42.73	817.96	516.53	102.72
GeForce GTX 560 Ti	101.33	77.81	35.97	521.21	360.81	140.18
Quadro 2000	38.50	31.24	13.79	196.61	143.93	47.54
Host PC	**37.46**	**27.95**	**2.21**	**79.65**	**64.88**	**4.23**
GeForce GT 650M	**21.15**	**19.27**	**8.45**	**171.68**	**53.27**	**32.65**
GeForce GT 430	21.17	17.34	7.09	106.37	56.57	20.67
Quadro 600	19.76	16.40	7.32	100.29	76.58	34.45

We see from the table that GPU single precision outperforms double precision, as noted above. We might be surprised that FFT performance as shown in this table is relatively low, even for high-end graphics cards: This can be explained by the fact that external GPU benchmarks measure a peak FFT performance that is achieved when a batch of small-sized FFTs is executed in parallel. Instead, *GPUBench* measures peak performance of a single-large FFT, giving a more practical estimate.

People often make the mistake of assuming that GPU performance is a factor of the Compute Capability. In fact, the Compute Capability just refers to the set of features that the GPU supports, not to its speed. GPU speed is a factor of the number of cores, the clock speed, the amount of device memory and cache, and the internal memory bandwidth.[1019]

When assembling or buying a new system, look not just at the processing power (GFlops), but no less importantly at the host-device bus throughput. PCIe 2.0 has double the data throughput as PCIe 1.x, and half of 3.0. It is also important to verify that you are using all possible 16 lanes. Of course, testing the actual data transfer rate (as shown above) is always good, even if impractical for a new system.

To ensure that our GPU dedicates itself fully to our computations, we should ensure that it does not do any external stuff (e.g., general display acceleration or browser support).[1020] On Windows, set the GPU to TCC (rather than WDDM) mode, and consider turning of its ECC feature (at the expense of some memory errors).

I recommend reading a study on optimization, which compares standard and GPU-accelerated MATLAB code with PCT or Jacket support.[1021] I also recommend reading Cornell's webpage for MATLAB GPU-coding best practices,[1022] and Boston University's PCT tutorial.[1023] In addition, Case Western Reserve University's HPC Cluster portal contains many interesting resources.[1024]

7

Explicit Parallelization by Other Means*

7.1 GPU Acceleration Using Jacket

Jacket is a library developed by AccelerEyes, which provides GPU acceleration of many MATLAB built-in functions and language constructs.[1025] The main design philosophy behind Jacket was to make the transition from a CPU-bound MATLAB program to a fully GPU-accelerated MATLAB program as transparent as possible to enable rapid development of GPU computing applications in MATLAB. In essence, this was achieved by introducing several GPU-specific data types in the M-language with overloaded operators, as well as by exploiting the inherent parallel nature of M-language constructs. As a result, existing MATLAB programs usually need no further modifications, except changing the data types, to run on GPU. This makes a notable distinction between Jacket and the GPU support in the Parallel Computing Toolbox (PCT). Unlike PCT, Jacket was supported on MATLAB releases as old as R2007b (7.5) and also some non-Nvidia GPUs such as AMD's ATI brand of GPUs.

On December 12, 2012, AccelerEyes announced[1026] that it discontinues the development and sale of Jacket products.† However, since there are still numerous active Jacket licenses, with a possible secondary market, it was decided to keep this section as it might still have some value. Naturally, the importance of Jacket rapidly dissolves with time.

7.1.1 Key Ideas of Jacket Design

Jacket extends the M-language with GPU support by defining a new set of typed array classes (called "garrays"), which mimic the behavior of the corresponding classes from the MATLAB standard interface. In other words, classes such as *double, single, uint16, zeros, ones,* and so on get replaced by *gdouble, gsingle, guint16, gzeros, gones,* and so on, respectively. These new classes provide exactly the same functionality as their CPU-based counterparts by means of *function overloading.*‡ More precisely, each standard method currently present in the standard classes is made available on the *garray* class via GPU-enabled MEX code. These standard methods include basic functionality, such as displaying the object in text, as well as much more sophisticated algorithms such as the fast Fourier transform (FFT), matrix multiplication, singular value decomposition,

---

* The majority of this chapter was authored by Pavel Emeliyanenko of Max-Planck Institute for Informatics (http://mpi-inf.mpg.de/~emeliyan).

† In late 2013, AccelerEyes was renamed ArrayFire, after the name of their low-level parallelization library (see §7.2.3).

‡ This basic idea was first used by MIT's MATLAB*P project (see §7.2.4), which was featured a 1999 MathWorks newsletter: http://mathworks.com/company/newsletters/articles/objectively-speaking.html (or: http://bit.ly/1cAQP2q).

and so on. As a result, large portions of already-existing MATLAB programs can be run on the GPU in this manner by only changing data types from their base class, such as *double* to the GPU base class equivalent *gdouble* (for which a set of overloaded built-in functions exists). This is very similar to PCT's use of *gpuArray*.

As noted above, Jacket employs MEX routines for accessing external system libraries from MATLAB and thereby interface with GPU hardware. When such an external routine is called, the interpreter makes input data available to that routine from the MATLAB environment, concedes control to the routine, and expects computation results immediately upon termination of that routine. However, the fact that each MEX call will force at least two GPU–host memory transfers means that large volumes of computations must be carried out *entirely* in one MEX routine in order to tolerate high memory access latencies. This fact, in turn, would make the GPU development in M-language prohibitively inefficient, if possible at all.

To overcome this problem, Jacket employs a *lazy compilation scheme*: operations on data are tallied and only executed when absolutely necessary, such as when a user requests to display data. More concretely, each standard method of the *garray* class eventually calls a single MEX file, which maintains static state and assumes the role of managing GPU memory, JIT (just-in-time) compilation, trace caching, garbage collection, and automatic thread extraction. Through this arrangement, the requirement for memory transfers is alleviated and the opportunity for end-to-end optimization of MATLAB programs for GPU execution is exposed.

For instance, consider the following piece of MATLAB code:

```
A = gones(3) + 3;
```

The events that ensue to build the computation tree, whose result is eventually stored in the variable A are as follows:

1. The instantiator, *gones*, calls the gateway function, which returns a GPU object of type *constant*.
2. Since *gones*(3) is replaced by a GPU object, the + operator calls the object's *plus* method, with the object and the value 3 as operands.
3. A new GPU object representing a plus operator is created and *gones*(3) and 3 are made its children. The dimensions of the new GPU object are computed based upon the children and the operation.

The actual computation is demand driven: when the user wants to print or visualize the data on the screen, Jacket examines and executes the recorded computations taking into account various considerations including memory availability, memory latencies, cached instructions, system load, opportunities for parallelism, and so on.

Jacket also features *automatic garbage collection* of the GPU data. Since all variables contained within the computation tree are actually MATLAB objects, when a tree is disconnected from its parents and there are no references to the root of that tree in the MATLAB workspace, then the tree is considered orphaned and returned to the free memory heap automatically by MATLAB. Thus, nodes within the computation tree need no extra cleanup or duplication checks in order to be cleared properly from memory after a computation has been performed. A method of reference counting is used to determine when it is safe to reuse GPU/CPU memory objects.

7.1.2 Jacket Interface to MATLAB

The following table lists a full set of GPU data types introduced by Jacket with the example usage:

Function	Description	Example Usage
gsingle	Casts a CPU matrix to a single-precision floating-point GPU matrix.	`A = gsingle(B);`
gdouble	Casts a CPU matrix to a double-precision floating-point GPU matrix.	`A = gdouble(B);`
glogical	Casts a CPU matrix to a binary GPU matrix. All nonzero values are set to '1'. The input matrix can be a GPU or CPU data type.	`A = glogical(B);` `A = glogical(0:4);`

gint8, guint8, gint32, guint32	Cast a CPU matrix to a signed, unsigned, 8-bit, or 32-bit integer GPU array.	`A = gint8(B);`
		`A = guint8(B);`
		`A = gint32(B);`
		`A = guint32(B);`

gzeros, zeros	Create a matrix of zeros on the GPU.	`A = gzeros(5,'double');` `A = zeros(2,6,gdouble);`
gones, ones	Create a matrix of ones on the GPU.	`A = gones(5,'double');` `A = ones([3 9], gdouble);`
geye	Creates an identity matrix on the GPU.	`A = geye(5);`
grand (grandn) or *rand (randn)*	Creates a random matrix on the GPU, with uniformly (normally) distributed pseudorandom numbers.	`A = grand(5,'double');` `A = rand(5,gdouble);`

Once a GPU data structure has been created, any operations on that GPU matrix are performed on the GPU rather than the CPU. To turn off GPU computation, simply cast the data back to the CPU using one of the CPU data types, for example, *double*:

```
A = gdouble(B); % to push B to the GPU from the CPU
B = double(A); % to pull A from the GPU back to the CPU
```

Other important Jacket-specific functions are summarized below:

Function	Description	Example Usage
gactivate	Manual activation of a Jacket license.	`gactivate;`

gfor	Executes FOR loop in parallel on GPU.	`gfor n = 1:10;`
		`% loop body`
		`gend`

gselect	Select or query which GPU is in use.	`gselect(1);`
timeit	Robust timing of CPU or GPU code.	`timeit(@grandn, 2000)`
Graphics Library	Functions contained in the Graphics Library.	`gplot(A);`

gcompile	Compile m-code directly into a single CUDA kernel.	`my_fn =`
		`gcompile('filename.m');`
		`[B C...] = my_fn(A)`

gprofile	Profile code to compare CPU versus GPU run-times.	`gprofile on; foo;`
		`gprofile off;`
		`gprofile report;`

gprofview	Visual representation of profiling data.	`gprofview;`
geval	Evaluate computation and leave results on GPU.	`geval;`
gsync	Block until all queued GPU computation is complete.	`gsync(A);`
glaunch	Prototype, execute, and benchmark CUDA kernels within M-files.	`glaunch;`
gmex	Compile GPU-enabled MEX files using the Jacket SDK (in Windows).	`gmex [options] source.cu;`

The following example approximates *pi* using a Monte-Carlo algorithm on the GPU:*

```
NSET = 1000000;
X = grand(1, NSET); % generate sets of random numbers
Y = grand(1, NSET);
distance_from_zero = sqrt(X.*X + Y.*Y);
inside_circle = (distance_from_zero <= 1);

% count the number of points within a unit circle
pi = 4 * sum(inside_circle)/NSET
```

⇨ pi = 3.1421

For a large set of tips and tricks on how to get the best performance from Jacket, visit Torben's Corner.[1027]

7.1.3 Using Parallel *gfor* Loops

Similar to *parfor* statements from PCT, Jacket supports parallel *for*-loops, called *gfor*, which are executed on the GPU. The *gfor*-loop construct may be used to simultaneously launch all of the iterations of a *for*-loop on the GPU at the same time, as long as the iterations are independent. Jacket does this by tiling out the values of all loop iterations and then performing computation on those tiles in one pass.

The following MATLAB functions and language constructs, among others, are supported within the body of *gfor*-loops:[1028]

- Element-wise arithmetic (addition, subtraction, multiplication, division, *power, exp*)
- *fft, fft2, fftn,* and their inverses *ifft, ifft2, ifftn*
- *transpose, ctranspose,* and *diag*
- Matrix–matrix, matrix–vector, vector–vector multiply (*mtimes*)
- Subscripted assignment/referencing
- Reductions (*sum, min, max, any, all*)

Similar to *parfor*-loops, each iteration of the *gfor*-loop body must be independent of the other iterations: accessing the result of a separate iteration produces undefined behavior. However, the *gfor* iterator range is *not* restricted to *consecutive integers* (unlike **parfor** iterators). The only requirement is that the iterator expression must be a row vector of *uniformly spaced* real values.

Also, keep in mind that *gfor*-loops do not support conditional statements: that is, no branching is allowed inside a loop body. Nevertheless, quite often, this restriction can be overcome by expressing the conditional statement as a multiplication by logical values.† Examples in the table below illustrate some *gfor*-loop usages:[1029]

---

* Also see §3.9.3.
† See §4.6.3.

Example	Description
| ```
A = gones(n);
B = gones(1,n);
gfor k = 1:n
  B(k) = A(k,:) * A(:,k);
gend
``` | Vector–vector multiply |
| ```
A = gones(n,n,m);
B = gones(n);
gfor k = 1:m
  A(:,:,k) = A(:,:,k) * B;
gend
``` | Matrix–matrix multiply |
| ```
A = gones(n,n,m);
B = gones(n,10);
gfor k = 1:2:m
  A(:,1:10,k) = k*B;
gend
``` | Complicated subscripting is supported |
| ```
[A B C] = deal(gones(n));
gfor k = 1:n
  A(:,k) = 4*B(:,k);
  C(:,k) = 4*A(:,k);% use it again
gend
``` | In-loop reuse: within the loop, we can use a result that we just computed |
| ```
A = gsingle(1:n);
gfor ii = 1:n
  B = local(ii, A);
  % each tile has its own B with a
  % different index zeroed out
  B(ii) = 0;
  % sum is different for each tile
  D(ii) = sum(B);
gend
``` | Assign results can be subscripted to a local variable where each *gfor*-iteration (tile) has its own uniquely modified copy of that variable |
| ```
A = gones(n,m);
gfor k = 1:n
  condition = k > 10;
  A(:,k) = ~condition*A(:,k) +...
      condition*(k + 1);
gend
``` | Removing conditional statements through multiplication by logical values |
| ```
gfor k = 1:n
  for j = 1:m
    %...
  end
gend
``` | Nesting *for*-loops is supported, as long as the *gfor* iterator is not used in the *for*-loop iterator |

7.1.4 Compiling M-Code to a CUDA Kernel with *gcompile*

Jacket enables to compile the m-code directly to CUDA kernels. Though this mechanism, Jacket can perform additional optimizations, which are not possible in standard "dynamic" MATLAB mode. This is especially useful for time-critical code that contains complex arithmetic expressions, small loops or conditionals, and constitutes the core of a more complicated algorithm. The embedded code is written as if operating on scalars: each input and output parameter is a scalar element of the larger array. Subscripting and indexing is not supported.
```

These static optimizations are available through two functions, which can be used interchangeably: *gcompile* and *arrayfun*. *gcompile* accepts both strings and function handles, while *arrayfun* requires function handles. In practice, using *gcompile* directly can yield a few percent performance improvement as it avoids some caching checks that *arrayfun* incurs. There are two ways for invoking *gcompile* on MATLAB code. The first is by using the M-file convention as follows:

```
my_fn = gcompile('filename.m'); % compile function to a CUDA kernel
[A, B,...] = my_fn(C,...)        % call the compiled function
```

Alternatively, one can inline function code directly to *gcompile* statement:

```
my_fn = gcompile(verbatim);
%{
function [x, y] = foo(z)
  x = z/4;
  y = z * 2;
%}
[A, B] = my_fn(C);
```

arrayfun can be used as follows:

```
A = arrayfun(fun, S)
A = arrayfun(fun, S, T,...)
[A, B,...] = arrayfun(fun, S,...)
```

where *fun* is a function handle. The *fun* function must be written in M language and stored in a file named *fun.m*. *my_fn* will be compiled by *gcompile* and then executed:

```
p = gsingle(rand(3));
m = arrayfun(@my_fn, p);
```

During the first *arrayfun* call, the function will be compiled, and then reused in all subsequent calls. Note that *gcompile* and *arrayfun* support conditional statements and loops (e.g., *for* and *while*). However, there are certain limitations on their usage:[1030]

- Some operations (*sum, min/max, any/all, find*, subscripting, indexing, etc.), are not supported.
- *gcompile* functions cannot call other user-defined functions.
- All input parameters must be either scalar or of the same type.
- All operations must be element-wise.
- At least one input parameter must be a Jacket data type (e.g., *gsingle, gdouble*).
- *for*-loops may only use constants in the loop and may only increment by ≥ 1.
- *switch* statements are not supported.

7.1.5 Multi-GPU Support

Perhaps one of the most impressive features of Jacket is the ability to seamlessly utilize multiple GPUs on a local machine or across a cluster. Note that while programming a single-threaded application with the GPU support can be done using CUDA/OpenCL only, distributed multithreaded multi-GPU programming is practically unmanageable with these low-level libraries alone as it requires enormous amount of work to set up communication and/or synchronization between different computing units. Jacket provides a universal solution to this problem, which can save many hours of tedious work.

The easiest way to utilize multiple GPU devices with Jacket is by using the *gselect* function. The following example generates random values and performs FFTs in parallel across all GPUs found in the system:

```
ngpu = getfield(ginfo, 'gpu_count');
for i = 1 : ngpu
    gselect(i) % switch device
    out{i} = fft(rand(4096, gsingle));
end
gsync('all') % wait for all devices to finish
```

We can use *gselect* to switch between the GPU devices at any time. However, keep in mind that Jacket does not support peer-to-peer GPU access through unified virtual addressing. Hence, variables created on one GPU *cannot* be used on a different GPU. As a result, data must be copied *from the host* to each GPU separately.

More advanced multi-GPU usage is available through integration with MathWorks' PCT. Namely, Jacket supports *parfor, pmode,* and *spmd* constructs to distribute computations among workers, which we discussed in §6.1. Note that to run computations across a multi-mode cluster, we need the Jacket High-Performance Computing (HPC) License, similar to the way that we would need the MathWorks Distributed Computing Server (MDCS, §6.3). Typically, the optimal performance is attained when there is a one-to-one correspondence between the number of workers (reserved with *parpool*) and the number of GPUs present in the system. To test a multi-GPU setup, we can run the *ginfo* command on each worker using *spmd* statement. For example, on our platform with 4 Tesla GPUs, we obtain the following results:

```
>> spmd(4), ginfo, end
Lab 1:
   Jacket v2.2 (build 9ebd1f8) by AccelerEyes (64-bit Linux)
   License:...
   Addons: MGL16, JMC, SDK, DLA, SLA
   CUDA toolkit 4.2, driver 4.2 (295.53)
   GPU1 Tesla S2050, 2688 MB, Compute 2.0 (single,double) (in use)
   GPU2 Tesla S2050, 2688 MB, Compute 2.0 (single,double)
   GPU3 Tesla S2050, 2688 MB, Compute 2.0 (single,double)
   GPU4 Tesla S2050, 2688 MB, Compute 2.0 (single,double)
   Memory Usage: 2018 MB free (2688 MB total)
Lab 2:
   Jacket v2.2 (build 9ebd1f8) by AccelerEyes (64-bit Linux)
   License:...
   Addons: MGL16, JMC, SDK, DLA, SLA
   CUDA toolkit 4.2, driver 4.2 (295.53)
   GPU1 Tesla S2050, 2688 MB, Compute 2.0 (single,double)
   GPU2 Tesla S2050, 2688 MB, Compute 2.0 (single,double) (in use)
   GPU3 Tesla S2050, 2688 MB, Compute 2.0 (single,double)
   GPU4 Tesla S2050, 2688 MB, Compute 2.0 (single,double)
   Memory Usage: 2018 MB free (2688 MB total)
Lab 3:
   ...
Lab 4:
   ...
```

Note that each lab (worker) is bound to a different GPU device. Jacket fully supports *spmd* and *pmode* parallel constructs. The *parfor*-loops are also supported but with some minor limitations:

- The loop iterator range must be specified with MATLAB basic data types and not with GPU data types such as *gsingle* or *gdouble*.
- Any variable created inside a parfor-loop will have a MATLAB type by default, and hence cannot have a GPU value assigned to it. This limitation can be resolved by preallocating the variable before the loop, as a GPU data type, which can then be used within the *parfor* loop. In some cases, it must also be referenced as a value within the loop, before the assignment.

Consider an example:

```
parfor i = 1 : 10
    n(i) = gsingle(i);
end
```

This will not work since n will be created as a CPU data type. Instead, we preallocate n before the loop starts:

```
n = gones(1,10); % preallocate the variable before the parfor loop
parfor i = 1 : 10
    n(i); % reference the value before the assignment
    n(i) = gsingle(i);
end
```

Let us now apply what we have learned so far on our matched filter running example.

7.1.6 Running Example: Using Parallel *gfor*-Loop

We again rewrite the vectorized version of the algorithm from §5.6.2 to enable GPU acceleration. The resulting algorithm looks *almost* the same as the CPU-bound version with the exception of a couple of *gzeros* preallocations and a *gfor*-loop construct used in the inner loop.*

```
function data = matched_filter_gfor(data)
    % define speed of light
    c = 299792458;

    % determine the size of the phase history data
    K = size(data.phdata,1); % number of frequency bins per pulse
    Np = size(data.phdata,2); % number of pulses

    % initialize the image with all zero values (on the GPU)
    data.im_final = gzeros(size(data.x_mat), 'double');

    % array to keep "slices" of an image for different frequencies
    im_slices = gzeros([K, size(data.x_mat)], 'double');
```

---

* This file can be downloaded from: http://UndocumentedMatlab.com/books/matlab-performance (or: http://bit.ly/1pKuUdM).

```
% loop through every pulse
for ii = 1 : Np

    % Calculate differential range for each image pixel (m)
    dR = sqrt((data.AntX(ii)-data.x_mat).^2 + ...
              (data.AntY(ii)-data.y_mat).^2 + ...
              (data.AntZ(ii)-data.z_mat).^2) - data.R0(ii);

    % Calculate the frequency of each sample in the pulse (Hz)
    freq = data.minF(ii) + gdouble(0:(K-1)) * data.deltaF;

    % Perform the Matched Filter operation
    gfor jj = 1 : K   % parallel for loop on the GPU
        im_slices(jj,:,:) = data.phdata(jj,ii) * ...
                            exp(1i*4*pi*freq(jj)/c*dR);
    gend

    % accumulate image data for different frequencies
    data.im_final = data.im_final + squeeze(sum(im_slices,1));
end   % for ii = 1:Np

    gsync('all');   % ensure all computations on the GPU are done
end
```

Note that the fields of the data structure must be *explicitly cast* to the GPU (using **gsingle** or **gdouble** constructors) before invoking this function. Otherwise, the execution will proceed on the host side. To make the inner loop eligible for parallel execution, we introduced an extra variable im_slices to store the image data for different frequencies. After the parallel loop, different "image slices" are accumulated to produce a final image data. im_final. The GPU-accelerated algorithm runs in **13.2 s** for an image of size 500×500.

7.1.7 Running Example: Using *gcompile*

Next, we try to improve performance by employing *gcompile*. From the Jacket profiler, activated with the *gprofile* command, we can see that the major part of the running time is spent in the inner *gfor*-loop. This is quite a natural thing to expect since the loop performs heavy computations with complex arithmetic across large 2D arrays. We shall try to compile it in a single GPU kernel with *gcompile* such that Jacket can apply additional low-level optimizations, which are not possible in "dynamic" mode. However, the compiled GPU code cannot be used inside *gfor*-loops: therefore, we again take the original vectorized algorithm from §5.6.2 and modify it accordingly. The resulting algorithm is shown below:*

```
function data = matched_filter_gcompile(data)
    % define speed of light
    c = 299792458;

    % determine the size of the phase history data
    K  = size(data.phdata,1); % number of frequency bins per pulse
    Np = size(data.phdata,2); % number of pulses
```

---

* This file can be downloaded from: http://UndocumentedMatlab.com/books/matlab-performance (or: http://bit.ly/1pKuUdM).

```
    % initialize the image with all zero values (on the GPU)
    data.im_final = gzeros(size(data.x_mat), 'double');

    % compile the algorithm's inner loop in a single GPU kernel
    mf_gcore = gcompile(verbatim);
    %{
    function im_slice = ff(freq, dR, phdata, im_final, c)
        im_slice =  im_final + phdata * exp(1i*4*pi*freq/c*dR);
    end
    %}

    % loop through every pulse
    for ii = 1 : Np
        % Calculate differential range for each image pixel (m)
        dR = sqrt((data.AntX(ii)-data.x_mat).^2 + ...
                  (data.AntY(ii)-data.y_mat).^2 + ...
                  (data.AntZ(ii)-data.z_mat).^2) - data.R0(ii);

        % Calculate the frequency of each sample in the pulse (Hz)
        freq = data.minF(ii) + gdouble(0:(K-1)) * data.deltaF;

        % Perform the Matched Filter operation
        for jj = 1 : K
            phdata = gdata.phdata(jj,ii);
            freqj = freq(jj);
            data.im_final = mf_gcore(freqj, dR, phdata, ...
                                     data.im_final, c);
        end
    end  % for ii = 1:Np
    gsync('all');  % ensure that all GPU computations are done
end
```

Note the usage of *verbatim* syntax with **gcompile**. The modified algorithm now takes **6.8 s** on our machine. Note that for our problem, **gcompile** yields better performance than a *gfor*-loop: this is, of course, not always the case. Therefore, it is always advisable to experiment with different constructs to find which is more suitable for a particular problem at hand.

7.1.8 Running Example: Using *spmd* and Multi-GPU Support

To span the computations across multiple GPUs, we'll use the **spmd** parallel construct. We set the number of workers to the number of GPUs available to achieve optimal performance. The multi-GPU version of the matched filter algorithm is as follows:*

```
function gdata = matched_filter_multigpu(data)
    % Define speed of light (m/s)
    c = 299792458;

    % Determine the size of the phase history data
    K  = size(data.phdata,1);  % number of frequency bins per pulse
    Np = size(data.phdata,2);  % number of pulses
```

---

* This file can be downloaded from: http://UndocumentedMatlab.com/books/matlab-performance (or: http://bit.ly/1pKuUdM).

```
gdata.im_final = zeros([size(data.x_mat)], 'double');
ngpu = getfield(ginfo, 'gpu_count');

% start parallel processing on ngpu labs
spmd (ngpu)
    % cast data explicitly to the GPU for each worker
    minF     = gdouble(data.minF);
    AntX     = gdouble(data.AntX);
    AntY     = gdouble(data.AntY);
    AntZ     = gdouble(data.AntZ);
    R0       = gdouble(data.R0);
    phdata   = gdouble(data.phdata);
    x_mat    = gdouble(data.x_mat);
    y_mat    = gdouble(data.y_mat);
    z_mat    = gdouble(data.z_mat);
    im_slices = gzeros(size(data.x_mat), 'double');

    % Loop through every pulse
    for ii = drange(1:Np)

        % Calculate differential range for each image pixel (m)
        dR = sqrt((AntX(ii)-x_mat).^2 + ...
                  (AntY(ii)-y_mat).^2 + ...
                  (AntZ(ii)-z_mat).^2) - R0(ii);

        % Calculate frequency of each sample in the pulse (Hz)
        freq = minF(ii) + gdouble(0:(K-1)) * data.deltaF;

        % Perform the Matched Filter operation
        for jj = 1:K
            im_slices = im_slices + phdata(jj,ii) * ...
                                    exp(1i*4*pi*freq(jj)/c*dR);
        end
    end
end   % spmd
gdata.im_final = zeros(size(data.x_mat), 'double');

% gather results computed by different workers
for ii = 1:numel(im_slices)
    gdata.im_final = gdata.im_final + double(im_slices{ii});
end
end
```

At the beginning of the *spmd* body, we cast data explicitly to the GPU because, as noted above, Jacket cannot access peer GPU memory directly. Finally, at the end, we need to sum up the individual image "slices" to obtain the final result. On our platform, the algorithm takes **4.6** s for an image of size 500×500. As a matter of fact, we might expect even higher speedup by using *gcompile* and *gfor* constructs inside the *spmd* statements, but this is unfortunately not supported by Jacket.

7.2 Alternative/Related Technologies

PCT is not distributed as part of MATLAB and most MATLAB users do not possess a separate license for the toolbox. Fortunately, there are many alternative libraries and techniques available to exploit parallelism in MATLAB in case PCT is not an option. Note that while PCT provides out-of-the-box support for both multi-processor and GPU parallelization, these two aspects are generally handled by separate products/libraries.

Multi-processor parallelization alternatives include a number of open-source libraries developed specifically for MATLAB, such as Multicore,[1031] MatlabMPI,[1032] pMATLAB (which uses MatlabMPI),[1033] and bcMPI (MatlabMPI-compatible).[1034] There are also several generic (non-MATLAB-specific) libraries that can be accessed via MEX interface (MPI, OpenMP, OpenCL,* TBB, PVM, etc.). Additional alternatives are listed in §7.2.4 and §7.3.

GPU parallelization alternatives developed specifically for MATLAB include Jacket (§7.1), GPUmat, and MATLAB plug-in for CUDA.[1035] There are also several generic GPU libraries, accessible via MEX interface (CUDA, OpenCL, CULA, ArrayFire, OpenACC, etc.).

For reasons of space, this section only covers the GPUmat, Multicore, and ArrayFire libraries, while OpenMP will be discussed separately in §7.3.2.

7.2.1 Using GPUmat

GPUmat is an open-source software that enables MATLAB code to run on CUDA-enabled GPUs.[1036] GPUmat is built on top of the CUBLAS and CUFFT libraries, which are provided in the standard CUDA toolkit.[1037] As with the Jacket library, discussed in §7.1, existing code can be ported and executed on GPUs with few modifications.[1038] Unlike Jacket, GPUmat is a freeware and still available.

To install the library, add the directory with unzipped files to MATLAB's path.[1039] Then, start GPUmat using the ***GPUstart*** command:

```
>> GPUstart
Starting GPU
- GPUmat version: 0.280
- Required CUDA version: 5.0
There is 1 device supporting CUDA
CUDA Driver Version:                    5.0
CUDA Runtime Version:                   5.0

Device 0: "GeForce GT 650M"
  CUDA Capability Major revision number:  3
  CUDA Capability Minor revision number:  0
  Total amount of global memory:          2147287040 bytes
  - CUDA compute capability 3.0
...done
- Loading module EXAMPLES_CODEOPT
```

---

* OpenCL is one of the rare frameworks that (like PCT) support both multi-processor and GPU parallelization. A comment made by MathWorker Sara Wait Zaranek a few years ago said that MathWorks has "chosen to support CUDA and not OpenCL due to the fact that CUDA currently has the only ecosystem with all of the libraries necessary for technical computing. This may change as open CL libraries advance and evolve": http:// blogs.mathworks.com/loren/2012/02/06/using-gpus-in-matlab#comment-32970 (or: http://bit.ly/1g2NNRC). With OpenCL support growing continuously (Intel even released an OpenCL SDK to access its on-die GPU on new CPUs), I expect that OpenCL support will indeed be added by MathWorks some day.

```
- Loading module EXAMPLES_NUMERICS
  -> numerics30.cubin
- Loading module NUMERICS
  -> numerics30.cubin
- Loading module RAND
```

If errors occur during GPUmat startup, we can use the diagnostic command *GPUmatSystemCheck* to investigate them. If the system check was successful, the command's output should look similar to the following:

```
>> GPUmatSystemCheck
*** GPUmat system diagnostics
* Running on            -> "glnxa64"
* Matlab ver.           -> "7.14.0.739 (R2012a)"
* GPUmat version        -> 0.280
* GPUmat build          -> 09-Dec-2012
* GPUmat architecture   -> "glnxa64"

*** ARCHITECTURE TEST
*** GPUmat architecture test -> passed.

*** CUDA TEST
*** CUDA CUBLAS -> installed (.../cuda/lib64/libcublas.so).
*** CUDA CUFFT  -> installed (.../cuda/lib64/libcufft.so).
*** CUDA CUDART -> installed (.../cuda/lib64/libcudart.so).
```

A frequently occurring error on Linux systems is when GPUmat cannot find CUDA library files because the CUDA toolkit is installed not in the default path. To fix this problem, add the CUDA library path to the `LD_LIBRARY_PATH` environment variable.

Another common error is *"invalid MEX file… specified module could not be found"*. In this case, updating to the latest CUDA toolkit could solve the issue.[1040]

GPUmat defines **GPUsingle** and **GPUdouble** data types, with associated overloaded MATLAB functions, to model operations with single- and double-precision floating-point data on the GPU. For example, the following code creates a single-precision MATLAB variable and then transfers its contents to the GPU:

```
A = GPUsingle(rand(100,100));   % copy the variable to the GPU
```

However, it is more efficient to create this variable directly on the GPU. For example, using the overloaded *rand* function:

```
A = rand(100,100,GPUsingle);    % create directly in GPU memory
```

Analogously, we can use the *colon* operator to create a vector directly on the GPU:

```
A = colon(0,1,1000,GPUsingle);  % create directly in GPU memory
```

Note that in contrast to PCT, GPUmat supports *all* CUDA-enabled graphics cards, even those that have no native support for double-precision arithmetic. To check if a card supports double precision, use the following command:

```
if GPUisDoublePrecision
   A = rand(100,100,GPUdouble);
end
```

A large subset of MATLAB built-in functions and operators (including matrix operations, *repmat*, *reshape*, FFT, and mathematical functions) are overloaded for **GPUsingle** and **GPUdouble** data types.* For example, the following code snippet multiplies two matrices on the GPU:

```
A = rand(1000);
B = rand(1000);

tic
dA = GPUdouble(A); % copy matrices to the GPU memory
dB = GPUdouble(B);
dC = dA*dB;  % perform matrix multiplication
C = double(dC); % copy the results back
toc
⇨ Elapsed time is 0.103097 seconds.
```

We can compare the execution time with that of **gpuArray**:

```
dA = gpuArray(A);
dB = gpuArray(B);
dC = dA*dB;
C = gather(dC);
⇨ Elapsed time is 0.108702 seconds.
```

As we can see, the run-time of both matrix multiplication examples is essentially the same. This suggests that both libraries use the same GPU routine for matrix multiplication (CUBLAS) internally.

GPUmat also overloads *subsref* and *subsasgn* commands. Therefore, an array of GPU data can be accessed just as any other MATLAB array. For example, the following assignments will execute on the GPU:

```
A = rand(50,GPUdouble); % A is on the GPU
B = A(2:end);
B = A(1,1:15);
B = A(:);
A(1:15) = A(21:35);
```

To access the contents of a GPU array, GPUmat additionally provides the functions *slice* and *assign*, which are usually faster than the overloaded MATLAB commands *subsref* and *subsasgn*. The following examples illustrate the usage of these functions:

```
A = rand(100);
Ad = GPUdouble(A);
B = A([2 3 1],:);                        % MATLAB syntax
Bd = slice(Ad,{[2 3 1]},':');  % equivalent slice syntax
B = A(1:10,:);                           % MATLAB syntax
Bd = slice(Ad, [1,1,10], ':');  % equivalent slice syntax
C = rand(4,10);
Cd = GPUdouble(C);
A([2 3 1 5],1:10) = C;                      % MATLAB syntax
assign(1, A, C, {[2 3 1 5]},[1,1,10]); % equivalent assign syntax
```

A very useful GPUmat feature is that it provides a source-to-source compiler, which translates GPU code into a single MATLAB function. The GPUmat compiler functions are listed below:

---

* Consult the *GPUmat User Guide* for a full set of overloaded functions and operators.

Function Name	Description
`GPUcompileStart(NAME,` `    OPTIONS, p1,p2,...,pn)`	Starts compilation of a function with specified NAME and compile OPTIONS
`GPUcompileStop(r1,r2,...,rn)`	Stops compilation of a function
`GPUcompileAbort`	Aborts compilation
`GPUfor it = a:b`	Starts a *for*-loop
`GPUend`	Ends a *for*-loop
`GPUcompileMEX`	Compiles a CPP file

For example, the following code:

```
GPUcompileStart(funcName,p1,p2,...,pn)
...
GPUcompileStop(r1,r2,...,rm)
```

generates a MEX function: `[r1,r2,...,rm] = funcName(p1,p2,...,pn)`. To check if the system is configured properly for compilation by GPUmat, use the following command:

```
GPUcompileCheck
```

The *for*-loops (including nested loops) can be integrated in a compiled GPU code using *GPUfor* and *GPUend* constructs:

```
N = 2000;
A = randn(N,GPUdouble)+1i*rand(N,GPUdouble);
tmp = complex(zeros(1,N,GPUdouble));
B   = complex(zeros(1,N,GPUdouble));

% compile GPU code to a single CUDA kernel
GPUcompileStart('myfun', '-f', A, B, tmp)
    C = exp(A.');
    GPUfor kk = 1:N
        assign(0,tmp,C,kk,':');   % tmp = C(kk,:)
        GPUtimes(tmp,3.0,tmp);    % tmp = tmp * 3
        GPUplus(tmp,B,tmp);       % tmp = B + tmp
        assign(0,B,tmp,':');      % B = tmp
    GPUend
GPUcompileStop(B)
```

Then, we can execute the function on the GPU as follows:

```
B = double(myfun(A,B,tmp)); % invoke the GPU function
```

This function runs in **0.18** s on our machine with a GT 650M graphics card. The analogous code fragment using PCT's *gpuArray* would look like this:

```
A = gpuArray(randn(N)+1i*rand(N));
B = gpuArray(zeros(1,N));
C = exp(A.');
for kk = 1:N
    tmp = C(kk,:);
    B = B + tmp*3;
end
```

which takes **0.31** s on our platform.

GPUmat compiler produces binary MEX files that can later be used in other applications. However, the compiler has a number of limitations, such as:

- Only GPU functions are recorded, MATLAB functions are not included in compilation.

- All formal parameters of a function being compiled must be defined before the compilation starts.

- Indexed references and assignments are only supported with *slice* and *assign* functions.

- Temporary variables used in indexed assignments must be defined before the compilation starts.

- Many arithmetic operations must be replaced with their low-level analogs (starting with GPU* prefix).

Another salient feature of GPUmat is that users can add functionality to the library by either directly modifying the source code or working with a *GPUmat User Module*. User modules are loaded using the GPUmat module manager as follows:

```
GPUuserModuleLoad(module_name, cubin_file);
```

which loads a cubin module **cubin_file** and assigns the name **module_name** to it. GPUmat internal types and function can be accessed via the MEX interface. Refer to *GPUmat Developers Guide* (included in the GPUmat ZIP) for additional information.

Many publicly available toolboxes and libraries use GPUmat. One example is Malcolm Lidierth's open-source *Waterloo* toolbox.[1041] Such toolboxes can be examined, and their source code can be adapted for our specific needs.

7.2.2 Multicore Library for Parallel Processing on Multiple Cores

Multicore[1042] open-source library by Markus Buehren adds the support for parallel processing in MATLAB on a single or on multiple machines that have access to a common directory. The library interface is quite straightforward and consists of two global functions: *startmulticoremaster* and *startmulticoreslave*. The first command initiates parallel processing:

```
out = startmulticoremaster(@fun, params, settings)
```

where **fun** denotes the handle of a function to be executed on different MATLAB workers, **params** is a cell array for each element of which the given function handle is to be evaluated, and **settings** is a structure containing additional options listed in the table below. The outputs of the function **fun** are returned in a cell array **out**. Only the first output argument of the function is returned. To get multiple outputs, we can modify the function to put the outputs in a single cell array.

Master Option Name	Description
multicoreDir	Directory to save temporary files (standard directory is used if empty)
nrOfEvalsAtOnce	Number of function evaluations gathered to a single job
maxEvalTimeSingle	Timeout for a single function evaluation (this parameter should be chosen carefully to get optimum performance)
masterIsWorker	If true, master process acts as a worker and coordinator, otherwise the master acts only as a coordinator
useWaitbar	If true, a waitbar (progress bar) is opened to inform about the overall progress

For example, a *for*-loop:

```
res = cell(size(params));
for k=1:numel(params)
    res{k} = myfun(params{k});
end
```

will be equivalent to the following call of ***startmulticoremaster***:

```
res = startmulticoremaster(@myfun, params);
```

The call to ***startmulticoremaster*** stores the function handle and input parameters in a temporary directory, which must be accessible by other MATLAB workers. The ***startmulticoreslave*** command starts a MATLAB worker as follows:

```
startmulticoreslave(multicoreDir, settings);
```

where **multicoreDir** specifies a shared directory with data files storing the function to be run on a slave process and its input parameters, and **settings** is a structure of additional options described in the library manual (in most cases, default values suffice). Note that this command is *blocking*, and hence it should be started from a separate MATLAB session. You can also use the following command to start a slave worker from your current MATLAB session:*

```
>> !matlab -nodisplay -r startmulticoreslave&
```

provided that MATLAB is on your executable search path. This will start a background screenless (headless) MATLAB session and execute ***startmulticoreslave*** command within it. To benchmark the performance of Multicore library, you can use ***multicoredemo*** script provided with the distribution. On our laptop, with **six** MATLAB slaves running, it produces the following results:

```
>> multicoredemo

Elapsed time running STARTMULTICOREMASTER: 6.21 seconds.
Elapsed time without slave support:        20.82 seconds.
Now opening Matlab pool with default configuration.
Starting matlabpool using the 'local' profile... connected to 6 labs.
Elapsed time running TESTFUN in parfor-loop: 3.92 seconds.
Elapsed time running TESTFUN directly:      19.83 seconds.
```

As one can see, Multicore performance is comparable to PCT *parfor*-loops, although PCT is noticeably more efficient. The Multicore performance improves as the number of jobs are more evenly distributed among the master and workers: In the run example above, using the default settings, evalTimeAll = 20, evalTimeSingle = 0.5 and nrOfEvalsAtOnce = 4, the workload of 40 (evalTimeAll/evalTimeSingle) function evaluations is divided into 10 jobs (40/nrOfEvalsAtOnce). In the first round, all seven workers (master plus 6 slaves) do one job each. As soon as the workers finished their first job, there are only three more jobs left, leaving four workers idle. If the number of workers was 5 rather than 7, or if evalTimeSingle was 0.36 rather than 0.5, then the distribution of jobs would be optimal and no worker would remain idle.

Altogether, the Multicore library makes an impression of a mature and easy-to-use product that allows us to take advantage of multicore processing when PCT is not an option.

---

* See §7.4.

But, certainly, it has a number of limitations: for example, as workers communicate by passing all tasks and the associated data through a shared file system, performance can degrade. Still, we would recommend this library as an open-source alternative to PCT.

7.2.3 Using ArrayFire Library via MEX Interface

Using ArrayFire[1043] we can relatively easily enable GPU acceleration in MATLAB. ArrayFire's design philosophy resembles that of Jacket library (see §7.1), developed by the same company. ArrayFire runs on both CUDA-enabled and OpenCL-enabled hardware. In other words, ArrayFire is not limited to Nvidia GPUs; it also runs on AMD. The main object of manipulations in ArrayFire is **array**, which represents an N-dimensional vector of data in GPU memory. An array object provides a set of overloaded arithmetic operations as well as many other functions very similar to MATLAB. These functions include linear algebra, image and signal processing, cumulative sums, sorting, FFTs, set operations, statistical functions, histograms, and so on. The ArrayFire library was started as a commercial venture, but is now fully open-source, under a BSD license.

Array objects can be initialized from host-based pointers or created directly in GPU memory. For instance, the following code fragment creates a 3D array of **doubles** from host data:

```
double *vol = new double[nx*ny*nz];
...                     //perform initialization of "volume" data
array a(nx, ny, nz, vol, afHost); //create a 3D GPU array
```

Note that data in `vol` should be stored in *column-major* order (same as in MATLAB). More initialization examples:

```
array a = randu(100,100,f32); //create a random matrix on GPU
array c = array(seq(1,100));  //a sequence 1..100 on GPU
```

Array objects also support subscripted indexing with the syntax borrowed again from MATLAB. Therefore, MATLAB users should find it easy to get used to ArrayFire functions. For example, the following constructs are equivalent:

```
A = B(seq(1,2,7),end);       // ArrayFire syntax
A = B(1:2:7, end);           // MATLAB syntax

A(seq(1,7),span,span) = B;   // ArrayFire syntax
A(1:7, :, :) = B;            // MATLAB syntax
```

Finally, **arrays** can be manipulated using arithmetic operations and a large set of mathematical functions, for example:

```
//compute a matrix product and take an elementwise cosine
C = cos(matmul(A,B.T()));

//sort matrix along 1st dimension and compute the FFT
D = fft2(sort(C, 1));
```

Further examples on ArrayFire usage, including parallel *for*-loops, data visualization, and multi-GPU support can be found in the online user's manual.[1044]

In what follows, we demonstrate how to invoke ArrayFire functions from within MATLAB via MEX interface using our matched filter running example. Note that ArrayFire does not use source-to-source compiling, unlike other GPU libraries, and thus there is no need to set up CUDA/OpenCL compilation in MEX files. However, certain compilation/runtime problems have been reported by the users.[1045] We use the following command to compile ArrayFire MEX files in Linux:

```
mex mf_arrayfire.cpp -I/path/to/AF/includes -L/path/to/AF/libs
    -lcuda -lcudart -laf -Wl,-rpath=/path/to/AF/libs
```

The realization of the matched filter algorithm with ArrayFire is given below:[1046]

```
#include <cuda.h>
#include <arrayfire.h>
#include <mex.h>

// matched filter realization using ArrayFire
void matched_filter_af(SAR_data& data) {
   double c = 299792458.0; // speed of light

   // Determine the size of the phase history data
   int K  = data.phdata.dims(0);  // # of frequency bins per pulse
   int Np = data.phdata.dims(1);  // # of pulses

   // Initialize the image with all zero values (complex)
   data.im_final = zeros(data.x_mat.dims(), c64);
   array im_slices = zeros(K, data.x_mat.dims(0),
                              data.x_mat.dims(1), c64);
   array fspan = array(seq(0.0, K-1)) * data.deltaF;
   for (int ii = 0; ii < Np; ii++) {
      // compute differential range for each image pixel (m)
      array dx = data.AntX(ii) - data.x_mat;
      array dy = data.AntY(ii) - data.y_mat;
      array dz = data.AntZ(ii) - data.z_mat;
      array dR = sqrt(dx*dx + dy*dy + dz*dz) - data.R0(ii);

      // calculate the frequency of each sample in the pulse (Hz)
      array freq = data.minF(ii) + fspan;
      array tt = data.phdata(span,ii);

      // perform the Matched Filter operation
      gfor (array jj, K) {
         im_slices(jj,span,span) =
            tt(jj)*exp(i*((double)(4.0*Pi/c)*freq(jj)*dR));
      }
      tt = sum(im_slices,0);
      data.im_final = data.im_final + reshape(tt, data.x_mat.dims());
   }
}

// initializes an GPU array from a given data field
void init_field(array& out, const mxArray *rhs,
                const char *field_name, bool is_complex = false) {
   mxArray *mxA = mxGetField(rhs, 0, field_name);
   const mwSize *dims = mxGetDimensions(mxA);
   array re(dims[0], dims[1], mxGetPr(mxA), afHost);
   if (!is_complex) {
```

```
      out = re;
    } else {
      // construct an array of complex data on the GPU
      array im(dims[0], dims[1], mxGetPi(mxA), afHost);
      out = complex(re, im);
    }
}

// the MEX gateway function
void mexFunction(int nlhs,        mxArray *plhs[],
                 int nrhs, const mxArray *prhs[]) {
  SAR_data data; // initialize the data structure
  init_field(data.x_mat, prhs[0], "x_mat");
  init_field(data.y_mat, prhs[0], "y_mat");
  init_field(data.z_mat, prhs[0], "z_mat");
  init_field(data.AntX,  prhs[0], "AntX");
  init_field(data.AntY,  prhs[0], "AntY");
  init_field(data.AntZ,  prhs[0], "AntZ");
  init_field(data.minF,  prhs[0], "minF");
  init_field(data.R0,    prhs[0], "R0");
  init_field(data.phdata, prhs[0], "phdata", true);
  data.deltaF = mxGetScalar(mxGetField(prhs[0], 0, "deltaF"));

  // invoke the algorithm
  matched_filter_af(data);
  int sx = data.im_final.dims(0);
  int sy = data.im_final.dims(1);

  // create the output data matrix
  plhs[0] = mxCreateNumericMatrix(sx, sy, mxDOUBLE_CLASS, mxCOMPLEX);

  // copy the results from im_final field
  double *pre = af::real(data.im_final).host<double>();
  memcpy(mxGetPr(plhs[0]), pre, sx*sy*sizeof(double));
  double *pim = af::imag(data.im_final).host<double>();
  memcpy(mxGetPi(plhs[0]), pim, sx*sy*sizeof(double));
  array::free(pre);  // free temporary storage
  array::free(pim);
}
```

The MEX function can be invoked directly with a MATLAB data structure containing all necessary data fields as described in §5.6, that is:

```
im_final = mf_arrayfire(data);
```

Again, we have run the benchmarks on our multicore cluster with 4 Tesla S2050 graphics cards (the specifications are given in §5.6). The algorithm takes **8.3** s to reconstruct an image of size 500×500, which we consider to be a good result. Note that we have only used a single GPU for computations. Interested readers may extend the algorithm to support multiple GPUs.

In conclusion, ArrayFire library is a mature commercial product with good online support and continuously added new features. The library design is targeted at MATLAB users. Owing to the support for OpenCL standard, the library can be utilized for a broad range of parallel architectures.

7.2.4 Additional Alternatives

So far in this section we saw an overview of three different libraries that can be used as a (partial) replacement for PCT. These libraries were chosen not based on some "importance" factor but to show that there are multiple alternatives for parallel MATLAB processing. The number of alternatives is large;[1047] some examples are

- Intel's SPMD compiler[1048] (ispc) can be used for code vectorization.

- Portland PGI compiler[1049] enables AVX vector instruction support and CUDA support with OpenACC directives.

- AMD Core Math Library (ACML)[1050] provides highly optimized threaded math routines for compute-intensive applications.

- OpenCL[1051] can access on-die GPU of Intel Ivy Bridge[1052] or AMD APUs.[1053]

- GPULib[1054] is a commercial library that enables GPU-accelerated computations by providing bindings for IDL (Interactive Data Language).[1055] A GPULib developer has recently experimented with OpenCL support.[1056]

- CULA[1057] is a set of commercial GPU-accelerated BLAS and LAPACK libraries for dense and sparse linear algebra, which can be called from C/C++, Fortran, MATLAB, and Python. They are free for individual academic use.

- PLASMA library[1058] is a linear algebra library for multicore architectures.

- NAG C library[1059] is a large commercially available collection of numerical algorithms for C with parallel support and interfaces to Excel, Java, MATLAB, .NET/C#, and so on. See §8.5.3 for additional details.

- ViennaCL[1060] is an open-source C++ interface for linear algebra routines on GPUs using CUDA, OpenCL, and/or OpenMP. It has a MATLAB interface and can also be used by C++ MEX files.

- MITMatlab/MATLAB*P[1061] was a research project at MIT led by Alan Edelman starting in 1998, which used MATLAB object polymorphism to interface MATLAB to external parallel math libraries using MPI.[1062] It was featured in a 1999 MathWorks newsletter.[1063] A MATLAB*P spin-off in 2004 was called *P (Star-P).[1064] In 2001, Jeremy Kepner developed MatlabMPI as the first parallel multi-process MATLAB, using MPI for inter-lab communication.[1065] This was extended in 2002 by Nadya Bliss using MATLAB*P polymorphism ideas, as *pMatlab*.[1066] These projects served as the basis for MathWorks' PCT and MDCS toolboxes.

- Ohio Supercomputing Center's bcMPI,[1067] was a separate project in 2007 that used MPI communication, and was mostly compatible with MatlabMPI.

- MPITB,[1068] ParaMat,[1069] NASA's MATPAR,[1070] Ohio State's GAMMA,[1071] Illinois University's DLab,[1072] UTexas' PLAPACK,[1073] Cornell's Multitasking toolbox,[1074] and MultiMATLAB[1075] were all MATLAB parallelization projects implemented over the past two decades. Most of them are no longer active.

- Techila[1076] offers a commercial pay-as-you-use plug-in (toolbox) for MATLAB, which can be used for distributed cloud/cluster computing.

Finally, it is worth looking at a survey article[1077] about code vectorization on modern CPUs, which contains a large number of links to other related resources.

7.3 Multithreading

One of the limitations of MATLAB already recognized by the community[1078] is that it does not provide the users *direct access* to threads, for example, letting some expensive computations to be run in the background without freezing the main application. Instead, in MATLAB, there is either *implicit multithreading* that relies on built-in threading support in some MATLAB functions, or *explicit multiprocessing* using PCT (note: PCT workers use heavyweight *processes*, not lightweight *threads*). So we can only achieve true multithreading in MATLAB via MEX, Java, or .Net.[1079] In a sense, this is a logical extension of the idea of explicit parallelism without PCT.

The alternatives that can be used to enable MATLAB multithreading include standard POSIX threads, native OS threads, OpenMP, MPI (*Message Passing Interface*), TBB (*Thread Building Blocks*),[1080] Cilk, OpenACC, OpenCL, or Boost. We can also use libraries targeting specific platforms/architectures: Intel MKL, C++ AMP, Bolt, and so on.

Note that we do not save any CPU cycles by running tasks in parallel. In the overall balance, we actually increase the amount of CPU processing, due to multithreading overhead. But in most cases, we are more interested in the responsivity of MATLAB's main processing thread than in reducing the computer's total energy consumption. In such cases, offloading work to asynchronous C++, Java, or .Net threads could remove bottlenecks from MATLAB's main thread, achieving significant speedup.

7.3.1 Using POSIX Threads

POSIX threads (Pthreads) is a standard API for multithreaded programming implemented natively on many Unix-like systems, and also supported on Windows. Pthreads includes functionality for creating and managing threads, and provides a set of synchronization primitives such as mutexes, conditional variables, semaphores, read/write locks, and barriers. POSIX has extensive offline and online documentation.[1081] We therefore proceed directly to the usage example.[1082]

The example code below demonstrates a basic usage of Pthreads from MATLAB via MEX interface. Our task will be to compute binomial coefficients[1083] "n choose k" $\binom{n}{k} = \dfrac{n!}{(n-k)! \cdot k!}$

in parallel.* Binomial coefficients can become large, so we shall use GNU MP library,[1084] then return the results as a cell array where each cell contains a vector of 32-bit words of a large integer. Note that the GNU MP library fully supports multithreading, so computations can be parallelized. The pseudocode is given below:

```
#include "mex.h"
#include <pthread.h>
#include <iostream.h>
#include <vector.h>
#include <stdlib.h>
#include <gmp.h>
unsigned *g_in;    // input parameters
mxArray *g_out;    // output cell array
pthread_mutex_t mex_mtx;
```

---

* Note the built-in function *nchoosek* and Jan Simon's much faster MEX-based *VChoosek* utility (http://mathworks. com/matlabcentral/fileexchange/26190-vchoosek or: http://bit.ly/1lAsKfp). Both of these are not parallelized.

```
/* thread compute function */
void *thread_compute(void *t)
{
   long tid = (long)t;          // get this thread's ID
   unsigned *in = g_in + tid*2; // and the inputs 'n' and 'k'
   mpz_t z;                     // initialize a GMP large integer
   mpz_init(z);

   // calculate binomial coefficient
   mpz_bin_uiui(z, (unsigned long)in[0], (unsigned long)in[1]);

   // estimate the size of the result (in 32-bit words)
   unsigned numb = sizeof(unsigned)*8;
   unsigned n_limbs = (mpz_sizeinbase(z,2) + numb-1) / numb;
   mxArray *res = mxCreateNumericMatrix(n_limbs, 1,
                                    mxUINT32_CLASS, mxREAL);

   // copy 32-bit words (limbs) to the output vector
   mpz_export(mxGetData(res), NULL, 1, sizeof(unsigned), 0, 0, z);

   // lock mutex and copy the result to the cell array
   pthread_mutex_lock(&mex_mtx);
   mxSetCell(g_out, tid, res);
   pthread_mutex_unlock(&mex_mtx);
   mpz_clear(z);          // release the large GMP integer and exit
   pthread_exit(NULL);
}

/* The MEX gateway function */
void mexFunction(int nlhs,        mxArray *plhs[],
                 int nrhs, const mxArray *prhs[]) {
   // number of threads is determined by the size of the inputs
   pthread_mutex_init(&mex_mtx, NULL);
   mwSize n_thids = mxGetNumberOfElements(prhs[0])/2;
   g_in = (unsigned *)mxGetData(prhs[0]);
   plhs[0] = mxCreateCellMatrix(n_thids, 1);
   g_out = plhs[0];
   std::vector< pthread_t > threads(n_thids);
   pthread_attr_t attr;

   // create attributes for joinable threads
   pthread_attr_init(&attr);
   pthread_attr_setdetachstate(&attr, PTHREAD_CREATE_JOINABLE);

   // launch the threads
   for (int i = 0; i < n_thids; i++) {
      pthread_create(&threads[i], &attr, thread_compute, (void *)i);
   }

   // wait until all threads finish processing
   for (int i = 0; i < n_thids; i++) {
      pthread_join(threads[i], NULL);
   }
   pthread_attr_destroy(&attr);
   pthread_mutex_destroy(&mex_mtx);
}
```

The MEX source file can be compiled as follows (on Macs/Linux[1085]):

```
mex pthreads_example.cpp -lpthread -lgmp
```

Here, we assume that GNU MP library is installed in the default location; otherwise, additional compile flags must be provided. Our function takes a vector of 32-bit integers as an input parameter where each pair of integers specify "n" and "k" values of a binomial coefficient to be computed. Correspondingly, the number of threads is set to the number of binomial coefficients. To run the MEX file from MATLAB, we use the following code snippet:

```
% initialize an array to compute 4 binomial coefficients
>> X = uint32([2000000 1000000   2050000 699999...
              2100011   800000   2366466 1655234]);
>> Y = pthreads_example(X);
>> disp(size(Z{1}))
      62500 1
```

The program runs in **34.6** s on an Intel Core i5 processor (see §6.4.1). The task manager shows that indeed, all four processor cores are utilized. As a side effect, we learn that the number "2000000 choose 1000000" has a length of 62500×32 bits.

Note: We call MEX functions from within the parallel portion of our code. This works well on recent MATLAB releases, since some MEX API functions have been made thread-safe. However, it might not work in earlier (or future) MATLAB versions. To make our code portable, it is therefore recommended to not interact with MATLAB at all during parallel blocks, or to protect MEX API calls by critical sections.

For comparison, we perform the same computation in MATLAB with PCT. Since MATLAB does not natively support large integer arithmetic, we shall use MuPAD (a part of the Symbolic Math Toolbox[1086]). MuPAD commands are evaluated in a separate process and can be used together with PCT parallel constructs as demonstrated below:

```
X = ([2000000 1000000   2050000 699999 ...
      2100011   800000   2366466 1655234]);
spmd(4)    % evaluate binomial coefficients in parallel
   idx = labindex;
   Z = feval(symengine,'binomial',X(2*idx-1),X(2*idx));
end
```

On our laptop, this ran in **21.8** s with a MATLAB pool of four workers. This is faster than the direct approach using Pthreads. Certainly, MuPAD is a highly optimized symbolic algebra package while GNU MP uses a rather basic algorithm to compute binomial coefficients.[1087] If you do not have Symbolic Toolbox, you can try third-party libraries that provide multiprecision arithmetic support in MATLAB.*

To complete the picture, on Windows we can use native (non-POSIX) threads,[1088] by #include-ing <process.h> and calling *_beginthread()*. All POSIX implementations on Windows are basically native Window thread wrappers. Using native threads directly is often faster, but is not portable to Macs/Linux, unlike POSIX-based code. Yuval Tassa's *mmx* utility employs both Pthreads (Mac/Linux) and Windows threads in its MEX file.[1089]

---

* See §8.5.5.

Readers are encouraged to review *mmx*'s code to see the specifics. Another example using Pthreads for parallel MATLAB I/O was recently posted.[1090]

7.3.2 Using OpenMP

A major disadvantage of Pthreads is that for many applications, its interface is too low-level. Manually creating threads and synchronizing shared resources can be very tedious and hard to debug. One alternative is to use the OpenMP API,[1091] which is built on top of Pthreads (or other native threads), and provides programmers with a simple parallel programming interface on shared memory architectures. The OpenMP standard is implemented on most major processor architectures and operating systems.[1092]

In C/C++, OpenMP parallel constructs are specified using *#pragma* directives. The programmer puts a code fragment to be run in parallel inside a pragma clause, then the library distributes the computations across a required number of threads, either specified at run-time or set by OMP_NUM_THREADS environment variable.[1093] Setting this variable to the number of effective cores (e.g., 4 on a hyperthreaded dual-core CPU) improves performance even if we do not directly use OpenMP MEX, since OpenMP is often used by other programs on our system, as well as by MATLAB itself. Do not set a value higher than the cores number: this will *degrade* performance.

An OpenMP program begins as a single execution thread, called the initial thread. When a thread encounters a **parallel** construct (directive), the thread creates a *team* (together with zero or more additional spawned threads) and becomes the master of the new team. Each thread within a team gets assigned a task, which is defined by the code inside the **parallel** construct. There is an implicit synchronization barrier at the end of the **parallel** construct, after which only the master thread continues execution.

OpenMP provides many directives, including those defining simple parallel regions, worksharing, tasking constructs, and various synchronization primitives. Let us consider some of them. The first one defines a parallel region executed by a group of threads distinguished by thread IDs (similar to PCT's *spmd* construct, see §6.1.2):

```
double data[1000];
#pragma omp parallel num_threads(8)
{
    int tid = omp_get_thread_num();//get this thread's ID
    do_compute(tid, data);       //do some work
}
```

which starts a parallel region with eight threads with a single copy of **data** shared between all threads. Another very useful directive is a *loop construct*, which specifies that the iterations of one or more associated loops will be executed in parallel by threads in a team (similar to *parfor*-loops, see §6.1.1). The following code fragment shows a loop construct with parallel reduction:

```
#pragma omp parallel for shared(X,n) private(i) reduction(+: Y)
for (i = 0; i < n; i++) {
    Y += X[i]*X[i] + 1;
}
```

Here, we specify that X and n are shared by all threads and Y is a reduction variable. The OpenMP directives are extensively described in OpenMP's documentation.[1094]

We can now rewrite the example code from §7.3.1 using OpenMP constructs. Notice how compact and easy-to-understand the new algorithm looks:

```c
#include <mex.h>
#include <omp.h>
#include <gmp.h>


/* parallel compute function */
void do_calculation(unsigned *g_in, mxArray *g_out, unsigned n_thids)
{
    // start parallel section with n_thids threads
    #pragma omp parallel num_threads(n_thids)
    {
        int tid = omp_get_thread_num();
        long tid = (long)t;           // get this thread's ID
        unsigned *in = g_in + tid*2; // and the inputs 'n' and 'k'
        mpz_t z;                      // initialize a GMP large integer
        mpz_init(z);

        // calculate binomial coefficient
        mpz_bin_uiui(z, (unsigned long)in[0], (unsigned long)in[1]);

        // estimate the size of the result (in 32-bit words)
        unsigned numb = sizeof(unsigned)*8;
        unsigned n_limbs = (mpz_sizeinbase(z, 2) + numb-1) / numb;
        mxArray *res = mxCreateNumericMatrix(n_limbs, 1,
                                        mxUINT32_CLASS, mxREAL);

        // copy 32-bit words (limbs) to the output vector
        mpz_export(mxGetData(res), NULL, 1, sizeof(unsigned),0,0,z);

        // protect access to g_out with critical section
        #pragma omp critical
        {
            mxSetCell(g_out, tid, res);
        }
        mpz_clear(z);  // release a large integer and exit
    } // end of parallel section
}


/* The MEX gateway function */
void mexFunction(int nlhs,         mxArray *plhs[],
                 int nrhs, const mxArray *prhs[])
{

    // number of threads is determined by the size of the inputs
    mwSize n_thids = mxGetNumberOfElements(prhs[0])/2;
    unsigned *in = (unsigned *)mxGetData(prhs[0]);
    plhs[0] = mxCreateCellMatrix(n_thids, 1);
    mxArray *out = plhs[0];
    do_calculation(in, out, n_thids);
}
```

To compile the MEX source file, we use the following command:

```
mex omp_example.cpp CXXFLAGS="\$CXXFLAGS -fopenmp"
               LDFLAGS="\$LDFLAGS -fopenmp" -lgmp
```

We need to explicitly add the -fopenmp flag to both CXXFLAGS (or CFLAGS if you compile a C source file) and LDFLAGS. Without this flag, the program will still compile and run, but OpenMP directives will simply be ignored as though we execute the program with only a single thread. An example for compiling OpenMP[1095] and CUDA[1096] MEX files were provided by Fang Liu on the MATLAB File Exchange.

Note that some compilers (e.g., LCC or Visual Studio Express) do not support OpenMP,[1097] so we would need to use either a commercial compiler or gcc.*

Using OpenMP does not ensure a performance speedup compared to using plain serial processing.[1098] Much depends on the overhead of setting up and finalizing the processing threads compared to the actual data processing time, compiler support, and hardware (CPU cores) availability. Optimizing OpenMP pragma directives and code rewriting for optimal performance can sometimes be a nontrivial art.

Parallelization with OpenMP is a very broad topic. Luckily, there is plenty of educational material available online.[1099] Once we master OpenMP directives, it is quite easy to switch to other "directive-driven" paradigms, such as OpenACC,[1100] which enables us to execute parallel programs on CUDA-enabled GPUs.

OpenMP is used by multiple MATLAB libraries and toolboxes, such as Alois Schloegl's NaN Toolbox,† and many others.[1101] Even the MATLAB Coder converts m-code into C/C++ code that uses OpenMP for multithreading.‡

To complete the picture of C++-based multithreading in MEX, MATLAB uses the Boost set of libraries[1102] internally, and bundles them in every MATLAB release.§ For our purposes, it is important to note the Boost threads library. Boost threads provide a middle ground between the low-level Pthreads and the high-level OpenMP pragmas. With Boost, we still need to create separate thread objects and manipulate them using their class methods, as with Pthreads. However, Boost's classes include internal mechanisms that enable simpler and safer thread manipulation and synchronization.

MATLAB does not include the Boost header files, but it is easy to view the documentation and download the relevant files from the Boost website.[1103] Be careful to download the version that fits your MATLAB's preinstalled version; otherwise, segmentation violations could occur.[1104] For example, on R2014a Windows, the Boost threading library file is called *boost_thread-vc100-mt-1_49.dll*, indicating version 1.49 dated February 2012,[1105] 2 years behind the latest Boost release. It is not advisable in this case to install a newer Boost and link our MEX file with it, since clashes might occur with the standard MATLAB libraries that use the older (preinstalled) Boost libs and are used internally by MEX.

7.3.3 Using Java Threads[1106]

MATLAB uses Java for numerous tasks, including networking, data-processing algorithms, and graphical user interface (GUI).¶ In order to use Java, MATLAB launches

---

* gcc is not officially supported on Windows/Macs, but it can be made to work (see §8.1.2.1); on Linux, gcc is fully supported.

† See §8.5.6.3.

‡ See §8.2.4.

§ For example, *C:\Program Files\Matlab\R2014a\bin\win64\boost*.dll* on a Win64 system.

¶ Under the hood, even MATLAB timers (§7.3.5) employ Java threads for their internal triggering mechanism.

its own dedicated JVM (*Java Virtual Machine*) when it starts.* Once started, Java can be directly used within MATLAB as a natural extension of the MATLAB language.[†][1107] In this section, we only discuss Java multithreading and its potential benefits for MATLAB users.[‡] Readers are assumed to know how to program Java code and how to compile Java classes.

To use Java threads in MATLAB, first create a class that implements the `Runnable` interface or extends `java.lang.Thread`. In either case, we need to implement at least the *run()* method, which runs the thread's processing core.

In the following example, we compute some data, save it to file on a relatively slow USB/network disk, and then proceed with another calculation. We start with a simple synchronous implementation in plain MATLAB:

```
data = rand(5e6,1); % pre-processing (5M elements, ~40MB)
fid = fopen('F:\test.data','w');
fwrite(fid,data,'double');
fclose(fid);
data = fft(data);  % post-processing
⇨ Elapsed time is 9.922366 seconds.
```

 Now let us replace the serial I/O with a very simple dedicated Java thread. Our second calculation (*fft*) will not need to wait for the I/O to complete, enabling much faster responsiveness on MATLAB's single processing thread (known as the *Main Thread, MATLAB Thread,* or simply *MT*). In this case, we get a 58× (!) speedup:

```
data = rand(5e6,1); % pre-processing (5M elements, ~40MB)
javaaddpath 'C:\Yair\Code\' % path to MyJavaThread.class
start(MyJavaThread('F:\test.data',data)); % start running in parallel
data = fft(data); % post-processing (Java I/O runs in parallel)
⇨ Elapsed time is 0.170722 seconds. % 58x speedup !!!
```

Note that the call to *javaaddpath* only needs to be done once in the entire MATLAB session, not repeatedly.

The definition of our Java thread class is very simple:[§]

```
import java.io.DataOutputStream;
import java.io.FileOutputStream;

public class MyJavaThread extends Thread
{
    String filename;
    double[] doubleData;
    public MyJavaThread(String filename, double[] data)
    {
        this.filename = filename;
        this.doubleData = data;
    }

    @Override
    public void run()
```

---

* Unless MATLAB is started with the -nojvm start-up option, in which case Java is not available; see §4.8.2.
† For additional information, refer to my book *Undocumented Secrets of MATLAB-Java Programming*, ISBN 9781439869031, http://undocumentedmatlab.com/matlab-java-book (or: http://bit.ly/Zqqojt).
‡ See §4.8.5, §8.5.2, §9.2.7, §9.7.2, and §11.7 for additional aspects of Java for MATLAB performance.
§ Real-life classes would not be as simplistic, but the purpose here is to show the basic concept, not to teach Java threading.

```
{
    try
    {
        DataOutputStream out = new DataOutputStream(
                                new FileOutputStream(filename));
        for (int i=0; i < doubleData.length; i++)
        {
            out.writeDouble(doubleData[i]);
        }
        out.close();
    } catch (Exception ex) {
        System.out.println(ex.toString());
    }
}
}
```

Note: When compiling a Java class that should be used within MATLAB, as above, ensure that you are compiling for a JVM version that is equal to, or lower than, MATLAB's JVM, as reported by MATLAB's *version* function:

```
% MATLAB R2014a uses JVM 1.7, so we can use JVMs up to 7, but not 8
>> version -java
ans =
Java 1.7.0_11-b21...
```

Java (and C++/.Net) threads are very effective when they can run entirely independently from MATLAB's main thread. But what if we need to synchronize the other thread with MATLAB's MT? For example, what if the Java code needs to run some MATLAB function, or access some MATLAB data? In MEX, this could be done using the dedicated and documented MEX functions; in Java, this can be done using the undocumented/unsupported JMI (*Java–MATLAB Interface*) package.[1108] Note that using standard Java threads without MATLAB synchronization is fully supported; it is only the JMI package that is undocumented and unsupported.*

Here is the relevant code snippet for evaluating MATLAB code within a Java thread:

```
import com.mathworks.jmi.Matlab; //in %matlabroot%/java/jar/jmi.jar
...
Matlab matlabEngine = new Matlab();
...
Matlab.whenMatlabReady(runnableClass);
```

where `runnableClass` is a class whose *run()* method includes calls to `com.mathworks.jmi.Matlab` methods such as:

```
matlabEngine.mtEval("plot(data)");
Double value = matlabEngine.mtFeval("min",{a,b},1); //2 inputs, 1 output
```

Unfortunately, we cannot directly call `matlabEngine`'s methods in our Java thread, since this is blocked in order to ensure synchronization. MATLAB only enables calling these methods from the MT,[1109] which is the reason for the `runnableClass`. Indeed, synchronizing Java code with MATLAB could be quite tricky, and can easily deadlock

---

* Calling MATLAB from Java used to be documented and supported in MATLAB 6, but has become undocumented and unsupported in MATLAB 6.5 (R13). A MathWorks newsletter article from 2002 describing the functionality has been removed from the Mathworks website some years ago and is now available only in archived version (http://bit.ly/1faxzUn).

MATLAB. To alleviate some of the risk, I advise not to use the JMI class directly: Instead, use Joshua Kaplan's *MatlabControl* class,[1110] a user-friendly JMI wrapper.

Note that Java's native *invokeAndWait()* method cannot be used to synchronize with MATLAB. M-code executes as a single uninterrupted thread (MT). Events are simply queued by MATLAB's interpreter and processed when we relinquish control by requesting **drawnow, pause, wait, waitfor,** and so on. MATLAB synchronization is robust and predictable, yet forces us to use the whenMatlabReady(runnableClass) mechanism to add to the event queue. The next time **drawnow,** and so on is called in m-code, the event queue is purged and our submitted code will be processed by MATLAB's interpreter.

Java threading can be quite tricky even without the MATLAB synchronization complexity.[1111] Deadlock, starvation, and race conditions are frequent problems with Java threads. Basic Java synchronization is relatively easy, using the synchronized keyword. But getting the synchronization to work *correctly* is much more difficult and requires Java programming expertise that is beyond most Java programmers. In fact, many Java programmers who use threads are not even aware that their threads synchronization is buggy and that their code is not *thread-safe*.

My general advice is to use Java threads just for simple independent tasks that require minimal interactions with other threads, the MATLAB engine, and/or shared resources.

Readers may find interest in Rodney Thomson's TCP client/server utility on the MATLAB File Exchange,[1112] which employs Java multithreading to handle multiple client connections with a server. The server frontend is in MATLAB, while the communication multithreading is done in Java.

In a related matter, Java has several libraries (packages) that wrap its basic concurrency into frameworks that are safer and easier to use, for example, JADE (*Java Agent Development Environment*).[1113] It might make sense to use one of these packages, rather than Java's basic concurrency classes.

7.3.4 Using .Net Threads[1114]

Dot-Net (.Net), like Java and C++, also enables multithreading. Unlike Java and C++, however, .Net is not platform-independent: it only works on Windows, and only on those systems where .Net is actually installed. .Net is commonly installed on Win7+, but typically not, for example, on XP. On Windows platforms where .Net is not installed, we need to download and install it before we can use any .Net features (duh!). To ensure .NET is supported on our current platform, use the ***NET.isNETSupported*** function:

```
>> NET.isNETSupported   % returns true/false
ans =
    1
```

.Net libraries (*assemblies*) are commonly distributed as DLL files, which can be loaded into MATLAB using the ***NET.addAssembly*** function, similarly to the ***javaaddpath*** function for Java classes. Using these assemblies in MATLAB is then as straightforward as in Java:

```
NET.addAssembly('C:\Yair\Code\NetThread.dll');
data = rand(5e6,1); % pre-processing (5M elements, ~40MB)
start(My.NetThread('F:\test.data',data)); % start running in parallel
data = fft(data); % post-processing (.Net I/O runs in parallel)
```

As with Java, the assembly only needs to be loaded once per MATLAB session, not repeatedly. The definition of the .Net class closely follows that of the Java class. Unfortunately,

.Net's `System.Threading.Thread` class[1115] is sealed (noninheritable), so we cannot extend it as we did in Java. However, we can instantiate an internal `Thread` object within our class and use it. Here is the corresponding C# source code:

```
using System;
using System.IO;
using System.Threading;

namespace My
{
    public class NetThread
    {
        string filename;
        double[] doubleData;
        Thread thread;

        public NetThread(string filename, double[] data)
        {
            this.filename = filename;
            this.doubleData = data;
            thread = new Thread(this.run);
        }

        public void start()
        {
            thread.Start();  // note the capital S in Start()
        }

        private void run()
        {
            try
            {
                BinaryWriter out = new BinaryWriter(
                            File.Open(filename,FileMode.Create))
                for (int i = 0; i < doubleData.Length; i++)
                {
                    out.Write(doubleData[i]);
                }
                out.Close();
            }
            catch (Exception ex)
            {
                Console.WriteLine(ex.Message);
            }
        }
    }
}
```

As with Java, when compiling a .Net class that should be used within MATLAB, we must ensure that we are compiling for a .Net Framework that is equal to, or lower than MATLAB's, as reported by .Net's `System.Environment.Version`:

```
% My system uses .Net 4.0, so we can use up to 4.0, but not 4.5
>> char(System.Environment.Version.ToString)
ans =
4.0.30319.18034
```

As with Java threads, .Net threads are most effective when they can run independently of MATLAB and of each other. There is no known mechanism for data synchronization between .NET threads and MATLAB. However, we can use one of the standard IPC mechanisms, as discussed in §7.4 below.

7.3.5 Using MATLAB Timers*[1116]

Multithreading helps application performance in two related but distinct ways:

- By allowing code to run *in parallel*, on different CPUs or cores
- By allowing code to run *asynchronously*, rather than in serial manner

C++, Java, and .Net threads, which we have seen in the preceding sections, can improve performance by both of these ways. MATLAB timers, on the other hand, only enable the second option, of running code asynchronously. The reason for this is that, as noted in §7.3.3, all m-code, including timer callback code, is executed by MATLAB's interpreter on a single processing thread (MT).

So, while a timer callback executes, no other m-code can run. This may seem on the face of it to be unhelpful. But in fact, the ability to schedule a MATLAB processing task for later (nonserial) invocation, could be very handy, if we can time it so that the timer callback is triggered when the application is idle, for example, waiting for user input, following complex GUI update, or during late hours of the night.

Here is an implementation of our asynchronous I/O example, this time using MATLAB timers. First, we define the timer's callback function, using pure m-code:

```
function timerCallback(hTimer,eventData,filename,data)
    try
        fid = fopen('F:\test.data','w');
        fwrite(fid,data,'double');
        fclose(fid);
    catch
        err = lasterror;
        fprintf(2,'Error saving to file:\n%s\n',err.message);
    end
end
```

We can now use this timer in MATLAB, similarly to our earlier Java/.Net examples:

```
data = rand(5e6,1); % pre-processing (5M elements, ~40MB)
timerFcn = {@timerCallback,'F:\test.data',data};
start(Timer('StartDelay',2, 'TimerFcn',timerFcn)); % start after 2s
data = fft(data); % post-processing (timer I/O will run later!)
```

The difference versus our earlier examples is that the timer code is not run in parallel to the *fft* postprocessing, but rather 2 s later, when MT is hopefully idle.

MATLAB timers have an advantage over Java/C++/.Net multithreading in their synchronization with MATLAB, since the m-code interpreter is single-threaded. We just need to handle cases where a timer callback might interrupt other m-code.[1117]

MATLAB timers run pure m-code (i.e., there is no need to know Java/C#/C++ or to use external compilers) and are easy to set up and use. They can be very effective when tasks can be postponed asynchronously to when the MT is idle.

---

* MATLAB timers are also discussed in §1.8.4, §2.1.6, §3.7, §3.8.1, §10.2.2, and §10.4.3.

7.4 Spawning External Processes[1118]

It is sometimes impractical to create threads. For example, an algorithm might only be available in executable binary format, or we might wish to use a separate memory space for the processing, to sandbox (isolate) it from the main application. In such cases, we can spawn heavyweight processes (as opposed to lightweight threads), either directly from within MATLAB, or externally.

The simplest way to spawn an external process in MATLAB is using the *system* function. This function accepts a string that will be evaluated in the OS prompt (shell), at MATLAB's current folder. By appending a '&' character to the end of the string, we let MATLAB return immediately, and the spawned process will run *asynchronously* (in parallel to MATLAB); otherwise, MATLAB will block until the spawned process ends (i.e., *synchronous* invocation of the process).

```
system('program arg1 arg2');    % blocking, synchronous
system('program arg1 arg2 &');  % non-blocking, asynchronous
```

MATLAB normally uses only a single core on a single CPU, except when using the PCT or when doing some implicit parallelization of vectorized code.* Therefore, on a quad-core dual-CPU hyperthreaded machine, we would normally see MATLAB's CPU usage at only $1/(2*2*4) = 6\%$. The simplest way to utilize the unused CPU cores without PCT is to spawn additional MATLAB processes. This can be done using the *system* function, as above.† The spawned MATLAB sessions can be made to run specific commands or functions. For example:

```
system('matlab -r "for idx = 1:100, doSomething(idx); end" &');
system(['matlab -r "processFile(' filename ');" &']);
```

At this point, we may wish to use processor affinity‡ to ensure that each process runs on a separate CPU core, to avoid inter-process contention and implicit parallelization overheads, thereby increasing overall throughput. Different OSes have different ways of setting CPU affinity; for example, on Windows, we can use Task Manager's processes context menu or Process Explorer's context menu.[1119]

When MATLAB spawns an external process, it passes to it the set of environment variables used in MATLAB. This may be different from the set that is normally used when running the same process from the OS's command prompt. This could lead to unexpected results, so care should be taken to update such environment variables in MATLAB before spawning the process, if they could affect its outcome.[1120]

Once an asynchronous (nonblocking) process is started, MATLAB does not provide a way to synchronize with it. We could of course employ external signals[1121] or the state or contents of some disk file, to let the MATLAB process know that one or more of the spawned processes has ended. When multiple processes are spawned, we might wish to employ some sort of *load balancing*[1122] for optimal throughput.

---

* See Chapter 5.
† We can also spawn background MATLAB processes using PCT's **batch** function: See http://mathworks.com/help/releases/R2013b/distcomp/introduction-to-parallel-solutions.html#brjw1fx-2 (or: http://bit.ly/1bQk5fQ); http://mathworks.com/help/releases/R2013b/distcomp/batch.html (or: http://bit.ly/1bQjr1O). Note that MATLAB's license terms may place a limit on the allowable number of concurrent MATLAB processes that may be spawned.
‡ See §2.1.6.2. We could also limit MATLAB to a single computational thread (see footnote in §5.1.1).

We can use OS commands to check if a spawned processId is still running. This ID is not provided by *system* so we need to determine it right after spawning the process. On Unix systems (Linux and Mac), both of these can be done using a *system* call to the OS's *ps* command; on Windows, we can use the *tasklist* or *wmic* commands.

An alternative is to use Java's built-in process synchronization mechanism, which enables more control over a spawned process. The idea is to spawn an external asynchronous process via Java, continue the MATLAB processing, and later (if and when needed) wait for the external process to complete:[1123]

```
runtime = java.lang.Runtime.getRuntime();
process = runtime.exec('program arg1 arg2'); % non-blocking
% Continue MATLAB processing in parallel to spawned process
```

When we need to collect scalar results, we could use the process' result code:

```
rc = process.waitFor(); % block MATLAB until external program ends
rc = process.exitValue(); % fetch an ended process' return code
```

Or, if we need to abandon the work, we could stop the spawned process:

```
process.destroy(); % force-kill the process (rc will be 1)
```

Another alternative, that enables output redirection and checking the progress of running processes, is Brian Lau's *MatlabProcessManager*.

While this mechanism enables synchronization of the MATLAB and external process at the basic *execution* level, it does not enable synchronization of the *data*. Doing this between processes (that have independent memory spaces) is much harder (and slower) than it is between threads (that share their memory) or MEX. For interprocess data synchronization (known as IPC, or *interprocess communication*[1124]), we can use shared memory, named pipes, data files, or network messages. There are various mechanisms and libraries that enable this using C++ and Java that could be used in MATLAB.[1125]

Examples of memory sharing are Joshua Dillon's *sharedmatrix*[1126] and Kevin Stone's *SharedMemory*[1127] utilities, which use POSIX shared-memory and the Boost IPC library.*[1128] Rice University's TreadMarks library[1129] is another example of a shared-memory approach that has been used with MATLAB, in the *MATmarks* package.[1130]

Another alternative is to use the open-source *memcached* library.[1131] *memcached* is typically used as a distributed memory caching mechanism, but we can also use it as a simple shared-memory mechanism that can easily be used by separate MATLAB sessions.[1132]

Named pipes[1133] can be used on Unix systems (Linux and Mac). In this case, the source process sends information to the pipe and the destination process reads from it. After setting up the pipe in the OS, it can be opened, updated, and closed just like any other data file. Unfortunately, this mechanism is not generally used on Windows.

MATLAB includes a dedicated doc page showing how to synchronize interprocess data using disk files.[1134] An approach that combines memory sharing and I/O is to use memory-mapped files† on R2008a or newer (*memmapfile* was buggy before then[1135]).

Oliver Woodford's *batch_job* utility uses data files to enable parallelism.[1136]

For network messages, we could use low-level sockets to send and receive messages. The MATLAB File Exchange includes several implementations of client/server communication,[1137] or we could use a separate C++/C#/Java class or library.

---

* See §9.5.2 for a discussion of these utilities. Also see §7.3.2 for a discussion of Boost multithreading.
† See §11.3.3.

Alternatively, we could employ more sophisticated communication frameworks based on the Message Passing Interface (MPI) standard. Several projects have used MPI for MATLAB parallelization, including MATLAB*P, pMatlab, MatlabMPI, Star-P, bcMPI and even MathWorks' PCT.*

Finally, we can use MATLAB's ability to serve as a COM/DCOM (*automation*) server to communicate with it from the external process via the COM interface.[1138] Data can be exchanged and MATLAB functionality can be invoked by the process.

---

* See §7.2.4.

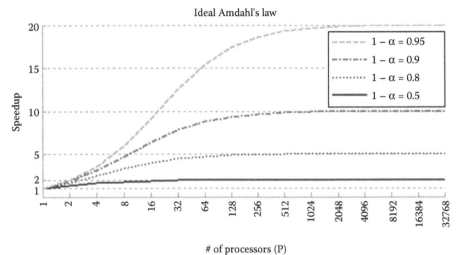

Ideal Amdahl's law of parallelization efficiency (see §1.7)

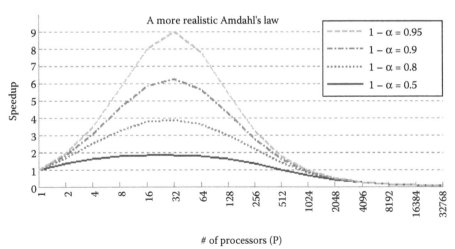

Realistic Amdahl's law of parallelization efficiency (see §1.7)

Profile Summary

Generated 06-Sep-2012 18:51:40 using cpu time.

Function Name	Calls	Total Time	Self Time*	Total Time Plot (dark band = self time)
surf	1	1.310 s	0.249 s	
newplot	1	0.873 s	0.499 s	
newplot>ObserveAxesNextPlot	1	0.359 s	0.032 s	
cla	1	0.327 s	0.046 s	
graphics\private\clo	1	0.281 s	0.108 s	
graph3d.surfaceplot.surfaceplot	1	0.141 s	0.078 s	
setdiff>setdifflegacy	2	0.095 s	0.064 s	
setdiff	2	0.095 s	0.000 s	
findall	1	0.079 s	0.063 s	
plotdoneevent	1	0.047 s	0.016 s	
peaks	1	0.031 s	0.031 s	

MATLAB Profiler summary report (see §2.1.1)

Function listing

Color highlight code according to [time ▾]

```
 time    calls line
                  1 function axReturn = newplot(hsave)
                  2 %NEWPLOT Prepares figure, axes for graphics according to NextPlot.
                  3 %    H = NEWPLOT returns the handle of the prepared axes.
                       H = NEWPLOT
```

```
           1 __ 60 if isempty(fig)
 0.514     1 __ 61     fig = getPlotFigure;
           1 __ 62 end
                63
           1 __ 64 fig = ObserveFigureNextPlot(fig, hsave);
                65 % for clay
           1 __ 66 set(fig,'nextplot','add');
                67
           1 __ 68 if isempty(ax)
           1 __ 69     ax = gca(fig);
                70 elseif ~any(ishghandle(ax))
                71     error(message('MATLAB:newplot:NoAxesParent'))
                72 end
                73
 0.359     1 __ 74 ax = ObserveAxesNextPlot(ax, hsave);
                75
           1 __ 76 if nargout
           1 __ 77     axReturn = ax;
           1 __ 78 end
```

MATLAB Profiler detailed report (see §2.1.1)

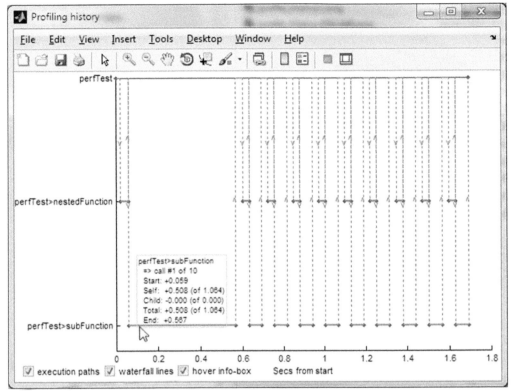

Function invocation timeline using the *profile_history* utility (see §2.1.4)

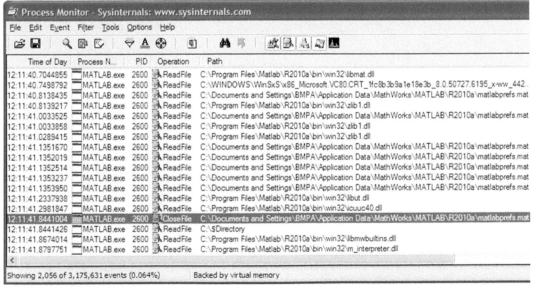

SysInternals *Process Monitor's* partial output for a MATLAB **getpref** command (see §2.4)

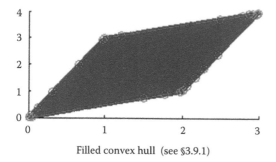

Filled convex hull (see §3.9.1)

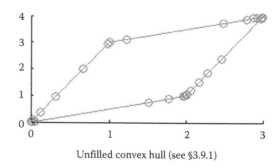

Unfilled convex hull (see §3.9.1)

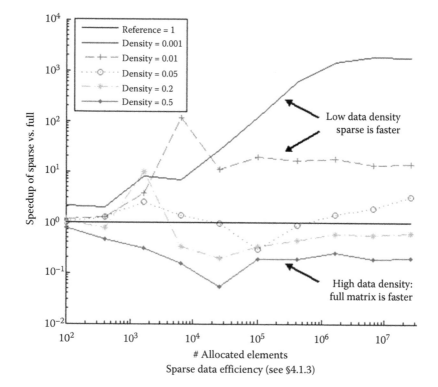

Sparse data efficiency (see §4.1.3)

```
c2={c{:} 123};
```
⚠ { A{:} B } can often be replaced by [A {B}], which can be much faster. [Details ▾]

Sample MLint (Code Analyzer) message (see §4.1.5, §4.2.3 and Appendix A.4)

```
v=version('-Release');
if str2num(v(1:end-1)) >= 2011
```
⚠ STR2DOUBLE is faster than STR2NUM; however, STR2DOUBLE operates only on scalars. Use the function that best suits your needs.

ST2NM MLint (Code Analyzer) message (see §4.1.5, §4.2.3, and Appendix A.4)

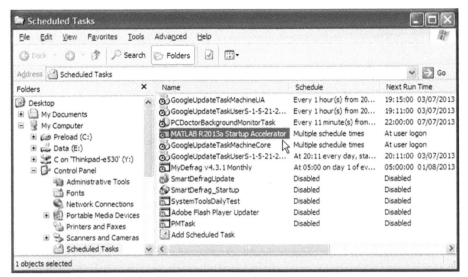

Windows XP Scheduled Tasks showing MATLAB Startup Accelerator (see §4.8.1)

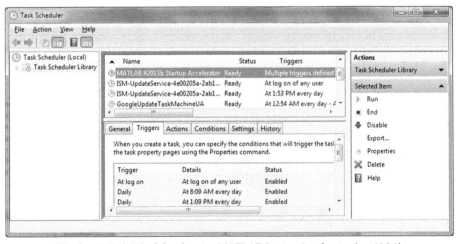

Windows7 Task Scheduler showing MATLAB Startup Accelerator (see §4.8.1)

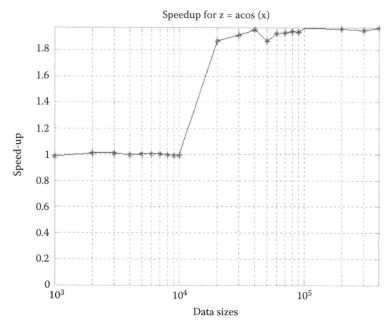

Implicit multi-threading speedup of the **acos** function (see §5.1.1)

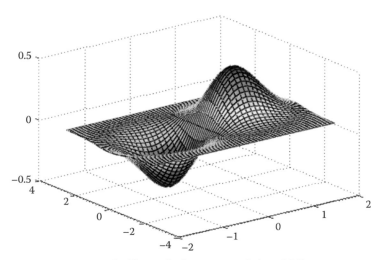

meshgrid vectorized usage example (see §5.3.7)

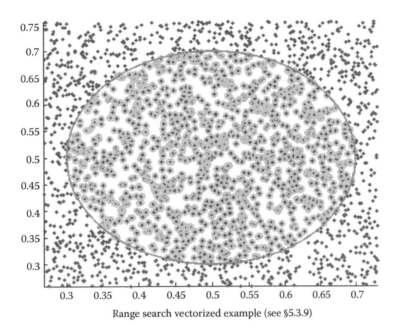
Range search vectorized example (see §5.3.9)

Visualization of the matched filter example (see §5.6.1)

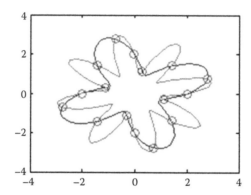

Visualization of the curves intersection example (see §5.7.17)

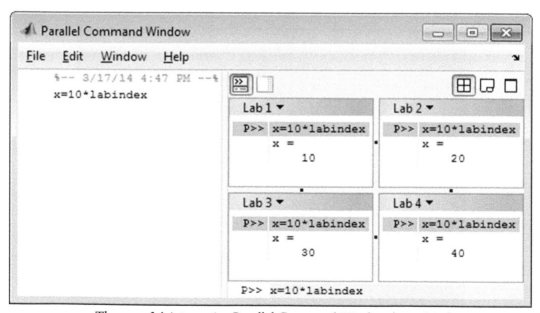

The *pmode's* interactive Parallel Command Window (see §6.1.4)

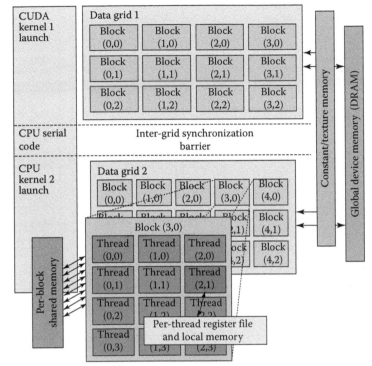

CUDA's basic architecture (see §6.2.1)

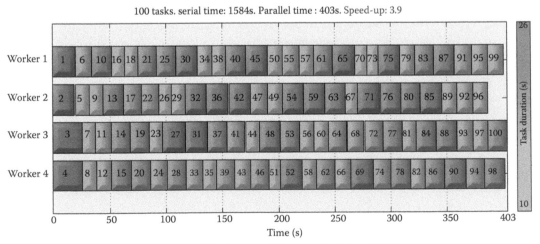

taskMap's report of distributed tasks (see §6.3.2)

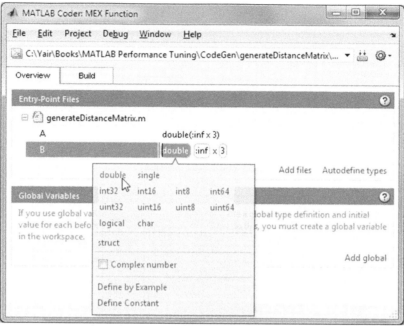

Defining function inputs in MATLAB Coder (see §8.2.2)

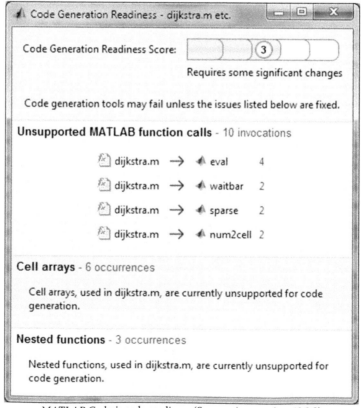

MATLAB Coder's code readiness (Screener) report (see §8.2.3)

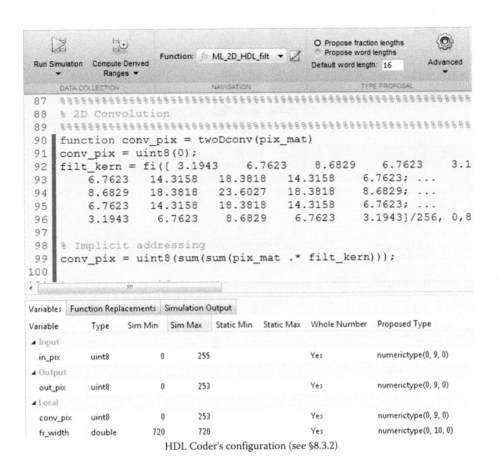

```
                                          O Propose fraction lengths
                                          O Propose word lengths
        Run Simulation  Compute Derived   Function: ʄx ML_2D_HDL_filt ▾ ✎   Default word length: 16     Advanced
                           Ranges ▾
             DATA COLLECTION                    NAVIGATION                    TYPE PROPOSAL

87  %%%%%%%%%%%%%%%%%%%%%%%%%%%%%%%%%%%%%%%%%%%%%%%%%%%%%%%%%%%%%%%%%%%
88  % 2D Convolution
89  %%%%%%%%%%%%%%%%%%%%%%%%%%%%%%%%%%%%%%%%%%%%%%%%%%%%%%%%%%%%%%%%%%%
90  function conv_pix = twoDconv(pix_mat)
91  conv_pix = uint8(0);
92  filt_kern = fi([ 3.1943    6.7623    8.6829    6.7623    3.1
93      6.7623   14.3158   18.3818   14.3158    6.7623; ...
94      8.6829   18.3818   23.6027   18.3818    8.6829; ...
95      6.7623   14.3158   18.3818   14.3158    6.7623; ...
96      3.1943    6.7623    8.6829    6.7623    3.1943]/256, 0,8
97
98  % Implicit addressing
99  conv_pix = uint8(sum(sum(pix_mat .* filt_kern)));
100
```

Variable	Type	Sim Min	Sim Max	Static Min	Static Max	Whole Number	Proposed Type
◢ Input							
in_pix	uint8	0	255			Yes	numerictype(0, 9, 0)
◢ Output							
out_pix	uint8	0	253			Yes	numerictype(0, 9, 0)
◢ Local							
conv_pix	uint8	0	253			Yes	numerictype(0, 9, 0)
fr_width	double	720	720			Yes	numerictype(0, 10, 0)

HDL Coder's configuration (see §8.3.2)

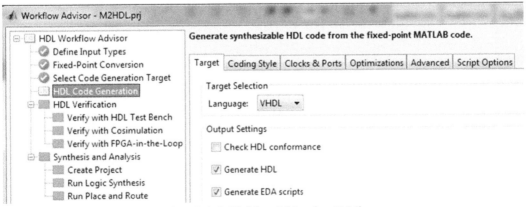

HDL Coder's Workflow Advisor (see §8.3.2)

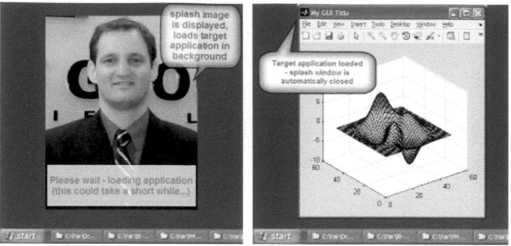

Splash image displaying while the MCR loads (left), until the main GUI appears (right) (see §8.4)

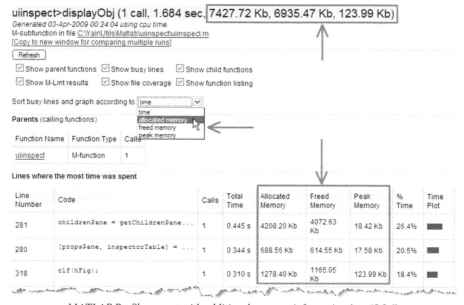

MATLAB Profiler report with additional memory information (see §9.2.6)

MATLAB Profiler report with additional memory information (see §9.2.6)

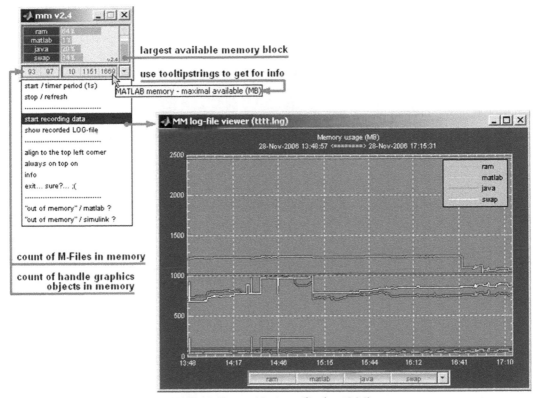

MATLAB Memory Monitor utility (see §9.2.8)

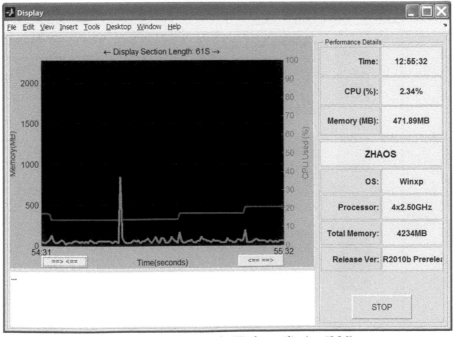

System Resource Monitor for Windows utility (see §9.2.8)

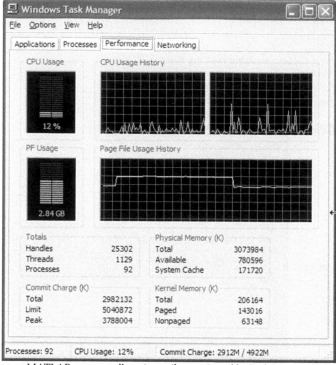

Temporary MATLAB memory allocation spike monitored by Task Manager (see §9.2.9)

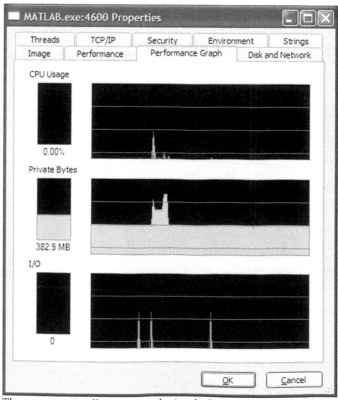

The *copy-on-write* effect monitored using the Process Explorer (see §9.5.1.1)

	A	B	C
1	3	-4	1
2	-2	0	2
3	-1	4	-3

An HTML-formatted MATLAB *uitable* (see §10.4.1.7)

Periods	Any period returns	Avg return signal	Avg gain	Avg draw	Gain/draw ratio	Max gain	Max draw	Random % pos	Signal % pos	Payout	% p-value
3	0.29	0.01	1.30	-1.91	-1.48	2.42	-2.12	58	60	0.01	76
5	0.49	-1.00	1.28	-1.57	-1.23	1.28	-3.63	58	20	-1.00	15
10	0.98	-2.39	4.04	-4.00	1.01	4.04	-4.87	62	20	-2.39	11
15	1.45	-6.39	0.12	-8.02	-67.49	0.12	-11.39	65	20	-6.39	2
20	1.89	-2.25	6.97	-8.39	-1.20	9.58	-11.24	67	40	-2.25	36
30	2.66	-6.54	7.50	-10.05	-1.34	7.50	-14.25	70	20	-6.54	10
40	3.40	-9.12	0.01	-9.12	-912.46	0.00	-15.71	71	0	-9.12	5

A *uitable* with custom cell-renderer (see §10.4.1.7)

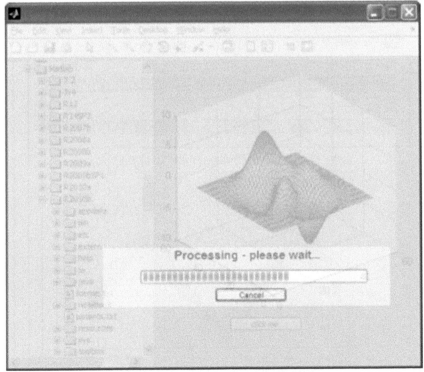

The *blurFigure* utility in action (see §10.4.2.4)

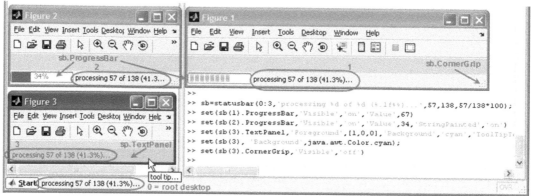

The *statusbar* utility, with customizable progressbars (see §10.4.2.5)

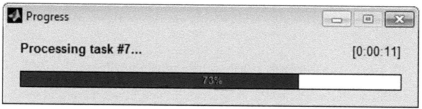

Good: a progressbar showing the overall (total) progress (see §10.4.2.5)

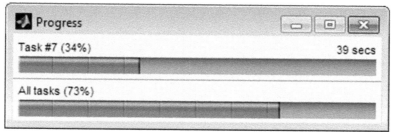

Excellent: Ben Tordoff's *multiWaitbar* (see §10.4.2.5)

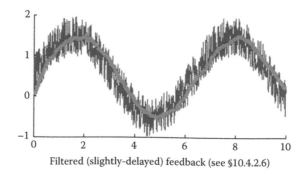

Filtered (slightly-delayed) feedback (see §10.4.2.6)

8

Using Compiled Code

MATLAB includes a variety of ways by which we could use the speed associated with native code in order to improve the performance of our MATLAB code. These mechanisms include:

- Using compiled C/C++/Fortran code in a specific (MEX) format
- Using MATLAB Coder Toolbox to convert m-code to compiled C/C++-code
- Running MATLAB code on dedicated FPGA hardware
- Improving the performance of compiled (deployed) code
- Integrating Java functionality in MATLAB
- Using external (third-party) libraries in MATLAB

8.1 Using MEX Code

8.1.1 Introduction

MEX (*MATLAB Executable*) files are written in C, C++, or Fortran in a certain standard format and then compiled using MATLAB's built-in *mex* function[1139] and one of several supported external compilers. The compilation result is a regular dynamic-link (or shared) library, which has a MATLAB-specific file extension (e.g., *.mexw64*).* This is a binary file that can be analyzed,† but cannot be edited in a text editor.

MATLAB code can directly call MEX functions on the MATLAB path, just like any other m-file or built-in function. The associated MEX file (shared/dynamic-link library) is dynamically loaded in MATLAB when first needed.‡ When MATLAB m-code invokes a MEX function, the library's gateway function (*mexFunction*) is called with the specified input args, and the output args are then passed back to the calling MATLAB workspace. To facilitate the transfer of control and data between the MATLAB workspace and the MEX file, MATLAB provides a wide range of support functions within the framework of the MEX API, which are discussed below. The original C/C++/Fortran source code is not necessary to run a MEX file, we only need the compiled binary library file (and possibly other libraries on which it depends).[1140]

MEX files can be very important for MATLAB performance, since they execute at nearly the speed of native compiled C/C++/Fortran code. In fact, MEX files *are* native-compiled

---

* Many years ago, MEX files used a standard *.dll* file extension, but this has been changed to MATLAB-specific extensions. This is merely a change in the file extension — the file format itself remained in DLL/SO format (depending on the specific platform).
† Using DLL analyzers such as Dependency Walker (http://dependencywalker.com) on Windows platforms.
‡ *[~,x] = inmem* returns the list of MEX files currently loaded in MATLAB memory; *clear(functionName)* clears from memory.

code and the only reason that they run a bit slower is the small overhead associated with the MATLAB interface and the MEX API calls.

In addition to the inherent speed of executing at compiled speed, MEX enables sophisticated programmers the extra speedup due to manipulating MATLAB and native memory in-place, bypassing MATLAB's copy-on-write (COW) mechanism* and the need to allocate and deallocate memory when manipulating data arrays.

MEX files also enable programmers to easily interface external C/C++/Fortran code, and statically link with external libraries. This avoids the need to use shared/dynamic libraries that needs special handling to be loaded and invoked in MATLAB.†

However, MEX also has drawbacks that should be understood before we happily start using it. First, MEX does not always provide a speedup compared to m-code. MEX is most effective at speeding up unvectorized algorithms having iterative loops, states and conditionals, and in heavy data processing that could benefit from in-place data access. In such cases, efficient MEX implementations can speed up the corresponding m-code by a factor of 2, 10, or even more. In other cases, MATLAB's optimizations, vectorization, implicit multithreading and use of highly optimized libraries and built-in functions could well cause m-code to run faster than the equivalent MEX function, even without explicit parallelization.‡

Second, writing MEX code is more difficult than equivalent m-code. MEX requires more low-level programming and run-time checks than m-code. MATLAB shields and protects the programmer from the need to handle memory management, null pointers, invalid data, and other similar aspects of real-life programming that must be handled explicitly in MEX. It is much easier to inadvertently cause memory leaks or segmentation faults that crash MATLAB using MEX code, than it is with m-code.

Each of the five MATLAB platform types (Linux 32/64 bit, Mac, and Win32/64) has a unique file format and extension; a file compiled on one platform type cannot be run on another.§ Moreover, some MATLAB versions have changed their internal representation of variables in memory (mxArray — see §8.1.4.1), so running a MEX file that was compiled on MATLAB release A might not work on MATLAB release B, even if they share the same platform (especially if B < A).[1141]

So while MEX promises improved performance in many cases, we should carefully consider its drawbacks when faced with the need to improve MATLAB performance. It is normally cost effective to MEX-convert only performance-critical processing sections that manipulate large data structures, where in-place data manipulation and faster native processing provide significant speedup. In other cases, the extra effort and potential problems associated with MEX may simply not be worth it.

8.1.2 Our First MEX Function

Note: In the following discussion and throughout §8.1, we shall discuss the C/C++ MEX variant.[1142] The Fortran variant is very similar, although naturally different in syntax. Readers interested in the Fortran variant are referred to the documentation for details.[1143] While the programming syntax is different, the underlying memory constructs (mxArray) and MEX API functions are the same for all variants.

---

* See §9.5.1.
† See §8.5.
‡ See §8.1.7.4.
§ The *mextext* function returns the MEX file extension that is associated with each platform type. See http://mathworks.com/help/matlab/matlab_external/platform-compatibility.html (or: http://bit.ly/L0Z6zJ).

8.1.2.1 Setting Up the Compiler

The first step is to set up a compiler. If one of the supported compilers[1144] is already installed on your computer, you're in luck and can skip the following two paragraphs.

If you have 32-bit Windows, the lcc compiler is already bundled with MATLAB. MATLAB does not include a compiler on other platforms, but free compilers can easily be installed:[1145] Windows SDK on Win64/Win32, Apple Xcode or GNU gfortran on Mac OS, and GNU gcc/g++/gfortran on Linux (gcc is often bundled with Linux). On Windows, you can also use the commercial products Microsoft Visual C++ or Intel C++/Fortran Composer XE.*[1146] In general, the commercial products generate code that is faster than code created by Windows SDK or lcc.† For example, MS Visual Studio 2012 includes an auto-vectorization feature that leverages Intel's SSE instructions.[1147] I find that for the vast majority of use cases, the free compilers are good enough. However, some people disagree.[1148] Still, even a simple compiler can do wonders to performance, compared to nonvectorized iterative MATLAB code.

An important aspect to note when installing a Windows compiler, is that most Windows installations already have a set of installed VC++ run-time redistributable files (e.g., *Microsoft Visual C++ 2005 Redistributable — x64 9.030729.17*). Unfortunately, they clash with the compiler's library files causing run-time errors when running *mex* to compile a MEX file.[1149] The solution is to uninstall the other library versions (some of which may actually be newer than the compiler's!), using the Control Panel (*Add/Remove Programs* on Windows XP; *Programs and Features* on Win7). After removing the libraries, reinstall the compiler. This will install the run-time files as required (other programs will also use the new run-time files).

Once you have at least one supported compiler installed, run the following:[1150]

```
>> mex -setup
```

Now, select the relevant compiler from the presented list, and you are done. This is a one-time process; we do not need to repeat it in any future MATLAB session on the same platform, unless we install a new compiler and wish to use it instead.

We can now use the *mex* function to compile any MEX-compliant source-code file:

```
>> mex myFunc.c
```

This command compiles the file *myFunc.c*, creating a file called *myFunc.mexw64* (or using some other file extension, depending on the platform). We can now call *myFunc* directly in our MATLAB m-code:

```
results = myFunc(input1,input2);
```

To link several source files and external libraries, simply specify them as extra arguments to the *mex* function. Just ensure that the gateway function is located in the first

---

* Older MATLAB releases supported other compilers (Borland, Watcom, etc.). It is possible to add settings for other compilers such as gcc (http://gnumex.sourceforge.net), Intel C++ XE (http://bit.ly/1n7PqQh), LF95 (http://lahey.com/miscfix.htm), or for unsupported combinations (e.g., http://mathworks.com /matlabcentral/answers/93013 or: http://bit.ly/1aAwGIO; http://mathworks.com/matlabcentral/fileexchange /index?term=mex+setup or: http://bit.ly/1gkcOWx). Note: using MEX with unsupported compilers is not officially supported by MathWorks and compiled code may not work in MATLAB in some cases.

† An exception to this rule was found by a user who reported the gcc-compiled MEX to be 3× faster than a VC++ compilation: http://mathworks.com/matlabcentral/newsreader/view_thread/325183 (or: http://bit.ly /LbwnYT).

specified file. The following example will create *myFunc.mexw64* into which the two support files and one external library will be statically linked:

```
>> mex myFunc.c supportFunc.c anotherFunc.c library.lib
```

The built-in function *mex.getCompilerConfigurations*[1151] returns a list of configuration settings for each of the supported, installed, or selected (default) compilers that can be used with MEX. This can be used to assess the options used by the compiler and linker when creating the MEX file. The options are stored separately for each compiler, in a corresponding XML file in the *%matlabroot%/bin/%arch%/mexopts/* folder.* Tweaking these options is not usually needed, but can be used to solve specific compiler issues, set the optimization level, and so on.[1152] Running *mex* with the -v switch displays a verbose output of all compilation options and steps. When we use the *mex* function to compile a MEX file, we can use the -f switch to specify a specific options file, overriding the compiler's defaults. We can also override specific compiler and linker flags as *mex* parameters, without modifying the options file.[1153]

We can test our MEX setup on any of the many example files that are provided in the *%matlabroot%/extern/examples/* folder and subfolders.[1154] Reviewing these examples is a great way to learn MEX, and we can use any of these examples as an adaptable baseline for our needs. MEX is a bit scary at first, but you'd be surprised how quickly you pick it up if you advance slowly, with small modifications to existing code.

The executable folder of MATLAB (*%matlabroot%/bin/*) contains operating-system script files (*mex.bat* for Windows; *mex.sh* for Mac/Linux) that enable compiling MEX files from outside MATLAB. This in turn enables using external make files and automated build processes. Both the internal *mex* function and the script files call the same *mex.pl* file (collocated with the OS script files in the *bin/* folder), which does the processing needed to run the compiler. Neither Perl nor the compilers depend on MATLAB, so they can be run using the external OS script files.

Once a MEX file is ready, it can be debugged. The specifics are different for each platform and supported compiler,[1155] but in general all platforms support source-code debugging, including setting breakpoints, inspecting variables, and code stepping.

8.1.2.2 *The Basic Format of a MEX File*

All MEX files share the same basic format. This standard format enables MATLAB to dynamically call MEX files in a consistent way regardless of the function contents.

The MEX file should #include "mex.h", which is a C/C++ header file located in the *%matlabroot%/extern/include/* folder,† followed by an implementation of the *mexFunction* function, which is defined exactly as follows:

```
#include "mex.h"
void mexFunction (
    int           nlhs,      /* number of expected outputs */
    mxArray       *plhs[],   /* array of pointers to output args */
    int           nrhs,      /* number of input args */
    const mxArray *prhs[])   /* array of pointers to input args */
{
    /* processing */
}
```

---

* For example, *C:\Program Files\Matlab\R2014a\bin\win64\mexopts\winsdk-7.1_cpp.xml* and *mssdk71opts.bat* on Windows.
† For example, C:\Program Files\Matlab\R2014a\extern\include\ on Windows.

This is the standard format of all MEX files. As can be seen, it is fairly simple. *mexFunction()*'s input args (nlhs etc.) define the MEX function's input and output args.* The processing itself is done within this gateway function. It could (and often does) call other functions, in the same source-code file or other files that are statically linked by the compiler. It could also call functions in external pre-compiled libraries, such as BLAS or LAPACK (see below).

MEX files should have exactly one *mexFunction* gateway function, and it should be defined exactly as above. Modifying the gateway function signature could cause run-time errors and MATLAB crashes.

8.1.2.3 "Hello, World!"

Our first MEX file is the ubiquitous function that simply outputs the string "Hello, world!" to the MATLAB console. The gateway *mexFunction* simply calls *printf()*:†

```
#include "mex.h"
void mexFunction (int nlhs,        mxArray *plhs[], /* outputs */
                  int nrhs, const mxArray *prhs[]) /* inputs */
{
    printf("Hello, world!\n");
}
```

Let us now compile this file:

```
>> mex hello.c
Building with 'Microsoft Windows SDK 7.1 (C)'.
MEX completed successfully.
```

This creates the file *hello.mexw64* in the current folder on my 64-bit Windows machine. I could have used a destination folder using the -outdir switch, but then I would need to ensure that this folder is on the MATLAB path in order to be able to run the MEX file.

```
>> which hello
C:\Yair\Books\MATLAB Performance\Mex\examples\hello.mexw64
```

We can now call this file directly in MATLAB:

```
>> hello
Hello, world!
```

Internal comments in MEX files are ignored by the compiler and are not stored within the generated binary file. MATLAB therefore cannot display any help message stored as source-code comments. This contrasts with m-files, where the help message is automatically generated from the file's top comment. As with p-coded files, we can specify the help message in a standalone file having the same name as the MEX file and a *.m* extension. This file should contain only the help comment (lines starting with %) and nothing else. Since MEX and p-code files have a higher precedence than m-files,[1156] when we invoke the function, the binary file would be run. But if we issue the ***help*** or ***doc*** commands (or their interactive GUI alternatives), then the m-file will be used.[1157]

---

* See §8.1.3 for details.

† The *mex.h* header file already #includes *stdio.h*, so we do not need to #include it ourselves. Moreover, *mex.h* #defines *printf()* as *mexPrintf()*, so we are actually not directly using stdio's *printf()* at all (this was not the case in old MATLAB releases). Be careful *NOT* to use C++'s *cout* but rather *printf*, since *cout* does not work properly in MEX files.

8.1.3 MEX Function Inputs and Outputs

As noted above, the MEX gateway function *mexFunction()* is defined as follows:

```
void mexFunction (
     int            nlhs,     /* number of expected outputs */
     mxArray        *plhs[],  /* array of pointers to output args */
     int            nrhs,     /* number of input args */
     const mxArray *prhs[])   /* array of pointers to input args */
```

The format is simple: a pair of gateway inputs that defines the MEX function's outputs (number of requested arguments and a corresponding array of pointers to them), and another pair that defines the MEX function's inputs (number of specified arguments and an array of pointers to them). MATLAB automatically sets *mexFunction*'s input args (nlhs etc.) based on the actual input and output args that were specified in the invoking m-code.

A compiled MEX function does not contain information about its external (invocable) function name. We can rename (or link to) the resulting binary file and simply call the function using its new file name. During run time, we can determine the actual name by which the function was called using *mexFunctionName()*, which is similar to the role of the *mfilename* function in m-code:

```
#include "mex.h"
void mexFunction (int nlhs,        mxArray *plhs[], /* outputs */
                  int nrhs, const mxArray *prhs[]) /* inputs */
{
    const char *name = mexFunctionName();
    printf("s() called with %d inputs,%d outputs\n",name,nrhs,nlhs);
}
```

After compiling, let us run this function:

```
>> hello
hello() called with 0 inputs, 0 outputs

>> hello(1,2,3)
hello() called with 3 inputs, 0 outputs

>> [a,b] = hello(1,2,3)
hello() called with 3 inputs, 2 outputs
One or more output arguments not assigned during call to "hello".
```

The last error, about unassigned outputs, is expected, since our MEX function has not assigned any output values (plhs). We shall shortly see how to do this.

Before using the input (prhs) and output (plhs) arguments, it is essential to check the number and type of these arguments that were actually passed to the function by MATLAB. This is more important than in regular m-code: accessing an unset argument in m-code causes a simple MATLAB exception, which can be trapped; doing so in MEX will generally lead to a memory segmentation violation causing an unrecoverable MATLAB crash. Similar effects can result from accessing a parameter using an incorrect type or size assumption. All this is second nature to C/C++ programmers, but it is clearly different from MATLAB m-code, so it is worth noting.

To check for the number of inputs/outputs, we can simply use `nrhs` and `nlhs`; to check the data type we can use the MEX helper functions *mxIsClass()* and its kin; to check the size we can use *mxGetN()*, *mxGetM()*, or *mxGetNumberOfElements()*. To raise an error back to the MATLAB caller, we can use *mexErrMsgTxt():**

```
if (nrhs < 1)
    mexErrMsgTxt("Input parameter must be specified.");
else if (mxIsSparse(prhs[0]) ||
        mxIsComplex(prhs[0]) ||
        mxIsClass(prhs[0],"char"))
    mexErrMsgTxt("Input must be full matrix of real values.");
else if (mxGetNumberOfElements(prhs[0]) > 10)
    mexErrMsgTxt("Input must be 10 or less data values.");
```

In addition to *mexErrMsgTxt()*, we could also use *mexErrMsgIdAndTxt()* or *mexWarnMsgIdAndTxt()* to generate errors and warnings having specific trappable/suppressible IDs and *sprintf*-like formatting:

```
mexErrMsgIdAndTxt("Yair:MEX:InvalidInput", "Invalid input %d.", data);
```

Checking `nrhs` and `nlhs` can improve performance by enabling the code to bypass costly calculations that may not be required in some cases (see §8.1.5 for example).

We can access the contents of input args using the standard MEX functions for accessing MATLAB variables (`mxArray`). See the following section for details.

8.1.4 Accessing MATLAB Data

8.1.4.1 mxArray

All MATLAB data is accessed via a common structure format (`mxArray`) that holds information on the data, such as its class, dimensions, and memory location. We do not normally need to access this structure directly, and in fact we should NOT do so. Instead, we should rely on the available MEX API functions provided in the *matrix.h* header file. This ensures that even if the internal `mxArray` format changes, as it often does, we do not need to modify our code. We can create functions using the fully supported MEX API functionality without ever needing to worry about `mxArray`'s internal format. The only possible exception is in-place data update, which can be important for performance (see §8.1.5).

Still, I believe that insight into MATLAB's underlying memory format improves programming effectiveness. It should be noted, however, that this information is not officially documented nor supported.[†] You can safely skip this section.

 Note: poking into MATLAB's memory internals is not condoned nor supported by MathWorks, and I highly discourage it for any regular usage. The memory format often changes without notice across MATLAB releases, which could lead to adverse effects and MATLAB crashes for any program that relies on it. So be EXTREMELY careful when relying on this information, and save your data often!

---

* Note the "e" in the "mex" prefix of the function name. Some MEX functions have a prefix of *mx*, while others use *mex*. The *mx* functions are defined in the *libmx* library, while the *mex* functions are defined in *libmex*.

[†] The `mxArray` structure was documented some years ago, but is no longer documented in recent MATLAB releases.

For recent versions of MATLAB, mxArray contains the following information:[1158]

Name or reverse crosslink pointer	Points to a string that contains the name of the variable in the workspace (if it has one) up through R2008b. NULL pointer from R2009a to R2010b. Points to the reverse crosslink variable (if any) in R2011a and later.
Class ID	Single, double, char, logical, etc.
Variable type	Normal, persistent, global, sub-element, temporary, property, etc.
Crosslink pointer	Points to the next shared data copy (SDC) of this variable in a circular linked list. If there are no SDCs of this variable then this pointer is NULL. For R2011a and later note that this is a double-linked list as stated above.
Number of dims	The number of dimensions.
Reference count	The number of reference copies of this variable contained in all variables, not including this one.
Bit flags	Flags for: Scalar double full, empty double full, temporary, sparse, constant, numeric, and user settable bits.
Number of rows	Data dimension #1
Number of columns	Product of all dimensions from 2:end.
Dimension pointer	Present for multidimensional variables >2D only, and occupies the same physical place in the structure as the M would have occupied. Typically known as dims.
Pointer to real data	Pointer to the real data memory. For empty variables this is NULL. Typically known as Pr. For nonnumeric variables (e.g., cells or structs) this points to whatever data type the variable is holding (e.g., mxArray *, or char, etc.).
Pointer to imaginary data	Pointer to the imaginary data memory, which is *not* contiguous with the real data memory. Typically known as Pi. For real variables or empty variables, this is NULL. Note: MATLAB's *isreal* function simply returns *true* if this pointer is NULL, and *false* if this pointer is not NULL, even if all of the imaginary values are zero.
Sparse row indexes pointer	Pointer to memory that contains the row indexes for the nonzero entries in a sparse matrix. Typically known as Ir.
User defined class ID (new)	For a *classdef* style (MCOS) object, this is the value that MATLAB has assigned to identify that class. It occupies the same memory that the sparse row indexes pointer uses.
Class name pointer (old)	For old style (pre-R2008a) class objects, this points to a string containing the class name. It also occupies the same memory that the sparse row indexes pointer uses.
Sparse column indexes pointer	Pointer to memory that contains the column indexes for the nonzero entries in a sparse matrix, typically known as Jc.
User defined class ID (old)	For an old style object, this is the value that MATLAB has assigned to identify the class. It occupies the same memory that the sparse column indexes pointer uses.
Size of sparse matrix memory	Maximum number of nonzero entries that the sparse matrix can hold based on the current memory allocations. Typically known as nzmax. This is *not* the actual number of nonzero entries that the matrix currently contains.

The C Language struct definition of all of this is defined as follows:

```
struct mxArray {
    void *name; /* Name of workspace or source variable */
    mxClassID classID;      /*  0 = unknown 10 = int16
                                1 = cell    11 = uint16
                                2 = struct  12 = int32
                                3 = logical 13 = uint32
                                4 = char    14 = int64
                                5 = void    15 = uint64
                                6 = double  16 = function_handle
                                7 = single  17 = opaque  (classdef)
                                8 = int8    18 = object  (old style)
                                9 = uint8   19 = index   (deprecated)
```

```
                                  10 = int16  20 = sparse (deprecated)        */
     int variableType;       /* 0 = normal
                                1 = persistent
                                2 = global
                                3 = sub-element (field or cell)
                                4 = temporary
                                5 = (unknown)
                                6 = property of opaque class object
                                7 = (unknown)                                 */
     mxArray *crossLink;     /* Address of next shared-data variable          */
     size_t ndim;
     unsigned int refCount;  /* Number of extra sub-element copies            */
     unsigned int flags;     /* bit  0 = is scalar double full
                                bit  2 = is empty double full
                                bit  4 = is temporary
                                bit  5 = is sparse
                                bit  8 = (unknown)
                                bit  9 = is numeric
                                bits 24 - 31 = User Bits                       */
     union {
        size_t M;            /* M = Row size for 2D matrices, or              */
        size_t *dims;        /* Pointer to dims array for nD > 2 arrays       */
     } Mdims;
     size_t N;               /* N = Product of dims 2:end                     */
     void *pr;               /* Real Data Pointer or cell/field elements      */
     void *pi;               /* Imag Datac Pointer (or field information)      */
     union {
        mwIndex *ir;         /* Pointer to row values of sparse arrays        */
        mxClassID classID;   /* New User Defined Class ID (classdef)          */
        char *className;     /* Pointer to Old User Defined Class Name        */
     } irClassNameID;
     union {
        mwIndex *jc;         /* Pointer to column values of sparse data       */
        mxClassID classID;   /* Old User Defined Class ID                     */
     } jcClassID;
     size_t nzmax;           /* Number of elements allocated for sparse       */
     /* unsigned int reserved; listed in h-files but NOT REALLY THERE!        */
};
```

When we create a new variable or function argument in MATLAB, a brand new structure is created in the above format. We can see the structure's address, as well as the real and imaginary data pointers Pr and Pi,* using the *format debug* command at MATLAB's prompt:[†]

```
>> format debug
>> X = [1 2 3]
X =
Structure address = 4a2eee0
m = 1
n = 3
pr = 181646a0
pi = 0
     1     2     3
```

---

* I will use Pr and Pi in the text instead of pr and pi, in order to avoid confusion with MATLAB's *pi* function.
[†] Also see §9.2.10.

In this example, X's memory structure starts at address 4a2eee0 (hex), the data size is 1×3, the memory for the real data (3 elements) starts at address 181646a0 (hex), and the imaginary data pointer Pi is NULL (0) since this is a real variable (i.e., there is no memory allocated for the imaginary part).

```
>> Y = [1 2 3] + [4 5 6]*1i
Y =
Structure address = 4a2dda0
m = 1
n = 3
pr = 1816aa00
pi = 18163ce0
   1.0000 + 4.0000i    2.0000 + 5.0000i    3.0000 + 6.0000i
```

In this example, the memory for the real data starts at address 1816aa00 (hex), and the memory for the imaginary data starts at address 18163ce0. Note that the imaginary part is not contiguous with the real part. That is, the real address Pr plus 3×8 bytes would be 1816aa00 (hex) + 18 (hex) = 1816aa18, which is different from the stated Pi address of 18163ce0. There is a gap between the real data and the imaginary data. This is typical of MATLAB complex variables, and differs from other languages where the real and imaginary data are interleaved in memory on a per-element basis. This becomes important if we call an external library function to operate on a complex variable, since it necessitates a data copy in both directions.

To revert back to the standard (non-debug) *format* mode, simply call *format*:

```
>> format   % revert back to standard (non-debug) display format
```

All full numeric variables are arranged as above, the only difference being the class type and the number of storage bytes per element. Logical and char variables are very similar, except that their *isNumeric* flag bit is reset to 0 in mxArray's flag field. For all nonnumeric data, the imaginary pointer Pi is NULL (0) since they hold no imaginary data.

Sparse matrices are more complicated: The Pr and Pi pointers point to the nonzero real and imaginary data, respectively, the indexes of the nonzero elements are stored in the memory pointed-to by the Ir and Jc pointers, and the size of the allocated data memory blocks is contained in the nzmax field.

Cell and struct arrays are a bit different: The basic variable structure is the same as outlined above, but the data of a cell or struct array is actually pointers to other MATLAB variables. That is, the Pr pointer points to a block of memory that holds other MATLAB variable pointers (mxArray*). For cell arrays it is pretty simple, each cell element is accessed via a separate mxArray. For structs there is the extra complication of field names, pointers to which are stored in Pi.

The specific format of the mxArray changes between MATLAB releases. Users should therefore be very careful about relying on its internal format. The *MxArrayDefinition* utility on the MATLAB File Exchange attempts to infer this internal format dynamically, on the specific platform and MATLAB release on which it runs.

8.1.4.2 Standard MEX API for MATLAB Data Access

MEX includes an extensive set of interface functions for accessing mxArray data.[1159] The interface functions use generic data types, in order for the code to be as independent as possible from the compilation platform. While we can always use the native C/Fortran data types (int, char*, etc.), it is better to use these generic data types. The generic types are basically typedef-ed synonyms, defined in *matrix.h* and *tmwtypes.h* (which is automatically #included by *matrix.h*). Here's the list for C/C++:

- Mapping of generic MEX data types to native C++ data types

Generic MEX Data Type	Default Native Type (C/C++)	Description
mwSize[1160]	int/size_t	Data size
mwIndex[1161]	int/size_t	Index values in arrays
mwSignedIndex[1162]	int/ptrdiff_t	Cross-platform indexing
mxChar[1163]	16-bits, compiler-dependent	Strings (16-bit Unicode characters)
mxLogical[1164]	bool	Logical (true/false) data
mxClassID[1165]	enum (int)	Data type (enumeration)
mxComplexity[1166]	enum (int)	Declare that data is real or complex

The two enumeration types (mxClassID and mxComplexity) are used in function signature declarations. When setting or comparing their values, we should use their enumeration values: mxDOUBLE_CLASS, mxINT8_CLASS etc. for mxClassID; mxREAL or mxCOMPLEX for mxComplexity.

To access existing mxArray data, we can use a multitude of predefined functions in *matrix.h*:[1167]

- MEX functions related to data size, dimensions, and type

Retrieval Function	Update Function	Description
mxGetNumberOfDimensions	—	Number of dimensions in data
mxGetElementSize	—	Number of bytes required to store each data element
mxGetDimensions	*mxSetDimensions*	Pointer to dimensions array
mxGetNumberOfElements	—	Number of elements in data
mxCalcSingleSubscript	—	Offset from first element to the desired element
mxGetM	*mxSetM*	Number of data rows
mxGetN	*mxSetN*	Number of data columns
mxGetClassID	—	Type of data (mxClassID enum)
mxGetClassName	—	Type of data (as a string)[a]

[a] Do not confuse *mxGetClassName* and *mxSetClassName*—despite their similar name, one is NOT the opposite of the other.

- MEX functions related to full (dense) numeric data arrays

Retrieval Function	Update Function	Description
mxGetScalar	—	Real component of first data element
mxGetPr	*mxSetPr*	Real data values of type double
mxGetPi	*mxSetPi*	Imaginary data values of type double
mxGetData	*mxSetData*	Real numeric data values
mxGetImagData	*mxSetImagData*	Imaginary numeric data values

- MEX functions related to sparse data

Retrieval Function	Update Function	Description
mxGetNzmax	*mxSetNzmax*	Number of elements in sparse matrix
mxGetIr	*mxSetIr*	Sparse matrix IR array
mxGetJc	*mxSetJc*	Sparse matrix JC array

- MEX functions related to data arrays of other types

Retrieval Function	Update Function	Description
mxGetChars	—	Pointer to character array data
mxGetString, mxArrayToString	—	Pointer to native `char` buffer
mxGetLogicals	—	Pointer to logical (bool) array data
mxGetCell	*mxSetCell*	Contents of a cell array
mxGetProperty	*mxSetProperty*	Public property of a class object

- MEX functions related to structs

Retrieval Function	Update Function	Description
mxGetNumberOfFields	—	Number of fields in the struct
mxGetFieldNameByNumber	—	Field name based on field index
mxGetFieldNumber	—	Field index based on field name
mxGetField	*mxSetField*	Field value of a specified field name
mxGetFieldByNumber	*mxSetFieldByNumber*	Field value of a specified field index
—	*mxAddField*	Add a new field to a struct
—	*mxRemoveField*	Remove a field from struct
—	*mxSetClassName*	Convert a struct into MATLAB class[a]

[a] Do not confuse *mxGetClassName* and *mxSetClassName*—despite their similar name, one is NOT the opposite of the other.

- MEX functions related to numeric constants

Retrieval Function	Update Function	Description
mxGetEps	—	Returns *eps* double-precision value
mxGetInf	—	Returns *inf* double-precision value
mxGetNaN	—	Returns *nan* double-precision value

MEX also defines multiple *mxIs<Xyz>* functions that can be used to validate the data type, emptiness, and values.[1168] The defined *<Xyz>* values are: *Double* (i.e., *mxIsDouble*), *Single, Complex, Numeric, Int64, Uint64, Int32, Uint32, Int16, Uint16, Int8, Uint8, Char, FunctionHandle,** *Logical, LogicalScalar, LogicalScalarTrue, Struct, Cell, Class, Inf, Finite, NaN, Empty, Sparse,* and *FromGlobalWS*.†

---

* For some unknown reason (probably a simple documentation oversight), *mxIsFunctionHandle* does not have a dedicated doc page. However, it is defined alongside the rest of the data-validation functions in *matrix.h*.

† *mxIsFromGlobalWs* in R2014a onward; *mxIsGlobal* in R2013b and earlier.

MEX also defines the validation functions *mxAssert* and *mxAssertS*. These are similar to C's *assert* function, but are only invoked in debug mode (*mex -g ...*), slightly improving run-time (non-debug) performance at the expense of reduced robustness.

The functions above are sufficient for reading the input parameters sent by MATLAB to our *mexFunction*. But any nontrivial algorithm would need to allocate memory for storing interim and output results in new `mxArray` objects. We also need to ensure that we deallocate any memory used by interim results, since C/C++, unlike MATLAB, does not automatically deallocate the memory of variables when they go out of scope.

MEX memory management is discussed below (§8.1.6). For the moment it is sufficient to note that we can allocate `mxArray` for a regular double-precision scalar using *mxCreateDoubleScalar*; a regular 2D double-precision full (dense) numeric array using *mxCreateDoubleMatrix*; and similarly *mxCreateCellArray*, *mxCreateString*. We can use *mxDuplicateArray* to deep-copy one `mxArray` into another. To free memory that was allocated with any of the *mxCreate** functions, we should use *mxDestroyArray*.* Refer to the MEX documentation for additional data types and invocation options.[1169]

Advanced MEX programmers sometimes modify the input data directly, to avoid the need for deep data copy.† This is easy to do in C/C++, using the pointers to the data provided by `mxArray`. However, such usage is not officially supported[1170] and could actually cause memory-violation crashes in certain cases.[1171] The safer and officially supported alternative is to create a duplicate of the data using *mxDuplicateArray* and then modify the copy rather than the original. An alternative to direct input manipulation is to create a shared copy of the data using undocumented MEX functions such as *mxCreateSharedDataCopy*[1172] or *mxGetPropertyShared*.[1173] We can similarly use *mxUnshareArray* to mimic MATLAB's Copy-on-Write (COW) mechanism,‡ which only copies the data if and when it is actually modified,[1174] avoiding deep copies when the data is only being read. While these functions could significantly speed up MEX code,[1175] they are not officially supported and their usage is discouraged. They bypass MATLAB's memory manager and might even confuse it when combined with `mxArray` data that was created in the "regular" manner.§

8.1.5 A Usage Example

Let's illustrate the above with a usage example, using the following m-code snippet:

```
data = zeros(1,1e6); % preallocate 1 million elements
for idx = 1 : numel(data)
    if mod(idx,2) ==0
        data(idx) = sin(x);
    else
        data(idx) = cos(x);
    end
end
⇨ Elapsed time is 0.127994 seconds.
```

---

* Naturally, we should only deallocate interim/temporary data — not data that is returned as output parameters (plhs).

† See §9.5.2 for a discussion of MATLAB's in-place data manipulation.

‡ See §9.5.1 for a discussion of MATLAB's Copy-on-Write (COW) mechanism.

§ In R2014a, the interface for some undocumented MEX functions (but not those mentioned here) has changed. For details and workarounds, see http://undocumentedmatlab.com/blog/serializing-deserializing-matlab-data#MEX (or: http://bit.ly/1mBFe21).

This could of course be vectorized as follows:

```
% Vectorized version: 5x faster
data(2:2:1e6) = sin(2:2:1e6);
data(1:2:1e6) = cos(1:2:1e6);
⇨ Elapsed time is 0.026597 seconds.
```

Now let us convert this simple algorithm into MEX. We create a file called *sincos.c*:

```c
#include "mex.h"
#include <math.h>

/* Core algorithm processing function */
void algorithm(const double *inData, double *outData,
               mwSize numElements)
{
    int idx;
    for (idx=0; idx<numElements; idx++)
        if (idx % 2)
            outData[idx] = sin(inData[idx]);
        else
            outData[idx] = cos(inData[idx]);
}

/* The MEX gateway function */
void mexFunction(int nlhs,       mxArray *plhs[],
                 int nrhs, const mxArray *prhs[])
{
    double *inData, *outData;
    mwSize mrows, ncols, numElements;

    /* early bail-out if no output was requested */
    if (nlhs < 1)
        return;

    /* sanity check: ensure we have input data */
    if (nrhs < 1)
        mexErrMsgTxt("Missing input data.");

    /* create a pointer to the input data */
    inData = mxGetPr(prhs[0]);

    /* get the # of elements in the input data */
    /* alternative: use mxGetNumberOfElements() */
    mrows = (mwSize) mxGetM(prhs[0]);
    ncols = (mwSize) mxGetN(prhs[0]);
    numElements = mrows*ncols;

    /* allocate the output data's memory area */
    plhs[0] = mxCreateDoubleMatrix(1, numElements, mxREAL);

    /* create a C pointer to the output matrix's data */
    outData = mxGetPr(plhs[0]);

    /* call a C function with core algorithm to update output data */
    algorithm(inData, outData, numElements);
}
```

We now compile and run this file:

```
mex sincos.c
data2 = sincos(1:1e6);
⇨ Elapsed time is 0.052400 seconds.

disp(max(abs(data-data2))) % ensure that the results are the same
⇨ 1.11022302462516e-16    % hurray!
```

We see from these results that although we compiled with the default options and the most basic compiler (Windows SDK 7.1 in this example), our simple MEX file was 2.5× faster than the original m-code. We also note that the vectorized version was twice as fast as the MEX code. This simple example illustrates the general point that MEX files are generally faster than regular CPU/memory intensive m-code, but often slower than highly vectorized m-code.

We can mimic the m-code vectorization by modifying the algorithm function's loop:

```
for (idx=1; idx<numElements; idx+=2)
    outData[idx] = sin(inData  [idx]);
for (idx=0; idx<numElements; idx+=2)
    outData[idx] = cos(inData  [idx]);
```

This results in a 10% speedup on my system (i.e., 47 ms run time). Such low speedup is rarely worth the extra effort, and often falls within the bounds of measurement inaccuracies.

Additional MEX speedup can be achieved using better compilers, more aggressive compiler optimizations (default: /O2), or using a vectorized library.* However, vectorized m-code will often out-perform even a highly tuned C-based MEX file, when numeric computation is a major part of the processing, as above.

Most real-life algorithms are *not* as numerically oriented as our example, and spend much of their time evaluating conditionals, processing strings, and other programming aspects that are not numeric and cannot be efficiently vectorized. In such cases, MEX often provides a better speedup alternative than an attempt to optimize the m-code.

Even when the m-code is vectorized, we can still achieve significant speedups in MEX, by avoiding intermediate arrays that need to be computed in m-code. For example, consider the case of summing the values of a section of a numeric matrix:[1176]

```
% Original m-code
data = magic(5000); % 5000x5000 values
total = sum(sum(data(100:4900,200:4800)));
⇨ Elapsed time is 0.096798 seconds.

% And now the MEX variant: 3.4x faster
total = mexSumSubmatrix(data,100,4900,200,4800);
⇨ Elapsed time is 0.028336 seconds.
```

Where the MEX function *mexSumSubmatrix.c* is defined as follows:†

```
#include "mex.h"
void mexFunction(int nlhs,         mxArray *plhs[],
                 int nrhs, const mxArray *prhs[])
{
```

---

* See §8.1.6.3.

† A natural improvement to this code would be to accept any number of data dimensions. See related: http://mathworks.com/matlabcentral/newsreader/view_thread/262036 (or: http://bit.ly/1epxbVv).

```
double *data, minX, maxX, minY, maxY; /* input data */
double total = 0;                         /* output result */
double *dataPtr;
int row, col, numRows;

/* Check for proper number of input and output arguments */
if (nrhs != 5)
   mexErrMsgIdAndTxt("YMA:MexSumSubmatrix:invalidNumInputs",
           "5 inputs required: data,minX,maxX,minY,maxY");
if (nlhs > 1)
    mexErrMsgIdAndTxt("YMA:MexSumSubmatrix:maxrhs",
           "Too many output arguments.");
if (!mxIsDouble(prhs[0]))
    mexErrMsgIdAndTxt("YMA:MexSumSubmatrix:invalidInput",
           "Input data must be of type double.");
if (!nlhs)
    return;

/* Get the input data */
data = mxGetPr(prhs[0]);
numRows = mxGetM(prhs[0]);
minX = mxGetScalar(prhs[1])-1;
maxX = mxGetScalar(prhs[2])-1;
minY = mxGetScalar(prhs[3])-1;
maxY = mxGetScalar(prhs[4])-1;

/* Compute the sum of the specified sub-matrix */
for (col=minY; col <=maxY; col++) {
    dataPtr = data + numRows*col + (int)minX;
    for (row=minX; row<=maxX; row++) {
        total += *dataPtr++;
    }
}

/* Return the number of elements written to the caller */
plhs[0] = mxCreateDoubleScalar(total);
}
```

8.1.6 Memory Management

C programming is notorious for its lack of automated memory management. Programmers need to explicitly allocate and deallocate memory for data, with memory leaks and/or segmentation violations commonplace due to bugs in the implementation (or lack of) these memory operations. Several successful memory debuggers have been developed to detect and report such memory issues, including Purify, Insure++, BoundsChecker, and Valgrind.[1177]

MATLAB takes the opposite route, of automatically managing all m-code memory. MATLAB m-code users do not need to worry about memory allocation and deallocation, it all happens automatically. As one of MathWorks' development managers put it,[1178] there are *"no knobs to turn"*.

MEX code provides a middle ground. On the one hand, it is native (C/C++/Fortran) code, and on the other it is often tightly integrated with MATLAB m-code. MATLAB programmers might be more likely than pure C/C++ programmers to encounter memory-related issues due to the opposing memory management schemes. For this reason, MathWorks included automated memory management in MEX.[1179]

While MATLAB's memory manager does not handle pure native memory, it does automatically handle any `mxArray` or data that was created using MEX's built-in functions (*mxCreateDoubleArray, mxCalloc*, etc.). Such data is automatically deallocated when the MEX function exits for any reason, naturally, due to user interrupt, or due to a raised exception. Only data that was returned as an output parameter (via `plhs`) or specifically made persistent (using *mexMakeArrayPersistent* for an mxArray created via *mxCreate**, or *mexMakeMemoryPersistent* for memory allocated using *mx*alloc*)[1180] will not be cleared.

Using persistent memory carries the promise of increased performance using various caching mechanisms, as described in §3.2. Properly used, persistent memory can help avoid the need to reallocate and/or recalculate data in subsequent invocations of the MEX function, thereby improving its performance.

Memory that was declared persistent needs to be managed by us. This means that whenever we update persistent memory, we need to first deallocate any existing data that it holds, before assigning the new data.

When *mx**-allocated memory needs to be deallocated, we can use *mxDestroyArray*[1181] (for any memory allocated with *mxCreate**), or *mxFree*[1182] (for any memory allocated with *mx*alloc*). We should *NOT* use the native *free()* to deallocate such memory, as this could confuse MATLAB's memory manager and potentially lead to a segmentation fault (MATLAB crash).[1183] In general, we should not use native memory and deallocation functions at all in MEX, only MEX's *mxMalloc, mxCalloc, mxRealloc*, and *mxFree*. This will ensure that all the MEX function's memory is properly and automatically managed by MATLAB's memory manager, reducing the likelihood of segmentation violations and memory leaks.[1184]

Another typical error that leads to segmentation faults is using *mxFree* to deallocate memory that was originally allocated using *mxCreate**,[1185] or conversely using *mxDestroyArray* to deallocate memory that was originally allocated using *mx*alloc*. Yet another error occurs when trying to *mxDestroyArray* or *mxFree* within a C++ destructor, since at that point the memory manager has already freed the data.[1186]

A different class of memory leaks occurs when using one of the *mx** functions (such as *mxArrayToString* or *mgGetProperty*[1187]) that return data. In many cases, these functions deep-copy the data internally, returning a pointer to the newly allocated copied data, rather than to the original data. Apparently, MATLAB's memory manager does not always automatically free this memory, so we should *mxDestroyArray* or *mxFree* it ourselves, to prevent memory leaks.[1188]

To prevent memory leaks when a MEX function is cleared, we should either prevent this function from being cleared using *mexLock*,*[1189] or set-up an exit handler function using *mexAtExit*,[1190] which would deallocate any persistent memory.

Some users have reported that while MATLAB's memory manager successfully frees *mx**-allocated memory in MEX (e.g., **.mexw32*) files, it does not do so when the MEX file is compiled into a dynamic library (e.g., **.dll*) that is then loaded into MATLAB using the **loadlibrary** function from within m-code.[1191] This naturally leads to a memory leak. In such cases, we need to either use *mxDestroyArray* or *mxFree* to free the memory, rather than rely on MATLAB's memory manager to do it for us, or use native allocation/deallocation routines rather than rely on the *mx** functions.

When using MEX-managed data, we should be careful to avoid memory leaks on the one hand, and deallocating data that is later referenced on the other. For example, we should be careful not to deallocate data that is returned to the MEX function's caller as an

---

* *mexLock* is similar to MATLAB's **mlock** function, which prevents an m-function from being cleared along with its internal persistent variables.

output argument — this could result in a segmentation violation. MathWorks has created a documentation page listing several other common memory issues in MEX.[1192] Readers are encouraged to review this page and avoid the listed programming patterns.* Similarly, readers are encouraged to read James Tursa's insightful discussion of MATLAB's memory manager and memory leak scenarios, in newsgroup postings.[1193]

When creating new `mxArrays` using one of MEX's *mxCreate** functions (e.g., *mxCreate-DoubleArray*), it is important to note that the memory is not actually being initialized, only allocated. This is actually good for performance. In cases where we need to initialize the allocated data, we can use C's *memset* function, which could take some time for large data arrays. The performance difference between merely allocating an `mxArray` and also initializing it, could explain the large performance differences between the various forms of MATLAB data preallocations.[1194]

Another performance-related MEX aspect is that modern MATLAB releases enable MEX functions to create and access huge data `mxArrays`, up to 2^48-1 data elements, on 64-bit machines running a 64-bit OS and a 64-bit MATLAB.[1195] We need to *mex*-compile using the *-largeArrayDims* parameter to benefit from such large arrays. This can help performance by enabling better vectorization and parallelization. On the other hand, we must ensure that our platform's hardware (physical RAM) and OS (virtual memory) support such large data, otherwise we may run into virtual-memory thrashing slowdowns† or out-of-memory errors.

Unlike MATLAB functions, MEX-file functions (binary MEX-files) do not have their own variable workspace. MEX-file functions operate in the caller's workspace. *mexEvalString*[1196] evaluates a string in the caller's workspace. In addition, you can use the *mexGetVariable*[1197] and *mexPutVariable*[1198] routines to get and put variables into the caller's or the base workspace. What this means, basically, is that our MEX file has direct access to all the variables defined in MATLAB. So even if a certain variable is not passed to the MEX function as input parameter, we can still access it in MEX code.

8.1.7 Additional Aspects

8.1.7.1 Profiling MEX Code

While professional profilers exist that could be used to profile our MEX code, a "poor-man's" solution, similar to MATLAB's *tic/toc*, is to measure the elapsed time between two measurements of the system clock. This can easily be done with the use of the *clock()* function, which is defined in *<time.h>*, which should be available on all platforms and C/C++ compilers:‡[1199]

```
#include <time.h>
#include "mex.h"


/* The MEX gateway function */
void mexFunction(int nlhs,        mxArray *plhs[],
                 int nrhs, const mxArray *prhs[])
{
    clock_t start, stop;
```

---

* See §9.7.1 for additional advice regarding MATLAB variables memory and avoiding memory leaks.
† See §9.1.
‡ This code contains C++-style comments (//). This compiles ok on some platforms but not others. Rather than modifying the MEX code to use standard C comments (/* ... */), we can modify the compilation flags to accept them: http://walkingrandomly.com/?p=2694 (or: http://bit.ly/1ijXjli).

```
        double elapsed;
        start = clock();
        ... //do something useful here...
        stop = clock();
        elapsed = (double)(stop-start)/CLOCKS_PER_SEC;  //[secs]
        mexPrintf("Time elapsed:%0.3f\n", elapsed);   //msec resolution
    }
```

The resolution of *clock()* is system dependent; it is typically 1 millisecond (i.e., CLOCKS_PER_SEC = 1000).

8.1.7.2 Calling MATLAB Functions

We can invoke MATLAB functions from within our MEX code using the *mexCallMATLAB* function.[1200]

One of the MEX examples, *sincall.c*, provides an example of a MEX function that fills a vector of data x, then call's MATLAB's *sin* function to compute a corresponding data vector y, and finally calls *plot* on x, y. The m-code for this is

```
MAX = 1000;
mm = MAX/2;
for idx = 1 : mm-1
    x(i) = idx * (4*3.14159/MAX);
end
y = sin(x);
plot(x,y)
```

while the core of the equivalent MEX code is as follows:

```
rhs[0] = mxCreateDoubleMatrix(max, 1, mxREAL);

/* pass the pointers and let fill() fill up data */
fill(mxGetPr(rhs[0]), &m, &n, MAX);
mxSetM(rhs[0], m);
mxSetN(rhs[0], n);

/* get the sin wave */
mexCallMATLAB(1, lhs, 1, rhs, "sin"); //lhs = sin(rhs)

/* plot(rhs, sin(rhs)) */
inplot[0] = mxDuplicateArray(rhs[0]);
inplot[1] = mxDuplicateArray(lhs[0]);
mexCallMATLAB(0, NULL, 2, inplot, "plot");//plot(inplot(1),inplot(2))

/* cleanup allocated memory that is not returned as outputs */
mxDestroyArray(rhs[0]);
mxDestroyArray(lhs[0]);
mxDestroyArray(inplot[0]);
mxDestroyArray(inplot[1]);
```

mexCallMATLAB can be used to invoke any of MATLAB's functions, including graphics and GUI. This has been used by Tim Davis' *waitmex* utility, which enables using a MATLAB *waitbar* within a MEX function.[1201]

A close relative of *mexCallMATLAB* is *mexCallMATLABWithTrap*.[1202] These functions have almost identical signatures, with the following differences:

- *mexCallMATLAB* returns an `int` value (0 if successful, nonzero if not); *mexCallMATLABWithTrap* returns a pointer (NULL if no error; `mxArray` of class `MException` in case an exception was raised by the invoked function).

- *mexCallMATLAB* raises an exception if the invoked MATLAB function raises one; *mexCallMATLABWithTrap* always traps the exception and proceeds to the next MEX source-code line.

We can use the corresponding *mexEvalString* and *mexEvalStringWithTrap* functions, which accept a string (`char*`) of MATLAB code to evaluate, just as if it was encountered by the MATLAB interpreter in the Command Window or in m-code.

Calling MATLAB functions from MEX enables fast development of working code. However, such calls have a nonnegligible overhead, so if we find that our MEX file spends much of its time in such calls, we should perhaps consolidate them into a single m-function call. As MathWorker Steve Lord put it:[1203]

> *Don't evaluate each individual command with separate engEvalString calls. Doing so is like grocery shopping by purchasing one item, driving home to put it on your shelf, driving back to the grocery store, buying another item, driving home, etc. Instead create a MATLAB script or function file and call that using ONE engEvalString call so that you only pay the overhead of switching between C++ and MATLAB twice (once C++ → MATLAB, once MATLAB → C++.)*

To improve performance it is often better to implement a pure-MEX solution. For instance, in the example above, we might consider replacing the call to MATLAB's *sin* function with an equivalent C loop that calls C's internal *sin()* function, as shown in §8.1.5. MATLAB's built-in functions often employ implicit parallelization and other optimizations, so this advice should be tested on a case-by-case basis. It may well turn out that MATLAB's internal optimizations outweigh the overheads of an engine function call in certain cases. However, we shall see below that MEX can provide significant speedups even for simple builtins such as *min/max*.

8.1.7.3 Calling Functions in External Libraries

External libraries often provide highly optimized implementations of specific functions. MATLAB itself uses the FFTW, MKL, BLAS, IPP, and LAPACK libraries, among others, and we can use these libraries in our MEX code. MEX code performance and development time is often improved by invoking functions in those external libraries, rather than hand coding in MEX or invoking the corresponding MATLAB functions via *mexCallMATLAB*. In fact, the need to invoke BLAS and LAPACK routines is so commonplace, that MATLAB's documentation includes a detailed doc page[1204] and sample code* that explain how to do this. The techniques outlined for BLAS and LAPACK are also applicable to other external libraries.

We may sometimes wish to use functions of existing libraries that may be faster than the versions used by MATLAB. For example, use MKL's *fft* rather than FFTW's,[1205] or use IPP functions for some algorithmic heavy lifting that is not necessarily imaging related.[1206]

---

* *%matlabroot%/extern/examples/refbook/matrixMultiply.c, matrixDivide.c* and *dotProductComplex.c.*

It's really quite simple: Within our MEX file we should `#include` the library's header (e.g., *"blas.h"*), so that we could reference its functions and data types in our MEX code. When the code is ready, add the library as a link parameter to the *mex* compilation command. For example, in the case of BLAS:*

```
mex -largeArrayDims mexFile.c -llibmwblas
```

MATLAB's default library folder, *%matlabroot%/extern/lib/...* is already included in the *mex* linker's default path, so for any library in that folder (e.g., *libmwblas.lib*) we do not normally need to specify a full path when linking.

Integrating a third-party library, that is, one that is not bundled in MATLAB, follows exactly the same route. The *mex* compiler and linker do not mind whether the requested linked library is pre-bundled or not, only that its path can be found. If we place the library in MATLAB's default libraries folder then it will be found by the linker, otherwise we need to provide the full path:

```
arch = computer('arch');
libFolder = fullfile(matlabroot,'extern','lib',arch,'Microsoft');
libPath = fullfile(libFolder,'libmwblas.lib');
mex('-largeArrayDims', 'mexFile.c', ['-L' libFolder], '-llibmwblas');
mex('-largeArrayDims', 'mexFile.c', libPath); % alternative
```

We can include and link several libraries at once:

```
#include "lapack.h"
#include "blas.h"

mex mexFile.c -lmwlapack -lmwblas
```

Note that we do not need to specify the MEX libraries (*libmx* and *libmex*), nor the compiler's standard runtime libraries — these libraries are automatically linked by the *mex* command.[1207] We just need to specify the nonstandard libraries.

When sending or deploying our MEX file to other computers, we need to ensure that any third-party *dynamic* library (**.dll* or **.ocx* on Windows; **.so* or **.dylib* on Linux/Macs) that is used by our MEX file is included in the distribution, otherwise the MEX code will fail in run time. We do not need to similarly deploy *static* libraries (**.lib* on windows; **.a* on Linux/Macs), since their code is statically inserted into the resulting MEX library file. The original C/C++/Fortran source code is not necessary to run a MEX file, we only need the compiled binary library file (and any dynamic libraries on which it may depend).[1208]

Some libraries may have specific requirements. For example, BLAS and LAPACK use 64-bit integers. This means that we should use `mwSignedIndex` rather than the native `int`,[1209] and that we should *mex*-compile with the -largeArrayDims flag.[1210]

In some cases, we may need to use a legacy 32-bit library on a 64-bit system, or vice versa. In this case, we cannot simply use `mwSignedIndex` or its kin, since they are defined differently in the library and in our MEX file. In such cases, the data itself needs to be converted to the library's specific representation, before we can use it.[1211]

Some libraries may be coded in a different language than our MEX file. For example, MATLAB's bundled BLAS, LAPACK, and MKL use Fortran.† Since Fortran uses a

---

* Note: the -largeArrayDims compilation flag is a specific requirement of BLAS and LAPACK, which use 64-bit integers, as discussed below. This flag is not generally required when linking external libraries.

† BLAS and LAPACK have some ports to other languages such as C, but these are not bundled in MATLAB; Intel's MKL also uses Fortran, but also provides interfaces for C/C++ (http://software.intel.com/en-us /articles/intel-math-kernel-library-intel-mkl-blas-cblas-and-lapack-compilinglinking-functions-fortran-and-cc-calls or: http://intel.ly/17FNTL4).

pass-by-reference mechanism, we should be careful to pass the data reference (&data) rather than the data value (data) as parameters to such functions.

Additional complications may arise due to platform inconsistencies. For example, Linux/ Mac functions append an underscore (_) to their names, for example: *dgemm()* → *dgemm_()*. The code could be made generic using a simple pre-processor define:

```
#if !defined(__OS2__) && !defined(__WINDOWS__) && !defined(WIN32) &&
    !defined(_WIN32) && !defined(WIN64) && !defined(_WIN64) &&
    !defined(_MSC_VER)
        #define dgemm dgemm_
#endif
```

For detailed examples of integrating BLAS/LAPACK routines, see James Tursa's *mtimesx* utility[1212] and compare with Yuval Tassa's corresponding *mmx*.[1213] Tim Toolan provided an m-code wrapper around MEX, to easily call BLAS/LAPACK from our own m-code, without having to mess around MEX C-code; reviewing the wrapper's MEX code could be a good starting point for integrating these external routines in your own MEX file.[1214] The File Exchange contains several additional utilities that interface with BLAS/LAPACK that could be used as reference.[1215]

For examples of integrating MEX with FFTW routines, see chapter 11 in Pascal Getreuer's guide *Writing MATLAB C/MEX Code*.[1216] Also of interest is Peter Carbonetto's discussion[1217] of the integration of GNU's Scientific Library (GSL).[1218]

MEX is sometimes used not for its speedup potential, but rather for its ability to interface to external systems (API, drivers, etc.), libraries and system calls. Examples of using MEX to interface to Operating System functions can be found in Jan Simon's *WindowAPI*[1219] and Bill York's *getScreenSize*.[1220] *getScreenSize* is so simple that it is included here in its entirety:

```
#include "mex.h"
#include <windows.h>

void mexFunction(int nlhs,        mxArray *plhs[],
                 int nrhs, const mxArray *prhs[])
{
    double sz[4];

    /* initialize size array */
    sz[0] = 1;
    sz[1] = 1;

    /* get width and height of the screen */
    sz[2] = GetSystemMetrics(SM_CXSCREEN);
    sz[3] = GetSystemMetrics(SM_CYSCREEN);

    /* initialise the returned mxArray */
    plhs[0] = mxCreateDoubleMatrix(1, 4, mxREAL);

    /* copy data into it */
    memcpy(mxGetPr(plhs[0]), sz, 4*sizeof(double));
}
```

8.1.7.4 Miscellaneous Aspects

MEX is not always guaranteed to run faster (although it usually does). In one reported case, a Coder-generated MEX file ran 10× slower than the equivalent m-function, since the m-code performed FFT analysis, which uses the highly optimized FFTW library, while the generated MEX code used a naïve (slow) C implementation.[1221] While we could obviously modify the MEX code to also use FFTW,* it does not seem to be worth the effort since there is not much overhead in MATLAB's implementation of the FFT wrapper code that can be optimized away in MEX. As discussed in detail for the MATLAB Coder below,† MEX code is not expected to provide much if any speedup for m-code that contains mainly built-in functions and operators or calls to highly tuned libraries (MKL, BLAS, LAPACK, IPP, and FFTW).

Similarly, no speedup is expected for functions for which a pure-MATLAB numeric vectorization and implicit parallelization is available.‡ In fact, it may well turn out that a MEX implementation would be slower than the MATLAB variant in such cases.§ MEX is similarly not effective when the MATLAB processing can be easily parallelized using explicit parallelization (e.g., *parfor* or GPU).¶

Speedup is mainly expected for m-code that spends much of its time in iterative loops, or evaluating strings, conditions, and states (as in runtime checks, or a state-machine engine). Similarly, since m-code is mostly tuned for full double-precision data, we may expect better performance from MEX code that is optimized different data types.** For best performance, we should try to use a professional (commercial) compiler and turn on all its compiler optimizations when we *mex*-compile our code.

Another use case that MEX can speed up is when we process the data several times independently. A typical example for this is when we need both the minimal and maximal data values. In this case we need to run both *min(data)* and *max(data)*. A MEX implementation would only need to scan the data once, returning both values at the same time. While this would suggest an upper bound of 2× speedup, it turns out that an efficient MEX implementation (Jan Simon's *MinMaxElem*)[1222] can actually achieve a 5× speedup for very large arrays. This utility also demonstrates the fact that MEX speedups can increase for non-*double* data types.

M-code that extensively uses matrix transpose and/or *permute*, can be a good candidate for speed-up using MEX. The reason is that these operations†† modify the layout of the data in memory. In such cases, MATLAB does not employ in-place optimizations but rather allocates a new memory block and moves the data there, in its new format. In such cases, employing C's ability to directly access the data via pointer arithmetic can do wonders for performance.

Examples showing how MEX can speed up MATLAB I/O are discussed in §11.7 and §11.8.2. Although MEX provides some speedup for I/O just as for regular algorithmic code, the speedup is often less than in the case of pure CPU-driven code, since I/O is normally bound by external factors such as memory and disk throughput, not CPU.

In general, it is not good practice to fine tune an algorithm to a specific platform, but in some specific cases we may need to do exactly this. In such cases, we could consider platform-specific aspects such as memory-access latency and CPU-level instruction parallelization.[1223] It is difficult to utilize platform-specific knowledge with pure m-code, but for example we can use the memory page size to calculate the optimal chunking size‡‡ or loop

---

* See §8.1.7.3.
† See §8.2.4.
‡ See Chapter 5.
§ It is always possible to write horribly vectorized code that could well be sped-up using MEX. But an efficient vectorization (which uses implicit multithreading) often outperforms efficient single-threaded C code.
¶ See Chapter 6.
** See §4.1. The "MATLAB to C Made Easy" webinar (https://mathworks.com/company/events/webinars/wbnr62736.html or: http://bit.ly/1bbKrxI) shows an example (~45:00 into the video) of accelerating a single-precision image compression algorithm using Coder-generated MEX, resulting in 37× speedup.
†† Unlike *reshape*, which leaves the memory intact, modifying only mxArray's M and N fields; this is done in-place.
‡‡ See §3.6.3 and §9.5.9.

size for best caching. In MEX files and Coder-generated C files, our abilities to leverage platform-specific information are larger.

An excellent example of trading accuracy for speed is an approximation of the *exp* function using MEX, which provides a 2–8× speedup at the expense of up to 4% inaccuracy.[1224] This detailed example also highlights a common performance pitfall when using MEX: while calling the generated *mexexp* function with vectorized data is faster than the MATLAB equivalent, it is much slower when called in a non-vectorized loop for scalar values, due to the MEX invocation overheads:

```
% Vectorized
x = 1 : 0.0001 : 70;
y1 = exp(x);
⇨ Elapsed time is 0.007243 seconds.

y2 = mexexp(x);
⇨ Elapsed time is 0.005824 seconds.

% Non-vectorized
for x = 1:0.0001:70, y1 = exp(x); end
⇨ Elapsed time is 0.040265 seconds.

for x = 1:0.0001:70, y2 = mexexp(x); end
⇨ Elapsed time is 1.194346 seconds.
```

A situation often occurs, whereby a C/C++/Fortran algorithm is already available. Rather than converting it to m-code in order to integrate into a larger MATLAB program, we can relatively easily transform it into a valid MEX file by simply adding the *mexFunction* wrapper and making a few minor modifications.[1225] Claude Tadonki provides an interesting tool where users can paste pure C or Fortran code and generate the corresponding MEX wrapper code online[1226] or offline.[1227]

The MATLAB Coder Toolbox can automatically convert m-code into valid readable MEX code.* The generated code is highly efficient, certainly for a first pass. Since the code is readable, complete with the original m-code comments, it is quite easy to use the generated MEX code as a baseline for additional customizations.

Starting with MATLAB release R2013a (8.1), we can manipulate data contained in *gpuArrays* directly within MEX functions, using the Parallel Computing Toolbox (PCT).†[1228] In other words, user-defined CUDA kernels can access *gpuArrays* without the need for memory transfer. This is, however, only supported on 64-bit platforms. MEX parallelization alternatives, which do not require PCT, are described in §7.2 and §7.3.

MEX has several drawbacks in its publicly exposed functionality. Some of these limitations can be overcome using a few undocumented MEX functions, while others have no known workarounds. A post by MEX heavyweight James Tursa, and follow-up comments by others, provide insight into those limitations.[1229] By understanding the boundaries of MEX's documented capabilities, I believe that we can better utilize its features, and understand what can and cannot be achieved.

There are plenty of MEX-based functions that were posted over the years, mainly in MATLAB's File Exchange. Reviewing these functions is a great way to learn and improve MEX skills. Of the engineers who have submitted MEX-based utilities and have provided

---

* See §8.2.
† Refer to Chapter 6 for a discussion of explicit parallelization in MATLAB.

advanced MEX support/answers in various forums over the years, three especially are of note: James Tursa,[1230] Bruno Luong,[1231] and Jan Simon.[1232] In addition to these MEX heavy-weights, Sebastien Paris has also submitted an extensive set of MEX utilities to the File Exchange.*[1233]

Ilias Konsoulas posted a set of small simple utility functions that illustrate how to do basic matrix operations (multiply, add, reshape) in MEX using pure C code (i.e., without using BLAS/LAPACK or advanced MEX functions).[1234]

A CSSM newsgroup thread provides many alternative implementations for counting "1" bits in a numeric value (aka, *bitcount*).[1235] It is instructional to review the different implementations and to compare their run speed. Similar MEX implementations can be used to speed up MATLAB's built-in bitwise functions (**bitor, bitand, bitxor, bitshift**).[1236]

Fang Liu has submitted a set of complete MEX examples to the MATLAB File Exchange,[1237] that use the *CMake* utility[1238] to compile MEX files. Using make files enables easier compilation and dependency-checks for large projects that include multiple files and inter-dependencies. Specific examples show how to integrate OpenMP-enabled[1239] and CUDA-enabled functions.[1240]

To avoid passing data by value across separate invocations of a MEX file from within MATLAB, we can persist the data in the MEX code and simply store the data's pointer in MATLAB, as a type-cast 64-bit double value. This value (pointer) could then be passed repeatedly to MEX functions, to access its underlying data. Oliver Woodford investigated several alternative methods for doing this,[1241] and finally posted a highly rated *MATLAB class wrapper for a C++ class* on the MATLAB File Exchange.[1242]

Debugging MEX files is not easy, due to the fact that they are basically dynamically-loaded (shared) libraries loaded into MATLAB's process at run-time. One method to bypass this limitation is to load the MEX file via a wrapper layer and starting a separate MATLAB process.[1243]

8.2 Using the MATLAB Coder Toolbox†

MATLAB Coder[1244] is a MathWorks toolbox that works with MATLAB and is available as a separate add-on. It enables the translation of MATLAB m-code into standard ANSI C/C++ code that can then be compiled into a variety of target binaries:

- MEX — for use by MATLAB m-code
- Dynamic (shared) libraries — DLLs or SO, depending on platform
- Static libraries (LIB)
- Standalone executable (EXE)
- FPGAs or DSPs (with the extra HDL/Embedded Coder Toolboxes — see §8.3)

The MATLAB Coder is a relatively new product. Formerly part of the Real-Time Workshop Toolbox, in R2011a (MATLAB 7.12) the Coder functionality was enhanced, becoming separate toolboxes — *MATLAB Coder* and *Embedded Coder*.[1245]

---

* While Tursa, Luong, and Simon are active on all fronts (File Exchange, CSSM newsgroup and Answers forum), Paris only focuses on FEX.

† Michael Donnenfeld, Application Engineer at Systematics Ltd. (michaeld@systematics.co.il) helped prepare this section.

Other tools existed in the past, which converted m-code into C code. The MATLAB Compiler used to do that, before changing paradigm to use the MCR in MATLAB 7. An independent product called Agility MCS (*MATLAB-to-C*) was acquired by MathWorks in 2009 and is no longer available.[1246] To the best of my knowledge, MATLAB Coder is currently the only available tool that converts m-code into C/C++.*

Coder-generated C code is quite efficient and does not contain much unnecessary overheads. Compiling the code by a good optimizing compiler has the potential for significant speedups compared to running the original m-code in MATLAB.

We can also gain significant speedups by using the Coder to compile the code into MEX or DLL files that can be used directly in MATLAB. This can be used to verify and tune C code, even when our final target platform is not MATLAB.†[1247]

In one example that is discussed below (§8.2.3), Dijkstra's shortest-path algorithm ran 18× faster when converted into MEX compared to exactly the same code running in m-code. MATLAB Coder is not free, but the cost could well be offset by the improved performance and the development and verification time that it saves.

When considering using the MATLAB Coder, we should note potential drawbacks:

- **Compatibility** — MATLAB Coder supports only a subset of the MATLAB language and toolboxes. Using the Coder often requires significant manual adaptation of the code in order to become Coder-compliant (see §8.2.1).

- **Performance** — MATLAB Coder generates C/C++ code that is not always faster than the original MATLAB code. If our original code is efficiently vectorized, parallelized, or JIT'ed, then it is highly possible that we will not see any significant speedups, and in some cases even slowdowns (see §8.2.4).

8.2.1 Code Adaptation

MATLAB Coder, unlike the MATLAB Compiler, does not use the MCR (MATLAB run time) engine, and so does not support the entire set of MATLAB functionality. While most of MATLAB's major functionality is supported,[1248] some aspects are not. As of R2014a (MATLAB 8.3), the following features are not supported:

- Anonymous functions
- Nested functions
- Scripts (as opposed to regular function files, which are supported)
- Cell arrays (except with *varargin* and *varargout*)
- Java/Dot-Net/COM integration
- Recursion
- Sparse data
- Exception handling (*try/catch*)
- Some internal MATLAB functions (e.g., *fscanf*‡)

---

* McGill University's Sable research group (http://www.sable.mcgill.ca/mclab) has several MATLAB converter projects, including one to Fortran.
† Starting in R2013b (MATLAB 8.2), the verification process became even easier, with the ability to use Embedded Coder Toolbox's SIL (*Software-in-the-Loop*) feature to automatically wrap static LIB files for direct use in the prototype m-code, having the ease-of-use of MEX files (unlike DLLs), along with the flexibility of using the actual target library file.
‡ *fread* was added to the list of supported functions in R2014a (MATLAB 8.3).

- Some features of MATLAB classes[1249]
- A few System Object features (most System Objects are supported)
- Most toolboxes
- Explicit parallelization (***parfor***)* — supported since R2012b (MATLAB 8.0)

To check whether your code is Coder-compliant, add the %#codegen pragma comment at the end of the line that defines the main function. For example:

```
% The following function is currently (R2013b) not Coder-compliant
function data = notCoderCompliant()   %#codegen
    data = {1, 2, 3};              % cell arrays are currently unsupported
    function nestedFunction()   % nested funcs are currently unsupported
    ○ Code generation does not support nested functions.
end
```

This directs MATLAB's MLint (Code Analyzer) tool to report as errors all instances within the source code that are not supported by the Coder. This can be done even if we do not have the Coder license, enabling us to check the estimated effort in converting our algorithm to become Coder-compliant. We could use this information as the basis of a decision whether an investment in the coder toolbox is worthwhile.

Since R2012b (MATLAB 8.0), Coder includes a code-readiness tool, which reports incompatibilities of the MATLAB code with the Coder, in a very easy-to-use GUI. We can launch this tool from the MATLAB Desktop's Current Folder panel, by right clicking the relevant .m file and selecting "Check Code Generation Readiness". We can also do it directly from the MATLAB Command Window (or a MATLAB script):

```
coder.screener('notCoderCompliant');
```

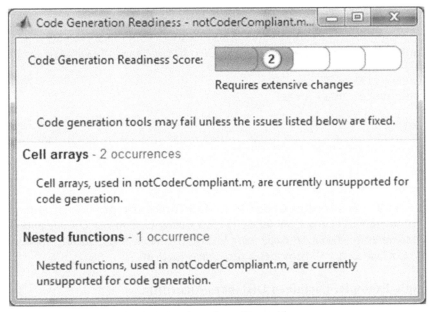

The Coder's code readiness (Screener) report

---

* See §6.1 and §8.2.4.

The Coder in its present state is a relatively new product. It improves and expands with each MATLAB release,[1250] so it is very possible that what was not available in R2013a will indeed be supported by the latest MATLAB release when you read this.

We can often bypass Coder incompatibilities by recoding our m-file program to become Coder-compliant. The Dijkstra example below will show this route.

In other cases, if our target platform is MATLAB (by compiling our code into MEX files that are then called from within the main MATLAB application), we can use the Coder's *extrinsic* command to declare MATLAB functions that should be called externally (i.e., not translated into C code). In other words, MATLAB syntax that cannot be *translated* into C code by the Coder can still be *invoked* by the C code, as long as it runs within a MATLAB process. Extrinsic functions support almost the entire MATLAB syntax, except (at least in R2013a) function handles.[1251]

Normally, MATLAB graphic visualization functions are automatically known to be extrinsic by the Coder, so there is no need to declare them as such. But if we have a MATLAB function that contains non-Compliant code then this could be an easy solution to enable translation of at least the rest of the code (that is Coder-compliant):

```
function mainFunction()
    % Declare notCoderCompliant() as extrinsic
    coder.extrinsic('notCoderCompliant');
    % Now use notCoderCompliant within the code
    notCoderCompliant();
end
```

To summarize, Coder's decision logic for resolving function names in m-code:

```
% Pseudo-code for resolving function names by the MATLAB Coder
if isExtrinsic(funcName)
    if isOnMatlabPath(funcName)
        dispatch to MATLAB engine to execute in run-time
    else
        error
    end
elseif isSubFunc(funcName)||isOnCoderPath(funcName)||isOnMatlabPath
(funcName)
    if isCompilable(funcName)
        generate C/C++ code
    else
        error
    end
else
    error
end
```

In addition to the official online Coder docs, MathWorks also provides useful quick-start guides:[1252] *Preparing MATLAB Code for MATLAB Coder, Generating C Code with MATLAB Coder,* and *Accelerating MATLAB Code with MATLAB Coder.* In addition, a webinar titled *"MATLAB to C Made Easy"*[1253] provides a nice introduction.

8.2.2 A Simple Example: Euclidean Distances Algorithm

Here is a simple example showing the Coder's performance speedup potential. In this case, we use the non-vectorized version of the Euclidean distances algorithm (§5.3.8) as the baseline. For this, we create the following standalone MATLAB function:

```matlab
function r = generateDistanceMatrix(A,B)
    sizeA = size(A,1); % # of points (rows) in A
    sizeB = size(B,1); % # of points (rows) in B
    r = zeros(sizeA,sizeB);
    for idxA = 1 : sizeA
        for idxB = 1 : sizeB
            coordDiff = A(idxA,:) - B(idxB,:); % 3-element array
            r(idxA,idxB) = sqrt(sum(coordDiff.^2));
        end
    end
end
```

We start the Coder GUI by launching it via the ***coder*** command, or interactively via the Apps tab in MATLAB Desktop's main toolstrip. Now, create a simple Coder project with this generateDistanceMatrix function as input: drag-and-drop into the "Entry-Points Files" section or click the "Add files" link. Then, specify the data type (double) and size (unlimited × 3) for each of the auto-detected input args:

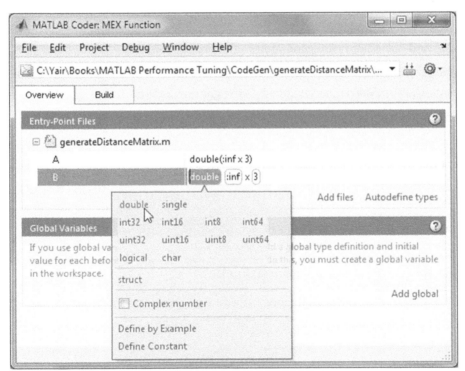

Defining function inputs in MATLAB Coder

Next, we build a MEX function target (which is the default) (see figure on page 418).

Note: when generating target files, we should ensure that they are not locked for use, otherwise the compilation (or rather, the linking phase) will fail with a rather cryptic and not-very-helpful message. This could happen, for example, if we have loaded our target library for testing purposes in MATLAB using ***loadlibrary***, and forgot to ***unloadlibrary*** before re-building.

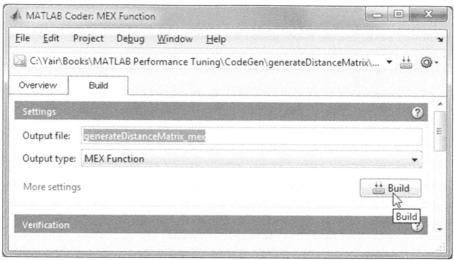

Defining the Coder's build target

Following a successful build, then assuming we have used the default settings, the following files and sub-folders are generated in the project folder:

- *generateDistanceMatrix.prj* — the Coder's project file
- *generateDistanceMatrix_mex.mexw64* — the generated MEX file*
- *codegen\mex\generateDistanceMatrix* — lots of .c and .h source-code files
- *codegen\mex\generateDistanceMatrix\html* — build log report

The generated C-code for the `generateDistanceMatrix` function resides in the new *generateDistanceMatrix.c* file, which has the following contents:

```
/* Include files */
#include "rt_nonfinite.h"
#include "generateDistanceMatrix.h"
#include "generateDistanceMatrix_emxutil.h"

/* function r = generateDistanceMatrix(A,B) */
void generateDistanceMatrix(const emxArray_real_T *A,
                            const emxArray_real_T *B,
                            emxArray_real_T *r)
{
  int32_T A_idx_0;
  int32_T i0;
  int32_T idxA;
  int32_T idxB;
  real_T coordDiff[3];
  real_T y[3];
  real_T b_y;

  /* 'generateDistanceMatrix:2' sizeA = size(A,1); */
  /* # of points (rows) in A */
```

---

* We have compiled on a Windows 64-bit platform, hence the .mexw64 extension.

```
/* 'generateDistanceMatrix:3'  sizeB = size(B,1);  */
/* # of points (rows) in B */
/* 'generateDistanceMatrix:4'  r = zeros(sizeA,sizeB);  */
A_idx_0 = A->size[0];
i0 = r->size[0] * r->size[1];
r->size[0] = A_idx_0;
emxEnsureCapacity((emxArray__common*)r,i0,(int32_T)sizeof(real_T));
A_idx_0 = B->size[0];
i0 = r->size[0] * r->size[1];
r->size[1] = A_idx_0;
emxEnsureCapacity((emxArray__common*)r,i0,(int32_T)sizeof(real_T));
A_idx_0 = A->size[0] * B->size[0];
for (i0 = 0; i0 < A_idx_0; i0++) {
  r->data[i0] = 0.0;
}

/* 'generateDistanceMatrix:5' for idxA = 1 : sizeA */
for (idxA = 0; idxA < A->size[0]; idxA++) {
  /* 'generateDistanceMatrix:6' for idxB = 1 : sizeB */
  for (idxB = 0; idxB < B->size[0]; idxB++) {
    /* 'generateDistanceMatrix:7' coordDiff=A(idxA,:)-B(idxB,:);  */
    for (i0 = 0; i0 < 3; i0++) {
      coordDiff[i0] = A->data[idxA + A->size[0] * i0] -
                      B->data[idxB + B->size[0] * i0];
    }
    /* 3-element array */
    /* 'generateDistanceMatrix:8'
       r(idxA,idxB) = sqrt(sum(coordDiff.^2));  */
    for (A_idx_0 = 0; A_idx_0 < 3; A_idx_0++) {
      y[A_idx_0] = coordDiff[A_idx_0] * coordDiff[A_idx_0];
    }
    b_y = y[0];
    for (A_idx_0 = 0; A_idx_0 < 2; A_idx_0++) {
      b_y += y[A_idx_0 + 1];
    }
    r->data[idxA + r->size[0] * idxB] = muDoubleScalarSqrt(b_y);
  }
}
}
```

As can be seen, the generated C-code is rather straightforward and easy to understand. The original MATLAB code is included as comments interlaced within the C-code to facilitate maintenance.* The Coder generates additional .c and .h files in the build folder to wrap the algorithm in a MEX interface, including numerous runtime checks that can be removed for improved performance (see §8.2.4).

To test our newly generated MEX file in MATLAB, we create a simple test script:

```
function perfTest
    A = rand(1000,3);
    B = rand(3000,3);
```

---

* This requires setting the "MATLAB source code as comments" in the "Code Appearance" tab of the project settings, which I highly recommend doing. Such comments help readability and maintainability without affecting the code performance.

```
% Slow, standard version
tic
sizeA = size(A,1); % # of points (rows) in A
sizeB = size(B,1); % # of points (rows) in B
r = zeros(sizeA,sizeB);
for idxA = 1 : sizeA
    for idxB = 1 : sizeB
        coordDiff = A(idxA,:) - B(idxB,:); % 3-element array
        r(idxA,idxB) = sqrt(sum(coordDiff.^2));
    end
end
non_vectorized_time = toc

% Test MEX (Coder C-code) version
tic, r2 = generateDistanceMatrix_mex(A,B); mex_time=toc
equal = isequal(r,r2)
speedup = non_vectorized_time/mex_time
end
```

In this script function, we have been careful to ensure both the correctness of the computation (isequal(r,r2)) and the performance. Running this script in MATLAB we see that the results are indeed identical, and that the MEX function provides a **33× speedup** (!) compared to the non-vectorized MATLAB version:

```
>> perfTest
non_vectorized_time =
     4.1413
mex_time =
     0.12401
equal =
     1 % == true
speedup =
     33.394
```

Comparing §5.3.8, we note that our MEX version is slower than *bsxfun*-vectorization of this algorithm. Multithreaded vectorization will always give a very good fight to sequential C code. Unfortunately, efficient vectorization is not always possible or easy to implement. In such cases, the Coder can provide speedups of 20× or even 100×, on par with handcrafted C code.[1254] Coder provides the speed of highly tuned C code, while preserving all the benefits of MATLAB's ease of development.

8.2.3 A More Realistic Example: Dijkstra's Shortest-Path Algorithm

In real life, converting an algorithm into C-code is seldom as simple as in the preceding example. Changes are often required in the algorithm to make it Coder-compliant. The situation has improved in recent years, as an increasing subset of MATLAB became supported by the Coder. But some important MATLAB features are still not supported. So, if we wish to use the MATLAB Coder effectively, we need to hand modify our code for it to become compliant.

In many cases we can simply rely on the Coder's extrinsic functions functionality to call the MATLAB engine to interpret those sub-functions that are noncompliant and that we did not yet modify (an example was shown in §8.2.1). However, these engine calls could result in significant overheads that degrade performance. Therefore, modifying the code to become Coder-compliant is advisable whenever possible, since that generates pure C code that is usually faster than the extrinsic function calls.

Our example here relies on Edsger Dijkstra's shortest-path algorithm[1255] from the 1950s, which is now extensively used as a building block in graph-related algorithms. A MATLAB implementation of this algorithm was created by Joseph Kirk in 2008 and posted on the MATLAB File Exchange.[1256] We shall now attempt to convert this algorithm into C code and check for a potential speedup (spoiler alert: it is worth the effort...).*

Kirk's implementation is a function `dijkstra` that received four inputs: a matrix A of logical true/false values indicating which nodes are directly connected; a matrix `xy` of special node positions; a vector `SID` of source node IDs; and a vector `FID` of destination nodes. In our tests below we shall use a single source and a single destination, but Kirk's function supports multiple nodes of each. In the discussion below, I use the May 1, 2009 version of the function, which is the latest at the time of this writing (July 2014).

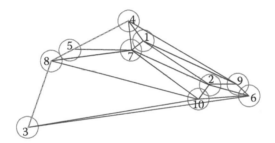

Dijkstra's graph algorithm (shortest path from 3 to 4 is highlighted)

8.2.3.1 Compatibility Check

The first step is to add the %#codegen pragma comment at the end of the line that defines the `dijkstra` function, or to use *coder.screener*, as explained in §8.2.1. This results in the same two types of errors shown there for *notCoderCompliant.m*, namely that cell arrays and nested functions are [currently] not supported by the Coder.

In addition, the Screener tool also reports several unsupported MATLAB function calls, which the Code Analyzer (via %#codegen) does not report for some reason (see figure on page 422).

We shall now fix these incompatibilities one by one.

8.2.3.2 Converting Nested Functions into Sub-Functions

This is actually the simplest adaptation: it requires three adaptation steps:

1. Move the nested functions out of the main function (i.e., following the main function's `end` keyword), to become sub-functions or m-functions.

---

* The source files for this running example can be downloaded from the book's webpage: http://UndocumentedMatlab.com/books/matlab-performance (or: http://bit.ly/1pKuUdM).

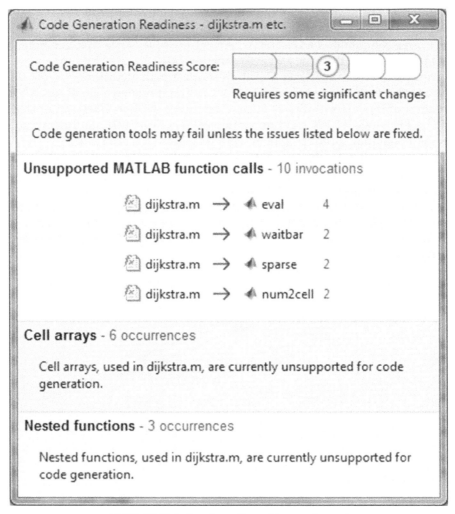

MATLAB Coder's code readiness (Screener) report

2. Ensure that any data declared by the parent function that is used within the nested function will be passed as an input parameter to the new functions.
3. Ensure that any data declared by the parent function that is updated within the nested function will be passed as an output parameter from the new functions.

In our *dijkstra.m* example, we have three nested functions:

The a2e, ve2d functions require no modification, since they already receive all data as input parameters, and modify no parent-function variable. They can safely be moved outside the parent function to become sub-functions or standalone m-functions.

The processInputs function is slightly more difficult, since it uses the m,n variables declared in the parent function: these now need to be passed as input params. The function already passes all its results as output params and does not directly modify parent-function data, so no changes are needed in this respect. The all_positive variable is modified, but not used, since Coder cannot use sparse data (see §8.2.3.4).

Old Code	New Code
132: `[n,nc] = size(AorV);`	`[n,nc] = size(AorV);`
133: `[m,mc] = size(xyCorE);`	`[m,mc] = size(xyCorE);`
134: `[E,cost] =` `processInputs(AorV,xyCorE);`	`[E,cost] =` `processInputs(AorV,xyCorE,`**`n,m,nc,mc`**`);`
233:	**`end`**
234: `% – – – – – – – – – – – – – -`	`% – – – – – – – – – – – – – -`
235: `    function [E,C] =` `processInputs(AorV,xyCorE)`	`    function [E,C] =` `processInputs(AorV,xyCorE,`**`n,m,nc,mc`**`)`
236: `        C = sparse(n,n);`	`        C = sparse(n,n);`
237: `        if n == nc`	`        if n == nc`
238: `            if m == n`	`            if m == n`
	`                ...`
	`                ...`
281: `        end`	`        end`
282:	
283: `    % Convert Adjacency Matrix...`	`    % Convert Adjacency Matrix...`
284: `    function E = a2e(A)`	`    function E = a2e(A)`
285: `        [I,J] = find(A);`	`        [I,J] = find(A);`
286: `        E = [I J];`	`        E = [I J];`
287: `    end`	`    end`
288:	
289: `    % Compute Euclidean Distance...`	`    % Compute Euclidean Distance...`
290: `    function D = ve2d(V,E)`	`    function D = ve2d(V,E)`
291: `        VI = V(E(:,1),:);`	`        VI = V(E(:,1),:);`
292: `        VJ = V(E(:,2),:);`	`        VJ = V(E(:,2),:);`
293: `        D = sqrt(sum((VI-VJ).^2,2));`	`        D = sqrt(sum((VI-VJ).^2,2));`
294: `    end`	`    end`
295: **`end`**	

8.2.3.3 Converting Cell Arrays to Numeric Arrays

Fixing the issue with cell arrays requires an understanding of the program's algorithm. In *dijkstra*'s case, the only cell array is `paths`: `paths{i,j}` stores the numeric IDs of all nodes that connect source node #i to destination node #j. Naturally, elements in the 2D cell array hold numeric arrays of varying lengths: some nodes are directly connected, some nodes are far apart with multiple interconnecting nodes.

We can bypass the need for the cell array by converting it into a 3D numeric array. The maximal path length is N (number of nodes), so the third dimension will have a size of N, and vacant path slots can be represented by the value NaN. For example:

```
paths{1,1} = [1 2]       →    paths(1,1,:) = [1,2,NaN,NaN]
paths{1,2} = [1 3 4]     →    paths(1,2,:) = [1,3,4,NaN]
```

Old Code	New Code
166: `paths = num2cell(nan(L,M));`	`paths = nan(L,M,n);`
175: `path = num2cell(nan(1,n));`	`path = nan(n,n);`
180: `path(I) = {I};`	`path(1,I) = I;`
195: `if isreversed`	`valid = ~isnan(path(:,I));`
196: `    path{J} = [J path{I}];`	`if isreversed`

(continued)

Old Code	New Code

197: else
198:     path{J} = [path{I} J];
199: end
	```path(:,J) = [J; path(valid,I);```
	``` nan(n-length(path(valid))-1,1)];```
	else
	``` path(:,J) = [path(valid,I); J;```
	``` nan(n-length(path(valid))-1,1)];```
	end

221: paths(k,:) = path(FID);	paths(k,:,:) = path(:,FID)';
228: paths = paths';	paths = permute(paths,[2 1 3]);
232: paths = paths{1};	paths = reshape(paths(1,1,:),1,[]);

The Coder Screener output now looks much cleaner, but some work still remains.

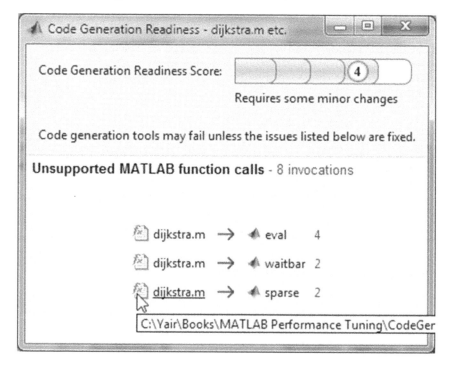

8.2.3.4 *Bypassing Unsupported MATLAB Calls*

As shown by the Screener output at the end of the previous section, the list of MATLAB functions in *dijkstra.m* that are not supported by the Coder includes three:

1. *eval* is used in four places within the code (lines 145, 149, 257, and 277) to display *dijkstra.m*'s help in the Command Window:

```
eval(['help' mfilename]);
```

   We cannot use the *help* function directly, because that function too is not supported. Instead, we elect to simply comment these four code lines. The reduced functionality (in this case, not displaying the help section in case of an error) is an acceptable price to pay for the potential Coder speedup.

2. *waitbar* is used in two locations within the code (lines 169 and 221). As above, we elect to simply comment out the relevant code lines.

In addition, we must also comment out line 223, which closes the ***waitbar*** handle: This line is not reported by %#codegen and the Screener, since *close* is a supported Coder function, but if we do not comment it then a runtime error will occur since the ***waitbar*** handle was never created.

3. ***sparse*** is also used in only two code locations (lines 172 and 236). But here we cannot simply comment out the code since the sparse data is used by the core algorithm. Instead, we convert the sparse data into a full (dense) numeric matrix. This requires additional code changes, to support the new data format:

Old Code	New Code
172: if all_positive, TBL = sparse(1,n); else TBL = NaN(1,n); end	TBL = NaN(1,n);
185: if all_positive, TBL(I) = 0; else TBL(I) = NaN; end	TBL(I) = NaN;
192: if all_positive, empty = ~TAB(J); else empty=isnan(TAB(J)); end	empty = isnan(TAB(J));
206: if all_positive, K=find(TBL); else K=find(~isnan(TBL)); end	K = find(~isnan(TBL));
236: C = sparse(n,n);	C = zeros(n);

The Screener output now looks perfect:

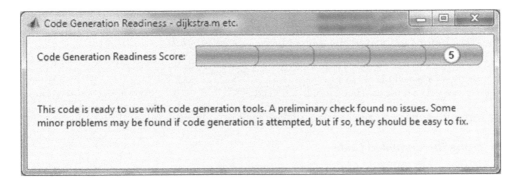

8.2.3.5 Compiling the Code

Before actually translating the code into C/C++ using the Coder, we must ensure that our manual changes have not changed the program's functionality and that we have not introduced any bugs. This important step should not be skipped. There is no point in performance tuning a buggy program. In our case, the result of running *dijkstra* is exactly the same, except for the expected change in the second output argument paths, which has extra NaNs due to our algorithm change in §8.2.3.3:

```
>> [costs,paths] = dijkstra(A,xy,3,4,0)
costs =
       108.717083270117
paths =
     3       8       5       4     NaN     NaN     NaN     NaN     NaN     NaN
```

To translate (compile) the m-code, we could use the Coder GUI as in §8.2.2. But instead, let's automate the build process using the Coder's programmatic interface:

```
>> cfg = coder.config('mex');
>> codegen -config cfg dijkstra
Preconditioning: No class precondition specified for input 'AorV' of
function 'dijkstra'. You must specify the class, size, and complexity
of function inputs. To specify class, use assert(isa(input, 'class_
name')).
```

The reason for this error is that we must specify the expected input type and size of the translated function's input arguments. We can do this interactively in the Coder GUI as shown in §8.2.2, or programmatically using *assert*[1257] or *coder.varsize*.[1258] Note that *coder. varsize* only sets the size — we must use *assert* in any case to set the type. The assertion calls must be placed at the top of our function, as the very first code lines following the function declaration. Combining assertions about the input data type and sizes is allowable and easy:

```
function [costs,paths] = dijkstra(AorV,xyCorE,SID,FID,iswaitbar)
% (main help comment section goes here)
    assert(all(size(AorV)<=10));
    assert(isa(AorV,'double'));
    assert(isa(xyCorE,'double') && size(xyCorE,1)<=10 &&
            size(xyCorE,2)==2);
    assert(isa(SID,'double') && isscalar(SID));
    assert(isa(FID,'double') && isscalar(FID));
    assert(isa(iswaitbar,'double'));
    coder.varsize('iswaitbar', [1,1]); % maxsize = [1,1]
```

codegen now generates a different error, due to our new *reshape(...,[])* on line #241 (which was line #232 before we added the *assert* lines — see §8.2.3.3):

```
Dimension 1 is fixed on the left-hand side but varies on the right
([1 x 1 x :?] ~= [:? x :?]).
Error in ==> dijkstra Line: 241 Column: 5
```

After removing the *reshape* command, *codegen* now completes without an error.

8.2.3.6 Using the Compiled Code

The compilation produces a MEX file called <mfilename>_mex.<mex_extension> in the current folder, in addition to C source code in a subfolder (see §8.2.2). In our case, the file *dijkstra_ mex.mexw64* was generated. We can run the generated MEX file just as the original m-file:

```
[costs,paths      ] = dijkstra_mex(A,xy,3,4,0);
```

We can remove the "_mex" part in the MEX filename, for as-is override of the original m-file. Whenever MATLAB detects a MEX file on the path, it automatically uses that file rather than the corresponding m-file:

```
>> which dijkstra -all
C:\Yair\Books\MATLAB Performance\CodeGen\dijkstra.mexw64
C:\Yair\Books\MATLAB Performance\CodeGen\dijkstra.m % Shadowed
```

The MEX file is about 6× faster than the original m-file:

```
m_time = 0;
mex_time = 0;
```

```
for i = 1 : 1000
    tic; dijkstra(A,xy,SID,FID,0);        m_time = m_time + toc;
    tic; dijkstra_mex(A,xy,SID,FID,0); mex_time = mex_time + toc;
end
speedup = m_time/mex_time
speedup =
            6.3211
```

We should note that we have used the default Coder configuration when compiling the m-code. We shall now see that the speedup can be improved by configuring the Coder in certain ways that take advantage of *a priori* information that we have on the compiled function.

In addition to the MEX usage within MATLAB, we can also compile the algorithm into a standalone dynamic library. This library can be used within MATLAB using *loadlibrary*, or outside MATLAB by dynamic linking to an external application.

Similarly, the code can be compiled into a static.LIB library. As with the dynamic library, the static library can be linked into a MEX file and used in MATLAB, or it can be linked into an external application.

The code can also be compiled into an executable (with optional ZIP packaging), if there is a main m-function that can be used as the executable's entry point.

We can use the generated C/C++-code within any external application that can incorporate and compile such code. If you compile in a Dot-Net application, be careful to mark the code as unmanaged.

Finally, it would be an interesting exercise to compare the generated C code with Sebastian Paris's MEX implementation of Dijkstra's algorithm.[1259]

8.2.4 Configuring the Coder for Maximal Performance

The MATLAB Compiler basically generates an encrypted version of the m-code, which is run by MCR using the same interpreter engine as in MATLAB Desktop, and therefore runs at the same speed as the original m-code. On the other hand, Coder-generated code compiles into native platform binaries and can therefore be faster (or slower) than the original, depending on our C compiler optimizations.

This lesson is worth repeating: Coder-generated C/C++ code indeed often runs faster than the original m-code, but this is NOT guaranteed. When I tested MEX for the data-binning algorithm of §5.3.5, I got a meager 15% speedup using MEX — much slower than the m-code vectorization speedups. Our time and energy would have been better spent on m-code vectorization in that case. In other cases, the code might even run **slower** (!) than the original m-code.[1260]

The speedup potential of Coder-generated code appears to depend on several factors:

- **Vectorization** — Vectorized code is very efficient in MATLAB, often employing internal multithreading (see Chapter 5). In many cases it can out-perform straight-forward (single-threaded) C loops on the same platform.

- **Built-in functions** — If most of our m-code time is spent in built-in functions, especially those calling optimized libraries such as BLAS or IPP, then we will likely not see any speedup. These functions often use highly tuned C-code, which autogenerated code will never be able to match.

- **Runtime checks** — The MATLAB Coder includes many options to increase the code speed at the expense of runtime checks (see below). In addition, we could always manually modify the generated C code to remove runtime checks and Coder-generated wrappers.

- **Extrinsic code** — The Coder automatically converts many MATLB functions into extrinsic calls that use the MATLAB interpreter from within the C code. These calls have a nonnegligible overhead (especially within loops) and should be avoided if possible. In fact, it is faster to leave them in the m-code and not "translate" them into C.

- **Data type** — MATLAB is highly tuned for double precision. If our data is single precision fixed point or integer, the expected speedup will be higher.*

- **Compiler optimization** — If our compiler is not set to optimize the code, the resulting binary might be less efficient than MATLAB's highly tuned JIT. Modern compilers employ a slew of performance optimizations (see Chapter 3); using a simple compiler or disabling compiler optimizations is negligent.

The runtime performance of the generated C/C++ code can be improved by employing *a priori* knowledge that we have on the compiled function and its usage. We can limit the range of possible sizes that the input arguments might have, we can remove runtime checks and we can disable dynamic runtime variable sizing.

The most effective optimization is to disable all checkboxes in the "Speed" tab of the Coder Settings window (click the "More settings" link in the Coder's Build tab).

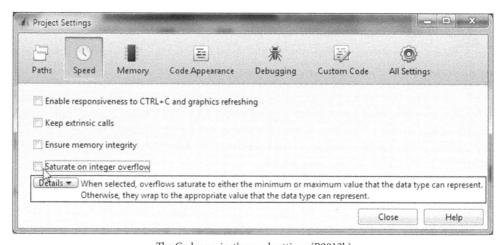

The Coder project's speed settings (R2013b)

These settings can also be set programmatically, using the coder configuration object:

```
cfg = coder.config;
cfg.ResponsivenessChecks = false;
cfg.ExtrinsicCalls = false;
cfg.IntegrityChecks = false;
```

---

* The "MATLAB to C Made Easy" webinar (https://mathworks.com/company/events/webinars/wbnr62736. html or: http://bit.ly/1bbKrxl) shows an example (~45:00 into the video) of accelerating a single-precision image compression algorithm using Coder-generated MEX, resulting in 37× speedup (or 67× using system objects — see §4.7.1).

```
cfg.SaturateOnIntegerOverflow = false;
codegen -config cfg dijkstra
```

All settings are selected (checked) by default. Disabling these checkboxes can have a negative effect on functionality. For example, without the Ctrl-C check, the program cannot be stopped in mid-run; without memory integrity checks the application might crash. But disabling these checks has a large impact on runtime performance: In our Dijkstra example (§8.2.3), the speedup versus original m-code increased from 6 to 16.

Note that some settings have changed between MATLAB R2013a and R2013b; some R2013a settings were removed in R2013b, while R2013b introduced some new ones.

All these settings, except integer-overflow saturation, are only applicable if the Coder output is MEX, not static/dynamic library or standalone executable. When our target is a library or executable, we can disable runtime checks for integer overflow saturation, and also support for nonfinite numeric values (NaN, Inf, -Inf). We also have the ability to specify in the Toolchain tab that the Build Configuration should be "Faster Runs", rather than the default "Faster Builds". This setting has the effect of compiling the C/C++ code with the /O2 optimization flag, or other corresponding flags, depending on our specific compiler. You can manually override these compilation flags (e.g., using a riskier but more effective /O3 flag).

In the Debugging tab, ensure that the "Disable C Compiler optimizations" checkbox is unselected. This is the default value, and it should remain unchecked for best performance. If you need to debug the generated C-code in the external compiler (e.g., by setting breakpoints), then for the debugging phase it might be useful to set this checkbox on, otherwise the code might appear to be misaligned with step-by-step debugging. But once the debugging phase is over, ensure to uncheck the checkbox. Note that this setting is only applicable if the Coder output is MEX, not static/dynamic library or standalone executable.

In the Memory tab, select the "Never" option from the *Dynamic memory allocation* pulldown menu, since dynamic allocation allocates memory in run time, as opposed to using static memory. This simple change, which for some reason is not set as default, can have a surprisingly large effect on performance.

Another checkbox option in the Memory tab, that of fixed versus variable size, is less useful for performance. Variable-sized arrays are required if we use functionality such as the *find* function or logical indexing. These are commonly used in MATLAB code, so disabling this checkbox should only be done in rare cases, typically of small self-contained numerical-computation functions. Such cases are also prime candidates for vectorization and parallelization so in practice disabling the checkbox is rarely useful.

Rather than disabling variable sizing configuration, it is more effective to limit the variable sizes explicitly, using *assert* or *coder.varsize* (see §8.2.3.5). Specifically for the function's input arguments, we can also do this interactively in the Coder project's GUI (see screenshot in §8.2.2). The types of variables, in decreasing efficacy order, are:

- **Fixed-size arrays** are the most efficient — They are allocated on the CPU stack (rather than the heap memory) when small. This is the fastest memory access option (excluding CPU registers). When the data is too large to fit on the CPU stack, then it is statically (i.e., at module load time) allocated on the heap memory with a fixed size.

- **Variable-sized arrays that have an upper size limit** are either statically or dynamically allocated, depending on the configuration for dynamic allocation, and on the variable size. Runtime computations are done (in MEX only) to check the actual size and ensure that it is within the upper size limit. When the size changes for dynamically allocated variables, memory is reallocated in run time, which can degrade performance (see §9.4.1).

- **Variable-sized arrays with unbounded maximal size** are the least efficient. They are always allocated on the heap at run time, using dedicated internal functions. The allocation and reallocation costs in this case are maximal.

We should tell the Coder that a variable has a fixed size, or at least never larger than some size — this will enable the Coder to use a more efficient memory allocation:

```
% Alternative #1: using coder.varsize
coder.varsize('local_data', [100, 1]);    % 100x1 array

% Alternative #2: using assert before the variable is created
assert(N==100);
local_data = zeros(N,1);

% Alternative #2: using assert after the variable is created
local_data = zeros(N,1);
assert(size(local_data,1)==100);
```

We should normally let the Coder figure out which variables are static/dynamic, and only fix the rare cases when the Coder is wrong. It is best for performance to use *coder.varsize* on all dynamic variables (variables that change their size in run time), if we know their maximal size in advance. Luckily, *coder.varsize* enables setting multiple variables at once:

```
coder.varsize('var1', 'var2', 'var3',..., [100, 1]);
```

Data preallocation is very important for m-code (see §9.4), but also for Coder-generated C code. By preallocating large matrices, we avoid the need for dynamic run-time memory reallocations.

Additional small benefits can be gained by disabling synchronization of *global* data in the Memory tab (MEX-target only), and by modifying the Advanced settings in the All Settings tab, such as setting the compiler optimization level, enabling OpenMP pragmas for *parfor* loops (see §7.3.2), and disabling initialization of allocated data to 0 (see §9.4.3). After these tweaks were done, the Dijkstra code ran **18× faster** in MEX than in the original m-code (in other words, 3× faster than by using the default Coder settings).

MATLAB Coder employs some built-in heuristics to increase the performance of the generated code. For example, in deciding when to inline code (see §3.1.2). These can sometimes be controlled via dedicated *coder.** functions:

- *coder.inline*[1261] enables controlling when the Coder will decide to inline a specific function. *coder.inline* only enables a selection between the values of 'never', 'always', and 'default'. Finer-grained control of the default heuristic's inlining thresholds can be achieved via the Advanced settings in the project's All Settings tab. Inlining can be controlled dynamically, based on the data:

```
function y = inline_division(dividend, divisor)
    if isscalar(dividend) && isscalar(divisor)
        % For scalar division, inlining produces smaller code
        % than the function call itself
        coder.inline('always');
```

```
      else
         % Vector division produces a for-loop
         % Prohibit inlining to reduce code size
         coder.inline('never');
      end
      y = dividend/divisor;
   end
```

- *coder.unroll*[1262] enables fine-grain control of loop-unrolling (see §3.1.5).

- *coder.nullcopy*[1263] enables pre-allocating memory for data, without incurring the overhead of initializing all its data elements (cf. §9.4.3). This is a very dangerous option that needs careful attention, since we need to ensure that we do not accidentally access any uninitialized data.

We can sometimes help the Coder generate faster and simpler C code by using simpler m-code. For example, logical indexing (§5.1.3) is highly efficient in m-code, but produces suboptimal C code. Instead, use a simple loop (note the maintenance drawback of supporting separate code segments for optimal m-code and c-code):

```
% Logical indexing - fast in m-code, slow C code
data(data>100) = 100;

% Alternative m-code for faster C code
for idx = 1 : numel(data)
    if data(idx) > 100
        data(idx) = 100;
    end
end
```

Numeric indexing can also be improved: instead of data(idx:idx + 5), use data(idx + (0:5)). This helps the Coder understand that the result has a fixed length of 6. At least in this case there is no maintenance disadvantage as before.

Many of the techniques presented in this book for improving m-code performance will also have an effect of improving the resulting compiled code's performance, although perhaps not at the same level of effectiveness as for the m-code. An example provided in the Coder's User Guide, is forcing in-place data manipulation (§9.5.2) by ensuring that the m-function's inputs and output both use the same variable. Similarly, reduce variable allocation by reusing existing variables (§9.5.3).

Starting in R2012b, we can use *parfor* to parallelize loops.[*][1264] MATLAB Coder automatically translates the *parfor* loop into a corresponding parallel C/C++ code using OpenMP pragmas (§7.3.2). MATLAB's Parallel Computing Toolbox (PCT) is <u>not</u> required for *parfor* translation by the Coder. Note that some compilers (e.g., LCC and Visual Studio Express) do not support OpenMP,[1265] so we would need to use either a commercial compiler, or gcc.[†] Also note that we should not use *parfor* when loop iterations inter-depend

---

[*] See §6.1 and §6.4.2 for additional details and advice on *parfor* usage. Note that *parfor* was only supported for a MEX target in R2012b (MATLAB 8.0) — support for other targets was added in R2013a (MATLAB 8.1).

[†] gcc is not supported for MEX on Windows/Macs, but it can still generate DLLs/LIBs via a Coder toolchain (requires Embedded Coder Toolbox). On Linux, gcc is natively supported (http://mathworks.com/support/compilers/current_release/?sec=glnxa64 or: http://bit.ly/1fk6Mce), so it compiles OpenMP pragmas also for MEX, and does not require a toolchain or Embedded Coder.

on their data or order, or when the loop is fast enough that the parallelization overhead outweighs the parallelization speedup. I have seen programs with *parfor* being sped-up by 200× when converted to OpenMP C-code by the Coder! One possible reason for the huge speedup is that *parfor* uses headless MATLAB worker processes, while OpenMP uses lightweight threads.

Additional performance improvement is achievable using a professional optimizing C compiler. MATLAB users often use the pre-bundled LCC compiler, Microsoft's free SDK,[1266] or the open-source gcc compiler. Unfortunately, their generated binaries are suboptimal compared to professional commercial compilers such as Microsoft's[1267] or Intel's.[1268] Naturally, there is little sense in purchasing expensive compilers if we do not utilize their internal optimizations. While some of these optimizations are turned on by defaults, others are located in custom settings that need to be enabled by the user. Different compilers have different ways to enable these optimizations — refer to your compiler's documentation for additional information.

Finally, performance improvements can sometimes be achieved by modifying the resulting C-code manually, before compiling. Unless you are a professional C programmer, such tweaks are not likely to generate significant speedups. In fact, you are more likely to introduce functional bugs into the program, so proceed with caution! If you do decide to modify autogenerated c-code, it is better to do it as part of the Coder flow using an automated m-code hook file (in the Custom Code tab).

When considering which m-function (or part of a function) to accelerate using hand-crafted MEX or the Coder, we should take the following advice under consideration:

- Only select m-code that has a significant portion of the overall run time. There is no point in optimizing a function that only takes 1% of the total run time.

- It is better to convert an entire loop, rather than the internal (looped) code. This will save the overhead of invoking MEX/DLL multiple times.

- It is best to select code that has minimal extrinsic functions or unsupported MATLAB features. This will reduce both run time and adaptation time.

- Only select non-vectorized code that does not spend much time in built-in library functions.

- If we happen to have handcrafted C code for a specific subpart of the code, we can integrate it directly using the *coder.ceval* function.

When considering whether to deploy a program using the MATLAB Compiler or Coder, note that the Compiler can only generate binaries for the same platform type as that on which it was compiled, and requires that platform's MCR distributable component to run. On the other hand, MATLAB Coder generates standard C code that can be compiled on, and cross-compiled for, any platform using any C compiler, without requiring MCR. This enables easy porting of MATLAB algorithms, even to target platforms that are not currently supported by MATLAB. In fact, the generated code is entirely independent of MATLAB and can be used stand alone, or integrated within any other software, at either the source code or library levels. In addition, the Compiler uses the MCR and so runs at the same speed as the m-code, while the Coder-generated code is usually faster than the original m-code. On the other hand, the Coder only supports a subset of MATLAB, while the Compiler supports all of it. For this reason, the Coder is often used for embedded targets, and the Compiler is used on desktops.

8.3 Porting MATLAB Algorithms to FPGA*

One of factors that limit the overall performance of a system is the number of the basic algebraic operations that the system can perform in any given instant of time. This limit is usually enforced by the hardware resource limitations, such as the number of CPU/GPU cores or the number of available algebraic units. MATLAB does possess some multicore capabilities.† But the limited number of cores and processors, which for most available computers does not exceed 16, puts unbreakable shackles on MATLAB's performance. A natural solution to the hardware resource limitation is therefore the use of some dedicated hardware designed to handle multiple algebraic calculations in parallel. One such alternative is the use of a GPU card;‡ another is the use of FPGA boards, which is the topic of this section.

FPGA stands for *Field-Programmable Gate Array*. It is an integrated circuit which, as its name implies, can be programmed by the user "in the field". Even the cheapest FPGA chips, produced for the last half a decade, have dozens of dedicated algebraic units, known as DSP (*Digital Signal Processing*) cells; more expensive chips have hundreds or thousands of DSP cells. Each DSP cell is capable of performing a basic algebraic operation (for example, sum or multiplication) completely independent of the other cells. Provided the input data flow for each cell is uninterrupted, the number of concurrent calculations is equal to the number of DSP cells.

To implement the seemingly nontrivial link of linking MATLAB, which is in essence interpreter-based highly abstract and hardly a "real-time" programming language/environment, to real-time FPGA hardware, a couple of *Hardware Description Language* (HDL) oriented MathWorks products can be used. Those products are the HDL Coder and HDL Verifier toolboxes.

HDL Coder[1269] is a MathWorks product designed to convert MATLAB functions, Simulink blocks, and StateFlow charts to an efficient, synthesizable, and IEEE-compliant HDL code (VHDL or Verilog), which can be used to program an FPGA chip (and APSoC chip's FPGA fabric) according to the desired functionality.

Originally, HDL Coder was designed for Simulink-based flow, it would support Verilog/VHDL code generation from Simulink models only, and this fact was reflected in its original name "Simulink HDL Coder". In release R2012a (MATLAB 7.14), the toolbox was extended with a dedicated GUI and workflow wizard that supports direct MATLAB to Verilog/VHDL conversion, the dependency on Simulink was removed, and the name was consequently changed to HDL Coder.

HDL Verifier[1270] is a MathWorks product designed to automate HDL design verification process by providing a clear and intuitive interface between the HDL design code and HDL simulators for example Mentor Graphics ModelSim/QuestaSim, Cadence Incisive, as well as FPGA-in-the-Loop (FIL) functionality for numerous FPGA boards from leading FPGA vendors such as Altera and Xilinx.

HDL Verifier started its life as three separate products, EDA Link MQ, EDA Link DS, and EDA Link IN. These products were responsible for integrating Mentor Graphics, Synopsys, and Cadence HDL simulation tools respectively, and lacked FIL capability. In release

---

* Igal Yaroslavski, senior application engineering team leader at Systematics Ltd. (igal@systematics.co.il) authored this section.
† See Chapter 5 for a discussion of MATLAB's multithreading and implicit parallelization.
‡ See §6.2.

R2009b, these products were unified under the name EDA Simulator Link. In R2012a, FIL functionality was added and the product was renamed HDL Verifier.

Note that neither HDL Coder nor HDL Verifier is currently supported for Macs, only Windows and Linux platforms, due to lacking vendor support on the Mac.[1271]

Using the HDL Coder and Verifier it is possible to convert computation-intense MATLAB functions to HDL code that could be downloaded onto an FPGA chip, which would function as de-facto math co-processor to the MATLAB environment. It is important to note some potential limitations of this workflow:

- **Algorithm type** — Not every algorithm will benefit from the FPGA in the Loop acceleration to the same degree. Low-level algorithms can be accelerated with great efficiency; abstract high-level algorithms will demand much adaptation just to run on FPGA hardware, and may eventually not be sped up at all.

- **Code compatibility** — As with the MATLAB Coder (§8.2), HDL Coder only supports a subset of the functions available in core MATLAB and toolboxes. Depending on the nature of the function to be accelerated, the amount of the adaptation needed to make the code HDL Coder-compatible may vary.

- **Feasibility and efficiency** — Since the target is an FPGA chip, that is, an integrated circuit, it imposes certain design and coding guidelines. Failure to comply with those guidelines may result in inefficient hardware implementation or even a complete lack of hardware compatibility.

8.3.1 Algorithm Adaptation

As noted, MATLAB algorithms need to be adapted to run on FPGA. These adaptations aim to make the code behave in a more "hardware-like" manner.

One example of such adaptation is the need for interface serialization. The need for interface serialization is that while generic MATLAB functions may support arguments of virtually infinite dimension and size, the actual FPGA has a limited number of hardware interface pins that must be taken into account.

Another example of a hardware-like adaptation is the need for a fixed point conversion: MATLAB's native data type is a 64-bit double-precision floating point, but most FPGAs are designed for fixed point or integer data-type calculations.*

In addition to these examples, functions that are not embedded- or hardware-oriented by nature, are unsupported for HDL code generation. An example of an unsupported functionality would be graphics output or file access, both of which have no meaning inside a hard-coded integrated circuit such as an FPGA. The list of MATLAB functionality that is supported for HDL code generation is available in HDL Coder's documentation.[1272]

The implementation of the full FGPA-in-the-Loop acceleration cycle is therefore similar to the following flow (see figure on page 435).

This flowchart may seem a bit confusing for a first-time user. Fortunately, the HDL Coder provides a clear and intuitive HDL Workflow Advisor, which is in fact an interactive wizard for MATLAB to HDL conversion and later FPGA implementation.

---

* Compare this to GPUs, which are typically designed for either single or double precision floating point data.

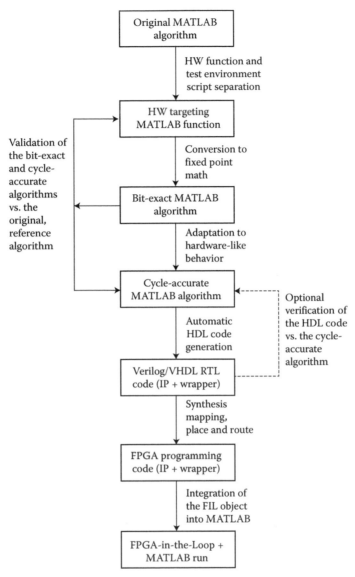

The FPGA-in-the-Loop MATLAB acceleration flow

8.3.2 HDL Workflow

Let us now walk through the full MATLAB to HDL workflow, based on a streaming high-definition (HD) video processing algorithm example.*

The flow begins by reading a video file, streamed pixel by pixel to a 2D filter function, *ML_2D_HDL_filt()*. Here is the relevant subsection within this function:

```
hVid      = VideoReader('Noisy_Ten_Sec_Timer.avi');
vidHeight = uint16(hVid.Height);
```

---

* The source code files for all variants used within §8.3 can be downloaded from this book's webpage: http://UndocumentedMatlab.com/books/matlab-performance (or: http://bit.ly/1pKuUdM).

```
vidWidth   = uint16(hVid.Width);
for hind = 1:vidWidth
    line = hind
    for vind = 1:vidHeight
        in_pix = in_frame(vind, hind);
        out_pix = ML_2D_HDL_filt(in_pix);
        out_frame(vind, hind) = out_pix;
    end
end
```

The function implements the 2D filter using a circular buffer, a section of which passes a 2D convolution with the filter kernel during each time instance. In our case the convolution kernel's dimensions are 5×5, which effectively means that each convolution requires 25 multiplication operations plus an accumulation operation.

```
function conv_pix = twoDconv(pix_mat)
  conv_pix = 0;
  filt_kern = ([3.1943   6.7623   8.6829   6.7623 3.1943;...
                6.7623 14.3158 18.3818 14.3158 6.7623;...
                8.6829 18.3818 23.6027 18.3818 8.6829;...
                6.7623 14.3158 18.3818 14.3158 6.7623;...
                3.1943   6.7623   8.6829   6.7623 3.1943]/256);%normalized
  conv_pix = sum(sum(pix_mat.* filt_kern));
```

Note the "manual" implementation of the convolution function. The convolution is intentionally implemented this way for better control of speed versus footprint simulation considerations, as well as actual HDL implementation efficiency that will play part in later stages of the conversion flow.

We can now start the MATLAB to HDL conversion flow:

```
>> coder
```

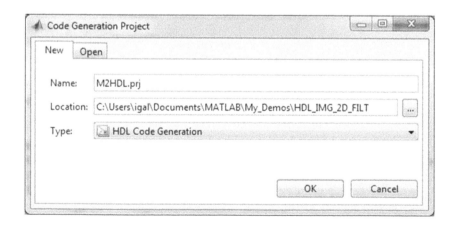

Once the name of the project and the type of the project are defined, an HDL Code Generation dialog opens. To get started you need to provide (1) a MATLAB function to be converted to the HDL language and the data type of its inputs, (2) a test-bench script capable of providing the function with the inputs it requires and analyzing the signals the function outputs. When done, we can start the HDL Workflow Advisor (3).

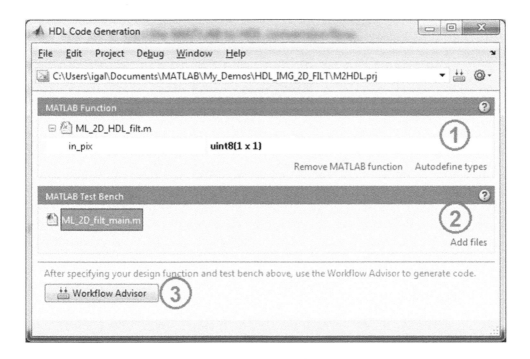

The Workflow Advisor consists of several consecutive steps. Each step is enabled only if the previous step has executed successfully. For example, automated fixed point conversion (2) is enabled only if the input types definition step (1) has passed:

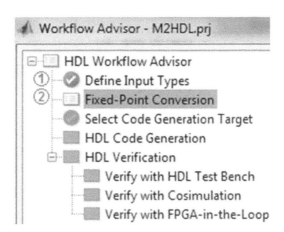

The automated fixed point conversion works by gathering all data values stored in each variable during the entire simulation. These data values are analyzed and the dynamic ranges, as well as spread statistics are concluded (see top figure on page 438).

On the basis of this information, for each variable a fixed point data type is proposed (see 2nd figure on page 438).

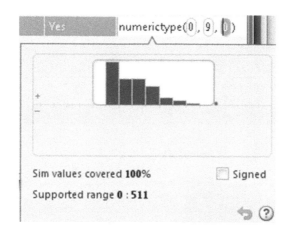

◢ Input					
in_pix	uint8	0	255	Yes	numerictype(0, 9, 0)
◢ Output					
out_pix	uint8	0	253	Yes	numerictype(0, 9, 0)
◢ Local					
conv_pix	uint8	0	253	Yes	numerictype(0, 9, 0)

The proposed data types are controlled by the "default word length" field and the "Advanced" menu, where fixed point aspects for the algorithm's mathematical behavior are configured. An example of a configuration is bit-width of multiplication operations, their precision (e.g., keep MSB) and rounding (e.g., nearest) mode (see figures on page 439).

In floating point to fixed point math conversion we need to note the following. First, the precision versus performance tradeoff* is ever present, that is, the choice of higher precision might result in slower hardware performance as well as higher resource utilization (design footprint), which in marginal cases may not even fit into the designated chip. One of the reasons for this tradeoff is the basic limitation of hardware resources that, for data types and operations outside their native support, demand a cascade of more than one resource for a single operation. An example for such tradeoff would be an attempt to multiply two 32-bit numbers on Xilinx Virtex 5 family FPGA, which has DSP cells with 25×18-bit multipliers. The synthesizer will therefore assign for this task more than one multiplier, which, in turn will result in larger footprint and slower implementation speeds. However, the reduction of fixed point data types and operations to the size compliant with the chip specifications will result in considerably smaller and faster implementation. It is therefore highly recommended to consider your target FPGA specifications prior to fine tuning the fixed point mathematics.

A second consideration is the fixed point calculation speed in MATLAB. Fixed point MATLAB simulation is based on objects, where every fixed point variable is a separate object with properties that include rounding methods (e.g., *floor*, *round*, etc.) algebraic operation properties (e.g., full precision multiply, word length limited multiply with "keep

---

* See §3.9.2.

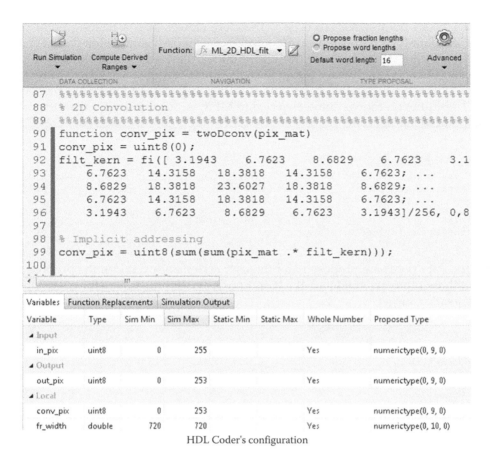

```
87  %%%%%%%%%%%%%%%%%%%%%%%%%%%%%%%%%%%%%%%%%%%%%%%%%%%%%%%%%%%%%%%%%%%%
88  % 2D Convolution
89  %%%%%%%%%%%%%%%%%%%%%%%%%%%%%%%%%%%%%%%%%%%%%%%%%%%%%%%%%%%%%%%%%%%%
90  function conv_pix = twoDconv(pix_mat)
91  conv_pix = uint8(0);
92  filt_kern = fi([ 3.1943      6.7623      8.6829      6.7623      3.1
93       6.7623     14.3158     18.3818     14.3158      6.7623; ...
94       8.6829     18.3818     23.6027     18.3818      8.6829; ...
95       6.7623     14.3158     18.3818     14.3158      6.7623; ...
96       3.1943      6.7623      8.6829      6.7623      3.1943]/256, 0,8
97
98  % Implicit addressing
99  conv_pix = uint8(sum(sum(pix_mat .* filt_kern)));
100
```

Variable	Type	Sim Min	Sim Max	Static Min	Static Max	Whole Number	Proposed Type
⊿ Input							
in_pix	uint8	0	255			Yes	numerictype(0, 9, 0)
⊿ Output							
out_pix	uint8	0	253			Yes	numerictype(0, 9, 0)
⊿ Local							
conv_pix	uint8	0	253			Yes	numerictype(0, 9, 0)
fr_width	double	720	720			Yes	numerictype(0, 10, 0)

HDL Coder's configuration

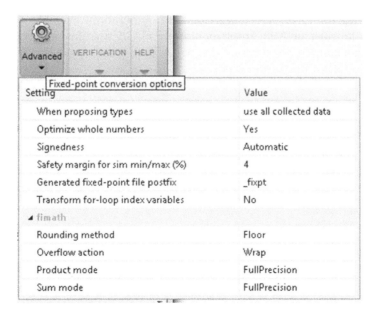

Fixed-point conversion options

Setting	Value
When proposing types	use all collected data
Optimize whole numbers	Yes
Signedness	Automatic
Safety margin for sim min/max (%)	4
Generated fixed-point file postfix	_fixpt
Transform for-loop index variables	No
⊿ fimath	
Rounding method	Floor
Overflow action	Wrap
Product mode	FullPrecision
Sum mode	FullPrecision

MSB" etc.). Fixed point MATLAB code is therefore slower than a floating point one. In addition, the data-type proposition is done based on empirical simulation data. Therefore, to have a good data-type estimation the wizard must collect substantial variable value statistics. Such collection might have forced longer simulation times. Luckily, to mitigate this seemingly considerable drawback, prior to running in the Fixed Point Designer wizard, all MATLAB function code is auto-converted to MEX, which runs faster than the original floating point MATLAB code.

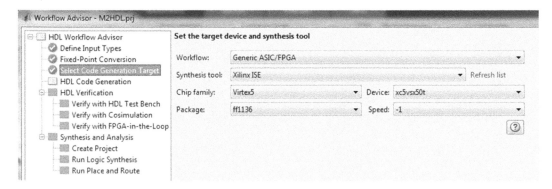

After the data types are proposed and the user accepts or overrides the proposed data type, and these types are validated, the Workflow Advisor proceeds to the next step: HDL Code Generation target:

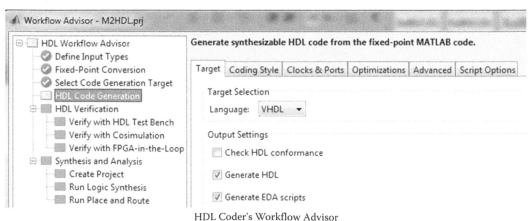

This step defines the hardware target for code generation. Possible options include numerous FPGA families or ASIC implementation, FPGA turn-key solutions and IP core generation. In this example, we choose "Generic ASIC/FPGA". Once the Target is defined we can proceed to HDL Code Generation:

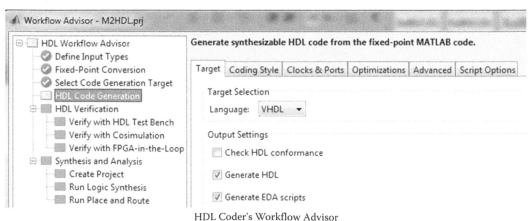

HDL Coder's Workflow Advisor

Upon completion of the Automatic HDL Code Generation step the user gets a fully synthesizable, IEEE standard compliant VHDL and Verilog languages, according to the user's preferences or company coding standards.

The HDL Generation step provides numerous HDL configurations and optimization options, such as automatic pipelining, loop unrolling (speed vs. footprint considerations*), and RAM usage threshold (use FPGA Block RAMs rather than cell registers for data structures larger than defined size). These configuration parameters enable users to fit their algorithms into an available hardware limitation, while ensuring the best possible performance.

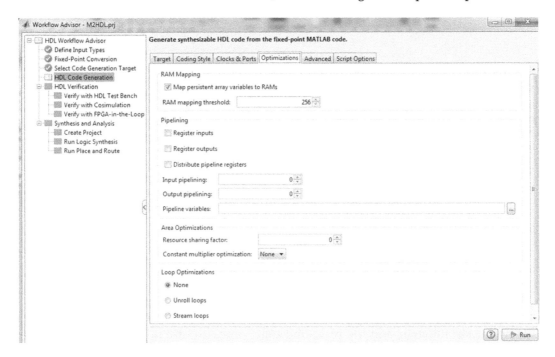

Next are the verification options. HDL Test Bench Verification and the co-simulation-based verification, allow the user to verify the generated (or manually written) HDL code with and without the MATLAB test bench, respectively. These verification variants are very useful for some tasks, but are entirely optional for the specific acceleration task at hand and will therefore be skipped here.

We next focus on FPGA-in-the-Loop project generation. As the name implies, this stage is all about the generation of the FPGA in the loop implementation and harness generation. Consequently, this step has three stages:

1. Generating the communication harness that will wrap the actual algorithm IP and provide connection of the IP interface with the FPGA board's Ethernet port (a few Altera boards also support JTAG communication).

2. The implementation of the FPGA in the loop, that is, the synthesis, the mapping, and the place-and-route processes for algorithm IP and Ethernet wrapper.

3. The generation of the MATLAB class and corresponding function responsible for the MATLAB to board linkage.

---

* See §3.1.5.

Fortunately, all three stages are performed automatically. Once the HDL Workflow Advisor is configured for the IP/MAC address of the FPGA board, we do not need to worry about the details of the underlying process.

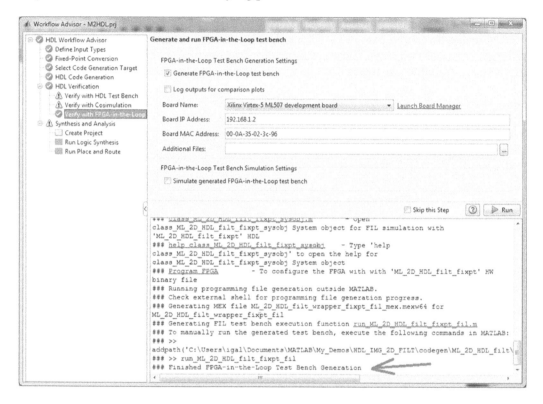

The synthesis, the map, and the place-and-route processes are performed by the dedicated FPGA vendor software that are auto-configured by the HDL Workflow Advisor and are run in batch mode:

Upon completion of the implementation stages the FPGA board is finally ready to run the FPGA in the loop.

At the same time, the MATLAB function responsible for the communication to the FPGA board is generated, using a MATLAB class (MCOS) object:

```
classdef ML_2D_HDL_filt_FixPt_fil < hdlverifier.FILSimulation
```

```
%ML_2D_HDL_filt_FixPt_fil is a filWizard generated class used for
%FPGA-In-the-Loop simulation with the 'ML_2D_HDL_filt_FixPt' DUT.
%ML_2D_HDL_filt_FixPt_fil connects MATLAB with a FPGA and co-simulate
%with it by writing inputs in the FPGA and reading outputs from FPGA.
    ...
    properties (Nontunable)
        DUTName = 'ML_2D_HDL_filt_FixPt';
    end
    methods
        function obj = ML_2D_HDL_filt_FixPt_fil

            %THE FOLLOWING PROTECTED PROPERTIES ARE SPECIFIC TO HW DUT
            %& MUST NOT BE EDITED; RERUN THE FIL WIZARD TO CHANGE THEM
            obj.InputSignals = char('in_pix');
            obj.InputBitWidths = [8];
            obj.OutputSignals = char('ce_out','out_pix');
            obj.OutputBitWidths = [1,8];
            obj.Connection = char('UDP','192.168.0.2',...
                                   '00-0A-35-02-3c-96');
            obj.FPGAVendor = 'Xilinx';
            obj.FPGABoard = 'Xilinx Virtex-5 ML507 development board';
            obj.ScanChainPosition = 5;

            %THE FOLLOWING PUBLIC PROPERTIES ARE RELATED TO SIMULATION
            %AND CAN BE EDITED WITHOUT RERUNING THE FIL WIZARD
            obj.OutputSigned = [false,false];
            obj.OutputDataTypes = char('fixedpoint','fixedpoint');
            obj.OutputFractionLengths = [0,0];
            obj.OutputDownsampling = [1,0];
            obj.OverclockingFactor = 1;
            obj.FPGAProgrammingFile =[pwd '\ML_2D_HDL_filt_FixPt_fil.bit'];
        end
    end
end
```

To invoke this from MATLAB we run the following:

```
MYFIL = ML_2D_HDL_filt_fixpt_fil
programFPGA(MYFIL)
```

And finally, the System Object that sends and receives the data to and from the board.

```
% System Object run, sampled mode
for hind = 1 : vidWidth %use column-major (not row) processing: §9.3.2
    line = hind
    for vind = 1:vidHeight
        in_pix = in_frame(vind, hind);
        [~, out_pix] = step(MYFIL,in_pix); % Sampled
        out_frame(vind, hind) = out_pix;
    end
end
```

In the code above, the data flow to and from the board is done serially, that is, a single sample at the time. While functionally correct, this option is not the most efficient as far as the Ethernet communication goes: the protocol overhead would be too large relative to the payload size. A much better option would be to run in *burst* mode,[1273] so that the sent and received data would be packed into packages.

```
% System Object run, burst mode
in_pix_vect = in_frame(:);
[~, out_pix_vect] = step(MYFIL,in_pix_vect);   % Vectorized
out_frame = uint8(reshape(out_pix_vect,size(in_frame)));
```

The usage of the burst mode may improve the FPGA in the loop run performance by two and more orders of magnitude! A technique that is well worth remembering.

8.3.3 Run-Time Measurements

Now that we have all components generated and ready, it is time to measure the run times for each of the options: behavioral MATLAB (pure m-code), bit-cycle accurate MATLAB, FPGA in the loop sampled mode and FPGA in the loop burst mode. Let us start with the original, detailed behavioral algorithm run:

```
% Detailed behavioral algorithm run time results
OPmode =
Beh_detailed
⇨ Elapsed time is 398.271270 seconds.
```

The processing of a single HD video frame takes nearly 7 min. While this might be sufficient in some cases, running this algorithm for extensive video processing is rather impractical. We therefore attempt the FPGA-in-the-Loop approach. According to the flow-chart, we first need to adapt the m-code to bit and cycle accuracy. For the sake of simplicity, we will skip bit adaptation, but not cycle accurate result:

```
% Bit and Cycle accurate algorithm run time results
OPmode =
Bit & Cycle accurate
⇨ Elapsed time is 1754.496107 seconds.
```

Now it takes ~30 min to finish a single frame! This result is not unexpected, as it is clear that the more elaborated the algorithm is, the slower it runs. This step, however, is essential for the FPGA implementation of the algorithm that we do next:

```
% Sampled mode FPGA in the Loop run time results
OPmode =
sampFIL
⇨ Elapsed time is 1045.781521 seconds.
```

Run time per frame: ~17 min. Not very accelerated… Why did this happen? Well, this time MATLAB's test bench generates the pixels that originate from the original video frame, the pixels are sent one-by-one to the target FPGA board, where the relevant image section is re-constructed and processed by the 2D image filter. The resulting pixel is then transmitted back to the test bench for result analysis. All this communication is done via Ethernet connection and data transition of single pixels. This imposes a large overhead on the relatively small "single 8-bit pixel" payload and clearly impairs our ability to benefit from FPGA's speed. But what if we package the pixels in packets, let's say the size of the original image frame? Then we will get FPGA in the Loop running in "burst mode", where each burst is a full video frame:

```
% Burst mode FPGA in the Loop run time results
OPmode =
framFIL
⇨ Elapsed time is 1.584631 seconds.
```

Burst mode of FPGA in the Loop only takes 1.6 s/frame, roughly 250× faster than the original MATLAB function. Not bad at all...

For additional advice and usage examples, visit the product pages of HDL Coder[1274] and HDL Verifier.[1275] There are also multiple MathWorks webinar videos, for novice to advanced users.[1276] There are also several dozen utilities and support packages for various FPGAs on the MATLAB File Exchange.[1277] Naturally, the MATLAB newsgroup[1278] and forum[1279] also contain hundreds of answered questions.

8.4 Deployed (Compiled) MATLAB Programs

As noted in §8.2, there is a major difference between the MATLAB Compiler and MATLAB Coder. The MATLAB Compiler basically generates an encrypted version of the m-code, which is run by MCR using the same interpreter engine as in MATLAB Desktop, and therefore runs at the same speed as the original m-code and only on the same platform as that of the originating compiler.*

On the other hand, Coder-generated code compiles into standard C source code and native platform binaries. Coder-generated code can therefore be faster (or slower) than the original, and can also be deployed to platforms different from that of the original, and in fact, also platforms (e.g., phones or tablets) that do not support MATLAB or its MCR. However, since the Coder-generated code does not use the MCR, it does not support the entire set of MATLAB functionality.

So, depending on the requirements, it may be possible that the MATLAB Coder may not provide a relevant solution and we would need to use the Compiler instead.

When using MATLAB Compiler, we should address two performance aspects:

- Our program's run-time performance
- MCR load and startup time

As noted, our program is expected to run at the same performance as the original m-code, assuming equal environmental aspects (hardware, running processes, I/O, etc.). Therefore, to improve the run-time program performance we should simply implement one of the performance-tuning techniques mentioned in other chapters of this book — these improvements will be directly reflected in the deployed application's performance. Run time might be different in rare cases, though.†

---

* This was not always the case: until R13 (MATLAB 6.5), the Compiler used to generate C/C++ source code like the modern Coder. The resulting code was often faster than the original, and did not require such a large MCR installation. This was changed in R14 (MATLAB 7.0), when the Compiler started using interpreted (rather than compiled) code. See technical solution 1-T6Y3P: http://mathworks.com/matlabcentral/answers/100607 (or: http://bit.ly/1b0H4TJ).

† For example, a user has discovered that automated graphics refresh does not run as often in the deployed application as in the MATLAB environment, causing a 10× slowdown in the deployed application compared to the original. The issue was solved by adding calls to *drawnow*. See http://stackoverflow.com/questions/16421560 (or: http://bit.ly/169MOKb); also see §10.1.5.

Since the run-time (deployment) environments are in most cases separate from the development environment, we could implement mechanisms in our code that bypass some checks or data processing that may only be relevant for the development version. For example, my *findjobj* utility on the File Exchange[1280] only checks for the availability of a newer version online in non-deployed mode (if ~isdeployed).

The main performance issue that is unique to the MATLAB Compiler is the MCR startup time. When running a compiled program for the first time on the target computer, the program might take 30 or even 60 seconds to load. The vast majority of this time is devoted to MCR load and initialization, and not at all to our program. In fact, even if our program would not have a single line of code, it would not affect the MCR startup time. MCR starts up before our program and loads the entire MATLAB (excluding desktop). Therefore, nothing that we might do to speed-up our program will have any effect on the MCR. Many users, when encountering the deployment startup delays, spend time trying to improve their program's performance, or to display a *msgbox* at the program's start. These attempts are entirely futile, since the program's startup time is usually insignificant compared to MCR startup time, and the *msgbox* will only appear once MCR has been loaded and our program starts to run.

MCR startup is much faster following its first invocation, often taking only a few (5–10 or even less) seconds. This is in fact not due to some smart caching by the MCR, but rather to standard OS mechanisms[1281] (cf. §4.8.1). While still a significant startup time, it is definitely much more acceptable than the initial MCR startup:

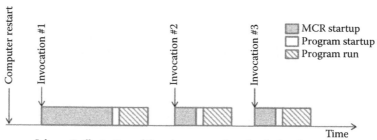

Schematic illustration of the relevant run times for deployed apps

During the prolonged MCR startup, no feedback is provided to the user about the progress of the startup, or even that it has indeed started. Due to this lack of feedback, I have seen many computer-savvy users, who double click the deployment app icon again (and again and again), thinking that perhaps their initial double click(s) did not register. This has a catastrophic effect on the startup time: Now, instead of a single MCR startup contending for CPU time and memory resources, there are several such MCR startup processes — all of them running at the same time!

Even when those users are told not to re-invoke the application in this manner, it becomes an exercise in self-restraint NOT to do this, after 30 or more seconds of total lack of feedback (and apparent inactivity) have passed.

To tackle the MCR's startup duration issue, we can use one of several strategies:

1. We can run the MCR from a local hard disk (preferably an SSD*), rather than from a network drive.† MCR weighs hundreds of MB and network I/O hurts!

---

* See §3.5.3.
† See §11.9.2.

2. We can modify our program so that it remains active for a long time, accepting inputs from time to time. In this manner we do not need to restart the application multiple times, saving at least the subsequent MCR startups (even if not the initial MCR startup). This is the official "best practice".[1282]

3. We can run MCR with reduced functionality (without graphics and Java), by specifying the -R -nojvm parameter to the Compiler (*mcc*) at build time,[1283] or to the *mclInitializeApplication* function at DLL run time.[1284] According to MathWorks, this should save 20%–30% of the MCR start-up time. Similarly, we can use other MATLAB startup parameters (e.g., -R -nojit).*

4. We can instruct the MCR to issue a short textual message to the console when it starts its startup, by specifying the -R -startmsg parameter to the Compiler (*mcc*) at build time.[1285] While these console messages are not highly visible, they do answer the need for immediate feedback, letting the user know that MCR and the deployed application are being started. Naturally, there is no point in using this feature if we also issue the parameter to suppress the MS-DOS console window (-e).[1286]

5. If we have the MATLAB Builder NE (for Dot-Net) toolbox, then we can specify the -S parameter to ensure that the MCR will only use a singleton instance of the MCR for all program instances. While this does not reduce the initial MCR startup time, it does eliminate this time for all subsequent concurrent program instances.[1287] We can do a similar thing with Java.[1288]

6. A blog reader has reported an undocumented trick that has been successful in reducing the MCR startup time in some cases:[1289]

Normally, the MCR and the standalone executable are unpacked upon every startup in the user's temp dir. Apparently, when the MCR_CACHE_ROOT environment variable[1290] is set, these files are only unpacked once and then kept for later reuse. This could significantly speed up the startup time of a compiled application in subsequent invocations.

On Linux:

```
export MCR_CACHE_ROOT=/tmp/mcr_cache_root_$USER #local to host
mkdir -p @MCR_CACHE_ROOT
./myExecutable
```

On Windows:

```
REM set MCR_CACHE_ROOT=%TEMP%
set MCR_CACHE_ROOT="C:\Documents and Settings\Yair\MATLAB Cache\"
myExecutable.exe
```

Setting MCR_CACHE_ROOT is especially important when running the executable from a network (NFS) location, since unpacking onto a network location could be quite slow.† If the executable is run in parallel on different machines (e.g., a computer cluster running a parallel program), then this might even cause lock-outs when different computers try to access the same network location. Using a network folder also has the risk that some users will not have write permission

---

* See §3.1.15 and §4.8.2.
† See §11.9.2.

in that folder, causing the application startup to fail. In all cases, setting MCR_CACHE_ROOT to a local folder (e.g., /tmp or %TEMP%) solves the problem.

If we plan to reuse the extracted files again, then perhaps we should not delete the extracted files but reuse them. Otherwise, simply delete the temporary folder after the executable ends. In the following example, $RANDOM is a bash function that returns a random number:

```
export MCR_CACHE_ROOT=/tmp/mcr$RANDOM
./matlab_executable
rm -rf $MCR_CACHE_ROOT
```

Setting MCR_CACHE_ROOT can also solve other performance bottlenecks in deployed applications, as explained in MathWorks technical solution 1-7RH0IV[1291] and a related article on the *MATLAB for Compbio* blog.[1292]

7. We can set the MCR_CACHE_SIZE environment variable to a higher value than the default (32M). MathWorks say that this can help performance, although details on this feature are a bit fuzzy.[1293]

8. A blog reader has reported that deleting the subfolders of the MATLAB preferences folder (*prefdir*)* before compiling significantly reduces the size of the executable and the time to unpack the executable in run-time. The Compiler includes files in these folders in the target executable, significantly bloating it. These subfolders are not mandatory for normal MATLAB usage, and deleting them may be a fair price to pay for the extra performance of the deployed application.

Here is a short script that can be run in MATLAB just prior to compilation, which deletes these sub-folders:

```
f = dir(prefdir);
for index = 1: numel(f)
    if f(index).isdir && ~any(strcmpi(f(index).name, {'.','..'}))
        subfolderName = fullfile(prefdir, f(index).name);
        fprintf('deleting %s...\n', subfolderName);
        rmdir(subfolderName, 's');
    end
end
```

Alternatively, these folders/files can be deleted from the generated CTF file (as well as the generated executable) directly, without affecting the actual preferences folder on disk. Both the CTF and executable files are simply a zip file having a nonstandard file extension. They can be opened and modified using any zip-file editor, such as WinZip or WinRAR.

9. We can tweak the compiled output to display a message-box before the MCR initialization (*mclInitializeApplication*) is called. While this does not improve the *actual* startup performance, it does improve its *perceived* performance (see §1.8). For a Windows application, the steps are as follows:[1294]

a. Compile the MATLAB program using *mcc* rather than the Deploy Tool:

```
mcc -m -v -C -e <programName>.m
```

---

* On my system: *C:\Users\Yair\AppData\Roaming\MathWorks\MATLAB\R2013b.*

b. Search the Command Window output of the *mcc* command for a command that starts with *"mbuild –O -v ..."*. It will include the name of the generated main (top-level) C file, typically *<programName>_main.c*.

c. Open this file in a text editor (MATLAB's Editor can be used for this).

d. Search for the *WinMain()/main()* function and at its very top (before the call to *mclInitializeApplication*) add the following line:[1295]

```
MessageBox(NULL, "Starting: click OK to continue", "title", MB_OK);
```

e. Save the file and recompile it using the *mbuild* –O -v ... from step b.

f. Ensure that you always deploy the **.ctf* file alongside the **.exe* file.

g. When the application is run, a modal MessageBox will appear immediately (before MCR initialization). Execution (MCR startup) will NOT proceed until the message box is dismissed. To make the box nonmodal, you will need to create a custom dialog window (this requires some C/C++ Windows GUI experience).[1296]

10. We can run our program on a MATLAB Production Server (MPS).[1297] MPS is a MATLAB toolbox that enables running MATLAB program on a webserver that is then accessible to numerous client terminals via a simple browser. The MCR is pre-loaded by the MPS and provides excellent responsivity, since it does not need to be reloaded with each program invocation.[1298] MPS is an expensive toolbox, so using it only for its performance benefit makes little sense. But if you deploy in a large organization that has dozens or more end users, then this option should definitely be explored.

11. We should unselect any unnecessary toolboxes from our compilation project's settings. The extra toolboxes unnecessarily bloat the resulting executable and prolong its load time.[1299]

12. If we have R2014a (MATLAB 8.3) or newer, we could set a splash screen image in the deployment tool's GUI.

13. Finally, we can wrap our deployed application in a splash-screen utility that presents immediate feedback when the application is invoked. This can be done in any programming language (e.g., C++ or Visual Basic).

It is possible that if you use a professional deployment installer,[1300] then it might have a built-in option for displaying a splash screen. If you are using such an installer, it is worth to check its documentation for this.

I have created a ready-made Windows utility called *splash.exe*, which presents a non-modal splash window with user-selected image. The MCR startup and application initialization continues in the background, while the splash window is displayed. Once the wrapper detects that the MATLAB figure window is displayed, or after a 60-second timeout, the splash image is automatically dismissed.[1301]

Here are the steps for using my splash wrapper:

a. Create a splash image, possibly with some overlaid textual instructions. The image format can be BMP, JPG, or GIF.

b. Determine the required splash window size. This is normally set to the image size, but if not then the image is automatically resized.

c. Get the title of the first MATLAB figure displayed by the target application.

d. Determine the target application's command, including any command-line input arguments.

e. Create a desktop shortcut icon with a target such as the following:

```
splash MySplashImage.jpg 600 450 "My GUI Title" 1 "C:\Program
Files\MyApps\MyTargetApplication.exe" param1 param2
```

f. Clicking the shortcut will launch the *splash* wrapper utility, loading the *MySplashImage.jpg* image into a 600×450 window (resizing if necessary), then runs *MyTargetApplication.exe* (which load the MCR, etc.). After 60 s have passed, or when a figure window whose title starts with "My GUI Title" appears, the splash window is automatically dismissed.

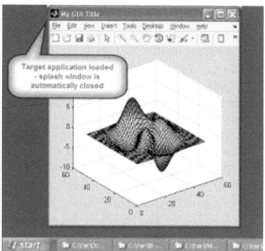

Splash image displaying while the MCR loads (left), until the main GUI appears (right)

8.5 Using External Libraries

8.5.1 Introduction

External libraries are often used in MATLAB for their functionality: interfacing to an external device, running some specific algorithm, and so on. In some cases, however, we can also use external libraries for the performance benefits that they may offer compared to a pure-MATLAB implementation.

We can integrate external libraries within MATLAB in various manners:

- As shared library objects (*.dll* or *.so*) using the ***loadlibrary*** and ***calllib*** functions[1302]
- As Java archives (*.jar*) in MATLAB's static or dynamic Java path*
- As a Dot-Net assembly in MATLAB workspace (Windows only)[1303]
- As a COM/DCOM client (ActiveX) or server (Windows only)[1304]

---

* See §8.5.2.

- As SOAP web services whose invoked functions will be run remotely[1305]
- As a direct MATLAB class replacement (e.g., Marcel Leutenegger's toolbox*)
- As direct replacement of MATLAB's bundled libraries†
- Called from within a MEX function‡

This extensive list highlights MATLAB's flexibility in terms of external integrations. In fact, MATLAB also enables external code to call its functions and code in a variety of manners. For the context of this text, we shall concentrate only on integration aspects that have performance implications.

As a general rule, at least as far as performance is concerned, we should try to avoid libraries that run in a separate process (e.g., COM servers), and certainly on a separate computer (e.g., SOAP web services).

Given a choice, dynamic (shared) libraries are slightly slower than the equivalent statically linked libraries, at the expense of a somewhat-larger compiled object size. Of course, not all libraries provide the option of a linkable static library. In fact, few libraries do. But if you have a choice of MEX-compiling with either a static or dynamic version of the same library, then prefer linking to the static version.

8.5.2 Java

The MATLAB programming environment uses Java for numerous tasks, including networking, data-processing algorithms, and graphical user interface (GUI). Many important Java libraries, typically packaged in ZIP-format files called *Java archive* (*.jar), are pre-bundled in every MATLAB installation and can be directly accessed within our m-code.§[1306]

Java has an extensive set of packages and classes dealing with I/O to external files, processes, systems, and hardware. It has extensive support for networking, from low-level TCP/IP to webpage URLs and XML documents. Java contains classes and libraries that provide wider functionality, more granular control and often improved performance than their MATLAB equivalents.

Section 3.2.6 and §4.9.6 provide a usage example of directly accessing the java.util. Hashtable class within our m-code. As that example has shown, Java classes can provide significant speedups (not to mention, expanded functionality) compared to the MATLAB counterparts. In addition to the Hashtable class, java.util.* includes a wide variety of predefined data structures (specifically Collections and Maps), which can easily be adapted to fit most programming tasks. It is unfortunate that MATLAB does not contain similar predefined collection types, apart from its basic cell, array and struct elements. MATLAB release R2008b (7.7) has added ***containers.Map***, which is a much-scaled-down version of the java.util.Map interface, but is a step in the right direction. Some MATLAB programmers prepared their own implementations of data structures, which can be found on the File Exchange.[1307]

Java collections have many advantages over hand-coded MATLAB equivalents, in addition to the obvious development time saving: Java's classes are extensively performance

---

* See §8.5.4.
† See §4.5.7.
‡ See §8.1.7.3.
§ For additional information, refer to my book *Undocumented Secrets of MATLAB-Java Programming*, ISBN 9781439869031, http://undocumentedmatlab.com/matlab-java-book (or: http://bit.ly/Zqqojt).

tuned, which is especially important when searching large collections. These classes provide a consistent interface, are highly configurable and extendable, enable easy cross-type interoperability and generally provide MATLAB users the full power of Java's collections without needing to program the nuts-and-bolts.

Most of the numerous standard Java classes under `java.*` and `javax.*` are pre-bundled in MATLAB and directly available as simply as in the `Hashtable` example.

MATLAB also enables easy access to external Java functionality, either third-party or user-created. Binary (compiled) *.class* files can be directly used in MATLAB if they are located in the current folder or one of the folders on the Java classpath (either static or dynamic). Package files (*.jar*) can be added to the dynamic classpath using the *javaaddpath* function, or to the static classpath by updating the *classpath.txt* file. Most classes and packages work well with the dynamic classpath. However, classes that use Java events, which need to be propagated to m-code callbacks as MATLAB events, need to use the static classpath. This includes, for example, JDBC's database connectivity library (§3.5.1.4). Once the classes have been successfully added to the Java path, they can be used within MATLAB m-code just as easily as the classes in the pre-bundled Java archives. Placing a Java class or archive on the static Java classpath may also slightly improve performance compared to the dynamic Java classpath.

Java's millions of programmers worldwide far outnumber MATLAB users. There are many active Java forums,[1308] blogs,[1309] articles,[1310] tutorials,[1311] and source code repositories,[1312] with traffic and content far beyond those available in MATLAB's. Therefore, there is a good likelihood that for any programming task, algorithm, or problem in a MATLAB application, somebody somewhere has already posted a Java solution which can relatively easily be integrated into our MATLAB program.

Performance aspects of using Java in MATLAB are also discussed elsewhere in this text:

- Automating GUI activities (§2.1.6.3 and §5.5)
- JDBC database connectivity (§3.5.1.4)
- Regular expression string parsing (§4.2.2)
- Converting Java strings into MATLAB (§4.2.5.3)
- Using Java's date/time functions (§4.4)
- Java startup performance issues (§4.8.5)
- Using Java threads for explicit multithreading (§7.3.3) and precise pause delays (§10.4.3.6)
- Profiling Java memory usage (§9.2.7)
- Performance-related recommendations for Java objects (§9.7.2)
- Java I/O (§11.4, §11.7, and §11.8.1)

8.5.3 NAG: Numerical Algorithms Group

NAG[1313] is a commercial collection of numerical, mathematical, and statistical algorithms, coded in Fortran. It is currently considered the de-facto mathematical library leader, and has been in constant evolution since 1970.[1314] Interfaces are provided to a wide variety of languages and platforms, including MATLAB, C/C++, C#, and Java. The MATLAB interface (*NAG Toolbox for MATLAB*[1315]) includes over 1500 wrapper functions to corresponding Fortran functions. Many of those functions provide direct alternatives to MATLAB's core or toolbox functions.[1316] Downloadable trials are available for MATLAB releases, from as old as R2007a (7.4) to the latest MATLAB release, on all supported MATLAB platforms.[1317]

NAG Toolbox for MATLAB comes as an executable installer that installs some ~400 MB of libraries and thousands of files on disk, fully integrating in MATLAB (MATLAB needs to be restarted following installation).

NAG's multitude of functions have both human-readable names and short names that follow a naming pattern based on semantic relationship and position within the relevant documentation chapters, which is easy to master after a short while. These function names are entirely equivalent, users can use either of them depending on personal preference.

For example, MATLAB's Optimization Toolbox's *fsolve*[1318] has several equivalent NAG functions (*c05qb*,[1319] *c05qc*, etc., or their equivalent names: *nag_roots_sys_func_easy*, *nag_roots_sys_func_expert*, etc.) that can be used under different circumstances. Mike Croucher reviewed these alternatives in 2010.[1320] In 2014, two NAG and several MATLAB releases later, the situation remains largely the same: MATLAB's *fsolve* has significantly improved its performance since then, but NAG still offers a speedup of 3–4x:*[1321]

```
% MATLAB version, using Optimization Toolbox fsolve()
function F = fsolve_obj_MATLAB(x)
    F = zeros(1,3);
    F(1) = exp(-x(1))  + sinh(2*x(2)) + tanh(2*x(3))  - 5.01;
    F(2) = exp(2*x(1)) + sinh(-x(2))  + tanh(2*x(3))  - 5.85;
    F(3) = exp(2*x(1)) + sinh(2*x(2)) + tanh(-x(3))   - 8.88;
end

options = optimset('Display','off'); % no output messages
startX = [0 0 0]; % our starting guess for the solution
X = fsolve(@fsolve_obj_MATLAB, startX, options);
⇨ Elapsed time is 0.005921 seconds.

% NAG version, using the NAG Toolbox c05nb() - 4x faster
function [F,user,iflag] = fsolve_obj_NAG(n,x,user,iflag)
    F = zeros(1,3);
    F(1) = exp(-x(1))  + sinh(2*x(2)) + tanh(2*x(3)) - 5.01;
    F(2) = exp(2*x(1)) + sinh(-x(2))  + tanh(2*x(3)) - 5.85;
    F(3) = exp(2*x(1)) + sinh(2*x(2)) + tanh(-x(3))  - 8.88;
end

startX = [0 0 0]; % our starting guess for the solution
X = c05qb(@fsolve_obj_NAG, startX);
⇨ Elapsed time is 0.001476 seconds.
```

Note that NAG's *c05qb* has a sibling function *c05qc* that solves the same optimization problem, but allows more user control via extra input parameters.[1322] Other NAG functions also have similar variants, for simple ("easy") and advanced ("expert") uses.

NAG results are by default accurate to within O($sqrt(eps)$),[1323] or around 1e-8. This tolerance value is also, as expected, the order of magnitude of the difference between MATLAB's *fsolve* and NAG's *c05nb()*. In most use-cases this is accurate enough, but if you need more accurate results, modify the tolerance in the NAG (via the optional xtol input parameter) or MATLAB (typically via *optimset* parameters) functions. This is yet another example

---

* The NAG Toolbox Mark 22 that Mike reviewed in 2010 had a limitation of not accepting MATLAB function handles and also transposing inputs. These limitations have since been removed.

where accuracy can be traded for speed.* In many cases, NAG is faster yet just as accurate as MATLAB. For example, in another Mike Croucher article,[1324] NAG's *e01bf()* was 9x faster and gave identical results to MATLAB's *interp1* function.†

Objective functions (such as `fsolve_obj_NAG` above) used by NAG routines sometimes require information in addition to the input parameters in the standard interface. MATLAB callbacks enable using the {} mechanism for passing such extra data, but this is currently not supported by NAG. We can use alternatives such as global variables, or data attached to GUI handles.‡ However, my favorite alternative is to use nested functions.§ For example, here we pass `externalData` to the objective function:

```
function X = nestedExample()
    ...
    externalData = magic(3);
    X = c05qb(@fsolve_obj_NAG, startX);
    function [F,user,iflag] = fsolve_obj_NAG(n,x,user,iflag)
        F = someFunctionOf(externalData);
    end
end
```

In some cases, NAG routines have dedicated parameters enabling users to attach extra data. For example, the *c05qb* function added the "user" argument (compared to its *c05nb* predecessor), for exactly this purpose.

NAG's wrapper functions are lightweight MEX functions, which do little more than pass the input args to the corresponding compiled Fortran library functions. For this reason, NAG typically outperforms those MATLAB functions that use m-code. MATLAB is expected to outperform NAG when explicit parallelization is used, since the NAG Toolbox only uses OpenMP multithreading (§7.3.2),[1325] and does not currently (Mark 24) support GPUs.¶ MATLAB may also be as fast as, or faster than NAG when our code spends most of its time in highly vectorized built-in functions such as linear algebra (which is implicitly multithreaded**) when using non-*double* data types,†† or when most of the time is spent evaluating m-code objective functions (such as `fsolve_obj_NAG` in the example above). Such cases need to be checked individually. NAG functions may still be faster if they are more efficient and require fewer steps.

NAG includes extensive detailed documentation, both online,[1326] in PDF format,[1327] and integrated in MATLAB, which I find as excellent. The documentation includes detailed help on all functions, advice on calling conventions and data passing,[1328] explanation of multithreading support,[1329] detailed release compatibility notes,[1330] and runnable MATLAB demos. The demos are explained in much detail in a series of highly readable articles on the NAG website.[1331]

---

* See §3.9.2 and §4.5.5.
† In 2014, the speedup was reduced to 5x and the results were the same to within *eps* (2e-16).
‡ See §9.5.5 and §10.3.7.
§ See §9.5.6.
¶ The core NAG library does support GPUs but this is not currently supported by NAG's MATLAB Toolbox. I assume that NAG will add such support in some future toolbox release.
** See Chapter 5.
†† Some NAG functions expect integers/logicals as inputs, but only double-precision floating-point values are supported, not single-precision. When calling NAG functions that expect integer/logical values, ensure that the data types match; use *int32*/*int64*/*logical* type-casting as needed. See http://nag.co.uk/numeric/MB/calling.asp#types (or: http://bit.ly/1gm8ERZ).

NAG's wrapper functions closely mirror the underlying Fortran functions. This enables rapid deployment of prototyped MATLAB applications that use NAG, to a production C/C++ version (for example). If we use the MATLAB Coder toolbox* to generate C/C++ code, the adaptation to use the direct NAG library functions is easy, since the function names and arguments remain unchanged; we just need to #include <nag.h>,[†] and take care of the issues presented in §8.1.7.3. The *nag.h* header file is not included in NAG's MATLAB toolbox, but you can get it from the NAG website,[1332] along with detailed explanations[1333] and FAQ.[1334] The libraries that need to be included for the link process are all available in the NAG install folder, as dynamic libraries.[‡] Static versions of these libraries are also not provided with the toolbox, but we can get them (with their corresponding header files) by downloading and installing NAG's C/Fortran versions. Mike Croucher, again, provided a detailed article explaining how to call NAG functions from MEX C code.[1335]

A few years ago, Marko Laine developed a generic utility (*genmex*) that created MEX C wrappers for Fortran libraries, with specific examples for integration with NAG. *genmex* is currently inaccessible,[1336] but for good reason: NAG's MATLAB Toolbox provides exactly such a MEX interface.

8.5.4 MATLAB Toolbox

The open-source MATLAB Toolbox[1337] by Marcel Leutenegger of EPFL, contains dozens of commonly used math functions that directly replace the built-in MATLAB functions, so no program modification is required. The toolbox includes function variants for single-precision,[1338] double-precision,[1339] and extended-precision,[1340] handling both real and complex data.[1341] It comes with pre-compiled MEX DLLs for Win32, but the source code can also be compiled for other 32-bit Intel platforms.

MATLAB Toolbox provides a very simple way of overriding MATLAB's built-in math functions, by providing MEX equivalents, which are typically based on performance-optimized assembler versions of these functions.[1342] The toolbox contains a benchmark utility that was used to display very impressive speedups on MATLAB Toolbox's webpage[1343] showing 2×–8× speedup for many functions and a staggering 25× speedup for *mod*, averaged over multiple data array sizes.[§] However, this benchmark was probably run long ago. On a newer R2013a the actual average speedups that I found were 1.3×–2.5× for most functions, and 3×–5× for *rem* and *mod*.

The reduced speedups are not surprising, in light of MathWorks' continuous effort to optimize its built-in function, and the integration of newer BLAS, LAPACK and MKL math libraries over the years. Even so, a 1.2×–2.5× speedup for core math functions should not be easily dismissed. It could significantly speed up numerically heavy computations. It is quite possible that a MEX compilation of the toolbox source code (which is provided) using the latest C and Assembler compilers will improve both the performance and the accuracy. Then again, it is also possible that the speedup is reduced on newer MATLAB releases, something that I have not tested.

---

* See §8.2.
† And possibly a few additional NAG header files, depending on the specific functions that we need to call.
‡ For example, *C:\Program Files\NAG\MB24\mbw6i24ddl\mex.w64\MBW6I24DD.dll* for Mark 24 installed on Win7 64bits.
§ Higher speedups for large data arrays and slowdowns for scalars and small arrays.

Note that MATLAB Toolbox was not updated since 2008, although the developer is still active in his research.[1344] Also note the toolbox only targets 32-bit Intel CPUs. Finally, some functions* produce slightly different results from MATLAB builtins. These are typically within numerically acceptable limits, but not always. It is actually often the MATLAB Toolbox that is more accurate, as the following example shows:

```
>> max(abs(sin(pow2(pi,0:50))))  % MATLAB builtin
ans =
        0.137446488227799
>> max(abs(sin(pow2(pi,0:50))))  % MATLAB Toolbox
ans =
     5.42101086242752e-20
```

To leverage the latest CPU vectorized instructions (SSE, AVX, FMA, and XOP), we can use assembler code integrated in MEX functions, as Marcel Leutenegger has done in his well-documented toolbox. An alternative is to use a C/C++ wrapper, such as Agner Fog's Vector Class library.[1345] This would benefit from Fog's work with vectorized Intel CPU instructions, which seems to be up-to-date, at least as of mid-2014.

8.5.5 MCT: Multi-Precision Computing Toolbox

MCT[1346] is a commercial library that provides high-performance quad- and arbitrary-precision computation functions for MATLAB. MCT leverages object-oriented techniques (MCOS classes and function overloading) to substitute default MATLAB functions with their multi-precision counterparts. As a result, existing MATLAB programs can be run with MCT with minimal code changes, basically only requiring conversion of their input/output data, similarly to PCT's GPU parallelization.† MCT includes analogs to hundreds of numeric functions, from simple arithmetic operations to advanced numerical algorithms (e.g., *fft, eig, svd, quad, ode45, fminsearch*), in addition to functionality not currently implemented in standard MATLAB.

Downloadable trials are available for MATLAB releases as old as R2009b (7.9), on all supported MATLAB platforms.[1347] Installation is straightforward, requiring only addition of the MCT folder to the MATLAB path. A short user guide[1348] and function reference[1349] are provided, and MCT's developer is very responsive for queries.

MCT is similar in concept to MathWorks' Symbolic Math Toolbox (SMT)'s *vpa* function, but reportedly outperforms[1350] and also expands *vpa*'s functionality, for example, to sparse matrix data (which is planned to be improved even further). It should be noted that *vpa* is only part of SMT: much of SMT operates at the *symbolic* level, for achieving analytic solutions. On the other hand, MCT operates at a *numeric* level, simply extending the default data precision from 64 bits to practically infinity.‡

MCT currently supports multithreading only in some of its functions. Development work is currently underway to extend multithreading support. There are also plans to add GPU support, which is currently not available for any MCP function.

For open source but slower and far less-capable alternatives, consider John D'Errico's Variable Precision Integer (VPI) Arithmetic,[1351] or HPF — Big Decimal Arithmetic.[1352] John D'Errico is working on a variant of VPI that is reportedly much faster (VPIJ),[1353] but it is still not available at the time of this writing (October 2014). You might also consider using

---

* Often *acos*, and some cases with *asin* and *atan*. Read the toolbox's *readme* file for detailed explanations.
† See §6.2.
‡ See §4.9.11 for some related advice on using symbolic arithmetic in MATLAB.

Ben Barrowes' Multiple Precision Toolbox (MPT),[1354] but it appears abandoned since 2008. Pavel Holoborodko (MCT's developer) posted a comparison of some of the alternatives.[1355]

Both MCT and MPT use the GNU Multi-Precision Arithmetic (GMP)[1356] and GNU Multi-Precision Floating-point and Rounding (MPFR, based on GMP)[1357] open-source libraries. Users who wish to integrate GMP in their MEX code, can review the example in §7.3.2, or Leonhard Asselborn's PPL4Matlab.[1358] Additional GMP-related C-based open-source libraries are GNU's Multi-Precision Complex (MPC)[1359] and the independent Multi-Precision Integer and Rationals (MPIR).[1360]

8.5.6 Additional Libraries

There are many other third-party libraries available, which are designed exclusively for, or interface with, MATLAB. Some of these libraries can offer significant performance advantages by the fact that in the most part they rely on highly tuned compiled binary code tackling specific domain functions. Some of these libraries that have received note for their performance aspects in MATLAB include:

8.5.6.1 LightSpeed

Lightspeed[1361] is an open-source library coded in C-MEX by Tom Minka that contains optimized versions of primitive functions such as *repmat*, set *intersect*, and **gammaln**. For many years and multiple MATLAB releases, Lightspeed's *repmat* was several times faster than MATLAB's m-coded version. Only in R2013b (MATLAB 8.2), when *repmat* became a built-in function implemented in efficient compiled C code, did it finally outperform Minka's MEX version.*

Lightspeed includes efficient random number generators and evaluation of common probability densities, as well as some other useful utilities such as filename globbing (wildcard filename expansion) and parsing of variable-length argument lists.

An accompanying library called Fastfit,[1362] which relies on Lightbox, implements efficient estimation of various statistical distributions.

8.5.6.2 Spiral

Spiral[1363] is an open-source system for automatically generating a library of mathematical functions based on specific attributes of the target platform's hardware. Spiral generates hardware-optimized source-code (C) functions for several signal transforms (FFT/DFT, cosine, wavelet), convolution/decoding, data sorting, and so on.

Spiral implements a variety of mathematical identities in order to automatically generate memory-based parallelization (SMP[1364]) and processor-based vectorization (SIMD[1365]) intrinsics, if supported by the target platform. OpenMP pragma directives (§7.3.2) are generated for multi-core threading. Algebraic simplifications, loop unrolling and other code optimization techniques are used to further optimize the source code.

It turns out that such platform-targeted code can significantly outperform the generic implementation. For example, Spiral includes an sFFT (§4.5.6) implementation that outperforms the generic baseline implementation by 2×–5×.†

---

* See §5.4.2 for additional details and other *repmat* alternatives.
† Note that the generic baseline sFFT is itself much faster than MATLAB's FFTW implementation — see §4.5.6.

Note that Spiral does not currently include a MATLAB interface: the Spiral-generated C source code needs to be compiled into a library that can be called from m-code (§8.5.1), or integrated within a callable MEX function (§8.1.7.3).

8.5.6.3 Additional Toolboxes

Additional libraries that were referenced with regard to MATLAB performance:

- **MATLAB Tensor Toolbox**,[1366] by Tamara (Tammy) Kolda et al. at Sandia National Labs, for multidimensional data analysis, discussed in §4.5.2.

- **NaN Toolbox**[1367] by Alois Schloegl of the Institute for Science and Technology, Austria — a toolbox for statistics and machine learning classification for data with and without NaN values.* This toolbox employs OpenMP multithreading (see §7.3.2) for improved performance.

- **TSA (Time-Series Analysis) Toolbox**,[1368] also by Alois Schloegl, is devoted to analyzing statistical properties of univariate and multivariate time series. A comparison with MATLAB's built-in *hist* and *roots* functions showed 10×–20× speedups in specific cases.[1369]

- **Wavelab**[1370] — An open-source wavelet analysis library from Stanford University containing hundreds of MEX-optimized functions.

- **LibLinear**[1371] — An open-source library for large linear data classification from National Taiwan University.

- **TOMLAB**[1372] — A commercial library of high-speed optimization functions, including solvers that are not available in MathWorks' Optimization, Global Optimization and Symbolic toolboxes. Many solvers (LP, QP, MIP, MIQP, etc.) and numerical derivatives benefit from multi-core multithreading.

- **FastRBF**[1373] — A commercial library for rapid interpolation and modeling of 2D and 3D data: contours, volumes, and surfaces. FastRBF was last updated in 2006 and my emails to them were not answered, so it may well be abandoned.

- **GNU's Scientific Library (GSL)**[1374] — An open-source repository of mathematical, statistical, optimization, and algorithmic functions. Peter Carbonetto explained how they can be integrated in MEX functions that can be used in MATLAB.[1375]

- Additional libraries were discussed in a dedicated WalkingRandomly post.[1376]

In a related matter, consider using the NIST *DLMF* (US National Institute of Standards and Technology's Digital Library of Mathematical Functions), in either digital[1377] or massive paperback book format.[1378] DLMF lists known mathematical identities and functions, as well as known implementations by various software libraries, including MATLAB, NAG, GSL, and others.[1379] If we use some esoteric function that is not implemented in MATLAB, or may be taking too long to execute, we might try integrating one of the other libraries that implement this function.

---

* See §4.5.1 for a discussion on the performance of NaN values in MATLAB.

9

Memory-Related Techniques

In many programming languages and platforms, memory considerations have an impact on performance. MATLAB is not different in this regard. In fact, since MATLAB's processing often involves large data objects, the memory effect is especially important in MATLAB.

This chapter provides some insight as to how MATLAB arranges its internal memory and how we can use this information to our performance advantage. The chapter will also show how to profile and isolate memory issues.

9.1 Why Memory Affects Performance

No program code can execute without accessing memory. This is necessary for storing variable data, which is the cornerstone of any computation. Each variable used in the program has a specific location in the computer's memory (RAM).* Using a variable in program code means that the program needs to allocate it some memory space, update the memory whenever we update the variable's value or contents, and read the memory location whenever we need to use the variable data in our code.

As a result, nearly every MATLAB command accesses memory, often numerous times. When this happens, the CPU (*central processing unit*) may need to wait until the memory completes its task; the longer the wait, the worse our code performance.

The effect of memory access on performance is exacerbated by the effect of *memory paging* (or *swapping*).[1380] Paging occurs when the computer's running processes require more memory than the physically installed RAM. In such a case, the operating system uses the hard disk as a virtual extension of the physical main memory (*virtual memory*).[1381] Memory is moved between RAM and disk in chunks called *pages*,[1382] which are typically 4KB or higher. The pages are kept on disk and when needed by the CPU they are recalled back to RAM at the expense of another RAM page that is offloaded to disk in its stead (*memory swap*).

During memory swapping, the CPU cannot access the memory and processing effectively stops. Needless to say, disk access is much slower (even for solid state disks (SSDs), but certainly for standard magnetic hard disks) than RAM access: RAM access time is measured in nanoseconds compared to milliseconds for hard disks, a million times slower.† Memory swapping is therefore disastrous to performance.

---

* In some specific cases, variables are stored elsewhere, but this technical detail does not really detract from the overall correctness of the text. There are several other technical simplifications in the text here that do not detract from the overall message. These inaccuracies seem to me to be of minor importance in this context.

† Modern CPUs have effective pipelining of RAM memory so that the memory is usually available at the CPU whenever it needs it, further reducing the access time for RAM. This mechanism is ineffective for swapped memory.

The operating system decides which memory pages swap to disk in an independent manner that is not really in our hands. The result is that when paging starts, we cannot ensure that MATLAB remains in RAM while the other processes get swapped. Moreover, once swapping starts, the CPU slows down, which also affects MATLAB.

When overall memory consumption gets high enough, the operating system constantly swaps memory pages, and the CPU spends more time waiting for memory than processing instructions. This situation is called *thrashing*.[1383] When it happens, the computer appears hung and unresponsive. Performance drops like a stone.

Code performance depends on the execution time of its slowest component. It is not enough to improve the performance of CPU-bound computations; at some point we would reach the limit imposed by memory performance. As MathWorker Stuart McGarrity has noted:[1384]

> *"memory performance has not increased at the same rate as CPU performance, [so] code today is often "memory-bound", its overall performance limited by the time it takes to access memory."*

Fortunately, we can utilize knowledge about how MATLAB interacts with the computer memory in order to improve the performance of the memory throughput, thereby increasing the performance of our entire MATLAB program.

This chapter explains several mechanisms for such improvements. It is hoped that by understanding the reasons underlying these improvements, readers will gain insight that will enable them to manage memory performance also in cases that are not directly addressed by the specific recommendations below.

As users, we are naturally thrilled at the ever-improving cost/performance ratio of both computing power (CPUs and GPUs) and storage (memory and disks). However, we should be aware that performance-tuning techniques related to memory are often trading one for the other. For example, data caching* increases memory usage for reduced CPU cycles (a *space-for-time* trade-off), whereas data compression† does the opposite, increasing CPU usage for reduced memory/storage size (a *time-for-space* trade-off). As technologies evolve, the relative costs of CPU and memory change, turning some of the suggested techniques on their heads.[1385] It may happen that CPU becomes so fast compared to memory cache misses that caching is no longer such a good idea and it would be faster to recompute a value than to look it up in a lookup table. Similarly, SSDs may render compression irrelevant since the extra I/O of the uncompressed data becomes faster than the time needed for compression and decompression. This may also change from one platform to another. In short, there is no alternative to actually testing on your target platform.

Readers who are interested in understanding the technical intricacies of memory's effect on software performance are referred to a few very detailed articles.[1386]

---

* See §3.2.
† See §11.4.

9.2 Profiling Memory Usage

Diagnosing memory-related performance bottlenecks can be quite tricky. Memory issues tend to be transient, depending on the specific conditions at a specific time. Running the same program again under different conditions may yield significantly different performance results.

For example, running the MATLAB program after closing a memory-hogging program such as a multitab browser might yield much faster performance. Conversely, running the program under a highly memory-fragmented MATLAB environment might be much slower than running it on a fresh MATLAB instance. In some cases, we might even run into out-of-memory errors in some of our runs.

Such behavior is often indicative of memory-related issues. The first step in fixing or circumventing these issues is to isolate the problem and identify its reason. MATLAB does not yet provide very sophisticated memory-profiling tools to enable us easy memory problem isolation. But a few available tools can certainly help us in this.

In the following discussion throughout this section, note that the process memory is discussed in its entirety. This includes, but is not limited to the memory space reserved by Java (mostly for GUI-related stuff).[1387] It is my personal opinion that the subset of Java memory affects overall system performance to a lesser degree than non-Java memory. For this reason, Java memory shall not be discussed separately.

9.2.1 Workspace Browser

The first tool is naturally MATLAB's Workspace Browser. This tool, which is part of the MATLAB Desktop, enables us to view the size of all objects in the current MATLAB workspace, both in terms of stored elements and in terms of size. To see this information, we need to modify the Workspace Browser's default displayed fields. This is done by right-clicking the workspace header and selecting the Bytes and Size fields (possibly replacing the Min and Max fields, if you are short in space):

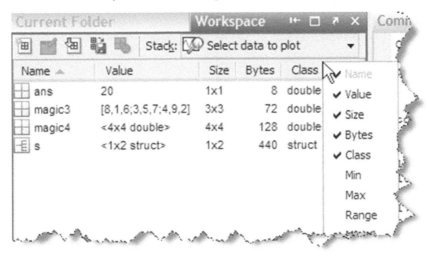

MATLAB's Workspace Browser

Note that the Workspace Browser displays only a small fraction of the actual space actually used by MATLAB: it only shows variables in the current workspace, and only non-global variables, and only the space used by the data (excluding the ~100 bytes overhead used by the mxArray descriptor — §8.1.4.1), and only user-created variables (excluding internal MATLAB variables and reserved memory).

Removing the statistical columns from the Workspace Browser can improve overall MATLAB performance when the workspace contains large data objects (arrays/matrices), since statistics are not updated whenever the workspace is refreshed.[1388]

If you depend on the statistical information for your work, at least limit the maximal array size for which stats are computed using the corresponding preference option in the Workspace preference panel. The default value (500000) is quite large, so you may find it useful to reduce this number.

The Workspace Browser enables us to immediately see such things as oversized variables. Once we identify such variables, we can decide what to do with them: perhaps we can *clear* them? Or possibly reuse the same variable rather than storing copies of it? Maybe change the variable type to a less memory-hogging type? Or perhaps we can reduce the variable size somehow?

The Workspace Browser changes its displayed contents based on the current MATLAB workspace. If we determine that we have a memory-related problem inside some program function, and place a breakpoint in the code within this function, then the Workspace Browser will display the variables in the function scope (workspace) when the code reaches the breakpoint.* Stepping through the code line-by-line will reveal to us how the variables change their size, and this usually provides very useful hints to help us reduce the memory usage.

Moreover, the Workspace Browser does not by default display global variables, although they are available for use within any MATLAB workspace. A global variable is only displayed after we declare it using the *global* directive:†

```
global myGlobalVar
```

9.2.2 *whos* Function

The same information seen in the Workspace Browser can also be accessed directly in the MATLAB Command Window, using MATLAB's built-in *whos* function:

```
>> whos
    Name          Size    Bytes    Class      Attributes
    ans           1x1         8    double
    magic3        3x3        72    double
    magic4        4x4       128    double
    myGlobalVar   0x0         0    double     global
    s             1x2       440    struct
```

The *whos* function can also be used programmatically, as follows:

```
>> vars = whos
vars =
5x1 struct array with fields:
    name
    size
    bytes
    class
    global
    sparse
```

---

* In nested functions, we will also see the variables from the containing functions.
† See §9.5.5 for more information on using global variables to improve performance.

```
        complex
        nesting
        persistent
>> vars(2)
ans =
          name: 'magic3'
          size: [3 3]
         bytes: 72
         class: 'double'
        global: 0
        sparse: 0
       complex: 0
       nesting: [1x1 struct]
    persistent: 0
```

We can use *whos'* programmatic form in our MATLAB program, as follows:

```
varInfo = whos('magic3');
if varInfo.bytes > 1e6
  msgbox('Variable is too large!', 'Memory issue', 'Warn');
  magic3 = []; % clear the variable, release memory
end
```

Such programmatic checks can help us detect (and possibly prevent) memory issues before they actually crash or hang our program.

Like the Workspace Browser, *whos* does not by default display global variables, until we declare them using the *global* directive. However, we can use *whos('global')* to display all global variables, even those that were not declared.

9.2.3 *memory* Function

While the Workspace Browser and *whos* can help us find individual memory-hogging variables, sometimes this information is not enough. For example, some memory-related problems in MATLAB are due to memory fragmentation, and this is not reported in the Workspace Browser or *whos*.

MATLAB does not provide us with a detailed memory map.* However, MATLAB releases since R2008a (7.6) provide the built-in *memory* function, which presents summary information that can be used to detect some memory issues.[1389] Note that as of MATLAB release R2014b (8.4), the *memory* function is only available on Windows platforms:

```
>> memory
Maximum possible array:         1022 MB (1.072e+09 bytes) *
Memory available for all arrays: 2351 MB (2.465e+09 bytes) **
Memory used by MATLAB:           496 MB (5.197e+08 bytes)
Physical Memory (RAM):          3002 MB (3.148e+09 bytes)

*  Limited by contiguous virtual address space available.
** Limited by virtual address space available.

>> [userData,systemData] = memory
```

---

* The *feature('dumpmem')* function provides something quite close, but it is undocumented, unsupported, and only available on Windows platforms. More information on this is presented in §9.2.4.

```
userData =
   MaxPossibleArrayBytes: 1072029696
   MemAvailableAllArrays: 2451574784
           MemUsedMATLAB: 520769536

systdspfemData =
   VirtualAddressSpace: [1x1 struct]
           SystemMemory: [1x1 struct]
         PhysicalMemory: [1x1 struct]

>> systemData.PhysicalMemory
ans =
   Available: 464371712
       Total: 3147759616

>> systemData.SystemMemory
ans =
   Available: 2451574784

>> systemData.VirtualAddressSpace
ans =
   Available: 2465599488
       Total: 3221094400
```

The data returned by *memory* provides us with the following information:

- Maximum possible array (`userData.MaxPossibleArrayBytes`) — This is the size in bytes of the maximal block of contiguous physical memory that MATLAB can use. Since MATLAB stores numeric data arrays in contiguous memory blocks, the largest block limits the size of any new MATLAB array. The number of elements stored in such an array can be computed by dividing this size by 8, which is the size of a standard *double* data element.*

- Memory available for all arrays (`userData.MemAvailableAllArrays` or `systemData.SystemMemory`) — This is the total memory that can be used by MATLAB for all new variables, subject to contiguous memory limitations.

- Memory currently used by MATLAB (`userData.MemUsedMATLAB`)

- Total physical memory (`systemData.PhysicalMemory.Total`) — This is the total physical memory (RAM) installed on the computer. The corresponding `systemData.PhysicalMemory.Available` indicates how many bytes of this total are actually currently free (i.e., unassigned).

- Virtual Address Space data (`systemData.VirtualAddressSpace`) includes information about the amount of available and total virtual memory (i.e., file-swapped memory) that is currently available for use by the MATLAB process. The amount of virtual memory space actually used by the MATLAB process is provided by the difference between these two values.

9.2.4 *feature memstats* and *feature dumpmem*

Detailed information about blocks of contiguous memory in the MATLAB process memory can be seen using the built-in unsupported *feature('memstats')* and *feature('dumpmem')*

---

* The ~100 bytes overhead used by the mxArray descriptor (§8.1.4.1) is negligible for large numeric arrays.

functions, as explained in MathWorks' official *Memory Management Guide* (note that this, again, is a Windows-only feature):[1390]

```
>> largestBlockSize = feature('memstats')
   Physical Memory (RAM):
          In Use:                       2488 MB  (9b84e000)
          Free:                          513 MB  (201a2000)
          Total:                        3001 MB  (bb9f0000)

   Page File (Swap space):
          In Use:                       2532 MB  (9e48f000)
          Free:                         2390 MB  (9562b000)
          Total:                        4922 MB  (133aba000)

   Virtual Memory (Address Space):
          In Use:                        721 MB  (2d17f000)
          Free:                         2350 MB  (92e61000)
          Total:                        3071 MB  (bffe0000)

Largest Contiguous Free Blocks:
          1.  [at 7ffe1000]            1022 MB  (3fe5b000)
          2.  [at 4d549000]             537 MB  (21917000)
          3.  [at 3e1bb000]             243 MB  ( f335000)
          4.  [at 31028000]             201 MB  ( c908000)
          5.  [at 28f8f000]             128 MB  ( 8071000)
          6.  [at 78e4b000]              30 MB  ( 1e55000)
          7.  [at 70539000]              21 MB  ( 1517000)
          8.  [at 6eee4000]              17 MB  ( 111c000)
          9.  [at 7f7f0000]               7 MB  (   7f0000)
         10.  [at ef40000]                7 MB  (   780000)
                                   = == = == = == = == = ==
                                    2216 MB  (8a818000)
largestBlockSize =
             1072017408
```

Where the returned `largestBlockSize` is provided in bytes (in this case, 1022.4MB, at memory address 7ffe1000). This is actually quite good. After several hours working intensively in MATLAB, the memory will be much more fragmented, and we would see many smaller contiguous blocks.

A much more detailed view of loaded modules in MATLAB process memory can be seen using the *feature('dumpmem')* function. It lists the start (base) address of each memory module (either loaded library or data), the amount of memory used by the module, and the amount of memory free in the module. An abridged listing is displayed below (the actual listing is quite long):

```
>> feature('dumpmem')
              Module                    Base       In Use       Free
== = == = == = == = == = == = == = == = ==   = == = ==   = = == = =   = = == = =
<anonymous>                            00010000   00002000   0000e000
<anonymous>                            00020000   00002000   0000e000
<anonymous>                            00030000   00003000   0000d000
<anonymous>                            00040000   00001000   0000f000
...
C:\MATLAB\bin\win32\libmwfl.dll        002c0000+  000d7000   00009000
...
C:\MATLAB\bin\win32\MATLAB.exe         00400000   0002b000   00005000
```

```
<anonymous>                             00430000    00800000*  00000000
C:\MATLAB\bin\win32\icudt44.dll         00c30000+   00e41000*  0000f000
<anonymous>                             01a80000    00001000   0000f000
C:\MATLAB\bin\win32\icuio44.dll         01a90000+   0000f000   00001000
...
C:\MATLAB\bin\win32\mcr.dll             7f270000    00126000*  0008a000
C:\MATLAB\bin\win32\hgbuiltins.dll      7f420000    00091000   0017f000*
C:\MATLAB\bin\win32\m_dispatcher.dll    7f630000    0007a000   00046000
<anonymous>                             7f6f0000    00100000*  007f0000*
<anonymous>                             7ffe0000    00001000   3fe5b000***
<anonymous>                             bfe3c000    00001000   00000000
...
<anonymous>                             bffdf000    00001000   00000000
<anonymous>                             bffe0000    00010000   00000000
                                                    ======     ======
Totals                                              2d1ee000   92e02000
```

```
Largest available memory block is 1072017408 bytes (1022.36 MB) located
at address 7ffe1000
```

Note that all the values are listed in hexadecimal format. We can convert these values to the standard decimal format using the *hex2dec* function:

```
>> value = hex2dec('3fe5b000')          % value in bytes
value =
                1072017408
>> value = hex2dec('3fe5b000')/2^20     % value in MB
value =
            1022.35546875
```

Note that a faster alternative to *hex2dec* is to use *sscanf*.*

A very useful utility for 32-bit Windows systems is *chkmem* by MathWorker Stuart McGarrity.[1391] *chkmem* uses *feature*('memstats') and *feature*('dumpmem'), analyzes MATLAB's memory map from their output, and then suggests specific solutions to several detected issues. *chkmem*'s analysis provides a big added value over using MATLAB's basic functions' raw data.

Another extension of the *feature*('memstats') function was provided by Bruno Luong.[1392] It encapsulates the information provided by *feature*('memstats') in an easy-to-use struct format, saving us the need to parse the displayed text.

9.2.5 *feature mtic/mtoc*

In MATLAB release R2008a (but not on newer releases), we could also use a nifty parameter of the undocumented *feature* function:[1393]

```
>> feature mtic; a=ones(100); feature mtoc
ans =
        TotalAllocated: 84216
            TotalFreed: 2584
     LargestAllocated: 80000
             NumAllocs: 56
              NumFrees: 43
                  Peak: 81640
```

---

* See §4.2.3.

As can easily be seen in this example, allocating 100^2 doubles requires 80000 bytes of allocation, plus some 4 KB others that were allocated (and 2 KB freed) within the function *ones*. Running the same code line again gives a very similar result, but now there are 80000 more bytes freed when the matrix a is overwritten:

```
>> feature mtic; a=ones(100); feature mtoc
ans =
       TotalAllocated: 84120
          TotalFreed: 82760
    LargestAllocated: 80000
           NumAllocs: 54
           NumFrees: 49
                Peak: 81328
```

This is pretty informative and very handy for debugging memory bottlenecks. Unfortunately, starting in R2008b, features mtic and mtoc are no longer supported "under the current memory manager".[1394] Sometime around 2010, the mtic and mtoc features were completely removed. Users of R2008b and newer releases therefore need to use the internal structs returned by the *memory* function, and/or use the Profiler's memory-monitoring feature, which is discussed in the next section.

9.2.6 Profiler's Memory-Monitoring Feature

By far the most useful tool in debugging MATLAB performance bottlenecks is the MATLAB Profiler. Luckily, the Profiler has a very useful undocumented/unsupported feature that enables monitoring memory usage on a line-by-line basis, in addition to profiling timing aspects.[1395] This can be used to detect memory leaks and to improve memory-related performance.

To turn on the Profiler's memory-monitoring feature, run the following in the MATLAB Command Window (this is only necessary once — it will be remembered for all future profiling runs):

```
profile -memory on;
profile('-memory','on'); % an alternative
```

We will now see in the profile report additional information on allocated, freed, and peak memory, as well as options to sort by allocated, freed, and peak memory:

Profile Summary
Generated 03-Apr-2009 00:33:49 using cpu time

Function Name	Calls	Total Time	Self Time*	Allocated Memory	Freed Memory	Self Memory	Peak Memory	Total Time Plot (dark band = self time)
uinspect	1	0.904 s	0.001 s	4410.86 Kb	4372.22 Kb	-0.26 Kb	23.34 Kb	
uinspect>displayObj	1	0.845 s	0.138 s	3563.64 Kb	3546.64 Kb	2.39 Kb	13.01 Kb	
com.jidesoft.grid.TableUtils (Java-method)	8	0.221 s	0.221 s	0.91 Kb	0.54 Kb	0.38 Kb	0.09 Kb	
uinspect>getMethodsPane	1	0.221 s	0.036 s	153.38 Kb	151.03 Kb	-0.22 Kb	6.40 Kb	
uinspect>getPropsPane	1	0.214 s	0.181 s	335.16 Kb	335.07 Kb	-3.69 Kb	9.01 Kb	

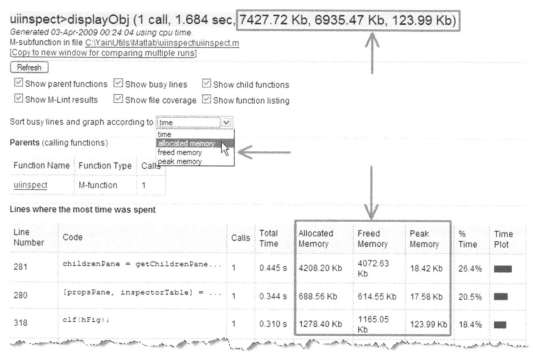

MATLAB Profiler report with additional memory information

Function listing
Color highlight code according to time

time	calls	mem	unjitted	line
				275 function hF
				276
				277 % Pre
0.21	1	432k/376k/14.4k	X	278 metho
0.06	1	400k/345k/20.5k	X	279 [call
0.34	1	689k/615k/17.6k	X	280 [prop
0.44	1	4.11m/3.98m/18.4k	X	281 child
				282
				283 % Pre
				284 impor
< 0.01	1	2.8k/2.26k/556b	X	285 right

MATLAB Profiler report with additional memory information

In the Profiler's memory report, we should take a careful look at any code line or function that has a large memory throughput. This could lead to loss of performance, and possibly also fragmentation that will eventually cause an "out-of-memory" error. Similarly, if we see that a code line or function does not free as much memory as it has allocated, it could indicate a potential memory-leak problem.

In addition to memory leaks, the report could also help diagnose memory-related performance hotspots due to extensive memory allocation and deallocation (see §9.5).

Finally, a large peak memory could be problematic if not enough physical memory is available. In this case, MATLAB might either issue an "out-of-memory" error, or start using the disk space (virtual memory), which would be disastrous for performance although at least the program will not have an out-of-memory error.

The nice thing about the Profiler is that once we have isolated potentially problematic lines or functions, it is very easy to modify our code and see the effect when rerunning the program under the Profiler.

Note that running the Profiler with this memory-monitoring feature has a drawback of degrading the overall run-time performance.[1396] The Profiler needs to track much more information, and this overhead is attributed to the profiled function. Depending on the specific situation, this might either be noticeable or not. It may cause the timing statistics reported by the Profiler to be misleading. For this reason, it is advisable to use memory profiling only in those cases where memory appears to be an issue worth an investigation. In other cases, it is better to run the regular Profiler, without memory monitoring. To turn the memory profiling off, simply run:

```
profile -nomemory
profile('-nomemory'); % an alternative
```

9.2.7 Profiling Java Memory

If our MATLAB application is heavily laden with GUI, it could mean extensive use of Java memory, in addition to memory used by MATLAB variables (both of these share MATLAB's process memory). The reason is that MATLAB's GUI is Java-based.* Profiling Java memory and solving issues related to it are different than for MATLAB variables. Java objects are not by default reported by MATLAB's *whos* function or the Workspace Browser, unless we allocate a specific MATLAB variable to reference them. Second, *whos* and the Workspace Browser always report Java references as having zero bytes, making it difficult to diagnose memory leaks.[1397]

A MathWorks technical article[1398] provided some assistance on using the *JConsole* utility to profile Java memory in MATLAB. We can also use the *JMap* and *JHat* utilities.[1399] These utilities are part of the free Java Development Kit (JDK) that can be downloaded online. Just ensure to use the same Java version used by MATLAB:

```
>> version -java
ans =
Java 1.7.0_11-b21 % i.e., Java 7 update 11
```

In addition to the JDK tools, I find the open-source *JVisualVM* utility[1400] informative and easy to use. We can also use JMP,†[1401] TIJMP,‡[1402] or other third-party tools.[1403]

Within MATLAB, we can use utilities such as *Classmexer* or *ObjectProfiler*[1404] to estimate a particular Java object's size (both shallow and deep referencing). We can also use `java.lang.Runtime.getRuntime()`'s methods (*maxMemory()*, *freeMemory()*, and *totalMemory()*) to monitor overall Java memory:[1405]

```
>> r = java.lang.Runtime.getRuntime
r =
java.lang.Runtime@5fb3b54

>> r.freeMemory % in bytes
```

---

* At least as of MATLAB R2014b (8.4).
† For JVM 1.5, that is, MATLAB R2007a (7.4) and earlier.
‡ For JVM 1.6, that is, MATLAB R2007b (7.5) and later.

```
ans =
   86147768

>> r.totalMemory
ans =
   268304384
>> usedMemory = r.totalMemory - r.freeMemory;
```

Jeff Gullet has suggested[1406] monitoring these values and programmatically activating a synchronous Java garbage-collection when the memory appears too "crowded":

```
if (r.freeMemory/r.totalMemory) < 0.1
   r.gc();
end
```

9.2.8 Using Third-Party Tools

Several third-party MATLAB utilities are available for monitoring memory usage, both the internal memory used for MATLAB variables as well as the Java memory. Note that some of these utilities only work on some platforms (e.g., Windows) and/or MATLAB releases. Consult the utilities' webpages for additional information.

Perhaps the most useful such tool (IMHO) is the *MATLAB Memory Monitor* by Elmar Tarajan, which enables continuous graphical monitoring of system and MATLAB memory, from within MATLAB itself:[1407]

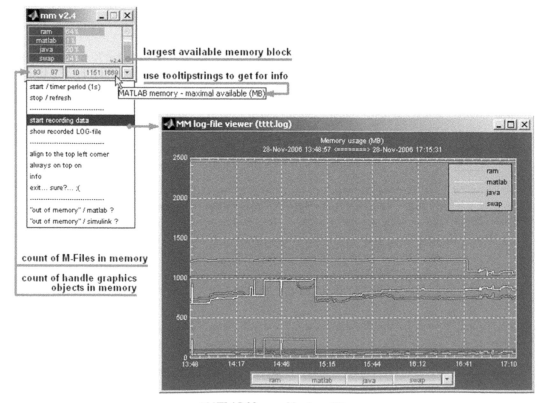

MATLAB Memory Monitor utility

A much simpler monitor by Michael Agostini uses the built-in ***memory*** function to plot ongoing MATLAB process memory in a MATLAB plot.[1408]

Another continuous graphic monitor is the *System Resource Monitor for Windows* utility by MathWorker Xin Zhao.[1409] While this utility is less detailed than Tarajan's monitor for memory, it does present other information that Tarajan's monitor does not (e.g., CPU load):

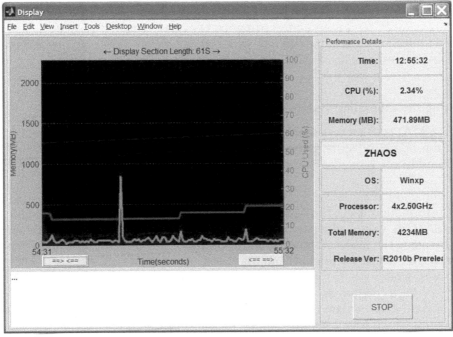

System Resource Monitor for Windows utility

Zhao's monitor uses an underlying *System Information Class* object,[1410] that can be used as a stand-alone GUI-less object in program code:

```
% Record performance data at 0.5 seconds intervals
startRecordPerformance;
... (do whatever)...
stopRecordAndDisplay; % stop collection timer, display usage graph


% Record performance manually, at user-controlled locations
>> perfData = SysInfoData
>> perfData.measure
>> disp(perfData)
SysInfoData with properties:

        TimeArray: 735662.542606273
     UsedCPUArray: 0.771395266056061
  UsedMemoryArray: 561.41796875
     UsedCPUUnits: '%'
  UsedMemoryUnits: 'MB'
```

```
     NumOfCPU: 4
  MachineName: 'THINKPAD-E530'
  TotalMemory: 7004
     CpuSpeed: '2.49GHz'
```

A variation of the built-in *whos* function (§9.2.2) was provided by Matt J on the File Exchange.[1411] This utility displays the memory size data in kilobytes rather than bytes, and displays the actual data dimensions even when they are 4D or higher (the built-in *whos* displays 4D data simply as "4-D double", rather than as 2×3×4×5).

The *totalmem* utility by MATLAB user Eric returns the sum of the sizes for all variables reported by *whos*.[1412] This is an underestimation of overall memory in use (e.g., it does not report any undeclared global variables or variables in parent scopes). However, it could be useful as a quick estimation of comparative sizes during program execution.

9.2.9 Using the Operating System's Tools

Monitoring MATLAB process memory can also be done from outside the MATLAB process. All the operating systems on which MATLAB runs offer basic monitoring capabilities. In addition, there is a wide variety of third-party tools that we can use for this purpose.

For example, on Windows machines, we can use the standard *Task Manager* to monitor basic information about the MATLAB process memory (this is the equivalent of the `MemUsedMATLAB` value, which was discussed above). Additional information can be gleaned by using the brilliant *Process Explorer* utility[1413] and its *VMMap* sibling[1414] — both of which are free utilities available from Microsoft.

Developers who wish to optimize their code to reduce memory paging can use Windows' *perfmon* (performance monitor) application.[1415] Note that paging is highly platform dependent and scenario dependent. Therefore, we should generally not invest time doing this unless we have a serious paging problem, or to discover some generally beneficial improvement (e.g., bunching together computations that affect a data matrix, rather than performing these computations in different places intermixed with other data). For tracking graphic handles and detecting GDI handle leaks, tools such as *Process Explorer* or *GDIView*[1416] can be used.

On Unix/Linux, we can get process snapshots using the *ps* command or continuous information using the *top* command (this may require administrator privileges to run).

Let us illustrate a very simple example, using Windows standard *Task Manager*. Start by launching the Task Manager by right-clicking the taskbar and selecting "Task Manager" or by clicking <Ctrl>-<Shift>-<Esc>, and select the "Performance" tab. Now run the following MATLAB code segment:

```
>> a = ones(10000,10000); % 100M elements = 800MB memory
(wait for MATLAB to be ready again - this could take a while...)
>> a = [];
```

Note that it could take MATLAB quite some time to allocate the 800 MB of memory. While the Task Manager shows the memory spike to be very fast, it takes a long time after the memory was allocated for MATLAB to be ready for additional commands. MATLAB appears to be "hung" in *busy* state for a few moments, until all the aftershocks of the memory allocation have subsided.

In the Windows *Task Manager*, we will be able to see the temporary 800 MB spike in system memory caused by the allocation of 100 million *double* elements in the matrix a. This spike will return to its previous level when we set the variable a to []:

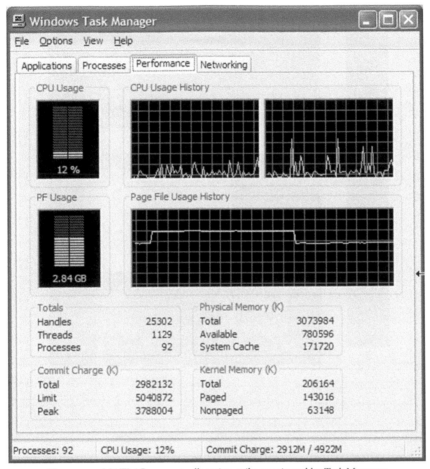

Temporary MATLAB memory allocation spike monitored by Task Manager

The same effect can be seen in the *Process Explorer* monitor. Here, we get much more detailed information, including the actual memory values specifically for the process (as opposed to the entire system), seeing the jump from 394 MB to 1.2 GB, as well as the short CPU spike when MATLAB needed to allocate the 800 MB memory space for the new matrix, and the I/O spikes for virtual memory use, during the allocation and deallocation (see figure in Section 9.2.10).

9.2.10 *format debug*

Using the tools mentioned in the preceding sections only gives us the overall picture, without details about the memory behavior of specific variables. In some very rare and specific cases, we may wish to gain fine-grained insight regarding memory allocation and

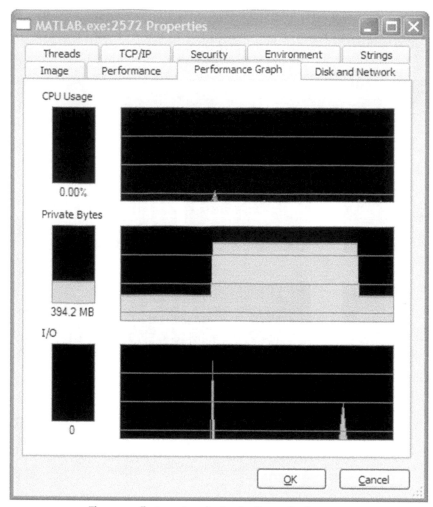

The same effect monitored using the *Process Explorer*

reallocation of specific variables or objects in our code. This can be done using an undocumented and unsupported parameter of the otherwise-fully-documented built-in *format* function:[1417]

```
>> format debug    % or: format('debug')
>> data = ones(2,3)
data =
Structure address = 71bb5a0 ⎤
m = 2                        ⎥
n = 3                        ⎬  Debug printout
pr = 26d031a0                ⎥
pi = 0                       ⎦
    1     1     1    ⎫  Standard printout
    1     1     1    ⎭
```

The debug printout on the command Window includes the following information:*

- Structure address — Pointer to an `mxArray` structure that describes the variable and contains its data.[1418] It is this pointer that is passed to MEX functions.†

- m,n — The number of data rows and columns, respectively.

- pr,pi — Pointers to memory blocks holding the real and imaginary data parts, respectively. $pi = 0$ (NULL) indicates the data has no imaginary component.

Whenever a variable is reallocated in memory, its *pr* and *pi* values will change to point to the new location in main memory. In general, the structure address will not change when a variable is reallocated, only when it is renamed. Using format debug in critical sections of the code, with debug printouts to the Command Window, could provide insight about the memory allocation behavior of the code variables.

To revert to regular (non-debug) mode, simply issue the *format* command without any parameter. You will then need to reissue any other format that you are normally using. For example:

```
>> format
>> format compact
>> format long g
```

Note: *format debug* is undocumented and unsupported. It reports internal information whose contents and interpretations may change or disappear without prior notice in any future MATLAB release. It has existed in its present form since at least 1999 through 2014, but could be removed at any time in the future.

9.3 MATLAB's Memory Storage and Looping Order

9.3.1 Memory Storage of MATLAB Array Data

MATLAB numeric array data‡ is organized in computer memory in *column-major order*.§[1419] This means that the first (leftmost) data dimension (namely, data rows) changes fastest and is stored in contiguous memory locations. Subsequent dimensions change progressively slower and are stored one after the other in serial order:

```
% 2D array:
>> a = magic(3)
a =
     8     1     6
     3     5     7
     4     9     2
>> a(:)'
```

---

* Numeric data only: not cell arrays, structs, an so on; if the data array is too long, use *evalc* to capture and process the information.
† See §8.1.3 and §8.1.4.1.
‡ As opposed to other types of MATLAB data types. See §4.1 for additional details.
§ This memory storage organization is a leftover from MATLAB's early days using Fortran.

```
ans =
    8       3       4       1       5       9       6       7       2


% 3D array:
>> b(:,:,1) = magic(3);
>> b(:,:,2) = -magic(3);
>> b
b(:,:,1) =
    8       1       6
    3       5       7
    4       9       2
b(:,:,2) =
   -8      -1      -6
   -3      -5      -7
   -4      -9      -2
>> b(:)'
ans =
 Columns 1 through 9
    8       3       4       1       5       9       6       7       2
 Columns 10 through 18
   -8      -3      -4      -1      -5      -9      -6      -7      -2


% 4D array:
>> c(:,:,1,1) =  magic(3);
>> c(:,:,1,2) = -magic(3);
>> c(:,:,2,1) =  magic(3)*10;
>> c(:,:,2,2) = -magic(3)*10;
>> c
c(:,:,1,1) =
    8       1       6
    3       5       7
    4       9       2
c(:,:,2,1) =
   80      10      60
   30      50      70
   40      90      20
c(:,:,1,2) =
   -8      -1      -6
   -3      -5      -7
   -4      -9      -2
c(:,:,2,2) =
  -80     -10     -60
  -30     -50     -70
  -40     -90     -20
>> c(:)'
ans =
Columns 1 through 9
    8       3       4       1       5       9       6       7       2
Columns 10 through 18
   80      30      40      10      50      90      60      70      20
Columns 19 through 27
   -8      -3      -4      -1      -5      -9      -6      -7      -2
Columns 28 through 36
  -80     -30     -40     -10     -50     -90     -60     -70     -20
```

9.3.2 Loop Down Columns Rather than Rows

This preceding memory-storage information may seem unimportant. However, it turns out to be critically important for performance: Since data is stored contiguously for the leftmost dimension (data rows), when looping over the data to perform some calculation, it is always better to loop over columns before looping over the rows. In general, in any N-dimensional array, it is always better to loop over a rightward dimension before a leftward dimension. In other words, for array A(w,x,y,z):

```
for zIdx = 1:size(data,4)          % right-most index in outer loop
    for yIdx = 1:size(data,3)
        for xIdx = 1:size(data,2)
            for wIdx = 1:size(data,1)  % left-most index in inner loop
                data = A(wIdx,xIdx,yIdx,zIdx);
                ...
            end
        end
    end
end
```

Looping down columns may seem somewhat unnatural: it is natural to loop according to the order of dimensions, that is, rows before columns. But performance dictates that we reverse the order of the loops and loop over any rightward dimension before looping over any leftward dimension. The effect may be nonexistent or negligible for small data arrays, since the entire array is likely stored in a single cached memory page and so no page swap will occur, regardless of the loop order.

However, if our data is larger than the platform's page size (typically 4 KB), then the effect could be significant.* 4 KB corresponds to 500 *double* data elements, so loops over larger arrays should loop over columns before looping over rows:

```
data = rand(5000,5000); % 25M elements, 200MB = 50 K memory pages

% Row-first loop (natural order, bad for performance)
total = 0;
for rowIdx = 1:size(data,1)
    for colIdx = 1:size(data,2)
        total = total + data(rowIdx,colIdx);
    end
end
⇨ Elapsed time is 0.556565 seconds.

% Column-first loop (less natural, but much better performance)
total = 0;
for colIdx = 1:size(data,2)
    for rowIdx = 1:size(data,1)
        total = total + data(rowIdx,colIdx);
    end
end
⇨ Elapsed time is 0.320137 seconds.
```

---

* A MathWorks newsletter article shows that speedups of up to 5× are possible, depending on the specific configuration.

The reason for the effect is that contiguous memory locations are likely to be stored in the same memory page cached by the CPU, a phenomenon called *spatial locality* or *locality of reference* principle.[1420] By looping over the rows in the internal (rather than the external) loop, we maximize CPU cache efficiency since the CPU does not need to access the main memory (RAM) to get the data; it can process the data directly.[1421]

Even if the memory page is not cached by the CPU, it is still more efficient to access data on the same memory page, thus reducing the number of required memory fetches due to *"cache misses"* and possibly even, heaven forbid, the number of swaps from virtual memory (system disk) due to *"page faults"*. Since this effect is general to all CPUs, the associated performance-tuning technique is not specific to MATLAB.[1422]

A computer's page size can sometimes be modified. In some specific cases, this could benefit performance for highly data-intensive programs. However, changing page size has wide repercussions[1423] and should not be done except in very rare specific cases. In general, careful loop ordering can reduce the need to modify the page size.

Different computers have different page and cache sizes. Owing to these differences, different machines may yield significantly different timing results for the same data. We should be careful not to tailor our code (e.g., number of rows, columns, or dimensions) to any particular profiled machine. Instead, we should adhere to the general suggestion of looping down rightward dimensions first.[1424]

If we cannot avoid looping over the left-most index before the right-most one, consider reshaping the data. A 2D matrix can simply be transposed, while 3+D data needs to use *permute* (be careful with complex data: §4.5.2). We can then switch the looping order accordingly, with the corresponding speedup.[1425] Once done, we should reshape the data back to its original shape. For example, to sum rows of 2D data:

```
a = rand(1000);
tic, for idx=1:100, b=sum(a,2); end, toc
⇨ Elapsed time is 0.137035 seconds.

tic, a=a'; for idx=1:100, c=sum(a,1)'; end, a=a'; toc % 2.3x faster
⇨ Elapsed time is 0.059988 seconds.
```

As an example for multidimensional data, if we have A(x,y,z) and we must loop over dimension x before y, then we could do the following:

```
data = rand(300,300,300); % 27M elements, 206MB = 52 K memory pages

% Naïve approach - slowest: loop over x,y,z
total = 0;
for xIdx = 1:size(data,1)
    for yIdx = 1:size(data,2)
        for zIdx = 1:size(data,3)
                total = total + data(xIdx,yIdx,zIdx);
        end
    end
end
⇨ Elapsed time is 0.697251 seconds.

% Slightly better - switch y,z loops: loop over x,z,y (5% faster)
for xIdx = 1:size(data,1)
    for zIdx = 1:size(data,3)
```

```
        for yIdx = 1:size(data,2)
               total = total + data(xIdx,yIdx,zIdx);
        end
    end
end
⇒ Elapsed time is 0.665435 seconds.

% Much better - permute dimensions: loop over z,y,x (5x faster)
data = permute(data, [3,2,1]); % in-place to prevent memory spikes
for zIdx = 1:size(data,3)
    for yIdx = 1:size(data,2)
        for xIdx = 1:size(data,1)
            total = total + data(xIdx,yIdx,zIdx);
        end
    end
end
data = permute(data, [3,2,1]); % reshape back to original
⇒ Elapsed time is 0.137673 seconds.
```

Loop order effects can also be seen with internal MATLAB code, even in cases where we might expect implicit parallelization to negate such effects. For example, (B'*A')' might be faster than the equivalent A*B., if A has many more rows than columns.[1426]

The effect of cache misses is made worse when the data size is a multiple of the CPU L2 cache's *critical stride*. In such cases, called *super alignment*, previously allocated cache elements are constantly evicted for new elements that share the same cache set and line.[1427] Looping over the rows in the innermost loop reduces the likelihood of super alignment and the rate of cache misses. If loop reversal cannot be done, we should at least ensure that our data size is not an integer multiple of the critical cache stride. It will be much faster to loop over 8191 or 8193 elements than over 8192:

```
function perfTest(N)
    data = rand(N,N);
    total = 0;
    for rowIdx = 1:size(data,1)
        for colIdx = 1:size(data,2)
            total = total + data(rowIdx,colIdx);
        end
    end
end

perfTest(8191)
⇒ Elapsed time is 3.581139 seconds.

perfTest(8192) % critical-stride multiple - much slower!
⇒ Elapsed time is 7.551756 seconds.

perfTest(8193)
⇒ Elapsed time is 3.650956 seconds.
```

Refer to §3.1.11 for a related technique of looping over the shorter dimension of multidimensional data.

9.3.3 Effect of Subindexing

When looping over data arrays, it is beneficial to reduce usage of subindexing wherever possible. It is sometimes even beneficial to loop over the entire data array dimension, rather than a subset of it, as funny as this may sound.[1428]

Consider the following example:

```
data  = rand(1000,1000);
data2 = rand(1000,1000);


% Test #1: Multiply entire data arrays, no indexing - fastest
data2 = data2 .* data;
⇨ Elapsed time is 0.004246 seconds.


% Test #2: Multiply entire data array, full indexing - 6x slower!
data2(:,:) = data2(:,:) .* data(:,:);
⇨ Elapsed time is 0.025267 seconds.


% Test #3: Multiply only part of the data arrays - even slower!
data2(:,1:999) = data2(:,1:999) .* data(:,1:999);
⇨ Elapsed time is 0.030754 seconds.
```

The reason for the slowness of the indexed operations (tests #2, #3) is that indexing is relatively expensive and may also hinder in-place data manipulation.* A possible explanation is that it creates a copy of the relevant data, rather than acting directly on the original data. If possible, avoid indexing altogether (as in test #1 or §3.1.1.2).

If this is not possible, then try to use the entire index range (as in test #2) rather than a subrange (test #3). Strange as it may seem, apparently this enables MATLAB's JIT to optimize the operation and run faster, possibly avoiding data copy.

Oddly enough, in such cases, another fast alternative is to loop over the necessary indices, rather than use direct subindexing in a vectorized manner:†

```
% Test #4: Multiply entire data arrays using loops: just slightly slow
for colIdx = 1 : 1000
   data2(:,colIdx) = data2(:,colIdx) .* data(:,colIdx);
end
⇨ Elapsed time is 0.005100 seconds.


% Test #5: Multiply only part of the data arrays - not much slower
for colIdx = 1 : 999
   data2(:,colIdx) = data2(:,colIdx) .* data(:,colIdx);
end
⇨ Elapsed time is 0.005474 seconds.
```

As a variation of this idea, it is faster to use *size(data,1)* rather than *length(data(:,1))*.

When we need to remove a subblock of data, it appears that keeping the remaining part (a=a(1:N)) is faster than deleting the unneeded part (a(N+1:end) = []).[1429]

---

* See §9.5.2 below.
† This is an example where looping is sometimes better than vectorized code. See §3.1 and Chapter 5 for additional details.

9.4 Array Memory Allocation

9.4.1 Dynamic Array Growth

Unlike some other programming languages (such as C, C++, C#, or Java) that use *static typing*,[1430] MATLAB uses *dynamic typing*. This means that it is natural and easy to modify array size dynamically during MATLAB program execution. For example:

```
fibonacci = [0, 1];
for idx = 3 : 100
    fibonacci(idx) = fibonacci(idx-1) + fibonacci(idx-2);
end
```

While this may be simple to program, it is not wise with regard to performance. The reason is that whenever an array is resized (in most cases, enlarged), MATLAB allocates an entirely new contiguous block of memory for the array, copying the old values from the previous block to the new, then releasing the old block for potential reuse.[1431] This operation takes time to execute. In some cases, this reallocation might require accessing virtual memory and page swaps, which are even slower. If the operation is done in a loop, then performance could quickly drop off a cliff.

The cost of such naïve array growth is theoretically quadratic. This means that multiplying the number of elements by N multiplies the execution time by about N^2. The reason for this is that MATLAB needs to reallocate and copy N times more than before, and each time takes N times longer due to the larger data size (the average block size multiplies by N):

```
% This is run on MATLAB 7.1 (R14 SP3):
f=[0,1]; for idx=3:10000, f(idx)=f(idx-1)+f(idx-2); end
⇒ Elapsed time is 0.149173 seconds.%baseline loop size & exec time

f=[0,1]; for idx=3:20000, f(idx)=f(idx-1)+f(idx-2); end
⇒ Elapsed time is 0.586088 seconds.%x2 loop size, 4x execution time

f=[0,1]; for idx=3:40000, f(idx)=f(idx-1)+f(idx-2); end
⇒ Elapsed time is 2.090217 seconds.%x4 loop size, 14x execution time
```

As can be seen from this MATLAB 7.1 (R14 SP3) run, as we multiply the loop size by N = 2, the execution time more-or-less quadruples ($N^2 = 4$), as expected from theory.

A very interesting discussion of this phenomenon and various solutions can be found in a newsgroup thread from 2005.[1432] Four main solutions were presented:

- Preallocate the arrays — This is the main topic of this section and shall be expanded below. It is always the fastest approach. The basic idea is to create a data array in its final expected size before actually starting the processing loop, thereby preventing the need for any reallocations within the loop. This solution is useful when the final size is known in advance; otherwise, we can use one of the other three solutions:

- Over-preallocate the array to some very large size. While wasteful in memory, this is normally very fast. Within the processing loop, keep track of the actual size of the intended array. Once the loop ends, remove the excess elements from the array, freeing memory.* Unless we run into memory barrier issues, the performance is nearly on par with that of precisely preallocating the array.

---

* See §9.4.3 for additional details and a sample implementation.

- Dynamically grow the array by a certain percentage factor each time. The idea is that when the array first needs to grow by a single element, we would in fact grow it by a larger chunk (say 50% of the current array size),* so that it would take the program some time before it needs to reallocate memory again. This has a theoretical cost of O(n·log(n)), which is nearly linear in n for most practical purposes. It is similar to preallocation in the sense that we are preparing a chunk of memory for future array use in advance. One might say that this is on-the-fly preallocation. Power-user Amro posted a sample implementation on Stack Exchange a few years ago.[1433] The optimal growth-step factor varies based on platform and usage, but most implementations use 50%–100%.[†1434]

- Use cell arrays to store and grow the data, then use ***cell2mat*** or *cell2vec*[1435] to convert the resulting cell array to a regular numeric array. Cell elements are implemented as references to distinct memory blocks,[1436] so concatenating to a cell array merely concatenates a reference. When a cell array is reallocated, only its internal references (not the referenced data) are moved.[‡] Like factor growth, using cell arrays is faster than quadratic behavior. Different situations may favor using either the cell arrays method or the factor growth mechanism.

The effect of preallocation shall be explored below; the effects of the other two solutions (factor growth and using cell arrays) is negligible for small data sizes and/or loop iterations (i.e., number of memory reallocations), but could be dramatic for large data arrays and/or a large number of memory reallocations. The difference could well mean the difference between a usable and an unusable ("hung") program.

John D'Errico has posted a well-researched utility called *growdata* that optimizes dynamic array growth for maximal performance.[1437] It is based in part on ideas mentioned in the newsgroup thread above, where *growdata* is discussed in detail.

As an interesting side note, John D'Errico has also posted[1438] an extremely fast implementation of the Fibonacci function.[§] The source code may seem complex, but the resulting performance gain is well worth the extra complexity. I believe that readers who will read this utility's source code and understand its underlying logic will gain insight into several performance tricks that could be very useful in general.

9.4.2 Effects of Incremental JIT Improvements[1439]

The introduction of JIT acceleration in MATLAB 6.5 (R13) caused a dramatic boost in performance.[¶] Over the years, MathWorks has consistently improved the efficacy of its computational engine and the JIT Accelerator in particular. JIT was consistently improved, giving a small improvement with each new MATLAB release. In MATLAB 7.11 (R2010b), the short snippet above executed at 0.098, 0.384, and 1.483 s, respectively (approximately 30% improvement compared to MATLAB 7.1 R14SP3). This is still

---

* For example, by using the ***repmat()*** function, or by concatenating a specified number of ***zeros()***, or by setting some way-forward index to 0 — see §9.4.3 for a comparison of these alternatives.
† There is a theoretical debate on the optimal value, without consensus. Different programming systems use different factors in practice.
‡ Note that this relies on the internal implementation of cell arrays in MATLAB, and may possibly change in some future release.
§ Fibonacci sequence generation is also discussed in §3.1.5, §3.2.4.2, §3.10.1.2, and §5.7.7.
¶ There is a distinction between the Accelerator and JIT: they are complementary speedup technologies used internally for interpreting m-code, and both have improved over the years. The distinction between the technologies has no practical impact on the discussion in the text, so for the sake of brevity, I refer to both technologies as "JIT".

quadratic in nature, and so in these releases, using the three solutions mentioned above could prove very beneficial.

In MATLAB 7.12 (R2011a), a major improvement was done in the MATLAB engine,* although the technical details of this improvement were never disclosed.[1440] The execution run-time of memory allocations improved significantly, and in addition have become linear in nature. This means that multiplying the array size by N only degrades performance by N, not N^2 — a very impressive achievement:

```
% This is run on MATLAB 7.12 (R2011a):
f=[0,1]; for idx=3:10000, f(idx)=f(idx-1)+f(idx-2); end
⇨ Elapsed time is 0.004924 seconds.%baseline loop size & exec time

f=[0,1]; for idx=3:20000, f(idx)=f(idx-1)+f(idx-2); end
⇨ Elapsed time is 0.009971 seconds.%x2 loop size, 2x execution time

f=[0,1]; for idx=3:40000, f(idx)=f(idx-1)+f(idx-2); end
⇨ Elapsed time is 0.019954 seconds.%x4 loop size, 4x execution time
```

In fact, it turns out that using either the cell arrays method or the factor growth mechanism is much slower in R2011a than using the naïve dynamic growth![1441]

This teaches us a very important lesson: It is not wise to program against a specific implementation of the engine, at least not in the long run. While this may yield performance benefits on some MATLAB releases, the situation may well be reversed on some future release. This might force us to retest, reprofile, and potentially rewrite significant portions of code for each new release. Obviously this is not a maintainable solution. In practice, most code that is written on some old MATLAB release would likely be carried over with minimal changes to the newer releases. If this code has release-specific tuning, we could be shooting ourselves in the leg in the long run.

MathWorks advises,[1442] and I strongly concur, to program in a natural manner, rather than in a way that is tailored to a particular MATLAB release (unless of course we can be certain that we shall only be using that release and none other). This will improve development time, maintainability, and in the long run also performance.

If you are determined to squeeze every last possible bit of performance from your code, at the possible expense of possibly programming in a less natural manner, then consider following Steve Eddins' advice[1443] to grow arrays along their last (trailing) dimension (i.e., in a 2D matrix, grow columns rather than the more natural rows).† However, note that I have not seen this particular technique to be helpful in my tests.

9.4.3 Preallocate Large Data Arrays

The only recommendation that has withstood the test of time well for all MATLAB releases is to preallocate large data arrays.[1444] The basic idea is to create a data array in the final expected size before actually starting the processing loop. This saves any reallocations within the loop, since all the data array elements are already available and can be accessed.

---

* Presumably in the JIT, but I am not sure about this: turning the JIT Accelerator off from the MATLAB Command Line (something that is NOT recommended under any circumstances) naturally degrades performance, but the behavior remains linear and not quadratic. So, apparently some deeper improvement, perhaps in the memory allocation module, was done. The release notes (http://mathworks.com/help/matlab/release-notes.html#bsy83rt-1 or: http://bit.ly/YYNzDI) only say that *"This release improves the performance of growing an array in the trailing dimension if that array has not been preallocated"*.

† Also see the related §9.31 and §9.3.2.

This solution is useful when the final size is known in advance, as the following snippet illustrates:

```
% This is run on MATLAB 7.12 (R2011a):
% Regular dynamic array growth:
f=[0,1]; for idx=3:40000, f(idx)=f(idx-1)+f(idx-2); end
⇨ Elapsed time is 0.019954 seconds.


% Now use preallocation - 5x faster than dynamic array growth:
f=zeros(40000,1); f(1)=0; f(2)=1;
for idx=3:40000, f(idx)=f(idx-1)+f(idx-2); end
⇨ Elapsed time is 0.004132 seconds.
```

On pre-R2011a releases the effect of preallocation is even more pronounced:

```
% This is run on MATLAB 7.1 (R14 SP3):
% Regular dynamic array growth:
f=[0,1]; for idx=3:40000, f(idx)=f(idx-1)+f(idx-2); end
⇨ Elapsed time is 2.090217 seconds.


% Now use preallocation - 35x faster than dynamic array growth:
f=zeros(40000,1); f(1)=0; f(2)=1;
for idx=3:40000, f(idx)=f(idx-1)+f(idx-2); end
⇨ Elapsed time is 0.057398 seconds.
```

Because the effect of preallocation is so dramatic on all MATLAB releases, it makes sense to utilize it even in cases where the data array's final size is not known in advance.* We can do this by estimating an upper bound to the array's size, preallocate this large size, and when we are done remove any excess elements (even if we do not remove the excess elements, it still makes sense to over-preallocate):[1445]

```
% The final array size is unknown - assume 1Kx3K upper bound (~23MB)
data = zeros(1000,3000); % estimated maximal size
numRows = 0;
numCols = 0;
while (someCondition)
    colIdx = someValue1; numCols = max(numCols,colIdx);
    rowIdx = someValue2; numRows = max(numRows,rowIdx);
    data(rowIdx,colIdx) = someOtherValue;
end


% Now remove any excess elements
data(:,numCols + 1:end) = []; % remove excess columns
data(numRows + 1:end, :) = []; % remove excess rows
```

In some cases, preallocating the maximal possible array size may require too much memory or time. In such cases, we could preallocate blocks of memory: We start by preallocating some significant memory block, and track the usage within the loop. When the memory block is exhausted (i.e., additional memory allocation is required), we allocate another block of memory.[1446] Refer to §9.4.1 for implementation variants.

---

* Note that this suggestion conflicts with MathWorks tech solution 1-18150, which advises *against* this, for performance reasons: http://mathworks.com/matlabcentral/answers/99124 (or: http://bit.ly/1b0IduJ). I believe that my suggestion is better.

Turning existing loops into preallocated loops is easy: simply reverse the loop direction.[1447] This enables straightforward performance improvement with only minor code changes (no need for a separate preallocation line). The very first loop iteration will access the largest data index, thereby automatically creating a data array of the specified size. Here are the three main equivalent variants for preallocation:

```
% Variant #1: explicit preallocation of data1
data1 = zeros(1000,3000);
for colIdx = 1 : 3000
    for rowIdx = 1 : 1000
        data1(rowIdx,colIdx) = someValue;
    end
end


% Variant #2: implicit preallocation of data2 (much faster, see below)
data2 = []; data2(1000,3000) = 0;
for colIdx = 1 : 3000
    for rowIdx = 1 : 1000
        data2(rowIdx,colIdx) = someValue;
    end
end


% Variant #3: implicit preallocation of data3 using loop reversal
for colIdx = 3000 : -1 : 1
    for rowIdx = 1000 : -1 : 1
        data3(rowIdx,colIdx) = someValue;
    end
end
```

There are some additional variants, but these are the main ones. One of the interesting variants, which does not improve performance compared to variants #2 and #3, is *zeros(1 000,0)*zeros(0,3000)*.[1448]

The effect of preallocation is negligible if the data size or the number of reallocations is not large (i.e., less than several hundred or perhaps thousands, depending on the situation). When preallocation is important, then a second order of importance could be given to the specific method of preallocation. There is no general rule here, so we need to test our program on the specific target platform: different variants are faster on different MATLAB configurations/platforms. Note that preallocations are usually done only once in the program, making the difference immaterial in most cases.

The reason for the performance difference appears to be that the *zeros* function does two things internally: allocates the required memory block, and then initializes all its elements to zero. On the other hand, variants #2 and #3 skip the initialization part. If an element's default value (zero, *false* or '') is needed, then a lazy-evaluation mechanism of some sort is apparently used to provide consistency.[1449]

Note that *zeros(1000,'double')* appears to bypass the initialization part and is therefore as fast as the fastest alternative, despite being exactly equivalent to *zeros(1000)*.[1450] I do not understand why MathWorks chose to include the unnecessary initialization stage in *zeros(1000)* but not *zeros(1000,'double')*. In any case, the result is that we should attempt to replace all unspecified-class calls to *zeros* by either adding the data type (...,'double') or using variant #2.

When we need to preallocate a specific (non-default) value into every data array element, we cannot use variant #2. The reason is that variant #2 only sets the very last data element, and all other array elements get assigned the default value (0, '', or *false*, depending on the array's data type). In this case, we can use one of the following alternatives (with their associated timings on R2013b):[1451]

```
scalar = 7; % for example...
data = []; tic, data = scalar(ones(1000,3000));        toc %A1: 42.8 ms
data = []; tic, data = scalar*ones(1000,3000);         toc %A2: 14.3 ms
data = []; tic, data(1:1000,1:3000) = scalar;          toc %B:   9.3 ms
data = []; tic, data = repmat(scalar,1000,3000);       toc %C:   7.1 ms
data = []; tic, data = scalar + zeros(1000,3000);      toc %D:   6.7 ms
data = []; tic, data(1000,3000) = 0; data = data + scalar; toc %E: 5.6 ms
```

Notice that Tony's trick (variant A1),* which for years was viewed as the faster index-replication method,† actually turns out to be — by far — the slowest of the tested variants. Even a direct comparison with variant A2 looks very badly for Tony's trick, although this was the exact example used to promote it in days of yore, when the situation was reversed.[1452] Once again this provides us with an example lesson that established common knowledge may well change with newer MATLAB releases.‡

As can be seen, variant B is about 3× faster than variant A1, and variants C–E are progressively faster.[1453] Again, in most cases, the differences are immaterial since the preallocation code would only run once in the program. In some cases, we may have a need to periodically refresh our data ("wipe the slate clean", so to speak) or to set large blocks of data to a certain value. In such cases, it would make sense to use variants C–E rather than A or B.

The difference between variants A2 and D is worthy of special note: I have seen many MATLAB programs that use variant A2 (namely, *ones()**scalar) rather than equivalent B (*zeros()* + scalar), which is usually faster[1454] (but not always[1455]). The amount of speedup varies depending on platform and allocation size. Much of the difference stems from the fact that *zeros* is faster than *ones*. This difference is more pronounced, in both absolute and relative terms, the larger the allocated size. Beyond 100 K elements, *zeros* is orders of magnitude faster than *ones*,[1456] possibly because multithreading kicks-in at that point for *zeros* but not for *ones* (see figure on page 487).[1457]

Note that in the case of *ones*, there is no noticeable difference between *ones(1000)* and *ones(1000,'double')*, unlike the case of *zeros* above. The reason is that when allocating memory for a numeric value, a value of 0 is automatically and implicitly set for all memory locations, making the initialization step unnecessary. For *ones* we cannot avoid initialization, in order to modify the initial memory values from 0 to 1.

The importance of specifying the data type in *ones* and *zeros* is that this prevents MATLAB from unnecessarily allocating interim data of a different size and type than the final result.§

Sometimes we are only concerned about allocating memory and not about the initial element values. This could happen, for example, if our algorithm is assured to set a valid value for each element, before it is used. In such cases, we should not spend time initializing all data elements using *ones* or *zeros*. Instead, we should use variant #2 (data(n,m) = 0).

---

* See §5.1.2.

† Even as recently as 2012: http://stackoverflow.com/a/10327626/233829 (or: http://bit.ly/1nzTFGs).

‡ Certainly after so many years and numerous MATLAB releases (Tony made the original suggestion to Loren Shure in 1990). A critical examination of Drea's report (that suggested Tony's trick, see endnote in the preceding sentence of the main text) versus modern MATLAB releases was provided by Matt Fig in 2009: http://mathworks.com/matlabcentral/newsreader/view_thread/261716#683236 (or: http://bit.ly/11hCi3G).

§ See §9.4.5.

We could also use James Tursa's *uninit* utility,[1458] which is a superfast MEX function that simply allocates the required memory without initializing it.

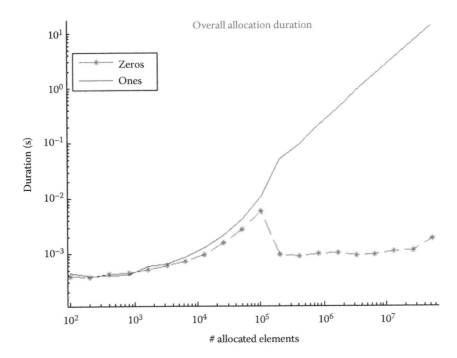

The MATLAB Editor indicates all cases of dynamic array growth with a Code Analyzer (M-Lint) warning: *"The variable a appears to change size on every loop iteration. Consider preallocating for speed".*

Sometimes we may wish to disregard this warning, for example, if our data array is small, or the loop is short, or if the dynamic growth within the loop only happens on certain rare conditions. Also, sometimes the Code Analyzer misidentifies preallocated arrays, and issues the warning although the array is indeed preallocated. In all such cases, we can disregard the warning on the specific code line by appending a %#ok<AGROW> comment to the end of the line:[†]

```
for idx = 1 : 1000
    data(idx) = someValue;  %#ok<AGROW>
end
```

When trying to preallocate a variable that is larger than the largest available contiguous block in physical memory (RAM), MATLAB needs to resort to using the much slower virtual memory. This is slow both because it uses the generic VM services of the operating system and slow disk swapping of memory pages. In such cases, we can use the trick of a memory-mapped file masquerading as a variable.[‡] This was implemented by Malcolm Lidierth of King's College London in the *vvar* class,[1459] as part of his *Waterloo*

---

* The complete list of MLint-reported performance-related warnings is presented in Appendix A.4.
† Since preallocation is a very important tool of performance tuning, be careful not to suppress the warning in the current file or worse, in all files.
‡ See §3.6.2 and §11.3.3 for a description of MATLAB's memory-mapped files and the *memmapfile* function.

utility of scientific graphics.[1460] *vvar* is naturally slower than real-memory variables when there is enough physical memory (e.g., variable size of 80 MB). But when we need a variable size of 800MB, *vvar* is 5–7× faster than using standard MATLAB variables, and 100× faster when the variable size is 8 GB. The speedup affects all types of variable access: preallocation, read, and update. It may seem strange to get such speedups when deliberately using file I/O rather than memory, but when we take into consideration the VM overheads, we can understand that direct memory-mapped I/O is more efficient than indirect VM I/O.

9.4.4 Preallocation Location within the Code

When preallocating large data arrays (as noted above, there is no real point in preallocating small arrays), we do not need to preallocate them immediately before the processing loop. We could preallocate anywhere preceding the processing loop. A typical location for such early preallocation is within an initialization function or code segment at the beginning of the program.

There are some conflicting trade-offs regarding the location of preallocated arrays, and readers should decide based on their particular program requirements:

- Preallocating an array immediately before its corresponding processing loop creates code that is easier to read and maintain, than programs where the preallocation occurs at the beginning of the code, far away from the processing loop. Program development and maintenance time is an important consideration, not to be easily dismissed.

- Placing preallocations at a different code location than the processing loop could well lead to a situation where the processing loop is moved to a different location in the code (for any of numerous possible reasons), where the preallocation is no longer visible in the execution scope. This will lead to degraded performance because the preallocation no longer comes into play. It will be very difficult for the developer to detect this, since the code will run correctly, without error, only slower.

- If the processing loop is removed altogether (e.g., if the program no longer needs it), then if the preallocation is done elsewhere, it would be easy for the developer to miss this and not delete the preallocation along with the loop. This will cause an unnecessary waste of memory.

- On the other hand, placing preallocation commands at the top of the program enables us to arrange them in such a way that larger arrays are preallocated before smaller ones.[1461] Since MATLAB arrays need contiguous memory blocks, this arrangement reduces the likelihood of running into memory allocation problems, and ultimately results in better performance.

9.4.5 Preallocating Nondouble Data

When preallocating an array of a type that is not *double*, we should take care to create it using the desired type, to prevent memory and performance inefficiencies. For example, if we need to process a large array of small integers (*int8*), it is inefficient to preallocate an array of *doubles* and type-convert to/from *int8* in every loop iteration. Similarly, it is inefficient to preallocate the array as a *double* type and then convert it to *int8*. Instead, we should create the array as an *int8* array in the first place:

```
% Bad idea: allocates 8 MB double array, then converts to 1MB int8 array
data = int8(zeros(1000,1000)); % 1M elements
⇨ Elapsed time is 0.000113 seconds.

% Better: directly allocate the array as a 1MB int8 array - faster
data = zeros(1000,1000,'int8');
⇨ Elapsed time is 0.000072 seconds.
```

For example, *int8(ones*(1000,1)) first allocates 8000 bytes for 1000 double-precision values, then allocates a separate 1000-bytes block for the converted data.* This is not only wasteful in CPU cycles and memory, but could also cause severe slowdowns when the data is large, possibly even thrashing and out-of-memory errors. Instead, directly allocate using the target data type: *ones(1000,1,'int8').*[1462]

Starting in MATLAB release R2013a (8.1), we can preallocate based on a specified input's dynamic run-time type, sparsity (sparse/dense) and complexity (real/complex). This is supported by the functions *zeros, ones, eye, inf, nan, true, false,* and *cast*:

```
data = zeros(1000,1000,'like',var);
```

Similarly, when reading binary data files, specify the target data precision directly:[1463]

```
fid = fopen('large_file_of_uint8s.bin', 'r');
a = fread(fid, 1e3, 'uint8');            % Requires 8KB
whos a
   Name        Size        Bytes    Class       Attributes
   a          1000x1        8000    double

a = fread(fid, 1e3, 'uint8=>uint8');     % Requires 1KB
whos a
   Name        Size        Bytes    Class       Attributes
   a          1000x1        1000    uint8
```

To preallocate a cell array, we can use the *cell* function (explicit preallocation), or the maximal cell index (implicit preallocation). These variants are functionally equivalent. On some releases, explicit preallocation is faster than implicit preallocation;[†] on other releases/platforms, the reverse is correct:[‡]

```
% Variant #1: Explicit preallocation of a 1Kx3K cell array
data = cell(1000,3000);

% Variant #2: Implicit preallocation
clear('data'), data{1000,3000} = [];
```

To preallocate an array of structs, we can use the *repmat* function to replicate copies of a single data element (explicit preallocation), or just use the maximal data index (implicit preallocation). In this case, unlike the case of cell arrays, implicit preallocation is much faster than explicit preallocation, since the single element does not actually need to be copied multiple times:[1464]

---

* In addition to the ~100 bytes overhead used by the mxArray descriptor (§8.1.4.1) of course.
† Note: this is contrary to the allocation of numeric arrays (discussed in §9.4.3) and other arrays (see below).
‡ Both variants were superfast until ~2011; in a release around then, variant #2 seemed to suffer major performance degradation.

```
% Variant #1: Explicit preallocation of a 100x300 struct array
element = struct('field1',magic(2), 'field2',{[]});
data = repmat(element, 100, 300);
⇨ Elapsed time is 0.001166 seconds.


% Variant #2: Implicit preallocation - 3x faster than explicit
clear('data'), data(100,300) = element;
⇨ Elapsed time is 0.000417 seconds.
```

When preallocating structs, we can also use a third variant, using the built-in feature of replicating the struct when the ***struct()*** function is passed a cell array. For example, struct('field1',cell(100,1), 'field2',5) will create 100 structs, each of them having the empty field 'field1' and another field called 'field2' with value 5.[1465] Unfortunately, this variant is slower than both of the previous variants.

```
% Variant #3: Explicit preallocation without using repmat
data = struct('field1',magic(2), 'field2', cell(100,300));
⇨ Elapsed time is 0.016966 seconds.
```

In a related matter, when creating a single struct object, rather than an array of structs, it is faster to use the direct assignment syntax than the built-in ***struct*** function:[1466]

```
% Variant #1: Direct field assignment
s1.a = pi; s1.b = 'hello'; s1.c = {1,2,'three'};
⇨ Elapsed time is 0.000022 seconds.


% Variant #2: using the builtin struct() function - 2.5x slower
s2 = struct('a',pi, 'b','hello', 'c',{{1,2,'three'}});
⇨ Elapsed time is 0.000054 seconds.
```

Implicit preallocation of class objects, like that of structs, is also faster than explicit preallocation, but for a different reason. When implicit expansion of class object arrays takes place, an abbreviated version of object instance creation takes place, which bypasses the constructor calls and just copies the instance properties.[1467] For example, array(9) = Widget creates an array of nine separate Widget objects, but the Widget constructor is only called for array(1) and array(9); array(1) is then expanded (copied over) to the remaining objects array(2:8).*

When preallocating, ensure that you are using the maximal expected array size. There is no point in preallocating an empty array or an array having a smaller size than the expected maximum, since dynamic memory reallocation will automatically kick-in within the processing loop. For this reason, avoid using the *empty()* method of class objects[1468] to preallocate, but rather ***repmat*** as explained above.[1469]

When using ***repmat*** to replicate class objects, always be careful to note whether you are replicating the object itself (this happens if your class is a *value class*, that is, does not derive from ***handle***) or its reference handle (which happens if you derive the class from ***handle*** or one of its subclasses). When replicating objects, we can safely update any of their properties independently of each other; but if we replicate references, we are merely using multiple copies of the same reference, so modifying any referenced object will also automatically affect all other objects referencing it.[1470] This may or may not be suitable for your particular program requirements, so be careful to check carefully. If you actually need to

---

* For some additional aspects of class object creation, see §4.7.1.

use independent object copies, you will need to call the class constructor multiple times, once for each new independent object.[1471]

A fast way to copy a handle object including its properties, but without calling any constructors or property setter methods, is via the *copy()* method provided by the ***matlab.mixin.Copyable*** superclass. This abstract class extends ***handle*** and is available on MATLAB R2011a (7.12) onward.[1472] We can use it as follows:

```
classdef MyClass < matlab.mixin.Copyable
    properties
        myScalar = 12.5;
        myValueObj = MyValueClass;
        myHandleObj = MyHandleClass;
    end
end


a = MyClass;
b = copy(a); % or: b = a.copy;
```

This is called *shallow copy*, since it copies the internal properties as is, without drilling down into contained objects. In other words, MyClass's myHandleObj property will be copied as a reference, without creating a new object having a mirror set of internal properties. If we now update b.myHandleObj's internal properties, we will see these changes reflected in a.myHandleObj. This is different from b.myScalar and b.myValueObj, which are entirely independent of their counterparts in a. We can override the *copyElement()* method to provide a customized copy behavior, for example, to implement a *deep copy* of handle properties (which would drill down into myHandleObj and create a true clone, not a reference), or to log the copy action.

Preallocation of class objects (class instances) can be used not just as a means of avoiding dynamic allocation, but also as a means of controlling the time of object initialization. Object initialization, typically done in the class's constructor method, could be lengthy (depending on your specific application). By preallocating the object, we can control the exact timing in which this initialization occurs, possibly at such a time that is less time-critical in the regular application time-flow.*

For preallocation of sparse data, refer to §4.1.3.

9.4.6 Alternatives for Enlarging Arrays[1473]

As mentioned above, it is sometimes impractical or impossible to preallocate data arrays. In such cases we may need to enlarge the array dynamically. There are several equivalent alternatives for doing this, having significantly different performances:

```
% This is run on MATLAB 7.12 (R2011a):

% Variant #1: direct assignment into a specific out-of-bounds index
data=[]; tic, for idx=1:100000; data(idx)=1; end, toc
⇨ Elapsed time is 0.075440 seconds.

% Variant #2: direct assignment into an index just outside the bounds
data=[]; tic, for idx=1:100000; data(end+1)=1; end, toc
⇨ Elapsed time is 0.241466 seconds. % 3x slower
```

---

* Compare *lazy initialization*, discussed in §1.8.3 and §3.8.1. Also see §4.7.1.

```
% Variant #3: concatenating a new value to the array
data=[]; tic, for idx=1:100000; data=[data,1]; end, toc
⇒ Elapsed time is 22.897688 seconds. % 300x slower!!!
```

As can be seen, it is much faster to directly index an out-of-bounds element as a means to force MATLAB to enlarge a data array, rather than using the end + 1 notation, which needs to recalculate the value of end each time.

In any case, try to avoid using the concatenation variant, which is significantly slower than either of the other two alternatives (300× slower in the above example!). In this respect, there is no discernible difference between using the [] operator or the *cat()* function for the concatenation.

Apparently, the JIT performance boost gained in MATLAB R2011a does not work for concatenation. Future JIT improvements may possibly also improve the performance of concatenations, but in the meantime it is better to use direct indexing instead.

The effect of the JIT performance boost is easily seen when we run the same variants on pre-R2011a MATLAB releases. The corresponding values are 30.9, 34.8, and 34.3 s. Using direct indexing is still the fastest approach, but concatenation is now only 10% slower, not 300× slower.

When we need to append a nonscalar element (e.g., a 2D matrix) to the end of an array, we might think that we have no choice but to use the slow concatenation method. This assumption is incorrect: we can still use the much faster direct-indexing method, as shown below (notice the nonlinear growth in execution time for the concatenation variant):

```
% This is run on MATLAB 7.12 (R2011a):
matrix = magic(3);

% Variant #1: direct assignment - fast and linear cost
data=[]; tic, for idx=1:10000; data(:,(idx*3-2):(idx*3))=matrix; end, toc
⇒ Elapsed time is 0.969262 seconds.

data=[]; tic, for idx=1:100000; data(:,(idx*3-2):(idx*3))=matrix; end, toc
⇒ Elapsed time is 9.558555 seconds.

% Variant #2: concatenation - much slower, quadratic cost
data=[]; tic, for idx=1:10000; data=[data,matrix]; end, toc
⇒ Elapsed time is 2.666223 seconds.

data=[]; tic, for idx=1:100000; data=[data,matrix]; end, toc
⇒ Elapsed time is 356.567582 seconds.
```

As the size of the array enlargement element (in this case, a 3×3 matrix) increases, the computer needs to allocate more memory space more frequently, thereby increasing execution time and the importance of preallocation. Even if the system has an internal memory-management mechanism that enables it to expand into adjacent (contiguous) empty memory space, as the size of the enlargement grows, the empty space will run out sooner and a new larger memory block will need to be allocated more frequently than in the case of small incremental enlargements of a single 8-byte *double*.

If preallocation is not possible, and JIT is not very helpful, consider using one of the following previously discussed alternatives:*

---

* See §9.4.1.

- Dynamically grow the array by a certain percentage factor each time the array runs out of space (on-the-fly preallocation).

- Use John D'Errico's *growdata* utility.

- Use cell arrays to store and grow the data, then use ***cell2mat*** or *cell2vec*[1474] to convert the resulting cell array to a regular numeric array.

- Wrap the data in a referential object (a class object that inherits from *handle*), then append the reference handle rather than the original data.[1475] Note that if your class object does not inherit from *handle*, it is not a referential object but rather a value object, and as such it will be appended in its entirety to the array data, losing any performance benefits. Of course, it may not always be possible to wrap our class objects as a *handle*.*

     References have a much small memory footprint than the objects that they reference. The objects themselves will remain somewhere in memory and will not need to be moved whenever the data array is enlarged and reallocated — only the small-footprint reference will be moved, which is much faster.†

- Reuse an existing data array that has the necessary storage space.‡

9.5 Minimizing Memory Allocations

As should be obvious to the reader by now, we should attempt to reduce the number of memory allocations and reallocations done by MATLAB. Memory reallocation was extensively discussed in the previous section, and this section will show different methods to minimize the need for initial allocation. If variables do not need to be allocated memory, then expensive memory operations (possibly requiring page swap with virtual memory) can be avoided, thereby improving performance.

9.5.1 MATLAB's Copy-on-Write Mechanism[1476]

MATLAB implements an automatic *copy-on-write* (sometimes called *copy-on-update* or *lazy copy*ing) mechanism,[1477] which transparently allocates a temporary copy of the data only when it sees that the input data is modified. This improves run-time performance by delaying actual memory block allocation until absolutely necessary, a variation of the *lazy loading* technique.§ The mechanism has two variants: during regular variable copy operations, and passing data as input parameters into a function.

9.5.1.1 Regular Variable Copies

When a variable is copied, as long as the data is not modified, both variables actually use the same shared memory block. The data is only copied onto a newly allocated memory

---

* See §9.4.5 for additional details about using arrays of class objects.
† Recall that this is also the reason that cell concatenation is faster than array concatenations for large objects (see §9.4.1).
‡ See §9.5.3.
§ See §3.8.1.

block when one of the variables is modified.* The modified variable is assigned the newly allocated block of memory, which is initialized with the values in the shared memory block before being updated:

```
data1 = magic(5000);    % 5Kx5K elements = 191 MB
data2 = data1;          % data1 & data2 share memory; no allocation done
data2(1,1) = 0;         % data2 allocated, copied and only then modified
```

If we profile our code using any of the means presented in §9.2, we will see that the copy operation *data2 = data1* takes negligible time to run and allocates no memory. On the other hand, the simple update operation *data2(1,1) = 0*, which we could otherwise have assumed to take minimal time and memory, actually takes a relatively long time and allocates 191 MB of memory[†]

```
time    calls                   mem line
                                     1 function perfTest
0.890     1     215m/23.9m/191m     2     data1 = magic(5000);
          1                         3     data2 = data1;
0.172     1     191m/0b/191m        4     data2(1,1) = 0;
          1                         5 end
```

The copy-on-write effect monitored using the MATLAB Profiler with the -memory option

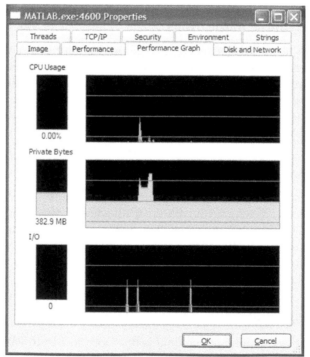

The *copy-on-write* effect monitored using the Process Explorer

---

* Note that this only occurs for variables that are non-referential (i.e., not a handle class object). When we modify a handle class object, all copies of this object automatically reflect this change, since they all reference the same memory location.

† In this Profiler report, the memory column lists the allocated, freed, and peak memory in MB units, for each executed code line.

We first see a memory spike (used during the computation of the *magic* square data), closely followed by a leveling off at 190.7 MB above the baseline (this is due to allocation of *data1*). Copying *data2 = data1* has no discernible effect on either CPU or memory. Only when we set *data2(1,1) = 0* does the CPU return, in order to allocate the extra 190MB for *data2*. When we exit the test function, *data1* and *data2* are both deallocated, returning the MATLAB process memory to its baseline level.*

There are several lessons that we can draw from this simple example. First, creating copies of data does not necessarily or immediately impact memory and performance. Rather, it is the update of these copies that may be problematic. If we can modify our code to use more read-only data and less updated data copies, then we would improve performance. The Profiler report will show us exactly where in our code we have memory and CPU hotspots — these are the places we should consider optimizing.

Second, when we see such odd behavior in our Profiler reports (i.e., memory and/or CPU spikes that occur on seemingly innocent code lines), we should be aware of the copy-on-write mechanism, which could be the cause for the behavior.[1478]

9.5.1.2 Function Input Parameters

The copy-on-write mechanism behaves similarly for input parameters in functions: whenever a function is invoked (called) with input data, the memory allocated for this data is used up until the point that one of its copies is modified. At that point, the copies diverge: a new memory block is allocated, populated with data from the shared memory block, and assigned to the modified variable. Only then is the update done on the new memory block.[1479]

```
data1 = magic(5000); % 5Kx5K elements = 191 MB
data2 = perfTest(data1);

function outData = perfTest(inData)
   outData = inData;   % inData & outData share memory; no allocation
   outData2(1,1) = 0; % outData allocated, copied and then modified
end
```

Readers should not be surprised by the corresponding Profiler report:

time	calls	mem	line	
			1	function outData = perfTest(inData)
	1		2	outData = inData;
0.140	1	191m/0b/191m	3	outData(1,1) = 0;
	1		4	end

The copy-on-write effect monitored using the MATLAB Profiler with the -memory option

One lesson that can be drawn from this is that whenever possible we should attempt to use functions that do not modify their input data. This is particularly true if the modified input data is very large. Read-only functions will be faster than functions that do even the simplest of data updates.[1480]

Another lesson is that perhaps counterintuitively, it does not make a difference from a performance standpoint to pass read-only data to functions as input parameters. We might think that passing large data objects around as function parameters will involve

---

* For the Process Monitor screenshot, I have inserted short *pauses* after the three code lines, to make the different stages distinct.

multiple memory allocations and deallocations of the data. In fact, it is only the data's reference* that is being passed around and placed on the function's call stack. Since this reference/structure is quite small in size, there are no real performance penalties. In fact, this only benefits code clarity and maintainability.

The only case where we may wish to use other means of passing data to functions is when a large data object needs to be updated. In such cases, the updated copy will be allocated to a new memory block with an associated performance cost. The following sections describe several means to reduce this cost.

9.5.2 In-Place Data Manipulations[1481]

MATLAB's interpreter, at least in recent releases, has a very sophisticated algorithm for using in-place data manipulation.[1482] Modifying data in-place means that the original data block is modified, rather than creating a new block with the modified data, thus saving any memory allocations and deallocations.

For example, let us manipulate a simple 4K×4K (122MB) numeric array:

```
>> m = magic(4000);     % 4Kx4K = 122MB
>> memory
Maximum possible array:             1022 MB (1.072e+09 bytes)
Memory available for all arrays:    1218 MB (1.278e+09 bytes)
Memory used by MATLAB:               709 MB (7.434e+08 bytes)
Physical Memory (RAM):              3002 MB (3.148e+09 bytes)

% In-place array data manipulation: no memory allocated
>> m = m * 0.5;
>> memory
Maximum possible array:             1022 MB (1.072e+09 bytes)
Memory available for all arrays:    1214 MB (1.273e+09 bytes)
Memory used by MATLAB:               709 MB (7.434e+08 bytes)
Physical Memory (RAM):              3002 MB (3.148e+09 bytes)

% New variable allocated, taking an extra 122MB of memory
>> m2 = m * 0.5;
>> memory
Maximum possible array:             1022 MB (1.072e+09 bytes)
Memory available for all arrays:    1092 MB (1.145e+09 bytes)
Memory used by MATLAB:               831 MB (8.714e+08 bytes)
Physical Memory (RAM):              3002 MB (3.148e+09 bytes)
```

The extra memory allocation of the not-in-place manipulation naturally translates into a performance loss:

```
% In-place data manipulation, no memory allocation
m = m * 0.5;
⇨ Elapsed time is 0.056464 seconds.

% Regular data manipulation (122MB allocation) - 50% slower
m2 = m * 0.5;
⇨ Elapsed time is 0.084770 seconds.
```

---

* Or more precisely, its `mxArray` structure — see §8.1.4.1.

The difference may not seem large, but placed in a loop it could become significant indeed, and might be much more important if virtual memory swapping comes into play, or when MATLAB's memory space is exhausted (out-of-memory error).

Similarly, when returning data from a function, try to update the original data variable whenever possible, thereby avoiding the need to allocate a new variable:[1483]

```
% In-place data manipulation, no memory allocation
>> d=0:1e-7:1; tic, d = sin(d); toc
⇨ Elapsed time is 0.083397 seconds.

% Regular data manipulation (76MB allocation) - 50% slower
>> clear d2, d=0:1e-7:1; tic, d2 = sin(d); toc
⇨ Elapsed time is 0.121415 seconds.
```

Within the function itself we should ensure that we return the modified input variable, and not assign the output to a new variable, so that in-place optimization can also be applied within the function. The in-place optimization mechanism is smart enough to override MATLAB's default copy-on-write mechanism, which automatically allocates a new copy of the data when it sees that the input data is modified:[1484]

```
% Suggested practice: use in-place optimization within functions
function x = function1(x)
   x = someOperationOn(x);   % temporary variable x is NOT allocated
end

% Standard practice: prevents future use of in-place optimizations
function y = function2(x)
   y = someOperationOn(x);   % new temporary variable y is allocated
end
```

To benefit from in-place optimizations of function results, we must (1) use the same variable name for the output and the input variables ($x = function1(x)$), in both the caller and the called workspaces, and also (2) ensure that the called function is optimizable. If any of these two requirements is not met then in-place function-call optimization is not performed.[1485]

Also, for the in-place optimization to work, we need to call the in-place function from within another function, not from a script or the MATLAB Command Window.[1486]

A related performance trick is to use masks on the original data rather than temporary data copies. For example, suppose we wish to get the result of a function that acts on only a portion of some large data. If we create a temporary variable that holds the data subset and then process it, it would create an unnecessary copy of the original data:

```
% Original data
data = 0 : 1e-7 : 1;              % 10^7 elements, 76MB allocated

% Unnecessary copy of data into data2 (extra 8MB allocated)
data2 = data(data<0.1);          % 10^6 elements, 7.6MB allocated
results = sin(data2);            % another 10^6 elements, 7.6MB allocated

% Use of data masks avoids the need for temporary variable data2:
results = sin(data(data<0.1));   % no need for the data2 allocation
```

In-place optimization is naturally used when assigning a scalar value to specific data elements (or array of indexes). We can use this to improve performance in cases where we need to filter out invalid or outlier data elements: rather than removing them from the original data, which requires reallocating the entire updated data, we could set the value to

Inf, NaN, or some other unique value. This is done in-place and is therefore much quicker than the removal (and reallocation) method. We just need to ensure that the marked values are properly handled by the algorithm.

For example, assume we need to get the value of the largest data element(s) smaller than 3 (i.e., closest to 3 from below):

```
% Original data
data = 10*rand(1,1e7); %10^7 elements between 0-10, 76MB allocated

% Alternative #1: in-place element marking
data(data>=3) = -Inf; maxVal = max(data);
⇨ Elapsed time is 0.170726 seconds.

% Alternative #2: remove elements >3 (60% slower)
data(data>=3) = []; maxVal = max(data);
⇨ Elapsed time is 0.280331 seconds.
```

In this case, we have set the marked elements to a value of -Inf, which is ignored by the subsequent algorithm (*max*).[1487]

In-place optimization is even employed for array slices (columns and rows):[1488]

```
v = data(:,1);
v = data*v;
data(:,2) = data * v;
data(2,:) = v' * data;
```

Of course, if the data changes its size (for example, by deleting rows or adding columns), then in-place optimization will not be applied. In such cases, the data will be reallocated in a new memory location. To limit the performance impact of such reallocations, we can try to consolidate the actions. For example, consider the task of deleting all rows and columns that contain only zeros in a data matrix:[1489]

```
% Initial test data
data = ones(8000,8000);  % =488 MB
data([2,3,7],:) = 0;
data(:,[5,8,9]) = 0;

% Naïve approach
data(~any(data,2),:) = [];
data(:,~any(data,1)) = [];
⇨ Elapsed time is 0.587138 seconds.

% Consolidated deletion – 2.5x faster
data = data(any(data,2), any(data,1));
⇨ Elapsed time is 0.233948 seconds.
```

A note of caution:[1490] we should not invest undue efforts to use in-place data manipulation if the overall benefits would be negligible. For example, writing *[L,A,P] = lu(A)* rather than the more natural *[L,U,P] = lu(A)* would probably not help much in performance due to *lu*'s relatively large processing time.* It would only help if we have a real memory limitation issue and the matrix *A* is very large.

---

* This is a theoretical example, because as of mid-2014 LU was not optimized for in-place operations... See http://blogs.mathworks.com/loren/2007/03/22/in-place-operations-on-data#comment-30551 (or: http://bit.ly/JuVJwE).

MATLAB in-place optimization is a topic of continuous development. Code that is not in-place optimized today (e.g., in-place manipulation on class object properties[1491]) may possibly be optimized in next year's release. For this reason, it is important to write the code in a way that would facilitate future optimization (e.g., $obj.x = 2*obj.x$ rather than $y = 2*obj.x$).

Some in-place optimizations were added to the JIT Accelerator as early as MATLAB 6.5 R13, but MATLAB 7.3 R2006b[1492] saw a major boost. As the JIT Accelerator improves from release to release, we should expect in-place data manipulations to be automatically applied in an increasingly larger number of code cases.

In some older MATLAB releases, and in complex data manipulations where the JIT Accelerator cannot implement in-place processing,[1493] a temporary storage is allocated that is assigned to the original variable when the computation is done. This naturally results in significant slowdown. To implement in-place data manipulations in such cases we could develop an external function (e.g., using MEX) that directly works on the original data block.[1494] Note that the officially supported MEX update method is to always create deep copies of the data using *mxDuplicateArray* and then modify the new array rather than the original; modifying the original data directly can be done using the undocumented *mxCreateSharedDataCopy*, which has exactly the same function signature as *mxDuplicateArray*:

```
/* Definition */
mxArray *mxCreateSharedDataCopy(const mxArray *data);


/* Sample usage */
plhs[0] = mxCreateSharedDataCopy(prhs[0]);
```

Using shared data is both discouraged[1495] and not officially supported,[1496] although it was featured in an official MathWorks support solution[1497] that demonstrated its usage with a sample MEX file[1498] (while stating the fact that this is unsupported):

```
# include "mex.h"
/* Add this declaration: it does not exist in the "mex.h" header */
extern mxArray *mxCreateSharedDataCopy(const mxArray *pr);

void mexFunction(int nlhs,        mxArray *plhs[],
                 int nrhs, const mxArray *prhs[])
{
   mxArray *copy1 = NULL;
   mxArray *copy2 = NULL;
  /* Check for proper number of input and output arguments */
   if (nrhs != 1)
       mexErrMsgTxt("One input argument required.");
   if (nlhs > 1)
       mexErrMsgTxt("Too many output arguments.");

   /* First make a regular deep copy of the input array */
   copy1 = mxDuplicateArray(prhs[0]);

   /* Then make a shared copy of the new array */
   copy2 = mxCreateSharedDataCopy(copy1);

   /* Print some information about the arrays */
   mexPrintf("Created shared data copy, and regular deep copy\n");
   mexPrintf("prhs[0] = %X, mxGetPr = %X, value = %lf\n",
             prhs[0], mxGetPr(prhs[0]), *mxGetPr(prhs[0]));
```

```
      mexPrintf("copy1 = %X, mxGetPr = %X, value = %lf\n",
                copy1, mxGetPr(copy1), *mxGetPr(copy1));
      mexPrintf("copy2 = %X, mxGetPr = %X, value = %lf\n",
                copy2, mxGetPr(copy2), *mxGetPr(copy2));

      /* Destroy the first copy */
      mxDestroyArray(copy1);
      copy1 = NULL;
      mexPrintf("\nFreed copy1\n");

      /* copy2 will still be valid */
      mexPrintf("copy2 = %X, mxGetPr = %X, value = %lf\n",
                copy2, mxGetPr(copy2), *mxGetPr(copy2));
}
```

Now back in MATLAB:

```
>> mex mxsharedcopy.c % compile the MEX file
>> mxsharedcopy(magic(3))
Created shared data copy, and regular deep copy
prhs[0] = AC82B70, mxGetPr = 701ECC00, value = 8.000000
copy1   = AC827F0, mxGetPr = 701EBD80, value = 8.000000
copy2   = AC805D0, mxGetPr = 701EBD80, value = 8.000000

Freed copy1
copy2 = AC805D0, mxGetPr = 701EBD80, value = 8.000000
```

Using a shared copy can be much faster than a deep copy, but be careful: using shared data incorrectly can easily crash MATLAB. If you do directly overwrite the original input data, at least ensure that you unshare any variables that share the same data memory block (using *mxUnshareArray*), thus mimicking the copy-on-write mechanism.[1499] If you do not unshare the data, then modifying one variable will affect another (the shared) variable, causing very hard-to-diagnose bugs.[1500]

Adventurous users with some MEX experience can utilize the ability to modify only a small portion of data via deep copy, while retaining the majority of the data shared. This can significantly improve performance when we need to access complex data structures (e.g., structs or cell arrays) that can share most of their internal data. The MEX function *mxUnreference* can be used in such cases.[1501] The degree of complexity here is higher than for regular shared arrays, and so is the related risk.

A similar mechanism for accessing MATLAB (MCOS) class properties in-place in MEX can be used: MATLAB's built-in *mxGetProperty* MEX function returns a copy of the relevant property, after allocating it in memory. James Tursa created MEX variant *mxGet-PropertyPtr* that returns the actual pointer to the property data.[1502] This is much faster, since it does not require memory allocation for the data copy. It also enables in-place property data updates, even for value classes.* On my system, computing sum((z.x-z.y).^2) where x,y are 100 MB each, using pure m-code took 44 ms, using a MEX file that uses *mxGetProperty* took 98 ms, and using *mxGetPropertyPtr* only took 16 ms.

Using MATLAB's internal in-place data manipulation is very useful, especially since it is done automatically without need for any major code changes on our part. But sometimes

---

* Accessing class object properties is further discussed in §4.7.2. Note that in-place property update should be used judiciously, to prevent updating read-only object properties.

we need certainty of actually processing the original data variable without having to guess or check whether the automated in-place mechanism will be activated or not. This can be achieved using several alternatives:

- Using global or persistent variable (see §9.5.5)
- Using a parent-scope variable within a nested function (see §9.5.6)[1503]
- Modifying a reference (class) object's internal properties (see §9.5.7)

Bruno Luong's *inplacearray* MEX utility can also help share array memory:[1504]

```
a = zeros(5000,5000);   % 5000x5000 = 200MB

% Without inplace - allocates extra COW memory for array b
b = a;           % create a copy of a and modify this copy
b(21:30) = pi;   % now do something with b
⇨ Elapsed time is 0.063786 seconds.

% Using inplacearray - 335x faster
a10 = inplacearray(a,20,10,1);   % 10 rows starting at offset 20
backupData = a10(1:end);         % copy data for backup
a10(1:10) = pi;
    % now do something with a
a10(1:end) = backupData(:);      % restore data from backup
releaseinplace(a10);
⇨ Elapsed time is 0.000190 seconds.
```

Note that when using *inplacearray* we must remember to release the reference using the associated *releaseinplace* function (included in Bruno's submission), before either the original or the reference variable is cleared, in order to avoid memory leaks and potential crashes due to invalid memory access.

Also note that *inplacearray* only works on full matrices, not sparse data, structs, or objects. Joshua Dillon has posted[1505] that he has a solution for sparse data, and suggested to contact him for details.[1506]

The same Josh Dillon also posted his *sharedmatrix* utility,[1507] which enables to use a shared copy (reference) of a variable across multiple MATLAB processes on the same computer (e.g., PCT workers), without having to resort to complex and lengthy interprocess communication (IPC) mechanisms.* The different MATLAB processes simply access the same memory location for the data, as simple as that. Note that *sharedmatrix* does not support class objects, although it does support structs and sparse data. Important usage notes appear in the utility's page as user comments that are worth reading.[1508]

Kevin Stone improved several aspects of *sharedmatrix* in his *SharedMemory* utility,[1509] which uses Tursa's aforementioned *mxGetProperty* by the way. Unlike *sharedmatrix*, which is apparently unsupported by Joshua since mid-2011, Kevin seems to continuously update his utility (at least as of end-2013). Both *sharedmatrix* and *SharedMemory* use POSIX shared memory and the Boost IPC library.†[1510] Refer to §7.4 for additional related IPC alternatives.

As a last resort, we can always modify the original data with C pointers in MEX. For example, the following function multiplies its input data by a specified multiplier:[1511]

```
#include "mex.h"
void mexFunction(int nlhs,        mxArray *plhs[],
```

---

* See §7.4.
† See §7.3.2 for a discussion of Boost multithreading.

```
                       int nrhs, const mxArray *prhs[])
{
    mwSize i, numel;
    double *pr, multiplier;
    if (nrhs != 2 || nlhs != 0) {
        mexErrMsgTxt("Need 2 inputs (data, multiplier) & 0 outputs");
    }
    if (!mxIsDouble(prhs[0]) || !mxIsDouble(prhs[1])) {
        mexErrMsgTxt("Both inputs must be double");
    }
    if (mxIsSparse(prhs[0])) {
        numel = *(mxGetJc(prhs[0])+ mxGetN(prhs[0]));
    } else {
        numel = mxGetNumberOfElements(prhs[0]);
    }
    pr = mxGetPr(prhs[0]);
    multiplier = mxGetPr(prhs[1]);
    for (i=0; i<numel; i++) {
        pr[i] *= multiplier;
    }
}
```

Usage would be as follows:

```
>> mex multBy.c     % compile the MEX file (one-time only)
>> multBy(data,3)   % data will have all its elements multiplied by 3
```

9.5.3 Reusing Variables (with Utmost Care)

Existing variables obviously do not need to be allocated, since they already have assigned memory blocks. Therefore, reusing variables in your code can improve performance by reducing the number of new variables that need memory allocation.

The practice of reusing existing variables should NOT be used often. In fact, reusing variables (as opposed to creating new ones) may be very detrimental to code readability and maintainability. Moreover, reusing variables can be an endless source of applicative bugs. Furthermore, unless the reused variable is a very large data object, the performance benefits will generally be negligible. If the variable data type is changed, then performance might even be degraded.

One scenario in which reusing variables may be useful and worth the risks is when our program carries out separate computations on large data objects. For example, a large image undergoing separate filter operations, or a large data set of financial data undergoing several stages of technical analysis.

In such cases, it makes sense to reuse the same large data objects for all calculations, rather than create new variables for each computational step. In these cases, the data object typically maintains its data type, dimensions, and size, making variable reuse efficient and less likely to cause bugs due to leftover data. It may assist the MATLAB engine to decide to use efficient in-place data manipulation.[1512]

Another scenario is when multiple class object instances need to use some common class element. In this case, it makes sense to make the common element a static (*constant*) object property,[1513] rather than a dynamic (per-instance) property. The static property will be shared by all the class instance objects, saving initialization, allocation, and garbage collection times:

```
classdef MyClass
    properties (Constant)
        ON = 'on';
        PI = 3.14159265358979;
        COMMON = someCommonClassObject;
    end
    ...
```

Note that constant properties are not the same as static fields of other object-oriented languages such as C++ or Java. In MATLAB, constant properties are indeed static in the sense that only a single instance is stored for all object instances, but unlike C++/Java, these properties cannot be modified in run-time. If you need to have this ability, you need to use a global variable, at least as of MATLAB R2012b (8.0).*

Using a constant class property enables direct (and fast) equality check, using the == (identity equals) operator, as opposed to implementing a dedicated *isequal* function. The reason is that since only a single instance of the property is stored by MATLAB application-wide, all its uses actually refer to the same instance (memory location). Such a technique is often called *canonicalization* and the constant property, which is typically specified in UPPERCASE, is called a *canonical object*. If we had not used a static property location, we would need to use our much slower *isequal* implementation, since separate instances of the property might actually be stored in different memory locations, even if their contents are exactly the same.

Constant properties, especially when used for canonicalization, are very similar in nature to enumerated properties.[1514] Depending on the situation, you may wish to use either of these close relatives. The main difference appears to be that while the constant properties have the type of their contents (e.g., PI is a double, COMMON is a class object), enumerated properties retain the type of their containing class and need to be type cast in order to get their underlying value. For example:

```
classdef MyClass < double
    enumeration
        PI (3.14159265358979)
    end
end

>> a = MyClass.PI
a =
  PI

>> b = double(a)
  b =
3.1415926535898

>> whos a b
  Name      Size      Bytes    Class        Attributes
   a        1x1          64    MyClass
   b        1x1           8    double
```

Class objects are sometimes used temporarily. It makes sense in such cases to keep these objects live, reusing them as needed. We need to ensure for each such reuse that we are clearing and setting all the relevant internal object fields, to prevent unexpected results.

---

* See §9.5.5 for additional information on global variables. Alternatively, we can use the *singleton* design pattern, described later in this section.

Reusing class objects prevent reallocations, deallocations, and object initialization (in the constructor), which may significantly help performance. We shall revisit this idea below, when we discuss object pools.

As an alternative to using constant properties, we might consider using the *singleton* design pattern.[1515] The basic idea is to have only a single class object available for use, and have all the class functionality be available via that singleton object. The pattern ensures that no more than a single instance of the class is ever created, preventing unnecessary memory allocations.

Singletons have several possible implementations. All implementations share the common aspect of making the object constructor private or protected, to prevent direct creation of class instances. The singleton instance is instead created by a static member method, which does have access to the constructor, on its first invocation. Here is the basic outline of one of the possible implementations:[1516]

```matlab
% Concrete singleton implementation
classdef Singleton < handle

    properties (Accesss=private)
        data
    end % private properties

    methods (Access=private)
        % The constructor is private, preventing external invocation.
        % Only a single instance of this class is created. This is
        % ensured by getInstance() calling the constructor only once.
        function newObj = Singleton()
            % Initialize - maybe load from disk?
        end % constructor
    end % private methods

    methods (Static)
        function setData(newData)
            obj = getInstance();
            obj.data = newData;
        end % setData()
        function data = getData()
            obj = getInstance();
            data = obj.data;
        end   % getData()
    end % static methods
end % classdef

% Note: this is deliberately placed *outside* the class, so that it
% ^^^^ is not exposed to the user. If we do not mind this, we could
% place getInstance() in the class's static methods group.
function obj = getInstance()
    persistent uniqueInstance
    if isempty(uniqueInstance)
        obj = Singleton();
        uniqueInstance = obj;
    else
        obj = uniqueInstance;
    end
end   % getInstance()
```

Usage of this Singleton class is very simple: There is only one class instance, which has no accessible member properties and only the two static methods *setData* and *getData*, which access the singleton's private data:

```
currentData = Singleton.getData();
Singleton.setData(newData);
```

Performance-wise, singletons improve performance by keeping just a single global class object in heap memory. Of course, the underlying private data might still be reallocated when it changes, but at least there is only a single copy in our application, and the encapsulating singleton object is not reallocated. We can also employ some tips from §9.5.1 in our static access methods, to try to use in-place manipulations.

Uses of singleton classes include loggers, configuration setups, and application-wide resource handling/pooling (which is described below). Singleton is a very basic design pattern, which is often used as a building block for other patterns (e.g., *object factory*, *state object*, and *façade*).

As a special case of reusing class objects, consider reusing exception objects. It turns out that throwing new exceptions is relatively costly in terms of performance, mainly because the function call stack needs to be populated. We often do not care about this call stack but just in the information that an exception has occurred. In such cases, we could rethrow a persistent error object instead of a new exception object (see §3.4).

One common way of handling objects is to use an *object pool*, typically employing a singleton access point (the pool object). An object pool is a preallocated array of objects that are ready for use (like a car pool). Whenever our code needs a new objects, it requests one from the pool (*obj = pool.pull()*). The pool returns one of the available pools, or creates a new one if the pool is exhausted. When the code finishes with the object, it returns it to the pool (*pool.recycle(obj)*) for possible later use by some other code segment. Naturally, effective use of the pool requires each code segment that *pull*s an object to also *recycle* it when it is no longer used. Otherwise, we end up with a memory leak (since the pool keeps a live reference) and the pool resource is not reused. See §4.7.1 for a simple implementation of such an object pool.

Still, even in these special cases, the performance gains from reusing the data variable is often negligible compared to the actual computational (data-processing) costs. Therefore, unless we truly run into an out-of-memory limitation, we should think twice before reusing variables in places where it would not be natural to code that way in the first place.

One specific type of variable reuse we should be especially careful of: reusing a variable while changing its type. For example, if the variable was previously a simple double and is now changed to be an object reference (or vice versa). This not only causes full memory reallocation, the exact thing that we were trying to avoid, but it also prevents JIT optimizations (§4.1.4) and is also a potential cause for myriad of bugs and a maintenance nightmare. Avoid this at all cost. If you must modify the variable type, allocate a new variable for this, using a new variable name.

9.5.4 Clearing Unused Workspace Variables

If our system has enough memory, we might think that the extra time used to clear data variables in our code is a simple waste of time. This is indeed the case most of the time, but not always.

Programs sometimes allocate large data objects (matrices, cell arrays, structures, class objects, etc.) during the course of their work. It could make sense to deallocate large objects as soon as they are no longer needed, rather than to wait for the function to end, at which time all local variables are automatically *garbage-collected*.*[1517] The reason is that this frees memory for other variables, making it easier to allocate contiguous memory blocks to new or reallocated variables. Moreover, it improves the behavior of the CPU's internal cache and reduces memory paging, since the remaining variables will not need to compete with the deallocated memory block for their place in the cache and the physical memory.

The easiest way to deallocate variables is using the *clear* function:

```
clear varName1 varname2
clear('varName1', 'varName2',...); % equivalent alternative
```

Starting in MATLAB R2008a (7.6), we can also use the *clearvars* function to clear variables, including the ability to clear all variables except a specific set. In R2007b (7.5) and earlier releases, we could use one of several utilities that were posted on the MATLAB File Exchange for this purpose.[1518] A related utility that allows some extra control over the kept and deallocated variables is *gcoll* by Urs (us) Schwarz.[1519]

Note that neither *clear* nor any of the abovementioned utilities release the memory immediately. This takes some nondeterministic OS-dependent time. On 32-bit Windows platforms, the cleared memory blocks are not even guaranteed to be returned to MATLAB for reuse.[1520] Therefore, we cannot rely that a *clear* command has immediately vacated the memory for use by new MATLAB variables.

It should be noted that the *clear all* command, which is an often-used variant of *clear*, also clears all functions from MATLAB's memory cache. These functions would need to be recompiled (re-JIT'ed) when next encountered. This might produce a performance slowdown when our application has a large body of MATLAB code.[1521] Another side effect is that all persistent variables that stored cached data are cleared, necessitating their containing function to repopulate the cache when they will be first called after then. This too could produce a slowdown.

If a large data object is repeatedly initialized or updated within our MATLAB program, then it is generally reallocated whenever this happens.† This means that a temporary object of the same size is allocated while the old object still exists, and only after being successfully created is the old memory block released and the variable is assigned to the newly allocated memory block. This results in short memory-usage spikes that may cause thrashing and even out-of-memory errors. To avoid such cases, simply *clear* the old variable before recreating it. This [almost] ensures that the memory is released before being reallocated, subject to the OS-dependent reservations in the preceding paragraph:

```
for idx = 1 : 100
    clear('largeData')
    largeData = rand(10000,10000); % 100M elements = 800MB
    ...
end
```

If large data objects are not needed for some time, and it is important to release memory for some memory-intensive computation, it could make sense to save the data objects to disk (e.g., using the *save* function), and then *clear* them to free memory space; once the

---

* Note that unlike Java's garbage collector (§9.2.7), we cannot control the timing or parameters of MATLAB's garbage collector, nor invoke it directly. As one MathWorker put it, *"there are no knobs to turn"*. Some unverified information can be found here: http://stackoverflow.com/questions/1446281/matlabs-garbage-collector (or: http://bit.ly/1h8WkTa).

† Unless in-place data manipulation (see §9.5.2) is used. But in practice, in-place data manipulation is rarely used in such cases.

memory-intensive computations are over and the data objects are needed again, they can be *load*ed back from disk onto memory. The extra time needed to save and load the data from disk may be less than the time needed for virtual-memory disk swaps during the memory-intensive computations, making the operation worthwhile:*

```
data = rand(10000,10000);            % Large data object computed here
save('data_storage.mat','data');     % Temporarily save to disk
clear('data');                       % Deallocate the memory

% Memory-intensive operation done here
load('data_storage.mat','data');     % Restore data from disk
```

Loading the data variable into the workspace in this manner is called *"poofing"* and is generally discouraged as a bad programming practice.[1522] An alternative approach is to use the following manner to load the data:

```
data = load('data_storage.mat','data');
data = data.data;
```

9.5.5 Global and Persistent Variables

Using global variables enable us to modify them within functions without needing to worry about the whether or not new copies will be made when they are modified (refer to the copy-on-write mechanism discussed in §9.5.1).[1523] This reduces the number of memory allocations and deallocations, and also simplifies the code since we do not need to pass these variables as function input arguments. However, keep in mind, especially in light of the plentiful drawbacks of using globals (see below), that in most cases we can use standard in-place data manipulations or nested functions to achieve the same memory benefits as globals.[1524]

Global variables are similar to references that we find in other languages (e.g., C++, C#, and Java), whereby the object behaves just like any other local variable, but in fact is a pointer into the main (*heap*) memory rather than a local stack allocation.[1525]

Unlike other variables, which have *local scope* (meaning that they are only accessible in the function that declared them and in its nested functions), global variables have *global scope*,[1526] meaning that they are accessible anywhere within the current MATLAB session, even outside their declaring scope.

Within any MATLAB function that needs to access global variables, we need to declare the variable to be global using the *global* directive:

```
global myGlobalVar
myGlobalVar(5) = pi; % use global variables just like regular vars
```

global declarations should appear at the top of their code block, that is, the beginning of the containing function. They can also appear anywhere later in the code, but in that case the JIT Accelerator apparently finds it harder to optimize the code.

The *global* directive, unlike most other MATLAB directives, does not have a functional form. This means that we cannot use a *global('myGlobalVar')* form in our code. As a direct consequence, we cannot dynamically declare variables (i.e., assemble global variable names on-the-fly, to be passed on to the *global* function) in MATLAB code. This is not

---

* Also read the related discussion of Malcolm Lidierth's vvar utility, at the end of §9.4.3.

necessarily a bad thing: having dynamic global variables can be a maintenance nightmare and a cause for hard-to-trace bugs.

In those rare cases where we need to use dynamic names globally, we could always create a global struct and dynamically manage its fields:

```
global myGlobalVar
dynamicFieldName = 'data';
myGlobalVar.(dynamicFieldName) = magic(3);
```

Using global variables can be very beneficial for performance, but has drawbacks:

- Using global variables is generally considered a bad programming practice.[1504] Using globals unwisely can lead to code maintenance difficulties, hard-to-trace bugs, unexpected behaviors, and so on (see details below). Extra care should therefore be taken with globals, to prevent such nastiness.

- Global variables can affect any part of the program, even parts that are entirely unrelated to the modification location. Doing so is an extremely bad programming practice (the *"Action at a Distance"* design antipattern[1527]) that often causes difficult-to-trace bugs and maintenance difficulties. If you must use globals variables, at least give them non-generic names and concentrate their updates at functionally relevant locations.

- We need to declare globals using the ***global*** directive wherever we need them (i.e., within every function). Aside from the associated nuisance, failing to do so may cause MATLAB to automatically assign a local variable by the same name that will *shadow* the global variable,[1529] causing hard-to-detect bugs.

- It can be very difficult to keep track of global variables usage and updates, especially if they occur in multiple functions and m-files.

- Using globals rather than passing parameters to functions improves the performance at the expense of code readability and maintainability, by hiding the functional dependencies that exist between different functions.[1530]

- The ***clear global*** directive clears all global variables, whereas ***clear global myVar*** only clears *myVar*. Failing to add the variable name to the directive can inadvertently clear all globals, causing hard-to-detect bugs.

- ***clear java***, ***javaclasspath***, and ***javaaddpath*** all have the same effect as ***clear global***, of clearing ALL global variables from memory. In this case, clearing the globals is actually an undesired side effect that is often difficult to trace since globals are not directly related to Java (at least in their usage).

- Global variables are persistent across program runs within the same MATLAB session, but not across different MATLAB sessions. This behavior is inconsistent and can be difficult to diagnose.

- Global variables are not protected from modification by other MATLAB programs, if these programs happen to access the same global variable names.

- Global variables are not protected from modification by users in the MATLAB workspace (Command Window) while the program is running.

- There can also be subtle inconsistencies between the behavior of globals in MATLAB versus deployed (compiled stand-alone) applications.[1531]

- The allocation of global variables on the heap ensures the persistence of the memory space, but also means that such variables have a somewhat slower read-access time than simple function input arguments or local variables, which typically use the stack or even CPU registers. So, if you are not using the persistence feature, and do not have a copy-on-write situation to avoid, then you may wish to avoid global variables.

- Global variables are generally unsafe in multithreaded environments. Since MATLAB's main computational engine is single threaded, this is one drawback that does not seem to apply in MATLAB's case.

As an alternative to using global variables, we could use MATLAB Desktop's **UserData** or **ApplicationData** properties to store global information (note that there are other GUI-based[1532] and class-based (singleton) alternatives):

```
set(0,'UserData',magic(3));
setappdata(0,dynamicFieldName,magic(3));
```

Both **UserData** and **ApplicationData** can store any MATLAB data type (scalars, multidimensional arrays, object handles, structs, cell arrays, etc.). However, the convention in MATLAB is to use **ApplicationData** to store only a single struct. In fact, this is assumed by many built-in MATLAB functions, so I strongly suggest sticking to this convention. This is actually easy to do, if we access **ApplicationData** only via the *getappdata, setappdata,* and *isappdata* built-in accessor functions.

Persistent variables[1533] are similar to global variables in the sense that they are allocated on the heap and act like references. Like globals, they are also preserved in memory throughout a MATLAB session, except if their containing function has been edited. Like globals, they are also declared at the top of the function (using the *persistent* directive) and are initialized to the empty array [].

However, unlike global variables, persistent variables have local scope, like any other local function variable. This means that while a persistent variable value is *preserved* across multiple invocations of its containing function, it is not accessible *outside* the function scope (unlike globals that are accessible anywhere in MATLAB).

Using persistent variables ensures that we are not reallocating memory each time the function is called. This way, we can reuse the previously allocated persistent memory block, even if we do not really need to use the previous values. Relying on in-place optimization we can minimize memory allocations without the drawbacks that accompany global variables.

Persistent variables are often used to implement caching, a widely used performance technique.* Persistent variables can also be used to prevent unnecessary running of code, for example, to prevent GUI callback re-entrancy.†

To prevent a persistent variable from being cleared by the *clear* command, use the *mlock* command within the function that uses the persistent variable.

9.5.6 Scoping Rules and Nested Functions

The scoping rules used for nested functions[1534] can provide a convenient and effective alternative to global and persistent variables.[1535] The basic idea is that nested functions share the workspace of their parent function, including all locally-defined variables and input

---

* See §3.2.
† See §10.4.3.3.

arguments. For this reason, there is no need to rely on the copy-on-write and in-place manipulation mechanisms, since the nested function works directly on the parent function's data.

For this reason, there is no need to pass data into a nested function as input arguments, nor to return them to the parent function as output arguments. This enables much simpler function interfaces than regular functions that might have numerous input and output arguments to achieve the same functionality.

Nested functions are very easy to use, since they do not require a declaration of any of the parent's variables/arguments in order to use and modify them within the nested function. This contrasts with global or persistent variables, which need to be declared in all functions in which they are used. This solves the data-copy problem without any specific code change required beyond placing the code within a nested function.

Using nested functions may create maintenance headaches, since the function's variables are not all defined in the scope of the immediate function, but may be declared in its parent, or some other ancestor function. If a variable is redeclared within a nested function, this may cause hard-to-trace bugs.

Also, by directly updating the parent workspace, side effects may be inadvertently introduced: nested function A can affect parent variable V although this is not directly evident from the code:

```
function V = nestedExample()
   V = magic(3);

   function nestedFunctionA()
      V = -1; % side-effect of changing the parent's V
   end
end
```

Therefore, while nested functions provide a powerful and easy way to facilitate in-place manipulation, care should be taken to reduce such side effects only to specific well-commented cases.

A limitation of nested functions is that all nested and parent functions must be enclosed by the end keyword, which simple functions and subfunctions do not require.

Another limitation is that nested functions must reside in the same file as their parent and all other ancestor functions. If the nested function is large, it may seem wise to split its contents into one or more separate files that contain a utility function. However, unless this is done carefully, we may run into the same problems of data-copying that we were trying to solve by using nested functions. We would then be faced with the trade-off dilemma of performance versus maintainability.

Using nested functions does carry a small performance penalty — the overhead of calling another function. Depending on the platform, this can amount to between a few to several hundred microseconds per invocation (on modern computers, the number is much closer to the lower limit). Therefore, depending on the situation (e.g., if the nested function is called many thousands of times), the benefit of sharing the parent's workspace may prove to be lower than the disadvantage of the extra performance overhead. Note, however, that such a situation is usually quite rare: in most cases the in-place data manipulation benefits outweigh the overhead cost.

Nested functions became available in MATLAB 7.0 (R14), and so this alternative is not available on earlier MATLAB releases.

One caveat of using nested functions is that on MATLAB release R2011b (7.13) and earlier, accessing parent cell arrays is significantly slower than accessing numeric arrays. This was apparently fixed in MATLAB R2012a (7.14):[1536]

```
% Nested function accessing a numeric array
function fh = perfTest
    A1 = zeros(5000,5000);
    fh = @nestedFunc;
    val = 1;
    function nestedFunc()
        val = val + 1;
        A1(1,1) = val;
    end
end
```

```
>> fh=perfTest; tic; for idx=1:100; fh(); end, toc
⇨ Elapsed time is 0.000551 seconds.
```

```
% Nested function accessing a cell array on R2011b
function fh = perfTest
    A1 = cell(5000,5000); % changed from numeric to cell array
    fh = @nestedFunc;
    val = 1;
    function nestedFunc()
        val = val + 1;
        A1{1,1} = val; % changed from numeric to cell array
    end
end
>> fh=perfTest; tic; for idx=1:100; fh(); end, toc
⇨ Elapsed time is 19.870618 seconds.        ← 36Kx slower(!)
```

9.5.7 Passing Handle References (Not Data) to Functions

An alternative means of ensuring that data is not copied in a called function is to encapsulate the data within a handle reference object. In MATLAB, this means to place the data as an internal member (class variable or property) of a class that derives from *handle*. This creates a *handle class*, as opposed to a *value class*.[1537]

Value classes use copy-on-write when passed as function arguments, whereas handle classes always update the original object (they use a *pass-by-reference* mechanism). This means that if we pass a value object to a function and then update some property, a copy is made of the entire object (not just the affected property). This could be quite costly, both during the creation time and during destruction (cleanup by MATLAB's garbage collector).[1538] On the other hand, updating a handle class always affects the original memory, so only the affected property is reallocated, not the entire object. In practice this means that using handle objects avoids the need for costly copies as in value objects.

The basic code structure for creating a handle class is as follows:

```
classdef MyClass <handle
    properties
        name = 'defaultName';
        dept = 'defaultDept';
        data = magic(4);
    end
end
```

Of course, if we set any of the class object's properties, that particular data might well be reallocated as necessary. But this will not affect or reallocate any of the other data members of the object. Moreover, MATLAB will not spend time to allocate a copy of the entire object at the top of the function, and deallocate it at the end.

As with nested functions,* updating the original data carries the risk of unintended side effects. However, in this case, the side effects are easier to spot, since the affected object is passed to the affecting function as an input parameter, so it is easy to identify the functions that affect an object. Using class objects also enables much better scope control than using *global* variables.

While being easier to understand when reading existing code, class objects (and object-oriented programming in general) are admittedly more cumbersome to develop than straightforward variables.

Note that reference (handle) classes are only available in MATLAB 7.6 (release R2008a) and newer.

9.5.8 Reducing Data Precision/Type

MATLAB's default data type is *double*, which uses 8 bytes (64 bits). Unless we specifically request using a different data type, 8 bytes will be allocated for every data element, even if this extra memory is not needed.

For example, if we only have low-resolution floating-point data (meaning that a data resolution of 10^{-7} is enough, and do not need 10^{-16}), then consider using the *single* data type (4 bytes), which uses half the memory as the same data in *double* resolution.

If our data contains integer values, we could perhaps use one of the integer data types: *int8* and *uint8* use only a single byte, useful for values –128 to +127, and 0 to 255, respectively; *int16* and *uint16* use 2 bytes for values between –32768 and +32767, or 0 to 65535, respectively. There are also corresponding *int32/uint32* and *int64/uint64* data types that use 4 and 8 bytes, respectively.

If our data only stores boolean (true/false, on/off) values, it would be much better to use the *logical* data type, which uses a single byte. Logical arrays are also much better for storing searchable array indices, as explained in §4.1. We can even pack logical arrays into bit arrays, storing elements in a single bit, such that each byte stores 8 logical values.[1539] This requires extra computations, but reduces memory usage by eightfold compared to the *logical* type, and 64-fold compared to *double*.

These alternatives are all better than using the default *double* data type. MATLAB will use much less memory for storing data in these data types, thereby requiring much less time to allocate, copy, and deallocate memory.

When creating data of a non-*double* data type, try to avoid temporary allocations of the default *double* data type, as explained in §9.4.5. For example:

```
% Bad idea: allocates 8MB double array, then converts to 1MB int8 array
data = int8(zeros(1000,1000)); % 1M elements
⇨ Elapsed time is 0.008170 seconds.

% Better: directly allocate the array as a 1MB int8 array - 80x faster
data = zeros(1000,1000,'int8');
⇨ Elapsed time is 0.000095 seconds.
```

---

* See §9.5.6.

Using non-*double* data type has some drawbacks: not all MATLAB functions support these data types (the exact list of supported functions changes between releases, each release adding non-*double* support to some additional functions); the performance of some MATLAB functions on non-*double* data is degraded compared to *double* data; care must be taken by the programmer to prevent underflows, overflows, and other similar issues that arise when mixing data types within a computation.

Still, the potential performance gain of using a reduced-precision data type is well worth the trouble of trying it. In one specific case, I have found that reducing the data type from *double* to *single* was all it took to make a nonresponsive application (due to disk thrashing, etc.) fast enough to work in real time versus a medical device. Refer to §4.1 for a more detailed discussion.

9.5.9 Devectorizing Huge Data Operations

Sometimes it makes sense to break up vectorized operations on huge data into smaller chunks, so internal memory allocations do not impede performance. For example, if we have Vector1, Vector2 both having 20 million elements (i.e., 152 MB) and try:[1540]

```
Vector1 = Vector1 + exp(param1+param2*rand(20e6,1)).*(param3<Vector2);
```

then we understandably get out-of-memory errors and memory-related performance issues. The code is indeed vectorized, but exactly because of this vectorization, all the data is processed at once, leading to memory bottlenecks.

Instead, split the data into smaller pieces that are easier for MATLAB to "chew":

```
numOfChunks = 10;
chunkSize = length(Vector1) / numOfChunks;
for idx = 1 : numOfChunks
    idxRange = (idx-1)*chunkSize+1 : idx*chunkSize;
    Vector1(idxRange) = Vector1(idxRange) + ...
        exp(param1+param2*rand(chunkSize,1)).*(param3<Vector2(idxRange));
end
```

Here, we have split up the code into 10 chunks, processing each chunk in a mini vectorized manner. Using indexing is less efficient than full vectorization,* and the resulting code is certainly harder-to-maintain, but at least this enables us to overcome the memory issues. In addition, chunking could also improve performance.† Depending on the situation, it should be fairly easy to determine an optimal chunk size for each specific problem, by simply playing around with different values for the number of chunks.

We do not need to avoid vectorization altogether. We can reshape the data using *mat-2cell* into a cell array where each cell has the requested subblock size. We can then process the subblocks in a vectorized manner using *cellfun*.[1541] For example:

```
% split data into sub-blocks of size N x M
cellRows = N*ones(size(data,1)/N,1);
cellCols = M*ones(size(data,2)/M,1);
dataCells = mat2cell(data, cellRows, cellCols);
results = cellfun(@someProcessing, dataCells); % process via cellfun
```

---

* See §9.3.3.
† See §3.6.3, §6.4.1, §11.3.4, and §11.3.5.

An alternative technique is to reshape the data using higher dimensions, such that the subblocks will occupy dimensions that can be operated on in a vectorized manner. For example, reshape data(X,Y) to data(N, X/N, M, Y/M) and then operate on the subblocks of size N×M:[1542]

```
data2 = reshape(A, N, size(A,1)/N, M, size(A,2)/M);
y = squeeze(sum(sum(data2,1),3))/(N*M); % mean values of all sub-blocks
```

Splitting up a large data set into separate independent data subblocks enables parallelization, if we have the relevant hardware and software support for this. Parallelizing such independent subblocks can significantly improve the overall processing speed, by simply replacing the *for*-loop with a *parfor*.*

MATLAB's Image Processing Toolbox (IPT) includes the **blockproc** function since R2009b (7.9), which simplifies processing image blocks in small blocks.[1543] This enables processing huge (multi-GB) images in small blocks without overflowing the physical memory. The function supports file-to-file workflow, meaning that each block is separately read from file, processed in memory, and then written to file. At no time does the entire image file need to be loaded/written. The **blockproc** function also includes a 'UseParallel' parameter,† which enables the processing to run in parallel using the Parallel Computing Toolbox (PCT) if available. A MathWorks newsletter from 2012 discusses this and other related IPT performance improvements, such as multithreading enhancements in R2007b (7.5) and R2010a (7.10).[1544]

Jan Simon's provided a fast and simple *BlockMean* MEX utility[1545] that does the same thing, namely, breaking up a large data array into N×M blocks and running a common function (in this case, **mean**, or rather its C equivalent) on each of the blocks. *BlockMean* is easy to use and its MEX code can easily be adapted to other processing functions rather than the mean. Importantly, it runs significantly faster than the corresponding pure m-code. However, if you have PCT, devectorizing using explicit parallelization might be even faster.

The idea of breaking up a large problem into smaller, more manageable chunks is further discussed in §3.6, §6.4.1, and §11.3. Readers are referred there for additional details and usage examples.[1546]

9.5.10 Assign Anonymous Functions in Dedicated Wrapper Functions

Anonymous functions are a great MATLAB construct, enabling us to assign functions on-the-fly (lambda constructs), to handle data in other functions that accept function handles. For example, we could use an anonymous function to test which members of a cell array contains some substring:

```
>> strings = {'1234','34536','234521'};
>> anonFunc = @(value) ~isempty(strfind(value,'23'));
>> cellfun(anonFunc,strings)
ans =
     1     0     1
```

Anonymous functions are great because they enable us to use functions and function handles without the overhead of setting up dedicated function code. That is:

```
anonFunc = @(value) ~isempty(strfind(value,'23'));
```

---

* See §6.1.
† We have already encountered UseParallel with the *optimset* function in §4.5.5.

is naturally simpler and easier to maintain than the otherwise equivalent

```
function result = anonFunc(value)
    result = ~isempty(strfind(value,'23'));
end
```

Unfortunately, anonymous functions entail a little-known (although fully documented[1547]) memory allocation overhead. In order to operate properly, the anonymous function stores a copy of the entire function workspace.*

If we only use the anonymous function within the same function in which it is defined (as is often the case), then the extra copy does not affect performance very much. The reason is that MATLAB's copy-on-write mechanism merely creates references to the data, and does not immediately copy the data itself. This changes, of course, if the code makes any modification to the data, and even if the data is not directly used by the anonymous function. For example:

```
function test()
    data = rand(5000);   % 5Kx5K = ~200MB
    anonFunc = @(value) value+5;
    data = rand(3000);   % 3Kx3K = ~72MB
    % do some other stuff here...
end
```

You might assume that when we reassign rand(3000) data, this should reduce the allocated function memory from ~200 MB to ~72 MB. Instead, since anonFunc remains in scope, it keeps a hold on the original 200 MB data, which cannot therefore be deallocated. Therefore, when data is reassigned, the 72 MB are added to function's memory, for a total of ~272 MB. We can easily see how this may be detrimental to performance, especially when the stored data is large.

The anonymous function only releases the stored workspace variables to the garbage collector, which can then deallocate the memory, when the containing function (test()) exits. Even clearing the anonymous function handle is of no use for the garbage collection.

We can see the stored workspace data using MATLAB's built-in *functions* function (the two workspaces apparently belong to the calling function and the container, which in this case is the MATLAB Desktop):

```
>> metaData = functions(anonFunc)
metaData =
     function: '@(value)value+5'
         type: 'anonymous'
         file: 'C:\Yair\Books\MATLAB Performance\Code\test.m'
    workspace: {2x1 cell}

>> metaData.workspace{:}
ans =
1x1 struct array with no fields.

ans =
        data: [3000x3000 double]
    anonFunc: @(value)value + 5
```

---

* Apparently, this is only done for anonymous functions created within a function; anonymous functions created in the MATLAB Desktop (Command Window) only store those variables that directly affect the function's inner calculations.

A further complication arises when we persist the function handle in a disk file.[1548] Naturally, storing a large amount of data on disk takes a long time for both the saving and the loading parts. This is especially painful when we consider the fact that quite often we do not actually need all the extra workspace baggage in the first place!

The solution to all these problems is to place the anonymous function assignment in its own helper (wrapper) function that has a clean workspace:[1549]

```
function test()
    data = rand(5000); % 5Kx5K = ~200MB
    anonFunc = assignAnonFunc();
    data = rand(3000); % 3Kx3K = ~72MB
    % do some other stuff here...
end


function anonFunc = assignAnonFunc()
    anonFunc = @(value) value+5;
end
```

In this case, anonFunc does <u>not</u> store a copy of data or any other variable, since they are not visible in the scope of the assignAnonFunc() function where anonFunc was created. In consequence, the second data reassignment deallocates the 200 MB, reusing part of the deallocated memory block for the 72 MB, as expected.

When this cannot be done, then at least *clear* any variable that is no longer needed prior to defining the anonymous function.

9.5.11 Represent Objects by Simpler Data Types

Whereas a normal engineering best practice is to abstract complex information (e.g., date/time information and machine status) in class objects, such class objects carry a performance penalty.

In rare cases, and only after you have verified that the creation of such class object instances is a performance bottleneck in your program, it could make sense to replace the use of the class object with a much simpler data type. For example, the date/time object could be represented by a single *double* value (like *datenum*), and machine status could be represented by another numeric value. Unfortunately, even simple *double*s are stored by MATLAB in heap memory rather than the function's stack. But even so, their allocation, initialization, and deallocation are faster than complex objects.

As a variant of the above suggestion, instead of using an array of class object instances, you may consider holding an array of simple variables, each of which holds a separate property. For example, rather than holding an array of 10 class instances

```
classdef MyClass
    properties
        width
        height
    end
end
...

classObject(10) = MyClass;
requestedHeight = classObject(10).height;
```

consider holding separate simple variable arrays as follows:

```
myClassWidths(1:10)  = 5;
myClassHeights(1:10) = 10;
...
requestedHeight = myClassHeights(10);
```

Along the same lines, if you have a hierarchy of class definitions, where each class holds properties that are specialized classes and so on, consider *flattening* out the class definition such that the base class only contains simple (primitive) data types. For example, instead of

```
classdef Employee
   properties
       name = NameClass('John','Doe');
       address = AddressClass(1500,'5th Ave.','New York','NY','USA');
       salary = 12345;
   end
end
```

consider using the following flattened class definition:

```
classdef Employee
   properties
       firstName = 'John';
       lastName = 'Doe';
       streetNumber = 1500;
       streetName = '5th Ave.';
       city = 'New York';
       state = 'NY';
       country = 'USA';
       salary = 12345;
   end
end
```

Of course, using these simpler data representations carries a nonnegligible maintenance cost. Your program is now harder to debug, understand, and in general maintain than before. But the effect of removing the allocation, initialization, and later deallocation of the class instances might just be the deciding factor for the performance of your program. I am certainly NOT recommending this as a general practice, but only in the rare cases when you really have no other way of achieving your desired performance goals.

9.6 Memory Packing

Memory packing refers to the process of rearranging MATLAB's process memory in such a manner as to pack all the used memory at the base of the MATLAB memory space. This increases the amount of *contiguous* memory available for MATLAB variables, thus avoiding out-of-memory errors during variable memory allocations.

Variables that are no longer needed can be *clear*ed.* This increases the total available memory and *may* also increase the size of the largest contiguous free memory block.

MATLAB has a built-in function *pack* that purports to pack memory. The basic idea is to *save* all workspace variables to a temporary MAT file, *clear* these variables from memory, and then *load* the variables back from the file. Unfortunately, *pack* only works in the MATLAB Command Prompt, and only affects the small portion of memory that stores base workspace variables. Memory used for storing internal, global, object, Java, and graphics data is unaffected by the *pack* mechanism. The bottom line is that *pack* is effective only in very rare circumstances, although it has starred in MATLAB's official memory management guide[1550] for years.

A trick that can be used to better pack MATLAB memory is to save all current data, restart MATLAB, then reload the data. This has the effect of loading all variables contiguously. The effect is only temporary of course, but could be useful in some cases, especially for long-running MATLAB sessions that have highly fragmented memory. It is certainly more effective than *pack*, since it also packs the internal data that is untouched by *pack*. Even better memory packing can be achieved by restarting the computer, since this also packs the memory used by other OS processes, which contend with MATLAB over the available system memory. Obviously, restarting MATLAB and/or the computer takes quite some time, and might be unacceptable. Restarting MATLAB can be done programmatically, as follows:

```
system('matlab.exe -r myFunctionName &'); exit;
```

Using *feature('dumpmem')*[†] we can check which memory modules are adjacent to the largest contiguous free memory block. If we see a library (DLL) loaded immediately adjacent, we can possibly move (*rebase*) the DLL to a different address, thereby increasing the contiguous memory block's size. This solution works only on Windows XP SP2 machines running MATLAB releases 7.0 (R14) through 7.3 (R2006b), and has some MATLAB functionality side effects.[1551] In some cases, it may be worthwhile to preallocate large memory blocks before loading user DLLs, in order to save these blocks for later use in MATLAB, rather than by the DLLs.[1552]

Additional hints on ways to enlarge the available memory are reported by the *memory* function.[‡]

---

* See §9.5.4.
† Stuart McGarrity's *chkmem* utility can be useful for analyzing its output—see §9.2.4 for details.
‡ See §9.2.3.

9.7 Additional Recommendations

9.7.1 MATLAB Variables

Some recommendations that appeared in earlier chapters are also related to memory management, and are therefore repeated here. Some of these can be found in MATLAB's official guide *Strategies for efficient use of memory.*[1553]

Load, process, and save only as much data as you really need, not all of it (see §3.6).

Avoid the form a(:) or data(:,:), but rather use a and data as is, without the (:) parts. Adding these full-index notations forces MATLAB to unnecessarily parse the data, which degrades performance:[*]

```
a = rand(1000);
b = rand(1000);
tic; c = a*b; toc
⇨ Elapsed time is 0.055416 seconds.

tic; c = a(:,:)*b(:,:); toc % 13% slower
⇨ Elapsed time is 0.062494 seconds.
```

Reduce the number and size of temporary nonprimitive objects, and large temporary data variables in general. For example, the simple operation *data = data + data* may run into memory issues (performance slowdown due to swapping, or even out-of-memory error) since it creates a temporary array holding the result before assigning back into *data*.[1554] In-place data manipulations (§9.5.2) are very useful in this regard, but they are not always available. It may be useful to employ data chunking (§9.5.9).

Use the ~ operator (R2009b and newer[1555]) to denote unused output arguments of function calls. In general, do not accept unneeded output args if they are not needed. For example, if we only need the first output arg of *f()*, then use *a = f()*; rather than *[a,b,c] = f()*. If we are lucky, the function *f()* will not spend any time computing the unrequested output variables. But even if it does compute them, MATLAB will not spend time in allocating them in the caller workspace.

Avoid creating objects in frequently called code segments (loops, functions, object methods). The objective here, again, is to reduce memory allocations. If necessary, reuse existing variables, use nested functions, and/or *global/persistent* variables.

Reduce the complexity of the structure that holds the data. *struct*s are generally more problematic than cell arrays, which are in turn generally more problematic than arrays (see §4.1). Using a *struct* of arrays is generally better than an array of *struct*s, both in terms of memory and performance, since arrays have minimal memory overhead and are stored contiguously, which in turn enables fast vectorization and parallelization.

On the other hand, *struct* elements are stored separately and have dedicated management overhead. Therefore, we would be better off whenever we can bypass *struct* overheads by directly accessing all the data within a simple array. If we do have the less efficient form (array of *struct*s), we can still use vectorization of constituent fields using the square brackets notation, as shown in §5.7.12.

Similarly, using a cell array of numeric data arrays is preferable (performance-wise) to using an array of cell arrays.

---

[*] See related §9.3.3.

In certain cases, sparse matrices can be orders of magnitude more memory-efficient than the corresponding regular (full) matrix. Refer to §4.1.3 for additional details.

Numeric data is often processed faster when using the native *double* precision. However, in certain specific cases it may be more efficient to use a *single* or integer data type. This becomes even more important for huge data arrays, where allocation performance and memory-limitation issues (i.e., swapping/thrashing) come into play.

The most compact data types (excluding bit-compressing algorithms) are integers, followed by single-precision data, double precision, cells, and structs. Class objects, datasets, tables, and categorical arrays (§4.1.6) use the most space, but provide metadata information and enable easier development and maintenance.

When performing a binary operation on data, it is often better and faster to use *bsxfun* (§5.2.4.2) than *repmat* or other alternatives (§5.4.2). Using *repmat* often requires data replications and interim matrices, sometimes enormous in size, which are avoided using *bsxfun*.

Increase your platform's physical memory (RAM) size (§3.5.3). The more physical memory is available, the less virtual memory swapping is needed, and the better the overall performance — not just of MATLAB but of all the computer's processes. Note that different operating systems have different limitations on the maximal amount of physical memory. For example, we cannot use more than 4 GB of memory on a 32-bit Windows running on an x86 32-bit system, and the maximal physical memory available to the MATLAB process is even lower than this.[1556] If you have a 32-bit MATLAB on Windows, it is advisable to use Windows' /3GB boot switch,[1557] in order to allow the MATLAB process access up to 3 GB of RAM (where available), rather than the default process limit of 2 GB. Note that this is a Windows (not MATLAB) startup switch, and should be added to the computer's *boot. ini* file.* The 3 GB mechanism is not a general panacea, otherwise Microsoft would have made it the default. Some odd drivers assume they have 2 GB of memory, where in fact the OS can only use 1 GB, with the other 3 GB reserved for user processes. This might cause Windows to misbehave when 3 GB is used.[1558]

To be able to use more physical memory, we need to use a 64-bit operating system running on a 64-bit CPU and running a 64-bit version of MATLAB. All three components must be 64-bit, otherwise MATLAB falls back to 32-bits.† The 64-bit MATLAB version is not inherently faster than the 32-bit version, but its access to the extra physical memory can significantly improve MATLAB application speed in R2011a (7.12) and newer. Unfortunately, on R2010b (7.11) and earlier, the 64-bit MATLAB might even be slower (!) than its 32-bit counterpart.[1559]

Increase the platform's allocated virtual memory (on Windows‡) or swap space (Macs or Linux§). This may improve performance by *reducing* memory thrashing, due to cross-effects with other OS processes that also use this virtual memory.

On Windows platforms, MATLAB uses Windows standard memory heap manager.[1560] As a result, memory allocation is not released to the general pool when cleared, but rather reserved for future use by new allocations. This behavior causes gradual fragmentation of the MATLAB process memory, preventing allocations of large memory arrays. The effect is

---

* On 32-bit Vista and Win7, the 3 GB mechanism is activated differently, via a *bcedit /set IncreaseUserVA 3072* command: http://msdn.microsoft.com/en-us/library/ff407021%28v=VS.100%29.aspx (or: http://bit.ly/1dkKO45).
† 64-bit OSes running 32-bit MATLAB do have one advantage over standard 32-bit MATLABs in that they can use up to 4 GB of process memory, rather than 2 or 3 GB. This could reduce memory swapping/thrashing for memory-hungry applications.
‡ Using *Control Panel / System / Advanced.*
§ Using the *ulimit* or *limit* command, depending on your *nix flavor.

normally limited to 512 KB data (a 64 K element double-precision array) as the documentation for the *memory* function implies, but can sometimes occur for larger allocations as well. As a result, it is better to allocate large arrays before smaller ones, to ensure that they have access to as much free contiguous memory as possible, before it gets fragmented.[1561] We can also use chunking (§3.6.3, §9.5.9) to reduce allocation sizes, thereby avoiding the need for large contiguous free memory.[1562] Cell and struct arrays allocate their elements in separate memory locations, so using them also avoids the need for large contiguous memory blocks.

MATLAB memory fragmentation occurs as early as during startup, when DLLs load into free memory slots (see §9.2.4). We can avoid some of this startup fragmentation using the *-shield* startup option on Windows platforms.[1563] Various levels of memory shielding are available, from 500MB up to 1.5 GB. See the documentation for details.

MATLAB is an extremely complex application, so it is not surprising that it has memory leaks in many different sections of its code.[1564] When reviewing the list of over 400 reported memory leaks on the CSSM newsgroup (many of which are actually user-caused, by forgetting to free MEX allocations),[1565] it appears that MathWorks has cleaned up most major leaks in recent releases, as the number of reports has dramatically reduced over the years, yet another reason to upgrade our MATLAB release. Still, when running MATLAB for any length of time, MATLAB's memory usage tends to creep upward. At some point in time, this causes performance degradation, when the MATLAB engine allocates new memory that requires page swapping and ultimately disk thrashing. Avoiding leaks by changing application code may be one solution. However, in practice, detecting and avoiding internal MATLAB memory leaks is very difficult. A much more cost-effective solution is to periodically restart the MATLAB process (or the compiled application, if it is deployed). This is also the most cost-effective solution to MATLAB memory fragmentation.

To reduce memory leaks in our application, we should ensure that all object references, especially those referenced by internal class fields, are released and freed. MEX functions are an endless source of memory leaks — we must ensure that we *mxFree()* and *mxDestroyArray()* all allocated memory constructs, especially if they are large and/or allocated numerous times (as in a loop or frequently called function).* When allocating memory in MEX code, ensure to use *mxAlloc()* rather than the standard C *alloc()*, in order to ensure that the memory is handled by MATLAB's memory manager and potentially garbage-collected when becoming unused.[1566] Similarly, use *mxFree()* rather than *free()*.

After some run time, an application might accumulate large amounts of data. It is a good idea to periodically clear old ("stale") data items from memory, thereby freeing space for new entries. If we have direct access to the data we can do this programmatically by removing some data elements, or by using *clear* on unused variables. For *persistent* cache variables, we can simply issue a *clear(functionName)* command, which will clear all the persistent variables in the specified function. The nice thing about this is that we can issue the *clear* command from outside the function, although we have no access there to the function's internal *persistent* data. This *clear(functionName)* mechanism can also be used to clear functions from MATLAB memory, after they have been processed and are no longer needed. This could prove useful if the functions include a large amount of internal memory (e.g., if data is loaded via human-editable m-files rather than from MAT files[1567]).

Some user-created objects remain in memory even when their original creation handle goes out of scope and may no longer be accessible.[1568] In this case, object scoping, which normally garbage-collects memory that goes out of scope, is not applicable. This is basically a user-generated memory leak, and *not* a MATLAB bug. Timers, figure windows, I/O handles,

---

* See §8.1.6 for details.

and loaded functions/libraries are all examples of such *user-managed objects*. The memory associated with these objects may be significant, and we should remember to delete them whenever they are no longer used. It is not sufficient to simply *clear* the handle variable, since the underlying object remains in memory even when the handle variable is cleared. Instead, we should specifically delete the objects from memory. Figures and timers can be deleted using *delete*; I/O handles using *fclose*; and functions using *clear*. Consider using *onCleanup* (§4.6.2) to ensure such objects are deleted even in case of an unexpected error.

I/O is often a limiting performance factor, so limiting the amount of data loaded from I/O is often beneficial. This can be done by partial reading of files (§11.3), processing sub-blocks (§3.6), and using System Objects for stream processing (§4.7.4).

9.7.2 Java Objects

When using Java object references in MATLAB, the objects are typically garbage-collected by the JVM (Java Virtual Machine) when the reference variable is cleared or updated. An exception to this rule may occur when MATLAB and Java code are interlaced in such a way that the JVM cannot garbage-collect an object since it is still referenced by an internal MATLAB object property, even after it has been deleted. This leads to a memory leak of the Java object, and multiple such leaks in run-time could result in a performance slowdown. To solve such potential memory leaks, we should specifically reassign the Java reference property in the MATLAB class object's destructor function (*delete*) to []. This reduces the Java object's reference count to zero, enabling JVM to garbage-collect it:[1569]

```
classdef MyClass
    properties
        jObject = java.util.Hashtable
    end
    methods
        function delete(obj)
            obj.jObject = [];
        end
    end
end
```

Tracking memory leaks is never easy. Tracking them in MATLAB is even more difficult than usual, due to lack of insight into MATLAB's actual memory map: the object seen in the Workspace represent only a small fraction of the objects that are actually allocated. Tracking Java memory leaks can be made somewhat easier using the tools and techniques presented in §9.2.7. A step-by-step tutorial on finding Java memory leaks was provided by Vladimir Sizikov some years ago.[1570]

One specific example of a Java memory leak in MATLAB is the fact that whenever a message is sent to the MATLAB Command Window, a small Java object is created to encapsulate this message. After many thousands of such messages, the Java heap space can get exhausted.[1571] The best solution appears to be a reduction in the amount of such messages, which is a good advice for performance in general (see §11.1).

To prevent memory leaks in complex GUIs, it is advisable to *get* and *set* callbacks using the *handle()* object, instead of directly using the so-called "naked" Java reference.[1572] Starting in MATLAB R2010b (7.11), setting callbacks on un-*handle*d Java references evokes a warning message, and in R2014a (8.3) setting callbacks on naked Java references was disabled altogether:

```
>> jb = javax.swing.JButton;
>> jbh = handle(jb,'CallbackProperties');
>> set(jbh,'ActionPerformedCallback',@myCallbackFcn)% ok!

% On R2010b-R2013b we can do the following, but we get a warning msg:
>> set(jb, 'ActionPerformedCallback',@myCallbackFcn)% bad! memory leak
Warning: Possible deprecated use of set on a Java callback.
(Type "warning off MATLAB:hg:JavaSetHGProperty" to suppress this
warning)
```

Similarly, MathWorks release notes for R2008b (7.7) warns that using *get* and *set* to access Java object properties might cause memory leaks:[1573]

```
% Bad usage - memory leaks
propertyValue = get(javaObject, 'PropertyName');
set(javaObject, 'PropertyName', newValue);

% Good usage - use accessor methods (no memory leaks)
propertyValue = javaObject.getPropertyName;
javaObject.setPropertyName(newValue);
```

Unfortunately, it appears that using the safer alternative (*getXYZ* and *setXYZ*) is slower than using *get* and *set*, or using dot-notation on the *handle*'d object:[1574]

```
jb = javax.swing.JButton;

% Using the standard set() function
for idx=1:10000, set(jb,'Text','testing'); end
⇨ Elapsed time is 0.278516 seconds.

% Using the HG handle() wrapper is about 35% faster (1.6x speedup):
jhb = handle(jb);
for idx=1:10000, set(jhb,'Text','testing'); end
⇨ Elapsed time is 0.175018 seconds.

% Using the HG handle() wrapper with dot notation is even faster
% (65% faster, 2.8x speedup):
for idx=1:10000, jhb.text='testing'; end
⇨ Elapsed time is 0.100239 seconds.

% Using the Java setText() function, is actually slower
% (faster with handle() wrapper, but still slower than dot-notation):
for idx=1:10000, jb.setText('testing'); end
⇨ Elapsed time is 0.587543 seconds.

for idx=1:10000, jhb.setText('testing'); end
⇨ Elapsed time is 0.201635 seconds.
```

The same holds true also for retrieving property values via the *get* function.

For efficient conversion between MATLAB and Java data, we can use the Lightspeed toolbox (§8.5.6.1)'s *toJava()* and *fromJava()* functions, as shown in §3.2.6.

 As a final Java-related recommendation, to convert a java.util.Vector (or one of its derived classes) into a MATLAB cell array, it turns out that the internal function *feature* is much faster than the standard *cell(jVector.toArray)*.[1575] Note that *feature* is an undocumented and unsupported function that can change without prior notice in new MATLAB releases:

```
% Prepare the data vector
v = java.util.Vector;
for idx=1:10000, v.add(idx); end    %10K elements

% Using the standard toArray() and cell() functions
data = reshape(cell(v.toArray),[],20);  % => 500x20 cell array
⇨ Elapsed time is 0.063515 seconds.

% Using the undocumented feature(44,...) function - 5x faster
data = feature(44,v,20);
⇨ Elapsed time is 0.012380 seconds.
```

10

Graphics and GUI

MATLAB has an extensive library of plotting functions, enabling data to be visualized in a variety of manners using plot charts and interactive controls (collectively called MATLAB *Handle Graphics*, or *HG*). Performance is usually not an issue when generating few simple graphs. However, performance may well be important in the following cases:

- When we generate complex charts, having numerous plotting elements
- When we need to generate multiple plots
- When we wish to update plotted data in a continuous flicker-less manner

In any of these situations, small performance improvements can add up to a significant effect on the overall timing, and may make the difference between an unusable flickering graph and a smoothly updating interactive plot.

The first part of this chapter is devoted to graphics performance, as it relates to data visualization in MATLAB plot axes. A didactic distinction is made between the initial creation of the plotted data (§10.1), and its subsequent update in real-time (§10.2). In practice, many of the presented ideas and techniques apply to both cases to some degree. Readers are therefore advised to keep both of these sections in mind when tuning graphics-laden programs.

Graphical user interface (GUI) is a generic name for any figure window having interactive user controls. Naturally, fast rendering of the GUI and timely interaction with its controls is very important for user-friendly GUIs. Particularly in GUIs, perceived performance* and visual feedback cues are of paramount importance. While a GUI may not in fact be faster, it would appear to the human brain as being much more responsive when animated visual feedback is presented and continuously updated. Although GUI appears superficially to be unrelated to plot axes, in reality they share many aspects and performance-tuning techniques, as we shall see.

The second part of this chapter discusses GUI: ways to speed up GUI rendering (actual performance), techniques to improve GUI interactivity and responsiveness, and mechanisms for continuous user feedback (perceived performance). As with plotting, GUI performance is also split between a discussion of figure-wide performance aspects (§10.3, which would also help speed up figures containing MATLAB plots), and performance aspects of specific GUI controls or other figure components (§10.4). The chapter concludes with a couple of performance pitfalls that apply equally to plotting and GUI.

---

* See §1.8.

10.1 Initial Graphs Generation

When generating a few simple graphs, it is normally not cost-effective to invest time in improving performance. However, if we need to generate complex plots that have numerous plotting elements, or when we need to generate multiple plots, then small performance improvements can add up to a large improvement in the overall timing.

Before describing specific tips, it is worth noting some common suggestions:

- Reduce the number of plotting elements (i.e., less graphic handles)
- Nonoverlapping plot elements plot faster than overlapping elements
- Use simple or no plot markers
- Plot vectorized data rather than plotting the elements one by one
- Use static axes properties
- Only use *drawnow* when you are finished plotting
- Use low-level graphic functions (*line, surface*) rather than high-level ones (*scatter, plot, surf*)
- Generate and customize plots while the figure is hidden
- Apply data reduction to plotted data to fit the display area
- Reuse plot axes whenever possible, rather than regenerating them
- Avoid using the *axes* function
- Use the painters figure **Renderer** with fast axes **DrawMode**
- Avoid *colorbar*s, or at least use a more efficient version of it
- Avoid *legend*s, or at least use a static one
- Reopen previously-saved axes and figures

10.1.1 Reduce the Number of Plotting Elements

The more plotting elements there are in a graph, the longer it takes MATLAB to prepare and display them. While this may not be meaningful in a plot having only a few lines and perhaps a few annotations (such as text objects), the effect starts to become noticeable when the number of elements rises over a few dozen.

MATLAB stores a reference to its plot elements in graphic handles. The more handles that MATLAB uses, the more memory it needs to allocate, renderings it needs to do, and ongoing management that it has to perform. All this translates into execution time, which adds up with the number of handles.

There are several ways in which we could reduce the number of graphic handles without affecting the visual output.

10.1.1.1 Do Not Plot Hidden Elements

In some cases, we may decide that we do not wish to display some data points, or plot lines, or text labels. It may appear simple from a programming standpoint to plot these graphic elements in any case, but just make them hidden (**Visible** = 'off'). However, creating these elements and then maintaining their handles carries a performance overhead. If we simply skip the generation of these elements altogether (rather than create them as hidden), the visual effect will be the same, while performance and memory will benefit. This may cause a bit of extra programming logic, to check whether the graphic elements need to be plotted or not, but the benefits might possibly outweigh this drawback.

```
% Standard plot having 100 hidden graphic elements
N=100;
x=rand(N,1);
y=rand(N,1);
cla; hold('on'); drawnow
tic
plot(0.5,0.5,'*r');
for idx = 1 : N
    plot(x(idx),y(idx)),'ob','Visible','off');
end
toc
⇨ Elapsed time is 0.085811 seconds.

% Same plot without the hidden graphic elements - 91x faster
cla; hold('on'); drawnow
tic, plot(0.5,0.5,'*r'); toc
⇨ Elapsed time is 0.000941 seconds.
```

Note that, as in the code snippet above, even if we do not store the graphic handles returned by the plotting functions, they are nonetheless created and can be accessible via the axes **Children** property. Therefore, we cannot rely on not storing the handles ourselves as a means for reducing the handles count.

A common programming practice when designing interactive GUIs is to plot all graphic elements at the beginning, but only make some of them visible. Depending on program conditions and user actions, some of the visible elements are made hidden, while some other previously hidden elements are made visible. This programming paradigm is relatively easy to set up, but its drawback is that at any point in time, all graphic elements (and their associate handles), even hidden ones, are actually plotted, taking up system resources and setup performance. It could prove beneficial to only plot the visible elements initially, adding visual elements as needed during run-time. This is admittedly more difficult to program. In some cases, it might even be detrimental to performance: setting an existing graphic handle's **Visible** property is much faster than actually plotting it from scratch, so online interactivity* would suffer. Users should decide whether or not to use this technique based on the question of how often the graphic elements need to change their visibility over the course of the program's execution.

It is sometimes possible to reuse existing graphic elements during the interactive GUI program execution stage. Instead of generating the new graphic elements, the existing elements' coordinates and other properties are updated. This improves the rendering performance. It also avoids the need to generate all possible elements from the onset, thereby improving the plot setup performance and overall resource utilization.

10.1.1.2 Do Not Plot Overlapped Elements

For similar reasons as above, if we have overlapping graphic elements, we should consider plotting only one of the elements. For example, if some of the data points have exactly the

---

* See §10.2 for a discussion of plot update performance techniques; see §3.8 for a related discussion of latency versus throughput.

same coordinates, then it would be beneficial for performance and memory to plot only one of the duplicate data points.*

The *histcompress* utility on the MATLAB File Exchange extends this idea and removes data points that are redundant within a specified tolerance.[1576] We can use this utility to remove data points that may not exactly overlap, but which do not really provide additional information and can therefore be removed safely.

Similarly, if we have an opaque *patch* object, then it would be beneficial not to plot any graphic element beneath the patch, since the patch would hide it in any case.

10.1.1.3 Use the Scatter Plot with More than 100 Data Points

The *scatter* plot has an undocumented side effect that when up to 100 data points are plotted, it stores a unique handle for each data point, whereas for 101 and more data points, it just stores a single handle.[1577] This naturally has a very beneficial effect for both performance and resource usage, which may explain why it was implemented in this manner.

We can use this side effect by simply recreating some data points such that we would have 101 or more, before calling the *scatter* function to plot our data:

```
% First plot our 70 data points in regular fashion
N=70; x=rand(N,1); y=rand(N,1);
cla; hold('on'); drawnow
tic, scatter(x,y); toc
⇨ Elapsed time is 0.041179 seconds.

% Now, ensure we have over 100 data points before plotting
% This is done by simply duplicating all 70 data points (total = 140)
tic, scatter([x;x],[y;y]); toc
⇨ Elapsed time is 0.005896 seconds.  ← 7x faster
```

A drawback to this trick is that since only a single handle is preserved for all data points, they can no longer be individually customized in terms of **Color** and **Marker**, as would be possible when 100 or less data points are plotted.

10.1.1.4 Use NaN or Inf Data Values to Reduce the Number of Line Segments

Rather than plotting separate lines in a single axes, we can concatenate the entire data into a single line that appears visually identical. To do this, we can use the fact that NaN and Inf data values appear as empty (transparent) data points that have no attached line connectors. Therefore, to concatenate different lines, all we need to do is to add a NaN or Inf data point in between the segments:

```
% Prepare the test data
t = 0 : 0.01 : 10;  % 1001 data points
y1a = sin(t);
y1b = cos(t);

% Standard (slower) alternative
line(t,y1a);
line(t,y1b);
drawnow
```

---

* See §10.1.1.5 for application of this technique. Also see §10.1.8.

```
% Faster alternative (visually identical)
t2 = [t,NaN,t];
y2 = [y1a,NaN,y1b];
line(t2,y2);
drawnow
```

In the particular example above, the performance difference is small (a few percent). However, the more line segments, the larger the performance boost. The difference can start to become significant in axes that contain dozens of separate line plots.

In addition to the cost of initial rendering, extra line segments carry the hidden cost of axes updates that occur in run-time. The more children objects that an axes has, the more the time MATLAB needs to spend to update them when the axes properties are updated in run-time (e.g., zoom in/out, panning, and resizing).

Additional performance aspects of MATLAB NaNs and Infs can be found in §4.5.1.

10.1.1.5 Use Only End-Points When Plotting Straight (Linear) Lines

We often need to plot straight lines, for example, to display upper/lower bounds, the X/Y axes, linear regression lines, and so on. In such cases, it may seem natural to use the same data size as the other plotted lines. For example, if our nonlinear plot has 100 K data points, we can easily create a linear plot line that also has 100 K data items, corresponding to the main data points.

This would be wrong from a practical standpoint. Instead, our linear plot line only needs to define its two end-points. Calculating the intermediate 99998 data points takes CPU time, extra memory, and extra graphic resources for plotting.

```
% Naïve approach: 100 K data points
plot(1:1e5,1:1e5); drawnow
⇨ Elapsed time is 0.029051 seconds.
```

```
% Smarter approach: only end-points - 4x faster
plot([1,1e5],[1,1e5]); drawnow
⇨ Elapsed time is 0.007346 seconds.
```

This technique is very effective in improving the performance and reducing memory usage of charts that display horizontal or vertical data, such as strip charts. In these cases, we should only plot the minimal and maximal data points (and the line connecting them) for each unique Y or X value, respectively.* Tucker McClure's *reduce_plot* utility[†] does this data reduction in run-time, but it is better to apply the data reduction at the source, before plotting, rather than to recompute it repeatedly in run-time. One way to do this is to use the *histcompress* utility.[‡]

For additional advice that relates to reducing the amount of plotted data, see §10.1.8.

---

* Strictly speaking, it would be even better to plot only two data points for each unique Y or X *pixel*, rather than *data value*. This will reduce the data even further, at the expense of extra computations and the need to recompute when the figure window is enlarged. For this reason I suggest to only merge unique *data values* and let the *reduce_plot* utility handle the pixel-reduction aspects in run-time.

† Discussed in §10.1.8 and §10.4.3.7.

‡ See §10.1.1.2.

10.1.2 Use Simple or No Plot Markers

Markers take extra time to render, so it makes sense to remove them if possible. However, removing markers may detract from the visual clarity of the graph, so this should be weighed against the performance benefits on a case-by-case basis.

Even if we cannot remove the markers altogether, we might at least consider using a simple marker (such as a dot or a plus) rather than a complex marker such as a star:

```
t = 0:0.01:10;
cla; hold('on'); drawnow
plot(t,sin(t)); drawnow          % no markers
⇨ Elapsed time is 0.004385 seconds.

plot(t,sin(t),'.r'); drawnow     % dot marker: 80% slower
⇨ Elapsed time is 0.007940 seconds.

plot(t,sin(t),'+r'); drawnow     % plus marker: 180% slower
⇨ Elapsed time is 0.012307 seconds.

plot(t,sin(t),'*r'); drawnow     % star marker: 400% slower
⇨ Elapsed time is 0.022105 seconds.
```

For the record, the relative performance of all available plot markers in the simple example above is as follows:

Marker	Single-Plot Marker Group	Elapsed Time	Relative Performance
None	None	0.004385	(baseline)
v	1	0.007286	65% slower
s (square)	1	0.007377	70% slower
<	1	0.007363	70% slower
^	1	0.007429	70% slower
>	1	0.007561	75% slower
.	1	0.007940	80% slower
h (hexagram)	1	0.008133	85% slower
d (diamond)	1	0.008020	80% slower
p (pentagram)	1	0.008223	90% slower
o	1	0.008264	90% slower
+	2	0.012307	180% slower
x	2	0.013485	210% slower
*	3	0.022105	400% slower

The differences between the first 10 markers listed in the table above, belonging to group 1, are miniscule and vary greatly across succeeding experiments. Different platforms may exhibit slightly different behavior (e.g., a pentagram marker may be slightly faster than a square). The important thing that can be learned from this table is that any of the markers in group 1 is faster than the + and x markers (group 2), which are faster in turn than the star (*) marker (group 3).

When we run the above experiments in a loop, the effect of marker performance is even more pronounced. In this case, all of the markers in group 1 above except the dot (.) and the circle (o) actually render more slowly than the star marker:

```
for idx=1:30; plot(t,sin(t),'r');   drawnow; end
⇨ Elapsed time is 0.266594 seconds.

for idx=1:30; plot(t,sin(t),'.r'); drawnow; end
⇨ Elapsed time is 1.131353 seconds.

for idx=1:30; plot(t,sin(t),'+r'); drawnow; end
⇨ Elapsed time is 1.279908 seconds.

for idx=1:30; plot(t,sin(t),'*r'); drawnow; end
⇨ Elapsed time is 1.978682 seconds.
```

Marker	Single-Plot Marker Group	Multi-Plot Marker Group	Elapsed Time	Relative Performance
None	None	None	0.266594	(baseline)
x	2	1	1.063220	300% slower
.	1	1	1.106585	310% slower
o	1	1	1.106785	310% slower
+	2	1	1.279908	380% slower
*	3	2	1.978682	650% slower
v	1	2	2.016563	650% slower
>	1	2	2.021466	650% slower
<	1	2	2.045129	660% slower
^	1	2	2.071235	680% slower
d (diamond)	1	2	2.233139	740% slower
s (square)	1	2	2.436108	800% slower
p (pentagram)	1	3	4.126895	1500% slower
h (hexagram)	1	3	4.593030	1600% slower

As can be seen, the only markers that have relatively good performance in both cases are the dot (.) and circle (o) markers.

One might think that a similar situation exists with the **LineStyle** property, such that there is a difference between a solid line (-), dotted (:), dashed (--), or dot-dashed (-.). However, it turns out that there is no noticeable performance difference between the various line styles. The only real difference is whether the plot has a marker, with or without a line. In other words, lines are much faster to plot than markers, so whenever possible we should use plots having lines (of any style) but no markers:

```
for idx=1:30; plot(t,sin(t),'.');   drawnow; end % marker only: slow
⇨ Elapsed time is 0.365715 seconds.

for idx=1:30; plot(t,sin(t),'.-'); drawnow; end % marker + line: slow
⇨ Elapsed time is 0.367715 seconds.

for idx=1:30; plot(t,sin(t),'-');   drawnow; end % line only: fastest
⇨ Elapsed time is 0.168918 seconds.
```

10.1.3 Use Vectorized Data for Plotting

MATLAB's plotting functions have a relatively large processing overhead. This means that calling such a function 100 times with a single data point is extremely inefficient compared to calling the same function once with an array containing all 100 data points. Most MATLAB plotting functions accept such vectorized inputs, and we should make use of this whenever possible.

Moreover, using the vectorized version usually means that only a single (vectorized) graphic handle is generally returned and maintained, which has an additional benefit in terms of system resources* and update performance.†

```
% First plot our 70 data points in regular fashion
N=70; x=rand(N,1); y=rand(N,1);
cla; hold('on'); drawnow
tic
for idx = 1 : N
    h(idx) = plot(x(idx),y(idx),'ob');
end
drawnow
toc
whos h
⇨ Elapsed time is 0.057370 seconds.
 Name        Size             Bytes  Class      Attributes
 h           70x1               560  double


% Now use the vectorized version: 12x faster
h = plot([x,x],[y,y],'ob','LineStyle','none');
drawnow
whos h
⇨ Elapsed time is 0.004957 seconds.
 Name        Size             Bytes  Class      Attributes
 h           2x1                 16  double
```

Similarly, use input triplets to *plot()*, rather than multiple *plot()* commands:[1578]

```
% Separate plot commands - slow
for idx = 1 : 5
    plot(x(idx,:),y(idx,:),'-');
end
drawnow
⇨ Elapsed time is 0.008221 seconds.


% Unified plot command using input triplets - 1.8x faster
plot(1:5, 12:16, '-b',...
     1:5, 22:26, '-r',...
     1:5, 32:36, '-g',...
     1:5, 42:46, '-c',...
     1:5, 52:56, '-m'); drawnow
⇨ Elapsed time is 0.004649 seconds.
```

---

* See §10.1.1.
† See §10.2.

10.1.4 Use Static Axes Properties

MATLAB invests a lot of time to ensure that all the graphic elements and properties automatically match each other. For example, the default axes limits are dynamic, so that whenever a plot is added to the axes or an existing plot's data is modified (using the **XData**, **YData**, and/or **ZData** properties), the axes recomputes its limits and updates the display accordingly. All this naturally takes time, both when initially setting up the plots and when updating them.

We often have a priori knowledge about the displayed data. This knowledge enables us to directly specify many axes properties, rather than letting MATLAB figure them out in a dynamic automated way. Setting properties in a way that disables MATLAB's automated computations can significantly improve plotting performance.[1579]

For axes, some of the more important properties in this respect are the limits (**XLim**, **YLim**, **ZLim**, **ALim**, and **CLim**) and tick labels (**XTick**, **YTick**, **ZTick** and their corresponding label strings **XTickLabel**, etc.).[1580] For 3D plots, the camera control properties are also relevant.

Whenever we set any of these properties (e.g., **XLim**) to any value, its corresponding dynamic computation mode (**XLimMode** in this case) is automatically changed from 'auto' to 'manual'. This means that automated recomputation will not be used for this particular property.

```
set(gca, 'XLim',[-5,5]); % XLimMode will now be 'manual', not 'auto'
```

We can toggle an axes property's mode between 'auto' and 'manual' during run-time, without affecting the actual axis limits. This can be used to temporarily disable auto-recomputation within a busy code section. For example:

```
set(gca, 'XLimMode','manual'); % temporarily disable recomputation
...% do plot updates without affecting Xlim
set(gca, 'XLimMode','auto');   % restore the default automated mode
```

Even if we cannot be certain in advance of the specific properties values, we can still benefit from this technique by plotting the initial graph (using MATLAB's default dynamic automated computation), and then set all the axes mode properties to 'manual'. This will ensure that all subsequent plotting and updates in the axes will not attempt to recompute these properties.

The relevant mode properties that can (should) be set to 'manual' are

Property	Description
ALimMode	Transparency limits
CLimMode	Color mapping limits
XLimMode, YLimMode, ZLimMode	Limits of the X, Y, and Z axes
DataAspectRatioMode	Plot data scaling and stretching
PlotBoxAspectRatioMode	Axes box scaling and stretching
CameraPositionMode, CameraTargetMode, CameraUpVectorMode, CameraViewAngleMode	Viewpoint, position, direction, stretching, and projection of 3D plots
TickDirMode,XTickMode, YTickMode, ZTickMode, XTickLabelMode, YTickLabelMode, ZTickLabelMode	Axes tick marks direction, position, and labels

In some cases, we may wish to keep using the default dynamic axes behavior, but just not let some specific plot line affect the axes and recompute its limits. This can be done using

the *LimInclude properties (**XLimInclude, YLimInclude, ZLimInclude, ALimInclude,** and **CLimInclude**) of the specific plot.[1581] These properties are normally 'on', which means that the respective axes property should be recomputed if its mode is 'auto'. Setting a LimInclude property to 'off' means that its corresponding axes property should not be recomputed even if its mode is 'auto'. For example:

```
set(gca, 'XLimMode','auto');            % 'auto' is actually the default mode
plot(0:10,0:10, 'XLimInclude','off'); % only Y axis is recomputed
plot(0:20,0:20);                        % both X,Y axes are recomputed
```

This feature is useful when plotting real-time (strip) data. If the data bounds are known in advance, we could set the axes **YLim** to these limits. But we often do not know the dynamic range a priori, so we need to use the axes' default dynamic Y-limit behavior. On the other hand, we may wish to display a vertical wavefront line that spans the entire height of the axes, even during zooming in/out. If we set the wavefront's height to [–5,5] (for example), then the axes **YLim** would automatically be set to [–5,5], even if the plotted data only spans [–1,1]. On the other hand, if we set the wavefront to [–1,1], then zooming out would show a cropped line! The solution is simply to set **YLimInclude**='off' for a wavefront line that spans [–100,100]:

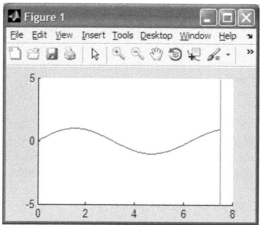
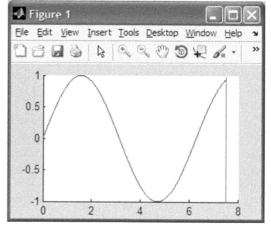

Regular plot (**YLimInclude**=on for both lines)   **YLimInclude**=off for the wavefront line (only)

10.1.5 Only Use *drawnow* when You Are Finished Plotting

MATLAB does not normally render graphics onscreen immediately, if it is in the midst of a processing block and not requested to do so explicitly using the ***drawnow*** function.* Rendering graphics takes time: the more complex the graphics, the more time is needed. So, if our MATLAB code contains multiple plotting commands, it makes sense to call ***drawnow*** only at the end of the block, thereby reducing overall graphics processing and rendering time. The effect can be dramatic:

---

* Graphics are also automatically updated when the user returns to the MATLAB command prompt; when MATLAB waits for user input (e.g., a modal dialog window); or when one of the ***pause, figure, getframe, input,*** or ***keyboard*** functions is invoked, directly or indirectly. Refer to the ***drawnow*** function documentation (http://mathworks.com/help/techdoc/ref/drawnow.html or: http://bit.ly/1uFtBgj) for additional information.

```
% Plot several lines, update the screen after each plot command
N=1000;
x=rand(N,1);
y=rand(N,1);
cla; hold('on'); drawnow
tic, for idx=1:N; plot(x(idx),y(idx)); drawnow; end; toc
⇨ Elapsed time is 33.895913 seconds.

% Plot the same lines, but update screen only at the end: 50x faster!
cla; hold('on'); drawnow
tic, for idx=1:N; plot(x(idx),y(idx)); end; drawnow; toc
⇨ Elapsed time is 0.694960 seconds.
```

In some cases involving highly interactive GUIs and nongraphic objects, it makes sense to use *drawnow('expose')* rather than simple *drawnow*. Adding the 'expose' parameter ensures that only the graphic objects are refreshed, and that callback events are not processed. This makes *drawnow* faster, and postpones the processing of callback events and refreshing of nongraphic objects to a later stage. On Windows and Linux (but not Macs), we can use *drawnow('update')* for even faster processing within a loop (as in animation): if the graphics renderer is busy, no refresh will be done.

While postponing the *drawnow* rendering improves overall performance, in some cases, we may actually wish to insert *drawnow* commands within our plotting commands block. There are three possible cases when we may wish to do so:

- **Debugging** — When plotting a complex chart, it sometimes helps to see the progressive graphic updates.

- **Java components** — Embedded Java GUI controls and their MATLAB containers need to be properly rendered before we can update them.

- **Perceived performance** — If it takes a long time to prepare the entire GUI, it sometimes makes sense to prepare and render graphics in subblocks, rather than waiting until the end to display everything.

I once consulted to a client who had a long-running MATLAB process consisting of dozens of GUI updates. Running *drawnow* after each update made the process too slow (several minutes). Using only a single *drawnow* at the end of the entire process made it run much faster (about 30 s), but the screen was blank and the application appeared frozen during this entire time. The solution was a compromise, running *drawnow* once every 0.5 s. The process was now slightly longer (35–40 s), but the screen was updated twice a second and the user received visual feedback during the entire process run-time. Here is the code skeleton:

```
persistent lastDrawnowTimestamp
lastDrawnowTimestamp = now; % initialize
ONE_SEC = 1/(24*60*60);

while someVeryLongRunnningLoop
    doSomeGraphicsProcessing();
    % Refresh the GUI only once every 0.5 seconds
    if now - lastDrawnowTimestamp > 0.5*ONE_SEC
        drawnow
        lastDrawnowTimestamp = now;
    end
end
```

Naturally, different applications may possibly require modifying the refresh parameter from 0.5 s to a higher or lower value, based on the specific use-case. The trade-off is quite simple: frequent calls to *drawnow* improve the screen's refresh rate at the expense of slower run-time.

In a related matter, note that a memory leak bug was fixed[1582] in MATLAB R2013a (8.1) that relates to using *drawnow* within a loop, yet another reason for avoiding it.

10.1.6 Use Low-Level Rather than High-Level Graphic Functions

MATLAB has several functions that can plot graphics, and they can often be used interchangeably. Some of these functions are categorized as *low-level* functions, and the others as *high-level* functions. Two well-known examples are *line/plot* for line plots (*line* being the low-level equivalent of *plot*), and *surface/surf* for 3D surfaces. The difference between the low-level and high-level functions is that the low-level functions only do the graphic plotting functionality, while the high-level functions also do some automatic cleanup work on the axes.[1583] Aside from this difference, low- and high-level functions can safely be used interchangeably. Importantly, using the low-level variants can provide significant speed-ups compared to their high-level counterparts.

For example, if we use *plot* or *plot3* function to plot a graph line, MATLAB looks at the axes **NextPlot** property to determine whether or not the axes contents should be cleared and its properties reset for the new line plot. In contrast, the *line* function ignores the **NextPlot** property and does not clear the axes nor reset its properties.

Obviously, if we do not need the extra cleanup work done by the high-level functions, it makes sense from a performance standpoint to use the low-level variants: *line* instead of *scatter, plot*, or *plot3*; and *surface* instead of *surf*.

Unlike *scatter*, the *line* function does not enable specific data-point marker customization, although the colors could be modified. On the other hand, *line* only uses a single handle object, saving memory and system resources compared to *scatter* keeping a separate handle for each data point.* So, if we just need to quickly plot a bunch of scattered points, then *line* could be a better choice than *scatter*.[1584]

```
% First plot our data points in regular fashion using scatter()
N=70;   x=rand(N,1);   y=rand(N,1);
cla; hold('on'); drawnow
tic; h=scatter(x,y); drawnow; toc
⇨ Elapsed time is 0.040249 seconds.

% Now use the line() function instead of scatter() - 9x faster
cla; hold('on'); drawnow
tic
props = {'LineStyle','none', 'Marker','o', 'MarkerEdge','b'};
h = line([x,x],[y,y],props{:}); drawnow; toc
⇨ Elapsed time is 0.004536 seconds.
```

...but if you must use *scatter*, then at least ensure that you plot more than 100 data points (as weird as this may sound...).†

---

* See §10.1.1.3 for a discussion of this.
† Refer again to §10.1.1.3.

Note that while the speedup effect (of using *line* rather than *scatter*) is less prominent when more than 100 data points are plotted, it is still evident:

```
% Scatter-plot 140 data points - line() is still 3x faster
cla; hold('on'); drawnow
tic, h=scatter([x;x],[y;y]); drawnow; toc
⇨ Elapsed time is 0.012150 seconds.
```

And similarly for 1000 data points:

```
% First plot our data points in regular fashion using scatter()
N=1000; x=rand(N,1); y=rand(N,1);
cla; hold('on'); drawnow
tic; h=scatter(x,y); drawnow; toc
⇨ Elapsed time is 0.037477 seconds.
```

```
% Now use the line() function instead of scatter() - 3x faster
cla; hold('on'); drawnow
tic
props = {'LineStyle','none', 'Marker','o', 'MarkerEdge','b'};
h = line([x,x],[y,y],props{:}); drawnow; toc
⇨ Elapsed time is 0.012776 seconds.
```

In general, the *line* function is also slightly faster than the *plot* function. In many cases, a call to *plot* may be replaced with an equivalent call to *line*. The speedup is usually small (up to 20%) compared to *plot*, but could be as high as 3× compared to *plot3*.[1585] Note that these functions have slightly different syntaxes, so be careful about how you specify your input arguments.[1586]

Other low-level plotting functions are *patch, rectangle, text, image, axes* and *light*.

10.1.7 Generate Plots while the Figure Is Hidden

When the containing figure window is hidden (**Visible**='off'), plotting is much faster. This is a natural and expected effect due to the fact that the system does not need to repaint the graphics after every plotting command. When the plot is ready, the figure can be made visible, and the overall time of the entire thing would generally be significantly less than plotting within a visible figure window.*

```
% Create a new visible figure and plot onto it
N=10000;
x=rand(N,1);
y=rand(N,1);

tic
f = figure;
h = scatter(x,y); drawnow
toc
⇨ Elapsed time is 0.465220 seconds.

% Now plot onto an invisible figure - 3x faster
f = figure('visible','off');
```

---

* See §10.4 for a similar advice regarding GUI components.

```
h = scatter(x,y); drawnow
set(f,'visible','on');
⇨ Elapsed time is 0.168826 seconds.
```

If we have plots with colorbars, legends, dynamic axes limits, and so on, then the effect of plotting onto an invisible figure becomes even more pronounced. This can make the difference between acceptable and unacceptable performance. A figure loading in 1–2 s might be acceptable, whereas loading in 4–10 s would not.

Many users use MATLAB's GUIDE (*Graphical User Interface Design Environment*) tool to create GUI figure windows. GUIDE automatically creates both a FIG file and a corresponding m-file. For example, if we call our GUI "MyGUI", then GUIDE will automatically create the files *MyGUI.fig* and *MyGUI.m*. The FIG file contains the definitions of all the figure components and their properties,[1587] while the m-file contains the code that controls the behavior of the figure and its components.

GUIDE-generated m-files always have a standard skeleton format, which has an *_OpeningFcn()* function and an *_OutputFcn()* function. In our MyGUI example, GUIDE would create *MyGUI.m* with both the *MyGUI_OpeningFcn()* and *MyGUI_OutputFcn()* functions. *_OpeningFcn()* is called before the figure is made visible, while *_OutputFcn()* is called afterwards. This means that for performance reasons it would be better to plot all graphs and update all GUI components in the *_OpeningFcn()* function, rather than in *_OutputFcn()*.

For debugging purposes, it is often useful to generate the plots and update the controls when they are visible, so that bugs in the code could easily be detected and fixed. This would imply that during the development phase we may want to place this code segment in *_OutputFcn()* rather than *_OpeningFcn()*. But as soon as our debugging is over, if performance is of any importance, we should move this code into the *_OpeningFcn()*. Since these two functions are syntactically equivalent, this migration is usually straightforward and entails no changes to the code.

When maintaining and debugging existing code, it sometimes makes sense to make the figure visible for the plotting code segment. We could of course migrate the code from *_OpeningFcn()* to *_OutputFcn()*, and back again to *_OpeningFcn()* once we are done with debugging. But a simpler approach is to make the figure temporarily visible within *_OpeningFcn()*. Just remember to remove or comment this line once you are done debugging, so that run-time performance will not suffer:

```
% The following makes the figure temporarily visible in *_OpeningFcn()
set(hObject,'Visible','on');%comment/remove this line after debugging
... (plotting/GUI commands here)
set(hObject,'Visible','off');
```

10.1.8 Apply Data Reduction to Plotted Data to Fit the Display Area

When plotting a hundred-point line plot on a full-screen figure axes, all the data points will naturally be plotted and visible. However, we often have more data points to plot than available screen real-estate (display pixels). In such cases, many data points in the plot will overlap one another without having any visual added value.

For performance reasons, it may be better to check whether or not this is the case, and if so then to apply some data reduction (also called *decimation* or *thinning*) so that we would only plot a small portion of the data, which would match the number of display pixels. In many cases, using a number that is twice the number of width pixels is a good choice.*

---

* Compare: Nyquist–Shannon sampling theorem.

For example, if our axes width is 400 pixels, then we might decimate the data to only 800 representative points. Highly volatile or scatterpoint data may need to use a higher multiple than 2, but we would rarely need to use 5 or higher.

To get the axes pixel width, we can use the built-in *getpixelposition* function:

```
positionInPixels = getpixelposition(hAxes);
widthInPixels = positionInPixels(3);
```

Several data reduction techniques could be used: The simplest is to just take every Nth data element for plotting. Some functions that could be handy for this are *linspace, logspace*, and *reducevolume*. We could of course also use direct indexing:

```
N = round(length(actualData)/(2*widthInPixels)); %use a multiple of 2
dataToPlot = actualData(1:N:end);
```

More complex data reduction techniques could take into account local data variability (see *diff, gradient*, and *del2*), or the data points proximity to each other, in order to smooth the data (see *filter, filter2*, and *reducepatch*). If we have the Signal Processing Toolbox, we could also use *filtfilt, decimate*, or *resample*.

Special care should be taken when the original data is not evenly distributed in the X axis, since the generated (data-reduced) midpoints may be too far apart, or (if the calculation was not done properly) even incorrect.

Specific plots may employ other dedicated techniques. For example, the *reducepatch* function reduces the number of patch vertices, which could be critical for the display performance of a complex multifaceted object.

It turns out that the extra time necessary for the data reduction is often smaller than the time savings, making the effort worthwhile:

```
% Prepare the original data
t = 0 : 0.001 : 10;    % 10,001 data points
originalYData = sin(t);

% Plot the data regularly, non-vectorized
for idx=1:length(t); plot(t(idx),originalYData(idx)); end; drawnow
⇨ Elapsed time is 7.356226 seconds.

% Now use data reduction - 8x faster:
pos = getpixelposition(gca);
pixels = pos(3);       % width of the axes in pixels
sampleRate = 2;        % data points per pixel
dataPointsToPlot = sampleRate * pixels;    % # of data points to plot
originalDataSize = length(originalYData);  % =10,001
dataReductionRatio = max(1,floor(originalDataSize/dataPointsToPlot));
dataIdxToPlot = 1 : dataReductionRatio : originalDataSize;
for idx = 1 : length(dataIdxToPlot)
    plot(t(dataIdxToPlot), originalYData(dataIdxToPlot));
end
drawnow
⇨ Elapsed time is 0.937279 seconds.
```

An example where data reduction techniques can be useful is loading an image from a file onto a plot axes. In such cases, it makes sense to programmatically reduce the image resolution to the axes size, before using the *image* function to plot the image.

In a related matter, refer to the technique of linear plot line data reduction, by using only the line's end-points and none of the interim data points.*

Note that data reduction is not useful in several cases:

- If our data can be plotted in a vectorized manner, then the extra performance penalty of data reduction is often higher than the possible performance gain from the smaller number of data points.

- If our initial axes size could be enlarged (e.g., by resizing or maximizing the figure window), then we would either need to take the largest possible pixel size in account when decimating, or hook onto the figure's **ResizeFcn** callback to dynamically add data points based on the updated axes size.

- If we wish to enable the user to zoom in on the plot, then we could increase our `sampleRate` from 2 to some higher value, to enable higher resolution when zooming in. If we wish to enable maximal zooming, then we would need to either use the original data (i.e., undecimated) from the onset, or to hook onto the zoom function's callback and dynamically add missing data points (or recompute `dataId-xToPlot` and update the plot's XData, YData properties to display the newly computed decimated data).

Instead of programming the data reduction code ourselves, we could simply use Tucker McClure's *reduce_plot* utility on the MATLAB File Exchange.[1588] This utility automatically selects the minimal set of data points to be shown, and only plots them. It also attaches listeners to the axes **XLim** and **Position** properties, so that changes to any of them (e.g., as a result of zoom/pan) will update the list of displayable data points. The end result is a much smoother and faster interactivity.†

10.1.9 Reuse Plot Axes

Creating and displaying GUI components, especially complex ones such as plot axes, might take some time. We can often reuse existing components that are no longer in use. This can improve performance, especially when reusing axes that require multiple nondefault customizations for properties such as labels, title, colors, and so on.‡

However, note that in general, axes creation is very quick, so we should not expect much speedup from this optimization. If the required code change is not immediate and natural, it is probably not worth the extra trouble.

10.1.10 Avoid Using the *axes* Function

MATLAB programmers often use the *axes* function in order to set a specified axes as active, for subsequent plotting. Unfortunately, *axes* has a nonnegligible overhead. Instead, we could use the fact that most plotting functions directly accept an axes handle as an optional input, thereby avoiding the need to use *axes*:

```
% Slower
axes(hAxes);
plot(xdata, ydata);
line(xdata, ydata);
```

---

* See §10.1.1.5.
† Also see §10.4.3.7.
‡ See also §10.2.1 and §10.4.1.2.

```
% Faster
plot(hAxes, xdata, ydata);
line('Parent',hAxes, 'xdata',xdata, 'ydata',ydata);   %note the syntax
```

Alternatively, we could set the figure's **CurrentAxes** property, then continue using the plotting function in their standard form, which plots on the currently active axes:

```
% Another fast alternative
set(hFig,'CurrentAxes',hAxes);
plot(xdata, ydata);
line(xdata, ydata);
```

Both of these techniques are faster than using the *axes* function.[1589] In past MATLAB releases, MATLAB's Code Analyzer (MLint) included a specific warning, alerting us when a call to *axes* was found in the code:

```
axes(ax(1));
```
⚠ Calling AXES(h) without an output can be slow. Include 'h' in plot function arguments.

However, there are cases where a single call to *axes* makes sense. After all, not all functions accept an axes handle as their first input parameter (e.g., *colorbar* and *line*).* Moreover, the overhead of a single call to *axes* is normally very small.

For these reasons, the MLint warning was misleading and it is a good thing that it was replaced with a newer warning variant (LAXES), which now only warns when *axes* is being called within a loop.†

```
axes(hAxes);
```
⚠ Calling AXES(h) in a loop can be slow. Consider moving the call to AXES outside the loop. [Details ▲]

Explanation

When you call axes(h) within a loop, MATLAB makes the specified axes the current axes. However, MATLAB also attempts to give focus to the axes and its parent figure with every loop iteration, which slows performance significantly.

Suggested Action

If you are calling axes because you have multiple axes and you need to specify in which axes to plot, pass the axes handle as an argument to the plot function instead. For example, replace code such as this:

a1 = subplot(1,2,1)

Still, calling *axes* in order to set the focus on a specific axes, even just once outside a loop, is slower than directly setting the figure's **CurrentAxes** property to the requested handle. The reason is that in addition to setting the active axes, *axes* also flushes (renders) the graphics queue (like *drawnow*), and then brings the containing figure into focus:

```
hFig = figure;
```

---

* Although in these cases we could specify the axes handle as an optional parameter, for example, *colorbar(...,'peer',hAxes)* or *line(...,'parent',hAxes)*.
† The complete list of MLint-reported performance-related warnings is presented in Appendix A.4.

```
hAxes1 = subplot(1,2,1);
hAxes2 = subplot(1,2,2);

tic, set(hFig,'CurrentAxes', hAxes1); toc
⇨ Elapsed time is 0.000038 seconds.

tic, axes(hAxes2); toc
⇨ Elapsed time is 0.052878 seconds.
```

Therefore, if all we need is to set the active axes, use the figure's **CurrentAxes** property rather than the *axes* function. Flushing the graphics queue can always be done by a direct call to *drawnow* (when we are finally ready, not before!*), and bringing the figure into focus can always be done using the *figure(hFig)* command.

10.1.11 Use the Painters Figure Renderer with Fast Axes DrawMode

When using 3D plots, and possibly also 2D plots having patches, MATLAB automatically checks which displayed graphic element hides (occults) other elements. If we are not worried about this (e.g., if we only have 2D line graphs, or simple 3D graphs), and wish to bypass these checks, we can use a combination of the figure's **Renderer** property and the axes **DrawMode** property, as follows:

```
set(gcf, 'Renderer','painters');  % Note: this is a figure property
set(gca, 'DrawMode','fast');       % 'normal' is the default axes value
```

The axes **DrawMode** property draws objects in the order in which they were created, regardless of their relative positions. This results in faster rendering because it requires no geometric calculations and object resorting, but can produce undesirable results if hidden elements are in fact present.

DrawMode is only relevant when the figure's **Renderer** property is painters.† If the **Renderer** is ZBuffer or OpenGL, MATLAB ignores **DrawMode** and always performs hidden surface elimination and object intersection handling. This is naturally slower.

MATLAB usually selects the painters renderer for figures that contain only simple plots. If a figure contains 3D surfaces, MATLAB may decide to use a ZBuffer or OpenGL renderer instead. In such a case, knowing that we do not need to check for element overlaps may improve rendering speed using the above combination of property values. If this is not an option, then perhaps we could modify our data and/or 3D viewing angle such that there will be minimal overlaps of graphical elements.

If we cannot prevent such overlaps, then consider the possibility of not plotting these elements (or parts of elements) that are hidden by the other elements. For example, if a 3D surface exists in such a way that it would always hide some interior object, then consider not plotting this interior object.‡

When considering whether or not to set the figure's **Renderer** to painters (in order to benefit from fast **DrawMode**), consider the possible alternative gains from using a hardware-accelerated OpenGL renderer, as explained in §10.3.1. In some cases, switching from hardware-assisted OpenGL to painters, possibly to improve some visualization aspects, may in fact significantly degrade performance.[1590]

---

* See §10.1.5.
† Note again that **DrawMode** is an axes property, whereas **Renderer** is a figure property. This combination is often confusing.
‡ See also §10.1.1.

Also be aware that using the painters renderer prevents using colormaps having more than 256 color shades. This is often insufficient for 3D visualizations with shading. If we need more shades, we would need to use either a ZBuffer or an OpenGL renderer.

In some cases, when painters is too slow and OpenGL cannot be used, ZBuffer might provide adequate visualization with better performance than painters.[1591]

10.1.12 Images

Image processing is a wide field encompassing numerous processing functionalities that have specific performance aspects. Readers are referred to dedicated image processing resources. In this section, I will just describe a few generic speedup techniques. Perhaps the most important suggestion is to apply data reduction techniques, as explained in §10.1.8. It may also be helpful to use the following tips:*

10.1.12.1 Reduce Image File Size

File I/O takes much longer than CPU processing in the vast majority of cases (SSD/flash disks and antique CPUs excluded). For this reason, it is more effective to store images in compressed files (lossy JPG or lossless PNG/GIF, e.g.) than in pure uncompressed bitmap files (BMP).† The extra CPU time needed to compress/decompress the image is much smaller than the I/O time saved by the smaller file size, in any file larger than a few KB.‡

Different file types employ different compression schemes that affect the resulting file size. For example, PNG uses a true color (RGB) format for each pixel whereas GIF uses a colormap index. For this reason, PNG typically compresses better for real-life images, whereas GIF is better for flat-colored images having a limited number of colors (cartoons).§ As before, use of a format that compresses better is important from a performance standpoint, due to the reduced associated I/O costs.

It is also sometimes beneficial to reduce the image's display size (width × height) to fit the window or monitor size. This can be done using cropping and/or resizing. The resulting image will contain much fewer pixels and will therefore weigh less.

The same compression format can often be applied more aggressively than typically used in imaging programs. Tools such as IrfanView[1592] or Yahoo!'s SmushIt[1593] enable drastic image size reduction without affecting image quality. They can reduce 20%–40% of the file size, even for already-compressed images. A one-time extra compression of large images that are loaded multiple times may be useful.

Shrinking image dimensions also decreases file size. We can use Image Processing Toolbox (IPT)'s sophisticated *imresize* function, Jan Simon's fast and simple *BlockMean* MEX utility,[1594] or external tools such as IrfanView. Better quality is achieved by resampling rather than resizing, but resampling is a bit slower.

10.1.12.2 Use Integer Image Data

By default, MATLAB uses double-precision numbers (64 bits, or 8 bytes) to store the RGB components of images. This translates into 24 bytes per pixel, which can add up to a lot of

---

* See §11.6 for performance aspects of loading and processing image files. See §10.2.2 for additional related aspects.
† The BMP format does enable compression, but compression is very rarely used with BMP files in practice.
‡ In smaller files, the I/O block size comes into play, making the benefit negligible or nil.
§ This is the reason that small computer and website icons are typically GIFs, while larger images are typically JPG/PNG.

memory for large images. For example, a 2K×2K image would need 96 MB. Such a large amount of memory requires extra time to allocate and reallocate,* and large files (with corresponding I/O costs) to store the data. Compressed file formats such as JPG or PNG reduce file size and I/O, but not memory footprint and allocation.

Instead of using double precision, images can often employ *uint8* (0-255) or *uint16* (0-65535) to store the color values.[1595] An image is typically considered to be "true color" if it has 24 bits of RGB data (i.e., 3 bytes) per pixel, which is equivalent to using a 1-byte *uint8* for each of the RGB components. Using *uint8* by itself reduces the image memory footprint eightfold compared to using the default *double*.[†]

Fortunately, modern MATLAB releases support *uint8* in many numeric and image processing functions. This was less true up to several years ago, so if you have a nonrecent MATLAB release, you should check this compatibility issue. On MATLAB releases that support *uint8*, using it for images speeds up processing times:[1596]

```
double_data = double(rand([1000 1000 3]));
uint8_data = uint8(255*double_data);

for idx=1:100, image(double_data); drawnow; end
⇨ Elapsed time is 2.172845 seconds.

for idx=1:100, h=image(uint8_data); drawnow; end
⇨ Elapsed time is 1.185261 seconds.
```

Image files can often be compressed without noticeable degradation by reducing their color depth to use fewer bits per pixel. A color-depth reduction followed by conversion of an RGB format (e.g., JPG or PNG) to indexed format (e.g., GIF) can lead to a large saving in file size. However, unless this is a one-time operation that is followed by multiple image usages, the potential speedup would be offset by the extra processing and reduced quality. Bit-depth reduction is better achieved by reducing the colormap size in a direct-indexed image (see the following section).

10.1.12.3 Use Direct Color Mapping (Indexed Images)

Direct color mapping, also known as indexed images, drastically reduce an image's memory and file footprint, since each pixel is only represented by a single data value (typically *uint8* or *uint16*, not even a full *double*), which is an index into a side table (colormap) that contains the actual RGB values. Storing such an image in a file format that supports indexed images (e.g., GIF) facilitates saving/loading the image.

This makes sense for large images that repeatedly reuse the same color shades, such as synthetically generated images. It is less useful for real-life pictures, which contain almost as many color shades as pixels (i.e., almost every pixel has a unique color).

To further reduce the image footprint, ensure to use a *uint8* or *uint16* class for the image data, rather than the default *double* class. Also consider reducing the colormap size to improve compression and to enable a smaller data type: *uint8* enables 256 distinct color shades in the image, whereas *uint16* enables 65536. Therefore, if we can rescale the colormap to only use 256 colors, then we could use *uint8*.

When using indexed images, always prefer to use direct color mapping (**CData-Mapping** = 'direct'), rather than scaled or intensity images, since these require extra runtime computations to convert from the pixel data value to the mapped color.

---

* See Chapter 9.
† Unfortunately, since images are typically compressed, the disk size is not reduced by a similar factor.

If the image uses transparency/translucency, scale all the image's **AlphaData** values between 0-1, then set the image's **AlphaDataMapping** property to 'none' to ensure that MATLAB does not spend time mapping the alpha values.

10.1.12.4 Cropping

MATLAB's Image Processing Toolbox (IPT) provides the well-known *imcrop* function for cropping images. This function is very useful but includes many internal checks (e.g., for partially overlapped pixels) that may be unnecessary in many applications. For performance purposes, it is faster to use simple subscript indexing rather than *imcrop* (the smaller the subimage, the larger the speedup):[1597]

```
img = imread('family.jpg');
I2 = imcrop(img,[75 68 130 112]);
⇨ Elapsed time is 0.000538 seconds.

I3 = img(68:180,75:205,:);              % 6x faster
⇨ Elapsed time is 0.000085 seconds.

disp(isequal(I2,I3))
⇨     1                                 % =true (i.e., equivalent)

I2 = imcrop(img,[200 300 1300 750]);
⇨ Elapsed time is 0. 003577 seconds.

I3 = img(300:1050,200:1500,:);          % only 30% faster
⇨ Elapsed time is 0. 002725 seconds.
```

10.1.13 Patches and Volume Surfaces

Several techniques that specifically improve patches and surfaces performance were discussed above:

- Avoid plotting any graphic elements that would be occluded (§10.1.1.2)
- Reduce the number of displayed facets/vertices (§10.1.8)

- Use the painters figure **Renderer** with fast axes **DrawMode** (§10.1.11)
- Use direct color mapping (**CDataMapping** = 'direct') and transparency mapping (**AlphaDataMapping** = 'none') — (§10.1.12.3)

For patches, modify all patches to use only three vertices (i.e., triangles), in order to remove the tessellation step from the patch rendering process.[1598] Also, reduce the number of rendered patch faces (preserving the overall shape) using *reducepatch*.[1599]

For surfaces, reduce the number of volume data points using *reducevolume*.[1600] In addition, set **EdgeColor** to 'none' and **FaceColor** to 'texture'.[1601] For example:

```
h = surf(peaks(1000));
shading('interp');
cd = get(h,'CData');
surf(peaks(24), 'FaceColor','Texture', 'EdgeColor','none', 'CData',cd)
```

Finally, consider using Tiangli Yu's 3D Offscreen Rendering Toolbox to render 3D objects using direct calls to OpenGL's advanced graphics-accelerated capabilities.[1602]

10.1.14 Colorbars

Colorbars are very expensive to create and update from a performance standpoint. Some suggestions that could help improve performance are

1. Do not use colorbars unless you really need them. Deleting the colorbar items from the standard toolbar and menu bar can be done as follows:[1603]

   ```
   delete(findall(gcf,'tag','Annotation.InsertColorbar'))
   delete(findall(gcf,'tag','figMenuInsertColorbar'))
   ```

2. Postpone the colorbar creation, or create it while the figure is still invisible. This is especially important in colorbars, since the addition of a colorbar to an axes forces the axes to change its position and size, causing numerous graphic updates. If the figure is invisible, all these updates are computed in memory rather than onscreen, which is much faster.

3. Use an older, simplified version of the *colorbar* function, from an older MATLAB release, downloadable from the MATLAB File Exchange.[1604] Unless you need the latest functionality and bug-fixes, this earlier version significantly improves the colorbar's creation time and appears to be a benign compatible in-place replacement. Note that mixing m-files from different MATLAB releases may cause unexpected bugs in your program.

10.1.15 Legends

MATLAB legends are notoriously slow; even simple legends can take noticeable time to create and update. There are several tips that may help reduce creation time:*

1. Avoid legends. This is the simplest advice, yet is often overlooked. Graphs are often self-explanatory, especially when having verbose titles and axes labels. Adding a legend could hurt performance as well as increase clutter.

   To prevent users from manually displaying legends, we could remove the legend entry from the figure's menu bar and toolbar. The easiest way to do so is to

---

* See §10.2.3 for techniques to improve legend *updates* performance, as opposed to legend *creation*, which is discussed here.

remove the standard figure menu bar and toolbar, and rely only on our custom menus and toolbar:

```
set(gcf,'MenuBar','none','ToolBar','none')
```

If we need to keep using the standard toolbar and menu bar, we can still remove just the legend items, as follows:[1605]

```
delete(findall(gcf,'tag','Annotation.InsertLegend'))
delete(findall(gcf,'tag','figMenuInsertLegend'))
```

2. Postpone legend creation: The initial plot would be displayed without a legend and therefore appear quicker. Users can then create a legend via the menu bar or toolbar. This action would be done in response to interactive user request, so it would not hamper the main flow's performance.

3. If a legend needs to be displayed at the onset, we can prepare it while the plot is still hidden. This saves a lot of time in legend position updates.*

4. The *legend* function takes some time to calculate the optimal legend-box position when the **Location** property/parameter is set to 'Best'.[1606] If possible, try to set a static location such as 'NorthEast'.†

5. Use a listbox or *uitable* to display the legend, rather than *legend*.[1607]

6. Reduce the number of components displayed in the legend. We can do this by specifying only the legendable graphic handles to the *legend* function:

```
legendableHandles = [hLine1, hSurface2, hPatch3];
legend(legendableHandles,'line #1','surface #2','patch #3');
```

7. If we cannot use this mechanism for any reason, use the following mechanism to tell *legend* to ignore specific handles:[1608]

```
hLegendInfo = get(get(hLine,'Annotation'),'LegendInformation');
set(hLegendInfo,'IconDisplayStyle','off');

hasbehavior(hLine,'legend',false);   %undocumented alternative
```

10.1.16 Reopen Presaved Figure and/or Plot Axes

In some cases, generating a graph or image can take a long time. If we regenerate the graph or image each time that we need it, we might need to wait a long time.[1609] Instead, we could cache the image data and even the entire figure window, and then simply display the cached version, which could be much faster.‡

```
% Save and load entire figure/GUI window
savefig(hFig, 'MyFigure');    % => MyFigure.fig in current folder
hFig = openfig('MyFigure');           % open in a new visible window
hFig = openfig('MyFigure','reuse');   % reuse existing window if open
```

We can reopen a saved figure even faster, by specifying the optional *openfig* 'invisible' parameter. Invisible figures load much faster, since no on-screen rendering is done. This

---

* See also §10.1.7.

† 'NorthEast' is the default location when the **Location** parameter is not specified for 2D plots; for other plots the default is 'NorthEastOutside'.

‡ For a discussion of information caching in MATLAB, see §3.2.

feature can be used, for example, to quickly loop over saved figures and save them as images:[1610]

```
files = dir('*.fig');
outputFormat = 'jpg';
for fileIdx = 1 : length(files)
    filename = files(fileIdx);
    disp(['Converting ', filename])
    hFig = openfig(filename,'new','invisible');
    saveas(hFig,filename,outputFormat);
    close(hFig);
end
```

Reusing previously saved axes or any other HG object* can be done using the *hgsave* and *hgload* functions:

```
% Save and load a specific axes
hgsave(hAxes, 'MyAxes');      % => MyAxes.fig in current folder
props = struct('parent',hFig, 'Units','norm', 'Position',[0,0,.7,.5]);
[hAxes,oldProps] = hgload('MyAxes',props);
```

Note: Be sure to specify a valid HG parent property value to *hgload*; otherwise, MATLAB might crash (at least as of R2013b)!

hgsave/hgload can be used to implement a dynamic configuration (preferences) GUI: The contents (plots, GUI controls) of different panels would be prestored in separate *.fig* files; selecting items in a listbox or tree control in a left panel would update the right panel:

```
delete(allchild(hPanel)); % clear the target panel
props = struct('parent',hPanel);
hgload(filename,props);   % load the relevant contents into the panel
```

10.1.17 Set Axes SortMethod to Childorder and Reduce Transparency

By default, MATLAB draws objects in a back-to-front order based on the current view. This means that objects (lines, patches etc.) which should appear "on top" of other objects are drawn last, overlapping the objects "beneath" them. Calculating the order of the objects, especially in complex plots having multiple overlapping segments, can take noticeable time.

We can improve performance by telling the renderer to draw objects in the order of the **Children** property, which is typically the order in which the objects were created. This can be done by setting the axes' **SortMethod** property to 'childorder' (default='depth').

Note that **SortMethod** became a fully documented and supported axes property in MATLAB R2014b (8.4), but has existed as a hidden (undocumented) property in earlier MATLAB releases as well. The functionality remained unchanged, so we can use this technique on the older MATLAB releases just as for R2014b or newer.

In a related matter, we should note that transparent/translucent patches and lines (having a fourth **Color** element value between 0.0 and 1.0) are slower to render for much the same reasons. Reducing the number of transparent/translucent objects will improve graphics rendering performance.

---

* Some HG objects (e.g., *javacomponents*) cannot be saved/loaded in this fashion, but all the mainstream HG objects can.

Of course, it would be even faster to avoid plotting overlapped (occluded) plot segments, as described in §10.1.1.2.

10.2 Updating Graphs and Images in Real Time

Once a graph has been plotted in a figure window, we often need to update it during the course of the program. If we do this often enough, or if the updates are extensive enough, noticeable flicker might result, degrading the user experience. This section details methods for improving the update performance and reducing plot flicker.

Note: MATLAB GUI cannot be updated from within a parallelized (***parfor***) loop.[1611] A possible workaround is to use the ***parfeval*** function (R2013b and newer).*

10.2.1 Axes Update

A general advice for axes update is to minimize the number of complete graphs regeneration. Axes are automatically regenerated when we use any high-level plotting function (e.g., ***plot***). Whenever an axes is cleared, noticeable flicker occurs since the entire axes contents need to be regenerated and displayed. All this takes time and degrades performance and user experience, especially when done in a loop:

```
for idx=1:N, [xdata,ydata]=getNewData(); plot(xdata,ydata); end
```

There are many ways to reduce the need for complete axes regeneration:

- If we use static axes limits,† then there is no need to redraw the axes when the contained data changes.
- If we modify a plot element's properties (especially the **XData, YData,** and **ZData** properties), then the element does not need to be redrawn from scratch and its update appears smooth and flicker-free. It is also much faster than regenerating the element using the original plotting function (see §10.2.2).

If we cannot avoid regenerating axes when replotting, at least try to reuse the axes' previous set of property values. This avoids the property reset time, as well as many automated checks and updates that would happen when we later add plots to the axes.

To clear an axes without resetting its properties, we can set the axes' **NextPlot** property to 'replacechildren', or to issue the *cla* function. **NextPlot**'s default value is 'replace', which resets the properties and corresponds to *cla('reset')*. As an alternative, we can set the axes hold state to 'on';[1612] then, when it is time to replace the existing plot lines with new ones, we simply delete the existing lines and plot the new ones:

```
% Initial plotting in the hAxes axes
hold(hAxes, 'on');
plot(hAxes,xdata,ydata);

% Subsequent plotting in the axes (no axes reset due to hold ON state
```

---

* See §6.1.2.
† See §10.1.4.

```
delete(get(hAxes,'Children'));    % delete old plot lines
plot(hAxes,xdata,ydata);          % display new lines
```

 ### 10.2.2 Plot and Image Update

It is much faster to update an existing graphic object's properties than to replace it with a new object. Plot elements (lines, markers, patches, etc.) are a specific example for this general advice. For example, instead of replotting the data, it is faster to update an existing plot line's data property values (**XData, YData,** and **ZData**):

```
% Prepare the original data
t = 0 : 0.001 : 10;    % 10,001 data points
originalYData = sin(t);

% Plot the data
hLine = plot(t,originalYData); drawnow

% Update the displayed data by replotting the data
for idx = 1 : 100
    hLine = plot(t,originalYData*(1+idx/100)); drawnow
end
⇨ Elapsed time is 0.888256 seconds.

% Update the data by updating the YData property - 2x faster
for idx = 1 : 100
    set(hLine,'YData',originalYData*(1+idx/100)); drawnow
end
⇨ Elapsed time is 0.464848 seconds.
```

In the sample code above, note how we have used a loop of 100 iterations to prove our point. In fact, if we only update our plot once or twice, the timing difference would be negligible. It is only meaningful when we need to repeatedly update our graph. This is the case, for example, when we continuously measure or calculate something and wish to display the results in "real time". In such cases, performance and flicker-free updates are very important. Updating **YData** results in flickerless update since the update is done using offscreen double-buffering. On the other hand, replacing a plot altogether causes flicker due to the time when the old plot is erased and its replacement has still not been rendered onscreen. So even if real performance is not important, we may still wish to update the data properties to avoid flicker.

In R2014b, we could also use the new *animatedline* function and object, which has a similar effect to updating the **Data** properties, but may perhaps be more readable.

Another property that affects performance is **EraseMode**, which exists for many (but not all) graphic objects. By default, this property has a value of 'normal', which means that whenever the object's data is updated, the entire object is deleted and then redrawn. By using any of **EraseMode**'s alternative values ('background', 'xor', or 'none'), we can further improve the rendering speed. However, these alternatives have functional implications, so we should consider each case separately:

- *normal* — the default **EraseMode** property value, normal plotting behavior
- *background* — erases the object by repainting it with the axes background color, thereby erasing any gridlines and overlapping objects. For one user, this sped up his plot updates from 12 Hz @50% CPU to 30 Hz @4% CPU.[1613]

- *xor* — erases only the object, without affecting overlapping elements, but the line color may be affected by overlapped objects. *xor* is typically used for graphic animations. Both *background* and *xor* are faster than *normal*.

- *none* — does not erase the old object at all. This is the fastest alternative (twice as fast as *normal*), but it leaves behind temporary artifacts of past versions of the object, until the figure is repainted.* This effect can be a good or bad feature, depending on the specific needs. For example, it can be a very efficient way to leave "comet trails" in radar plots of airplane or ship tracks.

When updating images, there is no dilemma about using **EraseMode**, since images are generally updated in-place, meaning that the new pixels are usually plotted right on top of the old ones, so the background is never displayed. It is therefore always advisable to set the image's **EraseMode** property to 'none', and then update the image's **CData** property, rather than redrawing the image from scratch using the *image* or *imagesc* functions, or keeping its default **EraseMode** of 'normal':

```
% Prepare the original 50x50 RGB image
hImg = image(rand(50,50,3)); drawnow

% Update the displayed data by replotting the image
for idx = 1 : 100
    hImg = image(rand(50,50,3)); drawnow
end
⇨ Elapsed time is 0.879869 seconds.

% Update the image by updating the CData property - 2.4x faster
for idx = 1 : 100
    set(hImg,'CData',rand(50,50,3)); drawnow
end
⇨ Elapsed time is 0.374022 seconds.

% Update the image by updating CData & EraseMode = none - 4x faster
set(hImg,'EraseMode','none');
for idx = 1 : 100
    set(hImg,'CData',rand(50,50,3)); drawnow
end
⇨ Elapsed time is 0.216366 seconds.
```

Note that non-normal **EraseMode** may conflict with the OpenGL renderer,[†] as well as with the **DoubleBuffer** functionality.[‡] Also note that **EraseMode** is deprecated in MATLAB's new graphics system (HG2, on R2014b), and should not be used there. **EraseMode** should only be used on R2014a and earlier releases, which do not use HG2.

Other properties can similarly be updated to improve performance and reduce flicker, for example, the position of annotations and text labels; the color, shape, and size of plot markers; the axes title string; and so on. In all these cases, it is generally advisable to update the existing object's property values rather than to recreate the object.

---

* We can force a figure repaint using a simple *set*(gcf, 'Visible', 'on') command, but we should really not do it after each update since that defeats the entire purpose of using **EraseMode** = none, and performance would degrade. Instead, force a figure repaint after a certain amount of time or updates have passed, or at the end of the updates loop.
† See §10.3.1.
‡ See §10.3.4.

When updating 3D surface plots, it is advisable to set the surface object's **EdgeColor** property to 'none' and **FaceColor** to 'texture', to enable faster surface rerendering.[1614]

When updating live (streaming) data in a GUI, it is often useful to employ buffering with a short start-delay. The idea is to constantly load the data into a data buffer on one side, and periodically read data from the buffer's other side. By using a short start delay, we ensure that the buffer always has some data for the retrieval function. In this manner, the display update rate remains constant even when the data rate fluctuates. In practice, users rarely notice a short start delay. It is certainly a negligible price to pay for achieving a constant update rate, with no temporary "hickups". This technique is often used to display streamed Internet movies*

In practical terms, the callback for new data will update the buffer (a matrix variable or some other data structure in MATLAB memory), rather than the GUI. A separate fixed-delay timer object will periodically be invoked, whose callback function will remove the oldest data element from the buffer and update the GUI with this data.†

When updating plots or images in a loop or callback function, avoid using the *axes* function to set the current axes, onto which the following plotting commands will be directed. Instead, use the ability of the plotting commands to specify a parent axes as input parameter, or set the figure's **CurrentAxes** property, as shown in §10.1.10.

When updating a graphic object's properties, we can improve performance by aggregating all property updates into a single *set* call, rather than updating properties one at a time. However, note that the resulting code may appear less readable:[1615]

```
h = plot(1:1e6); drawnow
markers = {'-.','-',':'};
colors = {'r','g','b'};

% Aggregating property updates - a bit faster
for idx = 1 : 1e3
    idx2 = mod(idx,3) +1;
    set(h,{'LineWidth', 'LineStyle', 'Color'},
          {idx2, markers{idx2}, colors{idx2}});
end
⇨ Elapsed time is 0.084513 seconds.

% Separating property updates - slightly slower but more readable
for idx = 1 : 1e3
    idx2 = mod(idx,3) +1;
    set(h,'LineWidth',idx2);
    set(h,'LineStyle',markers{idx2});
    set(h,'Color',colors{idx2});
end
⇨ Elapsed time is 0.109592 seconds.
```

10.2.3 Legends and Colorbars

Some graphic components update dynamically based on the plotted data. These include the *legend* and *colorbar* components. For performance purposes, it is better not to use

---

* See related aspects in §10.1.12 (processing images) and §10.2.6 (trading accuracy for speed to achieve a constant frame rate).

† MATLAB timers are also discussed in §1.8.4, §2.1.6, §3.7, §3.8.1, §7.3.5, and §10.4.3.

these components, which take some time to prepare and display.* However, if we need to display these graphic components, we could at least prepare the legend and/or colorbar on a different hidden axes, to prevent the need for MATLAB to update them whenever the plot data changes.

The setup for this is very easy: copy the relevant axes (including all its plot lines) onto another axes, attach a legend/colorbar to the new (second) axes, then hide the second axes. The legend/colorbar remains in place, and since the second axes is never updated, neither is the legend/colorbar, making updates to the first axes much faster:

```
% Prepare the original data
t = 0 : 0.001 : 10; % 10,001 data points
originalYData = sin(t);
updatedYData  = cos(t);

% Plot the data, then add a legend in the normal way
hLine = plot(t,originalYData);
legend(hLine); drawnow

% Now update the displayed data
tic, set(hLine,'YData',updatedYData); drawnow; toc
⇨ Elapsed time is 0.044120 seconds.

% Now try the suggested alternative: attach legend to a hidden axes
clf
hLine = plot(t,originalYData);
hAxes2 = copyobj(gca, gcf);
legend(hAxes2,hLine);
set(hAxes2,'Visible','off');
drawnow

% Now update the displayed data - 2x faster than before
tic, set(hLine,'YData',updatedYData); drawnow; toc
⇨ Elapsed time is 0.024725 seconds.
```

Another useful suggestion for legends is to limit the number of graphic handles included in the legend. The fewer the handles to update, the faster the update.†

 If the suggestion of creating a separate axes cannot be used for any reason, we can force the legend and colorbar to be static and to not update in real time.[1616] This relies on undocumented features and the technical details are MATLAB-release dependent, and is therefore discouraged. Since we have the separate-axes solution, I advise using it, rather than the unsupported techniques of static legends/colorbars.

10.2.4 Accessing Object Properties[1617]

There are several ways to access (read or update) HG object properties. The simplest is to use the built-in *get* and *set* functions on the HG object's handle. However, this is not the fastest method: a significant speedup is possible (see below).

Accessing individual properties is so fast that this speedup may not seem important. Individual properties are accessed in submilliseconds, which is a very short time for most

---

MATLAB programs. Indeed, if our application seldom accesses properties, it is probably not worth the effort to optimize this particular aspect of the program:

```
% Individual property access is extremely fast:
hFig = figure;
tic, figName = get(hFig,'name'); toc
⇨ Elapsed time is 0.000229 seconds.

tic, set(hFig,'name','testing'); toc
⇨ Elapsed time is 0.000270 seconds.
```

But if we have thousands of reads and/or updates, this could possibly become an important factor (*drawnow* calls are intentionally omitted to illustrate the effect):

```
for idx=1:10000, set(hFig,'name','testing'); end
⇨ Elapsed time is 0.229772 seconds.

% Using the HG handle() wrapper is 1.3x faster:
hFig = handle(hFig);
for idx=1:10000, set(hFig,'name','testing'); end
⇨ Elapsed time is 0.170205 seconds.

% Using the HG handle() wrapper with dot notation is 2.7x faster:
hFig = handle(hFig);
for idx=1:10000, hFig.name='testing'; end
⇨ Elapsed time is 0.083762 seconds.
```

We learn from this that using the HG *handle()* wrapper is useful for improving performance. This has the benefit of improving our code performance with minimal changes to our code, by simply updating our handle to be a *handle()* wrapper:

```
hFig = handle(hFig);
```

Using the *handle*'d handle enables further speedup using the dot notation (`handle. propertyName`), rather than the familiar *get/set* functions. This conclusion also holds for non-GUI handle properties (Java objects, MATLAB user classes, etc.*). In my opinion, the dot notation is not only faster, but also more readable/maintainable.

Note that with MATLAB's new graphics engine (HG2, R2014b), the *handle* notation became the default. However, it is too early to tell the performance implications of HG2 at the time that this text is written (mid-2014).[†]

10.2.5 Listeners and Callbacks

Listeners on graphic properties or GUI control events, as well as regular MATLAB callbacks take time to execute. Whenever we update a graphic object (either a GUI control or an axes plot), we should take into consideration the possibility that some callback(s) will be invoked. Depending on what the callback(s) actually do, and how efficiently, this may impose a significant penalty on the update performance.

A simple solution is to temporarily disable callbacks before important updates:

---

* See §4.7.2.
[†] These implications will be collected and updated over time in http://UndocumentedMatlab.com /?s=hg2+performance (or: http://bit.ly/1sROV0B).

```
oldCallback=get(hControl,'Callback');    % store callback for later use
set(hControl,'Callback',[]);             % temporarily disable callback
drawnow; pause(0.05);
set(hControl,...);                       % perform the update
drawnow; pause(0.05);
set(hControl,'Callback',oldCallback);    % restore the stored callback
```

and similarly for listeners:[1618]

```
hListener = addlistener(hAxes, 'XLim', 'PostSet', @callbackFunction);
  try
    % R2014a or older:
    set(hListener,'Enable','off');       % temporarily disable listener
    set(hAxes,'xlim',[minDate maxDate]); % perform the update
    set(hListener,'Enable','on');        % restore the listener
  catch
    % R2014b or newer:
    hListener.Enabled = false;           % temporarily disable listener
    set(hAxes,'xlim',[minDate maxDate]); % perform the update
    hListener.Enabled = true;            % restore the listener
  end
  drawnow
```

Remember to add a short *pause()* after clearing the callback and also before restoring it, to ensure that the update is correctly done without the callback present. In some cases (primarily GUI updates, not plot updates), the update action is performed on a separate processing thread than the main program, and so might occur before the callback removal has taken place, or after it has been restored. This may lead to unexpected results since the callback will be invoked although the callback property appears to be empty. By adding the intentional *drawnow* and *pause*, we ensure that the threads are in-sync at that point, so that the graphics update will occur without the callback.

Depending on the specific situation, we may omit one or both of these intentional *pause*s. To ensure correct behavior, it is safer to add them at first, and then test the behavior without each of the *pause*s separately. For additional performance aspects of such intentional pauses, see §10.5.1.

It is not always possible to disable listeners and regular callbacks, due to functional requirements. In such cases, we should at least attempt to improve the performance of the callbacks, so that they execute as fast as possible. For example, perhaps an early bail-out mechanism could be employed (see §10.4.3.3).

Another aspect to avoid is *callback chaining*, whereby the callback affects some property that in turn invokes a new callback and so on. Instead, disable all other callbacks and run the bunch of updates together in the parent callback.

10.2.6 Trading Accuracy for Speed

When displaying real-time data in a GUI, it is often more important to provide the appearance of a smooth flickerless update than it is to be highly accurate in the presented data. This is a variant of the accuracy versus performance trade-off.* It is particularly important when presenting rapidly changing graphic data (graphs and images), since in these cases

---

* See §1.6 and §3.9.2.

the human eye is more affected by update smoothness and the overall visual imagery, than by data accuracy.

For example, updating some graph to a 0.001 data precision might take 0.25 s, while updating the same graph to an accuracy of only 0.01 might take just 50 ms. The former only enables four updates per second, whereas the latter enables 20. At first sight, the 10-fold accuracy increase may seem well worth the 5-fold performance penalty. However, since human eyes can only perceive 10–12 individual frames (images) per second, the 50 ms update would appear to be in continuous motion, whereas the 250 ms updates would appear as a jittery series of still images. To the user, the end result is that the 0.001 variant is noticeably slow and flickery, while the 0.01 variant appears to run continuously, in real time.

In a rapidly changing graph/image, the human eye focuses its attention on the moving parts, at the expense of the static ones and of pixel position accuracy (i.e., data precision). After all, if we need to present data in a highly accurate manner, we would not be using graphs in the first place: we would display and update a table or text label with the actual data value. The eye has no problem immediately identifying a text-label change from 12.34 to 12.45, but this same change in a graph* would probably be overlooked by the user. By using a graph or image, we are taking the conscious decision to sacrifice perceived accuracy for clarity. There is no reason not to continue with a conscious decision to sacrifice actual accuracy for performance.

Moreover, the extra accuracy may not even be noticeable: graphs can only display a certain number of pixels;[†] similar data values possibly share the same onscreen pixels. This is used by some utilities in order to improve plot pan/zoom responsiveness.[‡] Similarly, JPG images can achieve 90% data compression with a barely noticeable impact on image accuracy. Since users may not even notice any visual difference between accuracy levels, the extra time taken to achieve higher precision is simply wasted.

If we need to handle interactive user zoom-in on the data, we could instrument the figure's zoom functionality with a dedicated user callback (see the **zoom** function): Our application would use the standard (lower) data precision by default, and use the zoom-in callback to temporarily increase computation accuracy when the user zooms in. In such cases, we could use the fact that only a small portion (the zoomed-in portion) of the graph or image is displayed, in order to reduce the computation cycle time by having to compute only a smaller subset of the data. In this manner, we could provide higher accuracy for the zoomed-in portion while still maintaining a real-time refresh rate.[§]

If the application might run on different platforms (monitor size, CPU speed, physical memory) or environments (OS processes, data, etc.), we could dynamically control its display accuracy based on the actual measured time[¶] that it takes to complete an update cycle. If the application sees that a cycle takes more than 100 ms (i.e., a frame rate less than 10 Hz), then it should reduce the accuracy; otherwise it could increase it.

In a complex dynamic application, we can continuously monitor the run-time performance and system load, and modify the target accuracy accordingly. In this manner, we dynamically throttle the accuracy in order to achieve a constant throughput/performance meeting the required QoS (*quality of service*) standard.

---

* Or even more so in an image, where the data value is often color-coded using color intensities.
† We can use the ***getpixelposition**(hAxes)* to determine the specific number of pixels for a particular graph/ image axes.
‡ See §10.4.3.7 and §10.1.8.
§ See §10.4.3.7 and §10.1.8.
¶ Wall-clock time, that is, *tic-toc* — see §2.2.1.

We can control the frame rate not just by controlling the display resolution, but also by controlling processing parameters, such as the filtering type or accuracy.[1619]

10.2.7 Avoid Update to the Same Value

Graphic programs are often dynamic in nature. As the application executes, the data changes and GUI controls change their state and properties. It is a common pitfall to update plot data and GUI controls to the same value as their current value. This does not cause any error or logical bug, and is therefore very difficult to detect during debugging and testing. It does however have an adverse effect on performance.

Whenever we update a graph or a GUI control, we should ask ourselves whether the update is actually required. We should only update if the value needs to change:

```
% Updating a graph
oldData = get(hLine,'YData');
if ~isequal(newData,oldData)
    set(hLine,'YData',newData);
end
```

MATLAB *uicontrols* generally have built-in protection against same-value updates. We can see this from the fact that the control's callback is generally not invoked in such cases. As long as we are aware of this behavior, it is a very useful protection. Unfortunately, this does not happen in all callbacks. For example, Java and ActiveX event callbacks are always triggered, even when the value remains unchanged. In such cases, we should implement a programmatic check, as above:

```
% Updating a GUI control
oldData = get(hControl,'String');
if ~isequal(newData,oldData)
    set(hControl,'String',newData);
end
```

10.2.8 Cache Graphic Handles

Storing handles in a persistent location can be very helpful. Figures created using GUIDE automatically use the `handles` struct to store handles of static GUI and graphic handles (components and axes). We can access these handles directly if we have access to the handles struct (typically as a function input parameter), or we can get it programmatically using the *guidata* function. We can use the *guidata* function to store and later retrieve additional (dynamic, run-time) handles in this struct.

For figures created programmatically, without GUIDE, we can store the handles in the figure or MATLAB desktop (0) handle. Mechanisms for attaching the data to such objects include setting their **UserData** property or using the built-in functions *getappdata/setappdata** (plus a few other GUI-based alternatives[1620]). For example:

```
% Using the UserData property
set(gcf,'UserData',hAxes);          % store cache data
hAxes = get(gcf,'UserData');        % load cache data
```

---

* Under the hood, *getappdata* and *setappdata* simply update a structure of user-defined fields that is placed in the object's hidden **ApplicationData** property. This mechanism is also used by the *guidata* and *guihandles* functions, which use the predefined field UsedByGUIData_m.

```
% Using getappdata/setappdata
setappdata(0,'hAxes',hAxes);        % store cache data
hAxes = getappdata(0,'hAxes');      % load cache data
```

The alternative to using cached handles is to search for them programmatically, using the *findobj, findall,* or *allchild* functions:

```
% Slow alternative, should be avoided where possible
hAxes = findobj(gcf,'Tag','mainAxes');
hLines = allchild(hAxes,'Color','b');
```

Such a search could be relatively costly in heavily laden figures, which have numerous graphic objects to be scanned. Even in simple figures, it could be too slow for rapidly triggered callbacks, such as mouse movements. In such cases it is much better to directly access the relevant cached handles.

If scanning for child handles cannot be avoided, then at least limit the number of scanned objects by specifying the lowest possible handle in the objects tree as the starting (ancestor) handle, for example, the containing axes or panel handle. We could also use the '-depth' parameter to limit the number of scanned handles.

Note that the *guihandles* function, which seems superficially very similar to *guidata*, uses a different mechanism: *guidata* directly accesses the data using *getappdata/setappdata*, and is therefore highly efficient. On the other hand, *guihandles* uses *findall* to process all handles in the figure, scan their **Tag** property and create the resulting handles struct. Given a choice, we should generally prefer *guidata* over *guihandles* for performance.

10.2.9 Avoid Interlacing Property *get* and *set*

A graphics rendering update is not just triggered by *drawnow*: it may also be triggered by a combination of property query (*get*) and update (*set*), since the engine might think it needs to update the graphics in order to retrieve correct property values. In fact, although the property query and update might be unrelated and would not affect each other, the engine would still update the graphics since *set* marks the entire graphics model as "dirty" (requiring a full rendering update), while *get* of certain properties causes an immediate update if the model is marked as "dirty".*

The bottom line for all this is that the following code snippet

```
% Multiple graphics rendering updates due to interlaced get,set
for idx = 1 : N
    propValue = get(hObject(idx), propName);
    set(hObject(idx),propName,newData);
end
```

would cause a rendering update in each loop iteration. It would therefore be faster to separate the query and update loops:

```
% Faster - Only a single rendering update (at the end)
for idx = 1 : N
    propValue(idx) = get(hObject(idx), propName);
end
```

---

* This only happens for auto-calculated properties such as **Position**, ***Lim**, ***Tick** etc. See http://mathworks. com/help/matlab/creating_plots/getting-and-setting-automatically-calculated-properties.html (or: http:// bit.ly/1nupMeG) for details.

```
for idx = 1 : N
    set(hObject(idx),propName,newData);
end
```

10.2.10 Use *hgtransform* to Transform Graphic Objects

For graphic objects that need to be translated (moved), scaled or rotated, we can use either basic MATLAB functions to effect the transformation, or use an *hgtransform*. It turns out that *hgtransform*, which is admittedly somewhat more complex to program, can be much faster. The reason is that only the transformation matrix needs to be sent to the graphics renderer upon each update, rather than the entire object data:

```
[x,y,z] = sphere(270);

% Using standard rotate() function - slower
s = surf(x,y,z,z, 'EdgeColor','none');
axis vis3d
for angle = 1 : 360
    rotate(s,[1,1,1],1);
    drawnow
end
⇨ Elapsed time is 21.367996 seconds.

% Using hgtransform() function - 4x faster
grp = hgtransform('Parent',gca);
s = surf(gca, x,y,z,z, 'Parent',grp, 'EdgeColor','none');
view(3)
axis vis3d
for angle = linspace(0,2*pi,360)
    tm = makehgtform('axisrotate',[1,1,1],angle);
    grp.Matrix = tm;
    drawnow;
end
⇨ Elapsed time is 5.600215 seconds.
```

10.3 Figure Window Performance Aspects

Several figure-wide mechanisms for improving performance were discussed in §10.1 above. Some of these mechanisms are also applicable for figure GUIs, not only for graphs. These mechanisms are

- Generate plots and GUI while the figure is hidden (§10.1.7)
- Prevent figure resizing to enable plot data reduction (§10.1.8)
- Set the figure's **CurrentAxes** property programmatically (§10.1.10)
- Set the figure's **Renderer** property to painters (§10.1.11)

In addition to these suggestions, there are several other suggestions that can improve the performance of entire MATLAB figures.

10.3.1 Use Hardware-Accelerated OpenGL Renderer and Functionality

Modern computers often have hardware graphics accelerators (GPUs). MATLAB can often use these accelerators in its OpenGL rendering implementation, in addition to the built-in software implementation of OpenGL.

Note that only Unix and Windows (but not Mac) platforms have an OpenGL software implementation; however, all MATLAB platforms support hardware acceleration in principle. Whether or not MATLAB recognizes a specific hardware accelerator depends on the specific case. Use the *opengl('info')* function to determine whether the hardware is recognized (in which case the Software flag will indicate `false`):

```
>> opengl('info')
Version         = 3.3.0 - Build 8.15.10.2778
Vendor          = Intel
Renderer        = Intel(R) HD Graphics 4000
MaxTextureSize  = 8192
Visual          = 04 (RGB 32 bits(08 08 08 08) zdepth 24, Hardware
Accelerated, Opengl, Double Buffered, Window)
Software        = false
...
```

Unfortunately, OpenGL's software variant is slower than the hardware variant. We can check whether MATLAB recognizes hardware-accelerated OpenGL on our system:

```
s = opengl('data');
if ~s.Software      % i.e., hardware-accelerated, not software variant
   set(gcf,'Renderer','opengl');
end
```

If our particular platform happens to not have a recognizable hardware accelerator, then MATLAB would use the slower OpenGL software-implementation variant if we set the figure's **Renderer** to OpenGL. We therefore need to be very careful about this.

MATLAB HG2, on R2014b, relies on advanced OpenGL hardware features much more than earlier releases. When MATLAB detects an incompatible display driver, it issues a warning and reverts to using the much slower software implementation. It is advisable to update the display driver to the very latest version, since this often solves such incompatibilities. Some computer vendors use custom OEM drivers whose latest version might still be incompatible. For example, on my Lenovo E530 laptop, the latest Intel HG Graphics 4000 driver (v.9.15) was still incompatible; I downloaded the latest generic driver (v.10.18) from Intel's website and this fixed the issue: HG2 now uses hardware-accelerated OpenGL on my laptop, which is much faster.

Using hardware-accelerated graphics rendering can significantly improve performance, despite the fact that using the OpenGL renderer involves additional computations that are skipped by the painters renderer (especially with **DrawMode** = 'fast'*). In such cases, it could make sense to programmatically use the OpenGL renderer, even if painters or ZBuffer would normally be selected.

The decision to modify the default figure **Renderer** should not be taken lightly. Using OpenGL, even hardware-accelerated, will not always improve performance over the painters renderer, although it would indeed outperform ZBuffer in most cases. Moreover, OpenGL has many nooks and crannies that we should be aware of. For example, OpenGL

---

* See §10.1.11.

does not support 8-bit color mode (only true-color displays); it does not support logarithmic axes, colormap color interpolation, and Phong lighting; it does not render complex patches correctly; and it is slow with complex GUIs and line markers. In such cases, MATLAB automatically reverts to the slow ZBuffer renderer, and we should consider switching to the painters renderer instead.

When using the OpenGL renderer, be aware that the OpenGL driver has several bugs that we can work around in software, using the *opengl* function*. For example, if graphics objects with the **EraseMode** property set to non-normal (xor, none, or background†) do not draw with OpenGL **Renderer**, we can run this:

```
opengl('OpenGLEraseModeBug',1);
```

The list of OpenGL driver bugs that can be worked around in this manner is listed in the *opengl('info')* output and in *opengl*'s function documentation:[1621]

- OpenGLBitmapZbufferBug — text and data-tips are not displayed
- OpenGLWobbleTesselatorBug — MATLAB crashes with complex patches
- OpenGLLineSmoothingBug — pixelizations on lines having **LineWidth** >3
- OpenGLDockingBug — MATLAB crashes when docking OpenGL figures
- OpenGLClippedImageBug — images and colorbars are not displayed
- OpenGLEraseModeBug — objects with **EraseMode** ~= normal are not displayed

OpenGL has some other bugs that have no workaround except switching renderer or playing with **WVisual** property.‡ One bug that is often encountered and can be fixed in this way is flipping text labels and legend entries upside down in some cases.[1622]

Also note that the axes **DrawMode** property is ignored in OpenGL. So, if we plan to use it (as suggested in §10.1.11), we need to set the figure's **Renderer** to painters.

Additional patch/image rendering speedups can be achieved by directly calling OpenGL's graphics-accelerated functions. For some examples of doing this, refer to Tiangli Yu's *3D Offscreen Rendering Toolbox*[1623] and the well-known *PsychToolbox* (PTB, widely used in computer-vision research).[1624] PTB includes *MOGL*,[1625] a collection of MATLAB wrapper functions for all OpenGL commands, which enables calling them directly from MATLAB without needing MEX or a C interface.

10.3.2 Set a Nondefault WVisual/XVisual Property Value

On Windows platforms, setting a figure's **Renderer** property automatically affects the related **WVisual** property value, if the corresponding **WVisualMode** value is 'auto' (which is the default value). The corresponding properties on non-Windows platforms (Mac and Linux) that use XServer, are **XVisual** and **XVisualMode**.

Adventurous users may wish to override MATTLAB's automatic selection of the **WVisual/XVisual** value to squeeze the performance lemon for a few more drops. The performance gain is not expected to be significant, and using an incompatible value may create rendering problems. So the general advice is in fact NOT to override this value. Refer to §10.3.4 and to the online doc for additional details.[1626]

---

* The ability to work-around OpenGL driver bugs is only available in R2014a and earlier, not on R2014b.
† See §10.2.2.
‡ See §10.3.2.

Note: The **WVisual**, **WVisualMode**, **XVisual**, and **XVisualMode** properties are not available in the upcoming new graphics system (HG2, on R2014b). They should only be used on R2014a and earlier releases, which do not use HG2.

10.3.3 Disable BackingStore

By default, MATLAB automatically keeps an off-screen buffer of each figure window's pixels. This buffer is used whenever a figure becomes obscured (fully or partially) by other windows, and then becomes unobscured. The buffer contents are then simply copied onto the unobscured window region. Without the buffer, MATLAB would be forced to perform a complete window redraw, which is more expensive in terms of performance.

If we can be certain that our figure window will not be obscured by other windows during run-time (for example, by setting the figure window to be "always on top"[1627]), then it could be efficient both in terms of memory and performance to set the figure's **BackingStore** property value to 'off'. This will be slower whenever a figure's region is uncovered by another window, since the entire figure will be redrawn, but at least the off-screen buffer will not be kept and maintained.

10.3.4 Disable DoubleBuffer

MATLAB has the ability to provide double-buffering functionality for figure updates. A copy of the entire figure is kept off-screen, and whenever an update is made to the figure contents, it is the off-screen copy that is updated, and copied onto the displayed figure window when the update is done. This reduces onscreen flicker and provides smooth animation. The disadvantage is that maintaining the off-screen buffer degrades performance and increases the memory usage.

Note: **DoubleBuffer** has been deprecated starting in MATLAB release 8.1 (R2013a). With earlier releases, double-buffering was disabled in MATLAB figures by default. On later releases, double-buffering is enabled by default and we cannot modify this setting. As of R2014b, the **DoubleBuffer** property is still accepted, but has no effect. The following discussion applies to MATLAB releases R2012b or earlier.

DoubleBuffer is similar in concept to the **BackingStore**.* The main difference is that figure updates are done to the off-screen **DoubleBuffer** before the displayed window is affected, whereas the order is reversed for **BackingStore** (i.e., the off-screen buffer is updated following the displayed window). Another difference is that by default, **DoubleBuffer** is 'off', whereas **BackingStore** is 'on'.

Double-buffering cannot work with the OpenGL renderer or nonnormal **EraseMode**. For it to work properly, the figure's **Renderer** must be set to 'painters' or 'ZBuffer', **DoubleBuffer** must be 'on', and the animated objects must have **EraseMode** 'normal'. Note that this conflicts with the suggestion to set **EraseMode** to a non-normal mode.† We may indeed need to choose between flicker-less animation and faster animation.

Note that the figure's **DoubleBuffer** property applies only to the painters renderer.

The ZBuffer renderer ignores the figure's **DoubleBuffer** property and automatically uses MATLAB's double-buffering. The user has no control over this behavior.

OpenGL renderer also ignores the figure's **DoubleBuffer** property and automatically uses internal (not MATLAB's) double-buffering. You can override OpenGL's double-buffering

---

* See §10.3.3.
† See §10.2.2.

mode by specifying a nondefault **WVisual/XVisual** format that does not contain internal double-buffering. **WVisual** is used on Windows platforms, whereas **XVisual** is used on non-Windows platforms. Refer to §10.3.2 above, and the official documentation for the **WVisual/XVisual** properties.[1628] For example:

```
>> set(gcf,'Renderer','opengl')
>> get(gcf,'WVisual')
ans =
02 (RGB 32 bits(08 08 08 08) zdepth 24, Hardware Accelerated, Opengl,
Double Buffered, Window)

>> set(gcf,'WVisual') % list the available (supported) modes
{ 00 (RGB 32 GDI, Bitmap, Window)}
  01 (RGB 32 bits(08 08 08 08) zdepth 24, Hardware Accelerated, Opengl,
  Window)
  02 (RGB 32 bits(08 08 08 08) zdepth 24, Hardware Accelerated, Opengl,
  Double Buffered, Window)
...

>> set(gcf,'WVisual','01')% HW-accelerated OpenGL, no double-buffering
```

Note that the ZBuffer renderer is not available in MATLAB's new graphics system (HG2, on R2014b). Also, in HG2 the **EraseMode** property is deprecated and should not be used. Finally, the **WVisual**, **WVisualMode**, **XVisual**, and **XVisualMode** properties are not available in HG2.

10.3.5 Set a Manual DitherMapMode on Old Platforms

Relatively old MATLAB releases supported screens that had low color resolution. Modern computer screens use true color, but on low-color displays setting the figure's **DitherMapMode** property to 'manual' could speed up the rendering.

Modern MATLAB releases have **DitherMapMode** set to manual by default and evoke the following warning when setting this property:

```
>> set(gcf,'Dithermapmode','manual')
Warning: figure DithermapMode is no longer useful with TrueColor
displays, and will be removed in a future release.
(Type "warning off MATLAB:HandleGraphics:SupersededProperty:DithermapMode"
to suppress this warning.)
```

Note: **DitherMapMode** is not available in MATLAB's new graphics system (HG2).

10.3.6 Reuse Figure Windows

When closing figure windows, MATLAB does not always delete them but sometimes clears and then hides the windows, making them inaccessible in MATLAB but still active in memory. Effectively, this means that not all graphics resources are released, which may cause a creep in memory consumption and graphic handles, leading to slowdowns after several thousand MATLAB figures have been created and closed.

```
>> jFrames = java.awt.Frame.getFrames % list of windows still in memory
```

Moreover, creating a new MATLAB figure takes some time. Behind the scenes, many computations are involved. It is often more efficient to hide a figure window than to close it. When we need to display a new window, clear the old (hidden) figure using *clf* and then redisplay it, rather than creating a new figure window from scratch.

10.3.7 Sharing Data between GUI Callback Functions

A GUI without callback functions is lifeless. The interactivity and usability of a GUI depends on callback functions being used to process interaction events. In many cases, a control's callback function needs to update some information that is used by another callback function, for example, the updated system state or configuration.

The standard way to do this is via the built-in *guidata* function. In heavily laden GUIs, this could be a performance hotspot and in such cases we might consider using one of the other alternatives for sharing data. These include (also see §9.5.5):[1629]

- Using *getappdata/setappdata* on specific control handles
- Setting the **UserData** or **ApplicationData** property of specific control handles
- Setting the callbacks as sibling nested functions that share parent variables
- Using *global* variables (this is highly discouraged!)

10.3.8 Disable Anti-Aliasing

On HG2 (MATLAB R2014b and newer), anti-aliasing is turned on by default, resulting in smooth plot lines, curves and patches. Anti-aliasing works by setting intermediate colors for pixels that are directly adjacent to the plotted lines, thereby presenting a visual illusion of smoothness. This technique is so widely used in graphics (not just in MATLAB) that it is now a standard feature of any renderer. However, the associated calculations take an extra bit of time.

By setting the figure's **GraphicsSmoothing** property to 'off' (rather than its default 'on'), we disable anti-aliasing, thereby improving rendering performance at the expense of jagged plot lines. Depending on the specific use-case, this could be a worthy tradeoff. Note that this property only has an effect for the painters and OpenGL renderers, not ZBuffer. In ZBuffer, anti-aliasing is not calculated and lines will always appear jagged. Disabling anti-aliasing is especially important if it is not natively supported by the graphics card or driver. This can be seen by issuing the *opengl* command, and checking the value of the SupportsGraphicsSmoothing field:

```
>> d=opengl('data'); d.SupportsGraphicsSmoothing
ans =
     1     % =true
```

The figure's **GraphicsSmoothing** property only affects certain types of graphic objects, such as axes and plot lines, but not others, such as text objects. For these objects, we can update their specific ***Smoothing** property (most objects have a dedicated **GraphicsSmoothing** or **LineSmoothing** or **FontSmoothing** property):

```
hText = text(xPosition, yPosition, string, 'FontSmoothing','off');
```

10.3.9 Use Smaller and Fewer Figure Windows

The larger the figure window, and the more figure windows exist, the more computations need to be done by the graphics renderer in order to render all the pixels. Reducing the size and number of the Matlab figures may help improve overall graphics performance. The performance would generally not improve significantly enough to offset the reduced usability, but this can be tested in cases where overall graphics performance is a limiting factor.

10.4 GUI Preparation and Responsiveness

Many of the ideas presented earlier in this chapter also apply to GUIs. When developing a GUI, responsiveness is one of the important concerns. GUI responsiveness includes the following topics:

- Creating the initial GUI within an "acceptable" time period
- Presenting user feedback during long operations (*perceived performance*)
- Performing asynchronous (user- or timer-triggered) actions in an "acceptable" time period

Let us tackle these topics one by one.

10.4.1 Creating the Initial GUI

There are several suggestions for improving the performance of initial GUI creation:

- Create and update GUI components in nonvisible mode
- Reuse existing figure windows and GUI components
- Reduce the number of property updates
- Prevent unnecessary callback invocations
- Postpone nongraphical setup and updates
- Simplify the GUI
- *uitable* considerations

10.4.1.1 *Create and Update GUI Components in Nonvisible Mode*

The most effective advice for improving performance in creating GUI components and figures is to create them in invisible mode. For standard MATLAB *uicontrols*, this means that the containing figure or panel is hidden (**Visible** property = 'off').

```
% Create a simple figure with a simple contained data table
f=figure; uitable('Parent',f, 'Data',magic(15)); drawnow;
⇨ Elapsed time is 0.161285 seconds.

% Create the data table while the figure is initially non-visible
f=figure('Visible','off');
uitable('Parent',f, 'Data',magic(15));
set(f,'Visible','on'); drawnow
⇨ Elapsed time is 0.135930 seconds. % 15% faster than before
```

The technique of using invisible containers was described above in detail.* As noted, many users use MATLAB's GUIDE tool to create GUI figure windows. GUIDE automatically creates both a FIG file and a corresponding m-file. GUIDE-generated m-files always have a standard skeleton format, which has an *_OpeningFcn()* function and an *_OutputFcn()* function. *_OpeningFcn()* is called before the figure is made visible, while *_OutputFcn()* is called afterwards. This means that for performance reasons it is better to generate and update all GUI components in the *OpeningFcn()* function, rather than in *OutputFcn()*.

---

* See §10.1.7.

A good example of the importance of updates in an invisible figure is a case when we need to resize containers (panels or figures). Such resizes often involve numerous updates and refreshes of the contained components and could be quite slow, and cause visibly disturbing flicker, if the figure is visible. When done in nonvisible mode, the graphic calculations and the display updates are much faster.

In some cases, we cannot avoid creating our GUI while the figure is visible. If the created component is complex enough (e.g., a large data table), it often makes sense to create and customize the component as a nonvisible component, and only make it visible when it is fully ready. This ensures that the component is not refreshed and redisplayed every time there is a minor update or customization done to it during its preparation phase. This, in turn, ensures that the component is created in the fastest time possible, without unnecessary interim updates/refreshes.

In some cases, a component needs to be visible to make changes (e.g., if we wish to incorporate and customize some ActiveX or Java components). We can still use the technique, by moving the figure to an off-screen position (e.g., [9999,9999,width,height]). The figure will not in fact be displayed,* but the component will be able to be updated as if it were.

In one project for which I consulted, preparation and display of a visible data table with several thousand cells took 20 s. By simply making the component invisible before the update and making it visible again afterwards, I got a 10× speedup: the update now took only 2 s.†

```
set(hComponent, 'Visible','off');
... % update the component here
set(hComponent, 'Visible','on');
```

Updating a nonvisible component is a general advice that is not limited to MATLAB components. In fact, it applied also to Java, C++ (ActiveX), and Dot-Net components that are integrated in our MATLAB GUI.

Similarly, this advice also applies to external applications. For example, updating a hidden instance of Excel is much faster than updating it in visible mode. Much of the slowdown is due to making the Excel application visible (it is nonvisible by default), but even later it is still much slower. This may not be noticeable for a single update, but if we make numerous data and/or format updates it becomes meaningful:

```
% Open an Excel file and update it in non-visible mode
xls = actxserver('Excel.Application');
xls.Workbooks.Open([pwd '\test.xls']); % non-visible by default

tic, xls.Range('A1:O15').Value = magic(15); toc
⇨ Elapsed time is 0.005367 seconds.

% Now update a visible Excel - much slower!
xls.Visible = true;
xls.Range('A1:O15').Value = magic(15);
⇨ Elapsed time is 0.259920 seconds.

% Once Excel is visible, updating is still much slower than in
```

---

* While the figure window will not be visible, it will indeed appear as a separate element in the operating system's task bar.

† In that particular case, due to application constraints, the update had to be done row-by-row. In general, it is a better practice to prepare the entire data in advance and only update the component with the final data (this is discussed further below in this section), but this was not possible in that particular case. Updating the hundreds of data rows one after another required hundreds of graphic refreshes when the table was visible.

```
% non-visible mode (this becomes important when updating in a loop)
xls.Range('A1:O15').Value = magic(15);
⇨ Elapsed time is 0.065930 seconds.
```

10.4.1.2 Reuse Existing Figure Windows and GUI Components

Creating and displaying new GUI components, especially complex component such as a MATLAB figure window or a *uitable*, can take some time. We can often reuse existing components that are no longer in use.* This can be very beneficial for performance, especially when reusing figure windows or *uitable*s.

In the following demonstration, we are only interested in the figure creation versus reuse time. Since data processing exists in both cases, it is excluded from the code snippets:

```
% Create 100 new figures
for idx = 1 : 100
    hFig = figure;
    % create the GUI here
    drawnow
    close(hFig);
end
⇨ Elapsed time is 12.705217 seconds.

% Create a single figure and update it 100 times - 40x faster!
hFig = figure;
for idx = 1 : 100
    clf; % clear the figure window in preparation for the new GUI
    % create the GUI here
    drawnow
end
close(hFig);
⇨ Elapsed time is 0.295012 seconds.
```

A similar technique can be used for reusing GUI components such as listboxes or *uitable*s. In these cases, we simply need to update some of the existing component's properties (**String** in the case of listboxes and editboxes; **Data** in the case of *uitable*s), rather than deleting and recreating the components from scratch:

```
% Create 100 new uitables
for idx = 1 : 100
    clf;
    uitable('Data',rand(3));
    drawnow
end
⇨ Elapsed time is 2.024527 seconds.

% Create a single uitable and update it 100 times – 5x faster!
clf;
hTable = uitable('Data',rand(3));
for idx = 1 : 100
    set(hTable,'Data',rand(3));
    drawnow
end
⇨ Elapsed time is 0.383843 seconds.
```

---

* See also §10.1.9.

10.4.1.3 Reduce the Number of Property Updates

A related advice is to create the property values in advance and only update the component properties when the property values are ready. For example, updating a data table's **Data** property repeatedly, each time adding a new row, could be quite slow. Instead, prepare the entire data in advance, in a numeric matrix or cell array, and then update the **Data** property just once with this data:

```
% Update the data table one row at a time
clf; hTable=uitable;
tic
for row = 1 : 100
    d = get(hTable,'Data');
    d(end+1,:) = [row, row*pi];
    set(hTable,'Data',d);
end
drawnow; toc
⇨ Elapsed time is 0.020205 seconds.

% Update data table only when entire data matrix is ready - 1.6x faster
clf; hTable=uitable; d=[];
tic
for row = 1 : 100
    % note: this can be vectorized but I am trying to make a point here
    d(end+1,:) = [row, row*pi];
end
set(hTable,'Data',d);
drawnow; toc
⇨ Elapsed time is 0.012713 seconds.
```

10.4.1.4 Prevent Unnecessary Callback Invocations

Whenever you create a component, ensure that you customize it <u>before</u> you set a callback property, otherwise the callback might be called unnecessarily, hampering performance. An extreme example would be setting a table's cell-edit callback before actually setting the table data — the callback would then be called unnecessarily for each of the table's data cells, which could bring the application (and MATLAB itself) into a frozen state for many minutes.

An important corollary is that GUI callbacks are only triggered after the control is actually rendered onscreen. We must therefore ensure not to set callbacks before the control is actually displayed; otherwise the callback may be triggered unnecessarily.

In one specific case, I investigated a case where table updates took many minutes, rather than the expected split second. It turned out that although the table update callback was turned off before the update, it was turned on immediately after the update. Since the MATLAB code executes on the main MATLAB thread while the GUI updates are done on a separate thread (Java's EDT: *Event Dispatch Thread*),[1630] it so happened that the MATLAB command to restore the table callback was executed before the table data has actually updated onscreen. This, in turn, caused the callback to be unnecessarily invoked hundreds of times, for each of the updated data cells. The solution was to simply force a screen refresh before restoring the callback:

```
% Initial data table set-up with an updates callback
h = uitable('v0',...);
set(h,'DataChangedCallback',@myCallbackFcn);
```

```
% Update the data table
set(h,'DataChangedCallback',[]); % temporarily disable callback
set(h,'Data',newData);

% The following appears innocent but is in fact very important:
% Ensure all updates are displayed onscreen before restoring callback
drawnow; pause(0.1);

% It is now safe to restore the regular data-updates callback
set(h,'DataChangedCallback',[]);
```

While the preceding example explained why we sometimes must insert seemingly unnecessary calls to *drawnow* and *pause* in our code, in many cases, it is actually better <u>NOT</u> to insert such calls except at the very end of our GUI preparation.* The reason is that the GUI refreshes that are triggered by these functions could take some time, and multiple refreshes could induce a flicker effect. By only updating the GUI after everything is ready, we minimize the setup time and reduce flicker. Inserting such intentional calls to *drawnow* and *pause* is only advisable in one of three cases:[†]

1. To control the display timing, as in the callback example above
2. To avoid a MATLAB hang following close of a modal dialog[1631]
3. To display immediate visual feedback, rather than wait until setup is complete

10.4.1.5 *Postpone Nongraphical Setup and Updates*

A general advice for GUI preparation performance is to update nonvisible items last. For example, there is no reason to populate the options of a drop-down control (also called a popup menu, or combo-box) early in GUI preparation. This takes some time and has no visual effect since the drop-down options are hidden until the user actually clicks the control. Therefore, we could postpone the population of the control's options until the entire GUI has been displayed (i.e., after the final call to *drawnow*).

Similarly, try to postpone all nongraphical back-end maintenance until after the GUI has been set up and displayed, for example, updating log files, reading information from a database or disk files, or accessing the Internet. While postponing these tasks will not improve the total *actual* setup time, it would significantly improve the *perceived* time (and the user experience), since the GUI will display faster.[‡]

As noted elsewhere in this book, Excel file I/O is often a significant performance hotspot.[§] Since reading data from an Excel file is often required for MATLAB GUIs and this is a nongraphical step, we can postpone reading the file until after the entire GUI is created and presented (e.g., in our *_OutputFcn()* function). The user will see the GUI (and possibly be able to interact with it) much faster than if the Excel file is read during the setup phase (*_OpeningFcn()*).[¶]

---

* See §10.1.5.
[†] Read more on intentional pauses in §10.4.3.6 and §10.5.1.
[‡] Compare *lazy initialization*, discussed in §3.8.1 and §1.8.3. Read more on perceived performance in §1.8.
[§] See §11.5 and §1.8.3, for example.
[¶] See §3.8.1 for details.

10.4.1.6 Make the GUI Simple to Use

Any GUI that is simple and streamlined to use appears to be faster than GUI that is more complex to operate (perceived performance).* The longer a user has to search for functionality in a multitude of complex menu items or figure panels, the longer that operation would take from end-to-end. Even if the actual time spent in operating the GUI is not large, a complex GUI would *appear* slower than a streamlined GUI.

It does not matter if a computational processing task is superfast if users take ages to activate it. In such a case, we could even argue that the perceived performance has turned into an actual performance issue.

The trade-off between wishing for GUI simplicity on the one hand, and providing a complex functionality on the other hand, is a long-standing design problem. Different applications have provided a variety of answers over the years. For example, Microsoft Office's ribbon (copied onto the MATLAB Desktop in R2012b) provides easy access to the functionality subset that is used by most users most of the time, while enabling the rest of the functionality to be accessed a few more clicks away.† Your specific application may well employ a different solution to the simplicity/functionality trade-off, but you must be aware of it from the onset of your design.

GUIs are often designed with a prominent "Calculate" or "Run" button that performs some important task based on the values of helper controls such as checkboxes, editboxes, drop-downs, and so on. While the need for such a centralized processing-trigger control may be unavoidable for long-running tasks, it is a mistake to use such a design when the task is fairly short (say, up to a few seconds in duration):

All MATLAB controls have callback properties that are automatically invoked (called) when the user interacts with the control. If the main processing task is quick enough, it is highly advisable to automatically recalculate the application's data and refresh the GUI display accordingly, upon every interaction with any of the GUI's controls. This produces immediate GUI responsiveness, which improves the user experience compared to the centralized processing design.

10.4.1.7 uitable Considerations

10.4.1.7.1 *Old versus New uitable*

MATLAB's **uitable** function has existed in all MATLAB 7 releases, but only became officially supported and documented in R2008a (MATLAB 7.6). In that release, the implementation and input parameters to **uitable** have changed dramatically.

The newly supported **uitable** is an entirely different "animal" than its unsupported predecessor: it is better integrated in MATLAB's GUI and has new features that were not available in its predecessor (e.g., row striping). Unfortunately, these new features made the underlying table implementation more complex. Each table data cell is now implemented as an independent class object that enables cell-specific formatting, data checks, and so on. As a corollary result, **uitable** data setup and update have become much slower in the new **uitable** compared to the old **uitable**. Updating a table that contains thousands of data cells can now take so much time as to become unacceptable.

---

* Read more on this in §1.8.5.
† Ribbons (called *toolstrip* in MATLAB) are highly contentious, with many users vehemently taking an opposing view. Both the Microsoft and the MATLAB forums are awash with criticism for the new design in terms of usability and user productivity.

If we can live without the extra features of the new *uitable*, and if we are willing to use an unsupported function that may well be removed at any future MATLAB release, then consider using the old *uitable* as an alternative. This is as simple as passing 'v0' as the first input parameter,* followed by an optional container handle, initial data values (a cell or numeric matrix), and a cell array of header strings:

```
>> hTable = uitable('v0', gcf, magic(3), {'a','b','c'})
hTable =
         javahandle_withcallbacks.com.mathworks.hg.peer.UitablePeer
```

Note that the returned handle is that of a Java object rather than a regular MATLAB HG handle. Additional properties such as **Position, Units, Visible, Editable,** and **GridColor** can be set on the returned handle. The old *uitable* does not have a cell selection callback like the new *uitable*; its data update callback property is called **DataChangedCallback.** Also note that the **Visible** and **Editable** properties accept logical (*true/false*) values, rather than 'on'/'off':

```
>> set(hTable,'Editable',false)
```

The difference in performance between the two versions of *uitable* is striking:

```
% Prepare the data (100x100 cells)
data = num2cell(magic(100));
headers=cellfun(@(c)sprintf('C%d',c),num2cell(1:100),'uniform',false);

% Old uitable version
hOld = uitable('v0',data,headers); drawnow
⇨ Elapsed time is 0.302210 seconds.

% New uitable version: 15x slower, unacceptable performance
hNew = uitable('data',data,'ColumnName',headers); drawnow
⇨ Elapsed time is 4.425472 seconds.
```

The difference in data update performance is comparably striking:
```
% Old uitable version
set(hOld,'Data',dc'); drawnow
⇨ Elapsed time is 0.157653 seconds.

% New uitable version: 14x slower, unacceptable performance
set(hNew,'Data',dc'); drawnow
⇨ Elapsed time is 2.185875 seconds.
```

Moreover, the old *uitable*, being based more directly on a Java JTable, enables direct access to and modification of specific table cells, without needing to update the entire **Data** as in the case of the new *uitable*. This is done via the underlying JTable's *setValueAt()* method that accepts the new value (string or number), row number (first row = 0), and column number (first column = 0). Updating specific table cells in this manner can be many times faster than updating the entire data:

```
% Update row 5 column 3 to the string value 'qwerty'
hOld.Table.setValueAt('qwerty',4,2); drawnow
⇨ Elapsed time is 0.018457 seconds.
```

---

* MATLAB releases R2007b (7.5) and older only have the old *uitable* version, so the "v0" input parameter should not be used on those releases.

Similarly, we can access specific data cells much quicker than reading the entire **Data** to read a single cell value (in this case, the new *uitable* is actually slightly faster than the old):

```
% Directly read old uitable's row 5 column 3
value = hOld.Table.getValueAt(4,2); drawnow
⇨ Elapsed time is 0.000437 seconds.

% Read entire data using the old uitable to get row 5 column 3:
data = cell(get(hOld,'Data')); value=data{5,3}; drawnow
⇨ Elapsed time is 0.115396 seconds.

% Read entire data using the new uitable to get row 5 column 3:
data = cell(get(hNew,'Data')); value=data{5,3}; drawnow
⇨ Elapsed time is 0.000315 seconds.
```

Using the old *uitable* may seem tempting. But remember that it uses unsupported functionality that may well be removed without any prior notice at some future MATLAB release. There are also usage quirks that are due to the different parameters format and the fact that it is basically a Java object and not a regular MATLAB HG one. We have seen above some effects of this: needing to specify 0-based row and column numbers for *set-ValueAt()* and *getValueAt()*; as well as needing to cast the result of get(hOld,'Data') into a cell array (matrix) using the *cell* function. All these quirks do not exist in the new supported *uitable*. Users should therefore carefully weigh the possible performance gains versus maintainability and development time.

For more information on using and customizing the new and old *uitable*, refer to Section 4.1 of my MATLAB-Java programming book.[1632]

10.4.1.7.2 *Using HTML and Cell-Renderers*

A common feature of Java Swing components is their acceptance of HTML and CSS for any of their JLabels.[1633] Since most MATLAB controls are based on Swing-derived components, this Swing feature automatically applies to MATLAB *uicontrols* and *uitable*.[1634] Whatever can be formatted in HTML (font, color, size, etc.) is inherently available in MATLAB controls, subject to some limitations on the internal Swing implementation of HTML and CSS.*

In *uitable* (both the old and the new version), we can HTML-format both the headers and the cells data. For example, let us start with a standard *uitable*:

```
labels = {'A';'B';'C'};
data = magic(3) - 5;
hTable = uitable('Data',data, 'ColumnName',labels);
```

	A	B	C
1	3	-4	1
2	-2	0	2
3	-1	4	-3

A standard MATLAB *uitable*

---

* Swing supports a wide subset of HTML 3.2 and CSS 2.0 , but not the entire specification. For example, HTML <div> tags are often ignored whereas <font>, <b>, <i>, <u>, <sup> and <sub> are honored; in CSS, the text-align directive appears to be ignored, while font directives (color/size, etc.) are honored.

Now let's add some HTML formatting. Naturally, the cells need to be changed from numeric to string for this:

```
>> dataCells = cellfun(@num2str,num2cell(data), 'uniform',false)
dataCells =
    '3'      '-4'     '1'
    '-2'     '0'      '2'
    '-1'     '4'      '-3'
```

Now let's update the headers to bold blue and the table cells with green/red colors based on whether cell values are positive or negative:

```
labels = strcat('<html><b><font color="blue">',labels);
dataCells(data>0) = strcat('<html><u><font color="#00FF00">',...
                    dataCells(data>0)); % positive cells: green
dataCells(data<0) = strcat('<html><b><font color="rgb(255,0,0)">',...
                    dataCells(data<0)); % negative cells: red
set(hTable, 'Data',dataCells, 'ColumnName',labels);
```

An HTML-formated MATLAB *uitable*

Note that HTML tags do not need to be closed (<tag>...</tag>), although it is good practice to close them properly. Also note the different ways in which HTML colors can be specified.

While such HTML formatting is very useful for visualization, it does carry a performance penalty for tables that have numerous cells. The reason for this is a suboptimal Java HTML processing engine, not MATLAB's fault at all. For this reason, if we need to apply formatting to more than a few dozen cells, it might be wise to use table cell-renderers rather than HTML formatting. In addition to performance benefits, cell-renderers also enable cell-specific tooltips, backgrounds, alignment, and so on, which cannot easily be achieved using the supported HTML/CSS subset.*

Programming table cell-renderers requires a little Java knowledge.[1635] On the other hand, using existing cell-renderers (many of which can be freely downloaded from Java forums) in a MATLAB *uitable* is quite easy and requires no Java knowledge. For example, consider using my *ColoredFieldCellRenderer* to modify cell foreground and background colors based on the cell values:[1636]

```
% Initialize our custom cell renderer class object
javaaddpath('ColoredFieldCellRenderer.zip');
cr = ColoredFieldCellRenderer(java.awt.Color.white);
cr.setDisabled(true); % to bg-color the entire column

% Set specific cell colors (background and/or foreground)
for rowIdx = 1 : size(data,1)
```

---

* HTML can create a background color, but it spans only the size of the cell text. To span the entire cell width, we need to either use a cell-renderer or widen the cell contents with multiple " ". For example, "text   ". For cell-specific tooltips, we can implement a mouse-movement callback function that checks the cell that the mouse cursor is currently on and then updates the table's tooltip accordingly. Both of these hacks are obviously lacking.

```
    % Red/greed foreground color for the numeric data
    for colIdx = 2 : 8
        if data(rowIdx,colIdx) < 0
            cr.setCellFgColor(rowIdx-1,colIdx-1,java.awt.Color.red)
        elseif data(rowIdx,colIdx) > 0
            cr.setCellFgColor(rowIdx-1,colIdx-1,java.awt.Color.green)
        end
    end

    % Yellow background for significant rows based on p-value
    if data(rowIdx,12) <= 5 && data(rowIdx,11) ~= 0
        for colIdx = 1 : length(headers)
            cr.setCellBgColor(rowIdx-1,colIdx-1,java.awt.Color.yellow)
        end
    end

    % Bold blue foreground for significant payouts
    if abs(data(rowIdx,11)) >= 2
        cr.setCellFgColor(rowIdx-1,10,java.awt.Color(0,0,1)) % blue
        % the following could also be done in the renderer, like colors
        boldPayoutStr = ['<html><b>' num2str(data(rowIdx,11),'%.2f')];
        dataCells{rowIdx,11} = boldPayoutStr;
    end
end

% Replace MATLAB's table model with something more renderer-friendly
jTable.setModel(javax.swing.table.DefaultTableModel(dataCells,headers))
set(hTable,'ColumnFormat',[]);

% Finally assign the renderer object to all the table columns
for colIdx = 1 : length(headers)
    jTable.getColumnModel.getColumn(colIdx-1).setCellRenderer(cr);
end
```

Periods	Any period returns	Avg return signal	Avg gain	Avg draw	Gain/draw ratio	Max gain	Max draw	Random % pos	Signal % pos	Payout	% p-value
3	0.29	0.01	1.30	-1.91	-1.48	2.42	-2.12	58	60	0.01	76
5	0.49	-1.00	1.28	-1.57	-1.23	1.28	-3.63	58	20	-1.00	15
10	0.98	-2.39	4.04	-4.00	1.01	4.04	-4.87	62	20	-2.39	11
15	1.45	-6.39	0.12	-8.02	-67.49	0.12	-11.39	65	20	-6.39	2
20	1.89	-2.25	6.97	-8.39	-1.20	9.58	-11.24	67	40	-2.25	36
30	2.66	-6.54	7.50	-10.05	-1.34	7.50	-14.25	70	20	-6.54	10
40	3.40	-9.12	0.01	-9.12	-912.46	0.00	-15.71	71	0	-9.12	5

A *uitable* with custom cell-renderer

Unlike *using* cell-renderers, which is quite simple as we just saw, *creating* cell-renderers is an advanced programming topic, which is more prone to problems and future support-ability issues than using simple HTML formatting.* So once again, we are faced with a performance-tuning trade-off to consider.

---

* Readers are referred to my book, *Undocumented Secrets of MATLAB-Java Programming*, CRC Press 2011, for details on creating and using cell-renderers, the differences between the *uitable* versions, and *uitable* customizations in general.

10.4.1.7.3 *Fixing the New **uitable**'s Performance*

Running the MATLAB Profiler when preparing a ***uitable***, we can easily see that the vast majority of the time is spent in the ***arrayviewfunc*** function. This function includes multiple helper functions, dispatched from the main function in switchyard format.* After drilling into the profiling report and the nontrivial code, the source of ***uitable***'s slowness appears to emanate from the fact that each table cell is considered separately, and wrapped within a MATLAB class object.

This implementation provides flexibility for cases when a table column might contain both numeric and nonnumeric data. But in the general and most common case, it simply adds enormous memory and performance overheads.

It is possible to modify ***arrayviewfunc***[†] in such a way that will significantly improve its performance, by avoiding the class object encapsulations, or at least by avoiding the numerous data-type run-time checks, if we are certain that it is real double, for example. After spending a few hours doing this for a client who had unacceptable table performance, I improved ***uitable***'s creation performance by 4×. The bad news is that ***arrayviewfunc*** changes across MATLAB releases, and the modified function caused errors and blank tables when we tried to use it with a newer MATLAB release.

So the bottom line is that while it is indeed *possible*, it is not in fact *advisable* to modify the internal ***arrayviewfunc***. Since (IMHO) the only major benefit of the new ***uitable*** compared to the old ***uitable*** appears to be row striping, it seems to me that it is much more cost-effective and maintainable to use the old ***uitable*** with a custom cell-renderer that takes care of the striping (based on the row index), than it is to fix the new ***uitable***.

I remain optimistic that MathWorks will fix ***uitable***'s performance in one of the upcoming MATLAB releases, thereby relieving us of the need to use such unsupported hacks.

10.4.2 Presenting User Feedback

10.4.2.1 Psychology of User Feedback

Presenting feedback to the user, while the application is performing lengthy tasks, is an important aspect of perceived performance in GUI design.

The classic example of the elevator is often used to illustrate this. When electric elevators were invented in the late 19th century, it was soon discovered that a simple mechanism that presents to the user the current elevator's position (floor number) causes the elevator users to be more relaxed, tolerant of delays, and willing to wait longer for the elevator to arrive. The more sophisticated the display, the longer users are willing to wait. Similarly, the simple act of placing a mirror in the elevator car makes the passengers more willing to accept longer travel times. After all, time seems to "fly" when we are not bored.

Another well-known example of these techniques is the phone. When phones were invented, it was soon realized that users are willing to accept much longer wait times if they hear a dial tone, which indicates that the phone is ringing on the other side. This dial tone is synthetic: we are not really hearing the other side ring. However, our mind subconsciously translates the dial tone into an image of the phone ringing on the other side, and this prompts us to wait longer for the other party to answer the phone. As in elevators, the more sophisticated the feedback, the longer users would be willing to wait: When we call a call center, we

---

* See §4.6.1.4.
† *%matlabroot%\toolbox\matlab\codetools\arrayviewfunc.m.*

are much more likely to wait if it uses an automated queuing system. We would never have waited several minutes with a simple dial tone, but we are willing to wait this long if we are constantly informed of our position in the queue and the expected response time.

Apparently, we humans are willing to accept longer wait periods when presented with a sensual feedback indicating that the system has "accepted" our request and is "working on it". When people get no feedback at all, they tend to "give up" much sooner. The phone's dial tone presents exactly this sort of feedback. Just think how much sooner we would hang up a phone if we did not hear a dial tone after dialing.

We can easily adapt these insights to MATLAB programs. Whenever we see that our program has a long-running task, we should add some feedback. The actual duration time for which it makes sense to present feedback naturally depends on the specific situation. In general, user actions should receive immediate (<0.1 s) acknowledgment (e.g., editbox becoming disabled, button appearing "unclicked") and any task running longer than 2–3 seconds should have some associated progress feedback.[1637]

Users who see no feedback for more than a few seconds perceive a slow system and may well abandon it before it's done. The feedback engages the user and delays abandonment to a much later phase. Even a simple thing such as an endless animated hourglass or progress bar can significantly improve the perceived performance.

10.4.2.2 Feedback in MATLAB Applications

Feedback features should be an integral part of our GUI design. Each task that takes more than a split second should optimally be accompanied by a corresponding feedback mechanism.

Applications should generally present their feedback in a manner consistent with the manner in which the user interacts with the application. If the application is primarily Command Window based, then the feedback should generally be presented in the Command Window, rather than in a popup window. Conversely, if the application is primarily GUI based, then the feedback should in general be presented within the GUI itself, or in a separate graphical popup window that looks consistent with the GUI, and not in the Command Window.

This feedback could be as simple as disabling a clicked button control, turning the mouse pointer into an hourglass symbol, or displaying a simple "please wait..." message in a *msgbox* popup or in the Command Window:

```
% Simple feedback mechanisms
set(hButton,'Enable','off');
set(gcf,'pointer','watch');
msgbox('Loading - please wait...'); % optional title, icon parameters
fprintf('Loading - please wait...\n');
```

10.4.2.3 Feedback in MATLAB Slider Controls

One particular GUI control has a behavior that affects its responsivity in MATLAB: The slider *uicontrol* (which is actually a scrollbar rather than a slider) has a single main callback property called **Callback**.* This callback is called whenever the slider's value is interactively changed, by clicking the scroll buttons or moving the knob.

---

* The slider control actually has a few other callback properties but these are of lesser importance and are rarely used in practice.

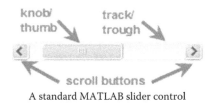

A standard MATLAB slider control

It so happens that while the callback is continuously called by MATLAB if a scroll button or the trough is continuously clicked, this is not the case when the knob is moved: In that case, the callback is only called when the knob is released in its final location. In order to force MATLAB to continuously call the callback during the knob movement, we can use the following code snippet:[1638]

```
try      % R2013b and older
   addlistener(hSlider,'ActionEvent',@myCallbackFcn);
catch   % R2014a and newer
   addlistener(hSlider,'ContinuousValueChange',@myCallbackFcn);
end
```

10.4.2.4 Temporarily Disabling User Interaction

If the processing task uses multiple GUI elements, it is prudent to disable all such elements until the task completes. Otherwise, the user might think that the task is done, and will interact with these elements while the original task is still processing in the background. This might cause problems in our application's logic.

For example, let's imagine a button control that starts a processing task that uses the state of a checkbox or editbox to perform some calculation. If we modify the checkbox state or editbox value while the task is still running, this could result in the first part of the calculation using different parameters than the second part, causing inconsistencies and errors. In such a case, it is advisable to disable the relevant GUI controls until the entire calculation task has ended.

We need to take into consideration the possibility that an error (exception) will occur in our processing task. This could result from programmatic errors (bugs), external factors (e.g., disconnected hardware or missing file), or even a user break with Ctrl-C. In order not to remain with a uselessly disabled GUI, we need to wrap our processing logic in an external *try-catch* block, and re-enable all controls once the block has completed, either successfully (in the *try* branch) or not (i.e., in the *catch* branch).

On the other hand, we should re-enable the disabled controls as soon as they are no longer used by the processing task and interaction with them would be safe. This could of course be done at the very end of the processing task, but faster perceived performance would result from earlier re-enabling.

For example, let's assume the last phase of a task is updating some Excel file with the calculation results. This subtask could take a long time and does not rely on, or affect, any GUI component. We could therefore restore all the disabled GUI controls to their normal state *before* updating Excel, rather than wait for this update to end. The file update will be performed in the background, while the user is able to once again interact with the GUI, resulting in a significant perceived performance improvement.

In complex GUIs, controls are often interrelated. Disabling one control during processing may not be sufficient, and several controls may need to be disabled at the same time (e.g., menu items, toolbar buttons, and GUI controls). It might be simpler and less error-prone to simply disable the entire figure during processing. This effect can be achieved using the figure's underlying Java window frame:[1639]

```
% Alternative #1: directly disable/enable the entire Java JFrame
jFigPeer = get(handle(gcf),'JavaFrame');
try
    jClient = jFigPeer.fFigureClient;   % This works up to R2011a
catch
    try
        jClient = jFigPeer.fHG1Client;  % This works on R2008b-R2014a
    catch
        jClient = jFigPeer.fHG2Client;  % This works on R2014b and up
    end
end
jClient.getWindow.setEnabled(false);        % disable the figure
 ... % do the processing here
jClient.getWindow.setEnabled(true);         % re-enable the figure
```

Alternatively, we can use my *enableDisableFig* utility,[1640] which wraps the Java-based code above in an easy-to-use MATLAB function:

```
% Alternative #2: using the enableDisableFig utility
oldState = enableDisableFig(gcf, 'on');   % disable the figure
 ... % do the processing here
enableDisableFig(oldState);                % re-enable the figure
enableDisableFig(true);                    % an alternative
```

Whenever we disable a major application functionality for a long-running task, we should enable users to stop the task in mid-run, using a dedicated GUI control that is made available when the task begins. For example, browsers normally enable users to stop a long-running download using a simple mouse-click.

Similarly, in MATLAB, we could present a control that would stop the task by raising a flag that would be checked in the main processing loop. An example of this concept is my *blur Figure* utility,[1641] which can be used to disable (and blur) a figure while presenting continuous progress bar feedback as well as a task-cancellation button (see figure on page 579).

Note that *blurFigure* only works on MATLAB R2013a (8.1) and earlier. It does not work on R2013b and newer releases, since those releases use Java 7, which does not support modifying window transparency in real time, in a way that can be used in MATLAB.

10.4.2.5 *Feedback for Long-Duration Tasks*

More sophisticated feedback is needed when the task duration is longer than a few seconds. In such cases, animation can often prolong the time that users will be willing to wait for the task to complete (which is obviously different from person to person).

For example, using MATLAB's built-in *ftp* function for downloading a very large file will appear to "hang" the application, when in fact it is still busy downloading the file; adding animation and/or download progress indication will make users much more receptive to a long download.[1642]

A very simple animation feedback can be achieved using a textual progress bar on MATLAB's Command Window. This is useful for applications that are Command Window based rather than GUI based. The idea is to use the backspace control character (BS, or *sprintf('\b')*, or *char(8)*) repeatedly in order to erase the preceding characters from the Command Window, before displaying the new text:[1643]

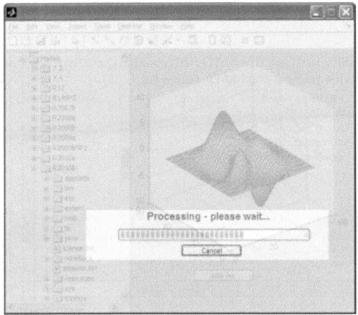

The *blurFigure* utility in action

```
reverseStr = '';
for idx = 1 : maxVal

    % Do some computation here...

    % Display the progress
    percentDone = 100 * idx/maxVal;
    msg = sprintf('Percent done:%3.1f', percentDone)
    fprintf([reverseStr, msg]);
    reverseStr = repmat(sprintf('\b'), 1, length(msg));
end
```

An extension of this principle has been implemented in Evgeny Prilepin's *ConsoleProgressBar* utility.[1644] This utility displays a customizable textual progress bar, as well as the elapsed and estimated remaining work time:

```
% Create a customized progress bar object
maxVal = 500;
cpb = ConsoleProgressBar();
cpb.setMinimum(0);
cpb.setMaximum(maxVal);
cpb.setElapsedTimeVisible(true);
cpb.setRemainedTimeVisible(true);
cpb.setElapsedTimePosition('left');
cpb.setRemainedTimePosition('right');
cpb.start();
for idx = 0 : maxVal
    % Update the progress bar and associated text
    cpb.setValue(idx);
    cpb.setText(sprintf('Progress:%d/%d', idx, maxVal));
    % Do some computation here...
```

```
end
cpb.stop();
⇒ E 00:00:11 68% [==============>......] R 00:00:05 Progress: 342/500
```

GUI-based applications can use MATLAB's standard *waitbar* function to present a popup window with a simple graphical progress bar:

```
h = waitbar(0,'Please wait...', 'Name','My GUI Application');
for idx = 1 : maxVal
    % Update the progress bar
    waitbar(idx/maxVal);
    % Do some computation here...
end
close(h);
```

A standard *waitbar* window

More sophisticated graphical progress bar utilities are available on the MATLAB File Exchange.[1645] These are improvements of the standard *waitbar* function, enabling a display of the percentage value, elapsed/remaining time, multiple progress bars, and so on. Like *waitbar*, most of them display a GUI popup window with progress bar:

A more sophisticated progress bar, with percentage value
and estimated remaining time

Several utilities are available that enable embedding the progress bar directly within the GUI.[1646] For example, my *statusbar* utility[1647] enables embedding a customizable progress bar in a figure window's or the MATLAB Desktop's status bar (see top figure on page 581).

Another utility, *waitmex*, enables using a wait-bar from within a MEX function.[1648]

While MATLAB GUI cannot be directly updated from within a parallelized (*parfor*) loop, we can use the *parfeval* function (R2013b and newer) to update the progress.*

Sometimes, when displaying progress data, the percentage value and remaining time cannot be calculated. In such cases, it is still useful to present information about the

---

* See §6.1.2.

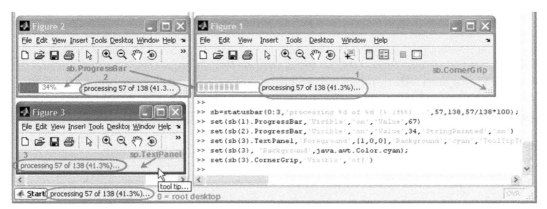

The *statusbar* utility, with customizable progress bars

number of items or subtasks completed, and the currently processing item. This tells the user relevant information about the relative processing position, and feedback that the system is "working". This is better than not displaying any information at all.

If even this information cannot be presented (perhaps the entire long task cannot be broken into subtasks), then it is still useful to present an animated visual cue that the system is "working". The typical method for doing this is to change the figure cursor:

```
set(gcf,'Pointer','watch');
```

We can also embed animated spinner icons within our GUI.[1649]

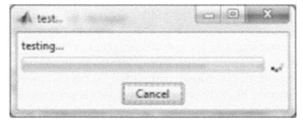

An animated spinner icon   An animated spinner embedded in a GUI

An alternative is to use a so-called "indeterminate" (endless looping/cyclic) progress bar.[1650] This can be embedded in MATLAB GUIs as follows:

```
jpb = javaObjectEDT(javax.swing.JProgressBar);
jpb.setIndeterminate(true);
[hpb,hContainer] = javacomponent(jpb,[10,10,100,20],gcf);
```

An indeterminate progress bar

When presenting feedback, we may need to deliberately slow-down the feedback, to prevent it from changing too rapidly for the human eye,[1651] or on the other extreme, to avoid excessive GUI updates that are indistinguishable by a human eye.

For example, there is no point to update a progress bar in steps smaller than 1%–2%: this would not be distinguishable, and would take unnecessary GUI processing time. Instead, we can display feedback only once every few steps.* The displayed feedback information could be filtered or averaged data, to avoid hyperactive fluctuations.† The reduced feedback processing and GUI updates will visibly speed up the run-time, while presenting a fluid continuous-looking feedback.[1652]

```
prevPercent = 0;
lastFeedbackTime = now;
ONE_SEC = 1 / (24*60*60);
for idx = 1 : N
    doSomeProcessing();

    % update feedback only once every 200 loop iterations or 0.5 sec
    if ~rem(idx,200) || now - lastFeedbackTime > 0.5*ONE_SEC
        presentFeedback();
        lastFeedbackTime = now;
    end

    % update waitbar only in 1% steps
    currentPercent = fix(100*idx/N);
    if currentPercent > prevPercent
        waitbar(idx/N);
        prevPercent = currentPercent;
    end
end
```

When using progress bars, we should report progress as percentage of the total work, not of something partial. For example, say that we have several long-running tasks: If we are using a multibar progress bar, then we could present the progress within the current task in the upper bar, and the overall progress in the lower bar (or vice versa).

Alternatively, we could display only a total progress bar, calculating our position in the current task (and all previously completed tasks) as a percentage of the overall.

What we should NOT do is to display a single progress bar that only shows the position within the *current* task, and once complete loops back to 0 at the beginning of the next task. This frustrates users and takes the point out of the feedback.

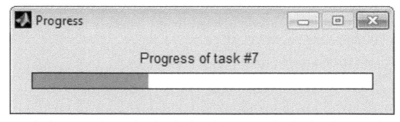

Misleading: a simple ***waitbar*** that displays the progress of each sub-task separately

---

* See §10.1.5 and §10.4.3.8 for similar implementations. To solve the rapid feedback issue, we could also insert intentional calls to *pause*. However, this will slow down overall run time; see §10.5.1.
† See §10.4.2.6.

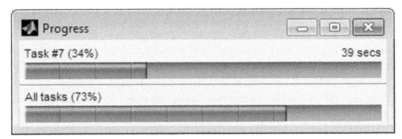

Good: a progress bar showing the overall (total) progress

Excellent: Ben Tordoff's *multiWaitbar*[1653]

10.4.2.6 Feedback for High-Volatility Data

When presenting feedback for tasks that change data quickly, it sometimes makes sense to smooth the data in order to present the user with an averaged, more relaxed information. Using a moving average (e.g., using the *filter* function, §5.7.7) is a typical solution that smoothes the data at the expense of a small delay in the feedback. To the user, the small delay may be unnoticeable, and yet the smoothed data might be a blessing, removing the high-frequency jitter. For example:

```
% Create and plot the noisy actual data
t = 0 : 0.01 : 10;
data = sin(t) + rand(size(t));
plot(t,data); hold on

% Create and plot the filtered feedback data
windowSize = 50;
dataFiltered = filter(ones(1,windowSize)/windowSize,1,data);
plot(t,dataFiltered,'-r','linewidth',5);
```

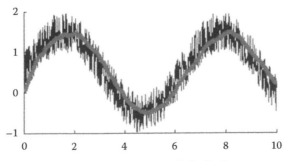

Filtered (slightly-delayed) feedback

When we use the feedback to manually control some program parameter, it is much easier for us to control a smoothed feedback. When feedback is unnecessary, the high-frequency information may or may not be advisable, depending on the specific case.

Two File Exchange utilities by Damien Garcia, *smoothn*[1654] and *smooth1q*,[1655] are of note for their fast and robust smoothing implementation to a wide range of data. John D'Errico's *SLM (shape language modeling)* utility[1656] provides excellent curve-fitting results, having far better accuracy than simple smoothing. The File Exchange has ~50 other smoothing-related utilities that may be useful for specific use-cases.[1657]

10.4.2.7 Reducing Intermediate Updates

Instead of computing, saving, and displaying intermediate results, it is faster to merge the intermediate steps, saving and displaying only the end results. Saving intermediate results is prime candidate for removal. Users are usually not interested in intermediate results — a simple progress bar is almost always sufficient for ongoing feedback.

In rapidly firing mouse callbacks, there is usually no need to recompute and update the display every single time (§10.4.3.8). This is achieved by dropping intermediate events, or by setting the callback's **Interruptible** property to 'off', and **BusyAction** to 'cancel',[1658] and in general by preventing callback re-entrancy (§10.4.3.3).

10.4.3 Performing Asynchronous Actions

In addition to the above-mentioned techniques for improving our GUI's actual and perceived performance in *synchronous* programming, we can do several additional things to improve the performance when *asynchronously* interacting with an active GUI via timer* or user-triggered actions.

10.4.3.1 Update GUI Selectively

The most important advice here is to only update and refresh the GUI elements that actually need to be updated. For example, there is no use in recreating an entire GUI window when we only need to update the state of a single checkbox. Similarly, we can simply update a specific graph rather than update all graphs. This technique is standard practice in modern GUI design. In fact, it underlies AJAX, which is used in modern websites to refresh selective webpage data in "real time".

There is a natural trade-off here: specific updates in response to specific user actions or application states require more coding. Careful attention to detail and all the possible application states (including edge cases) is important. In fact, programming and debugging this might take much longer than justified by the performance gain.

10.4.3.2 Reduce Callback Execution Time

GUI callbacks should be designed to be as short as possible. The quicker the callback, the more responsive the GUI. Users may lose patience with long callback execution.†

Callbacks sometimes need to perform a lengthy computation or update (e.g., updating an Excel file). In such a case, consider delegating the lengthy update task to a timer object

---

* Also see §7.3.5.
† See §10.4.3.8 for additional aspects of callback execution time.

that will execute asynchronously, enabling the synchronous callback to complete its work sooner. This will ensure better GUI responsiveness, without affecting the program logic. Here is a simple example (we may wish to specify some more timer properties — the snippet below is only for illustration):

```
% Regular callback function - takes a long time to complete
function myCallbackFcn(varagin)
    % Call the utility function directly (synchronously)
    utilityFcn();
end

% A better callback - completes much faster, using asynchronous timer
function myCallbackFcn(varagin)
    % Start an asynchronous timer to perform the lengthy update
    start(timer('StartDelay',0.5, 'TimerFcn',@utilityFcn));
end

% A utility function that performs a lengthy calculation/update
function utilityFcn()
    % some lengthy calculation/update done here (e.g., update Excel)
end
```

10.4.3.3 Prevent Callback Re-entrancy

GUI callbacks sometimes have a bad habit of being called several times for a single user action. There are several reasons for this, and the specific cases when this happens depend on the particular control and user action. For example, continuously clicking an arrow button of a MATLAB slider uicontrol (which is actually a scrollbar GUI control) causes its corresponding callback to be invoked continuously, until the user releases the arrow button.* A similar situation occurs when the user continuously presses a keyboard key, or moves the mouse.

The end result is similar: our callback function is called multiple times, rather than just once. The callback may be successively invoked before it has time to complete its callback functionality, even if we make our callback function superfast. This would not only affect the overall callback responsiveness, but also cause programmatic errors. The reason is that the callbacks are invoked so close to each other that they could actually execute the same code at the same time, causing inconsistent results and programmatic/ algorithmic errors.†

Since it is generally not possible to control the callback invocations, we should at least prevent the callback code from being executed by more than a single invocation instance at any point in time. Doing this is called *preventing callback re-entrancy.*

The issue of callback re-entrancy is well researched in software engineering literature. Different programming languages and computing platforms have different built-in mechanisms to facilitate handling such cases (e.g., *mutexes* or *critical sections*).

In MATLAB, we could set the callback's **Interruptible** property to 'off' (the default value is 'on'), and **BusyAction** to 'cancel' (the default is 'queue').[1659] This will ensure that none of

---

* See §10.4.2.3.
† Strictly speaking, this is incorrect. In fact, only a single callback invocation thread is executed at any single point in time. However, at specific commands in the callback, the running invocation may yield CPU execution to another invocation in the events queue, thereby causing both invocations to effectively run parts of the callback function "at the same time".

the callbacks that honor these properties will have re-entrancy issues. Any callback events that occur while a previous callback is still being executed, will be discarded and will not be processed. In many cases, this is okay, since we do not care about processing each individual event. In cases where each and every event is indeed important, we should keep **BusyAction** as 'queue', causing new events to pile up in an event queue for processing one after the other. Note, however, that keeping a queue of unhandled events may cause errors if the callbacks are invoked too quickly, causing the event queue to explode.

Unfortunately, the **Interruptible** and **BusyAction** properties are only honored by a few standard MATLAB callbacks: **ButtonDownFcn, KeyPressFcn, KeyReleaseFcn, WindowButtonDownFcn,** **WindowButtonMotionFcn,** **WindowButtonUpFcn, WindowKeyPressFcn, WindowKeyReleaseFcn,** and **WindowScrollWheelFcn** for figure windows, and **ButtonDownFcn, Callback, CellSelectionCallback, KeyPressFcn, SelectionChangeFcn, ClickedCallback, OffCallback,** and **OnCallback** for GUI controls.[1660]

For the other callbacks, which do not honor the **Interruptible** and **BusyAction** properties, we need to use an applicative solution for quick "bail-out" at the beginning of the callback. Here are two possible variants of this idea:[1661]

```
% Variant1
function myCallbackFcn1(hObject, eventData, varargin)
    persistent inCallback
    if ~isempty(inCallback), return; end
    inCallback = true;
    try
        % do something useful here
    catch
        % error trapping here
    end
    inCallback = [];
end     % myCallbackFcn1

% Variant2
function myCallbackFcn2(hObject, eventData, varargin)
    inCallback = getappdata(hObject,'inCallback');
    if ~isempty(inCallback), return; end
    setappdata(hObject,'inCallback',true);
    try
        % do something useful here
    catch
        % error trapping here
    end
    setappdata(hObject,'inCallback',[]);
end     % myCallbackFcn2
```

10.4.3.4 Disable Control Callbacks during Automated Updates

An additional callback-related advice, which was already mentioned above* but is worth repeating, is to temporarily disable the relevant control callback(s) before any large automated (noninteractive) update operation. The importance of this seemingly simple advice cannot be overstated. Its effect on performance can be very significant.

---

* See §10.4.1.4.

10.4.3.5 *Prefer Asynchronous Callbacks to Synchronous Polling*

As noted above,* code should be asynchronous whenever possible, reducing the time spent in synchronous execution. This is true for callbacks in general: rather than churn CPU cycles until a condition is met (a bad programming pattern called *polling*), it is better to use callbacks that use CPU only when the condition actually occurs.

For example, assume we are continuously plotting data from some external device connected to our MATLAB, and wish to trigger some alarm when the plotted value exceeds some predefined limit. We could continuously poll the axes limits for this condition, and trigger an alarm when the limit exceeds our threshold. A better solution would be to use a property listener to invoke a callback only when the limits change:[1662]

```
% Polling synchronously- a very bad choice
while true
    limits = get(hAxes,'YLim');
    if limits(2) > 5, processLimitsChanged(); break; end
end


% Using an asynchronous property listener - a much better choice here
addlistener(hAxes, 'YLim', 'PostSet', {@checkLimits,hAxes});

function checkLimits (hProp, eventData, hAxes)
    limits = get(hAxes, 'YLim');
    if limits(2) > 5, processLimitsChanged(); end
end
```

As a variant of the asynchronous callback mechanism, consider using timer callbacks to periodically poll the monitored condition.† If the timer is set up to periodically trigger, a minimum of CPU cycles is wasted on the polling, freeing the main program to do useful work in the time-spans between timer callback invocations.

```
start(timer('Period',0.5, 'TimerFcn',{@checkLimits,hAxes})); % every 0.5s
```

When using MATLAB timers, we should take their inherent timing inaccuracies into consideration. Submillisecond accuracy seems to be impractical, at least on Windows platforms.[1663] In practice, the timing precision can reach multiple (even tens of) milliseconds, depending on system load and duration amount. A similar inherent limitation exists for intentional synchronous delays using the *pause* function.[1664] So if you need an ultra-high-frequency trading (UHFT) system that relies on periodic microsecond samples, consider using MEX (§8.1) and Coder-compiled (§8.2) C-code. But if your system can sustain millisecond-order timing inaccuracies, then you may well be able to use pure MATLAB m-code, which is simpler to develop and maintain.

For mechanisms to stop long-running and/or asynchronous tasks, refer to §3.7.

10.4.3.6 *Optimize Polling Performance*

Replacing polling with listener callbacks or asynchronous timer callbacks is not always possible. In such cases, we may have no choice but to endlessly poll the condition until it is met. There are performance considerations even in these cases: Looping over the condition

---

check without any delay causes the CPU to be used almost entirely by the looping code. This may prevent the condition-affecting code to execute. We may end up with an endless loop and a "hung" application simply because the monitored condition did not get CPU cycles to be updated!

Instead, it is better to add short delays in the polling loop, to enable execution to yield the CPU to other processes waiting to execute. This is done using the *pause* function:

```
% Polling with internal delays
while someCondition()
    pause(0.1); % delay a bit before retrying
end
```

It is often advantageous to set a timeout value, so that the polling loop does not block forever. The above code can easily be modified to accommodate this:

```
% Polling with internal delays and a timeout
timeout = 30;      % 30 second timeout
pauseDelay = 0.1;
for idx = 1 : (timeout/pauseDelay)
    if ~someCondition(), break; end
    pause(pauseDelay); % delay a bit before retrying
end
```

Setting the *pause* value (0.1 s in the above example) is more of an art than an exact science. Multiple small *pause*s with checks in between are generally better than using a large *pause*. If the delay is too large, we might respond too slowly when the condition is actually met. On the other hand, if the delay is too small, this could defeat the purpose of using the delay, the CPU will churn continuously, and the condition could actually be delayed simply because the other code segments did not have a chance to execute.[1665]

The *pause* delay is inherently inaccurate. To control the delay duration more precisely, use java.lang.Thread.sleep(millisecs) instead.[1666] For example:

```
java.lang.Thread.sleep(100);% more accurate than equivalent pause(0.1)
```

By knowing the expected condition update rate in our specific application and using the MATLAB Profiler, we can fine-tune the pause delay to a value that is suited for our particular application. Read §10.5.1 for additional advice on intentional *pause*s.

An alternative to *pause* is to use *timer* objects. However, timers are apparently less efficient and consume more CPU than the corresponding *pause*.[1667]

10.4.3.7 Zoom and Pan

An effect that many people notice with MATLAB graphs is that while MATLAB is quite efficient at quickly plotting millions of data points, the entire MATLAB appears to be stuck when we then try to pan or zoom within the plot.[1668]

We can tackle this in several manners: we could set the **HitTest** property of plotted elements to 'off';[1669] we could implement our own zoom and pan callback functions to handle the plot updates; we could down-sample the data to only display some of the data elements;* we could prevent plotting invisible (or out-of-limits) data points.†

---

* See §10.1.8.
† See §10.1.1.

Instead of programming the data-processing code ourselves, we could simply use ready-made utilities on the MATLAB File Exchange. An excellent example is former MathWorker Tucker McClure's *reduce_plot*.[1670] This utility automatically selects the minimal set of data points to be shown, and only plots them. For each X value, a line is plotted between the maximal and minimal Y values, ignoring intermediate values.* Listeners are attached to the axes **XLim** and **Position** properties, so that changes to any of them (e.g., as a result of zoom/pan) will update the list of displayable data points. The end result is a much smoother and faster interactivity.†

Another excellent utility is MathWorker Jiro Doke's *dsplot*, which takes a different approach, of instrumenting the zoom and pan (and data-cursor) callback functions to display a down-sample of the data.[1671] *dsplot* is more limited than *reduce_plot*, since it only works on line plots, whereas *reduce_plot* also works on many other chart types.

Jake Reimer's *jplot* utility[1672] also instruments zoom and pan (although not data-cursor), but uses a more sophisticated algorithm than *dsplot*, of both downsampling (decimation) and out-of-limits data cropping (only **XLim** is used for this). For very large data sets, downsampling is more useful at low zoom factors, while cropping is more effective at high zoom factors, but in fact both are always being applied.

A similar approach is taken by Yuval Cohen's *tplot*,[1673] which instruments only the zoom functionality. *tplot* crops the displayed data to the axes **XLim**, as well as downsamples the data if more than $w*10$ data elements are detected within these limits (w being the width in pixels of the axes).

Any of these utilities can solve the zoom/pan reactivity performance problem without requiring complex coding on the user's part. The utilities use different algorithms that have different performance and support for different types of plots and usage scenarios (e.g., panning and data-cursor), so I suggest to compare them for your specific needs. Most of these utilities seem to assume 2D plots, and may need adaptation for 3D. Additional adaptation is required if we wish to crop data that is outside the **YLim** (in addition to **XLim**, which they all seem to check).

10.4.3.8 Mouse Movement and Hover

Many programmers are not aware of the performance impact of suboptimal mouse-movement callbacks, such as the MATLAB figure's **WindowButtonMotionFcn** callback property. Remember that mouse-movement events and their associated callbacks are continuously invoked whenever we move the mouse pointer. The longer such a callback takes to execute, the more "sluggish" our application will appear. Callback execution times of 200 ms or longer will give the user a distinct taste of delayed response. If we reduce the callback execution time below the visual perception rate of 40 ms (25 Hz), the application will appear to be instantly reactive, without any perceptible delay. Mid-range execution times (50–150 ms) will cause the application to appear responsive and fluid, but users will experience a slight delay.

Reducing the time spent in mouse-movement callbacks is an art, involving trade-off between functionality and performance no less than optimizing the source code to run faster. Indeed, we may well decide to not update some graphic element during mouse movements, for the sake of improved callback performance.‡

---

* See §10.1.1.5.
† Also see §10.1.8.
‡ Also see §10.4.3.2.

A technique that I have found useful is to only update the GUI/graphs if a certain amount of time has passed since the last update, thereby ensuring that only one in N callbacks will be slow. The user will see the graphs update at a rate of 2–3 times a second, without any perceptible delay in the other aspects of the mouse callback:

```
function mouseCallbackFcn(varargin)
    persistent lastTime
    if isempty(lastTime), lastTime=-1; end

    ONE_SEC = 1/(24*60*60);
    if now - lastTime > 0.3*ONE_SEC
       % do some expensive GUI/graphs update
    end
    % do some regular inexpensive update
end
```

There is a slight delay between the time the mouse enters a control's area and when the control's tooltip actually appears. The tooltip appears for a certain duration, and then disappears. Sometimes the default delays are too slow or fast for our application. They can be controlled using javax.swing.ToolTipManager.[1674] ToolTipManager sets these parameters globally, including for MATLAB desktop components. Some examples using the ToolTipManager:[1675]

```
btn = uicontrol('String','Button','Tooltip',...
          'This is a button.','Pos',[100,100,75,25]);
txt = uicontrol('Style','edit','String','Edit Text',...
          'Tooltip','This is editable text','Pos',[100,50,75,25]);

% Use a static method to get ToolTipManager object
tm = javax.swing.ToolTipManager.sharedInstance;

% Get the delay before display in milliseconds (=750ms on my system)
initialDelay = javaMethodEDT('getInitialDelay',tm);

% Set tooltips to appear immediately
javaMethodEDT('setInitialDelay',tm,0);

% Get delay before tooltip disappears (=10000ms =10sec on my system)
dismissDelay = javaMethodEDT('getDismissDelay',tm);

% Set the dismiss delay to 2 seconds
javaMethodEDT('setDismissDelay',tm,2000);

% Turn off all tooltips in system (including the MATLAB desktop)
javaMethodEDT('setEnabled',tm,false);
javaMethodEDT('setEnabled',tm,true);  %...now turn them back on

javaMethodEDT('setInitialDelay',tm,initialDelay);
javaMethodEDT('setDismissDelay',tm,dismissDelay);
```

Note that these settings are not persistent between MATLAB sessions. Also note the extensive use of the *javaMethodEDT* function to execute Java Swing methods on the Swing *Event Dispatch Thread* (EDT).[1676]

10.5 Avoiding Common Pitfalls

There are several common pitfalls that affect all types of graphic code in MATLAB, either plotting in axes or using a GUI.

10.5.1 Minimize Intentional Pauses

We sometimes need to insert an intentional delay into our code. This can be used, for example, to control display timing* or during a polling loop† or to avoid a MATLAB hang following close of a modal dialog.[1677] We should minimize such intentional pauses so that our program does not delay more than the minimal duration necessary for its logic to function correctly. Even short delays may adversely affect performance when placed within a loop, or in a callback function that should be highly interactive.

One way to minimize intentional pauses is to move a later code segment so that it would run before the *pause*. The time that would take this code segment to execute could be enough to remove the need for the *pause*, or at least to reduce its duration.

As explained above, setting the correct *pause* duration is more of an art than an exact science, so we should spend a bit of time to fine-tune the value. Both low <u>and</u> high duration value could cause unnecessary delays. In some cases, when developing a graphic application, the logic changes and code segments are moved or copied. In such cases we should revisit any *pause* in the affected code segments to ensure that it is still needed — perhaps there is no longer a need for *pause* at all? If it is still needed, perhaps its duration should be modified?

When fine-tuning *pause* durations, we should consider the fact that the application might be run on both slower and faster computers, so it would not be cost-effective to overly fine-tune the value. We should use common sense in deciding when to stop tuning.

As noted above,‡ we should also take into consideration the inherent inaccuracies of the *pause* function, and consider using `java.lang.Thread.sleep(millisecs)` instead.

In some cases, the delay value might be dependent on the application state. It is tempting to simply use a long duration in such cases, to cover all possible states. However, this would cause unnecessary delays in the majority of the states in which this long delay is unwarranted. If this is important in our specific application, then we could control the delay programmatically. For example:

```
if condition1
    pause(0.1); % delay a bit (the more common program state)
elseif condition2
    pause(0.5); % a longer pause in some other program state
else
                % no pause at all in yet another program state
end
```

---

* See §10.4.1.4.
† See §10.4.3.6.
‡ See §10.4.3.6.

10.5.2 Delete Unused Graphic Objects

Graphic components use system resources: memory and graphic handles, as well as CPU cycles when updating the graphics content area. When we're done with a graphic object, if we do not intend to reuse it later,* then it is better to *delete* it, rather than to make it invisible. This may sound like a trivial advice, but it is actually surprisingly common in GUI applications to find a large number of nonvisible HG objects still being preserved by MATLAB for no apparent reason. We can track the number of all non-visible HG objects as follows:

```
hHiddenObjects = findall(0,'Visible','off');
```

Note that this excludes objects that are visible by property but not in practice, either because they are occluded by another object, or because they are positioned outside the visible area.

When using context menus linked to HG objects, we should be aware that the menus are actually attached to the parent figure, and not to the HG object. If we delete the HG objects, their linked context menus are not deleted with them. This causes hard-to-diagnose memory and handle leaks, and severe performance degradation over time. The solution to this problem is to delete any associated context menu before deleting the HG object:[1678]

```
delete(findobj(hObject,'type','uicontextmenu'))
```

A related aspect is that reportedly leaving multiple figure windows open degrades overall MATLAB performance, even when those figures are not directly accessed.[1679] Closing figures as soon as possible could improve the overall program speed.

---

* See §10.2.2.

11

I/O Techniques

Input/output (I/O) processing is notoriously slow compared to CPU processing. For example, memory (RAM) access time is measured in nanoseconds, compared to milliseconds for typical (non-SSD) hard disks, a million times slower. A simple read or write operation from an Excel file in MATLAB can take many seconds.*

These examples illustrate the reason why in many applications I/O is the performance-limiting factor. In such application, there is usually no use spending time to tune the data processing (CPU-intensive) or data (memory-intensive), since the I/O tends to be the dominating factor. Owing to I/O's relative slowness, even simple improvements can have a dramatic effect on the overall program performance.

Unfortunately, despite the fact that I/O rather than CPU performance is quite often the limiting factor in application performance, I/O monitoring and profiling tools have lagged behind the traditional CPU profilers.[1680] This highlights the importance of using correct coding practices when coding I/O.

I/O is a generic term, which refers to devices and the MATLAB console as much as to disks and networks. These all share common traits and even some MATLAB functions (*fopen, fprintf* etc.). In this chapter we use the term I/O loosely. We discuss mainly disk and network I/O, but console I/O could also be an important performance factor for some MATLAB programs.

11.1 Reducing the Amount of I/O

As noted, I/O processing is normally much slower than CPU processing. For this reason, reducing the amount of I/O, *even at the expense of increased CPU processing*, would normally improve performance.

There are many ways to reduce I/O bandwidth, which should be adapted differently and separately to each application. For example:

- We should reduce console output, especially high-frequency outputs. Remember to add the semi-colon character (;) at the end of each assignment and function call, to suppress output of the result to the command prompt.[†1681]

    In existing MATLAB code, seek (and disable where possible) usages of the *disp* and *fprintf* functions.[‡] If we need such outputs in certain cases, then it is best to

---

* Mostly due to the time it takes the Excel process to load, in order to service the request. See §11.5 below.

† The Editor's MLint / Code Analyzer will alert about this (warning id NOPRT) for assignments, but not for function calls.

‡ This refers mainly to *fprintf* outputs to the Command Window, but if we can also disable output to files then all the better.

limit the output only to these cases. We can do this, for example, by wrapping the output in a dedicated utility function such as this:

```
function printf_util(varargin)
    if someCondition
        fprintf(varargin{:});
    end
end
```

In some cases, we cannot effectively remove console output. For example, when using p-coded functions, or when there are numerous output locations in an invoked function. In such cases, consider wrapping the function call with *evalc*. This sends all console output (excluding errors) to a string variable, which can then be parsed. Although the console output is still prepared by the invoked function, the fact that it is not actually printed significantly improves the run-time performance:[1682]

```
str = evalc('invokedFunction(arg1,arg2);');
```

If our code generates a significant number of MATLAB warnings (e.g., about numeric inaccuracies, etc.), consider suppressing these messages using the *warning('off')* command.[*]

As noted in §9.7.2, reducing console outputs avoids possible memory issues, in addition to the performance benefits.

- We could turn detailed logging off by default and only turn it on if specifically requested by the user. Detailed logging is very useful during development, but there is usually no reason to leave it on once the development has finished.

- We could use caching to prevent I/O duplication.[†] Caching is especially useful for I/O, where data changes are often slow compared to program run time.

- We could read only part of a data file using smart data seek.[‡]

- We could filter a database query's result set using smart SQL.[§]

- We could combine I/O requests to reduce round-trip network latencies and server-side processing. For example, we could implement a webpage that returns aggregated data to a caller (via *urlget*), avoiding multiple requests.

  Similarly, we could implement a database stored-procedure that returns aggregated data in a single call, thereby avoiding the need for multiple separate SQL queries.[¶]

  Similarly, when reading or writing a disk file, we should process as much as possible in one go, thus saving disk seek time and reducing the number of fetches, rather than use multiple I/O requests.

- We could save data in binary rather than text files[**]

- We could save or transmit compressed data.[††]

---

[*] Also see §4.9.4.
[†] See §3.2.
[‡] See §3.6.2.
[§] See §3.5.1.5 and §3.6.1
[¶] Using database stored procedures has additional performance benefits. See §3.5.1.5 for additional details.
[**] See §11.3.1.
[††] See §11.4.

11.2 Avoiding Repeated File Access

A major factor in file I/O slowness can often be attributed to repeated cycles of open-read/write-close to or from the same disk file. Whenever this happens, the operating system needs to seek the relevant file, which on standard hard disks means mechanically spinning the disk and moving the read/write head assembly into position on top of the file. These mechanical movements naturally take considerable time compared to the electronic CPU processing speed.

A simple way to improve I/O performance is to simply keep the file open for as long as we need to read or update it, only closing the file when we are done or when an error occurs. For example:*

```
% Reopening and closing the file for each data update - slow
fname = 'test.txt';
delete(fname);
tic
for idx = 1 : 1000
    fid = fopen(fname,'at');
    fprintf(fid,'.');
    fclose(fid);
end
toc
⇨ Elapsed time is 0.225754 seconds.

% Keeping the file open during the updates - 12x faster
delete(fname);
tic
fid = fopen(fname,'wt');  % no need to use append mode in this case
for idx = 1 : 1000
    fprintf(fid,'.');
end
fclose(fid);
toc
⇨ Elapsed time is 0.019267 seconds.
```

The advice of keeping files open during repetitive data updates or reads is especially important with Excel files, as shown in §11.5.

This technique can easily be extended: we could open a file during program initialization, cache its file ID or handle (using *global*, *persistent* or some object's property), and then use this open file throughout our entire program run duration.

However, we should note the tradeoff: keeping files open conflicts with the requirement for robustness. If the application crashes or errors for any reason, the file may remain open, potentially causing its data to become corrupt or otherwise unusable. So in a real-world application, we should balance the performance-robustness tradeoff and possibly not leave a file open for the entire program duration.

---

* Some readers might speculate that the reason for the speedup may be due in some part to an output-stream flush being done in the first case (in *fclose*) for each loop iteration, but not in the second case. However, as we shall see in §11.3.5, MATLAB automatically flushes the buffer in each *fprintf* call, so there is no difference between the cases in this respect.

A related outcome of the desire to reduce repetitive file open/close cycles is to minimize the usage of MATLAB's ***getpref*** and ***setpref*** built-in functions. These functions enable users to get and set specific user preferences. For example:

```
dataMode    = getpref('myApp','dataMode');
dataSize    = getpref('myApp','dataSize');
isDebugMode = getpref('myApp','isDebugMode');

setpref('myApp','dataMode',    'binary');
setpref('myApp','dataSize',    [10,5]);
setpref('myApp','isDebugMode', false);
```

However, under the hood both ***getpref*** and ***setpref*** use the single preferences file [***prefdir***, '\matlabprefs.mat'].* Therefore, the above ***getpref*** and ***setpref*** sequence opens, parses, updates,† and closes this mat file six separate times. It would be faster to update the preferences struct tree data in memory, and save to disk only when ready:

```
% Get some specific application preference in the program
Preferences = getpref(); % get entire prefs struct tree from file
dataMode    = Preferences.myApp.dataMode;
dataSize    = Preferences.myApp.dataSize;
isDebugMode = Preferences.myApp.isDebugMode;

% Update the Preferences struct tree with updated application prefs
Preferences.myApp.dataMode = 'binary';
Preferences.myApp.dataSize = [10,5];
Preferences.myApp.isDebugMode = false;

% Update the preferences on disk
prefsFileName = [prefdir(1),'/matlabprefs.mat'];
save(prefsFileName, Preferences);
```

By extension, we can read the preferences file when the application starts, store the preferences in memory throughout the program's duration (as with the file ID above), and only update the preferences file when some preference is updated by the user.

Caching preferences data is especially important when accessing preference values in a loop, since repeated I/O access might severely degrade the loop performance. Instead, we should access the relevant preferences just once, outside the loop.‡

A related aspect of minimizing file access is merging multiple ***fread/fwrite*** calls.§

---

* See the screenshot in §2.4.
† Updates are naturally done only by ***setpref***, but parsing is done by both ***getpref*** and ***setpref***.
‡ See §3.1.1.2.
§ See §11.3.4 and §11.3.5.

11.3 Reading and Writing Files

11.3.1 Text versus Binary Format

Reading and writing binary data is generally faster than text, due to reduced I/O. Storing double-precision floating-point values in text format requires 15–20 characters compared to only 8 bytes in binary format. Moreover, text files require field separators (delimiters) and newline characters that take up additional space. For these reasons, text files are usually 2–3 larger than the corresponding binary files even without data packing or compression.* In addition to the extra I/O's performance cost, text files require parsing, whereas binary files can load data directly into memory.

The end result is that it is usually much faster to save and load data in binary rather than text format.[1683] For example, when saving data:

```
data = rand(1e6,1); % 1M double-precision values

% Text format save, 18MB
fid = fopen('test.txt','wt');
fprintf(fid,'%.15f ',d);
fclose(fid);
⇨ Elapsed time is 1.317067 seconds.

% Binary format save - 4x faster, 8MB
fid = fopen('test.bin','wb');
fwrite(fid,d,'double');
fclose(fid);
⇨ Elapsed time is 0.293236 seconds.
```

The speedup is even more pronounced when reading binary data:

```
% Text format load
fid = fopen('test.txt','rt');
data = fscanf(fid,'%f');
fclose(fid);
⇨ Elapsed time is 2.399103 seconds.

% Binary format load - 140x faster
fid = fopen('test.bin','rb');
data = fread(fid,inf,'double');
fclose(fid);
⇨ Elapsed time is 0.016699 seconds.
```

If a text file (e.g., parameters configuration) is read often, it might make sense to spend a bit of time to convert it into binary format, to improve run-time load speed.

Regardless of the specific binary format in which we save the data (MAT file, data stream, or some custom format), in general it will always outperform any text format. The question of whether this speedup is worth the reduced file readability and maintainability, is naturally application dependent.

---

* Unicode (non-ASCII) files are even larger. See §11.4 for details on the nonintuitive effects of compression on performance.

11.3.2 Text File Pre-Processing

In some cases, we may wish to pre-process a textual data file so that we could more-easily use functions such as *textscan* to parse the data. In almost all cases, the fastest way to do this is to read the entire file into MATLAB memory as a long string, preprocess this string using *strrep* and *regexprep*,* and then pass the resulting string to *textscan* for parsing. This is preferable to externally pre-processing the file.

For example, to process a space-delimited text file as CSV, we might be tempted to pre-process it with an external utility such as *perl* (bundled with MATLAB†):[1684]

```
% Pre-process 18MB text file using perl
perlCmd = 'perl -p -e "s/ /,/g" test.txt > test.csv'; % Linux/Mac
if ispc % the following is required on Windows
    perlPath = fullfile(matlabroot, 'sys\perl\win32\bin\');
    perlCmd = ['set PATH =',perlPath, ';%PATH%&' perlCmd];
end
[status, result] = system(perlCmd);
⇨ Elapsed time is 39.734800 seconds.
```

In fact, processing the file in MATLAB is 45× faster, and would be even faster if we continue to parse the `str` string in memory, without bothering to save *test.csv*:

```
% Process in MATLAB memory - 45x faster
fid = fopen('test.txt','rt');
str = fread(fid,inf,'*char')';
fclose(fid);
str = strrep(str,' ',',');

% The following step could possibly be skipped
fid = fopen('test.csv','wt'); fwrite(fid,str); fclose(fid);
⇨ Elapsed time is 0.894013 seconds.
```

Similarly, rather than count the number of lines in a text file using an external script,[1685] we can simply count the number of newlines in `str`, which is much faster:

```
fid = fopen('test.txt','rt');
str = fread(fid,inf,'*char')';
fclose(fid);
numLines = sum(str==10);
```

The drawback to using this method as-is, is that it reads the entire file into memory at once. So if the file is 1 GB text file, we would end up with a 2 GB string (MATLAB *chars* use 2 bytes). Even if this were possible in terms of memory and OS capabilities, it would obviously not be performant. Parsing such huge files in chunks would be better and faster.‡ Doing it in MATLAB memory would still outperform external scripts.

Also note that we can also use *textscan*'s ability to skip unneeded fields (see §11.3.4).

11.3.3 Memory-Mapped Files

Memory-mapped files[1686] are memory segments that are directly mapped, at the operating system's kernel level, to device memory or disk files. This mechanism enables faster access

---

* See §4.2 for performance recommendations for string processing.
† Unfortunately, MATLAB's built-in *perl* function prohibits running direct perl commands, only perl files (*.pl). This is easily overcome by directly accessing the perl executable from within MATLAB, as shown in the code snippet.
‡ See §3.6, §9.5.9, and §11.3.4.

(both read and write) to disk files than even low-level functions such as *fread* and *fwrite*, in almost all use cases. In addition to performance benefits, memory mapping provides an efficient mechanism for sharing data between different processes (MATLAB and an external application, or two MATLAB instances[1687]).

In addition to its performance benefits in direct data access, memory-mapped files enable easy implementation of the *lazy loading* paradigm, in which we load segments of a large data file only as needed, rather than all in advance.* This improves both performance and memory usage. It can also be considered a variant of the *Copy-on-Write* (COW) mechanism: the file is only updated when the data buffer is modified.†

MATLAB supports memory-mapped files since R14 SP2 (7.0.4), using the built-in *memmapfile* function.[1688] *memmapfile* can be very useful for accessing large data files, and is also much simpler to use than the corresponding low-level I/O functions.

Unfortunately, MATLAB's memory-mapping implementation is less efficient than it could be, in the sense that MATLAB wraps the functionality in a class object (which has some overhead) and by default pre-allocates buffer memory for the entire file size rather than allocating as-needed.‡[1689] MATLAB's use of a memory buffer voids many standard memory-mapping performance benefits, such as avoiding data copy between user and kernel space, or the need to pre-allocate and de-allocate I/O buffers.

The mapping buffer pre-allocation may result in out-of-memory errors if the mapped file is too large. Moreover, the MATLAB overhead may result in less efficient performance than directly using low-level I/O:§

```
% Memory-map an updatable binary data file
mObj = memmapfile('test.dat', 'Format','double', 'Writable',true);
⇨ Elapsed time is 0.001697 seconds.

% Read a specific segment from mid-file
partialData = mObj.Data(1001:1400);
⇨ Elapsed time is 0.000504 seconds.

% Update a specific segment in mid-file
mObj.Data(8001:8500) = 101:600;
⇨ Elapsed time is 0.001495 seconds.

% Un-map the file
clear mObj
⇨ Elapsed time is 0.000281 seconds.

% Corresponding low-level I/O: less easy to use but faster
fid = fopen('test.dat','r+b');
⇨ Elapsed time is 0.000446 seconds.

fseek(fid,1000*8,'bof'); partialData = fread(fid,400,'double');
⇨ Elapsed time is 0.000295 seconds.
```

---

* See §3.8.1 for a general discussion, and §3.6.2 for specific details on *memmapfile* usage.
† See §9.5.1 for the related COW mechanism used by MATLAB's memory management.
‡ Buffer memory usage reduction can be achieved via the **Offset** and **Repeat** parameters.
§ This phenomenon of memory-mapped I/O is not unique to MATLAB, but also occurs in other programming environments. For example: http://java.dzone.com/articles/memory-mapped-files-file-io (or: http://bit.ly/1d7Qz9d) shows this for C#.

```
fseek(fid,8000*8,'bof'); fwrite(fid,101:600,'double');
⇨ Elapsed time is 0.000330 seconds.

fclose(fid);
⇨ Elapsed time is 0.000305 seconds.
```

This example shows that, contrary to widespread belief among *memmapfile* users,* MATLAB's memory mapping is not a general panacea to I/O slowness. It is *possible*, but far from *certain*, that *memmapfile* will improve I/O performance. This needs to be tested in the context of the specific application, data and file-system being used.

In some cases, using *memmapfile* could actually severely degrade performance, for example, if we unmap a file when there are existing copies of the mapped data.[1690] Another limitation is that the file cannot be deleted or updated by nonmapping mechanisms (e.g., *fwrite* or *save*) as long as there is an active memory mapping.

Memory-mapped files are less efficient than standard low-level I/O when the file is small (several KB). In this case, the performance costs associated with setting up the mapping can outweigh the minor benefits of data access. MATLAB's documentation says that memory mapping can help with small files that we want to read once and access frequently.[1691] However, I find that a simple cache† of the read data may be just as simple to set up, require less resources and have better performance.

Naturally, memory mapping is more efficient for binary data than text files. While mapping text files is certainly possible, the larger file size means more memory is pre-allocated for MATLAB's internal data buffer. Moreover, additional processing and memory will be required to convert the textual data to the final data format used by the program. Storing the data in binary format avoids these issues.

Memory mapping is also less efficient when large sections of the data are updated.[1692] In this case, repeated memory paging can significantly degrade the overall performance.‡

Finally, mapping is only effective for relatively simple files whose data format can easily be described as a simple (possibly multidimensional) array of some primitive data type. More complex files can indeed be described by combining different repeated primitive formats[1693] and creating multiple *memmapfile* objects having different file offset values, but the associated development and maintenance complexity may outweigh the performance gains.

11.3.4 Reading Files Efficiently

There are several built-in MATLAB functions for reading general-format files (as opposed to specific formats such as MAT, HDF, or XLS that have dedicated functions). Binary files can be read using *fread* or *memmapfile* (see previous section); text files can be read using *textscan* or *fscanf*. Both *textscan* and *fscanf* require a file descriptor of an *fopen*'ed file, and have very similar format specifiers. Performance-wise, *textscan* is generally faster than *fscanf* (there are two alternatives that are even slower: *importdata*[1694] and *textread*,[1695] that even MathWorks suggests avoiding[1696]):

```
% data.txt: 15MB file with 100000x10 space-separated numeric values
fid = fopen('data.txt','rt');
```

---

* Unfortunately, I have found that few MATLAB programmers are aware of *memmapfile* or the memory-mapping mechanism. This is regrettable since there are situations in which it could bring important value.
† See §3.2.
‡ See §9.1.

```
% Read the data using fscanf
fseek(fid,0,'bof');
data = fscanf(fid,'%f');
data = reshape(data,10,[])';
⇒ Elapsed time is 1.727475 seconds.

% Read the data using textscan - 2x faster
fseek(fid,0,'bof');
data = textscan(fid,'%f');
data = reshape(data{1},10,[])';
⇒ Elapsed time is 0.863646 seconds.

% Close the file
fclose(fid);
```

Note how we have used *reshape* and *transpose* to convert the data back to its original format. This is done since the vectorized form of *fscanf* and *textscan* that we have used reads the entire data into a single long column array. In *textscan's* case this is also placed within a cell array. We can save a bit more processing time by telling *textscan* that the data consists of 10 numbers that can be merged into a numeric matrix:

```
fseek(fid,0,'bof');
data = textscan(fid,'%f %f %f %f %f %f %f %f %f %f','CollectOutput',1); data = data{1};
⇒ Elapsed time is 0.790411 seconds.
```

The *textscan* function has several optional parameters that we can use to skip parts of the input file without having to resort to lengthy iterative loops. These parameters include CommentStyle, HeaderLines, Delimiter, Whitespace, and the %* field specifier. One user reported that using the CommentStyle parameter resulted in a 2.5× speedup (60% run-time decrease) when processing a large text file having several identical header lines distributed throughout the file.[1697]

If the text file contains numeric data in a rectangular format (i.e., exactly the same number of numbers in each line), then we can also use *load*. *load* is simpler to use but slower and less versatile than either of the low-level functions *fscanf* and *textscan*:

```
data = load('data.txt');
⇒ Elapsed time is 5.219745 seconds.
```

If the text file contains rectangular numeric data <u>and</u> uses a single delimiter character (such as space, comma, or tab), we can also use *dlmread*. *dlmread* is faster than *load*, but unlike *load* it cannot read files that have multiple delimiters:

```
data = dlmread('data.txt','\t');
⇒ Elapsed time is 3.212678 seconds.
```

When reading enormous data files, it can significantly improve performance to include a short header at the top of the file (or perhaps in a separate attached file) that contains something like a "table-of-contents". Then, if we wish to read only a certain item, we will not have to read the enormous file all the way down to the requested data. Instead, we would read only the short header and extract the offset location of the requested item: For binary file we would *fseek* directly to that item's location, while for text files we could skip the number of specified text lines (e.g., using the 'HeaderLines' parameter of the *textscan* function). We could further optimize performance by caching the header data in MATLAB memory.

Reading huge files using *textscan* may be extremely slow due to memory concerns, once the MATLAB process starts thrashing due to insufficient physical memory.[1698] In such cases we need to revert to alternatives (such as the simpler and less capable *dlmread*), or processing the file in manageable sub-blocks.

Reading large chunks of I/O requires extra memory, so for huge files (tens of MB or more) it may be more efficient to read and process large chunks of (say) 10 MB at a time.* This would also help the perceived performance, since the program will "hang" for a much shorter amount of time and an iterative feedback will be able to display much smoother update progress. In fact, for improved feedback smoothness we may actually wish to *reduce* the chunk size, so that the progress could be updated more frequently, balancing between perceived and actual performance for our specific case.

I once needed to process a 300 MB text file, using multiple *regexp* calls. The process ran all night at full CPU, before I killed it in the morning. Modifying the process such that it worked in 10 K-line blocks enabled the process to complete within a couple of minutes. Here is the basic code snippet that broke the text into blocks:

```
% Save the modified string in the target file
[fid,errMsg] = fopen(newFileName,'Wt');

% Process 10 K-line text blocks one at a time
blockSize = 10000; % max # lines per block
allEolIdx = [0, find(text==10)];
nEOLs = length(allEolIdx);
eolIdx = 1;
while eolIdx < nEOLs
    % Prepare a block of up to 10 K text lines for processing
    startIdx = allEolIdx(eolIdx) + 1;
    endIdx = allEolIdx(min(nEOLs,eolIdx + blockSize-1));
    subStr = text(startIdx:endIdx); % text is HUGE...

    % Process the text block
    subStr = processStr(subStr);

    % Save the modified string in the target file
    fwrite(fid,subStr);
    fprintf('=> %d lines saved in%s\n', sum(subStr==10), newFileName);
    eolIdx = eolIdx + blockSize - 1;
end % loop over all blocks

% Close the file
fclose(fid);
```

When reading data iteratively in a loop, one element at a time, especially with large files that could take a long time to read and process, it is very important to provide a visual feedback to the user about the I/O progress.† In addition, we could consolidate the I/O by reading the entire data (or large blocks of it) before processing. This would make more efficient use of the operating system's I/O buffers and streamline the I/O.

```
data = zeros(1e6,1); % preallocate the memory

% Element by element I/O (slow)
fid = fopen('test.dat');
for idx = 1 : 1e6
    data(idx) = sqrt(fread(fid,1,'double'));
```

---

* See §3.6.2 and §9.5.9.
† See §10.4.2.

```
end
fclose(fid);
⇨ Elapsed time is 8.685103 seconds.

% Using bulk I/O - 170x faster (non-vectorized)
fid = fopen('test.dat');
data = fread(fid,inf,'double');
for idx = 1 : 1e6
    data(idx) = sqrt(data(idx));
end
fclose(fid);
⇨ Elapsed time is 0.050718 seconds.
```

When reading data from an external hardware interface (e.g., an instrument or hardware device connected via USB or serial port), rather than a file, it makes sense to use buffering.[1699] Using data buffers help to process data in chunks, rather than each data element separately as it arrives. In essence, data buffering mimics the I/O consolidation recommendation above.

In many cases, it is more efficient to load the entire file into MATLAB memory and then process it in memory, than to process the file directly (via its ID).[1700] Naturally, this will fail if the file is so large that it does not fit into physical memory. Even then, we are likely to get a performance speedup by reading entire file sub-blocks into memory and parsing it there, rather than doing the same directly on the file. The underlying reason is that I/O is more efficient when it is all done at once (the hard-disk spindle is already in place, the disk cache is primed, etc.), than at multiple times:

```
% Variant #1: process data directly from a 15MB (1Mx25) file
fid = fopen('test.data','rt');
data = textscan(fid, format, 'CollectOutput',true);
fclose(fid);
data = data{1};
⇨ Elapsed time is 1.200251 seconds.

% Variant #2: read data file into memory, then process it in memory
fid = fopen('test.data','rt');
str = fread(fid,'*char');
fclose(fid);
data = textscan(str, format, 'CollectOutput',true);
data = data{1};
⇨ Elapsed time is 0.849768 seconds.    ← 1.4x faster
```

When reading binary data files, it is better to specify the target data precision directly, rather than type casting the results:*[1701]

```
% test.dat is a binary files containing 1M uint8s
fid = fopen('test.dat','rb');

data = uint8(fread(fid,'uint8'));
⇨ Elapsed time is 0.080544 seconds.

data = typecast(fread(fid,'double'),'uint8');
⇨ Elapsed time is 0.044824 seconds.

data = fread(fid,'*uint8'); %equivalent to fread(fid,'uint8 = >uint8')
⇨ Elapsed time is 0.032347 seconds.
```

---

* See §9.4.5.

Similarly for textual data, *fread(fid,'*char')* is faster than *char(fread(...))*. This specific case is actually detected and reported by MLint/Code Analyzer (message ID: FREAD):

```
data = char(fread(fid));
```
⚠ FREAD(FID,...,'*char') is more efficient than CHAR(FREAD(...)).

In this regard, it is worth noting another MLint/Code Analyzer warning (N2UNI), which notes that *fread* no longer requires *native2unicode* in R2006a (7.2) and later releases; the conversion is done automatically internally.

When reading an entire text file into a MATLAB string, it is easier (but just as fast) to replace the three commands *fopen-fread-fclose* with a single call to *fileread*:*

```
str = fileread('test.data');
```

When reading files of a different endianness than our platform, it is faster to read data normally and then use *swapbytes*, than to use *fopen*'s ability to set the endianness.[1702]

For additional recommendations about reading data files in small blocks, refer to §3.6.2; also see the data compression section (§11.4).

Accessing files in folders that contain many thousands of files may cause slowdown.[1703] In such cases, it may be faster to copy the files to smaller folders before processing.

Once we have read and processed the input file efficiently, it could make sense to save it in a more compact (binary) format, for later reuse. Subsequent reads of the data will then be much more efficient than the first processing round. This brings us to the next section, which explains how to save data files efficiently.

11.3.5 Writing Files Efficiently

When writing data files, we have a choice of using one of several formats:

- One of MATLAB's MAT formats, using the *save* function
- Numeric data in text format using the *save* function
- General text format using the *fprintf* or *dlmwrite* functions
- General binary format using the low-level *fwrite* function

The decision about which format to use depends on how the data will later be read, no less than on aspects of simplicity and performance of the saving process.

From a performance standpoint, it can be said that in general saving general-format binary files, and MAT files in the older -v6 format, are the fastest of the saving options. Saving text files is somewhat slower, followed by MATs in the -v7 format. Saving -v7.3 MAT files is generally the slowest option.[†]

As noted, performance is only one of the considerations. For example, -v7.3 MAT files can save MATLAB class objects; this is not natively supported by any of the other options without dedicated serialization support (and deserialization upon read).

---

* *fileread* is a very simple m-file, which basically just has a few sanity checks of the inputs, and calls *fopen-fread-fclose*.
† See §11.4.

Another consideration is that developing a MATLAB program based on simple *save* and *load* functions can be easier and faster than developing a faster program using dedicated low-level *fwrite* and *fread*.

Another aspect is that the *save* function is atomic in the sense that MATLAB hangs while it is processing and we have no feedback about the saving progress, which in some cases could take several hours (!).[1704] On the other hand, saving data in general text or binary format enables us to update the user with a feedback on the writing progress. This contributes to perceived performance and can be as important as the actual overall performance.*

Yet another aspect is that the data-read performance does not necessarily correspond to the relative data-save performance. For example, while *save* -v7.3 may be the slowest data-save option, its corresponding *load* is actually one of the fastest. Saving a general text format is one of the faster options (following the general binary format), but loading this format and parsing the data is one of the slowest alternatives.

We should also consider the number of times that a file will be read. Sometimes it is worth waiting an extra minute to save the data, if it will save several seconds whenever reading a file hundreds of times over the course of its lifetime.

Many of the suggestions mentioned in the previous section, regarding reading data, are also applicable when writing data. This includes saving a file header for easier read access to specific data elements, providing feedback to the user during the I/O progress, and saving data in large chunks rather than numerous small elements.

Unlike C/C++'s implementation, MATLAB's *fprintf* and *fwrite* functions automatically flush the output buffer whenever they are called, even when the newline character ('\n', or rather *char(10)*) is not present in the output stream.[1705] The only exceptions to this rule are when the file was *fopen*'ed with the 'W' or 'A' specifiers, or when outputting to the MATLAB's Command Window.† Writing data without buffering in this manner severely degrades I/O performance:[1706]

```
data = randi(250,1e6,1); % 1M integer values between 1-250

% Standard unbuffered writing - slow
fid = fopen('test.dat', 'wb');
for idx = 1:length(data), fwrite(fid,data(idx)); end
fclose(fid);
⇨ Elapsed time is 14.006194 seconds.

% Buffered writing - 4x faster
fid = fopen('test.dat', 'Wb');
for idx = 1:length(data), fwrite(fid,data(idx)); end
fclose(fid);
⇨ Elapsed time is 3.471557 seconds.
```

In other words, MATLAB's default low-level I/O is <u>NOT</u> buffered, resulting in lousy performance. By simply opening the file with the 'W' or 'A' specifiers (rather than the default 'w' or 'a' equivalents), we add buffering to the I/O, speeding it significantly.

When we have an antivirus program running in the background, every update to the output file causes the antivirus to kick in and scan the entire file. Moreover, if we are using a standard hard-disk (rather than an SSD), each file update necessitates a separate disk-head seek of the appropriate file position. Since this is a mechanical process, it is relatively

---

* See §10.4.2.
† Or more precisely, to STDOUT (fid = 1) and STDERR (fid = 2).

slow. So, reducing the number of file update actions, by using buffering and consolidating *fprintf* calls, significantly improves the overall performance.

Consolidating the I/O, by assembling the entire data in memory within a long numeric or char array, and then using a single *fprintf* or *fwrite* to save this data to file, is even faster, almost as fast as we can expect to get.[1707] Further improvement lies in optimizing the array assembly (which is CPU and memory-intensive) rather than the I/O itself.

```
% Combined I/O - 1000x faster (!)
fid = fopen('test.dat', 'wb');
tic, fwrite(fid,data); toc
fclose(fid);
⇨ Elapsed time is 0.014025 seconds.
```

It might seem at first glance that since the same data is output to disk, no speed-up should be expected. However, a single large I/O update is much more efficient than multiple small ones, as explained in the previous section.

In this example, the I/O was so fast (14 ms) that it makes sense to write everything at once. But this requires the entire data to be assembled in memory, which might be impractical for huge arrays.* Even after data assembly, for huge data and slow disks,† writing the entire data to file in this manner can take long minutes or even hours. As explained in the previous section, we can break up the data into smaller chunks and *fwrite* them separately in a loop, all the time providing progress feedback to the user, to improve the perceived performance:

```
% Divide the data into chunks (last chunk is smaller than the rest)
h = waitbar(0, 'Saving data...', 'Name','Saving data...');
cN = 100; % number of steps/chunks
dN = length(data);
dataIdx = [1:round(dN/cN):dN, dN+1]; % cN+1 chunk location indexes

% Save the data
fid = fopen('test.dat', 'Wb');
for chunkIdx = 0 : cN-1
    % Update the progress bar
    fraction = chunkIdx/cN;
    msg = sprintf('Saving data... (%d%% done)', round(100*fraction));
    waitbar(fraction, h, msg);

    % Save the next data chunk
    chunkData = data(dataIdx(chunkIdx+1) : dataIdx(chunkIdx+2)-1);
    fwrite(fid,chunkData);
end

fclose(fid);
close(h);
```

In some cases, such as *dataset* export,[1708] we are unaware that MATLAB uses multiple unconsolidated I/O operations internally. If an export action takes too long, serializing the object (see below) and saving the serialized data might speed up I/O.

As an extension of the basic idea of combining I/O operations, consider combining multiple small files into a single large one. Multiple small preferences or data files may

---

* Memory-related performance is discussed in Chapter 9.
† I use a local SSD; using network hard disks is much slower. Network I/O is further discussed in §11.9.2.

be useful from an operational aspect. But if multiple such files need to be updated each time, then *combining* them into a single file can improve performance, by consolidating the I/O. Then again, if the small data segments are entirely independent and are accessed separately by different parts of the program, then actually *splitting* them into different independent small files would improve the performance. Once again, there is no general answer — it depends on the specific application.

11.4 Data Compression and the *save* Function

Compression[1709] and decompression are highly CPU-intensive, but the saved I/O sometimes outweighs their processing time. A typical example is using the built-in *save* function to store data in binary (compressed) MAT format rather than in a textual (uncompressed) format.* Similarly, using compressed GIF/JPG/PNG formats to store images rather than uncompressed BMP.†

> Note: the following discussion is mainly relevant for standard magnetic disks. Solidstate disks (SSD) are much faster, so the processing costs associated with compression and decompression may well outweigh the I/O benefits.

MAT is MATLAB's default data format for the *save* function. MAT's format is publicly available[1710] and adaptors are available for other programming languages.[1711] MATLAB 6 and earlier did not employ automatic data compression; MATLAB versions 7.0 (R14) through 7.2 (R2006a) use GZIP compression; MATLAB 7.3 (R2006b) and newer can use an HDF5-variant format,[1712] which uses SZIP compression[1713] but can also use GZIP.[1714] Note that MATLAB's 7.3 format is not a pure HDF5 file,[1715] but rather a HDF5 variant using an undocumented internal format.

The following table summarizes options for saving data using the *save* function:[1716]

save Option	Available Since	Data Format	Compression	Major Functionality
-v7.3	R2006b (7.3)	Binary (HDF5)	SZIP	2 GB files, class objects
-v7	R14 (7.0)	Binary (MAT)	GZIP	Compression, Unicode
-v6	R8 (5.0)	Binary (MAT)	None	N-D arrays, cell arrays, structs
-v4	All MATLABs	Binary (MAT)	None	2D numeric data
-ascii	All MATLABs	Text	None	Tab/space delimited

HDF5 uses a generic format to store data of any conceivable type, and has a nonnegligible storage overhead in order to describe the file's contents. Moreover, MATLAB's HDF5 implementation does not by default compress nonnumeric data (struct and cell arrays).[1717] For this reason, HDF5 files are typically larger and slower than non-HDF5 MAT files, especially if the data contains cell arrays or structs.‡ This holds true for both pure-HDF

---

* As we shall see below, compression is not necessarily better than uncompressed binary format, but it is indeed often faster than uncompressed text format.

† BMP actually has a little-known compression option, but it is not supported by MATLAB. See §11.6 for additional details.

‡ This may be different in specific cases due to idiosyncrasies between the different GZIP and SZIP compression algorithms.

files (saved via the *hdf* and *hdf5* set of functions, for HDF4 and HDF5 formats, respectively), and v7.3-format MAT files. Note that MATLAB's HDF5 implementation is generally suboptimal, both in terms of performance[1718] and compression.[1719]

For these reasons, I suggest to ensure that the default preference is for *save* to use -v7, even on new releases that support -v7.3. This preference can be changed in MATLAB's Preferences/General window, or we could always specify the -v7/-v7.3 switch directly when using *save*:

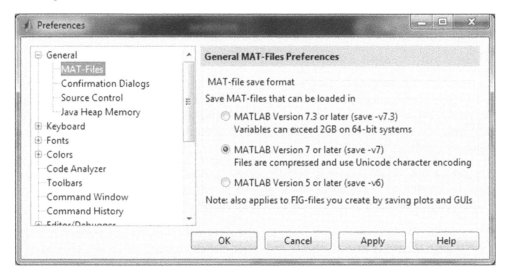

Over the years, MathWorks has fixed several inefficiencies when reading HDF5 files. Some of these fixes include patches for older releases, and readers are advised to download and install the appropriate patches.[1720]

One might think that due to the generic descriptive file header and the increased I/O, as well as the inefficiency bugs, the -v7.3 (HDF5) format would always be slower than -v7 (MAT) format in *save* and *load*. This is indeed often the case, but not always:

```
A = randi(20,1000,1200,40,'int32');     % 48M int32s → 184 MB
B = randn(500,1000,20);                  % 10M doubles → 78 MB
ops.algo = 'test';                       % non-numeric

tic, save('test1.mat','-v7','ops','A','B'); toc
⇒ Elapsed time is 11.940455 seconds.     % file size: 114 MB

tic, save('test2.mat','-v7.3','ops','A','B'); toc
⇒ Elapsed time is 6.963135 seconds.      % file size: 116 MB
```

In this case, the HDF5 (-v7.3) format was much faster than MAT, offsetting the benefits of the MAT's reduced I/O. This example shows that we need to check our specific application's data files on a case-by-case basis. For some files -v7 may be better; for others -v7.3 would be best. The widely accepted conventional wisdom of only using the new -v7.3 format for enormous (>2 GB) files[1721] is inappropriate. In general, if the data contains many nonnumeric elements, the resulting -v7.3 HDF5 file would be larger and slower than a -v7 MAT file.[1722] On the other hand, if the data is mostly numeric, then -v7.3 would be faster and comparable in size.

Surprisingly, we can often sacrifice compression to (paradoxically) achieve better performance, for both *save* and *load*, at the expense of a much larger file size. This is done by

saving the numeric data in uncompressed HDF5 format, using the *savefast* utility on the
MATLAB File Exchange, using the same syntax as *save*:[1723]

```
savefast('test3.mat','ops','A','B');
⇨ Elapsed time is 3.164903 seconds.    % file size: 259 MB
```

Even better performance, and similar or somewhat lower file size, can be achieved by
using *save*'s uncompressed format -v6. The -v6 format is consistently faster than both -v7
and -v7.3, at the expense of a larger file size.[1724] However, *save* -v6 is limited to <2 GB file
sizes and cannot save Unicode or class objects. A workaround to the latter problem is to
serialize the saved data, as shown below.

The astute reader may have noticed that the above analysis used data that was mainly
random, hence less suitable for compression. One might contend that the situation is different with data that is more compressible. So let us rerun the previous analysis using
highly compressible data, namely the same number of zero values:

```
A = zeros(1000,1200,40,'int32');        % 48M int32s → 184 MB
B = zeros(500,1000,20);                 % 10M doubles → 78 MB
ops.algo = 'test';                      % non-numeric

tic, save('test1.mat','-v7','ops','A','B'); toc
⇨ Elapsed time is 1.282518 seconds. % file size: 0.7 MB

tic, save('test2.mat','-v7.3','ops','A','B'); toc
⇨ Elapsed time is 1.900017 seconds. % file size: 2.1 MB

tic, save('test3.mat','-v6','ops','A','B'); toc
⇨ Elapsed time is 0.514884 seconds. % file size: 193 MB

tic, savefast('test4.mat','ops','A','B'); toc
⇨ Elapsed time is 0.531771 seconds. % file size: 259 MB
```

We see that the overall conclusions remain valid: *save* -v6 is faster than -v7 or -v7.3, even
for highly compressible data. The only exception to this rule might be on slow I/O (e.g.,
network drive or slow hard disk), where the reduced I/O for the compressed files may
outweigh the extra processing overheads. For this reason, we need to retest the situation
on the particular target platform, using representative data. For example, I have seen some
data saved faster with -v6 compared to -v7 on a 32-bit WinXP machine with a slow hard
disk, and the reverse on a 64-bit WinXP having a fast disk!

Another lesson here is that depending on the relative size of the numeric and nonnumeric data being saved, different data formats may be advisable. As the application evolves
and the saved data's size and mixture change, we might need to revisit the format decision.
Here is a summary on one specific computer, using the numeric variable A (184 MB) above,
together with a cell array of varying size:

```
B = num2cell(randn(1,dataSize)); % dataSize = 1e3, 1e4, 1e5, 1e6
```

Numeric Data (MB)	Nonnumeric Data (MB)	*save* -v7.3	*save* -v7	*save* -v6	*savefast*
184	0.114	3.8 s, 43 MB	9.3 s, 40 MB	1.6 s, 183 MB	2.1 s, 183 MB
184	1.14	4.3 s, 46 MB	9.5 s, 40 MB	1.6 s, 184 MB	2.1 s, 186 MB
184	11.4	12.6 s, 78 MB	9.9 s, 41 MB	2.7 s, 189 MB	11.1 s, 219 MB
184	114	87.5 s, 402 MB	13.9 s, 50 MB	5.8 s, 244 MB	85.2 s, 544 MB

As noted, and as can be clearly seen in the table, compression is not enabled for nonnumeric data in the *save* -v7.3 (HDF5) option and *savefast*. However, we can implement our own *save* variant that does compress, by using low-level HDF5 primitives in m-code[1725] or MEX c-code.[1726]

A general conclusion that can be drawn from all this is that in the specific case of *save*, the additional time for compression is often NOT offset by the reduced I/O. So the general rule is to use -v6 whenever possible.

Although *save* -v6 does not compress its data, it does store data in a more compact manner than MATLAB memory.* So, while our test set's cell array held 114 MB of MATLAB memory, on disk it was only stored within 61 MB (= 244 MB – 183 MB).

 The performance of saving nonnumeric data can be dramatically improved (and the file size reduced correspondingly) by manually serializing the data into a series of *uint8* bytes that can easily be saved very compactly. When loading the files, we would simply deserialize the loaded data. We could use Christian Kothe's excellent *Fast serialize/deserialize (hlp_serialize, hlp_deserialize)* utility on MATLAB's File Exchange,[1727] which does this programmatically for certain MATLAB data types (but not all). This utility uses pure MATLAB m-code that is fully documented and supported, although it does not support all MATLAB data types (e.g., class objects).

 An alternative is to use MATLAB's builtin ***getByteStreamFromArray*** function to serialize any MATLAB data into a byte stream, and ***getArrayFromByteStream*** to deserialize the byte stream back into the original MATLAB data.[1728] This pair of internal built-in functions is faster than Kothe's utility and supports all MATLAB data types. Unfortunately, these functions are undocumented, unsupported, and may be removed in a future MATLAB release:

```
byteStream = getByteStreamFromArray(data);        % Serialize saved data
save('test4.mat', '-v6', 'byteStream');           % Save the serialized data

dataS = load('test4.mat');                         % Load the serialized data
data = getArrayFromByteStream(dataS.byteStream);  % Deserialize the data
```

The huge gain in performance and file size when using serialized data, using any of these alternatives, is absolutely amazing:

```
B = num2cell(randn(1,1e6));  % 1M cell array, 114 MB in MATLAB memory
B_ser = hlp_serialize(B);    % or: B_ser=getByteStreamFromArray(B);
```

Saved Variable	MATLAB Memory (MB)	*save* -v7.3	*save* -v7	*save* -v6	*savefast*
B	114	83 s, 361 MB	4.5 s, 9.2 MB	3.5 s, 61 MB	83 s, 361 MB
B_ser	7.6	1.21 s, 7.4 MB	1.17 s, 7.4 MB	0.93 s, 7.6 MB	0.94 s, 7.6 MB

Serializing data in this manner enables *save* -v6 to be used even for Unicode and class objects (that would otherwise require -v7), as well as huge data (that would otherwise require >2 GB files, and the usage of -v7.3). One user has reported[1729] that the run time for saving a 2.5 GB cell array of structs was reduced from hours to a single minute using serialization, although he was using a simple nonoptimized serialization,[1730] not the faster Kothe utility or the built-in functions.

---

* It is not the data itself which is compacted, but rather the associated ~100-byte mxArray descriptor (§8.1.4.1).

In addition to the performance benefits, saving class objects in a serialized manner solves a memory leak bug that occurs when saving objects to MAT files on MATLAB releases R2011b–R2012b (7.13–8.0).[1731] In this case, under certain circumstances, switching to a different MAT-file format may also solve the problem.

When the data is purely numeric, as in serialized data, we could use *hdf5write** or *h5create* + *h5write*, as an alternative to *save* and *savefast*.[1732] Here are the corresponding results for 184 MB of numeric data on a standard 5400 RPM hard disk and an SSD:

	hdf5write	*h5create* + *h5write* (*Deflate = 0*)	*h5create* + *h5write* (*Deflate = 1*)	*save* - v7.3	*save* -v7	*save* -v6	*savefast*
File size (MB)	183	366	55	42	40	183	183
Time (hard disk) (s)	4.4	14.3	7.4	6.1	10.8	4.2	4.3
Time (SSD) (s)	2.1	0.2	4.5	4.1	9.5	1.5	1.6

As noted, MATLAB's HDF5 implementation is generally suboptimal. Better performance can be achieved by using the low-level HDF5 functions, rather than the high-level *hdf5write, hdf5read* functions.[1733] In recent releases the *hdf5write, hdf5read* functions have been deprecated in favor of the newer *h5write* and *h5read*. The multitude of low-level HDF5 functions[1734] enables much greater control and flexibility over the file format and properties than is possible using the simpler high-level functions.

In addition to HDF5, MATLAB also supports the HDF4 standard, using a separate set of built-in *hdf* functions.[1735] Despite their common name and origin, HDF4 and HDF5 are incompatible; use different data formats; and employ different designs, APIs, and MATLAB access functions. HDF4 is generally much slower than HDF5.[1736]

As noted, *save*'s -v7.3 format is usually significantly slower than the alternatives for storing entire data elements. But one specific case where -v7.3 should be considered is when we need to update or load only a small part of the data, on R2011b (MATLAB 7.13) or newer. In this case we could use the new *matfile* function, which uses the -v7.3 MAT-file format (see §3.6.2). This could potentially save a lot of I/O, especially for large MAT files where only a small part is updated or loaded. The potential savings in both memory and performance in such cases could be large indeed.

Readers should note an inherent problem in the HDF5 format that underlies the -v7.3 format: enlarging stored data via *save -append* or *matfile* does not compress the data. This could result in extremely large data files, increased I/O, and lousy performance. This problem was classified as a bug from R2008a (7.6) until it was reportedly fixed in R2012a (7.14).[1737] Even after the R2012a fix, the enlarged data is apparently still not compressed to the maximal possible extent.

When we need to store data in a format readable outside MATLAB, we might wish to save using a standard compression format such as ZIP rather than HDF or MAT. Unfortunately, the built-in compression functions *zip, gzip,* and *tar* do not really help run-time performance, but rather hurt it. The reason is that we would be paying the I/O costs three times: first to write the original (uncompressed) file, then to have *zip* or its counterparts read it,

---

* Note that *hdf5write* will be phased out in a future MATLAB release; MathWorks advises to use *h5create* + *h5write* instead.

and finally to save the compressed file.* Using *zip/gzip/tar* only makes sense if we need to pass the data file to some external program on some remote server, whereby compressing the file could save transfer time. But as far as our MATLAB program's performance is concerned, these functions bring little value.

In contrast to file-system compression, which is what *zip/gzip/tar* do, on-the-fly (memory) compression makes more sense and can indeed help performance. In this case, we are compressing the data in memory, and directly saving to file the resulting (compressed) binary data. The following example compresses data that is easily convertible to bytes (*int8* data type), such as strings or small integers (or any other data that was previously serialized into a *uint8* byte array):[1738]

```
fos = java.io.FileOutputStream('data.zip');
zos = java.util.zip.ZipOutputStream(fos);
  % or: org.apache.tools.zip.ZipOutputStream as used by MATLAB's zip.m
ze  = java.util.zip.ZipEntry('data.dat');
  % or: org.apache.tools.zip.ZipEntry as used by MATLAB's zip.m

ze.setSize(numel(data));
zos.setLevel(9); % set the compression level (0=none, 9=max)
zos.putNextEntry(ze);
dataInBytes = int8(data); % or: getByteStreamFromArray(data)
zos.write(dataInBytes, 0, numel(dataInBytes));
zos.finish;
zos.close;
```

This will directly create a zip archive file called *data.zip* in the current folder. The archive will contain a single entry (*data.dat*) that contains our original data. Note that *data.dat* is entirely virtual: it was never actually created, saving us its associated I/O costs. In fact, we could have called it simply *data*, or whatever other valid file name.

Saving a gzip file is faster and simpler, since GZIP files have single file entries. There is no need for a ZipEntry[1739] as in zip archives that may contain multiple file entries:

```
fos = java.io.FileOutputStream('data.gz');
zos = java.util.zip.GZIPOutputStream(fos); % note the capitalization
dataInBytes = int8(data); % or: getByteStreamFromArray(data)
zos.write(dataInBytes,0,numel(dataInBytes));
zos.finish;
zos.close;
```

The limitation on the data being convertible to *int8* is due to ZipOutputStream's limitation in accepting only byte arrays for its *write()* method.[1740] Therefore, if we have complex data that we wish to save, we should devise some way to serialize the data into an int8 array (and then deserialize it back when we read the data). This is all done in memory, so the saved I/O costs could be worth the trouble.

If instead of saving to a file we wish to transmit the compressed data to a remote process (or to save it ourselves later), we can simply wrap our ZipOutputStream with a ByteArrayOutputStream[1741] rather than a FileOutputStream:[1742]

```
% Zip variant:
baos = java.io.ByteArrayOutputStream;
zos  = java.util.zip.ZipOutputStream(baos);
ze   = java.util.zip.ZipEntry('data.dat');
```

---

* *tar* is worst in this respect, since it does both a GZIP compression and a simple tar concatenation to get a standard *tar.gz* file.

```
ze.setSize(numel(data));
zos.setLevel(9);
zos.putNextEntry(ze);
dataInBytes = int8(data); % or: getByteStreamFromArray(data)
zos.write(dataInBytes,0,numel(dataInBytes));
zos.finish;
zos.close;
compressedDataArray = baos.toByteArray; % array of MATLAB int8

% Gzip variant:
baos = java.io.ByteArrayOutputStream;
zos = java.util.zip.GZIPOutputStream(baos);
dataInBytes = int8(data); % or: getByteStreamFromArray(data)
zos.write(dataInBytes,0,numel(dataInBytes));
zos.finish;
zos.close;
compressedDataArray = baos.toByteArray; % array of MATLAB int8
```

Similar logic applies to reading compressed data: We could indeed use ***unzip/gunzip/untar***, but these would increase the I/O costs by reading the compressed file, saving the uncompressed version, and then reading that uncompressed file into MATLAB.

A better solution would be to read the compressed file directly into MATLAB. Unfortunately, the corresponding input-stream classes do not have a *read()* method that returns a byte array.* We therefore use a small hack to copy the input stream into a ByteArrayOutputStream, using MATLAB's own stream-copier class that is used within all of MATLAB's compression and decompression functions:

```
import com.mathworks.mlwidgets.io.*
streamCopier = InterruptibleStreamCopier.getInterruptibleStreamCopier;

% Zip variant:
baos = java.io.ByteArrayOutputStream;
fis = java.io.FileInputStream('data.zip');
zis = java.util.zip.ZipInputStream(fis);
% Note: although the ze & fileName variables are unused in the Matlab
% ^^^^  code below, they are essential in order to read the ZIP!
ze   = zis.getNextEntry;
fileName = char(ze.getName); % => 'data.dat' (virtual data file)
streamCopier.copyStream(zis,baos);
fis.close;
uncompressedDataArray = baos.toByteArray; % array of MATLAB int8
originalData = getArrayFromByteStream(uncompressedDataArray);

% Gzip variant:
baos = java.io.ByteArrayOutputStream;
fis = java.io.FileInputStream('data.gz');
zis = java.util.zip.GZIPInputStream(fis);
streamCopier.copyStream(zis,baos);
fis.close;
uncompressedDataArray = baos.toByteArray; % array of MATLAB int8
originalData = getArrayFromByteStream(uncompressedDataArray);
```

---

* They do have a *read()* method, but it works by modifying a specified data buffer in-place—this is not supported in MATLAB.

I have created a pair of utilities *savezip* and *loadzip* that implement these functionalities.[1743] Their usage is quite simple:

```
savezip('myData', magic(4)) %save data to myData.zip in current folder
savezip('myData', 'myVar')  %save myVar to myData.zip in current folder
savezip('myData.gz', 'myVar') %save data to myData.gz in current folder
savezip('data\myData', magic(4))    %save data to .\data\myData.zip
savezip('data\myData.gz', magic(4)) %save data to .\data\myData.gz
myData = loadzip('myData');
myData = loadzip('myData.zip');
myData = loadzip('data\myData');
myData = loadzip('data\myData.gz');
```

11.5 Excel Files (and Microsoft Office Files in General)

Excel files have such a prominent place in industry, engineering, and science, and their performance impact is so large that they deserve a dedicated discussion. MATLAB has built-in support for reading and writing Microsoft Excel files, using the *xlsread* and *xlswrite* functions, respectively. Unfortunately, these functions do not by default process the files directly. Instead, they start a dedicated Excel process as a nonvisible (GUI-less) COM server and process the read or update request using this server.

This mechanism is extremely detrimental to the MATLAB program's performance, since launching an Excel process can take a long time, and opening a multi-MB file in Excel can take many additional seconds.* Moreover, MATLAB does not reuse the Excel server for subsequent operations: Each call to *xlsread* and *xlswrite* starts its own dedicated Excel process and reopens the file. Multiple calls to *xlsread* and *xlswrite* can definitely grind the MATLAB program to a standstill.

Being aware of this internal implementation, several workarounds are apparent. First, caching† is a great way to reuse the Excel data without re-reading the file. Due to the high cost of Excel file operations, caching can bring significant speedups.

If we really need to read or update some disk file, we should ask ourselves whether we actually need to use a native Excel format. Instead, we can often use another, much faster, data file format. Some possibilities include:

- MAT format — Using the built-in *load* and *save* functions
- TXT format (tab or space delimited) — Using *dlmread/load* and *dlmwrite/save*
- CSV format (comma-separated values) — Using *csvread* and *csvwrite*‡

The CSV format could be considered a "light-XLS", since it is often associated with Excel. Double clicking a *.csv* file in Windows will automatically open it in Excel. CSV has the additional benefit of being fully portable across all MATLAB platforms, while XLS is a native-Windows format. CSV can also be read/updated outside Excel.

The following table lists the earliest MATLAB release that supports various Excel file formats on different platforms:

---

* The specific amount of time depends on many factors such as system load, available (free) RAM, and the file format: XLSX is slowest; XLSB and XLS are faster (see discussion below).

† See §3.2.

‡ Under the hood, *csvread* and *csvwrite* simply call *dlmread* and *dlmwrite* with a ',' delimiter parameter.

File Format	Windows *xlsread*	Windows *xlswrite*	Non-Windows *xlsread*	Non-Windows *xlswrite*
XLS	6.0 (R12)	7.0 (R14)	6.0 (R12)	–
XLSX	7.8 (R2009a)	7.8 (R2009a)	7.14 (R2012a)	–
XLSB	7.8 (R2009a)	7.8 (R2009a)	–	–
XLSM	7.8 (R2009a)	7.8 (R2009a)	8.0 (R2012b)	–

As can be seen, the support on non-Windows platform is only partial. In fact, even where supported, the functionality is limited (so-called *'basic'* mode) compared to Windows. The conclusion is that in addition to performance benefits, using non-Excel formats enables better cross-platform compatibility.

If for some reason we have to use the XLS format, we can combine calls to *xlsread* and *xlswrite*. Instead of writing separately to different worksheet locations, we can create the entire worksheet in MATLAB memory, and then use a single *xlswrite* call to update the Excel file. Owing to the I/O's dominance, the speedup factor is almost the same as the number of calls to *xlswrite* that are being avoided:*

```
parameters = {'Min',100; 'Max',200; 'Factor1',10; 'Factor2',-20};
tableHeaders = {'Row','data1','data2','data3'};

% Separate xlswrite calls - slow
xlswrite('data.xls',parameters, 'Sheet1','A1');
xlswrite('data.xls',tableHeaders, 'Sheet1','A6');
for row = 1 : 10
    xlswrite('data.xls',someFunc(1,4), 'Sheet1',sprintf('A%d',6+row));
end
⇨ Elapsed time is 20.786891 seconds.

% Unified xlswrite call - 10.6x faster
data = parameters;                       % 4x2 cell array
data(end+2, 1:4) = tableHeaders;         % 6x4 cell array
for row = 1 : 10
    data(end+1, :) = num2cell(someFunc(1,4));
end
xlswrite('data.xls',data, 'Sheet1','A1'); % 16x4 cell array
⇨ Elapsed time is 1.963062 seconds.
```

⊿	A	B	C	D
1	Min	100		
2	Max	200		
3	Factor1	10		
4	Factor2	-20		
5				
6	Row	data1	data2	data3
7	0.821721	0.429921	0.887771	0.391183
8	0.769114	0.396792	0.808514	0.755077
9	0.377396	0.216019	0.790407	0.949304
10	0.327565	0.671264	0.438645	0.833501
11	0.768854	0.167254	0.86198	0.969872
1?	0 ⁵·	⁻3 0.8842⁻	0.588⁰⁻⁻	0.1⁵47⁻⁻

---

* In the following example, 12 calls to *xlswrite* are replaced by one (avoiding 11), and the speedup is 10.6.

Similarly, when reading XLS files, we can read the entire file en-bulk and process it in MATLAB memory, rather than make separate reads from different locations.

Updating an Excel worksheet in a double *for*-loop, over the rows and columns of a data matrix, is such a common (and performance-wise disastrous) use case that it is worth repeating the lesson:

```
% Update I/O inside the loop - bad for performance
data = magic(3); % 3x3 magic square
for col = 1 : size(data,2)
    for row = 1 : size(data,1)
        cellAddr = sprintf('%s%d', 'A'+col-1, row);
        range = [cellAddr ':' cellAddr];
        xlswrite('test.xls',data(row,col),range);
    end
end
⇨ Elapsed time is 40.801892 seconds.

% Taking the update I/O outside the loop - 8x faster
lastCellAddr = sprintf('%s%d', 'A'+size(data,1)-1, size(data,2));
range = ['A1:' lastCellAddr];
xlswrite('test.xls',data,range);
⇨ Elapsed time is 5.049598 seconds.
```

As an alternative, we can use the fact that on Windows both *xlsread* and *xlswrite* use the *actxserver* function to get the Excel COM server's handle for their processing. As implemented, *actxserver* always starts a new process instance, which causes the above slowdown. We can easily modify this function to reuse an existing COM server handle if such a process is already active,[1744] relying on the ***actxGetRunningServer*** function, available on MATLAB 7.4 (R2007a) and newer. The change to the *actxserver.m* file (*%matlabroot%/toolbox/matlab/winfun/ actxserver.m*) is simply to check for a running server at the very top of the file, as follows:

```
function h = actxserver(progID, varargin)
%ACTXSERVER Creates a COM Automation server.
...
error(nargchk(1, 5, nargin, 'struct'));

% Yair: Try to reuse an existing COM server instance if possible
try
    h = actxGetRunningServer(progID);
    return; % no crash so probably succeeded - return
catch
    % Never mind - continue normally to start the COM server and
    % connect to it
end
% End of Yair's modifications

machinename = '';
interface = 'IDispatch';
...
```

This simple change ensures that any exiting code that uses *axtxserver*, including *xlsread* and *xlswrite* functions, will now try to reuse a running COM server process if possible, starting a new one only if this fails. The code is fault-tolerant in that it also works on old MATLAB releases (R2006b and earlier) where the ***actxGetRunningServer*** function is not available. This seems like such a natural and simple fix that I fail to understand why it was not implemented in *actxserver* since R2007a.

Instead of modifying MATLAB's installation file, which could be problematic when deploying applications to end users, I created a copy of *actxserver.m* somewhere in my user folders that is high on the MATLAB path. This way I can modify the file and bundle it with any application that I send to clients. When creating this copy, we also need to include a copy of a private function called *newprogid()*.* The relevant code snippet to include (at the bottom of the *actxserver.m* file) is:

```
% From: [matlabroot '\toolbox\matlab\winfun\private\newprogid.m']
function convertedProgID = newprogid(progID)
    convertedProgID = lower(progID);
    convertedProgID = regexprep(convertedProgID, '_', '__');
    convertedProgID = regexprep(convertedProgID, '-', '___');
    convertedProgID = regexprep(convertedProgID, '\.', '_');
    convertedProgID = regexprep(convertedProgID, ' ', '____');
    convertedProgID = regexprep(convertedProgID, '&', '_____');
```

When using this code, all we need to do is to ensure that we have a reference handle to an open (active) COM server before calling *xlsread* or *xlswrite*:

```
% 4x faster than the original code (but still slower than unified code)
Excel = actxserver('excel.application'); % New addition
xlswrite('data.xls',parameters,    'Sheet1','A1');
xlswrite('data.xls',tableHeaders, 'Sheet1','A6');
for row = 1 : 10
    xlswrite('data.xls',someFunc(1,4),'Sheet1',sprintf('A%d',6+row));
end
⇨ Elapsed time is 5.039221 seconds.
```

We should keep the `Excel` reference handle in MATLAB memory for as long as we need it in *xlsread/xlswrite* calls. When this handle is deleted, the Excel process exits.

A variant of this technique, of reusing an existing Excel process, is implemented by the *xlswrite1*[1745] and *officedoc*[1746] utilities on the MATLAB File Exchange. In both cases, ***actxserver*** is called once to start a new process, the Excel file is then opened, followed by one or more calls to the utility function to update, finally closing the file. In essence, *xlswrite1* is simply ***xlswrite*** with the ***actxserver*** part taken out and replaced with a reference handle called "Excel" in the base workspace:

```
% Start the Excel COM server process
% Note: to use xlswrite1, the handle *MUST* be called "Excel"
Excel = actxserver('excel.application');

% Load the workbook
filename = [pwd,'/data.xls'];
if ~exist(fileName,'file')
    ExcelWorkbook = Excel.workbooks.Add;
    ExcelWorkbook.SaveAs(fileName,1);
    ExcelWorkbook.Close(false);
end
Excel.Workbooks.Open(fileName);

% Update the workbook
xlswrite1(fileName, parameters, 'Sheet1','A1');
```

---

* In MATLAB 7.11 (R2010b) and earlier, *newprogid()* had a bug, which is solved by the code snippet presented here. See http://undocumentedmatlab.com/blog/fixing-matlabs-actxserver#fix (or: http://bit.ly/XAvFYH) for additional details.

```
xlswrite1(filename, tableHeaders, 'Sheet1','A6');
for row = 1 : 10
    xlswrite1(fileName,someFunc(1,4),'Sheet1',sprintf('A%d',6+row));
end

% Close the workbook and the Excel process
Excel.Quit;
Excel.delete;
clear Excel
```

Similar utilities exist for fast update of PowerPoint (*.ppt) files.[1747] In PowerPoint and other Microsoft Office files (e.g., *.doc), the standard way of accessing the file is via *actx-server*, where the techniques presented above (for Excel) are directly applicable.

Using the COM server handle directly can be very powerful, enabling MATLAB programs to format data (colors, fonts, borders, shading, and any other formatting supported by Office), add images and charts, and the like. This idea is utilized by the *officedoc* utility, which is much richer in functionality than the standard *xlsread* or *xlswrite*. *officedoc* enables both reading, writing, and formatting several Microsoft Office file formats (Excel, Word, and PowerPoint) using a similar interface. Under the hood, it uses the mechanism of reusing an existing open reference handle to the COM server:

```
% Open the workbook
hFile = officedoc('data.xls', 'open', 'mode','append');

% Update the workbook
officedoc(hFile, 'write', 'data',parameters);
officedoc(hFile, 'write', 'data',tableHeaders,...
                         'bold','on', 'fgcolor',[1,0,0]);%formatting
for row = 1 : 10
    officedoc(hFile, 'write', 'data',someFunc(1,4));
end

% Close the workbook and the Excel process
officedoc(hFile, 'close', 'release',true);
```

When reading Excel files, we can similarly reuse the COM server process handle. Alternatively, we could use *xlsread*'s so-called 'basic' mode. This mode is used on non-Windows platforms, as well as Windows platforms that do not have Excel. It is more limited in functionality compared to the standard (nonbasic) *xlsread*, in that we cannot specify the input cell range. However, basic mode is much faster than standard mode, since it uses mex* to directly parse the file, without starting the Excel process or its COM server interface at all. The performance speedup is striking:

```
% Standard usage - without reusing the COM server process handle
d = xlsread('data.xls','Sheet1',);
⇨ Elapsed time is 1.717196 seconds.
```

---

* *%matlabroot%/toolbox/matlab/iofun/private/biffparse.*.* **biffparse** works on the BIFF (Binary Interchange File Format) data that is read directly from the file in the *biffread.m* function. BIFF is the binary data format used by *.xls* files. (http://download.microsoft.com/download/0/B/E/0BE8BDD7-E5E8-422A-ABFD-4342ED7AD886/ Excel97-2007BinaryFileFormat(xls)Specification.xps or: http://bit.ly/ZuSN7r). Starting in Office 2007, the default format has changed to Office Open XML (http://en.wikipedia.org/wiki/Office_Open_XML or: http:// bit.ly/14nVv7r), with the main subtype being *.xlsx*, which is basically a ZIP file containing XML documents. This file can also be read directly in *xlsread's* basic mode.

```
% Reusing an active COM server process handle - 10x faster
d = xlsread('data.xls','Sheet1');
⇨ Elapsed time is 0.171422 seconds.

% Using basic mode - 300x faster (30x faster than reusing COM handle)
d = xlsread('data.xls','Sheet1','','basic');
⇨ Elapsed time is 0.005669 seconds.
```

I received reports that in some cases basic mode fails to correctly load/parse the file. In such cases, we can simply revert to using the slower standard *xlsread* mode.

When given a choice, it is faster and more I/O-efficient to save the data in XLSB format, rather than XLSX or XLS. The effect is visible even for relatively small files, and increases with worksheet size. The following snippet demonstrates this for a relatively small file with 250×250 data cells:*

```
% XLS format (1136 KB)
xlswrite('data.xls',rand(250));
⇨ Elapsed time is 2.624264 seconds.

data = xlsread('data.xls');
⇨ Elapsed time is 2.330853 seconds.

% XLSX format (894 KB)
xlswrite('data.xlsx',rand(250));
⇨ Elapsed time is 3.345375 seconds.

data = xlsread('data.xlsx');
⇨ Elapsed time is 2.438249 seconds.

% XLSB format (643 KB) - smallest and fastest
xlswrite('data.xlsb',rand(250));
⇨ Elapsed time is 2.480065 seconds.

data = xlsread('data.xlsb');
⇨ Elapsed time is 1.871067 seconds.
```

As a final note on Excel and other Microsoft Office applications (Word, PowerPoint), we may wish to send the COM server multiple separate commands for data update, formatting, and so on. This is often relatively straightforward to do in MATLAB. However, a simpler and faster alternative is to place all these commands as VBA code in a macro within the file. In many cases we can simply record a macro that records our mouse/keyboard actions, without having to know anything about VBA. Once the file contains a macro that does what we need, we can simply run it from MATLAB:[1748]

```
Excel = actxserver('excel.application');  % connect to Excel COM server
Excel.Workbooks.Open([pwd,'/data.xls']);  % open XLS workbook

Excel.Range('B3:D5').Select;      % select a specific cell-range
Excel.Run('Macro1');              % run Macro1 on this selection

Excel.ActiveWorkbook.Save;        % save the file after updating
Excel.Quit;
clear Excel
```

---

* Sending larger data to the Excel COM server in one go causes an Excel exception to be returned; 256×256 (64 K elements) is the accepted limit for the XLS format; larger data can be saved using XLSB/XLSX or using multiple XLS writes.

In Office 2003 and earlier the macro recorder is accessible via the Tools-Macros menu. The macro recording feature was removed in PowerPoint 2007 for some unknown reason, but we can still use this feature in PowerPoint 2003 and earlier and most of these macros will still work on PowerPoint 2007 and newer. In Word and Excel 2007 and newer, customize the Ribbon to display the Developer tab, which contains the macro recording and editing features. Viewing and editing the existing macros can be done using the Tools-Macros (Office 2003), the Developer Tab (Office 2007+), or by clicking Alt-F8. When moving VBA code to MATLAB, we often need to make some small modifications, to make the commands MATLAB-compliant, as shown in the code snippet examples above.

11.6 Image Files

Many image file formats have dedicated MATLAB reader functions, which are accessible via the switchyard function *imread*.[1749] Some of these readers have optional input parameters that enable direct access to part of the file, based on the file's format specification. However, we should use these optional parameters with care, since different readers have different behavior with regard to partial file access.

For example, CUR, ICO, TIFF, and HDF4 files can store multiple images (*multiframe*) in a single file.* By default, the first image stored in the file can be read. If a second optional index argument is specified, then the image at the specified index location will be read instead, that is, *imread(filename,3)* will directly read the third stored image, without reading the rest of the file. This is important for HDF4 files that can be huge.[†]

The GIF format[1750] also supports multiple images in a single file, enabling animation. But *imread* has a different behavior for GIF than for CUR. By default, it reads all file images into a 4D N×M×1×F array of *uint8* values, which represent the color index at each pixel for each of the F frames of size N×M. We can request a specific frame, but this would be inefficient: it reads the entire file and then simply discards the unneeded frames.[1751] So if we need to process GIF frames one by one, it is faster to load the entire file once into memory and then separate the frames in memory:

```
% Read the GIF image file one frame at a time
for frameIdx = 1 : 9
    frameData = imread('animated.gif', frameIdx);
    process(frameData);
end
⇨ Elapsed time is 2.583558 seconds.

% Read entire GIF file once then process frames in memory - 9x faster
allData = imread('animated.gif');
for frameIdx = 1 : 9
    frameData = allData(:,:,:,frameIdx); % separate frame in memory
    process(frameData);
end
⇨ Elapsed time is 0.274761 seconds.
```

---

* For CUR and ICO, this is typically used for storing multiple image sizes (e.g., 16×16, 32×32, and 48×48), and not for animation. Animated cursors use ANI format, which is not supported by *imread* as of R2014b (MATLAB 8.4); animated icons use GIF.

† HDF image files can also be read using the dedicated HDF functions for the HDF4 and HDF5 formats; see §11.4.

The JPEG 2000 format (*.jp2* or *.jpx*) does not enable storing multiple images, but its reader enables specifying the PixelRegion parameter. If we are only interested in a small and predefined part of the image, then this can improve performance since the requested image region is loaded directly, reducing the I/O on irrelevant image parts:

```
data = imread('image.jp2');
⇨ Elapsed time is 0.310153 seconds.

data = imread('image.jp2', 'PixelRegion',{[20,30],[40,50]});
⇨ Elapsed time is 0.187535 seconds.
```

Note that using *imread* with JPEG 2000 files on MATLAB releases R2010b–R2012b (7.11–8.0) has a memory leak bug, which can cause memory and performance issues when used within a loop. An official patch fix is available for all these releases.[1752]

TIFF (*.tif* or *.tiff*) is a favorite format for storing multi-image files. Multi-image TIFFs can reach multi-GB file sizes, making performance an important consideration. MATLAB's TIFF reader and writer functions were significantly improved for performance and stability in the past few releases, especially in R2011b (7.13)[1753] and in R2013a (8.1).[1754] MathWorks have released patches to improve TIFF-handling performance for MATLAB releases as far back as R2009b (7.9).[1755] I highly recommend installing the appropriate patch on MATLAB releases older than R2013a.

Like the JPEG 2000 reader, the TIFF reader also supports the PixelRegion parameter. Additional performance gains can be achieved by passing the output of the *imfinfo* function as input to *imread* (via the Info parameter).[1756] Depending on the specific configuration, the speedup can be anywhere between nothing and a factor of 5:

```
fileName = 'movie.tif';
tiffInfo = imfinfo(fileName);      % Get the TIFF file information
for frameIdx = 1 : numel(tiffInfo) % loop over all images in the file
    movie{frameIdx} = imread(fileName, frameIdx, 'Info',tiffInfo);
end
```

 Additional speedup can be achieved by directly using the *tifflib* mex function.[1757] This gives an order-of-magnitude speedup for MATLAB releases that do not include the R2013a fix/patch, but apparently only a 10%–20% speedup when these fixes are in place.[1758] Note that directly using *tifflib* may break in a future MATLAB release and bypasses some checks in the MATLAB wrapper function.* I recommend using *tifflib* directly only on MATLAB installations that do not have the R2013a fix/patch (e.g., when deploying to machines where installing the patch is not feasible).

When given a choice, we should use compressed images, rather than uncompressed bitmaps (BMP).† Most of today's widely used formats (including JPG, GIF, PNG, and TIF) support compressing, which can save up 80% or more of I/O costs, depending on the specific image and compression settings. Some formats are more suitable than others for specific image types,‡ so try several formats rather than just a default JPG.

---

* For additional information about using internal helper functions and bypassing sanity checks, see §3.3 and §4.3.
† BMP actually has a little-known compression option, but it is rarely used and is not supported by MATLAB.
‡ GIF is best for small images with few distinct colors; PNG is good for larger images and screenshots; JPG is good for real-life pictures; TIF is good for storing multi-image files and can use a variety of compression algorithms (both lossy and lossless). To save PNG efficiently use savepng: http://mathworks.com/matlabcentral /fileexchange/40384-savepng (or: http://bit.ly/Z5ogVl).

If we need to read BIP (*Band-Interleaved Pixels*) files, we should be aware that a patch is available for MATLAB releases R2006a to R2012a (7.2–7.14) that significantly improves performance,[1759] by consolidating internal *fread* commands (see §11.3.4). This fix was incorporated into the R2012b (8.0) release.

Use the **blockproc** function (§9.5.9) to devectorize image processing in small blocks. See §10.1.12 for additional performance aspects of image processing in MATLAB.

For videos, try saving using M-JPEG format and to use external preprocessors.[1760]

11.7 Using Java and C I/O

Many of MATLAB's built-in I/O functions have direct equivalents in standard Java I/O functions,[1761] which are included in every MATLAB installation out-of-the-box. As we shall see in §11.8.1, the Java equivalents are often more efficient than the corresponding built-in MATLAB functions. Some examples include:

- *dir* — Can be replaced by `java.io.File.list()`
- *exist* — Can be replaced by `java.io.File.exists()` or `java.nio.file.Files.exists()`
- *copyfile* — Can be replaced by `java.nio.file.Files.copy()`
- *movefile* — Can be replaced by `java.io.File.renameTo()` or `java.nio.file.Files.move()`
- *delete* or *rmdir* — Can be replaced by `java.io.File.delete()` or `java.nio.file.Files.delete()` and `deleteIfExists()` methods
- *mkdir* — Can be replaced by `java.io.File.mkdir()` or `java.nio.file.Files.createDirectory()` and `createDirectories()` methods
- *fileattribs* — Can be replaced by `java.io.File.is*()` and `set*()` methods, or `java.nio.file.Files.is*()`, `getAttribute()` and `setAttribute()` methods
- *fileparts* — Can be replaced by `java.io.File.getPath()`

Note that `java.nio.file.Files`[1762] is included in Java's New I/O (NIO) library,[1763] part of Java 7, which is available in MATLAB R2013b (8.2) or newer. Older MATLAB releases use earlier Java versions that only have `java.io.File`.[1764] In most cases, `java.nio` is faster than `java.io`. For example, NIO uses the Zero-copy mechanism where available.[1765] However, NIO is not always faster.* Moreover, `java.nio` is harder to program and will not work on MATLAB R2013a (8.1) or earlier.

Using Java requires learning the `java.nio` and/or `java.io` interfaces, which is not very difficult. In any case, the performance versus development time and maintainability tradeoff should indeed be considered here, since I/O operations can be such a significant performance bottleneck. Still, it does not make sense to replace the entire MATLAB I/O with Java-based I/O, except in those performance-critical hotspots where the MATLAB I/O performance is determined to be the limiting factor (e.g., within an inner loop or a function that executes numerous times).

Note: §7.3.3 showed an example of using multithreaded Java I/O to significantly improve the performance of MATLAB's main processing thread.

---

* An example for this is provided in §11.8.1.

For MAT-file I/O in Java, consider using the open-source *JMATIO* library.[1766]

Instead of using Java I/O, we might consider using low-level MEX C-code.* MATLAB's m-code I/O is closely modeled after the standard C I/O, helping make the transition from m-code to C-code relatively painless. MATLAB includes a dedicated doc page on using C/C++ I/O in MEX,[1767] with a couple of corresponding working examples.† Here is a simple example (a multi-threaded version is also available[1768]):

```
% The original m-code
data = rand(1000);  % 1000x1000 double values = ~8MB
fid = fopen('data.bin','w');
len = fwrite(fid,data,'double');
fclose(fid);
```

And now the corresponding MEX file (*mexIO.c*):

```
#include <stdio.h>
#include <string.h>
#include "mex.h"

void mexFunction(int nlhs,        mxArray *plhs[],
                 int nrhs, const mxArray *prhs[])
{
    FILE *fp = NULL;
    char *filename = NULL;
    char *str = NULL;
    size_t numElementsExpected = 0;
    size_t numElementsWritten = 0;

    /* Check for proper number of input and output arguments */
    if (nrhs != 2)
        mexErrMsgIdAndTxt("YMA:MexIO:invalidNumInputs",
                "Two input arguments required: filename, data");
    if (nlhs > 1)
        mexErrMsgIdAndTxt("YMA:MexIO:maxlhs",
                "Too many output arguments");
    if (!mxIsChar(prhs[0]))
        mexErrMsgIdAndTxt("YMA:MexIO:invalidInput",
                "Input filename must be of type string");

    /* Open the file for binary output */
    filename = mxArrayToString(prhs[0]);
    fp = fopen(filename, "wb");
    if (fp == NULL)
        mexErrMsgIdAndTxt("YMA:MexIO:errorOpeningFile",
                "Could not open file %s", filename);

    /* Write the data to file */
    if (mxIsChar(prhs[1])) {
        str = mxArrayToString(prhs[1]);
        numElementsExpected = strlen(str);
        numElementsWritten = (size_t) fwrite(str, sizeof(char),
                                        numElementsExpected, fp);
        mxFree(str);
```

---

* See §8.1 for a discussion of MEX.

† *%matlabroot%/extern/examples/mex/mexatexit.c* and *mexatexit.cpp*.

```
    } else if (mxIsDouble(prhs[1])) {
        numElementsExpected = mxGetNumberOfElements(prhs[1]);
        numElementsWritten = (size_t) fwrite(mxGetPr(prhs[1]),
                                             sizeof(double),
                                             numElementsExpected, fp);
    }
    fclose(fp);

    /* Ensure that the data was correctly written */
    if (numElementsWritten != numElementsExpected)
        mexErrMsgIdAndTxt("YMA:MexIO:errorWritingFile",
                "Error writing data to %s: wrote %d, expected %d\n",
                filename, numElementsWritten, numElementsExpected);

    /* Return the number of elements written to the caller */
    if (nlhs > 0)
        plhs[0] = mxCreateDoubleScalar(numElementsWritten);
}
```

We now *mex*-compile and use the new function:

```
mex mexIO.c
len = mexIO('data.bin',data);
```

Some users have reported that using MEX has significantly improved their I/O performance compared to standard m-code.[1769] The larger the data chunks being written to file, the more will the performance be bound by the underlying hardware and driver, rather than by MATLAB's CPU overheads. Therefore, for data size of 8 MB (as in the example above), I see only ~10% speedup of MEX versus the original m-code, on my specific system running R2014a on Win7 64-bit using SSD. This is hardly worth the effort to use MEX. However, when the data size is reduced to 100×100 (~80 KB), MATLAB's CPU overheads start having a more noticeable effect and the speedup becomes 30%–40%. Naturally, the specific speedup depends on the running platform. Issues such as disk and memory caching, latency, and throughput might have more impact than the question of whether we use m-code or MEX.

In some specific cases, we can increase the overall application I/O throughput by parallelizing independent I/O. For example, while saving data in an Excel file, we could also store data in a database. MATLAB is not a real multi-threading environment, and so real parallelization is not possible using MATLAB threads. However, if we delegate the I/O to separate Java, C++, or C# threads,* we could indeed achieve true parallelization, if (and this is a very important *"if"*) we are careful for the thread(s) not to require any MATLAB computation. We could also spawn an external process for dedicated I/O,† but this would be far less efficient than threads.

---

* See §7.3.
† See §7.4.

11.8 Searching, Parsing, and Comparing Files

11.8.1 Searching for Files

When searching for files, we can either use MATLAB's built-in *dir* function, or one of the functions in the `java.io` (or `java.nio`) package, or call an external operating system command via *system* (or *!* or *dos*). MATLAB's built-in *dir* outperforms the *system* alternative, but the Java-based approach seems to be the fastest by far, even when we use a vectorized approach with *dir*'s returned struct-array:

```
fileNames = strsplit(evalc('system dir/b/d;'));
⇨ Elapsed time is 0.094225 seconds.

files = dir; fileNames = {files.name};
⇨ Elapsed time is 0.008869 seconds.

fileNames = cell(java.io.File(pwd).list);
⇨ Elapsed time is 0.001504 seconds.
```

An alternative is to use Jonathan Sullivan's *dir2* MEX utility on the MATLAB File Exchange.[1770] In addition to replicating *dir*'s functionality, *dir2* also enables recursive directory search and specifying wildcard-filters. *dir2* is much faster than *dir*, although still slower than the Java variant:*

```
files = dir2; fileNames = {files.name};
⇨ Elapsed time is 0.002527 seconds.
```

When checking whether a file or a directory exists, we can similarly use the built-in *exist* function or the Java *exists()* methods. Here are the different variants, followed by a table listing the timing results for both a successful and unsuccessful match:[1771]

```
% Variant 1: MATLAB's built-in exist function
existsType = exist(filename);

% Variant 2: MATLAB's exist function with 'file'/'dir' spec: §4.9.9
existsType = exist(filename, 'file');

% Variant 3: java.io.File's exists() method
existsFlag = java.io.File(filename).exists;

% Variant 4: java.nio.file.Files's exists() method
lo = javaArray('java.nio.file.LinkOption',1);
lo(1) = java.nio.file.LinkOption.NOFOLLOW_LINKS;
filepath = java.io.File([pwd '/' filepath]).toPath;
existsFlag = java.nio.file.Files.exists(filepath, lo);
```

Variant	File Exists	File Does Not Exist	Directory Exists	Directory Does Not Exist
exist	486 µs	571 µs	348 µs	668 µs
exist(name,'...')	141 µs	359 µs	250 µs	585 µs
java.io.File.*exists*	209 µs	202 µs	212 µs	174 µs
java.nio...*exists*	361 µs	314 µs	390 µs	360 µs

---

* Unfortunately, the MEX's C-code uses Windows-specific function calls and so only works on Windows. Linux and Macs fall-back to an m-file implementation that is significantly slower, although still providing more functionality than the built-in *dir*.

As can be seen from the results, `java.io.File.exists` appears to be a clear winner.

When comparing the different variants, note that MATLAB's *exist* returns an index (0 if the item was not found, 2 if this file was found, 7 if a directory was found). On the other hand, the Java functions return a logical *true/false* flag.

Also note the conclusion from §4.9.9 that specifying the optional-type parameter is always preferable when using MATLAB's *exist* function.

Jan Simon has created a C-based MEX utility called *GetFullPath* that returns the full path for file names that possibly include wildcards or relative path.[1772] Processing is multiple times faster than the corresponding functionality via MATLAB's *which* function, Java's `java.io.File`, or Dot-NET's `System.IO.FileInfo`. A similar MEX function called *exist-file* can be found as part of Markus Buehren's *Multicore* utility.*

MATLAB's *dir* function can be used to return timestamp information of a file. A faster implementation was provided by Jan Simon in his *FileTime* utility.[1773] *FileTime* provides more detailed timestamps (creation and last access times, not just the latest save time that *dir* reports), and can also be used to update either of these file timestamps, a useful feature that no built-in MATLAB function offers.

11.8.2 Parsing and Scanning Files

MATLAB includes the built-in *depfun* function for parsing MATLAB code files in order to extract the list of its dependencies (invoked functions). This is useful when we need to bundle all dependent function files for distribution. Unfortunately, *depfun* is quite slow in its work, very often taking long minutes (or even hours) to run.

A far better alternative is to use Urs (us) Schwartz's *fdep* utility,[1774] which is much faster (often 20–50× faster!), and provides greater detail, than *depfun*.† *fdep* even provides the option to display its report in a GUI, in addition to programmatic access to the results via the function's output. The utility includes an HTML document with detailed usage instructions.[1775] *fdep* was selected as a MATLAB-Central Pick-of-the-Week (POTW) for its usefulness and robustness.[1776] *fdep* is actually a wrapper to *depfun*, but intelligently uses *depfun*'s nondefault parameters and smart tree pruning to remove unnecessary duplicate processing of the scanned files.

Another alternative is to use Malcolm Wood's *exportToZip* utility,[1777] which provides speedup by skipping built-in MATLAB functions. In addition to its inherent parsing speedup, *exportToZip* also packages all the dependent user (non-MATLAB) m-files into a single zip file, enabling simple code distribution. *exportToZip* was also selected as POTW.[1778] It was extended by Mark Morehead to support both relative and absolute paths, to handle cases where dependencies have no common root folder.[1779]

Readers interested in scanning and parsing m-code files, may also be interested in the related utility *farg*,[1780] also by Urs Schwartz. *farg* parses the specified m-file and returns the calling syntax, included the expected input and output arguments, for each of its included functions of various types (main, nested, sub-, anonymous, *eval*, etc.).

We often need to scan files for some string token or expression. For a single file we can easily *fopen* the file, *fread* its contents into a large string, and then use *strfind* or *regexp* to search for the requested string. But when we need to search multiple files, this becomes

---

* See §7.2.2 for details on this very useful utility.

† *depfun*'s performance was even further degraded in MATLAB releases R2011b–R2012b (7.13–8.0). It was somewhat improved in R2013a (8.1), but is still significantly slower than alternatives such as Urs' *fdep* utility. See http://www.mathworks.com/support/bugreports/759496 (or: http://bit.ly/18GGo5u).

a bit more complicated. If we are using the MATLAB Desktop interactively, we could use the Editor's built-in multi-file search functionality for simple static strings (regular expressions and wildcards are not supported). We could also use (using MATLAB's *system* command) external utilities such as *grep*,[1781] which is either pre-installed in the computer, or can easily be downloaded. *grep* is also included as a built-in function of Perl, which is included with the MATLAB installation.* Of all these alternatives, I find that, once again, Urs Schwartz provides the fastest and most comprehensive solution, with his *grep* utility on the MATLAB File Exchange,[1782] which was also chosen as POTW.[1783]

When comparing text files, I normally use an external utility such as the excellent open-source WinMerge.[1784] MATLAB includes a built-in comparison utility in its Desktop, although I find it lacking in features and speed compared to WinMerge. Still, both of these utilities are GUI-based, not command-line-based. In order to compare files from the MATLAB command-line and possibly integrate the results in a MATLAB application, I suggest using Urs's *cmp* utility.[1785]

When scanning the contents of MAT files, we can use the built-in *whos* function, or Ben Kraus' faster MEX-based *matwho* utility on the File Exchange.[1786] The difference is especially striking for large MAT files, since *who* and *whos* scan the entire file, whereas *matwho* (or actually, MEX's *matGetDir* function) only scans the header:

```
% Standard m-code (100MB data file)
varNames = who('-file','largeData.mat');
⇨ Elapsed time is 10.188956 seconds.

% Using matwho: 14000x faster (!)
varNames = matwho('largeData.mat');
⇨ Elapsed time is 0.000723 seconds.
```

11.9 Additional Aspects

11.9.1 Using p-Code

There is a common misconception among MATLAB users that a major part of MATLAB run time is spent interpreting the m-file source code into platform-independent *byte-code*,[1787] also called *p-code* ("portable code"). Byte codes are not unique to MATLAB and were actually invented in the 1960s, long before MATLAB. They are binary instructions that are then interpreted by the Virtual Machine (e.g., Java's JVM) into platform-specific executable *machine code*.[1788]

In MATLAB, the source-code to byte-code interpretation is done on-the-fly, by the internal MATLAB JIT.† The common misconception is that the run-time performance can be significantly improved if we pre-interpret the entire application. When the JIT interprets m-files, it does so in memory, without leaving the generated byte-code for later reuse. However, we can manually interpret m-files and store the generated byte-codes by using the *pcode* function. This saves the p-code in files that have the *.p* extension, thereby called *p-files*. These files, when present, are used directly by the MATLAB Virtual Machine saving the run time cost of the JIT interpretation:

---

* See §11.3.2 for related information.
† See §3.1.15.

```
pcode('myFunction.m');  % p-code a specific m-file
pcode('*.m');           % p-code multiple m-files in the current folder
```

The reality is that the MATLAB source code is only interpreted once per each MATLAB session, when the code is first encountered. The interpreted (p-code) version is then kept in MATLAB memory for possible later reuse. The full list of MATLAB files that are currently loaded can be seen with the built-in *inmem* function. We can use the *clear all* command to clear these functions from memory (along with the workspace variables). But in the vast majority of cases, our m-files will only be interpreted once (if at all) during the entire application run time.

Interpreting m-code is done at extremely fast rates (thousands of lines of code per second). Therefore, unless your application has hundreds of thousands of code lines, the extra time taken by the JIT will be entirely negligible.

Moreover, when using p-code files, we need to ensure that we re-*pcode* any m-file that is ever modified. If we fail to do so, the older p-files will still be used, rather than the newer m-files. This carries a nonnegligible maintenance overhead, and can lead to difficult-to-diagnose bugs. In my opinion, p-files are more important for IP protection (code obfuscation) than for any possible performance gain.

Still, if a MATLAB project has many thousands of MATLAB m-code lines, we can shave off several seconds from the initial startup time by using pre-interpreted p-files. Whether these precious seconds are worth the trouble or not, is up to you to decide.

11.9.2 Network I/O

 As already noted, by far the largest factor in file I/O's relative slowness is the physical disk access. In a nutshell, the faster the disk, the faster the I/O. Keeping this in mind, it is obvious why we should attempt to run our MATLAB application directly from a local disk, rather than a remote network drive. In fact, it could very well be faster to copy the entire application folder from the remote drive to the local disk, before running the program and then run on this local disk, than to run directly on the network, even if it is a fast local LAN.

I was able to speed up a client's application by a factor of 10 simply by moving it from network to local disk. In such cases, it does not matter that both the remote server and the local computer are super-fast high-performance platforms — the mere presence of the network traffic is the limiting factor. Each MATLAB function can possibly call dozens or hundreds of other functions, and each of these could potentially need to be fetched (and re-fetched ad infinitum) from the remote server. In many cases, we have multiple very small m-files, and in such cases the networking performance overhead might be much larger than the actual data transfer time!

In addition to networking overheads, reading from a remote server prevents usage of built-in cache and read-ahead mechanisms that are available when using a local disk.

When the application runs locally but processes remote files, it is similarly beneficial to first copy the files locally (e.g., via *copyfile*) and then process it locally. One user has reported that doing this improved his file read time from 196 s to 5 s.[1789]

Network storage does have important benefits for security, maintainability, and teamwork. Using a pure local disk should be considered versus these other factors. For this reason, I suggest that we keep using network storage to keep the latest files, but modify our program to *copyfile* the latest version from the remote server to our local disk before we actually start running the program. Alternatively, we could use a code-versioning system (CVS), which synchronizes between the local files and the server's master version.

Exactly the same logic applies to data files as to the application code files. If possible, we should strive to move them locally and then access them on our local disk, rather than directly on a remote server. We need to be careful to always copy the latest file, and also to always update the master (server) version whenever we modify the local copy. In many respects we can view our local disk as a local cache of the server's version.*

Similarly, when using HTML to display images in MATLAB GUI,[1790] we should use a local rather than remote image resource:

```
% Remote resource location
logoLocation = 'http://UndocumentedMatlab.com/images/logo_68x60.png';
htmlStr = ['<html><b>Logo</b>: <img src="' logoLocation '"/>'];
hButton = uicontrol('Pos',[10,10,120,70],'String',htmlStr,'Back','w');
drawnow
⇨ Elapsed time is 0.804578 seconds.

% Local resource location - 36x faster!
logoLocation = ['file :///' pwd '/logo_68x60.png'];
htmlStr = ['<html><b>Logo</b>: <img src="' logoLocation '"/>'];
hButton = uicontrol('Pos',[10,10,120,70],'String',htmlStr,'Back','w');
drawnow
⇨ Elapsed time is 0.022098 seconds).
```

MATLAB *uicontrol* with an HTML image

Given a choice, we should use a solid-state disk (SSD) rather than a traditional hard disk. SSDs are relatively inexpensive nowadays and are significantly faster than traditional hard disks.[1791] The performance gap is so significant that even low-end SSDs generally outperform high-end hard disks.[†] Using an SSD should not be done at the expense of networking — a remote SSD often provides slower overall I/O than a local traditional hard disk. But on a local system we should always prefer an SSD.

The tradeoff between latency and throughput[‡] is especially important in I/O, where the relative costs are high for both the latency (network traffic round-trip delays or disk seek time) and throughput.[1792]

If we cannot avoid using network I/O, we should try to minimize the networking round trips by combining resources. For example, by placing helper functions as sub-functions within a single large m-file, rather than as separate small m-files. Similarly, by placing our data in a single large data file, rather than in separate small files. If multiple I/O requests are unavoidable, we should at least use caching extensively.[§]

Another network I/O aspect is that we should avoid encrypted connections (TLS/SSL) whenever possible, since they can be significantly slower than unencrypted ones.

---

* See §3.2 for a discussion of caching in MATLAB.
† See §3.5.3.
‡ See §3.8.
§ See §3.2.

11.9.3 Miscellaneous

- On R2013a (MATLAB 8.1), using *fopen* and *finfo* with certain files is much slower than on previous MATLAB releases. A patch for this bug is available.[1793]

- Windows has traditionally encountered slowdowns when accessing files in folders having many hundreds of files or more. Things really slow down when a folder has several thousand files.[1794] In such cases, consider creating sub-folders to store files.

- Several suggestions have been posted for speeding up *saveas*[1795] and *movefile*.[1796]

- Jan Simon posted several super-fast file-access utilities that are worth perusing.[1797]

- Finally, refer to §3.5.2 for hardware-related aspects that affect I/O performance.

Appendix A: Additional Resources

A.1 Online Resources

A.1.1 MATLAB

A.1.1.1 Reference Guides

- *MATLAB 7 Programming official MathWorks guide,* Chapters 25 + 26
  http://mathworks.com/help/releases/R2014a/pdf_doc/matlab/matlab_prog.pdf
  (or: http://bit.ly/1mfXwKz)
- MathWorks' *Code Performance* section in the online documentation
  http://mathworks.com/help/matlab/code-performance.html (or: http://bit.ly/1bc6ejj)
- MathWorks *Graphics Performance* section in the online documentation
  http://mathworks.com/help/matlab/graphics-performance.html (or: http://bit.ly
  /1wUyClH)
- MathWorks' *Accelerating MATLAB Algorithms and Applications* article
  http://mathworks.com/company/newsletters/articles/accelerating-matlab-
  algorithms-and-applications.html (or: http://bit.ly/1kSUB7P)
- MathWorks' *MATLAB Acceleration* portal
  http://mathworks.com/discovery/matlab-acceleration.html (or: http://bit.ly /1knyCny)
- Peter Acklam: *MATLAB array manipulation tips and tricks* (see §5.7.15)
  http://home.online.no/~pjacklam/matlab/doc/mtt/doc/mtt.pdf (or: http://bit.ly/
  14Wq3dG)
- Pascal Getreuer: *Writing Fast MATLAB Code*
  - http://mathworks.com/matlabcentral/fileexchange/5685-writing-fast-matlab-
    code (or: http://bit.ly/15qLGnS)
  - http://blogs.mathworks.com/pick/2011/01/14/good-matlab-coding-practices
    (or: http://bit.ly/UfDlxA)
- Pascal Getreuer: *Writing MATLAB C/MEX Code* (see §8.1.7.3)
  http://mathworks.com/matlabcentral/fileexchange/27151-writing-matlab-
  cmex-code (or: http://bit.ly/15qLF39)
- Nico Schlömer: *Guidelines for writing clean and fast code in MATLAB* http://
  mathworks.com/matlabcentral/fileexchange/22943-guidelines-for-writing-
  clean-and-fast-code-in-matlab (or: http://bit.ly/15qLzIM)
- Marios Athineos: *MATLAB tips and tricks*
  http://www.ee.columbia.edu/~marios/matlab/matlab_tricks.html (or: http://bit.
  ly/15qLwwA)
- Reza Sameni: *Writing Efficient MATLAB Codes*
  http://sameni.info/Publications/Lecture%20Notes/Matlab_Tutorial.pdf (or: http: //
  bit.ly/1lAFBhz). Note that this guide is from 2006 — much has changed since then
  in the MATLAB performance arena!

- Boston University's research computing resources
  - http://scv.bu.edu/matlab — main portal page
  - http://bu.edu/tech/about/research/training/online-tutorials/matlab-pct (or: http://bit.ly/NohK5v)
  - http://bu.edu/tech/about/research/training/live-tutorials#HPC (or: http://bit.ly/15qKdxY)
- Cambridge University Dept. of Engineering: *MATLAB — Faster Scripts*
  http://www-h.eng.cam.ac.uk/help/tpl/programs/Matlab/faster_scripts.html (or: http://bit.ly/1i15LDm)
- Iain Murray, University of Edinburgh: *Efficient Matlab and Octave*
  http://homepages.inf.ed.ac.uk/imurray2/compnotes/matlab_octave_efficiency.html (or: http://bit.ly/1lAGcjq)
- MATLAB Versions Compatibility list
  http://www.dynare.org/DynareWiki/MatlabVersionsCompatibility (or: http://bit.ly/1dKeYUZ)

A.1.1.2 Blogs

- Undocumented MATLAB (performance-related articles)
  http://undocumentedmatlab.com/blog/tag/performance (or: http://bit.ly/17YllJ)
- Loren on the Art of MATLAB
  http://blogs.mathworks.com/loren (or: http://bit.ly/17FYs0D)
- MathWorks blogs (performance-related articles)
  http://google.com/search?q=site:blogs.mathworks.com+performance (or: http://bit.ly/15NNesf)
- Doug Hull's short video tutorials (performance-related videos)
  http://google.com/search?q=site:blogs.mathworks.com%2Fvideos+performance (or: http://bit.ly/15NMxiK)
- Walking Randomly (performance-related articles)
  http://walkingrandomly.com/?s=Matlab+performance (or: http://bit.ly/11gA052)
- MATLAB Tips (performance-related articles)
  http://matlabtips.com/?s=performance (or: http://bit.ly/11gAmZi)

A.1.1.3 Webinars (and associated files, where available)

- List of performance-related MATLAB webinars by MathWorks (note that some webinars discuss a different type of "performance" than "speed")
  http://mathworks.com/company/events/webinars/?q=performance (or: http://bit.ly/1mfVw4Q)
- *MATLAB to C Made Easy* (MATLAB Coder)
  https://mathworks.com/wbnr62736 (or: http://bit.ly/1bbKrxl)
- *Speeding up MATLAB Applications*
  http://mathworks.com/wbnr60898 (or: http://bit.ly/17FVIk2)
- *Speeding up MATLAB Applications* (a different webinar)
  http://mathworks.com/wbnr49643 (or: http://bit.ly/17FVM3c)

- *Generate C Code from MATLAB Using Embedded MATLAB*
  https://mathworks.com/wbnr40173 (or: http://bit.ly/17FVQ2N)
- *Large Data Sets in MATLAB*
  http://mathworks.com/wbnr33692 (or: http://bit.ly/17FWTQh)
  - http://mathworks.com/matlabcentral/fileexchange/29916-large-data-sets-in-matlab-webinar-examples (or: http://bit.ly/17FWY6E)
- *Parallel Computing with MATLAB*
  http://mathworks.com/videos/parallel-computing-with-matlab-81694.html   (or: http://bit.ly/Uh3f7W)
  - http://mathworks.com/matlabcentral/fileexchange/31336-demo-files-for-parallel-computing-webinar (or: http://bit.ly/17FV4Tz)
- *GPU Computing with MATLAB*
  http://mathworks.com/wbnr59816 (or: http://bit.ly/17FVFES)
- *MATLAB for CUDA Programmers*
  http://mathworks.com/videos/matlab-for-cuda-programmers-81971.html   (or: http://bit.ly/Uh3pwd)
  - http://mathworks.com/matlabcentral/fileexchange/38401-matlab-for-cuda-programmers (or: http://bit.ly/12mBL3a)
- *Accelerating Signal Processing and Communications Algorithms using GPU*
  http://mathworks.com/company/events/webinars/wbnr73569.html
- *Accelerating your MATLAB Code* (workshop)
  http://mathworks.com/company/events/conferences/matlab-computational-finance-conference-nyc/2013/proceedings/accelerating-your-matlab-code.pdf (or: http://bit.ly/GJAt99)
  - http://mathworks.com/company/events/conferences/matlab-computational-finance-conference-nyc/2013/proceedings/accelerating-your-matlab-code.zip (or: http://bit.ly/1crS4jc)

A.1.2 Non-MATLAB

- *Optimizing software in C++*
  (http://agner.org/optimize/optimizing_cpp.pdf or: http://bit.ly/1fJFAQd) — an excellent detailed tutorial about performance optimization in C++ by Agner Fog, Technical University of Denmark. Additional resources available at http://agner.org/optimize.
- *Linux Multicore Performance Analysis and Optimization in a Nutshell*
  (http://icl.cs.utk.edu/~mucci/latest/pubs/Notur2009-new.pdf   or: http://bit.ly/1mHtsZy) — a very detailed theoretical and practical introduction to multi-core performance tuning by Philip Mucci of Tennessee University.
- http://c2.com/cgi/wiki?CategoryOptimization — *Portland Pattern Repository* consisting of numerous performance-tuning design and coding patterns.
- *How to Write Fast Numerical Code: A Small Introduction*
  (http://users.ece.cmu.edu/~franzf/papers/gttse07.pdf or: http://bit.ly/1ijrARz) — a detailed paper from Carnegie Mellon University, explaining how to optimize numerical algorithms using a variety of techniques.

- *What Every Programmer Should Know About Memory* (http://akkadia.org/drepper/cpumemory.pdf or: http://bit.ly/1kMxjAt) — an *extremely* detailed technical report by Ulrich Drepper that explains the structure of modern memory subsystems and how to utilize them efficiently in software programs.

- *Memory Optimization* (http://research.scea.com/research/pdfs/GDC2003_Memory_Optimization_18Mar03.pdf or: http://bit.ly/OAKtFz) — a detailed presentation on cache/memory performance intricacies by Christer Ericson. A bit outdated but still very interesting.

- http://msdn.microsoft.com/en-us/library/aa292152(v=vs.71).aspx — Microsoft's Performance portal.

- http://howtospeed.com/speed-up-your-computer (or: http://bit.ly/1fJOCNd) — tips for general computer tuning; Windows-based but some tips also applicable to Linux/Mac.

- http://articlecity.com/articles/computers_and_internet/article_3255.shtml (or: http://bit.ly/PUEXyz) — general tuning tips for Windows-based computers.

- http://JavaPerformanceTuning.com/tips (or: http://bit.ly/1fJMvJ3) — tips

- http://performancetroubleshooting.com/blog (or: http://bit.ly/1fJNw47) — tips.

- http://performancewiki.com — general tuning tips that are easy to understand and use.

- http://stackoverflow.com/questions/761204/what-resources-exist-for-database-performance-tuning (or: http://bit.ly/PUJbWY) — extensive list of resources related to database performance.

- http://en.wikibooks.org/wiki/The_Performance_Guide/Database_and_Performance (or: http://bit.ly/OF9wa7) — a concise list of database tuning tips.

A.2 Books

A.2.1 MATLAB

- John Mathews, Kurtis Fink: *Numerical Methods Using MATLAB*, Pearson, 2004, ISBN 0130652482.

- Sergiy Butenko, Panos Pardalos: *Numerical Methods and Optimization*, CRC Press, 2014, ISBN 1466577770.

- P. Venkataraman: *Applied Optimization with MATLAB Programming*, John Wiley, 2009, ISBN 047008488X.

- Jung Suh, Youngmin Kim: *Accelerating MATLAB with GPU Computing*, Morgan Kaufmann, 2013, ISBN 0124080804.

- Jeremy Kepner: *Parallel MATLAB for Multicore and Multinode Computers*, SIAM, 2009, ISBN 089871673X.

- Richard Johnson: *The Elements of MATLAB Style*, Cambridge University Press, 2010, ISBN 0521732581. A synopsis of the main ideas is available at: http://datatool.com/downloads/matlab_style_guidelines.pdf (or: http://bit.ly/1mfXXEL)

- Yair Altman: *Undocumented Secrets of MATLAB-Java Programming*, CRC Press, 2011, ISBN 1439869030. http://UndocumentedMatlab.com/matlab-java-book (or: http://bit.ly/Zqqojt)

A.2.2 Non-MATLAB

- Jack Shirazi: *Java Performance Tuning*, O'Reilly, 2003, ISBN 0596003773.
- Charlie Hunt: *Java Performance*, Prentice-Hall, 2011, ISBN 0137142528.
- Georg Hager, Gerhard Wellein: *Introduction to High Performance Computing for Scientists and Engineers*, CRC Press, 2010, ISBN 143981192X. http://rrze.fau.de/dienste/arbeiten-rechnen/hpc/HPC4SE (or: http://bit.ly/1mfZiLC)
- Stefan Goedecker, Adolfy Hoisie: *Performance Optimization of Numerically Intensive Codes*, SIAM, 2001, ISBN 0898714842.
- Jon Bentley: *Writing Efficient Programs*, Prentice-Hall, 1982, ISBN 013970244X — an out-of-print classic.
- Jon Bentley: *Programming Pearls*, Addison-Wesley Professional, 1999, ISBN 0201657880 — another classic, strong focus on performance tuning but not just.
- Jason Sanders, Edward Kandrot: *CUDA by Example*, Addison Wesley Professional, 2010, ISBN 0131387685.
- Eric Raymond: *The Art of Unix Programming*, Addison Wesley Professional, 2003, ISBN 0131429019. See especially §1.6 and the entire Chapter 12. http://faqs.org/docs/artu (or: http://bit.ly/1hkGARl)

A.2.3 Database

A.2.3.1 General SQL Tuning

- Dan Tow: *SQL Tuning*, O'Reilly, 2003, ISBN 0596005733.
- Markus Winand: *SQL Performance Explained*, 2012, ISBN 3950307826. http://sql-performance-explained.com (or: http://bit.ly/PUNpha)
- Peter Gulutzan, Trudy Pelzer: *SQL Performance Tuning*, Addison-Wesley Professional, 2002, ISBN 0201791692.

A.2.3.2 Oracle

- R. Shee, K. Deshpande, K. Gopalakrishnan: *Oracle Wait Interface: A Practical Guide to Performance Diagnostics & Tuning*, Oracle Press, 2004, ISBN 007222729X.
- G. Vaidyanatha, K. Deshpande, J. Kostelac: *Oracle Performance Tuning 101*, Oracle Press, 2001, ISBN 0072131454.
- Cary Millsap, Jeff Holt: *Optimizing Oracle Performance*, O'Reilly, 2003, ISBN 059600527X.
- Mark Gurry, Peter Corrigan: *Oracle Performance Tuning*, O'Reilly, 1996, ISBN 1565922379.

A.2.3.3 SQL Server

- Sajal Dam, Grant Fritchey: *SQL Server 2008 Query Performance Tuning Distilled*, Apress, 2009, ISBN 1430219025.

- Steven Wort et al.: *Professional SQL Server 2005 Performance Tuning*, Wrox, 2008, ISBN 0470176393.
- Brad McGehee: *Mastering SQL Server Profiler*, Simple Talk Publishing, 2009, ISBN 1906434158.

A.2.3.4 MySQL, PostgreSQL, and DB2

- Baron Schwartz, Peter Zaitsev, Vadim Tkachenko: *High Performance MySQL*, O'Reilly, 2012, ISBN 1449314287.
- Ronald Bradford: *Effective MySQL: Optimizing SQL Statements*, Oracle, 2011, ISBN 0071782796.
- Robert Schneider: *MySQL Database Design and Tuning*, MySQL Press, 2005, ISBN 0672327651.
- Gregory Smith: *PostgreSQL 9.0 High Performance*, Packt Publishing, 2010, ISBN 184951030X.
- Tony Andrews: *DB2 SQL: 75+ Tuning Tips for Developers*, P+T Solutions, 2012, ISBN 0615264972.

A.3 Performance-Related Official MATLAB Bugs

In the following tables, all the bugs IDs can be accessed online using a URL of the format http://www.mathworks.com/support/bugreports/≤BugID≥. For example, Bug #12345 would be described in http://www.mathworks.com/support/bugreports/12345.

To check the very latest status list of performance-related MATLAB bugs, use the following link:

http://www.mathworks.com/support/bugreports/search_results?search_executed=1&keyword=performance&release_filter=Exists+in&release=0&commit=Search (or: http://bit.ly/1lUxfyn).

Bug ID	Description	First Affected Release	Fixed in Release	Work-around Available?	Discussed In
759496	*depfun* runs more slowly in R2011b than in R2011a	R2011b	—	—	§11.8.2
929447	Numerous Workspace Browser variables cause MATLAB slowdown	R14SP2	R2014a	—	§4.8.3
468781	Scripts that use variables defined outside the script are much slower than functions	R2006a	—	—	§4.6.1
255850	Transparent patches sometimes render incorrectly when using OpenGL renderer	R14SP1	—	Yes (partial)	—
331250	Using *hdf5read* to read compound data types from HDF5 files is slow	R14	—	—	§11.4
433624	JIT is unavailable on SELinux platforms	R13	—	Yes	§4.8.3
951125	*parfor* loops can suffer performance degradation	R2013a	R2013b	Yes (patch)	—
938249	Loading certain types of HDF5 files is slow	R2013a	R2013b	Yes (patch)	§11.4

(continued)

Bug ID	Description	First Affected Release	Fixed in Release	Work-around Available?	Discussed In
905821	*open* and *finfo* are much slower when opening some types of files	R2013a	R2013b	Yes (patch)	§11.9.3
914792	*imread* fails to read certain TIFF files	R2012b	R2013a	Yes (patch)	§11.6
872110	*imread* fails to read TIFF with a zero-count tag	R2012b	R2013a	Yes (patch)	§11.6
861335	*imfinfo* returns incorrect default value for ResolutionUnit TIFF tag	R2008b	R2013a	Yes (patch)	§11.6
857319	Storing objects in MAT-files can cause a memory leak and prevent the object class from being cleared	R2011b	R2013a	Yes (partial)	§11.4
793155	Memory leak when reading JPEG 2000 files with *imread*	R2010b	R2013a	Yes (patch)	§11.6
863941	Performance improvements for *Tiff* class	R2009b	R2013a	Yes (patch)	§11.6
846754	*publish* takes long time to complete	R2012a	R2012b	Yes	—
None[1798]	Memory leak using plot annotations	R2012a (?)	R2012b	—	—
792204	Performance degradation converting large cell arrays to .NET System.String arrays	R2011b	R2012b	Yes	—
799101	*multibandread* is too slow when reading all data from BIP files	R2006a	R2012b	Yes (patch)	§11.6
768825	Slow performance of certain indexed assignments	R2011b	R2012a	Yes	§4.8.2
819115	Memory leak using *drawnow* in a loop	R2011a	R2013a	Yes (patch)	§10.1.5
736830	*h5info* is slow when reading files with a very large number of objects	R2011a	R2012a	Yes (patch)	§11.4
535814	Performance degradation of linear algebra operations involving sparse matrices or small full matrices on 8-core machines running 32-bit Windows XP	R2008b	R2012a	Yes	§4.1.3
784028	*save-append* and growing a variable in a *matfile* object does not compress the variable in a V7.3 MAT-file	R2008a	R2012a	—	§11.4
688983	Poor performance saving and loading large cell arrays in V7.3 MAT-file format	R2009b	R2011b	—	—
732450	Startup hangs on Windows platforms having 32 or more logical CPUs	R2008b	R2011b	Yes (partial patch)	§4.8.3
534993	Large data arrays do not display or display very slowly in the Variable Editor	R2007a	R2011a	Yes (partial)	—
581959	Cannot start license manager via *lmstart* utility on systems running Mac OS 10.6	R2009b	R2011a	Yes	§4.9.13
641369	Publishing: Sometimes MATLAB becomes unresponsive	R2008a	R2011a	—	—
686102	Navigating certain XML files in the Current Folder browser is extremely slow	R2010b	R2011a	Yes (patch)	—
636170	Selecting Workspace Browser variables in debug state when Signal Processing Toolbox is installed can hang MATLAB	R2010a	R2010b	Yes (patch)	—
582905	*save* and *load* of containers in v7.3 MAT-files exhibits memory leak	R2009a	R2010b	—	—

(continued)

Bug ID	Description	First Affected Release	Fixed in Release	Work-around Available?	Discussed In
565962	Saving and loading very large container arrays in v7.3 MAT-files is slow	R2007b	R2010a	—	—
611334	Slow performance in *regexpi*	R2009a	R2010b	Yes	—
624798	*xlsread* performance regression for certain files	R2010a	R2010b	Yes	—
573283	*ldl* and *inv* may hang MATLAB under certain conditions	R2008b	R2010a	—	—
448843	*svd* with one output argument has performance regression	R2008a	R2010a	—	—
574807	Memory leak in MATLAB COM Automation client interface may cause degraded performance	R14SP2	R2010a	—	§9.7.1
425632	Memory leak in MATLAB COM Automation client interface may cause degraded performance	R2006b	R2008b	—	—
412219	Mac OSX 10.5 slow scrolling in MATLAB Editor and some GUIs	R2007a	R2008b	Yes (patch)	§4.9.13
405619	Memory leak using COM Automation Server *PutWorkspaceData* function in MATLAB degrades performance	R14SP3	R2008a	—	—
None[1799]	*memmapfile* cannot be used for IPC	R2007a (?)	R2008a	—	§7.4
368993	ActiveX *invoke* commands using incorrect case take much longer in MATLAB 7.4 (R2007a) than earlier	R2007a	R2007b	Yes (patch)	—
558989	MATLAB enters infinite loop when setting or getting value of a dynamic property	R2008b	R2010a	—	—
558698	Memory leak when destroying *libpointer* objects of type stringPtrPtr	R2006a	R2010a	Yes	—
577083	*conv2* has a performance regression for matrices that contain mostly zeros	R2008b	R2010a	—	—
523801	The MuPAD kernel can take several minutes to initialize on Windows	R2008b	R2009b	Yes (patch)	§4.8.3, §4.9.11
503572	*mldivide* on a nearly singular real symmetric indefinite matrix may sometimes cause MATLAB to hang	R2008a	R2009b	—	—
508142	Memory leak calling a nested function	R2008a	R2009b	—	—
491164	Value class scalar numeric property access slow on 32-bit platforms	R2008a	R2009a	—	—
500221	Memory leak in *contourf* with the Zbuffer figure renderer	R12.1	R2009a	—	—
486393	Calling *clear all* when the workspace contains MATLAB class objects slows down subsequent calls to class constructors	R2008a	R2008b	Yes	—
476589	Functions with ambiguous return types may cause a performance problem	R2006a	R2008b	Yes	—
463914	Calling a function with multiple output arguments may be slower when one of the outputs is subscripted	R2006b	R2008b	Yes	—

(continued)

Bug ID	Description	First Affected Release	Fixed in Release	Work-around Available?	Discussed In
470886	Performance regression in *waitbar*	R2008a	R2008b	—	—
461619	Performance regression in *num2str*	R2008a	R2008b	Yes (patch)	—
446891	*interp2* may run more slowly than expected for data type single	R14	R2008b	—	—
331335	Using *get* or *set* functions on Java objects and some Handle Graphics objects results in memory and performance problems	R14	R2008b	Yes	§9.7.2
250986	Serial port *fprintf* sporadically times out	R14SP1	R2008a	Yes (partial patch)	—
563214	Out of memory when reading multi-page TIFF containing thousands of images	R14SP3	R2008a	Yes (patch)	—
417149	*hdf5write* results in a memory leak	R2007b	R2008a	Yes (patch)	—
390945	Delay when resizing GUI windows	R2007a	R2008a	—	—
366601	MATLAB startup hangs when trying to access My Documents\MATLAB\ folder on Windows	R2007a	R2007b	Yes	§4.8.3
346884	*fscanf* takes considerably longer when using a scanset in the format string	R2006b	R2007a	Yes	—
342880	*cdfread* very slow reading large datasets	R14	R2007a	—	—
277100	*help* function runs slowly	R14	R2007a	Yes (partial)	—
313223	Creating an object causes a deep copy of input argument	R14	R2007a	—	—
335557	*input* command can cause MATLAB to use 100% of CPU time	R2006b	R2007a	Yes (partial)	—
306972	Scalar times sparse matrix yields performance degradation	R2006a	R2006b	Yes	§4.1.3
299539	Current Directory browser: sorting by Last Modified column is sometimes slow	R2006a	R2006b	—	—
279325	MATLAB may hang or terminate if the *libtiff.dylib* file is not removed on MacOS	R14SP3	R2006a	Yes	§4.9.13
205255	Inaccuracies in *cputime* on Pentium 4 with Hyperthreading Running Windows	R14	R2006a	Yes	§2.2.3
252672	GUI listbox performance issue	R14SP2	R14SP3	Yes	—
299698	must call *tic* before calling *toc* message	R12.1	R14SP2	Yes (partial)	—

A.4 Performance-Related Code Analyzer (MLint) Warnings

MATLAB's Code Analyzer (MLint), available in the MATLAB Editor, includes a wide range of warning messages about possible problems in an analyzed m-file, including a large number of warnings pertaining to performance. These warnings can be seen in the Profiler results (see §2.1.2), as well as in the MATLAB Editor:

```
c2={c{:} 123};
```
⚠ { A{:} B } can often be replaced by [A {B}], which can be much faster. [Details ▼]

There are over 550 different messages, of which about 50 are performance related. The complete list of these messages can be seen in the MATLAB Preferences (Code Analyzer → "Potential Performance Problems" group) or by the following command:
```
>> mlint -allmsg sin
```

Here is the list of the performance-related messages in MATLAB R2013b (8.2):

ID	Description	Discussed In
WNON	*warning('on',msgID)* is faster than *warning('on')*	§4.9.4
WNOFF	*warning('off',msgID)* is faster than *warning('off')*	§4.9.4
TLEV	<reserved word> (*global/persistent*) could be very inefficient unless it is a top-level statement in its function	§4.9.14
TRIM1	Use *strtrim(str)* instead of nesting *fliplr* and *deblank* calls	§4.2.5.1
TRIM2	Use *strtrim(str)* instead of *deblank(strjust(str,'left'))*	§4.2.5.1
RGXP1	Using *regexp(string, pattern, 'once')* is faster in this case	§4.2.2
STTOK	Use one call to *textscan* instead of calling *strtok* in a loop	§4.2.2
STCCS	*strcmpi/strncmpi* can be replaced by a faster, case-sensitive compare	§4.2.2
STNCI	Use *strncmpi(str1,str2)* instead of *upper/lower* in a call to *strncmp*	§4.2.4
MATCH2	*strmatch* is not recommended. Use *strncmp* or *validatestring* instead	§4.2.4
FLUDLR	Replace *flipud(fliplr(x))* or *fliplr(flipud(x))* by a faster *rot90(x,2)*	§4.9.14
RPMT0	Replace *repmat(0,x,y)* with *zeros(x,y)* for better performance	§5.4.2.2
RPMT1	Replace *repmat(1,x,y)* with *ones(x,y)* for better performance	§5.4.2.2
RPMTN	Replace *repmat(NaN,x,y)* with *NaN(x,y)* for better performance	§5.4.2.2
RPMTI	Replace *repmat(Inf,x,y)* with *Inf(x,y)* for better performance	§5.4.2.2
RPMTF	Replace *repmat(false,x,y)* with *false(x,y)* for better performance	§5.4.2.2
RPMTT	Replace *repmat(true,x,y)* with *true(x,y)* for better performance	§5.4.2.2
FNDSB	To improve performance, use logical indexing instead of *find*	§5.1.3
MXFND	Use *find* with the 'first' or 'last' option	§5.1.3
EFIND	Replace *isempty(find(X))* with *isempty(find(X,1))* for performance	§5.1.3
FLPST	For better performance in some cases, use *sort* with 'descend' option	§4.9.14
VCAT	Constructing a cell array is faster than using *strcat*	§4.2.5.2
CCAT	{A{:} B} can often be replaced by [A {B}] which can be much faster	§4.1.5
CCAT1	{A{I}} can be replaced by A(I) or A(I)', which can be much faster	§4.1.5
MMTC	*mat2cell* should be replaced by a simpler, faster call to *num2cell*	§4.9.5
AGROW	The variable <name> appears to change size on every loop iteration. Consider preallocating for speed.	§2.1.2, §9.4.3

(continued)

ID	Description	Discussed In
SAGROW	The variable <name> appears to change size on every loop iteration (within a script). Consider preallocating for speed.	§2.1.2, §9.4.3
NASGU	The value assigned to variable <name> might be unused.	§3.10.1.3
PSIZE	*numel(x)* is usually faster than *prod(size(x))*	§4.9.14
ISMT	Using *isempty* is usually faster than comparing *length()* to 0	§4.9.8
LOGL	Use *true* or *false* instead of *logical(1)* or *logical(0)*	§4.1.7
ST2NM	*str2double* is faster than *str2num*; however, *str2double* operates only on scalars. Use the function that best suits your needs.	§4.2.3
EXIST	*exist* with two input arguments is generally faster and clearer than with one input argument	§4.9.9
UDIM	Instead of using transpose ('), consider using a different DIMENSION input argument to <FUNCNAME>	§5.3.2
TRSRT	Transposing the input to <FUNCNAME> is often unnecessary	§5.3.2
ISCEL	Use *iscell* instead of comparing the class to 'cell'	§4.9.8
ISCHR	Use *ischar* instead of comparing the class to 'char'	§4.9.8
ISLOG	Use *islogical* instead of comparing the class to 'logical'	§4.9.8
ISSTR	Use *isstruct* instead of comparing the class to 'struct'	§4.9.8
ISMAT	When checking if a variable is a matrix consider using *ismatrix*	§4.9.8
ISROW	When checking if a variable is a row vector consider using *isrow*	§4.9.8
ISCOL	When checking if a variable is a column vector consider using *iscolumn*	§4.9.8
MIPC1	On Windows platforms, calling *computer* with an argument returns 'win32' or 'win64', but never 'PCWIN'	§4.9.14
SFLD	Use dynamic fieldnames with structures instead of *setfield*	§4.9.3
GFLD	Use dynamic fieldnames with structures instead of *getfield*	§4.9.3
FREAD	*fread(FID,...,'*char')* is more efficient than *char(fread(...))*	§9.4.5, §11.3.4
N2UNI	*fread* no longer requires *native2unicode* in R2006A and later releases	§11.3.4
MINV	*inv(A)*b* can be slower and less accurate than $A\backslash b$. Consider using $A\backslash b$ for *inv(A)*b* or *b/A* for *b*inv(A)*	§4.5.2
IJCL	Replace complex i and j by 1i for speed and improved robustness	§4.5.3
TNMLP	Consider moving toolbox function <FUNCNAME> out of the loop for better performance	§3.1.1.5
LAXES	Calling *axes(h)* in a loop can be slow. Consider moving outside the loop.	§3.1.1.5, §10.2.2
MRPBW	To use less memory, replace *bwlabel(bw)* by *logical(bw)* in a call of *regionprops*	§4.9.14
SPRIX	This sparse indexing expression is likely to be slow	§4.1.3
SPEIG	*eig* function is called in an invalid manner with a sparse argument	§4.9.14
GRIDD	Consider replacing *griddata* with *scatteredInterpolant/TriScatteredInterp* for better performance. Note: *TriScatteredInterp* will be removed in a future release; use *scatteredInterpolant* instead.	§4.9.14

Appendix B: Performance Tuning Checklist

As noted often in the text, an application's performance heavily depends on numerous factors, including the hardware, OS, MATLAB release, data, and application type. Different applications require different tuning paths and the relative importance of the tuning techniques that were presented in this book vary widely.

Still, there are several basic ideas that can prove helpful for optimizing most programs. Readers should note that the following list is intentionally partial. For more tuning tips and additional details the reader is referred to the main text.

Keeping the list concise was a challenge. After all, many other tips that are mentioned in the main text can be enormously effective for some applications. However, for most MATLAB applications, the guidelines in the following list should cover the majority of the low-hanging fruits that bear the largest speedup potential.

Note that many of the items have significant tradeoffs that must be considered. The list only refers to the performance aspects, but as explained in §1.6, there are many other sides to the story. For example, avoiding MATLAB classes (OOP) may be good for performance but bad for maintainability; using MATLAB toolboxes may not be within the budget; and avoiding certain MATLAB constructs often requires extra development effort. The tips in the list below, and throughout this book, cannot be used without also considering such tradeoffs, which are unique to each specific case.

1. Decide when *not* to performance-tune (§1.2).
2. Decide on realistic performance targets (§1.3).
3. Measure application performance before deciding to tune, and repeatedly during the tuning process. Stop when the performance target is reached (§1.3).
4. Improve the application's perceived performance in addition to the actual performance (§1.8, §10.4.2).
5. Use profiling and wall-clock timing measurements (*tic/toc*) to determine performance hotspots (Chapter 2).
6. Optimize loops by hoisting invariants outside and simplifying the loop contents (§3.1).
7. Heavily use caching (§3.2).
8. Process smaller data blocks (§3.6).
9. Trade latency for throughput (§3.8).
10. Reduce the problem's algorithmic complexity and target accuracy (§3.9).
11. Use appropriate data types, preferably numeric; use sparse data were relevant; do not modify data type in run-time (§4.1).
12. Avoid datasets, tables, and categorical arrays — use alternatives (§4.1.6).
13. Use numeric data, rather than strings (§4.2), especially for date/time (§4.4).
14. Provide MATLAB functions with optional input data where relevant (§4.5).
15. Avoid anonymous functions if regular function handles can be used (§4.6.1).
16. Use procedural programming rather than MATLAB's OOP (§4.7).

17. Disable MATLAB's Startup Accelerator (§4.8.1).

18. Start MATLAB in no-desktop mode when using it in batch mode (§4.8.2).

19. Use the latest MATLAB release (§4.9.7).

20. Heavily employ vectorization and bulk indexing (Chapter 5).

21. Use Parallel Computing Toolbox for multicore and GPU programming (Chapter 6).

22. Use Distributed Computing Server for grid/cluster/cloud programming (Chapter 6).

23. Use MEX to create C code that can be invoked in MATLAB functions (§8.1).

24. Use MATLAB Coder to translate MATLAB code into C automatically (§8.2).

25. Use a splash screen with deployed (compiled) MATLAB applications and reduce the executable size (§8.4).

26. Use third-party libraries for specific computational purposes (§4.5.7, §8.5).

27. Loop down columns rather than rows (§9.3).

28. Preallocate large data matrices (§9.4).

29. Use in-place data manipulation wherever possible (§9.5).

30. Reuse graphic objects, updating their properties rather than recreating (§10.1).

31. Use static or no legends and colorbars (§10.1, §10.2.3).

32. Only use *drawnow* when you are finished updating the graphics (§10.1.5).

33. Create and update graphic components in non-visible mode (§10.4.1.1).

34. Use the old (unsupported) rather than the new (supported) *uitable* (§10.4.1.7).

35. Present continuous feedback to the user for long-running tasks (§10.4.2).

36. Defer nongraphical tasks to the end of the display update (§10.4.3).

37. Minimize intentional *pause*s in the application (§10.5.1).

38. Suppress console output by appending a semi-colon (;) to statements (§11.1).

39. Merge file access operations (§11.2).

40. Use binary data files rather than text files wherever possible (§11.3).

41. Use *textscan* rather than alternatives when reading text files (§11.3).

42. Use serialization with *save* -v6 rather than the newer -v7 or -v7.3 (§11.4).

43. Keep Excel open and invisible when it is being read or updated (§11.5).

44. Use Java or MEX-based I/O for certain file operations (§11.7, §11.8).

45. Minimize networking; run MATLAB and access data from local disk (§11.9).

46. Use utilities on the MATLAB File Exchange for fast implementations of numerous functionalities, in some cases replacing builtin MATLAB functions.

References and Notes

Note: Due to the transient nature of URLs, some of the references below might no longer work when you try to use them. However, in many cases, the resources can still be found in web archives long years after their removal. For example, cached versions of http://undocumentedmatlab.com/blog/tag/performance can be found at http://web.archive.org/web/*/http://undocumentedmatlab.com/blog/tag/performance.

1. For example, http://mathworks.com/help/matlab/code-performance.html (or: http://bit.ly/1bc6ejj)
2. http://mathworks.com/company/jobs/opportunities/search?keywords=performance (or: http://bit.ly/1vIojoA)
3. http://www.famfamfam.com/lab/icons/silk (or: http://bit.ly/VtWQOT)
4. http://en.wikipedia.org/wiki/Performance_tuning (or: http://bit.ly/107ecZu)
5. For example: http://www.slideshare.net/joshfraz/sept-2012rumtalk (or: http://slidesha.re/1agYv4a); http://seogadget.com/improving-site-speed-talk-about-the-business-benefit (or: http://bit.ly/18MBTd2); http://www.strangeloopnetworks.com/resources/infographics/web-performance-and-ecommerce/impact-of-1-second-delay (or: http://bit.ly/18MAh2Y)
6. http://www.oraclejavamagazine-digital.com/javamagazine/july_august_2013#pg8 (or: http://bit.ly/1c4fRWE); https://home.java.net/poll/how-critical-performancescalability-success-apps-you-develop (or: http://bit.ly/1c4fKul). Java is a registered trademark of Javasoft, now an Oracle company (http://java.sun.com).
7. http://faqs.org/docs/artu/ch12s01.html (or: http://bit.ly/1elsJ7w) — Eric Raymond, *The Art of Unix Programming*, ISBN 0131429019, §12.1.
8. 1974 Turing Award Lecture, Communications of the ACM **17** (12), December 1974, p. 671. Archived version: http://bit.ly/1mmCbNw. Other variants by Knuth: http://en.wikiquote.org/wiki/Donald_Knuth (or: http://bit.ly/1mmCp76).
9. http://www.hans-eric.com/2007/11/23/optimize-late-benchmark-early (or: http://bit.ly/18bav4e)
10. http://thinkingparallel.com/2006/08/07/my-views-on-high-level-optimization (or: http://bit.ly/1ijzaeW)
11. Adapted from Lewis Carroll, *Alice's Adventures in Wonderland*, Chapter 6 — dialog between Alice and the Cheshire Cat.
12. http://en.wikipedia.org/wiki/Pareto_principle
13. http://en.wikipedia.org/wiki/Hot_spot_(computer_programming) (or: http://bit.ly/Zt6Al5)
14. http://en.wikipedia.org/wiki/Just-in-time_compilation (or: http://bit.ly/1gVbTBI)
15. http://www.youtube.com/watch?v=7gtf47D_bu0#t=341 (or: http://bit.ly/18MYgPB) and associated slides http://www.igvita.com/slides/2013/fluent-perfcourse.pdf (or: http://www.igvita.com/slides/2013/fluent-perfcourse.pdf)
16. http://undocumentedmatlab.com/ib-matlab (or: http://bit.ly/QYcgBl)
17. http://interactivebrokers.com
18. Also see http://en.wikipedia.org/wiki/Computer_performance#Aspects_of_performance (or: http://bit.ly/1gTJPyY)
19. An excellent resource is http://queue.acm.org/detail.cfm?id=1117403 (or: http://bit.ly/U7Gix3). A more humoristic article on this topic, with additional insight, is http://www.oraclejavamagazine-digital.com/javamagazine/20131112#pg32 (or: http://bit.ly/I8H0Lt).
20. Some R&D teams define a *"performance hawk"* position, whose job is to identify and reject performance-degrading fixes or features, as a counter-measure to programmers' natural

tendency to prefer functionality over performance (Jeremy Kepner, *Parallel MATLAB for Multicore and Multinode Computers*, SIAM 2009, ISBN 089871673X, §5.11.1). This has some advantages, but could lead to belated and ineffective tuning, as well as to a general feeling among the developers that performance is "somebody else's problem".

21. A related (although not directly relevant) list of pitfalls in using irrelevant metrics or benchmarks, is provided in http://en.wikipedia.org/wiki/Benchmark_(computing)#Challenges (or: http://bit.ly/18ZwU5N)

22. http://queue.acm.org/detail.cfm?id=2413037 (or: http://bit.ly/UkgGzL)

23. Rob Pike, *Notes on Programming in C*, http://www.lysator.liu.se/c/pikestyle.html

24. http://www.infoq.com/articles/9_Fallacies_Java_Performance (item #4) (or: http://bit.ly/1dryB3m)

25. http://blogs.mathworks.com/videos/2013/11/18/knowing-when-to-optimize-your-graphics-in-matlab (or: http://bit.ly/1biHVVN)

26. http://en.wikibooks.org/wiki/Optimizing_Code_for_Speed/Optimizing_by_Reducing_Feature-Set (or: http://bit.ly/1ggOS6r)

27. http://mathworks.com/matlabcentral/newsreader/view_thread/321539 (or: http://bit.ly/10dLAy3)

28. http://blogs.mathworks.com/loren/2012/01/13/best-practices-for-programming-matlab#comment-32891 (or: http://bit.ly/RAd3Y5)

29. "A Case Against the GOTO", Proceedings of the 25th National ACM Conference, August 1972, pp. 791-797, quoted in http://en.wikiquote.org/wiki/William_Wulf (or: http://bit.ly/VnAMWh)

30. See for example: http://blogs.mathworks.com/videos/2012/08/01/naive-vs-complex-solutions (or: http://bit.ly/RAcxcz)

31. http://en.wikipedia.org/wiki/Scalability (or: http://bit.ly/12CQROm)

32. http://en.wikipedia.org/wiki/Amdahl's_law (or: http://bit.ly/YqJASr)

33. http://en.wikipedia.org/wiki/Fallacies_of_Distributed_Computing (or: http://bit.ly/TrLXmx)

34. http://en.wikipedia.org/wiki/Parallel_slowdown (or: http://bit.ly/TaTpik)

35. http://en.wikipedia.org/wiki/Gustafson's_law (or: http://bit.ly/1gstkJv); http://blogs.mathworks.com/cleve/2013/11/12/the-intel-hypercube-part-2-reposted#aa5c77c4-80eb-4546-8752-6d0f820545b8 (or: http://bit.ly/1biDY3d)

36. http://en.wikipedia.org/wiki/Perceived_performance (or: http://bit.ly/VYMt6B)

37. Refer for example to the issue of the speed of MATLAB's *comet* demo animation, http://blogs.mathworks.com/pick/2013/11/01/comet3-with-speed-control (or: http://bit.ly/1bivTeT)

38. Numerous resources are available for optimizing website performance. Google (http://developers.google.com/speed) and Yahoo! (http://developer.yahoo.com/yslow) have great content on this. I also highly recommend Google's Ilya Grigorik YouTube videos.

39. http://en.wikipedia.org/wiki/Interlacing_(bitmaps) (or: http://bit.ly/1ggB8so)

40. http://en.wikipedia.org/wiki/1-Click (or: http://bit.ly/VYTdkS)

41. http://en.wikipedia.org/wiki/Worst-case_execution_time (or: http://bit.ly/Vnx7HP)

42. http://java.dzone.com/articles/little-knowledge-and-profiler (or: http://bit.ly/1eJ5aa8)

43. http://java.dzone.com/articles/percentiles-made-easy (or: http://bit.ly/1cdD4Gk)

44. See http://buytaert.net/files/oopsla07-georges.pdf (or: http://bit.ly/1epsyJZ) for a detailed treatment of statistically rigorous performance evaluation. While the paper discusses Java performance, its general treatment and main conclusions are also relevant to MATLAB.

45. http://en.wikipedia.org/wiki/Real-time_computing (or: http://bit.ly/18Zzs3W)

46. http://en.wikipedia.org/wiki/Profiling_(computer_programming) (or: http://bit.ly/14ITty1)

47. http://mathworks.com/help/matlab/matlab_prog/profiling-for-improving-performance.html (or: http://bit.ly/VbX018)

48. e.g., http://blogs.mathworks.com/videos/2012/07/11/using-the-matlab-profiler-to-speed-your-code (or: http://bit.ly/NXfuSQ); http://blogs.mathworks.com/pick/2006/10/19

/profiler-to-find-code-bottlenecks (or: http://bit.ly/NXfL8b); http://blogs.mathworks.com /videos/2009/04/17/speeding-up-dougs-solution-with-the-profiler (or: http://bit.ly /NXfXEx)

49. e.g., http://blogs.mathworks.com/community/2011/06/28/speed-up-your-guis-with-profiling (or: http://bit.ly/NXfQsu); http://blogs.mathworks.com/community/2010/02/01/speeding-up-your-program-through-profiling (or: http://bit.ly/NXg3fj); http://blogs.mathworks.com /pick/2008/09/03/how-our-developers-make-matlab-faster (or: http://bit.ly/U5Skqa)

50. e.g., http://mathworks.com/company/newsletters/news_notes/may03/profiler.html (or: http://bit.ly/U5SrlD). This newsletter article *"Picking up the Pace with the MATLAB Profiler"* from 2003 was removed from the MathWorks website in 2010 as some of its suggestions are no longer relevant; archived versions of this article can be found here: http://web.archive.org /web/20091025140723/http://mathworks.com/company/newsletters/news_notes/may03 /profiler.html (or: http://bit.ly/1c3ynLQ) and here: http://www.ee.columbia.edu/~marios /matlab/Picking%20up%20the%20Pace%20with%20the%20MATLAB%20Profiler.pdf (or: http://bit.ly/19oh1Zy).

51. e.g., www.mathworks.com/company/events/webinars/wbnr60898.html (or: http://bit.ly /NXgRAV)

52. http://UndocumentedMatlab.com/blog/undocumented-profiler-options-part-4 (or: http:// bit.ly/14ISs99). While the *–detail builtin* option is unsupported, it is indeed mentioned in the official documentation: http://mathworks.com/help/matlab/matlab_prog/profiling-for-improving-performance.html#f9-17518 (or: http://bit.ly/15vfO3f)

53. http://mathworks.com/help/techdoc/ref/profile.html (or: http://bit.ly/QIELuX)

54. http://mathworks.com/help/matlab/matlab_prog/profiling-for-improving-performance. html (or: http://bit.ly/VbX018)

55. http://UndocumentedMatlab.com/blog/undocumented-profiler-options-part-2 (or: http:// bit.ly/PdWrk1)

56. http://blogs.mathworks.com/contest/2002/11/13/using-the-profiler-and-jit-to-accelerate-a-fractal-calculation (or: http://bit.ly/QzggI5)

57. http://mathworks.com/help/distcomp/examples/illustrating-three-approaches-to-gpu-computing-the-mandelbrot-set.html (or: http://bit.ly/1g2GF7E); http://blogs.mathworks. com/loren/2011/07/18/a-mandelbrot-set-on-the-gpu (or: http://bit.ly/18UE9i3) and http://blogs.mathworks.com/loren/2011/04/21/deploying-multiple-c-shared-libraries (or: http://bit.ly/18UEili); also see the related *mandelbrotViewer* demo — http://mathworks. com/matlabcentral/fileexchange/30988-a-gpu-mandelbrot-set (or: http://bit.ly/18UEn8C), http://mathworks.com/matlabcentral/fileexchange/33201-gpu-julia-set-explorer (or: http://bit.ly/18UEqBp)

58. http://en.wikipedia.org/wiki/Call_graph (or: http://bit.ly/14ITUYW)

59. http://UndocumentedMatlab.com/blog/undocumented-profiler-options-part-3 (or: http:// bit.ly/QWtoPZ)

60. http://mathworks.com/matlabcentral/fileexchange/46976-profile-history (or: http://bit.ly /1ycnJya); http://UndocumentedMatlab.com/blog/function-call-timeline-profiling (or: http://bit.ly/1ycnOlg)

61. http://UndocumentedMatlab.com/blog/buggy-profiler-option (or: http://bit.ly/QPrwLP)

62. See Bill MacKeeman's whitepaper "Measuring MATLAB Performance" (http://mathworks. com/matlabcentral/fileexchange/18510-matlab-performance-measurement or: http://bit.ly /IDpc9c); also see http://mathworks.com/matlabcentral/answers/39788-how-does-warm-up-overhead-scale-with-data-size-or-iteration-count (or: http://bit.ly/LRHwuk) and http:// mathworks.com/matlabcentral/answers/100271 (or: http://bit.ly/1drRG5F)

63. http://mathworks.com/matlabcentral/newsreader/view_thread/323928#890059 (or: http:// bit.ly/10fiiz1)

64. http://mathworks.com/help/matlab/matlab_prog/profiling-for-improving-performance. html#brgp8gx (or: http://bit.ly/10fiaiM); http://walkingrandomly.com/?p=736 (or: http:// bit.ly/10fiIWa); http://stackoverflow.com/questions/9778287

/set-processor-affinity-for-matlab (or: http://bit.ly/1dLflIb); http://mathworks.com/matlabcentral/answers/94049 (or: http://bit.ly/1eSMFTR).

65. http://software.intel.com/en-us/articles/de-mystifying-software-performance-optimization (or: http://intel.ly/1gXvHVg).

66. http://mathworks.com/matlabcentral/newsreader/view_thread/321539 (or: http://bit.ly/10dLAy3)

67. http://UndocumentedMatlab.com/blog/gui-automation-robot (or: http://bit.ly/9Hj3D9); http://UndocumentedMatlab.com/blog/gui-automation-utilities (or: http://bit.ly/cWodk2)

68. http://blogs.mathworks.com/videos/2013/05/03/speeding-up-user-interfaces-in-matlab-with-profiler (or: http://bit.ly/1247NO7)

69. Read a related general discourse on profiling limitations here: http://faqs.org/docs/artu/ch12s02.html (or: http://bit.ly/1eluzoF) — Eric Raymond, *The Art of Unix Programming*, ISBN 0131429019, §12.2.

70. http://stackoverflow.com/questions/18952552/timing-code-in-matlab#18955502 (or: http://bit.ly/15qVlef); http://mathworks.com/matlabcentral/answers/18576-profiler-biases-execution-time-of-optimized-code#answer_25053 (or: http://bit.ly/12NjgkQ)

71. http://timsalimans.com/the-power-of-jit-compilation (or: http://bit.ly/1bp3qys); http://mathworks.com/matlabcentral/answers/29899-implausible-execution-times (or: http://bit.ly/18L4LF9)

72. http://www.oraclejavamagazine-digital.com/javamagazine/july_august_2013#pg31 (or: http://bit.ly/K1qM7d)

73. http://stackoverflow.com/questions/18952552/timing-code-in-matlab#18955502 (or: http://bit.ly/15qVlef); http://mathworks.com/matlabcentral/answers/18576-profiler-biases-execution-time-of-optimized-code#answer_25053 (or: http://bit.ly/12NjgkQ)

74. For example: http://mathworks.com/matlabcentral/answers/18576-profiler-biases-execution-time-of-optimized-code (or: http://bit.ly/12NjiJo)

75. http://UndocumentedMatlab.com/blog/undocumented-profiler-options#comment-64 (or: http://bit.ly/1eaHRVH)

76. http://mathworks.com/matlabcentral/newsreader/view_thread/248045#640002 (or: http://bit.ly/1eaIa2 V)

77. http://UndocumentedMatlab.com/blog/undocumented-profiler-options (or: http://bit.ly/1eaHNFp)

78. http://UndocumentedMatlab.com/blog/tic-toc-undocumented-option (or: http://bit.ly/OkmXqv); http://mathworks.com/company/newsletters/articles/improvements-to-tic-and-toc-functions-for-measuring-absolute-elapsed-time-performance-in-matlab.html (or: http://bit.ly/NmNDtL)

79. http://mathworks.com/company/newsletters/articles/improvements-to-tic-and-toc-functions-for-measuring-absolute-elapsed-time-performance-in-matlab.html (or: http://bit.ly/NmNDtL) provides an interesting technical description of *tic/toc's* underlying mechanism, differences between platforms, and evolution over the years. MathWorker Bill McKeeman, who coauthored this technical article, also authored the interesting 2008 whitepaper *"Measuring MATLAB Performance"*: http://mathworks.com/matlabcentral/fileexchange/18510-matlab-performance-measurement (or: http://bit.ly/IDpc9b).

80. http://www.luxford.com/high-performance-windows-timers (or: http://bit.ly/116KnWD)

81. http://undocumentedmatlab.com/blog/pause-for-the-better (or: http://bit.ly/116KHoh)

82. A nice presentation on the difference between structural and sampling profilers was presented by Google's Ilya Grigorik and John McCutchan in http://www.youtube.com/watch?v=nxXkquTPng8 (or: http://bit.ly/1ee4hWj). MATLAB's profiler seems to incorporate both aspects: 1 KHz sampling as well as automated function-call instrumentation, however this is speculative (official information on MATLAB's profiler is unavailable).

83. http://en.wikipedia.org/wiki/Heisenbug (or: http://bit.ly/1drW8Br)

84. http://mathworks.com/help/matlab/ref/cputime.html (or: http://bit.ly/13IZAzn)

85. http://mathworks.com/matlabcentral/answers/82622#answer_92297 (or: http://bit.ly /13VkiQG)

86. http://mathworks.com/help/matlab/matlab_prog/analyzing-your-programs-performance. html#bqo2pc_ (or: http://bit.ly/13J05cI). Note the underscore (_) at the end of the original link.

87. http://www.mathworks.com/support/bugreports/205255 (or: http://bit.ly/13J20Ol). This bug report contains a detailed technical explanation and test case.

88. http://UndocumentedMatlab.com/blog/more-undocumented-timing-features#feature (or: http://bit.ly/NKDmZP)

89. http://mathworks.com/matlabcentral/fileexchange/18798-timeit-benchmarking-function (or: http://bit.ly/1fvOs1r). *timeit* was discussed in http://blogs.mathworks.com/steve/ 2008/02/29/timing-code-in-matlab (or: http://bit.ly/OkoBIR) and: http://blogs.mathworks. com/steve/2008/12/31/another-timeit-update (or: http://bit.ly/OkoDAl).

90. http://mathworks.com/matlabcentral/fileexchange/33532-log4matlab (or: http://bit.ly /Pdmmxc)

91. http://mathworks.com/matlabcentral/fileexchange/37701-log4 m-a-powerful-and-simple- logger-for-matlab (or: http://bit.ly/PdmP2n)

92. http://logging.apache.org; http://en.wikipedia.org/wiki/Log4j (or: http://bit.ly/PdnbWL)

93. http://technet.microsoft.com/en-us/sysinternals/bb896653 (or: http://bit.ly/RhGBmb)

94. http://github.com/jezhumble/javasysmon (or: http://bit.ly/1ee0IPV)

95. http://mathworks.com/matlabcentral/newsreader/view_thread/269202#903119 (or: http:// bit.ly/1ee0yIk)

96. http://technet.microsoft.com/en-us/sysinternals/bb896645 (or: http://bit.ly/RhGJSD)

97. http://www.wireshark.org (or: http://bit.ly/TDAiRF)

98. http://www.fiddler2.com/Fiddler2/version.asp (or: http://bit.ly/SBbpnY)

99. http://technet.microsoft.com/en-us/sysinternals/dd535533 (or: http://bit.ly/IRC8mn)

100. http://en.wikipedia.org/wiki/List_of_performance_analysis_tools (or: http://bit.ly/VSmUbo)

101. http://en.wikipedia.org/wiki/VTune (or: http://bit.ly/VSmXE3)

102. http://en.wikipedia.org/wiki/Valgrind (or: http://bit.ly/VSn96e)

103. http://en.wikipedia.org/wiki/Shark_(application)#Shark (or: http://bit.ly/VSnE08)

104. http://en.wikipedia.org/wiki/Instruments_(application) (or: http://bit.ly/VSnPsh)

105. http://en.wikipedia.org/wiki/GlowCode (or: http://bit.ly/VSo1rE)

106. http://en.wikipedia.org/wiki/IBM_Rational_Purify (or: http://bit.ly/13J0LGC); http:// www-03.ibm.com/software/products/us/en/purifyplus (or: http://ibm.co/13J0mnI)

107. http://en.wikipedia.org/wiki/Insure%2B%2B (or: http://bit.ly/13J0ALy); http://www. parasoft.com/jsp/products/insure.jsp?itemId=63 (or: http://bit.ly/13J0weR)

108. http://www.eclipse.org/tptp (or: http://bit.ly/VSoeLx)

109. http://mathworks.com/company/jobs/opportunities/10433 (or: http://bit.ly/VSmlyq) is a job listing for a performance-tuning engineer (also see http://linkd.in/16ISMAy) that mentions these tools. Similarly, http://bit.ly/13j0bZC, a listing for a QA engineer. Unfortunately, these specific job listing are no longer available (apparently the position has been filled), but similar requirements can be found in other MathWorks job listings. These tools can also be inferred from the CVs of MathWorks engineers (e.g., Murali Gopinathan who uses GlowCode — http:// linkd.in/13iZwqU).

110. A somewhat different categorization can be found here: http://en.wikipedia.org/wiki /Compiler_optimization (or: http://bit.ly/1gLdG7e)

111. http://en.wikipedia.org/wiki/Fast_path (or: http://bit.ly/1d9Z1qR)

112. http://en.wikipedia.org/wiki/Loop_optimization (or: http://bit.ly/YIoj12). Note that several of the techniques listed in this resource are irrelevant for MATLAB programmers.

113. http://en.wikipedia.org/wiki/Loop-invariant_code_motion (or: http://bit.ly/VXU2i5)

114. http://blogs.mathworks.com/loren/2013/05/04/recent-question-about-speed-with-subar- ray-calculations (or: http://bit.ly/102Ojt9). Also see related §9.3.3.

115. http://mathworks.com/matlabcentral/answers/119149 (or: http://bit.ly/1pq0hrH)

116. http://en.wikipedia.org/wiki/Loop_unswitching (or: http://bit.ly/WpKU88)

117. http://stackoverflow.com/questions/15919553/memory-and-excecution-speed-in-matlab #15922754 (or: http://bit.ly/ZmOv1K)

118. http://en.wikipedia.org/wiki/Matryoshka_doll (or: http://bit.ly/1oqxQHs). This is sometimes erroneously referred-to as a babushka doll.

119. http://mathworks.com/matlabcentral/newsreader/view_thread/325681#895209 (or: http://bit.ly/YT9j2J)

120. http://www.matlabtips.com/inline-your-lines (or: http://bit.ly/WGaeBG); http://stackoverflow.com/questions/15919553/memory-and-excecution-speed-in-matlab#15922754 (or: http://bit.ly/ZmOv1K) mentions a specific case where this simple technique caused a 1200× speedup, reducing the execution time from days to minutes!

121. http://mathworks.com/matlabcentral/newsreader/view_thread/323890 (or: http://bit.ly/Ys6Dd0)

122. http://mathworks.com/matlabcentral/newsreader/view_thread/330868#909178 (or: http://bit.ly/1cDd6uN)

123. http://en.wikipedia.org/wiki/Instruction_pipeline (or: http://bit.ly/1gcTMHu)

124. http://www.new-npac.org/projects/cdroms/cewes-1999-06-vol1/nhse/hpccsurvey/orgs /sgi/bentley.html#loop2 (or: http://bit.ly/1gySu9u)

125. http://en.wikipedia.org/wiki/Branch_predictor (or: http://bit.ly/13yC0r5). An excellent discussion of the benefits, usages and pitfalls of branch prediction vis-à-vis conditional expressions within loops can be found in http://stackoverflow.com/questions/11227809/why-is-processing-a-sorted-array-faster-than-an-unsorted-array (or: http://bit.ly/13yC9Lb).

126. http://blogs.mathworks.com/videos/2013/10/17/when-only-small-speed-improvements-are-possible-in-matlab (or: http://bit.ly/176fxoY)

127. http://en.wikipedia.org/wiki/Loop_unwinding (or: http://bit.ly/Vr3A6d)

128. http://faqs.org/docs/artu/ch12s03.html (or: http://bit.ly/Vr7mfR) — Eric Raymond, *The Art of Unix Programming*, ISBN 0131429019, §12.3.

129. http://en.wikipedia.org/wiki/Loop_splitting (or: http://bit.ly/WpKheR)

130. http://en.wikipedia.org/wiki/Inner_loop (or: http://bit.ly/193Oo72)

131. http://mathworks.com/matlabcentral/newsreader/view_thread/329351 (or: http://bit.ly/11h2o2w)

132. http://mathworks.com/matlabcentral/answers/120913-speed-up-3-nested-for-loops (or: http://bit.ly/1ppPYnB)

133. http://en.wikipedia.org/wiki/Loop_interchange (or: http://bit.ly/XDQsJp)

134. See for example http://stackoverflow.com/questions/11227809/why-is-processing-a-sorted-array-faster-than-an-unsorted-array#comment14985551_11227902 (or: http://bit.ly/XDQ9yu) and http://stackoverflow.com/questions/11227809/why-is-processing-a-sorted-array-faster-than-an-unsorted-array#11303693 (or: http://bit.ly/Z9G8qH); while this example actually involves LIH optimization more than loop-inversion optimization (something that the commenters on that page have failed to note), the discussion on this page about loop inversion and its importance is in fact interesting and well worth reading.

135. For example: http://mathworks.com/matlabcentral/newsreader/view_thread/333365 (or: http://bit.ly/P8bjVW). See related: http://mathworks.com/matlabcentral/newsreader /view_thread/332835 (or: http://bit.ly/1bvM1Z6).

136. http://mathworks.com/matlabcentral/newsreader/view_thread/328199 (or: http://bit.ly /11RMaRn)

137. http://en.wikipedia.org/wiki/Loop_fusion (or: http://bit.ly/WpLumy)

138. http://en.wikipedia.org/wiki/Instruction_pipeline (or: http://bit.ly/1gcTMHu)

139. http://en.wikipedia.org/wiki/Loop_fission (or: http://bit.ly/YInwx4)

140. http://stackoverflow.com/questions/8547778/why-is-one-loop-so-much-slower-than-two-loops (or: http://bit.ly/1eDsb1L)

141. http://codereview.stackexchange.com/questions/20980/naive-c-matrix-multiplication-100-times-slower-than-blas/20983#20983 (or: http://bit.ly/17Fn1L3)

142. http://mathworks.com/matlabcentral/newsreader/view_thread/325206#893738 (or: http://bit.ly/1lpatRG)

143. This technique could be considered a variant of *run-time algorithm specialization*: http://en.wikipedia.org/wiki/Run-time_algorithm_specialisation (or: http://bit.ly/18IUZxO), although it is not really the same.

144. http://en.wikipedia.org/wiki/Just-in-time_compilation (or: http://bit.ly/1gVbTBI)

145. A technical explanation of MATLAB's JIT was provided with release R13 (6.5) in a 2002 newsletter. It is no longer available on the MathWorks website, probably because the JIT has evolved quite a bit from then. But I find that it still provides interesting insight and is worth reading. Archived versions can be found here: https://web.archive.org/web/20110302034013/http://mathworks.com/company/newsletters/digest/sept02/accel_matlab.pdf (or: http://bit.ly/1bp72AE) and here: http://www.ee.columbia.edu/~marios/matlab/accel_matlab.pdf (or: http://bit.ly/1bp7tuP). A discussion of JIT can be found in Ned Gulley's 2003 newsletter article *"Picking up the Pace with the MATLAB Profiler"*, which has also been removed from the MathWorks website in 2010; archived versions can be found here: http://web.archive.org/web/20091025140723/http://mathworks.com/company/newsletters/news_notes/may03/profiler.html (or: http://bit.ly/1c3ynLQ) and here: http://www.ee.columbia.edu/~marios/matlab/Picking%20up%20the%20Pace%20with%20the%20MATLAB%20Profiler.pdf (or: http://bit.ly/19oh1Zy). While detailed insight into MATLAB's JIT is not available, a description of the underlying technologies can be found in publicly-available papers on Java's HotSpot JIT: http://www.oraclejavamagazine-digital.com/javamagazine_open/20120506#pg49 (or: http://bit.ly/18L3rC5); http://www.oraclejavamagazine-digital.com/javamagazine_open/20120910#pg46 (or: http://bit.ly/18L3qOA); http://www.oraclejavamagazine-digital.com/javamagazine_open/20130708#pg42 (or: http://bit.ly/18L3yxG)

146. http://blogs.mathworks.com/loren/2008/06/25/speeding-up-matlab-applications#comment-29607 (or: http://bit.ly/1bp0oKO)

147. http://mathworks.com/matlabcentral/newsreader/view_thread/328427 (or: http://bit.ly/K4mWuM); http://mathworks.com/matlabcentral/newsreader/view_thread/325638#895100 (or: http://bit.ly/K4n0ui); http://blogs.mathworks.com/loren/2012/01/13/best-practices-for-programming-matlab#comment-32911 (or: http://bit.ly/184Z66R)

148. http://www.matlabtips.com/matlab-is-no-longer-slow-at-for-loops (or: http://bit.ly/1dTm7MD)

149. http://blog.accelereyes.com/blog/2010/04/05/converting-matlab-loops-to-gpu-code (or: http://bit.ly/UfDAsF)

150. http://blog.accelereyes.com/blog/2010/04/05/converting-matlab-loops-to-gpu-code (or: http://bit.ly/UfDAsF). See also §5.2.4.2.

151. http://stackoverflow.com/questions/17809077/performance-of-index-creation (or: http://bit.ly/17FHzDg)

152. Technical solution 1-OV2IV: http://mathworks.com/matlabcentral/answers/99056 (or: http://bit.ly/1b0OkPA)

153. http://mathworks.com/matlabcentral/fileexchange/18510-matlab-performance-measurement (or: http://bit.ly/1eDp4qz)

154. http://stackoverflow.com/a/24675978/233829 (or: http://bit.ly/1vNqas2). Also see http://mathworks.com/matlabcentral/newsreader/view_thread/325638#895124 (or: http://bit.ly/1eDp8Xw)

155. http://www.mathworks.com/support/bugreports/433624 (or: http://bit.ly/GZuQDD)

156. http://www.akkadia.org/drepper/selinux-mem.html (or: http://bit.ly/1eaLnPY); https://bugzilla.redhat.com/show_bug.cgi?id=218653 (or: http://red.ht/1eaLuen) is a specific discussion of the MATLAB issue in SELinux.

157. For example, http://mathworks.com/matlabcentral/answers/29899-implausible-execution-times (or: http://bit.ly/18L4LF9); http://mathworks.com/matlabcentral/newsreader/view_thread/332811 (or: http://bit.ly/18L7hLA); http://stackoverflow.com/questions/18898698/doing-something-is-faster-than-doing-nothing-with-a-variable-in-matlab-loop (or: http://bit.ly/19f68a2); https://web.archive.org/web/20110707092709/http://www.ugcs.caltech.edu/~srbecker/computer.shtml (or: http://bit.ly/1cowgno)

158. http://UndocumentedMatlab.com/blog/undocumented-profiler-options#comment-64 (or: http://bit.ly/1eaHRVH); http://blogs.mathworks.com/loren/2008/06/25/speeding-up-matlab-applications#comment-29607 (or: http://bit.ly/1bpOoKO); http://mathworks.com/matlabcentral/newsreader/view_thread/284759#755654 (or: http://bit.ly/18L63jA)

159. http://timsalimans.com/the-power-of-jit-compilation (or: http://bit.ly/1bp3qys)

160. http://polaris.cs.uiuc.edu/~galmasi/majic/majic.html (or: http://bit.ly/1BhRyyO). Main paper: http://forge.scilab.org/index.php/p/richelieu/source/tree/master/Papers/MaJIC%3A%20Compiling%20MATLAB%20for%20Speed%20and%20Responsiveness.pdf (or: http://bit.ly/Xl4lBl) or an alternate archive: http://dch360.com/read/6d935a0 (or: http://bit.ly/1BhR9wl)

161. http://www.sable.mcgill.ca/publications/papers/2010-3/mcvmcc2010.pdf (or: http://bit.ly/1bp3Uoc)

162. http://en.wikipedia.org/wiki/Cache_(computing) (or: http://bit.ly/Z9I6HC)

163. http://en.wikipedia.org/wiki/Space-time_tradeoff (or: http://bit.ly/18IUmEJ)

164. http://en.wikipedia.org/wiki/Locality_of_reference (or: http://bit.ly/1gyJYr3)

165. Butler Lampson, after David Wheeler; http://en.wikipedia.org/wiki/Indirection (or: http://bit.ly/1drApJV)

166. http://www.infoq.com/articles/9_Fallacies_Java_Performance (item #5) (or: http://bit.ly/1dryB3m)

167. Adapted from http://mathworks.com/matlabcentral/newsreader/view_thread/324859#892815 (or: http://bit.ly/12HAaQg)

168. http://en.wikipedia.org/wiki/Common_subexpression_elimination (or: http://bit.ly/YnFgAG)

169. http://matlab.wikia.com/wiki/FAQ#How_can_I_share_data_between_callback_functions_in_my_GUI.28s.29.3F (or: http://bit.ly/KW7IRu)

170. http://faqs.org/docs/artu/ch12s04.html#binary_caches (or: http://bit.ly/14Warc9) — Eric Raymond, *The Art of Unix Programming*, ISBN 0131429019, §12.4.

171. http://en.wikipedia.org/wiki/Cache_algorithms (or: http://bit.ly/WE8Mjq)

172. http://en.wikipedia.org/wiki/Memoization (or: http://bit.ly/XGLR9x)

173. http://mathworks.com/matlabcentral/newsreader/view_thread/311639#850249 (or: http://bit.ly/1cd9nFp)

174. http://mathworks.com/matlabcentral/newsreader/view_thread/314928#860425 (or: http://bit.ly/1fUIdUj); http://mathworks.com/matlabcentral/newsreader/view_thread/279056#735008 (or: http://bit.ly/1be6HCk); http://mathworks.com/matlabcentral/answers/2126-the-efficiency-of-recursion-in-matlab (or: http://bit.ly/1bdXLNc)

175. http://blogs.mathworks.com/loren/2006/02/08/use-nested-functions-to-memoize-costly-functions (or: http://bit.ly/1fyq75a)

176. http://mathworks.com/matlabcentral/fileexchange/34766-the-fibonacci-sequence (or: http://bit.ly/1chuENs)

177. http://en.wikipedia.org/wiki/Lookup_table (or: http://bit.ly/Z3suZn) provides many examples of using LUTs in computing; http://mathworks.com/matlabcentral/fileexchange/44371-bitcount-bitwise-hamming-distance (or: http://bit.ly/1hOaiLi) provides a MATLAB example of using a lookup table to efficiently compute the bit-count of any number; http://blogs.mathworks.com/pick/2012/11/25/converting-images-from-grayscale-to-color#comment-16102 (or: http://bit.ly/Yksp0q) provides a MATLAB example of precomputing a lookup table of colors to colorize an image with many more pixels than LUT

elements. The same technique is used by indexed images (such as GIFs) to store pixel colors efficiently. Also see: http://en.wikipedia.org/wiki/Precomputation (or: http://bit.ly/IWbT5c).

178. http://mathworks.com/matlabcentral/fileexchange/34565-micro-cache-memoize-cache-function-results (or: http://bit.ly/Z3nTX6)

179. http://mathworks.com/matlabcentral/fileexchange/21352-disk-memory-caching (or: http://bit.ly/Wrdia1)

180. http://mathworks.com/matlabcentral/fileexchange/37465-cacheresults (or: http://bit.ly/WEbbus)

181. http://UndocumentedMatlab.com/blog/datestr-performance (or: http://bit.ly/Z9K8aG)

182. http://en.wikipedia.org/wiki/Hashtable (or: http://bit.ly/102NjEV)

183. http://docs.oracle.com/javase/6/docs/technotes/guides/collections (or: http://bit.ly/13CanOg); http://docs.oracle.com/javase/tutorial/collections (or: http://bit.ly/ZWXGIT)

184. http://UndocumentedMatlab.com/blog/using-java-collections-in-matlab (or: http://bit.ly/ZWYQnE)

185. http://undocumentedmatlab.com/blog/serializing-deserializing-matlab-data (or: http://bit.ly/1ilyapo)

186. http://blogs.mathworks.com/loren/2006/01/18/think-about-your-algorithm (or: http://bit.ly/1jrBWQo)

187. http://mathworks.com/help/matlab/ref/interp1q.html (or: http://bit.ly/1f45KOa)

188. http://mathworks.com/matlabcentral/newsreader/view_thread/258413#672827 (or: http://bit.ly/1f46JOi)

189. http://mathworks.com/matlabcentral/fileexchange/25463-ScaleTime (or: http://bit.ly/1f46XVS)

190. http://mathworks.com/matlabcentral/newsreader/view_thread/330868#909132 (or: http://bit.ly/14nUhJT)

191. http://www.new-npac.org/projects/cdroms/cewes-1999-06-vol1/nhse/hpccsurvey/orgs/sgi/bentley.html#procedure2 (or: http://bit.ly/19M0z2O)

192. http://undocumentedmatlab.com/blog/trapping-warnings-efficiently#comment-214877 (or: http://bit.ly/NGNT93)

193. http://mathworks.com/matlabcentral/fileexchange/22597-programming-patterns (or: http://bit.ly/189UfjA)

194. http://en.wikipedia.org/wiki/Online_transaction_processing (or: http://bit.ly/13DFVPg)

195. http://en.wikipedia.org/wiki/Online_analytical_processing (or: http://bit.ly/13DFYdQ)

196. http://en.wikipedia.org/wiki/Entity-relationship_model (or: http://bit.ly/13DFLYn)

197. http://en.wikipedia.org/wiki/Database_normalization (or: http://bit.ly/13DGtoo)

198. http://en.wikipedia.org/wiki/Third_normal_form (or: http://bit.ly/13DGzw6)

199. http://en.wikipedia.org/wiki/Denormalization (or: http://bit.ly/13DGYyT)

200. http://en.wikipedia.org/wiki/Materialized_view (or: http://bit.ly/13DKtFg)

201. http://en.wikipedia.org/wiki/Star_schema (or: http://bit.ly/148A8EV)

202. http://en.wikipedia.org/wiki/Snowflake_schema (or: http://bit.ly/148zZRN)

203. http://en.wikipedia.org/wiki/Batch_processing (or: http://bit.ly/148AOtR)

204. http://en.wikipedia.org/wiki/Pivot_table (or: http://bit.ly/16ru8Yw)

205. For example, for SQL Server: http://www.mssqltips.com/sqlservertip/1582/implementing-data-compression-in-sql-server-2008 (or: http://bit.ly/WMpv9I) and http://msdn.microsoft.com/en-us/library/dd894051(v=sql.100).aspx (or: http://bit.ly/WMpD9h); for MySQL InnoDB: http://dev.mysql.com/doc/refman/5.5/en/innodb-compression.html (or: http://bit.ly/WMqKFT) and for MySQL MyISAM: http://dev.mysql.com/doc/refman/5.5/en/myisampack.html (or: http://bit.ly/WMrmeP)

206. http://www.techrepublic.com/article/save-disk-space-by-compressing-mysql-tables/5852557 (or: http://tek.io/WMshvT)

207. http://en.wikipedia.org/wiki/Partition_(database) (or: http://bit.ly/XR48BT)

208. For example: http://server.dzone.com/articles/inexpensive-ssds-database (or: http://bit.ly/18b6GMf)

209. http://en.wikipedia.org/wiki/B-tree (or: http://bit.ly/148pWvV)

210. http://en.wikipedia.org/wiki/Database_index (or: http://bit.ly/WMvhZ7)

211. http://sqlmag.com/blog/what-fill-factor-index-fill-factor-and-performance-part-1 (or: http://bit.ly/1815CwO); http://blog.sqlauthority.com/2009/12/16/sql-server-fillfactor-index-and-in-depth-look-at-effect-on-performance (or: http://bit.ly/1815Dkg); http://msdn.microsoft.com/en-us/library/ms177459.aspx (or: http://bit.ly/1815HjR); http://www.sqlskills.com/blogs/kimberly/database-maintenance-best-practices-part-ii-setting-fillfactor (or: http://bit.ly/1815IUU)

212. http://en.wikipedia.org/wiki/Unique_key (or: http://bit.ly/WMyWpF)

213. http://architects.dzone.com/articles/innodb-scalability-issues-due (or: http://bit.ly/1cd51OG)

214. http://en.wikipedia.org/wiki/Surrogate_key (or: http://bit.ly/148oPMZ)

215. http://en.wikipedia.org/wiki/Natural_key (or: http://bit.ly/148raY6). Read the contained references (External Links section) for a discussion of the disadvantages of surrogate keys versus natural keys. Also read http://en.wikipedia.org/wiki/Surrogate_key#Disadvantages (or: http://bit.ly/148rjLl).

216. http://en.wikipedia.org/wiki/Database_index#Covering_index (or: http://bit.ly/13DMjWP)

217. http://en.wikipedia.org/wiki/Virtual_column (or: http://bit.ly/16rrUIJ)

218. Compare the concept of *dependent property* in MATLAB classes: http://blogs.mathworks.com/loren/2011/03/08/common-design-considerations-for-object-properties#4 (or: http://bit.ly/16rsfv3); http://mathworks.com/help/matlab/matlab_oop/example-representing-structured-data.html#f2-85430 (or: http://bit.ly/16rshTI)

219. http://docs.oracle.com/cd/B10501_01/appdev.920/a96590/adg07iot.htm (or: http://bit.ly/1hJAvf1)

220. For example, SQLServer: http://technet.microsoft.com/en-us/library/ms190457.aspx (or: http://bit.ly/1hJAFTt); http://technet.microsoft.com/en-us/library/aa933131(v=sql.80).aspx (or: http://bit.ly/1hJAN5i); http://technet.microsoft.com/en-us/library/aa174523(v=sql.80).aspx (or: http://bit.ly/1hJAXK1). Note that the first reference is for SQLServer 2012, the others for 2000.

221. Technical solution 1-CL94AW: http://mathworks.com/matlabcentral/answers/101206 (or: http://bit.ly/1b0IqOg)

222. *Undocumented Secrets of MATLAB-Java Programming*, CRC Press 2011; http://UndocumentedMatlab.com/matlab-java-book (or: http://bit.ly/HNpg5X)

223. http://docs.oracle.com/javase/6/docs/api/java/sql/Connection.html (or: http://bit.ly/GZj1gO)

224. http://sqlite.org

225. http://sourceforge.net/projects/mksqlite (or: http://bit.ly/1fhCGlv)

226. For example, http://java.dzone.com/articles/ssl-performance-overhead-mysql (or: http://bit.ly/1cd5OPr)

227. http://www.correlsense.com/it-ops/windows-functions-make-for-better-sql (or: http://bit.ly/HmPePi)

228. http://en.wikipedia.org/wiki/Query_plan (or: http://bit.ly/Zm8Qpv)

229. http://mathworks.com/matlabcentral/fileexchange/36367-fetch-big (or: http://bit.ly/1fyvwiG)

230. http://stackoverflow.com/questions/23244179/how-to-speed-up-table-retrieval-with-matlab-and-jdbc (or: http://bit.ly/1rsePr5). In addition to explaining the advice on creating a thin Java connector layer to JDBC, this post by Andrew Janke also contains other important advice on optimizing JDBC query performance.

231. http://en.wikipedia.org/wiki/ACID (or: http://bit.ly/1fywOKp)

232. http://mssqltips.com/sqlservertip/2470/understanding-the-sql-server-nolock-hint (or: http://bit.ly/1pOSKE2)

233. http://en.wikipedia.org/wiki/Query_optimizer (or: http://bit.ly/XR5L2s)

234. http://en.wikipedia.org/wiki/Query_optimization (or: http://bit.ly/Zm92Fr)

235. http://stackoverflow.com/questions/2577174/join-vs-subquery (or: http://bit.ly/17zPXYF),
     http://stackoverflow.com/questions/141278/subqueries-vs-joins (or: http://bit.ly/17zQ32B),
     http://datacharmer.blogspot.com/2008/09/drizzling-mysql.html (or: http://bit.ly/17zR0rA),
     http://social.msdn.microsoft.com/Forums/en-US/transactsql/thread/34f119aa-5dd1-4729-
     b104-2c47f2f46ba4 (or: http://bit.ly/17zVdvy)

236. http://blog.jooq.org/2013/08/21/the-myth-about-slow-sql-join-operations (or: http://bit.ly
     /1cdRt5n)

237. http://www.dbspecialists.com/files/presentations/semijoins.html (or: http://bit.ly
     /1cdRhTS)

238. http://en.wikipedia.org/wiki/Hint_(SQL) (or: http://bit.ly/Zm9lQn)

239. http://en.wikipedia.org/wiki/Stored_procedure (or: http://bit.ly/XR4IQf)

240. http://en.wikipedia.org/wiki/Cursor_(databases) (or: http://bit.ly/16rwUx5)

241. http://en.wikipedia.org/wiki/Database_trigger (or: http://bit.ly/XR4k4d)

242. http://en.wikipedia.org/wiki/Database_transaction (or: http://bit.ly/XR4Wqq)

243. http://en.wikipedia.org/wiki/Integrity_constraints (or: http://bit.ly/13DECzS); http://
     en.wikipedia.org/wiki/Check_constraint (or: http://bit.ly/13DEDnk)

244. http://stackoverflow.com/questions/8153674/do-foreign-key-constraints-influence-query-
     transformations-in-oracle (or: http://bit.ly/HmNHst); http://asktom.oracle.com/pls/apex
     /z?p_url=ASKTOM.download_file?p_file=8322231124282761811&p_cat=MetadataMatters.
     ppt (or: http://bit.ly/16Nie8Z)

245. http://en.wikipedia.org/wiki/Extract,_transform,_load (or: http://bit.ly/Ck1hE)

246. http://en.wikipedia.org/wiki/Database_tuning (or: http://bit.ly/Zm9UK6)

247. Here's another case: http://www.mathworks.com/matlabcentral/answers/108068# answer_
     116767 (or: http://bit.ly/1lkUiFe)

248. Reviews of disk performance are abundant online, and relative results constantly change
     as new models are released. Here is a typical review from 2013 (newer disks are faster and
     cheaper by the time you read this): http://java.dzone.com/articles/testing-intel-samsung-and
     (or: http://bit.ly/1mBkVnn)

249. http://server.dzone.com/articles/inexpensive-ssds-database (or: http://bit.ly/18b6GMf)

250. http://dropbox.com

251. http://java.dzone.com/articles/your-storage-probably-effects (or: http://bit.ly/1mBuBhG)

252. Here is a related article that I found interesting and informative, although it's rather outdated
     by now: http://queue.acm.org/detail.cfm?id=1117404 (or: http://bit.ly/1bMeVa8)

253. http://mathworks.com/matlabcentral/newsreader/view_thread/323933#890434 (or: http://
     bit.ly/1jAwz0G)

254. http://en.wikipedia.org/wiki/MTU_(networking) (or: http://bit.ly/1fspWhA)

255. https://web.archive.org/web/20080510152112/http://www.daemon.be/maarten/ipperf.html
     (or: http://bit.ly/1bMdY1s). I do not consider myself a networking expert by any means, and
     cannot comment on the validity of his advice. Moreover, I am quite certain that there are better
     and more recent resources that can be found on this.

256. http://mathworks.com/matlabcentral/answers/72264 (or: http://bit.ly/1jBsfOH)

257. http://stackoverflow.com/questions/2052019/net-why-cant-i-get-more-than-11gb-of-allocated-
     memory-in-a-x64-process (or: http://bit.ly/1jBomcD)

258. http://mathworks.com/products/matlab/choosing_hardware.html (or: http://bit.ly/1puXBtO)

259. http://en.wikipedia.org/wiki/Hyper-Threading (or: http://bit.ly/1pv0FWR)

260. For example, the following reports a 20% speedup when running 24 local parallel MATLAB
     workers on a 12-core CPU, compared to running 12 workers on the same setup: http://math-
     works.com/matlabcentral/newsreader/view_thread/334485#919894 (or: http://bit.ly
     /1nybyZF).

261. http://software.intel.com/en-us/articles/parallelism-in-the-intel-math-kernel-library (or: http://
     intel.ly/1lhzIDu); http://walkingrandomly.com/?p=3736#comment-86371 (or: http://bit.ly
     /PxbISf).

262. Technical solution 1-18C2A: http://mathworks.com/matlabcentral/answers/92044 (or: http://bit.ly/1puWFWm). This resource is quite outdated, so I advise caution if you decide to follow it.

263. http://howtospeed.com/speed-up-your-computer (or: http://bit.ly/1fJOCNd). This resource is Windows-oriented, but some of the tips are also applicable to Mac/Linux.

264. http://mathworks.com/matlabcentral/newsreader/view_thread/302977#918415 (or: http://bit.ly/1qjel6W)

265. http://mathworks.com/help/matlab/matlab_prog/strategies-for-efficient-use-of-memory.html#brh72ex-27 (or: http://bit.ly/YcoKF5)

266. http://mathworks.com/matlabcentral/newsreader/view_thread/281500 (or: http://bit.ly/10Xrx4c)

267. http://mathworks.com/matlabcentral/fileexchange/36367-fetch-big (or: http://bit.ly/140k5d9)

268. http://mathworks.com/help/matlab/release-notes.html#bs10ply-2 (or: http://bit.ly/11sz5xa); http://blogs.mathworks.com/loren/2011/10/14/new-mat-file-functionality-in-r2011b (or: http://bit.ly/N6W25Z)

269. http://mathworks.com/help/matlab/import_export/import-large-text-files.html (or: http://bit.ly/Z87Gz0)

270. http://blogs.mathworks.com/loren/2013/07/09/using-memmapfile-to-navigate-through-big-data-binary-files (or: http://bit.ly/1kLV7qx)

271. http://mathworks.com/matlabcentral/fileexchange/35227-easy-memory-mapped-arrays-for-large-datasets (or: http://bit.ly/PtjJaL)

272. http://mathworks.com/help/matlab/import_export/supported-file-formats.html (or: http:// bit.ly/Z8kbuG)

273. http://www.bic.mni.mcgill.ca/users/wolforth/Matlab_Help/matlab_memory.html (or: http://bit.ly/Z7MPfb)

274. http://UndocumentedMatlab.com/blog/datenum-performance#comment-43265 (or: http://bit.ly/Z7NahZ)

275. http://en.wikipedia.org/wiki/Locality_of_reference (or: http://bit.ly/1gyJYr3)

276. http://en.wikipedia.org/wiki/CPU_cache (or: http://bit.ly/1evHZ6C)

277. http://en.wikipedia.org/wiki/Page_cache (or: http://bit.ly/1evI36A)

278. http://en.wikipedia.org/wiki/Loop_tiling (or: http://bit.ly/1mfT8LG)

279. http://en.wikipedia.org/wiki/CPU_cache#Cache_miss (or: http://bit.ly/1evIuOf)

280. http://faqs.org/docs/artu/ch12s03.html (or: http://bit.ly/1evHj15) — Eric Raymond, *The Art of Unix Programming*, ISBN 0131429019, §12.3.

281. http://mathworks.com/matlabcentral/answers/89737-imfilter-speed-for-volume (or: http://bit.ly/1dhNPs0)

282. http://UndocumentedMatlab.com/blog/waiting-for-asynchronous-events (or: http://bit.ly/QdHcvj)

283. Latency and throughput are often confused. See http://www.cadence.com/Community/blogs/sd/archive/2010/09/13/understanding-latency-vs-throughput.aspx (or: http://bit.ly/ZGkxWo) for a clear definition of the difference between them; see http://faqs.org/docs/artu/ch12s04.html (or: http://bit.ly/ZGkFVU) for usage examples in software programming (the examples are Unix-oriented but the discussion is general enough that it can be adapted to MATLAB) — Eric Raymond, *The Art of Unix Programming*, ISBN 0131429019, §12.4.

284. http://en.wikipedia.org/wiki/Lazy_initialization (or: http://bit.ly/W1vSze)

285. http://en.wikipedia.org/wiki/Lazy_loading (or: http://bit.ly/W1wp3Y)

286. http://en.wikipedia.org/wiki/Lazy_evaluation (or: http://bit.ly/TQCSAq)

287. http://www.devx.com/tips/Tip/18007 (or: http://bit.ly/ZNds8q)

288. http://www.javaworld.com/javaworld/javatips/jw-javatip67.html (or: http://bit.ly/WqgtxS)

289. http://en.wikipedia.org/wiki/Instruction_prefetch (or: http://bit.ly/10Xi1OJ), http://en.wikipedia.org/wiki/Prefetch_input_queue (or: http://bit.ly/140gk7y)

290. http://en.wikipedia.org/wiki/Speculative_execution (or: http://bit.ly/10Xho7Q)

291. For example, the Windows Prefetcher, which improves the OS and applications start-up time (http://en.wikipedia.org/wiki/Prefetcher or: http://bit.ly/10Xjbd8) or anticipatory paging to improve page swapping performance (http://en.wikipedia.org/wiki/Page_replacement_algorithm#Anticipatory_paging or: http://bit.ly/140fYgZ).

292. http://projecteuler.net

293. http://projecteuler.net/languages

294. http://blogs.mathworks.com/videos/2012/08/01/naive-vs-complex-solutions (or: http://bit.ly/RAcxcz)

295. http://en.wikipedia.org/wiki/Sorting_algorithm (or: http://bit.ly/1o96Z2K). A MathWorks newsletter from 2004 discussed performance aspects of various sorting algorithms used by the sort function: http://mathworks.com/company/newsletters/articles/an-adventure-of-sortsbehind-the-scenes-of-a-matlab-upgrade.html (or: http://bit.ly/1ghiJvv). For a truly spell-bounding visualization, see AlgoRythmics' set of videos that illustrate the major sorting algorithms using various European folk dances: http://youtube.com/user/AlgoRythmics (or: http://bit.ly/1o97iuB) — hilarious and informative at the same time!

296. http://blogs.mathworks.com/videos/2013/07/23/knowing-when-to-optimize-code-in-matlab (or: http://bit.ly/1fvW8ka)

297. For example, http://stackoverflow.com/questions/20461972/fast-searching-for-the-lowest-value-greater-than-x-in-a-sorted-vector (or: http://bit.ly/1bvIYQE). Also see §4.3.1.

298. A very nice utility by John D'Errico computes the minimal bounding shape for data, for various 2D and 3D geometric shapes (rectangles, circles, spheres, etc.): http://mathworks.com/matlabcentral/fileexchange/34767-a-suite-of-minimal-bounding-objects (or: http://bit.ly/1mxnhlF). The example shown in the text uses the builtin *convhull* function instead.

299. http://mathworks.com/matlabcentral/newsreader/view_thread/334591#919217 (or: http://bit.ly/P7OgL9). Also see related http://mathworks.com/matlabcentral/newsreader/view_thread/333670 (or: http://bit.ly/1ppKv09).

300. http://blogs.mathworks.com/loren/2013/12/26/double-integration-in-matlab-understanding-tolerances (or: http://bit.ly/JAcnOS)

301. http://mathworks.com/matlabcentral/newsreader/view_thread/323853#889796 (or: http://bit.ly/1buvaBC)

302. http://mathworks.com/matlabcentral/fileexchange/19345-efficient-k-nearest-neighbor-search-using-jit (or: http://bit.ly/11tp5HB)

303. John D'Errico, private communication.

304. http://en.wikipedia.org/wiki/Fast_inverse_square_root (or: http://bit.ly/1eMs0uz)

305. http://mathworks.com/matlabcentral/newsreader/view_thread/318467 (or: http://bit.ly/T2N74 g). A related vectorized implementation of numeric integration that uses Simpson's parabolic rule, which is much more accurate than *trapz*: http://mathworks.com/matlabcentral/fileexchange/25754-simpsons-rule-for-numerical-integration (or: http://bit.ly/ 1erBdgD).

306. http://www-h.eng.cam.ac.uk/help/tpl/programs/Matlab/mex.html (or: http://bit.ly/1fJmbAK). See discussion in §8.1.7.4.

307. http://en.wikipedia.org/wiki/Strength_reduction (or: http://bit.ly/1kFhqeU); http://en.wikibooks.org/wiki/Optimizing_Code_for_Speed/Order_of_Complexity_Optimizations (or: http://bit.ly/1gHLo1G)

308. http://en.wikipedia.org/wiki/Computational_complexity_theory (or: http://bit.ly/1dtrqDw); http://en.wikipedia.org/wiki/Algorithmic_efficiency (or: http://bit.ly/1e4jIiS)

309. http://walkingrandomly.com/?p=5377 (or: http://bit.ly/1cQThQd); http://mathworks.com/matlabcentral/newsreader/view_thread/287247 (or: http://bit.ly/1mqQkeF)

310. http://mathworks.com/matlabcentral/fileexchange/36534-hpf-a-big-decimal-class (or: http://bit.ly/1bJitdx)

311. http://en.wikipedia.org/wiki/Taylor_series (or: http://bit.ly/1jDfDqa)

312. http://en.wikipedia.org/wiki/Newton's_method_in_optimization (or: http://bit.ly/1gX5NgO)

313. http://blogs.mathworks.com/loren/2012/01/13/best-practices-for-programming-matlab#comment-32907 (or: http://bit.ly/1dthE4s)

314. http://en.wikipedia.org/wiki/List_of_common_coordinate_transformations (or: http://bit.ly/1f9RZin)

315. See for example, http://mathworks.com/matlabcentral/newsreader/view_thread/317390#874787 (or: http://bit.ly/1gsxGAg)

316. http://en.wikipedia.org/wiki/Lagrange_multiplier (or: http://bit.ly/1kAXGa4); http://mathworks.com/matlabcentral/newsreader/view_thread/320310 (or: http://bit.ly/1kAYJqJ)

317. http://en.wikipedia.org/wiki/Separation_of_variables (or: http://bit.ly/1f9T1L7)

318. http://mathworks.com/matlabcentral/newsreader/view_thread/323973 (or: http://bit.ly/1gQQpmy) is a related (although not exact) example

319. http://mathworks.com/matlabcentral/newsreader/view_thread/319660#875446 (or: http://bit.ly/1gsyvcn)

320. http://mathworks.com/company/newsletters/articles/computing-pi.html (or: http://bit.ly/1gh9zzb); http://en.wikipedia.org/wiki/Approximations_of_pi (or: http://bit.ly/1ghaeAC). One of the other approximations is shown in §7.1.2.

321. http://en.wikipedia.org/wiki/Sieve_of_Eratosthenes (or: http://bit.ly/Js1VtF)

322. http://stackoverflow.com/questions/7640010/optimizing-matlab-code (or: http://bit.ly/1bFikjV)

323. http://mathworks.com/matlabcentral/newsreader/view_thread/320196 (or: http://bit.ly/1gQOuhM)

324. http://mathworks.com/matlabcentral/newsreader/view_thread/330582 (or: http://bit.ly/1gQMx5d). Additional suggestions for improving ODE performance can be found here: http://stackoverflow.com/questions/25314561/how-to-speed-up-matlab-integration (or: http://bit.ly/VnZPjT).

325. http://mathworks.com/matlabcentral/fileexchange/authors/22524 (or: http://bit.ly/1mVvmWD)

326. For example, http://mathworks.com/matlabcentral/fileexchange/20652-hungarian-algorithm-for-linear-assignment (or: http://bit.ly/1gRELah)

327. http://mathworks.com/matlabcentral/fileexchange/authors/679 (or: http://bit.ly/1gX22HY)

328. http://mathworks.com/matlabcentral/fileexchange/authors (or: http://bit.ly/1mVx4qS)

329. For example, Sebastian Paris: http://mathworks.com/matlabcentral/fileexchange/authors/13308 (or: http://bit.ly/1dYYq4z)

330. http://en.wikipedia.org/wiki/Recursion_(computer_science) (or: http://bit.ly/103QP1S)

331. http://en.wikipedia.org/wiki/Tail_call (or: http://bit.ly/103QHQ0)

332. http://en.wikipedia.org/wiki/Fibonacci_number (or: http://bit.ly/1e1qhnv)

333. http://mathworld.wolfram.com/BinetsFibonacciNumberFormula.html (or: http://bit.ly/1e1sFut)

334. http://en.wikipedia.org/wiki/Dead_code_elimination (or: http://bit.ly/1mfQBBa)

335. http://en.wikipedia.org/wiki/Fast_path (or: http://bit.ly/1d9Z1qR)

336. http://www.new-npac.org/projects/cdroms/cewes-1999-06-vol1/nhse/hpccsurvey/orgs/sgi/bentley.html#logic3 (or: http://bit.ly/19LZGak)

337. http://mathworks.com/matlabcentral/newsreader/view_thread/330579#908136 (or: http://bit.ly/1aHpA4L)

338. http://en.wikipedia.org/wiki/Short-circuit_evaluation (or: http://bit.ly/YAgbjf); http://mathworks.com/help/matlab/ref/logicaloperatorsshortcircuit.html (or: http://bit.ly/YAi6EE); http://mathworks.com/help/matlab/matlab_prog/operators.html#f0-39129 (or: http://bit.ly/YAi9Af)

339. http://mathworks.com/help/matlab/matlab_prog/techniques-for-improving-performance.html#f8-790490 (or: http://bit.ly/12zhfag)

340. Specifically, using || and && requires more CPU cycles and can also hurt branch prediction. See http://en.wikipedia.org/wiki/Short-circuit_evaluation#Code_efficiency (or: http://bit.ly/YAhmPW)

341. http://www.new-npac.org/projects/cdroms/cewes-1999-06-vol1/nhse/hpccsurvey/orgs /sgi/bentley.html#logic2 (or: http://bit.ly/19LZ7NL)

342. http://en.wikipedia.org/wiki/Huffman_coding (or: http://bit.ly/102MUSV)

343. http://en.wikipedia.org/wiki/Data_compression (or: http://bit.ly/102FAa0)

344. For example: http://apple.stackexchange.com/questions/126081/os-x-10-9-2-has-slowed-down-matlab (or: http://bit.ly/1hdIsoO). Process priority demotion is a feature in all modern Operating Systems, and is not specific to Mac OS.

345. http://en.wikipedia.org/wiki/Self-tuning (or: http://bit.ly/YkoIHY)

346. http://www.fftw.org/fftw-paper-ieee.pdf (or: http://bit.ly/YkpUeo)

347. Jon Bentley: *Writing Efficient Programs, Prentice Hall* 1982, ISBN 013970244X

348. e.g., http://www.physics.ohio-state.edu/~prewett/writings/BookReviews/WritingEfficientPrograms.html (or: http://bit.ly/1cL0aCa); http://www.new-npac.org /projects/cdroms/cewes-1999-06-vol1/nhse/hpccsurvey/orgs/sgi/bentley.html#expression1 (or: http://bit.ly/ 1cL0dhh)

349. http://mathworks.com/matlabcentral/answers/8781-does-matlab-2010b-provide-jit-acceleration-support-for-structures-and-cell-arrays#answer_12113 (or: http://bit.ly /ZB2Smf)

350. http://UndocumentedMatlab.com/blog/matlabs-internal-memory-representation (or: http:// bit.ly/14Qjjhk); http://mathworks.com/help/matlab/matlab_prog/strategies-for-efficient-use-of-memory.html#brh72ex-38 (or: http://bit.ly/StZNxr)

351. http://mathworks.com/help/matlab/matlab_prog/memory-allocation.html#brh72ex-9 (or: http://bit.ly/191c1rl); http://web.archive.org/web/20111229195941/http://mathworks.com /help/techdoc/matlab_prog/brh72ex-2.html (or: http://bit.ly/1dkNDSL)

352. http://mathworks.com/matlabcentral/newsreader/view_thread/323801#889723 (or: http:// bit.ly/16Jxl5N); http://blogs.mathworks.com/pick/2008/04/22/matlab-basics-array-of-structures- vs-structures-of-arrays (or: http://bit.ly/Xq9WiV); http://web.archive.org/web /20111229195941/http://mathworks.com/help/techdoc/matlab_prog/brh72ex-2.html (or: http://bit.ly/1dkNDSL)

353. For example: http://mathworks.com/matlabcentral/newsreader/view_thread/328291#902351 (or: http://bit.ly/1177osk); http://stackoverflow.com/questions/19505916/improve-performance-of-nested-loops-matlab#19507194 (or: http://bit.ly/1gUJTOV); http://stackoverflow.com/questions/21402031/how-to-speed-up-this-triple-loop-in-matlab (or: http://bit. ly/1gUKiAM)

354. http://mathworks.com/help/matlab/matlab_prog/techniques-for-improving-performance. html#btimcn8 (or: http://bit.ly/17AlTvK); http://mathworks.com/matlabcentral /newsreader/view_thread/328291 (or: http://bit.ly/1176XhD)

355. http://www-h.eng.cam.ac.uk/help/tpl/programs/Matlab/faster_scripts.html#Sparse (or: http://bit.ly/14WgFqs)

356. http://mathworks.com/matlabcentral/newsreader/view_thread/323596 (or: http://bit.ly /19akI3F)

357. http://blogs.mathworks.com/loren/2007/03/01/creating-sparse-finite-element-matrices-in-matlab (or: http://bit.ly/19aiEZc). A related example by Nico Schlömer in his *Guidelines for writing clean and fast code in MATLAB* (http://mathworks.com/matlabcentral /fileexchange/22943-guidelines-for-writing-clean-and-fast-code-in-matlab or: http://bit.ly /15qLzIM), p. 26 showed a 120x speedup when creating a tri-diagonal matrix using **spdiags** rather than a full matrix using a loop; solving a set of linear equations was subsequently shown to be 150,000x (!!!) faster with the sparse matrix compared to the full one.

358. http://stackoverflow.com/questions/17030993/matlab-construction-of-very-large-sparse-band-matrix (or: http://bit.ly/19aljlL)

359. http://research.microsoft.com/en-us/um/people/minka/software/matlab.html (or: http:// bit.ly/1iAqbt1). This webpage author (Tom Minka) is also the author of the Lightspeed toolbox (§8.5.6.1).

360. http://stackoverflow.com/a/19618275/233829 (or: http://bit.ly/O1vt2Y)

361. http://mathworks.com/matlabcentral/newsreader/view_thread/324487#891637 (or: http://bit.ly/1o9pILD)

362. http://www.mathworks.com/support/bugreports/306972 (or: http://bit.ly/19ao46x)

363. http://www.mathworks.com/support/bugreports/535814 (or: http://bit.ly/19ao8TP)

364. http://mathworks.com/matlabcentral/fileexchange/44980-packed (or: http://bit.ly/O1FEEz); http://en.wikipedia.org/wiki/Packed_storage_matrix (or: http://bit.ly/O1FXiK)

365. http://mathworks.com/help/matlab/matlab_prog/techniques-for-improving-performance.html#f8-791752 (or: http://bit.ly/12zgFcv); http://mathworks.com/matlabcentral/fileexchange/22597-programming-patterns (or: http://bit.ly/189UfjA)

366. http://mathworks.com/matlabcentral/fileexchange/17476-typecast-and-typecastx-c-mex-functions (or: http://bit.ly/1e0Zrh1); http://mathworks.com/matlabcentral/answers/106903-which-matlab-operations-functions-need-speeding-up#answer_115925 (or: http://bit.ly/1koKHJZ)

367. http://mathworks.com/help/stats/dataset-arrays.html (or: http://bit.ly/1ac4HyL)

368. http://mathworks.com/matlabcentral/newsreader/view_thread/268923 (or: http://bit.ly/1ac5obj); http://blogs.mathworks.com/loren/2012/03/26/considering-performance-in-object-oriented-matlab-code#comment-33023 (or: http://bit.ly/1eEDrHc); http://stackoverflow.com/questions/16243523/matlab-data-structure-for-mixed-type-whats-time-space-efficient (or: http://bit.ly/1cHp0A6); http://mathworks.com/matlabcentral/newsreader/view_thread/329704#906473 (or: http://bit.ly/1ysuU2r)

369. http://mathworks.com/support/2013b/matlab/8.2/demos/tables-and-categorical-arrays-in-release-2013b.html (or: http://bit.ly/P8aOLG); http://blogs.mathworks.com/loren/2013/09/10/introduction-to-the-new-matlab-data-types-in-r2013b (or: http://bit.ly/P8aWux)

370. For example: http://mathworks.com/matlabcentral/newsreader/view_thread/333365 (or: http://bit.ly/P8bjVW)

371. http://mathworks.com/videos/large-data-sets-in-matlab-81614.html (or: http://bit.ly/1kbMdjp). This webinar describes categorical arrays as part of the Statistics toolbox; they were incorporated in core MATLAB in R2013b, a year after the webinar was created. Note that categorical array performance was improved in R2014b for several important functions (*sort, readtable*).

372. http://stackoverflow.com/questions/19178633/matlab-speed-optimisation#comment-28377707 (or: http://bit.ly/1c2Ry8F)

373. http://stackoverflow.com/questions/17809077/performance-of-index-creation (or: http://bit.ly/17FHzDg)

374. http://stackoverflow.com/questions/16527212/slower-to-mix-logical-variables-with-double (or: http://bit.ly/17Fwfae)

375. http://software.intel.com/en-us/intel-ipp (or: http://intel.ly/168Dp4x); http://en.wikipedia.org/wiki/Integrated_Performance_Primitives (or: http://bit.ly/168DfdA)

376. http://mathworks.com/help/matlab/matlab_prog/command-vs-function-syntax.html (or: http://bit.ly/18a0Lqu); http://blogs.mathworks.com/loren/2013/05/22/duality-between-function-and-command-syntax (or: http://bit.ly/18a0ROX)

377. http://blogs.mathworks.com/loren/2008/09/08/finding-patterns-in-arrays (or: http://bit.ly/11yuE2T); http://sites.google.com/site/jimhokanson/onmatlab/2011_10_31_strfind (or: http://bit.ly/Yp6if2)

378. http://mathworks.com/matlabcentral/newsreader/view_thread/32602#82610 (or: http://bit.ly/Yp6uLz)

379. http://mathworks.com/matlabcentral/fileexchange/1518-strpat-a-pedestrian-exactly-matching-pattern-finder-replacer (or: http://bit.ly/YpcrrG)

380. http://mathworks.com/matlabcentral/fileexchange/6070-fpat-a-fuzzy-pattern-detector (or: http://bit.ly/105jIXS)

381. http://mathworks.com/matlabcentral/fileexchange/7212-asort-a-pedestrian-alphanumeric-string-sorter (or: http://bit.ly/105keFv)

382. http://mathworks.com/matlabcentral/fileexchange/6436-rude-a-pedestrian-run-length-decoder-encoder (or: http://bit.ly/105kBjk)

383. http://mathworks.com/matlabcentral/fileexchange/9647-grep-a-pedestrian-very-fast-grep-utility (or: http://bit.ly/105lrwx); http://blogs.mathworks.com/pick/2012/05/25/grep-text-searching-utility (or: http://bit.ly/105lt7I)

384. http://mathworks.com/help/matlab/matlab_prog/regular-expressions.html (or: http://bit.ly/11zQ6Zc); http://blogs.mathworks.com/loren/2006/04/05/regexp-how-tos (or: http://bit.ly/Yotg3Q); http://en.wikipedia.org/wiki/Regular_expression (or: http://bit.ly/16pMWGt)

385. http://mathworks.com/matlabcentral/newsreader/view_thread/323460#888470 (or: http://bit.ly/16WsHRT)

386. http://mathworks.com/matlabcentral/newsreader/view_thread/101189 (or: http://bit.ly/182MPSe)

387. http://mathworks.com/help/matlab/matlab_prog/dynamic-regular-expressions.html#bqvj29s (or: http://bit.ly/123ZGzb)

388. http://www.mit.edu/~pwb/matlab/splitstr.c (or: http://bit.ly/O1Idql)

389. http://docs.oracle.com/javase/tutorial/essential/regex (or: http://bit.ly/16pM8kR)

390. http://mathworks.com/matlabcentral/fileexchange/24380-cstrainbp (or: http://bit.ly/1lAtL78)

391. http://mathworks.com/matlabcentral/answers/47214-matlabs-jit-engine (or: http://bit.ly/1dhVg2k)

392. http://undocumentedmatlab.com/blog/sprintfc-undocumented-helper-function (or: http://bit.ly/1eAFTNQ). Also see §4.3.

393. http://mathworks.com/matlabcentral/fileexchange/27032-strncmpr (or: http://bit.ly/1qffr4A)

394. http://mathworks.com/matlabcentral/fileexchange/24341-cstrcatstr (or: http://bit.ly/1mxjVjT)

395. http://mathworks.com/matlabcentral/fileexchange/26077-cstr2string (or: http://bit.ly/1lAv0mO); http://mathworks.com/matlabcentral/newsreader/view_thread/329128#904997 (or: http://bit.ly/1yspbJZ)

396. http://docs.oracle.com/javase/tutorial/i18n/text (or: http://bit.ly/XTAgCb); http://docs.oracle.com/javase/1.5.0/docs/api/java/text/Collator.html (or: http://bit.ly/179GPqY)

397. Adapted from http://UndocumentedMatlab.com/blog/ismembc-undocumented-helper-function (or: http://bit.ly/XsYaEv)

398. http://www.matlabtips.com/how-to-load-tiff-stacks-fast-really-fast (or: http://bit.ly/Zk8jT4)

399. http://mathworks.com/matlabcentral/newsreader/view_thread/83149 (or: http://bit.ly/1kEE7m9); http://mathworks.com/matlabcentral/newsreader/search_results?dur=all&search_string=sortrowsc (or: http://bit.ly/1kEEih2)

400. http://mathworks.com/help/finance/working-with-financial-time-series-objects.html (or: http://bit.ly/13y79e0)

401. http://mathworks.com/matlabcentral/answers/67454-financial-time-series-object-performance (or: http://bit.ly/13y6WYi)

402. http://mathworks.com/matlabcentral/fileexchange/25594-dateconvert (or: http://bit.ly/13m6QEm)

403. http://mathworks.com/matlabcentral/fileexchange/25594-dateconvert/content/DateConvert.m (or: http://bit.ly/13m7vWx)

404. http://UndocumentedMatlab.com/blog/datenum-performance (or: http://bit.ly/YKejEN)

405. http://stackoverflow.com/questions/5818583/faster-function-than-datenum-in-matlab (or: http://bit.ly/168nxkk); http://stackoverflow.com/questions/5880242/matlab-datenum-generation#comment-6768527 (or: http://bit.ly/168pCNa)

406. Examples: http://stackoverflow.com/questions/5831563/matlab-date-format (or: http://bit.ly/YKfYdt); http://stackoverflow.com/questions/5880242/matlab-datenum-generation (or: http://bit.ly/YKg4lv)

407. http://mathworks.com/matlabcentral/fileexchange/28093-datestr2num (or: http://bit.ly/13m3PUN)

408. http://UndocumentedMatlab.com/blog/datenum-performance#comment-42624 (or: http://bit.ly/X3pkH0)

409. http://mathworks.com/matlabcentral/fileexchange/28093-datestr2num?download=true (or: http://bit.ly/11x61Yn)

410. http://n-simon.de/mex

411. http://undocumentedmatlab.com/blog/datenum-performance#comment-43249 (or: http://bit.ly/Xy3sP2)

412. http://mathworks.com/matlabcentral/answers/7997-how-to-show-miliseconds-in-a-date-string (or: http://bit.ly/1i9FGXU)

413. http://mathworks.com/matlabcentral/answers/112772-help-optimize-slow-datestr-function-performance (or: http://bit.ly/1e1xq54)

414. A short tutorial on using Java date formats is provided in http://docs.oracle.com/javase/tutorial/i18n/format/dateintro.html (or: http://bit.ly/Yxnz35)

415. http://mathworks.com/matlabcentral/fileexchange/25594-dateconvert (or: http://bit.ly/1n69jwr)

416. http://mathworks.com/help/matlab/matlab_prog/infinity-and-nan.html (or: http://bit.ly/14S2aWd)

417. http://blogs.mathworks.com/loren/2011/09/23/another-possible-surprise-ignored-nan-values (or: http://bit.ly/14SbZU4)

418. http://mathworks.com/matlabcentral/newsreader/view_thread/45565 (or: http://bit.ly/14S9LEb); http://mathworks.com/matlabcentral/newsreader/view_thread/328199#902119 (or: http://bit.ly/1ftIHw7); http://blogs.mathworks.com/loren/2011/09/23/another-possible-surprise-ignored-nan-values#comment-32513 (or: http://bit.ly/14Sb71A); http://blogs.mathworks.com/loren/2011/09/23/another-possible-surprise-ignored-nan-values#comment-32799 (or: http://bit.ly/1a05qPP); http://blogs.mathworks.com/loren/2006/01/18/think-about-your-algorithm#comment-23 (or: http://bit.ly/1jryFjW)

419. http://stackoverflow.com/questions/19036809/matlab-relational-operator-performance-in-presence-of-nan-values (or: http://bit.ly/1gUFdbL); http://randomascii.wordpress.com/2012/05/20/thats-not-normalthe-performance-of-odd-floats (or: http://bit.ly/1gUG7oH)

420. http://mathworks.com/matlabcentral/answers/111478-are-infs-still-faster-than-nans (or: http://bit.ly/1rorI5k); http://mathworks.com/matlabcentral/newsreader/view_thread/256238#665904 (or: http://bit.ly/1rorQ4E)

421. http://blogs.mathworks.com/loren/2011/09/23/another-possible-surprise-ignored-nan-values#comment-32511 (or: http://bit.ly/14Sd0LU)

422. http://mathworks.com/matlabcentral/newsreader/view_thread/324344#891213 (or: http://bit.ly/15mAWDv)

423. http://walkingrandomly.com/?p=4912 (or: http://bit.ly/15mBQQw)

424. http://en.wikipedia.org/wiki/General_Matrix_Multiply (or: http://bit.ly/15mEjKM); http://mathworks.com/matlabcentral/newsreader/view_thread/45565#115725 (or: http://bit.ly/15mEB4a)

425. http://www.math.utah.edu/software/lapack/lapack-blas/dsyrk.html (or: http://bit.ly/15mJgD6)

426. http://walkingrandomly.com/?p=4912#comment-553041 (or: http://bit.ly/15mJpGS)

427. http://math.stackexchange.com/questions/28395/is-it-faster-to-multiply-a-matrix-by-its-transpose-than-ordinary-matrix-multipli (or: http://bit.ly/15mFN7E). Also read the related interesting discussion by James Tursa here: http://mathworks.com/matlabcentral/answers/6411-matrix-multiplication-optimal-speed-and-memory (or: http://bit.ly/1iOZ3Eh).

428. http://walkingrandomly.com/?p=4912#comment-552366 (or: http://bit.ly/15mKA9b)

429. http://en.wikipedia.org/wiki/Matrix_multiplication#Algorithms_for_efficient_matrix_multiplication (or: http://bit.ly/1dIFYmX)

430. http://www.sdsc.edu/~allans/cs260/lectures/matmul.ppt (or: http://bit.ly/15QMSRb); archived copy from May 2010: http://bit.ly/15tl4Fm

431. http://mm-matrixmultiplicationtool.googlecode.com/files/mm.pdf (or: http://bit.ly/17zTwtL); archived version: http://bit.ly/173Lbzv

432. http://mathworks.com/help/matlab/matlab_external/calling-lapack-and-blas-functions-from-mex-files.html (or: http://bit.ly/19y1m5Y)

433. http://mathworks.com/matlabcentral/fileexchange/16777-lapack (or: http://bit.ly/1fXGzlN)

434. http://mathworks.com/matlabcentral/fileexchange/index?term=tag:blas+OR+tag:lapack (or: http://bit.ly/1fXHcfd)

435. http://mathworks.com/matlabcentral/fileexchange/25977-mtimesx (or: http://bit.ly/1dB7Icd)

436. http://stackoverflow.com/questions/19322595/how-to-organize-data-to-maximize-speed-and-functionality-in-matlab (or: http://bit.ly/17y1kdD) claims a speedup of 1000×s (!) in her specific case

437. In general, I found the source code readable and informative, despite Tursa's comment to the contrary: http://mathworks.com/matlabcentral/newsreader/view_thread/322295#888894 (or: http://bit.ly/Lbkl1D)

438. http://mathworks.com/matlabcentral/fileexchange/37515-mmx-multithreaded-matrix-operations-on-n-d-matrices (or: http://bit.ly/1j7WGOO); http://www.cs.washington.edu/people/postdocs/tassa/code (or: http://bit.ly/1n7O8od)

439. http://mathworks.com/matlabcentral/fileexchange/37515-mmx/content/mmx_package/html/mmx_web.html (or: http://bit.ly/LIUQ8T)

440. http://mathworks.com/matlabcentral/newsreader/view_thread/324487#891637 (or: http://bit.ly/1o9pILD)

441. http://www.mit.edu/~pwb/matlab (or: http://bit.ly/1f3V9bt)

442. http://undocumentedmatlab.com/blog/matlabs-internal-memory-representation (or: http://bit.ly/14Qjjhk)

443. http://documents.epfl.ch/users/l/le/leuteneg/www/MATLABToolbox/VectorFunctions.html (or: http://bit.ly/1nWIDcs)

444. http://documents.epfl.ch/users/l/le/leuteneg/www/MATLABToolbox/FPUasmHints.html (or: http://bit.ly/1nWDW2t). The toolbox uses the open-source Netwide Assembler (http://nasm.us) for the assembler functionality.

445. http://mathworks.com/matlabcentral/fileexchange/8773-multiple-matrix-multiplications (or: http://bit.ly/18oSNvA)

446. http://www.sandia.gov/~tgkolda/TensorToolbox (or: http://1.usa.gov/15p3tIz)

447. B. W. Bader and T. G. Kolda, *Efficient MATLAB computations with sparse and factored tensors*, SIAM Journal on Scientific Computing 30(1):205–231, December 2007, http://dx.doi.org/10.1137/060676489 (or: http://bit.ly/18Q7uH8).

448. Tammy Kolda, private communication.

449. http://www.sandia.gov/~tgkolda/pubs/pubfiles/ACM-TOMS-TensorToolbox.pdf (or: http://1.usa.gov/15vBJYJ); archived: http://bit.ly/15vBPzF

450. http://www.sandia.gov/~tgkolda (or: http://1.usa.gov/15vBqgA)

451. http://mathworks.com/matlabcentral/fileexchange/24576-inplacearray (or: http://bit.ly/1dB7T7w)

452. http://mathworks.com/matlabcentral/newsreader/view_thread/324344#891282 (or: http://bit.ly/1aXpsgM)

453. Nico Schlömer's *Guidelines for writing clean and fast code in MATLAB* (http://mathworks.com/matlabcentral/fileexchange/22943-guidelines-for-writing-clean-and-fast-code-in-matlab or: http://bit.ly/15qLzIM), p. 27-29. Be careful with LU-decomposition techniques, since incorrect usage could easily lead to numerical instabilities and inaccuracies.

454. http://mathworks.com/help/matlab/ref/inv.html (or: http://bit.ly/1aOKHh0)

455. http://mathworks.com/help/matlab/ref/inv.html (or: http://bit.ly/1aOKHh0). Also see Nico Schlömer's *Guidelines for writing clean and fast code in MATLAB* (http://mathworks.com/matlabcentral/fileexchange/22943-guidelines-for-writing-clean-and-fast-code-in-matlab or: http://bit.ly/15qLzIM), p. 24–25.

456. Additional alternatives are analyzed in the following newsgroup thread, where it is shown that LU-decomposition may be the fastest alternative of all: http://mathworks.com /matlabcentral/newsreader/view_thread/334687 (or: http://bit.ly/N72yto).

457. http://undocumentedmatlab.com/blog/matlabs-internal-memory-representation (or: http:// bit.ly/14Qjjhk)

458. http://stackoverflow.com/questions/19178633/matlab-speed-optimisation#19179030 (or: http://bit.ly/1c3xJhq). A discussion of this idea can be found in Ned Gulley's 2003 newsletter article *"Picking up the Pace with the MATLAB Profiler"*, which has been removed from the MathWorks website in 2010 as some of its suggestions are no longer relevant; archived versions of this article can be found here: http://web.archive.org/web/20091025140723/http:// mathworks.com/company/newsletters/news_notes/may03/profiler.html (or: http://bit.ly /1c3ynLQ) and here: http://www.ee.columbia.edu/~marios/matlab/Picking%20up%20 the%20Pace%20with%20the%20MATLAB%20Profiler.pdf (or: http://bit.ly/19oh1Zy).

459. http://mathworks.com/matlabcentral/newsreader/view_thread/325336#894141 (or: http:// bit.ly/1c3xVgJ)

460. http://mathworks.com/matlabcentral/newsreader/view_thread/96268#907202 (or: http:// bit.ly/1bI1EPv)

461. http://stackoverflow.com/questions/18958231/fast-computation-of-a-gradient-of-an-image-in-matlab (or: http://bit.ly/15qUx95); http://regularize.wordpress.com/2013/06/19/how-fast-can-you-calculate-the-gradient-of-an-image-in-matlab (or: http://bit.ly/15qT9Du)

462. http://mathworks.com/matlabcentral/fileexchange/29887-dgradient (or: http://bit.ly /1bI1LKR)

463. http://mathworks.com/help/optim/ug/fmincon.html (or: http://bit.ly/1m20IqD)

464. http://en.wikipedia.org/wiki/Newton's_method_in_optimization (or: http://bit.ly/1gX5NgO)

465. http://mathworks.com/company/newsletters/articles/solving-large-scale-optimization-problems-with-matlab-a-hydroelectric-flow-example.html (or: http://bit.ly/1eVbNq5)

466. http://mathworks.com/matlabcentral/fileexchange/33597-improving-matlab-performance-when-solving-financial-optimization-problems (or: http://bit.ly/1gX7J8U); http://wilmott.com /magazine1105.cfm (or: http://bit.ly/1gX8mzq)

467. http://arc.vt.edu/resources/software/matlab/docs/fdi_2013_opt_tool.pdf (or: http://bit.ly /1e1fGGV)

468. http://mathworks.com/matlabcentral/newsreader/view_thread/271215 (or: http://bit.ly /1gXbvze)

469. http://mathworks.com/matlabcentral/newsreader/view_thread/269936 (or: http://bit.ly /1eVK1Kg)

470. A typical example: http://mathworks.com/matlabcentral/newsreader/view_thread/318718 (or: http://bit.ly/1gXaoQ1)

471. http://mathworks.com/help/optim/ug/writing-constraints.html (or: http://bit.ly/VnY9GX)

472. http://mathworks.com/matlabcentral/newsreader/view_thread/158607#399348 (or: http:// bit.ly/1eq3gx7)

473. See related: http://mathworks.com/help/optim/examples/using-quadratic-programming-on-portfolio-optimization-problems.html (or: http://bit.ly/1eVKMmD)

474. http://mathworks.com/help/optim/ug/choosing-a-solver.html (or: http://bit.ly/189Pnlr)

475. http://www-03.ibm.com/software/products/en/ibmilogcpleoptistud (or: http://ibm. co/189HZGI) — commercial; the MATLAB interface is at http://control.ee.ethz.ch/~hybrid/ cplexint.msql (or: http://bit.ly/189I7G8)

476. http://www.omegacomputer.com/staff/tadonki/using_cplex_with_matlab.htm (or: http:// bit.ly/1iYM30s)

477. http://control.ee.ethz.ch/~mpt/3 (or: http://bit.ly/1eVQtRm); open-source, GNU GPL license. Read: http://control.ee.ethz.ch/index.cgi?page=publications&action=details&id=4438 (or: http:// bit.ly/1aCGmvJ).

478. http://sigpromu.org/quadprog (or: http://bit.ly/1eVMa8H); free for academic/non-commercial use.

479. http://mathworks.com/matlabcentral/fileexchange/43097-newton-raphson-solver (or: http://bit.ly/1g7OiNI)

480. http://projects.coin-or.org/Ipopt (or: http://bit.ly/1g7MvrT). A detailed example of using IPOPT in MATLAB was provided by Mark Mikofski: http://poquitopicante.blogspot.com/2013/09/ipopt-as-non-linear-solver-for-matlab.html (or: http://bit.ly/1g7N2dp)

481. http://users.isy.liu.se/johanl/yalmip/pmwiki.php?n=Main.WhatIsYALMIP (or: http://bit.ly/1g7Pb93); open-source, free for non-commercial use

482. http://www.sztaki.hu/~meszaros/bpmpd (or: http://bit.ly/189IEbf); free for non-commercial use. The MATLAB interface is available at http://www.pserc.cornell.edu/bpmpd (or: http://bit.ly/189IL6D).

483. http://tbxmanager.com

484. http://mathworks.com/matlabcentral/newsreader/view_thread/328201 (or: http://bit.ly/1gQLp1s). This post discusses neural-network convergence, but in essence this is also an optimization problem and many of the insights are applicable to general optimization tasks.

485. http://mathworks.com/matlabcentral/newsreader/view_thread/119565#920415 (or: http://bit.ly/Qspbep) discusses this idea in the context of an ODE solver, but it can be generalized to solvers in general.

486. http://mathworks.com/help/optim/ug/optimization-options-reference.html (or: http://bit.ly/QspLJj)

487. An overview of this series: http://blogs.mathworks.com/pick/2010/10/29/optimize-quickly (or: http://bit.ly/1gWV45Q)

488. http://mathworks.com/videos/tips-and-tricks-getting-started-using-optimization-with-matlab-81594.html (or: http://bit.ly/1gWVut9). Files used in the webinar: http://mathworks.com/matlabcentral/fileexchange/21239-tips-tricks-getting-started-using-optimization-with-matlab (or: http://bit.ly/1gWWFsr)

489. http://mathworks.com/videos/global-optimization-with-matlab-products-81716.html (or: http://bit.ly/1gWVRDW). Files used in the webinar: http://mathworks.com/matlabcentral/fileexchange/27178-global-optimization-with-matlab (or: http://bit.ly/1gWWVaP)

490. http://mathworks.com/videos/speeding-up-optimization-problems-using-parallel-computing-81753.html (or: http://bit.ly/1gWWa1q). Files used in the webinar: http://mathworks.com/matlabcentral/fileexchange/28487-speeding-up-optimization-problems-with-parallel-computing (or: http://bit.ly/1gWXa5S)

491. http://mathworks.com/matlabcentral/fileexchange/8553-optimization-tips-and-tricks (or: http://bit.ly/1gWYaXz); http://blogs.mathworks.com/pick/2007/06/15/optimization-tips-and-tricks (or: http://bit.ly/1gWYfuk).

492. http://mathworks.com/matlabcentral/fileexchange/10093-fminspleas (or: http://bit.ly/1gX1jH0)

493. http://mathworks.com/matlabcentral/fileexchange/8277-fminsearchbnd-fminsearchcon (or: http://bit.ly/1gX1x0C)

494. http://mathworks.com/matlabcentral/fileexchange/authors/679 (or: http://bit.ly/1gX22HY)

495. As a typical example, consider John's comment on Apr 27th, 2010 on the *fminspleas* File Exchange page (http://mathworks.com/matlabcentral/fileexchange/10093-fminspleas or: http://bit.ly/1gX1jH0), explaining the deficiencies of simulated annealing for optimization problems. Also see http://mathworks.com/matlabcentral/newsreader/author/45419 (or: http://bit.ly/N68VgY); http://mathworks.com/matlabcentral/answers/contributors/869215-john-d-errico (or: http://bit.ly/N69c3r)

496. For example, http://mathworks.com/matlabcentral/fileexchange/23245-fminlbfgs-fast-limited-memory-optimizer (or: http://bit.ly/N68wuM)

497. http://mathtools.net/MATLAB/Optimization (or: http://bit.ly/1jLSMeM)

498. http://en.wikipedia.org/wiki/Fast_Fourier_transform (or: http://bit.ly/1cYexBH)

499. http://fftw.org; http://en.wikipedia.org/wiki/FFTW (or: http://bit.ly/1cYfuKk)

500. *Cleve's Corner*, MathWorks newsletter, winter 2001. This was removed online; here is an archived version: http://web.archive.org/web/20120214101107/http://mathworks.com/company /newsletters/news_notes/clevescorner/winter01_cleve.html (or: http://bit.ly/1cYhrX9)

501. http://UndocumentedMatlab.com/blog/math-libraries-version-info-upgrade (or: http://bit. ly/13lAsja)

502. http://mathworks.com/help/matlab/ref/ifftn.html#f86-998401 (or: http://bit.ly/12EKWWz). Also see the discussion in http://blogs.mathworks.com/steve/2014/04/07/timing-the-fft (or: http://bit.ly/1iZTyVf).

503. http://mathworks.com/matlabcentral/newsreader/view_thread/304045 (or: http://bit.ly /1jAF76U)

504. http://mathworks.com/help/matlab/ref/fftw.html (or: http://bit.ly/13LkIsD); http:// www.fftw.org/fftw-paper-ieee.pdf (or: http://bit.ly/YkpUeo)

505. http://blogs.mathworks.com/steve/2013/06/25/homomorphic-filtering-part-1 (or: http:// bit.ly/189EE3A)

506. http://mathworks.com/matlabcentral/fileexchange/24504-fft-based-convolution (or: http:// bit.ly/12EIxLq)

507. http://en.wikipedia.org/wiki/Convolution_theorem (or: http://bit.ly/1bdamUy)

508. http://mathworks.com/matlabcentral/answers/38066 (or: http://bit.ly/1eMpkNE)

509. http://mathworks.com/matlabcentral/newsreader/view_thread/324520 (or: http://bit.ly /M8XWCA)

510. http://web.mit.edu/newsoffice/2012/faster-fourier-transforms-0118.html (or: http://bit.ly /1jLW81l); http://spectrum.ieee.org/computing/software/a-faster-fast-fourier-transform (or: http://bit.ly/1jLWbdr)

511. http://groups.csail.mit.edu/netmit/sFFT (or: http://bit.ly/1gXpHIw); http://groups.csail. mit.edu/netmit/sFFT/Documentation.pdf (or: http://bit.ly/1e3jVSk)

512. http://spiral.net/software/sfft.html (or: http://bit.ly/1bhg7So); http://spiral.ece.cmu. edu:8080/pub-spiral/pubfile/Final_Thesis_SFFT_169.pdf (or: http://bit.ly/1bhgLPZ) — Jörn Schumacher's Master's thesis at ETH, Zurich.

513. http://spiral.net; http://spiral.ece.cmu.edu:8080/pub-spiral/pubfile/paper_146.pdf (or: http://bit.ly/1bhh9hi). See §8.5.6.2

514. http://groups.csail.mit.edu/netmit/sFFT/results.html (or: http://bit.ly/1e3hZZV); http:// groups.csail.mit.edu/netmit/sFFT/paper.html (or: http://bit.ly/1e3ihjD)

515. http://stackoverflow.com/a/22407882/233829 (or: http://bit.ly/1ggT8CV)

516. http://mathworks.com/matlabcentral/newsreader/view_thread/325336 (or: http://bit.ly/ 1jNGRNA)

517. http://icl.cs.utk.edu/magma (or: http://bit.ly/1dTmN4O); http://stackoverflow.com /questions/6058139/why-is-matlab-so-fast-in-matrix-multiplication/6058811#6058811 (or: http://bit.ly/1dTmMh9)

518. http://software.intel.com/en-us/intel-ipp (or: http://intel.ly/168Dp4x); http:// en.wikipedia.org/wiki/Integrated_Performance_Primitives (or: http://bit.ly/168DfdA)

519. http://software.intel.com/en-us/intel-mkl (or: http://intel.ly/13lAmYI); http:// en.wikipedia.org/wiki/Math_Kernel_Library (or: http://bit.ly/1atUMAU)

520. http://undocumentedmatlab.com/blog/math-libraries-version-info-upgrade (or: http://bit. ly/13lAsja)

521. http://software.intel.com/en-us/articles/intel-ipp-71-library-release-notes (or: http://intel. ly/19y2fvl)

522. http://mathworks.com/matlabcentral/newsreader/view_thread/129077 (or: http://bit.ly /19y4rmJ)

523. http://software.intel.com/en-us/articles/intel-math-kernel-library-intel-mkl-for-windows- using-intel-mkl-in-matlab-executable-mex-files (or: http://intel.ly/15TT8Kx); https:// software.intel.com/en-us/articles/using-intel-math-kernel-library-with-mathworks-matlab- on-intel-xeon-phi-coprocessor-system (or: http://intel.ly/1yEEaAx)

524. http://martin-thoma.com/matrix-multiplication-python-java-cpp#comment-1203201 (or: http://bit.ly/19zy9Md)

525. http://en.wikipedia.org/wiki/Automatically_Tuned_Linear_Algebra_Software (or: http://bit.ly/19zACGm); http://sourceforge.net/projects/math-atlas (or: http://bit.ly/19zB4EA). A 2000 newsletter article by Cleve Moler (http://mathworks.com/company/newsletters/articles/matlab-incorporates-lapack.html or: http://bit.ly/14fltcr) hints that MATLAB once used ATLAS as its BLAS implementation on some platforms.

526. http://cs.utexas.edu/~flame (or: http://bit.ly/1rbhqKw)

527. http://www.stanford.edu/~echu508/matlab.html (or: http://stanford.io/15TSzk2); http://mathworks.com/matlabcentral/newsreader/view_thread/129077 (or: http://bit.ly/19y4rmJ); http://en.wikipedia.org/wiki/GotoBLAS (or: http://bit.ly/19zCcbh); http://tacc.utexas.edu/tacc-projects/gotoblas2 (or: http://bit.ly/1rbh0Uv); http://open-blas.net

528. http://mathworks.com/matlabcentral/newsreader/view_thread/129077 (or: http://bit.ly/19y4rmJ); http://www.stanford.edu/~echu508/matlab.html (or: http://stanford.io/15TSzk2); http://software.intel.com/en-us/articles/using-intel-mkl-with-matlab (or: http://intel.ly/15TSAEG)

529. http://mathworks.com/help/matlab/release-notes-older.html#bq6a0cg-1 (or: http://bit.ly/1e7Vq8X)

530. http://mathworks.com/help/matlab/release-notes-older.html#bq6a0cg-1 (or: http://bit.ly/1e7Vq8X); http://mathworks.com/help/releases/R2007a/techdoc/matlab_env/f1-94809.html#bq35sqs (or: http://bit.ly/MglrKC)

531. http://software.intel.com/en-us/articles/intel-math-kernel-library-intel-mkl-intel-mkl-version-compatibility (or: http://intel.ly/15TV0D1)

532. http://mathworks.com/help/matlab/release-notes-older.html#bq17fgj-1 (or: http://bit.ly/1fZ4y4f); http://en.wikipedia.org/wiki/Mersenne_twister (or: http://bit.ly/1fZ6sBJ); http://www.math.sci.hiroshima-u.ac.jp/~m-mat/MT/emt.html (or: http://bit.ly/1fZ6vh9). Also see this related posting about a performance degradation of some RNG functions in R2013a compared to earlier releases: http://mathworks.com/matlabcentral/answers/67667-performance-degradation-of-random-number-generation-in-matlab-2013a (or: http://bit.ly/1iP0gLV).

533. http://mathworks.com/matlabcentral/newsreader/view_thread/328139#901809 (or: http://bit.ly/1fZeomK). See related: http://mathworks.com/matlabcentral/newsreader/view_thread/327817#902446 (or: http://bit.ly/1fZeESx)

534. http://mathworks.com/help/matlab/math/creating-and-controlling-a-random-number-stream.html#brvku_2 (or: http://bit.ly/1fZfLla); http://blogs.mathworks.com/loren/2011/07/07/simpler-control-of-random-number-generation-in-matlab#comment-32949 (or: http://bit.ly/GNBY5u); http://walkingrandomly.com/?p=4870 (or: http://bit.ly/17W1zDC). See related: http://mathworks.com/help/distcomp/establish-arrays-on-a-gpu.html#bs_td05 (or: http://bit.ly/17W2W5n) and http://walkingrandomly.com/?p=2755 (or: http://bit.ly/17W1Lmj)

535. http://www.nag.co.uk/IndustryArticles/usingtoolboxmatlabpart3.asp#g05randomnumber (or: http://bit.ly/GNCD6V)

536. http://mathworks.com/matlabcentral/fileexchange/33004-fasterparallel-random-number-generator (or: http://bit.ly/1bZ9ocv)

537. http://www-h.eng.cam.ac.uk/help/tpl/programs/matlab.html#Optimisation (or: http://bit.ly/1hIlNVG)

538. http://blogs.mathworks.com/loren/2012/01/13/best-practices-for-programming-matlab#comment-32911 (or: http://bit.ly/184Z66R)

539. http://mathworks.com/help/matlab/matlab_prog/techniques-for-improving-performance.html#btimcn8 (or: http://bit.ly/17AlTvK)

540. http://www.mathworks.com/support/bugreports/468781 (or: http://bit.ly/1851Hhh)

541. http://stackoverflow.com/questions/17645932/reduce-assignment-time-in-matlab (or: http://bit.ly/1e17b1b)

542. http://mathworks.com/company/newsletters/articles/programming-patterns-some-common-matlab-programming-pitfalls-and-how-to-avoid-them.html (or: http://bit.ly/1e1igzh)

543. http://undocumentedmatlab.com/blog/ishghandle-undocumented-input-parameter (or: http://bit.ly/19jJWdu)

544. http://mathworks.com/help/matlab/matlab_prog/support-variable-number-of-outputs.html (or: http://bit.ly/1burrIm)

545. http://mathworks.com/help/matlab/release-notes-older.html#br65zmd-1 (or: http://bit.ly/1b utTym)

546. http://blogs.mathworks.com/loren/2008/03/10/keeping-things-tidy (or: http://bit.ly/1iITHvK)

547. http://blogs.mathworks.com/pick/2013/01/11/functional-programming-constructs (or: http://bit.ly/1a20FWT)

548. http://blogs.mathworks.com/pick/2013/01/11/functional-programming-constructs #comment-18249 (or: http://bit.ly/1d7GeGu)

549. http://blogs.mathworks.com/pick/2013/01/11/functional-programming-constructs #comment-18425 (or: http://bit.ly/1a23ebF)

550. http://UndocumentedMatlab.com/blog/function-definition-meta-info (or: http://bit.ly/19l4gQp)

551. 1-6MBFQL, recently reposted as http://mathworks.com/matlabcentral/answers/103193 (or: http://bit.ly/19l3lzA)

552. http://mathworks.com/help/matlab/matlab_prog/techniques-for-improving-performance.html#btimcn8 (or: http://bit.ly/17AlTvK)

553. http://mathworks.com/matlabcentral/answers/103056 (or: http://bit.ly/1b0GchV)

554. http://mathworks.com/matlabcentral/newsreader/view_thread/328551#903460 (or: http://bit.ly/15WTi2C)

555. http://undocumentedmatlab.com/blog/datestr-performance#comment-60515 (or: http://bit.ly/15vdmJz)

556. http://mathworks.com/help/matlab/matlab_prog/function-precedence-order.html (or: http://bit.ly/1e1p53P)

557. http://mathworks.com/matlabcentral/answers/42675 (or: http://bit.ly/16gOKjE); http://mathworks.com/matlabcentral/newsreader/view_thread/327677 (or: http://bit.ly/H5zBuQ)

558. http://mathworks.com/help/matlab/matlab_prog/techniques-for-improving-performance.html#btimcn8 (or: http://bit.ly/17AlTvK)

559. http://mathworks.com/matlabcentral/answers/77843-what-functions-can-benefit-from-simple-patching#answer_87648 (or: http://bit.ly/1jUZ78C)

560. http://mathworks.com/help/releases/R2012a/techdoc/matlab_oop/brdqi79.html (or: http://bit.ly/1gUIyrg). The relevant section on overloading MATLAB's built-in functions is missing from the current online documentation webpage: http://mathworks.com/help/matlab/matlab_oop/overloading-functions-for-your-class.html (or: http://bit.ly/18fKQIR)

561. http://mathworks.com/help/matlab/matlab_prog/techniques-for-improving-performance.html#btimcn8 (or: http://bit.ly/17AlTvK); http://mathworks.com/help/releases/R2012a/techdoc/matlab_prog/f8-784135.html#f8-790494 (or: http://bit.ly/1gUIckq) provides a bit more information. It is unclear why MathWorks chose to remove this extra information from the latest webpage.

562. http://mathworks.com/help/matlab/object-oriented-programming.html (or: http://bit.ly/1dJeNbU); http://en.wikipedia.org/wiki/Object-oriented_programming (or: http://bit.ly/1dJeNJc)

563. http://blogs.mathworks.com/loren/2012/03/26/considering-performance-in-object-oriented-matlab-code (or: http://bit.ly/IyGQwc)

564. http://undocumentedmatlab.com/blog/class-object-creation-performance (or: http://bit.ly/1aYjbMJ)
565. http://en.wikipedia.org/wiki/Object_pool_pattern (or: http://bit.ly/IyHIkv)
566. http://en.wikipedia.org/wiki/Singleton_pattern (or: http://bit.ly/IyIbDb)
567. http://en.wikipedia.org/wiki/Factory_method_pattern (or: http://bit.ly/IyI5LJ)
568. http://mathworks.com/help/matlab/matlab_oop/comparing-handle-and-value-classes. html (or: http://bit.ly/IyJNga)
569. Inspired by MathWorker Bobby Nedelkovski's Singleton class implementation: http:// mathworks.com/matlabcentral/fileexchange/24911-design-pattern-singleton-creational (or: http://bit.ly/13aCP6Q)
570. Also see http://mathworks.com/matlabcentral/answers/96960 (or: http://bit.ly/InztrY)
571. http://mathworks.com/help/matlab/ref/handle.html (or: http://bit.ly/1dJkabg)
572. http://mathworks.com/help/matlab/ref/hgsetget.html (or: http://bit.ly/1dJk57 m)
573. http://mathworks.com/help/matlab/matlab_oop/creating-object-arrays.html#bru6o00 (or: http://bit.ly/1cYFNze)
574. http://advanpix.com/2013/08/02/performance-of-array-manipulation-operations (or: http:// bit.ly/1k5vgJY). See related §8.5.5.
575. http://UndocumentedMatlab.com/blog/performance-accessing-handle-properties#Class (or: http://bit.ly/1dJh4E3)
576. http://mathworks.com/matlabcentral/fileexchange/30672-mxgetpropertyptr-c-mex-function (or: http://bit.ly/Su78jQ)
577. http://mathworks.com/help/matlab/matlab_oop/implementing-a-set-get-interface-for-properties.html (or: http://bit.ly/1dJkmXX)
578. http://UndocumentedMatlab.com/blog/setting-class-property-types (or: http://bit.ly/1eDER4S). It is a pity that while Dave Foti himself admitted the usefulness of this construct (http://blogs.mathworks.com/loren/2012/03/26/considering-performance-in-object-oriented-matlab-code#comment-33027 or: http://bit.ly/184Vk38), he never mentioned that it already existed at that time. Also read the related http://UndocumentedMatlab.com/blog /class-object-tab-completion-and-improper-field-names#related (or: http://bit.ly/1tjb7Cd) that described a mechanism for limiting the specific values that a class property is allowed to accept.
579. http://UndocumentedMatlab.com/blog/setting-class-property-types#comment-243356 (or: http://bit.ly/1eDF1sS)
580. http://mathworks.com/help/matlab/matlab_oop/example-representing-structured-data. html#f2-85430 (or: http://mathworks.com/help/matlab/matlab_oop/example-representing-structured-data.html#f2-85430); http://mathworks.com/help/matlab/matlab_oop/property-access-methods.html#bsxanmy (or: http://bit.ly/1dJiCxV)
581. http://mathworks.com/help/matlab/ref/handle.addlistener.html (or: http://bit.ly/1dJkgzs); http://mathworks.com/help/matlab/ref/event.proplistener.html (or: http://bit. ly/1dJjdiY)
582. http://blogs.mathworks.com/loren/2012/03/26/considering-performance-in-object-oriented-matlab-code#3 (or: http://bit.ly/1dJzbd4)
583. http://mathworks.com/matlabcentral/newsreader/view_thread/328883 (or: http://bit.ly /1eEFL0R); http://blogs.mathworks.com/loren/2012/03/26/considering-performance-in-object-oriented-matlab-code#3 (or: http://bit.ly/1dJzbd4)
584. http://blogs.mathworks.com/loren/2012/03/26/considering-performance-in-object-oriented-matlab-code#comment-33025 (or: http://bit.ly/184UL9f)
585. http://blogs.mathworks.com/loren/2012/03/26/considering-performance-in-object-oriented-matlab-code#comment-33050 (or: http://bit.ly/1jYwn9u)
586. http://blogs.mathworks.com/loren/2012/03/26/considering-performance-in-object-oriented-matlab-code#11 (or: http://bit.ly/1dJB2P4)
587. http://stackoverflow.com/questions/1693429/is-matlab-oop-slow-or-am-i-doing-something-wrong#1745686 (or: http://bit.ly/1b6Gqbw)

588. https://github.com/apjanke/matlab-bench (or: http://bit.ly/1ggIOLk)

589. This is corroborated by http://mathworks.com/matlabcentral/answers/71633 (or: http://bit.ly/1eEGE9J) and http://mathworks.com/matlabcentral/answers/57794-why-does-adding-a-class-to-a-package-significant-slow-down-performance (or: http://bit.ly/1eEGZt6)

590. http://mathworks.com/matlabcentral/fileexchange/41349-performance-in-object-oriented-matlab-code (or: http://bit.ly/188Lv1S)

591. http://mathworks.com/discovery/stream-processing.html (or: http://bit.ly/19cWLHH)

592. http://mathworks.com/company/newsletters/articles/accelerating-matlab-algorithms-and-applications.html#SystemObjects (or: http://bit.ly/1dL8apu)

593. http://mathworks.com/help/comm/examples/simulation-acceleration-using-system-objects-matlab-coder-and-parallel-computing-toolbox.html (or: http://bit.ly/1mn3aug)

594. Unfortunately, this webinar by Bill Chou (#62736) has been removed and is no longer accessible online.

595. http://blogs.mathworks.com/community/2011/12/19/matlab-startup-accelerator (or: http://bit.ly/17TqOZp); http://mathworks.com/help/install/license/post-installation-tasks.html (or: http://bit.ly/17Tr5eW); Technical solution 1-FBGXHZ: http://mathworks.com/matlabcentral/answers/92426 (or: http://bit.ly/1b0ICxe)

596. http://mathworks.com/products/matlab/whatsnew.html (or: http://bit.ly/165S68g)

597. http://mathworks.com/help/matlab/release-notes.html#R2011b (or: http://bit.ly/17TrBJY)

598. http://mathworks.com/matlabcentral/answers/68644#answer_80291 (or: http://bit.ly/17TuRFd)

599. I suspect this is also the root cause for this slowdown report: http://mathworks.com/matlabcentral/answers/64262 (or: http://bit.ly/16gPs0m)

600. See also: http://undocumentedmatlab.com/blog/matlab-installation-take-2#comment-112457 (or: http://bit.ly/15g6OfO)

601. http://undocumentedmatlab.com/blog/more-undocumented-timing-features#comment-50584 (or: http://bit.ly/15gavCg)

602. Technical solution 1-FBGXHZ: http://mathworks.com/matlabcentral/answers/92426 (or: http://bit.ly/1b0ICxe)

603. http://mathworks.com/help/install/license/installing-the-software-in-a-concurrent-license.html#bs69rfo-1 (or: http://bit.ly/15g6Fcb)

604. Technical solution 1-18CD3: http://mathworks.com/matlabcentral/answers/93674 (or: http://bit.ly/1b0Mcau)

605. http://www.mathworks.com/support/bugreports/768825 (or: http://bit.ly/1etyYau)

606. Technical solution 1-2Z18MA: http://mathworks.com/matlabcentral/answers/92566 (or: http://bit.ly/1b0MrlR)

607. Technical solution 1-CQM2PT: http://mathworks.com/matlabcentral/answers/94362 (or: http://bit.ly/1b0MFta)

608. Technical solution 1-17VEB: http://mathworks.com/matlabcentral/answers/93345 (or: http://bit.ly/1b0MHkU); Technical solution 1-2Z18MA: http://mathworks.com/matlabcentral/answers/92566 (or: http://bit.ly/1b0MrlR)

609. http://mathworks.com/matlabcentral/newsreader/view_thread/270948 (or: http://bit.ly/1gM7gYx); http://mathworks.com/matlabcentral/answers/50660 (or: http://bit.ly/16gP020)

610. Technical solution 1-17VEB: http://mathworks.com/matlabcentral/answers/93345 (or: http://bit.ly/1b0MHkU); Technical solution 1-PAT1H: http://mathworks.com/matlabcentral/answers/95275 (or: http://bit.ly/1b0MX3q)

611. Technical solution 1-2Z18MA: http://mathworks.com/matlabcentral/answers/92566 (or: http://bit.ly/1b0MrlR)

612. http://mathworks.com/matlabcentral/newsreader/view_thread/297938#801552 (or: http://bit.ly/1gM3Hlh)

613. http://mathworks.com/matlabcentral/newsreader/view_thread/285865#765189 (or: http://bit.ly/1gM4Mth)

614. http://www.mathworks.com/support/bugreports/929447 (or: http://bit.ly/1b5eLWR)

615. http://mathworks.com/matlabcentral/answers/42675 (or: http://bit.ly/16gOKjE); http://mathworks.com/matlabcentral/newsreader/view_thread/327677 (or: http://bit.ly/H5zBuQ)

616. Technical solution 1-186XH: http://mathworks.com/matlabcentral/answers/101387 (or: http://bit.ly/1b0N7aV)

617. http://www.mathworks.com/support/bugreports/366601 (or: http://bit.ly/1aOVI21)

618. http://mathworks.com/matlabcentral/newsreader/view_thread/297938#801339 (or: http://bit.ly/1gM40N0)

619. http://mathworks.com/matlabcentral/newsreader/view_thread/160387 (or: http://bit.ly/1c16HtX)

620. http://mathworks.com/matlabcentral/newsreader/view_thread/51961 (or: http://bit.ly/16gOm4I)

621. http://mathworks.com/help/matlab/ref/matlabwindows.html (or: http://bit.ly/17FftYL). Memory reservation can be further fine-tuned using the MATLAB_RESERVE_LO and MATLAB_RESERVE_HI environment variables (http://kb.wisc.edu/cae/page.php?id=7188, http://www.mathworks.com/support/bugreports/398525, http://mathworks.com/matlabcentral/newsreader/view_thread/166620, http://mathworks.com/matlabcentral/newsreader/view_thread/244472#653535).Some additional memory-related startup options were removed in R2009b: http://mathworks.com/help/matlab/release-notes-older.html#br5ktrh-3 (or: http://bit.ly/19Zinb3)

622. http://mathworks.com/help/matlab/matlab_env/startup-options.html (or: http://bit.ly/19I5Wnj); http://mathworks.com/help/matlab/ref/matlabwindows.html (or: http://bit.ly/17FftYL); http://mathworks.com/help/matlab/ref/matlabunix.html (or: http://bit.ly/19I65Hr); and numerous resources online that refer to the various specific startup parameters

623. Technical solution 1-2Z18MA: http://mathworks.com/matlabcentral/answers/92566 (or: http://bit.ly/1b0MrlR)

624. http://mathworks.com/help/matlab/matlab_env/toolbox-path-caching-in-the-matlab-program.html (or: http://bit.ly/1gM6Kd4); Technical solution 1-186EP: http://mathworks.com/matlabcentral/answers/91507 (or: http://bit.ly/1b0GOEq); http://mathworks.com/matlabcentral/newsreader/view_thread/327677#900400 (or: http://bit.ly/H5zu2l)

625. Technical solution 1-186XH: http://mathworks.com/matlabcentral/answers/101387 (or: http://bit.ly/1b0N7aV); Technical solution 1-18CJ5: http://mathworks.com/matlabcentral/answers/97798 (or: http://bit.ly/1b0NtOY)

626. Technical solution 1-18CD3: http://mathworks.com/matlabcentral/answers/93674 (or: http://bit.ly/1b0Mcau)

627. Technical solution 1-6MBFQL: http://mathworks.com/matlabcentral/answers/103193 (or: http://bit.ly/1iq7LYC)

628. http://www.mathworks.com/support/bugreports/433624 (or: http://bit.ly/GZuQDD)

629. Technical solution 1-2Z18MA: http://mathworks.com/matlabcentral/answers/92566 (or: http://bit.ly/1b0MrlR); Technical solution 1-186XH: http://mathworks.com/matlabcentral/answers/101387 (or: http://bit.ly/1b0N7aV); Technical solution 1-PAT1H: http://mathworks.com/matlabcentral/answers/95275 (or: http://bit.ly/1b0MX3q); http://mathworks.com/matlabcentral/answers/14868#comment_34332 (or: http://bit.ly/1gM5iYm); http://mathworks.com/matlabcentral/answers/22384-matlab-startup-delay-linked-to-initsunvm (or: http://bit.ly/1bBGJaL); http://mathworks.com/matlabcentral/answers/106727-matlab-requires-15-minutes-to-start-up (or: http://bit.ly/1bBGCvV)

630. http://mathworks.com/matlabcentral/answers/14868#comment_168253 (or: http://bit.ly/17uHBfT)

631. http://mathworks.com/matlabcentral/newsreader/view_thread/73885#192568 (or: http://bit.ly/1gMaFa3)

632. Technical solution 1-2Z18MA: http://mathworks.com/matlabcentral/answers/92566 (or: http://bit.ly/1b0MrlR); Technical solution 1-PAT1H: http://mathworks.com/matlabcentral

/answers/95275 (or: http://bit.ly/1b0MX3q); http://mathworks.com/matlabcentral/newsreader/view_thread/72010 (or: http://bit.ly/16gObq7)

633. http://www.mathworks.com/support/bugreports/732450 (or: http://bit.ly/1gM2vya)

634. Technical solution 1-18CF6: http://mathworks.com/matlabcentral/answers/102499 (or: http://bit.ly/1b0NyC4); Technical solution 1-186XH: http://mathworks.com/matlabcentral/answers/101387 (or: http://bit.ly/1b0N7aV)

635. http://www.mathworks.com/support/bugreports/523801 (or: http://bit.ly/1gLXNAE); http://walkingrandomly.com/?p=1341 (or: http://bit.ly/1aOTfV7)

636. Technical solution 1-186XH: http://mathworks.com/matlabcentral/answers/101387 (or: http://bit.ly/1b0N7aV)

637. http://undocumentedmatlab.com/blog/more-undocumented-timing-features#startup (or: http://bit.ly/1aOQCml)

638. http://mathworks.com/help/releases/R2009a/techdoc/ref/matlabunix.html (or: http://bit.ly/1gLX6Y6)

639. Technical solution 1-18I2C: http://mathworks.com/matlabcentral/answers/92813 (or: http://bit.ly/1kulUWn)

640. http://blogs.mathworks.com/community/2010/04/26/controlling-the-java-heap-size (or: http://bit.ly/1duCoZd)

641. http://mathworks.com/matlabcentral/newsreader/view_thread/281500#875159 (or: http://bit.ly/1duBGLA); http://mathworks.com/matlabcentral/newsreader/view_thread/294336 (or: http://bit.ly/1duBOuo)

642. http://www.mathworks.com/support/bugreports/273783 (or: http://bit.ly/197M6hI)

643. http://www.mathworks.com/support/bugreports/398525 (or: http://bit.ly/1duAN5E)

644. See http://docs.oracle.com/javase/7/docs/technotes/tools/windows/java.html (or: http://bit.ly/197n6ac) and http://www.oracle.com/technetwork/java/javase/tech/vmoptions-jsp-140102.html (or: http://bit.ly/197mf9B) — especially note the performance tuning section in both of these webpages. A detailed article explaining the subset of options that can be used to performance-tune Java's garbage collection: http://www.oracle.com/technetwork/java/gc-tuning-5-138395.html (or: http://bit.ly/18HNcG2). Various additional JVM performance-tuning docs can be found here: http://www.oracle.com/technetwork/java/javase/tech/index-jsp-136373.html (or: http://bit.ly/18HNpJk)

645. Technical solution 1-333DU4: http://mathworks.com/matlabcentral/answers/101904 (or: http://bit.ly/1b0NFNX); http://www.mathworks.com/support/bugreports/929447 (or: http://bit.ly/1b5eLWR)

646. http://undocumentedmatlab.com/blog/matlab-installation-take-2#comment-114575 (or: http://bit.ly/1b5dFuj)

647. http://UndocumentedMatlab.com/blog/using-java-collections-in-matlab (or: http://bit.ly/Ir1IWR)

648. http://java.dzone.com/articles/performance-java-collections (or: http://bit.ly/1ilAa0U)

649. E.g., http://mathworks.com/matlabcentral/fileexchange/6514-simle-hashtable (or: http://bit.ly/18kHep7); http://mathworks.com/matlabcentral/fileexchange/28586-simple-hashtable-repackaged (or: http://bit.ly/18kHh4i); http://mathworks.com/matlabcentral/fileexchange/26778-hash-table-declaration (or: http://bit.ly/18kHiVJ); http://mathworks.com/matlabcentral/fileexchange/15831-hashtable-class (or: http://bit.ly/18kHlRy); http://mathworks.com/matlabcentral/fileexchange/19381-lookuptable (or: http://bit.ly/18kHpR6); http://mathworks.com/matlabcentral/fileexchange/33068-a-multidimensional-map-class (or: http://bit.ly/18kHsMZ); http://mathworks.com/matlabcentral/fileexchange/20876-use-a-hash-table (or: http://bit.ly/18kHBjo) ; http://mathworks.com/matlabcentral/fileexchange/19647-dict (or: http://bit.ly/18kHEf5); http://mathworks.com/matlabcentral/fileexchange/33901-dictionary-data-structure (or: http://bit.ly/18kHHr6)

650. http://www.mathworks.com/support/bugreports/search_results?search_executed=1&keyword=performance&release_filter=Exists+in&release=0&commit=Search (or: http://bit.ly/1bDXveO)

651. http://mathworks.com/help/pde/examples/poisson-s-equation-on-rectangular-domain-using-a-fast-poisson-solver.html (or: http://bit.ly/1b9Cw3b). Also see http://mathworks.com/matlabcentral/answers/94247-how-do-i-define-the-right-side-of-poisson (or: http://bit.ly/1vIeMxC). Also see the related comparison between the standard *eigs* and PDE Toolbox's *pdeeig* in http://mathworks.com/help/pde/examples/vibration-of-a-circular-membrane-using-the-matlab-eigs-function.html (or: http://bit.ly/1b9DfkP).

652. http://en.wikipedia.org/wiki/Poisson's_equation (or: http://bit.ly/1b9C7xA)

653. http://mathworks.com/matlabcentral/answers/89828-speed-differences-in-sym-and-vpa (or: http://bit.ly/15YpGTQ)

654. http://mathworks.com/matlabcentral/newsreader/view_thread/311639#850249 (or: http://bit.ly/1cd9nFp)

655. http://mathworks.com/matlabcentral/newsreader/view_thread/311639#913056 (or: http://bit.ly/1cd8vk0); http://mathworks.com/matlabcentral/answers/112270-is-there-a-way-to-increase-the-speed-of-this-code (or: http://bit.ly/1vI8dv3)

656. http://www.mathworks.com/support/bugreports/523801 (or: http://bit.ly/1gLXNAE); http://walkingrandomly.com/?p=1341 (or: http://bit.ly/1aOTfV7); http://matlabician.wordpress.com/2009/12/18/complex-magic-part-2-differentiation-via-integration (or: http://bit.ly/Holj9P)

657. http://www.mathworks.com/support/bugreports/license/accept_license/2063?fname=win32_inprocess.zip&geck_id=523801 (or: http://bit.ly/1butSIF)

658. http://mathworks.com/help/symbolic/code-performance.html (or: http://bit.ly/1lnAUFe)

659. http://mathworks.com/matlabcentral/answers/108155-symbolic-toolbox-speed-comparisons-on-2010a-versus-2012b-for-mac-os (or: http://bit.ly/1bvLeax)

660. http://undocumentedmatlab.com/blog/improving-simulink-performance (or: http://bit.ly/1jreynM)

661. http://mathworks.com/help/simulink#performance (or: http://bit.ly/15YrWug)

662. http://mathworks.com/company/newsletters/articles/improving-simulation-performance-in-simulink.html (or: http://bit.ly/164v6az); http://mathworks.com/company/newsletters/articles/improving-simulink-design-optimization-performance-using-parallel-computing.html (or: http://bit.ly/1bLoK7Y)

663. http://mathworks.com/videos/speeding-up-simulink-applications-81795.html (or: http://bit.ly/1evZu39); http://mathworks.com/videos/speeding-up-simulations-with-parallel-computing-81752.html (or: http://bit.ly/1evZqQR); http://mathworks.com/videos/speeding-up-simulink-for-control-systems-applications-81736.html (or: http://bit.ly/1evZoIL); http://mathworks.com/videos/speeding-up-simulink-for-signal-processing-applications-81748.html (or: http://bit.ly/1evZqQR)

664. http://blogs.mathworks.com/seth/category/performance (or: http://bit.ly/15NNQOz)

665. http://mathworks.com/help/simulink/acceleration.html (or: http://bit.ly/1bLpMk9); http://mathworks.com/help/simulink/ug/comparing-performance.html (or: http://bit.ly/1ew3aSp)

666. http://mathworks.com/help/simulink/gui/optimization-pane-general.html (or: http://bit.ly/164vSUU)

667. http://blogs.mathworks.com/seth/2010/10/28/tips-for-simulation-performance (or: http://bit.ly/1ew1Hvm)

668. http://mathworks.com/help/simulink/gui/simulation-target-pane-general.html (or: http://bit.ly/164Az1d)

669. http://mathworks.com/help/stateflow/ug/speeding-up-simulation.html (or: http://bit.ly/1gvEWtn)

670. http://mathworks.com/help/simulink/ug/scope-signal-viewer-characteristics.html (or: http://bit.ly/1gvFqjb)

671. http://mathworks.com/help/dsp/ug/sample-and-frame-based-concepts.html (or: http://bit.ly/1kRItU9); http://mathworks.com/help/dsp/release-notes.html#bs1rpr_-1 (or: http://bit.ly/1dpW2Uw)

672. http://blogs.mathworks.com/seth/2010/10/28/tips-for-simulation-performance (or: http://bit.ly/1ew1Hvm)

673. http://mathworks.com/help/simulink/gui/mask-editor-overview.html (or: http://bit.ly/164zQwS)

674. http://mathworks.com/help/matlab/file-opening-loading-and-saving.html (or: http://bit.ly/164v6az)

675. http://www.linkedin.com/groups/Hi-Im-using-simulink-process-109866.S.192318533?view=&gid=109866&item=192318533 (or: http://linkd.in/1ew0uV1)

676. http://mathworks.com/help/simulink/ug/capturing-performance-data.html (or: http://bit.ly/164wMB1)

677. http://mathworks.com/help/simulink/automatic-performance-optimization.html (or: http://bit.ly/15Ys78S); http://blogs.mathworks.com/seth/2012/11/21/automatically-improving-model-performance (or: http://bit.ly/176dNfj); http://mathworks.com/videos/simulation-analysis-and-performance-74976.html (or: http://bit.ly/1lMWL7h)

678. http://mathworks.com/help/gads/choose-a-solver.html (or: http://bit.ly/1ed5jSQ); http://mathworks.com/help/releases/R2012a/toolbox/simulink/ug/f11-69449.html (or: http://bit.ly/1ew2oEV)

679. http://blogs.mathworks.com/seth/2012/06/04/the-most-useful-command-for-debugging-variable-step-solver-performance (or: http://bit.ly/176elBQ)

680. http://blogs.mathworks.com/seth/2012/06/04/the-most-useful-command-for-debugging-variable-step-solver-performance#comment-1850 (or: http://bit.ly/176e55V)

681. http://blogs.mathworks.com/seth/2013/04/26/zero-crossing-detection-what-are-your-options (or: http://bit.ly/176dCRb); http://mathworks.com/help/simulink/ug/simulating-dynamic-systems.html#bridiag-5 (or: http://bit.ly/1bLo6XY); http://mathworks.com/help/stateflow/ug/when-to-enable-zero-crossing-detection.html (or: http://bit.ly/1bLo5TL)

682. http://mathworks.com/help/simulink/ug/saving-and-restoring-the-simulation-state-as-the-simstate.html (or: http://bit.ly/1bLokhH)

683. http://mathworks.com/help/simulink/ug/running-parallel-simulations.html (or: http://bit.ly/1bLoGoK); http://mathworks.com/company/newsletters/articles/improving-simulink-design-optimization-performance-using-parallel-computing.html (or: http://bit.ly/1bLoK7Y); http://blogs.mathworks.com/seth/2009/03/31/parallel-computing-with-simulink-model-reference-builds (or: http://bit.ly/176f1Hq); http://blogs.mathworks.com/seth/2010/10/17/parallel-computing-with-simulink-running-thousands-of-simulations (or: http://bit.ly/1ew3EYE)

684. http://blogs.mathworks.com/seth/2010/10/28/tips-for-simulation-performance (or: http://bit.ly/1ew1Hvm)

685. http://www.linkedin.com/groups/Hi-Im-using-simulink-process-109866.S.192318533?view=&gid=109866&item=192318533 (or: http://linkd.in/1ew0uV1)

686. http://www.linkedin.com/groups/Hi-Im-using-simulink-process-109866.S.192318533?view=&gid=109866&item=192318533 (or: http://linkd.in/1ew0uV1)

687. http://mathworks.com/help/simulink/sfg/error-handling.html#f4-84705 (or: http://bit.ly/1lpLy1E)

688. http://mathworks.com/company/newsletters/articles/improving-simulation-performance-in-simulink.html (or: http://bit.ly/164v6az)

689. https://www.ices.utexas.edu/sysdocs/commercial/matlab.html (or: http://bit.ly/1c0VN7t); also note the detailed discussion in http://mathworks.com/matlabcentral/newsreader/view_thread/144169 (or: http://bit.ly/1c10M87)

690. http://www.mathworks.com/support/bugreports/412219 (or: http://bit.ly/1c0XvFK); Also see technical solution 1-31CIOM: http://mathworks.com/matlabcentral/answers/96635 (or: http://bit.ly/1c16erB)

691. http://mathworks.com/matlabcentral/newsreader/view_thread/297938#801339 (or: http://bit.ly/1gM40N0)

692. http://mathworks.com/matlabcentral/newsreader/view_thread/160387 (or: http://bit.ly/1c16HtX)

693. http://mathworks.com/matlabcentral/newsreader/view_thread/160387#700723 (or: http://bit.ly/1k15Tag)

694. M.L. — private communication.

695. http://mathworks.com/matlabcentral/answers/104567-macbook-pro-mid-2012-retina-display-terrible-benchmark-performance-for-graphics (or: http://bit.ly/1k16kkR)

696. http://mathworks.com/matlabcentral/newsreader/view_thread/299444 (or: http://bit.ly/1mxyp3f)

697. http://mathworks.com/matlabcentral/newsreader/view_thread/158424#400835 (or: http://bit.ly/1bCGkHW); http://www.mathworks.com/support/bugreports/279325 (or: http://bit.ly/1bCGDma)

698. http://www.mathworks.com/support/bugreports/581959 (or: http://bit.ly/1c12xC8)

699. http://mathworks.com/help/matlab/ref/triscatteredinterp.html (or: http://bit.ly/1bh1knn)

700. http://mathworks.com/matlabcentral/fileexchange/8998-surface-fitting-using-gridfit (or: http://bit.ly/1fgnl4J)

701. http://blogs.mathworks.com/loren/2011/04/21/deploying-multiple-c-shared-libraries#comment-32227 (or: http://bit.ly/1bQx1Cp)

702. http://mathworks.com/matlabcentral/newsreader/view_thread/297938#801552 (or: http://bit.ly/1gM3Hlh)

703. http://mathworks.com/matlabcentral/newsreader/view_thread/285865#765189 (or: http://bit.ly/1gM4Mth)

704. Technical solution 1-CQM2PT: http://mathworks.com/matlabcentral/answers/94362 (or: http://bit.ly/1b0MFta)

705. http://mathworks.com/matlabcentral/newsreader/view_thread/297938#801339 (or: http://bit.ly/1gM40N0)

706. http://en.wikipedia.org/wiki/Automatic_parallelization (or: http://bit.ly/13vcaF1)

707. http://en.wikipedia.org/wiki/Embarrassingly_parallel (or: http://bit.ly/13vd6t7). MATLAB's co-founder Cleve Moler claims to have invented this term: http://blogs.mathworks.com/cleve/2013/11/12/the-intel-hypercube-part-2-reposted#096367ea-045e-4f28-8fa2-9f7db8fb7b01 (or: http://bit.ly/1bizLwv).

708. This design goal was outlined by Roy Lurie, MathWorks VP Engineering, in a 2007 paper: http://archive.hpcwire.com/hpcwire/2008-08-28/ecosystems_are_messy-1.html (or: http://bit.ly/1h8lHrz).

709. http://mathworks.com/help/matlab/matlab_prog/vectorization.html (or: http://bit.ly/13vgoMR)

710. An article by MathWorker Sarah Wait Zaranek described a use-case where vectorization achieved a 500x-1000x speedup (!): http://blogs.mathworks.com/loren/2008/06/25/speeding-up-matlab-applications (or: http://bit.ly/1jro8W2).

711. Technical solution 1-4PG4AN: http://mathworks.com/matlabcentral/answers/95958 (or: http://bit.ly/19FvAFE)

712. For a bit dated but very detailed theoretical and practical introduction to multi-core performance tuning by Philip (Phil) Mucci see http://icl.cs.utk.edu/~mucci/latest/pubs/Notur2009-new.pdf (or: http://bit.ly/1mHtsZy).

713. Technical solution 1-WNXI8: http://mathworks.com/matlabcentral/answers/94417 (or: http://bit.ly/1b0NXEF)

714. http://mathworks.com/help/releases/R2007a/techdoc/matlab_env/f1-94809.html#bq35sqs (or: http://bit.ly/MglrKC)

715. Technical solution 1-46OY0H: http://mathworks.com/matlabcentral/answers/94591 (or: http://bit.ly/1eSIjw1)

716. http://mathworks.com/help/matlab/release-notes.html (or: http://bit.ly/153Dt6l); http://mathworks.com/help/matlab/release-notes-older.html (or: http://bit.ly/153Dvv3)

717. Technical solution 1-4PG4AN: http://mathworks.com/matlabcentral/answers/95958 (or: http://bit.ly/19FvAFE); http://mathworks.com/matlabcentral/answers/95958 (or: http://bit.ly/19FvAFE)

718. http://walkingrandomly.com/?p=1894 (or: http://bit.ly/1adXKcx)

719. http://mathworks.com/matlabcentral/answers/23157-multithreaded-filter (or: http://bit.ly/1hYYskk); http://mathworks.com/matlabcentral/newsreader/view_thread/324169#890739 (or: http://bit.ly/1iLESWZ)

720. http://mathworks.com/matlabcentral/answers/uploaded_files/1624/DataSizes_vs_SpeedUp.doc (or: http://bit.ly/18oJZLt)

721. http://mathworks.com/company/newsletters/articles/the-origins-of-matlab.html (or: http://bit.ly/14fmcdL); http://mathworks.com/company/newsletters/articles/matlab-incorporates-lapack.html (or: http://bit.ly/14fltcr)

722. For example: http://en.wikipedia.org/wiki/General_Matrix_Multiply (or: http://bit.ly/1dTpCTc), used in matrix multiplications

723. http://stackoverflow.com/questions/6058139/why-is-matlab-so-fast-in-matrix-multiplication (or: http://bit.ly/1dTmJ4T)

724. http://software.intel.com/en-us/intel-ipp (or: http://intel.ly/168Dp4x); http://en.wikipedia.org/wiki/Integrated_Performance_Primitives (or: http://bit.ly/168DfdA)

725. http://software.intel.com/en-us/intel-mkl (or: http://intel.ly/13lAmYI); http://en.wikipedia.org/wiki/Math_Kernel_Library (or: http://bit.ly/1atUMAU)

726. http://undocumentedmatlab.com/blog/math-libraries-version-info-upgrade (or: http://bit.ly/13lAsja)

727. http://en.wikipedia.org/wiki/Streaming_SIMD_Extensions (or: http://bit.ly/12HhC2r)

728. http://en.wikipedia.org/wiki/Advanced_Vector_Extensions (or: http://bit.ly/12HhNLk)

729. http://en.wikipedia.org/wiki/SIMD (or: http://bit.ly/12HhYpZ)

730. See for example: http://stackoverflow.com/questions/8389648/how-to-achieve-4-flops-per-cycle (or: http://bit.ly/17FMiVB). However, see http://mathworks.com/matlabcentral/answers/114039-flop-rate-of-matlab-code (or: http://bit.ly/1mESF5T) for a counter-example. It is very difficult to infer exactly when and how MATLAB operations employ SIMD processing.

731. See for example http://codereview.stackexchange.com/questions/20980/naive-c-matrix-multiplication-100-times-slower-than-blas (or: http://bit.ly/17Fito6)

732. http://developer.amd.com/tools-and-sdks/cpu-development/amd-core-math-library-acml (or: http://bit.ly/17FxiqQ); http://en.wikipedia.org/wiki/AMD_Core_Math_Library (or: http://bit.ly/17FxfLD)

733. http://mathworks.com/help/matlab/release-notes.html#btz639l (or: http://bit.ly/18WpCjC)

734. http://www.matlabtips.com/matlab-is-no-longer-slow-at-for-loops (or: http://bit.ly/1dTm7MD). See §3.1.15 and §9.4.2 for additional detail.

735. http://mathworks.com/company/newsletters/articles/matrix-indexing-in-matlab.html (or: http://bit.ly/1bc1Cd9); http://mathworks.com/help/matlab/learn_matlab/array-indexing.html (or: http://bit.ly/137YsBN); http://mathworks.com/help/matlab/math/matrix-indexing.html (or: http://bit.ly/137YgTf); http://mathworks.com/help/matlab/indexing.html (or: http://bit.ly/137YfPc); http://mathworks.com/help/matlab/matlab_prog/vectorization.html#btisqhq (or: http://bit.ly/12k6EUa); http://blogs.mathworks.com/loren/2006/08/09/essence-of-indexing (or: http://bit.ly/1jrEAp8)

736. http://mathworks.com/help/matlab/matlab_prog/operators.html#f0-38155 (or: http://bit.ly/1380vG9)

737. http://blogs.mathworks.com/steve/2008/01/28/logical-indexing (or: http://bit.ly/WVIU7X); http://mathworks.com/help/matlab/math/matrix-indexing.html#bq7egb6-1 (or: http://bit.ly/WVIYol); http://mathworks.com/company/newsletters/articles/matrix-indexing-in-matlab.html (or: http://bit.ly/1bc1Cd9)

738. http://www-h.eng.cam.ac.uk/help/tpl/programs/Matlab/tricks.html#Examples (or: http://bit.ly/10Y91uU)

739. http://www.cs.umd.edu/class/sum2003/cmsc311/Notes/BitOp/pointer.html (or: http://bit.ly/14cHY1B); http://en.wikipedia.org/wiki/Pointer_(computer_programming) (or: http://bit.ly/14cHZ5t)

740. http://stackoverflow.com/questions/13382155/is-indexing-vectors-in-matlab-inefficient (or: http://bit.ly/1itf3OP); http://mathworks.com/matlabcentral/answers/54522-why-is-indexing-vectors-matrices-in-matlab-very-inefficient (or: http://bit.ly/1fjFZJ2)

741. http://stackoverflow.com/a/17585725/233829 (or: http://bit.ly/1lJDM1A). In that particular case, additional speedup could be achieved using sparse matrices (http://stackoverflow.com/a/17607738/233829 or: http://bit.ly/1lJE08X) — see §4.1.3 for additional details.

742. http://stackoverflow.com/questions/13382155/is-indexing-vectors-in-matlab-inefficient (or: http://bit.ly/1itf3OP)

743. http://blogs.mathworks.com/videos/2013/01/17/example-speed-up-matlab-code-by-profiling/ (or: http://bit.ly/18ZWnvY)

744. For example, http://mathworks.com/matlabcentral/newsreader/view_thread/326037 (or: http://bit.ly/18zr1eJ)

745. http://ubcmatlabguide.github.io/html/speedup.html#21 (or: http://bit.ly/1ggXcmW) or http://www.cs.ubc.ca/~murphyk/Software/matlabTutorial/html/speedup.html#20 (or: http://bit.ly/14cEqfA); another example of such a confusion: http://stackoverflow.com/questions/5589251/why-is-this-so-much-faster (or: http://bit.ly/YYLRST)

746. See for example http://blogs.mathworks.com/loren/2009/01/20/more-ways-to-find-matching-data (or: http://bit.ly/YYDRkM)

747. http://mathworks.com/help/matlab/release-notes.html#btz639a (or: http://bit.ly/18WolsX)

748. http://mathworks.com/help/matlab/release-notes.html#btz131f-1 (or: http://bit.ly/18Wozjx)

749. http://blogs.mathworks.com/loren/2006/06/02/structures-and-comma-separated-lists (or: http://bit.ly/1jrHxWJ)

750. A typical example: http://mathworks.com/matlabcentral/newsreader/view_thread/325929#895939 (or: http://bit.ly/16aPl5C)

751. http://undocumentedmatlab.com/blog/matlabs-internal-memory-representation (or: http://bit.ly/14Qjjhk); http://mathworks.com/matlabcentral/newsreader/view_thread/81509#207407 (or: http://bit.ly/1el95gK)

752. http://undocumentedmatlab.com/blog/matrix-processing-performance#comment-51189 (or: http://bit.ly/12wwSD5)

753. http://mathworks.com/matlabcentral/fileexchange/25346-ntimes (or: http://bit.ly/17PjCqP)

754. For example: http://stackoverflow.com/questions/21715047/how-to-vectorize-the-for-loop-in-matlab (or: http://bit.ly/1hWC2ym)

755. http://mathworks.com/matlabcentral/newsreader/view_thread/327687#900516 (or: http://bit.ly/1fbkvyN)

756. http://undocumentedmatlab.com/blog/cellfun-undocumented-performance-boost (or: http://bit.ly/10KEQHn)

757. http://mathworks.com/help/matlab/ref/bsxfun.html (or: http://bit.ly/UfBxoy); http://blogs.mathworks.com/loren/2008/08/04/comparing-repmat-and-bsxfun-performance (or: http://bit.ly/12ffKy5)

758. http://blogs.mathworks.com/loren/2008/08/04/comparing-repmat-and-bsxfun-performance#comment-29656 (or: http://bit.ly/18KIh61); http://mathworks.com/matlabcentral/answers/5883#answer_8215 (or: http://bit.ly/16nbeyt)

759. http://mathworks.com/matlabcentral/answers/106903-which-matlab-operations-functions-need-speeding-up#answer_115814 (or: http://bit.ly/1dKl5IU). This thread is well worth reading: it discusses potential future MATLAB speedups.

760. http://en.wikipedia.org/wiki/Outer_product (or: http://bit.ly/151odHa)

761. http://mathworks.com/matlabcentral/newsreader/view_thread/327653 (or: http://bit.ly/151p7Du)

762. http://stackoverflow.com/questions/17111331/ho-do-i-make-this-code-segment-faster-in-matlab (or: http://bit.ly/151s2fi); http://mathworks.com/matlabcentral/newsreader/view_thread/330514#908777 (or: http://bit.ly/1jrLZVA); http://stackoverflow.com/questions/20585252/efficient-way-to-subtract-each-vector-element-from-matrix-matlab (or: http://bit.ly/18IrzA0)

763. http://en.wikipedia.org/wiki/Kronecker_product (or: http://bit.ly/1aXnYDs)

764. http://mathworks.com/matlabcentral/fileexchange/24499-kronecker (or: http://bit.ly/15ytsAf)

765. http://mathworks.com/matlabcentral/fileexchange/28889-kronecker-product (or: http://bit.ly/1aXnGfO)

766. http://mathworks.com/matlabcentral/fileexchange/24557-kronecker-tensor-product (or: http://bit.ly/1aXmU2n). It is interesting to note that this implementation was helped in part by Bruno Luong.

767. http://mathworks.com/matlabcentral/fileexchange/25969-efficient-object-oriented-kronecker (or: http://bit.ly/1aXmZ62)

768. http://mathworks.com/matlabcentral/fileexchange/32578-superkron (or: http://bit.ly/1aXn1uT)

769. http://mathworks.com/matlabcentral/fileexchange/23606-fast-and-efficient-kronecker (or: http://bit.ly/1aXn6i8)

770. For example: http://blogs.mathworks.com/loren/2012/10/10/when-is-a-number-perfect#comment-33201 (or: http://bit.ly/1gQKcaf)

771. http://mathworks.com/matlabcentral/fileexchange/23084-binary-array-expansion-function (or: http://bit.ly/18LSH49)

772. For example, http://stackoverflow.com/questions/18782234/how-to-improve-performance-in-matlab-operation-with-3-for-loops (or: http://bit.ly/1fdK03F); http://stackoverflow.com/questions/17598345/increase-speed-of-array-population-when-using-nested-for-loops-matlab (or: http://bit.ly/1hYVlZO); http://mathworks.com/matlabcentral/newsreader/view_thread/330966#909350 (or: http://bit.ly/1fdK17 W)

773. http://mathworks.com/matlabcentral/newsreader/view_thread/324431#891492 (or: http://bit.ly/12kgx49)

774. http://blogs.mathworks.com/pick/2012/11/25/converting-images-from-grayscale-to-color#comment-16040 (or: http://bit.ly/151q3rp); http://mathworks.com/matlabcentral/newsreader/view_thread/324523 (or: http://bit.ly/151slqA)

775. http://mathworks.com/matlabcentral/fileexchange/23005-bsxfun-substitute (or: http://bit.ly/14WvGbG); http://mathworks.com/matlabcentral/fileexchange/10333-generalized-array-operations (or: http://bit.ly/14Wwttc) potentially provides a solution for even older releases.

776. http://mathworks.com/matlabcentral/fileexchange/18685-bsxfun (or: http://bit.ly/14WvVDL)

777. http://mathworks.com/matlabcentral/fileexchange/18686-bsxarg (or: http://bit.ly/18LRmdE)

778. http://mathworks.com/matlabcentral/fileexchange/32681-bsxcat (or: http://bit.ly/18LPnGt)

779. http://mathworks.com/matlabcentral/fileexchange/23821-bsxops (or: http://bit.ly/18LQBBx)

780. http://mathworks.com/matlabcentral/newsreader/view_thread/93652#237933 (or: http://bit.ly/1mqUvHs)

781. http://mathworks.com/matlabcentral/fileexchange/24380-cstrainbp (or: http://bit.ly/1lAtL78)

782. http://www-h.eng.cam.ac.uk/help/tpl/programs/Matlab/tricks.html (or: http://bit.ly/1dC99CR); older cached version with some different material: http://www.ee.columbia.edu/~marios/matlab/Matlab%20Tricks.pdf (or: http://bit.ly/15DvHjF)

783. http://stackoverflow.com/questions/17600835/matlab-removing-rows-when-there-are-repeated-values-in-columns (or: http://bit.ly/11Wrlau)

784. http://mathworks.com/matlabcentral/newsreader/view_thread/331920#912196 (or: http://bit.ly/19lbFMT)

785. http://mathworks.com/matlabcentral/newsreader/view_thread/327563 (or: http://bit.ly/153yYJ6)

786. http://stackoverflow.com/questions/7640010/optimizing-matlab-code (or: http://bit.ly/1bFikjV). Note that the original code in this reference was modified for illustration purposes.

787. http://www.linkedin.com/groups/Loop-Matrix-Operation-134533.S.224145319?gid=134533#commentID_126924600 (or: http://linkd.in/1boMlYV). The original posted looped version was highly unoptimized — it was simplified and somewhat optimized in the text here, keeping the original "look-and-feel" of the double-loop algorithm.

788. Apparently, I am not the only one who has this feeling: http://blogs.mathworks.com/loren/2008/02/20/under-appreciated-accumarray (or: http://bit.ly/1abdNbh)

789. http://mathworks.com/matlabcentral/newsreader/view_thread/330428 (or: http://bit.ly/10Mp9iV)

790. http://mathworks.com/matlabcentral/newsreader/view_thread/324859#892831 (or: http://bit.ly/12HAL4f); http://mathworks.com/matlabcentral/newsreader/view_thread/323437#888649 (or: http://bit.ly/18QFWX2); http://mathworks.com/matlabcentral/newsreader/view_thread/164470#921205 (or: http://bit.ly/1pgHhxq)

791. http://mathworks.com/matlabcentral/newsreader/view_thread/320297#877581 (or: http://bit.ly/18QFBnc)

792. http://mathworks.com/matlabcentral/newsreader/view_thread/324487#891637 (or: http://bit.ly/1o9pILD)

793. http://stackoverflow.com/questions/24091457/trying-to-improve-the-efficency-of-a-triple-for-loop-in-matlab (or: http://bit.ly/1sINLZ0)

794. http://mathworks.com/company/newsletters/articles/using-matlabs-meshgrid-command-and-array-operators-to-implement-one-and-two-variable-functions.html (or: http://bit.ly/16gFmvE). This article first appeared in the Sepember 2000 issue of the MATLAB newsletter.

795. http://mathworks.com/matlabcentral/newsreader/view_thread/264698#914122 (or: http://bit.ly/1brHm6G)

796. http://mathworks.com/matlabcentral/newsreader/view_thread/264698#914124 (or: http://bit.ly/HUbrES)

797. http://mathworks.com/matlabcentral/fileexchange/19480-fast-range-search-through-jit-ver-2 (or: http://bit.ly/14Wua8m)

798. http://stackoverflow.com/questions/19504719/using-matrix-structure-to-speed-up-matlab (or: http://bit.ly/HYNhZW)

799. Adapted from http://mathworks.com/matlabcentral/newsreader/view_thread/323890 (or: http://bit.ly/1agYrSp). Also see http://matlabtricks.com/post-40/speed-measurement-of-repeating-a-row-multiple-times (or: http://bit.ly/1elYPoS).

800. http://mathworks.com/matlabcentral/newsreader/view_thread/332406#913480 (or: http://bit.ly/1cnXM3P)

801. http://mathworks.com/matlabcentral/newsreader/view_thread/326075#896307 (or: http://bit.ly/1bxQV7o)

802. http://mathworks.com/matlabcentral/newsreader/view_thread/326075#896340 (or: http://bit.ly/1bxO5PR)

803. http://mathworks.com/matlabcentral/newsreader/view_thread/326075#896328 (or: http://bit.ly/1bxPk1g)

804. http://mathworks.com/matlabcentral/newsreader/view_thread/326075#896307 (or: http://bit.ly/1bxQV7o)

805. http://home.online.no/~pjacklam/matlab/doc/mtt/doc/mtt.pdf#page=22 (or: http://bit.ly/1fuNAWA)

806. http://mathworks.com/matlabcentral/fileexchange/24536-expand (or: http://bit.ly/1bxS0vY)

807. http://i217.photobucket.com/albums/cc229/spamanon/tester_output-2.png (or: http://bit.ly/14QaQut)

808. http://mathworks.com/matlabcentral/fileexchange/22319-replicate (or: http://bit.ly/1bxS7Yu)

809. For example, http://mathworks.com/matlabcentral/newsreader/view_thread/317005 (or: http://bit.ly/XWlSi0); http://mathworks.com/matlabcentral/newsreader/view_thread/311475#848866 (or: http://bit.ly/WDx0NT)

810. http://mathworks.com/matlabcentral/fileexchange/24323-ScreenCapture (or: http://bit.ly/1a37yGn); http://undocumentedmatlab.com/blog/screencapture-utility (or: http://bit.ly/1eALoML)

811. http://download.oracle.com/javase/6/docs/api/java/awt/Robot.html (or: http://bit.ly/1a3a2V2)

812. Technical solution 1-2WPAYR: http://mathworks.com/matlabcentral/answers/100155 (or: http://bit.ly/1b0O03t)

813. http://mathworks.com/matlabcentral/fileexchange/17476-typecast-and-typecastx-c-mex-functions (or: http://bit.ly/1e0Zrh1)

814. http://mathworks.com/matlabcentral/newsreader/search_results?dur=all&search_string=bruno+reshape (or: http://bit.ly/12hoX7E)

815. http://mathworks.com/matlabcentral/newsreader/view_thread/311475#848866 (or: http://bit.ly/WDx0NT)

816. I could not find an online copy on MIT's website, but a cached copy is available here: http://ee.columbia.edu/~marios/matlab/Fast%20manipulation%20of%20multi-dimensional%20arrays%20in%20Matlab%20(2002).pdf (or: http://bit.ly/1jrKMNV)

817. http://en.wikipedia.org/wiki/Synthetic_aperture_radar (or: http://bit.ly/1aJ9mHd)

818. https://osubp.googlecode.com/hg-history/60095eb90ef2ea48858f3fca37ac09ab41577704/SPIE10toolbox.pdf (or: http://bit.ly/ToG2wK)

819. See http://blog.accelereyes.com/blog/2012/08/28/sar-image-formation-algorithms-on-the-gpu (or: http://bit.ly/VQ27Rq) for more detailed algorithm description and recent benchmarks.

820. http://mathworks.com/matlabcentral/newsreader/view_thread/278246 (or: http://bit.ly/16gD5k1)

821. http://mathworks.com/matlabcentral/newsreader/view_thread/278246#732229 (or: http://bit.ly/181TT5d)

822. http://mathworks.com/matlabcentral/newsreader/view_thread/278246#732236 (or: http://bit.ly/16gEpmP) — both the vectorized and non-vectorized solutions given by Matt Fig.

823. http://mathworks.com/matlabcentral/fileexchange/19344-efficient-k-means-clustering-using-jit (or: http://bit.ly/11tp2v6), http://mathworks.com/matlabcentral/fileexchange/19345-efficient-k-nearest-neighbor-search-using-jit (or: http://bit.ly/11tp5HB)

824. http://stackoverflow.com/questions/17572332/why-in-matlab-this-code-is-faster#comment-25596038 (or: http://bit.ly/14Qr9aC)

825. http://stackoverflow.com/questions/16219233/quickest-way-to-get-elements-given-matrix-of-indices-in-matlab#16219256 (or: http://bit.ly/125sMOs)

826. http://mathworks.com/matlabcentral/newsreader/view_thread/336167 (or: http://bit.ly/1sITyxs)

827. http://mathworks.com/matlabcentral/newsreader/view_thread/330421 (or: http://bit.ly/12OnvK9)

828. http://mathworks.com/matlabcentral/newsreader/view_thread/330421#907750 (or: http://bit.ly/12OnTbl)

829. http://mathworks.com/matlabcentral/newsreader/view_thread/330076 (or: http://bit.ly/12hmDxr)

830. http://mathworks.com/matlabcentral/newsreader/view_thread/330076#906973 (or: http://bit.ly/12hnCOl)

831. http://mathworks.com/matlabcentral/newsreader/view_thread/81509#207542 (or: http://bit.ly/1eli69u)

832. http://mathworks.com/matlabcentral/newsreader/view_thread/324344#891641 (or: http://bit.ly/15ybuhm)

833. http://blogs.mathworks.com/community/2012/08/06/matlab-central-community-freshens-up-mathworks-support-site#comment-9822 (or: http://bit.ly/1d7yLXS)

834. http://mathworks.com/matlabcentral/answers/107170-how-to-replace-a-for-loop-with-something-faster (or: http://bit.ly/1kSTl4s)

835. http://blogs.mathworks.com/loren/2006/05/17/fibonacci-and-filter (or: http://bit.ly/1bejJ2J); http://blogs.mathworks.com/loren/2013/09/22/timing-code (or: http://bit.ly/1lp8AET)

836. http://mathworks.com/matlabcentral/answers/28396-speed-up-recursive-loop (or: http://bit.ly/1lBxjFQ)

837. http://mathworks.com/matlabcentral/fileexchange/32261-filterm (or: http://bit.ly/1l6wtOs); Additional recommendations for improving *filter*'s performance: http://stackoverflow.com/a/24418028/233829 (or: http://bit.ly/Vo3dv1).

838. http://mathworks.com/matlabcentral/answers/5883#answer_8697 (or: http://bit.ly/12faDha); http://mathworks.com/matlabcentral/answers/9900#answer_13623 (or: http://bit.ly/1e5H87f)

839. http://mathworks.com/help/signal/ref/buffer.html (or: http://bit.ly/1e5IMG3)

840. http://mathworks.com/help/signal/ref/filtfilt.html (or: http://bit.ly/1e5IFKx). In general, *filtfilt* removes the time-lag distortion produced by *filter* (the window size). A technical comparison between *filter* and *filtfilt* can be found here: http://dsp.stackexchange.com/questions/9467/what-is-the-advantage-of-matlabs-filtfilt (or: http://bit.ly/1e5IlLS).

841. http://blogs.mathworks.com/videos/2012/04/17/using-convolution-to-smooth-data-with-a-moving-average-in-matlab (or: http://bit.ly/14fKLIo)

842. http://mathworks.com/matlabcentral/fileexchange/24504-fft-based-convolution (or: http://bit.ly/12EIxLq)

843. http://en.wikipedia.org/wiki/Convolution_theorem (or: http://bit.ly/1bdamUy)

844. http://mathworks.com/matlabcentral/fileexchange/24504-fft-based-convolution#feedback (or: http://bit.ly/1lBV40F) — scroll down to the comment by Romesh on Feb 12, 2012; also read other comments which echo this statement

845. http://mathworks.com/matlabcentral/fileexchange/44466-beating-matlabs-convolution-function-conv-m-for-long-real-sequences (or: http://bit.ly/IfUdBT)

846. http://mathworks.com/matlabcentral/newsreader/view_thread/326037#896234 (or: http://bit.ly/18ztWUK). A similar application by Bruno Luong of conv2 as an efficient replacement for a double loop can be found here: http://mathworks.com/matlabcentral/newsreader/view_thread/336920#925287 (or: http://bit.ly/Y8QyOe).

847. http://mathworks.com/matlabcentral/newsreader/view_thread/261037#720879 (or: http://bit.ly/1e9ZmqE) and subsequent posts on the same thread, which provide a very interesting discussion by Matt and Bruno of the intricacies and performance-optimization of *conv2*.

848. http://mathworks.com/matlabcentral/newsreader/view_thread/261037#897426 (or: http://bit.ly/1ea09YC)

849. http://mathworks.com/matlabcentral/newsreader/view_thread/324985 (or: http://bit.ly/1o90BIU)

850. http://mathworks.com/matlabcentral/answers/109753-integralfilter-vs-conv2-speed#answer_118374 (or: http://bit.ly/18wTOa5). IPT's *imdilate* function was also suggested as a possible substitute: http://mathworks.com/matlabcentral/newsreader/view_thread/336920#925299 (http://bit.ly/Y8PRod).

851. http://mathworks.com/matlabcentral/fileexchange/29648-gpuconv2 (or: http://bit.ly/1dKpzLW)

852. http://blogs.mathworks.com/pick/2013/01/11/functional-programming-constructs#comment-18249 (or: http://bit.ly/1d7GeGu)

853. http://stackoverflow.com/questions/24466284/vectorization-of-nested-loops-and-if-statements-in-matlab (or: http://bit.ly/VnKwHN)

854. http://walkingrandomly.com/?p=5043 (or: http://bit.ly/1919QW5)

855. http://mathworks.com/matlabcentral/newsreader/view_thread/325881#895740 (or: http://bit.ly/12EC96Z)

856. http://stackoverflow.com/a/16092668/233829 (or: http://bit.ly/12ECzdl)

857. http://www.ee.columbia.edu/~marios/matlab/What's%20the%20big%20deal.pdf (or: http://bit.ly/15DrwUZ); http://mathworks.com/company/newsletters/articles/whats-the-big-deal.html (or: http://bit.ly/15DrLPS)

858. http://stackoverflow.com/questions/20071950/matlab-function-improving-speed-performance-of-code (or: http://bit.ly/1cHvbom)

859. http://mathworks.com/company/newsletters/articles/enhancing-multi-core-system-performance-using-parallel-computing-with-matlab.html (or: http://bit.ly/140hz10)

860. http://mathworks.com/matlabcentral/newsreader/view_thread/331015 (or: http://bit.ly/19aAxrE)

861. http://mathworks.com/matlabcentral/newsreader/view_thread/331015#909522 (or: http://bit.ly/19aAj3Q), slightly edited for clarity

862. http://home.online.no/~pjacklam (or: http://bit.ly/13gyD71); http://mathworks.com/matlabcentral/fileexchange/authors/1392 (or: http://bit.ly/13gz5SN)

863. http://home.online.no/~pjacklam/matlab/doc/mtt/doc/mtt.pdf (or: http://bit.ly/14Wq3dG)

864. http://mathworks.com/matlabcentral/newsreader/view_thread/22983 (or: http://bit.ly/13gBYmI)

865. http://matlab.wikia.com/wiki/FAQ (or: http://bit.ly/13gDEwy)

866. Description: http://home.online.no/~pjacklam/matlab/doc/mtt (or: http://bit.ly/14WrkkP); Download: http://home.online.no/~pjacklam/matlab/doc/mtt/mtt (or: http://bit.ly/14Wqtk0)

867. http://www.ee.columbia.edu/~marios/matlab/mtt.tar.gz (or: http://bit.ly/1aMPWlX)

868. http://mathworks.com/matlabcentral/newsreader/view_thread/323824 (or: http://bit.ly/12HuvcK)

869. http://mathworks.com/matlabcentral/newsreader/view_thread/326112 (or: http://bit.ly/18QCOus)

870. http://mathworks.com/matlabcentral/newsreader/view_thread/326491 (or: http://bit.ly/18o1Ehf)

871. http://mathworks.com/help/matlab/matlab_prog/vectorization.html#btjk58d-1 (or: http://bit.ly/18o1NBm)

872. http://mathworks.com/matlabcentral/newsreader/view_thread/329079 (or: http://bit.ly/18o2Xg2)

873. http://mathworks.com/matlabcentral/fileexchange/22441-curve-intersections (or: http://bit.ly/1jVaYRL)

874. http://mathworks.com/matlabcentral/fileexchange/11837-fast-and-robust-curve-intersections (or: http://bit.ly/1jVdvvc). This utility uses a different vectorized approach that is worth understanding and comparing to *InterX*'s. For the provided example, both utilities produce the same results (to within floating-point numerical accuracy), at the same run-time. Also see related http://mathworks.com/matlabcentral/newsreader/view_thread/332277 (or: http://bit.ly/1rogrBU).

875. http://mathworks.com/company/newsletters/articles/matlab-and-simulink-in-the-world-parallel-computing.html (or: http://bit.ly/1bc2zlF). Readers might find interest in LLNL's Introduction to Parallel Computing: http://computing.llnl.gov/tutorials/parallel_comp (or: http://1.usa.gov/1tupqVC).

876. https://developer.nvidia.com/category/zone/cuda-zone(or: http://bit.ly/1iFLgm2); http://en.wikipedia.org/wiki/CUDA (or: http://bit.ly/1fusisl)

877. http://khronos.org/opencl (or: http://bit.ly/1iFKANx); http://en.wikipedia.org/wiki/OpenCL (or: http://bit.ly/1iFKDJf)

878. A concise and highly-readable history of CPU/GPU evolution can be found the following series of articles: http://blog.accelereyes.com/blog/2013/03/07/cpu-processing-trends-for-dummies (or: http://bit.ly/NjyISw), http://blog.accelereyes.com/blog/2013/03/17/heterogeneous-computing-trends-for-dummies (or: http://bit.ly/NjyYRD); http://blog.accelereyes.com/blog/2013/04/08/parallel-software-development-trends-for-dummies (or: http://bit.ly/Njz1gl). A detailed overview of GPU historical evolution can be found here: http://blog.accelereyes.com/blog/2013/04/23/history-of-the-modern-gpu-series (or: http://bit.ly/NjAwLp).

879. http://mathworks.com/tagteam/72903_92027v00Cleve_Why_No_Parallel_MATLAB_Spr_1995.pdf (or: http://bit.ly/1bc4Opd)

880. As explained by Cleve Moler in his June 2007 follow-up article *"Parallel MATLAB"*: http://mathworks.com/company/newsletters/articles/parallel-matlab-multiple-processors-and-multiple-cores.html (or: http://bit.ly/10NmZMt)

881. http://mathworks.com/company/newsletters/articles/parallel-matlab-multiple-processors-and-multiple-cores.html (or: http://bit.ly/10NmZMt)

882. http://mathworks.com/products/parallel-computing/examples.html (or: http://bit.ly/18UArET)

883. http://mathworks.com/products/parallel-computing/builtin-parallel-support.html (or: http://bit.ly/12E4q0m)

884. For example: http://walkingrandomly.com/?p=2856 (or: http://bit.ly/1fNKrD8); Compiler started supporting parallel code in R2011b; Coder started generating parallel MEX code for *parfor* loops in R2012b; etc.

885. http://link.springer.com/article/10.1007/s10766-008-0082-5 (or: http://bit.ly/1ghexMt)

886. http://mathworks.com/discovery/matlab-multicore.html (or: http://bit.ly/1it3Awb)

887. http://mathworks.com/help/distcomp/troubleshooting-and-debugging.html#f5-15798 (or: http://bit.ly/1gr0XWE)

888. http://web.eecs.utk.edu/~luszczek/pubs/parallelmatlab.pdf (or: http://bit.ly/1nCnTGP)

889. See additional *parfor* examples at http://mathworks.com/company/newsletters/articles/enhancing-multi-core-system-performance-using-parallel-computing-with-matlab.html, http://mathworks.com/matlabcentral/newsreader/view_thread/321142#880479, and http://blogs.mathworks.com/loren/2009/10/02/using-parfor-loops-getting-up-and-running/

890. http://mathworks.com/help/distcomp/troubleshooting-and-debugging.html#f5-15586 (or: http://bit.ly/1gr00gP)

891. http://mathworks.com/matlabcentral/newsreader/view_thread/325235 (or: http://bit.ly/1dJgDqf)

892. http://mathworks.com/help/distcomp/getting-started-with-parfor.html#brdqn6p-1 (or: http:// bit.ly/1btbk0T)

893. http://stackoverflow.com/questions/4495000/multi-threading-in-matlab (or: http://bit.ly/18UCuJ5)

894. http://stackoverflow.com/questions/22297083/matlab-parfor-cannot-run-due-to-the-way-p-is-used (or: http://bit.ly/1d3VCdi)

895. http://stackoverflow.com/questions/21989725/optimising-multidimensional-array-performance-matlab (or: http://bit.ly/QsR7ik)

896. http://mathworks.com/matlabcentral/answers/109358-puzzling-performance-trends-in-for-parfor-for-drange (or: http://bit.ly/1dJj0Jz)

897. http://mathworks.com/matlabcentral/newsreader/view_thread/323737 (or: http://bit.ly/106oNFu)

898. http://mathworks.com/matlabcentral/fileexchange/31673-parfor-progress-monitor-v2 (or: http://bit.ly/13ci76q); http://mathworks.com/matlabcentral/fileexchange/35609-matlab-parforprogress2 (or: http://bit.ly/13cibmF); http://mathworks.com/matlabcentral/fileexchange/32101-progress-monitor-progress-bar-that-works-with-parfor (or: http://bit.ly/13ciazj)

899. http://mathworks.com/matlabcentral/newsreader/view_thread/251091#649117 (or: http://bit.ly/1tJgOtj)

900. http://mathworks.com/matlabcentral/newsreader/view_thread/306658 (or: http://bit.ly/1bQmQOe)

901. For example: http://web.eecs.utk.edu/~luszczek/pubs/parallelmatlab.pdf (or: http://bit.ly/1nCnTGP)

902. For more information on initializing random numbers in PCT workers, see http://mathworks.com/help/distcomp/control-random-number-streams.html (or: http://bit.ly/1mqWCer) http://mathworks.com/matlabcentral/newsreader/view_thread/327817 (or: http://bit.ly/1mqWJq9); http://walkingrandomly.com/?p=2755 (or: http://bit.ly/17W1Lmj).

903. The seminal 2009 paper by Gaurav Sharma and Jos Martin (http://link.springer.com/article/10.1007/s10766-008-0082-5 or: http://bit.ly/1ghexMt, §5.1.2 p. 9) on the inner workings of PCT and DCS, states a limit of 256 KB. In practice, due to serialization overheads, the actual limit is a bit smaller than 64 KB.

904. Adapted from http://alenblog.wordpress.com/2011/04/21/run-two-matlab-functions-simutaneously-in-parallel (or: http://bit.ly/18UDfCa)

905. http://blogs.mathworks.com/loren/2013/12/09/getting-data-from-a-web-api-in-parallel (or: http://bit.ly/1gNuCOV)

906. http://www.mathworks.com/help/distcomp/working-with-codistributed-arrays.html#bqjuynt (or: http://bit.ly/1j0bNrp)

907. http://mathworks.com/matlabcentral/newsreader/view_thread/316347#889130 (or: http://bit.ly/13cjZvV)

908. http://mathworks.com/help/distcomp/profiling-parallel-code.html (or: http://bit.ly/1dJymhn)

909. http://mathworks.com/help/distcomp/mpiprofile.html (or: http://bit.ly/1dJEOoy)

910. http://mathworks.com/help/distcomp/profiling-parallel-code.html (or: http://bit.ly/1dJymhn); http://mathworks.com/help/distcomp/mpiprofile.html (or: http://bit.ly/1dJEOoy)

911. http://mathworks.com/matlabcentral/newsreader/view_thread/331120#910812 (or: http://bit.ly/1p2HOCV)

912. http://mathworks.com/matlabcentral/fileexchange/27472-partictoc (or: http://bit.ly/Nju3jy)

913. PCT's GPU-computing capabilities are listed at http://mathworks.com/help/distcomp/gpu-capabilities-and-performance.html (or: http://bit.ly/1je8kr8)

914. http://mathworks.com/help/distcomp/release-notes.html (or: http://bit.ly/MD6tyD). One specific example: http://mathworks.com/matlabcentral/newsreader/view_thread/330362#907594 (or: http://bit.ly/1gvoF66).

915. http://mathworks.com/products/parallel-computing/tutorials.html (or: http://bit.ly/1fNPzqO)

916. http://mathworks.com/matlabcentral/fileexchange/31336-demo-files-for-parallel-computing-with-matlab-on-multicore-desktops-and-gpus-webinar (or: http://bit.ly/17FV4Tz), which are the files of the webinar recorded here: http://mathworks.com/wbnr56334 (or: http://bit.ly/17FVj0T); also see http://mathworks.com/wbnr59816

917. http://mathworks.com/products/parallel-computing/code-examples.html (or: http://bit.ly/1fNQ9oE); http://blogs.mathworks.com/steve/2013/12/10/image-processing-with-a-gpu (or: http://bit.ly/NjAOll)

918. http://mathworks.com/discovery/matlab-gpu.html (or: http://bit.ly/1g2Zk3r); http://mathworks.com/discovery/seismic-data-processing.html (or: http://bit.ly/1lFW6Xp)

919. http://mathworks.com/matlabcentral/fileexchange/?term=gpu (or: http://bit.ly/18UDPj6)

920. http://mathworks.com/help/distcomp/examples/illustrating-three-approaches-to-gpu-computing-the-mandelbrot-set.html (or: http://bit.ly/1g2GF7E); http://blogs.mathworks.com/loren/2011/07/18/a-mandelbrot-set-on-the-gpu (or: http://bit.ly/18UE9i3) and http://blogs.mathworks.com/loren/2011/04/21/deploying-multiple-c-shared-libraries (or: http://bit.ly/18UEili); also see the related *mandelbrotViewer* demo — http://

mathworks.com/matlabcentral/fileexchange/30988-a-gpu-mandelbrot-set (or: http://bit. ly/18UEn8C), http://mathworks.com/matlabcentral/fileexchange/33201-gpu-julia-set-explorer (or: http://bit.ly/18UEqBp)

921. Consult CUDA's C Programming Guide (http://docs.nvidia.com/cuda/cuda-c-programming-guide or: http://bit.ly/18UErW9) for additional details

922. http://en.wikipedia.org/wiki/General-purpose_computing_on_graphics_processing_units (or: http://bit.ly/1fuuGzd)

923. See also http://blogs.mathworks.com/loren/2012/02/06/using-gpus-in-matlab (or: http:// bit.ly/10iEfZI)

924. See http://mathworks.com/matlabcentral/fileexchange/42779-gpu-vs-cpu-speed-test-of-finite-difference-equation (or: http://bit.ly/1je8uPh) for a performance comparison of the solution of a finite difference problem using GPU and CPU in MATLAB

925. http://mathworks.com/company/events/conferences/automotive-conference-stutt-gart/2012/proceedings/primer-parallel-computing-with-matlab-and-simulink.pdf#page=12 (or: http://bit.ly/1fuN23f). A similar analysis was performed for another example here: http://blogs.mathworks.com/loren/2012/02/06/using-gpus-in-matlab (or: http://bit. ly/10iEfZI)

926. See introductory video at http://mathworks.com/videos/introduction-to-gpu-computing-with-matlab-68770.html (or: http://bit.ly/XAt0j7)

927. http://en.wikipedia.org/wiki/Comparison_of_Nvidia_graphics_processing_units (or: http://bit.ly/1fNN1Jn); http://en.wikipedia.org/wiki/CUDA#Supported_GPUs (or: http:// bit.ly/1fut1ty), which also has a good listing of the functional differences between the various SM levels.

928. http://mathworks.com/help/distcomp/examples/illustrating-three-approaches-to-gpu-computing-the-mandelbrot-set.html (or: http://bit.ly/1g2GF7E)

929. http://mathworks.com/help/distcomp/using-gpuarray.html (or: http://bit.ly/1g2wmAF)

930. http://mathworks.com/help/distcomp/run-built-in-functions-on-a-gpu.html (or: http://bit. ly/1g2vPyD); http://linkedin.com/groups/Benchmarking-new-NVIDIA-Drivers-through-134533.S.254310232 (or: http://linkd.in/NjBI1d).

931. http://icl.cs.utk.edu/magma (or: http://bit.ly/1dTmN4O); http://stackoverflow.com /questions/6058139/why-is-matlab-so-fast-in-matrix-multiplication/6058811#6058811 (or: http://bit.ly/1dTmMh9)

932. See also http://walkingrandomly.com/?p=3537 (or: http://bit.ly/10bEhTu)

933. http://mathworks.com/help/distcomp/examples/improve-performance-of-element-wise-matlab-functions-on-the-gpu-using-arrayfun.html (or: http://bit.ly/1g2AWig). A more sophisticated usage example is provided here: http://mathworks.com/help/distcomp /examples/using-gpu-arrayfun-for-monte-carlo-simulations.html (or: http://bit.ly/1g2CVDd)

934. http://walkingrandomly.com/?p=3634 (or: http://bit.ly/1kmO9FT)

935. See http://mathworks.com/help/matlab/ref/bsxfun.html (or: http://bit.ly/UfBxoy) for the list of built-in functions supported by *bsxfun*.

936. A recent MathWorks newsletter article explains how a geometric scene-rendering problem was modified using *arrayfun* and GPUs, achieving a 15Kx (!) speedup compared to iterative code, and 10x compared with vectorized code: http://mathworks.com/company/newsletters /articles/solving-large-geometric-and-visualization-problems-with-gpu-computing-in-mat-lab.html (or: http://bit.ly/1sRMqxp); source-code: http://mathworks.com/matlabcentral /fileexchange/46502-rayshapearticle (or: http://bit.ly/1sRMVYe)

937. http://quantsupport.com/1/post/2013/09/generating-correlated-random-numbers-on-a-gpu.html (or: http://bit.ly/1aCpXvS)

938. http://mathworks.com/help/distcomp/pagefun.html (or: http://bit.ly/1jef0Wf)

939. http://mathworks.com/help/distcomp/gputimeit.html (or: http://bit.ly/1jeeRC5); http:// mathworks.com/help/distcomp/measure-and-improve-gpu-performance.html (or: http:// bit.ly/1dKaL03)

940. http://developer.nvidia.com/cuda-toolkit (or: http://bit.ly/1fXV3zk)

941. http://mathworks.com/matlabcentral/newsreader/view_thread/326686 (or: http://bit.ly/18UFmpx)

942. http://mathworks.com/company/newsletters/articles/prototyping-algorithms-and-testing-cuda-kernels-in-matlab.html (or: http://bit.ly/1dKgEKq)

943. http://mathworks.com/videos/matlab-for-cuda-programmers-81971.html (or: http://bit.ly/1gBGaQN). The source-code files can be found here: http://mathworks.com/matlabcentral/fileexchange/38401-matlab-for-cuda-programmers (or: http://bit.ly/NjxM0A) and http://mathworks.com/matlabcentral/fileexchange/41089-color-balance-demo-with-gpu-computing (or: http://bit.ly/NjxUgB).

944. http://devblogs.nvidia.com/parallelforall/calling-cuda-accelerated-libraries-matlab-computer-vision-example (or: http://bit.ly/1thNCcw)

945. http://developer.nvidia.com/nvidia-visual-profiler (or: http://bit.ly/1pEkYSP)

946. http://nvidia.com/object/nsight.html (or: http://bit.ly/1gvtxKX)

947. http://developer.nvidia.com/cuda

948. Jason Sanders, Edward Kandrot: *CUDA by Example, Addison Wesley Professional* 2010, ISBN 0131387685

949. For CUDA Toolkit libraries, visit http://developer.nvidia.com

950. An example of this was shown in the second half of the following MathWorks webinar: http://mathworks.com/videos/matlab-for-cuda-programmers-81971.html (or: http://bit.ly/1gBGaQN). The source-code files can be found here: http://mathworks.com/matlabcentral/fileexchange/38401-matlab-for-cuda-programmers (or: http://bit.ly/NjxM0A) and http://mathworks.com/matlabcentral/fileexchange/41089-color-balance-demo-with-gpu-computing (or: http://bit.ly/NjxUgB)

951. http://thrust.github.io; https://developer.nvidia.com/Thrust (or: http://bit.ly/NjwAu7)

952. http://accelereyes.com/arrayfire/c/; see §7.2.3.

953. http://developer.nvidia.com/openacc (or: http://bit.ly/1o5byhx); http://openacc-standard.org (or: http://bit.ly/1o5bQ89)

954. http://culatools.com

955. http://mathworks.com/help/distcomp/run-mex-functions-containing-cuda-code.html (or: http://bit.ly/1dKl3x1)

956. http://mathworks.com/help/distcomp/examples/accessing-advanced-cuda-features-using-mex.html (or: http://bit.ly/1g2Hk98)

957. See, for instance, http://mathworks.com/matlabcentral/fileexchange/25314-cuda-mex (or: http://bit.ly/XAxKFy)

958. http://cs.ucf.edu/~janaka

959. http://cs.ucf.edu/~janaka/gpu/using_nvmex.htm. The tool can be downloaded from http://cs.ucf.edu/~janaka/gpu/nvmex_tool.zip (or: http://bit.ly/UMCAeE) or from http://UndocumentedMatlab.com/files/nvmex_tool.zip (or: http://bit.ly/1bp9rwq)

960. Janaka used the *cutil* library in his compilation of the Simulink S-function block that uses CUDA: http://cs.ucf.edu/~janaka/gpu/simulink_cuda.htm (or: http://bit.ly/MvWU3W)

961. http://cs.ucf.edu/~janaka/gpu/issues.htm (or: http://bit.ly/UMIQD1)

962. http://thrust.github.io; https://developer.nvidia.com/Thrust (or: http://bit.ly/NjwAu7)

963. http://developer.nvidia.com/cuda-toolkit (or: http://bit.ly/1fXV3zk)

964. http://mathworks.com/help/distcomp/run-mex-functions-containing-cuda-code.html (or: http://bit.ly/1dKl3x1)

965. http://mathworks.com/matlabcentral/fileexchange/29648-gpuconv2 (or: http://bit.ly/1dKpzLW)

966. An archived version of the webpage from 2012: https://web.archive.org/web/20120228181441/http://developer.nvidia.com/matlab-cuda (or: http://bit.ly/1fY1y5 k). Installations and usage instructions for the plugins can be found here: http://cs.smith.edu/classwiki/index.php/Matlab_and_CUDA (or: http://bit.ly/1aAxyx1) and http://cs.ucf.edu/~janaka/gpu (or: http://bit.ly/1aAxUE3) and http://mathworks.com/matlabcentral/newsreader/view_thread/292928 (or: http://bit.ly/1gcSaeR).

967. http://www.smast.umassd.edu/CMLAB/notebook/wp-content/uploads/2010/09/cuda_matlab.pdf (or: http://bit.ly/1dKoow9)

968. http://www.systematics.co.il/MathWorks/News/Pdf/2009/nvidia.pdf (or: http://bit.ly/18UG1qS) or alternatively https://www.ljll.math.upmc.fr/groupes/gpgpu/tutorial/Accelerating_Matlab_with_CUDA.pdf (or: http://bit.ly/19pdZaM)

969. Technical solution 1-DOIZVT: http://mathworks.com/matlabcentral/answers/98849 (or: http://bit.ly/MdjUoL)

970. For example: http://mathworks.com/matlabcentral/newsreader/view_thread/324086 (or: http://bit.ly/1kDdKKO)

971. http://www.cac.cornell.edu/matlab/TechDocs/Examples/BestPracticesGPU.aspx (or: http://bit.ly/1qMxQG6)

972. http://en.wikipedia.org/wiki/Distributed_computing (or: http://bit.ly/NocKxT)

973. http://mathworks.com/products/distriben/supported (or: http://bit.ly/10MROX1)

974. http://mathworks.com/matlabcentral/newsreader/view_thread/336485#924135 (or: http://bit.ly/1pwUgZD)

975. http://mathworks.com/help/distcomp/use-a-job-manager.html (or: http://bit.ly/XAzCOB)

976. http://www.hpc.maths.unsw.edu.au/tensor/matlab#distributed-computing-server (or: http://bit.ly/18UGrgU)

977. http://mathworks.com/help/distcomp/job-monitor.html#bs6wi5t-1 (or: http://bit.ly/18UGI3t)

978. http://mathworks.com/matlabcentral/fileexchange/23728-tcpip-distributed-waitbar (or: http://bit.ly/18UGKby)

979. http://mathworks.com/matlabcentral/fileexchange/7751-visual-timing-report-for-distributed-tasks (or: http://bit.ly/17xpOUJ)

980. http://mathworks.com/matlabcentral/fileexchange/27509-taskmap (or: http://bit.ly/17xpKEG)

981. http://mathworks.com/help/distcomp/parallel.cluster.html#btdlfig-4 (or: http://bit.ly/1h8cQWK). Usage examples: https://csguide.cs.princeton.edu/software/matlab (or: http://bit.ly/1h8bRpy)

982. See http://mathworks.com/help/distcomp/parallel.job.html (or: http://bit.ly/XAzIFO) for job object properties

983. http://mathworks.com/matlabcentral/newsreader/view_thread/331001#909477 (or: http://bit.ly/1g32nZs)

984. http://mathworks.com/help/distcomp/share-code-with-the-workers.html#bqur7ev-11 (or: http://bit.ly/1g31XCa)

985. http://mathworks.com/help/distcomp/troubleshooting-and-debugging.html (or: http://bit.ly/18UH227); http://mathworks.com/matlabcentral/newsreader/view_thread/323497 (or: http://bit.ly/18UHai6); http://www.cac.cornell.edu/wiki/index.php?title=Red_Cloud_with_MATLAB/Tips (or: http://bit.ly/18UHeP8)

986. See http://blogs.mathworks.com/loren/2012/04/20/running-scripts-on-a-cluster-using-the-batch-command-in-parallel-computing-toolbox (or: http://bit.ly/10MS6Nx) for examples on using the **batch** command

987. See http://mathworks.com/help/distcomp/batch.html (or: http://bit.ly/UAzolY) for more details

988. See http://mathworks.com/matlabcentral/newsreader/view_thread/330615 (or: http://bit.ly/1jeeAio) for a discussion of the **CommandWindowOutput** property

989. http://scv.bu.edu/matlab

990. http://bu.edu/tech/about/research/training/scv-software-packages/matlab/matlab-batch (or: http://bit.ly/1jIRsXW)

991. http://mathworks.com/help/distcomp/program-independent-jobs-for-a-supported-scheduler.html (or: http://bit.ly/1kTKIJz)

992. See also http://blogs.mathworks.com/pick/2011/01/14/good-matlab-coding-practices (or: http://bit.ly/UfDlxA) and http://blogs.mathworks.com/loren/2012/01/13/best-practices-for-programming-matlab (or: http://bit.ly/UfDnWu)

993. Jeremy Kepner, *Parallel MATLAB for Multicore and Multinode Computers*, SIAM 2009, ISBN 089871673X, §5.11

994. http://stackoverflow.com/questions/5041328/cuda-coalesced-memory (or: http://bit.ly /QYxwa5)

995. http://mathworks.com/matlabcentral/newsreader/view_thread/328054#901610 (or: http:// bit.ly/1kiqREk)

996. http://stackoverflow.com/questions/19477224/matlab-convolution-using-gpu (or: http:// bit.ly/H9j1LF). Also see related http://stackoverflow.com/questions/22305113/the-perfor- mance-of-convn-using-gpu-in-matlab (or: http://bit.ly/1ppOREk)

997. http://mathworks.com/company/events/conferences/matlab-computational-finance-vir- tual-conference/2013/proceedings/videos/speeding-up-algorithms-when-parallel-comput- ing-and-gpus-do-and-dont-accelerate.html (or: http://bit.ly/OtiRlf)

998. http://walkingrandomly.com/?p=3736 (or: http://bit.ly/1kmHxaq); http://csgillespie. wordpress.com/2011/07/12/how-to-review-a-gpu-statistics-paper (or: http://bit.ly/1iFHjxI)

999. http://mathworks.com/help/distcomp/advanced-topics.html#brdsl35-1 (or: http://bit.ly /1nS6Os7)

1000. http://blogs.mathworks.com/cleve/2012/11/26/magic-squares-meet-supercomputing (or: http://bit.ly/1ggZlio)

1001. See http://mathworks.com/products/parallel-computing/builtin-parallel-support.html (or: http://bit.ly/12E4q0 m) for the full list of toolboxes with built-in parallel support. See also example usage at http://mathworks.com/company/newsletters/articles/improving-optimi- zation-performance-with-parallel-computing.html (or: http://bit.ly/12E4qNZ).

1002. http://walkingrandomly.com/?p=3667 (or: http://bit.ly/12E4wFw)

1003. Read the explanation by Edric Ellis, senior PCT developer: http://stackoverflow. com/a/10546979/233829 (or: http://bit.ly/18lFGww)

1004. http://walkingrandomly.com/?p=3736 (or: http://bit.ly/1kmHxaq)

1005. http://mathworks.com/matlabcentral/newsreader/view_thread/324377 (or: http://bit. ly/12E4PQw)

1006. A complex GPU parallelization example can be found at http://mathworks.com/company /newsletters/articles/gpu-programming-in-matlab.html (or: http://bit.ly/12E4GfZ)

1007. http://blogs.mathworks.com/loren/2012/02/06/using-gpus-in-matlab#comment-32991 (or: http://bit.ly/12E4Jsc)

1008. http://mathworks.com/matlabcentral/newsreader/view_thread/329028#904570 (or: http:// bit.ly/Mdnq2i)

1009. http://mathworks.com/help/distcomp/measure-and-improve-gpu-performance. html#bt2g5ac-1 (or: http://bit.ly/N3Gfpz)

1010. http://mathworks.com/matlabcentral/newsreader/view_thread/329028#904570 (or: http:// bit.ly/Mdnq2i)

1011. http://mathworks.com/matlabcentral/newsreader/view_thread/330362 (or: http://bit.ly /MdiFWt); http://www.slideshare.net/vdimitris/hybrid-cpu-gpu-matlab-image-processing- benchmarking (or: http://slidesha.re/NjD8Jb)

1012. http://mathworks.com/help/distcomp/measure-and-improve-gpu-performance.html (or: http://bit.ly/1dKaL03)

1013. http://mathworks.com/company/events/conferences/matlab-computational-finance-vir- tual-conference/2013/proceedings/videos/speeding-up-algorithms-when-parallel-comput- ing-and-gpus-do-and-dont-accelerate.html (or: http://bit.ly/OtiRlf)

1014. http://mathworks.com/matlabcentral/newsreader/view_thread/333296 (or: http://bit.ly /1kilwNh)

1015. http://walkingrandomly.com/?p=3634 (or: http://bit.ly/1kmO9FT)

1016. Technical solution 1-FHLD0F: http://mathworks.com/matlabcentral/answers/99300 (or: http://bit.ly/1bO9DZ)

1017. http://blogs.mathworks.com/loren/2012/12/14/measuring-gpu-performance (or: http:// bit.ly/10iXVfZ); also: http://mathworks.com/help/distcomp/examples

/measuring-gpu-performance.html (or: http://bit.ly/N3H3e4). Both of these resources provide a detailed analysis that is well worth reading (they are not exactly the same).

1018. http://mathworks.com/matlabcentral/fileexchange/34080-GPUBench (or: http://bit.ly /UfDJw7); http://blogs.mathworks.com/pick/2013/05/17/benchmarking-your-gpu (or: http:// bit.ly/19tW9AY)

1019. http://stackoverflow.com/questions/25355495/unenhanced-performance-of-matlab-gpu-computing (or: http://bit.ly/1qtOQ3l)

1020. http://mathworks.com/help/distcomp/measure-and-improve-gpu-performance. html#bt2g515 (or: http://bit.ly/N3EzMC)

1021. The three parts are available at: http://walkingrandomly.com/?p=3604 (or: http://bit.ly /1iFHQjc), http://walkingrandomly.com/?p=3978 (or: http://bit.ly/1iFHVDw) and http:// walkingrandomly.com/?p=4062 (or: http://bit.ly/1dK3OvQ).

1022. http://www.cac.cornell.edu/matlab/TechDocs/Examples/BestPracticesGPU.aspx (or: http:// bit.ly/1qMxQG6)

1023. http://bu.edu/tech/about/research/training/online-tutorials/matlab-pct (or: http://bit.ly /NohK5v). Additional resources from Boston University are referenced in Appendix §A.1.1.

1024. https://sites.google.com/a/case.edu/hpc-upgraded-cluster (or: http://bit.ly/1mU1MwV); https://sites.google.com/a/case.edu/hpc-upgraded-cluster/hpcc-training (or: http://bit. ly/1mU1R3H)

1025. https://web.archive.org/web/20130203134931/http://www.nvidia.com/object/tesla-jacket-gpu-acceleration (or: http://bit.ly/1aApGM8)

1026. http://blog.accelereyes.com/blog/2012/12/12/exciting-updates-from-accelereyes (or: http:// bit.ly/VQpp9Y)

1027. http://wiki.accelereyes.com/wiki/index.php/Torben%27s_Corner (or: http://bit.ly /18UHMnT)

1028. See http://wiki.accelereyes.com/wiki/index.php/Category:GFOR (or: http://bit.ly/UfCZqO) for the full list of supported functions.

1029. See http://wiki.accelereyes.com/wiki/index.php/GFOR_Usage (or: http://bit.ly/UfD65E) for more comprehensive discussion of *gfor*-loops and example code.

1030. The full set of supported operations and the example usage of both constructs can be found at: http://wiki.accelereyes.com/wiki/index.php/Category:GCOMPILE (or: http://bit.ly/UfDcu1)

1031. http://mathworks.com/matlabcentral/fileexchange/13775-multicore (or: http://bit.ly /1cAWENf)

1032. http://www.ll.mit.edu/mission/isr/matlabmpi/matlabmpi.html (or: http://bit.ly /1cAWllA)

1033. http://www.ll.mit.edu/mission/isr/pmatlab/pmatlab.html (or: http://bit.ly/1cAVVM0)

1034. http://archive.osc.edu/bluecollarcomputing/applications/bcMPI/index.shtml (or: http:// bit.ly/QAIwcX)

1035. https://developer.nvidia.com/matlab-cuda (or: http://bit.ly/1cAVXDQ)

1036. http://gp-you.org or http://sourceforge.net/projects/gpumat (or: http://bit.ly/NkG4pZ)

1037. http://developer.nvidia.com/cuda-toolkit (or: http://bit.ly/1fXV3zk). Also see §6.2.5 and §6.2.7.

1038. http://sccn.ucsd.edu/wiki/GPU_and_EEGLAB (or: http://bit.ly/MBkhcs) shows how an early version of GPUmat from 2010 provided a 20x speedup for a single-precision power operation on a 112-core GPU, compared to standard vectorized MATLAB on a 16-core CPU.

1039. There are separate download ZIP files for separate architectures, with precompiled MEX files for Win32/64 and Linux 32/54: http://sourceforge.net/projects/gpumat/files (or: http://bit. ly/1d6DbCD)

1040. http://sourceforge.net/p/gpumat/discussion/help/thread/9f59b5b6 (or: http://bit.ly /1d6BTY4)

1041. http://waterloo.sourceforge.net (or: http://bit.ly/1kM1Ocd), http://undocumentedmatlab. com/blog/waterloo-graphics (or: http://bit.ly/1jnmAfl). Also see Malcolm's related *sig-TOOL* project: http://sourceforge.net/projects/sigtool (or: http://bit.ly/1o5fgaY).

1042. http://mathworks.com/matlabcentral/fileexchange/13775-multicore (or: http://bit.ly /12Sg1Hh); http://tech.groups.yahoo.com/group/multicore_for_matlab (or: http://bit.ly /12SgjOd)

1043. http://accelereyes.com/products/arrayfire (or: http://bit.ly/LzeFje); http://developer. nvidia.com/accelereyes-arrayfire (or: http://bit.ly/1fNZLzJ)

1044. http://accelereyes.com/arrayfire/c (or: http://bit.ly/LzeQuQ)

1045. See discussion at http://forums.accelereyes.com/forums/viewtopic.php?f=17&t=2656 (or: http:// bit.ly/Lzf19p)

1046. Adapted from http://blog.accelereyes.com/blog/2012/08/28/sar-image-formation-algo-rithms-on-the-gpu (or: http://bit.ly/VQ27Rq). This file can be downloaded from: http:// UndocumentedMatlab.com/books/matlab-performance (or: http://bit.ly/1pKuUdM).

1047. Surveys of some of the mentioned alternatives and several others can be found here: https:// web.archive.org/web/20070315200101/http:/www.interactivesupercomputing.com/ downloads/pmatlab.pdf (or: http://bit.ly/1lBYvBK) and here: https://web.archive.org/ web/20090222164513/http:/www.interactivesupercomputing.com /reference/parallelMatlabsurvey.php (or: http://bit.ly/1lBYw8 J)

1048. http://ispc.github.com, http://walkingrandomly.com/?p=3988 (or: http://bit.ly/1cB47Mu)

1049. http://pgroup.com, http://walkingrandomly.com/?p=4064 (or: http://bit.ly/1cB4syy), http://walkingrandomly.com/?p=3378 (or: http://bit.ly/1cB3XVd)

1050. http://developer.amd.com/tools-and-sdks/cpu-development/amd-core-math-library-acml (or: http://bit.ly/17FxiqQ); http://en.wikipedia.org/wiki/AMD_Core_Math_Library (or: http://bit.ly/17FxfLD)

1051. http://khronos.org/opencl (or: http://bit.ly/1iFKANx); http://en.wikipedia.org/wiki /OpenCL (or: http://bit.ly/1iFKDJf)

1052. http://software.intel.com/en-us/vcsource/tools/opencl-sdk (or: http://intel.ly/1cB3rXn), http://walkingrandomly.com/?p=4255 (or: http://bit.ly/1cB3xOA)

1053. http://makeuseof.com/tag/apu-technology-explained (or: http://bit.ly/NjzDCD)

1054. http://txcorp.com/home/gpulib (or: http://bit.ly/1cB2WMV)

1055. http://exelisvis.com/ProductsServices/IDL.aspx (or: http://bit.ly/1cB3a6M)

1056. http://gpulib.blogspot.com/2012/08/experiments-with-opencl.html (or: http://bit.ly /1cB3mCU)

1057. http://culatools.com, http://walkingrandomly.com/?p=4460 (or: http://bit.ly/1cB3PVT)

1058. http://icl.cs.utk.edu/plasma (or: http://bit.ly/1fO11CT)

1059. http://nag.com/numeric/CL/CLdescription.asp (or: http://bit.ly/1cB2D4L)

1060. http://sourceforge.net/projects/viennacl (or: http://bit.ly/LITXwq)

1061. MITMatlab was the PhD thesis of Parry Husbands, in Edelman's group: http://dspace. mit.edu/handle/1721.1/79973 (or: http://bit.ly/1uWhXO3). This evolved in 2001 into MATLAB*P, which was redesigned by Ron Choy in 2002 using MatlabMPI: http://dspace. mit.edu/handle/1721.1/3687 (or: http://bit.ly/1cAR9hy); http://dspace.mit.edu /handle/1721.1/87313 (or: http://bit.ly/1uWjjZj).

1062. http://mcs.anl.gov/research/projects/mpi (or: http://1.usa.gov/1uWoDvD); http:// en.wikipedia.org/wiki/Message_Passing_Interface (or: http://bit.ly/1k886AS); http://com-puting.llnl.gov/tutorials/mpi (or: http://1.usa.gov/1turyfQ)

1063. http://mathworks.com/company/newsletters/articles/objectively-speaking.html (or: http://bit.ly/1cAQP2q)

1064. http://www.ll.mit.edu/HPEC/agendas/proc04/abstracts/choy_ron.pdf (or: http://bit.ly /1cB9fQB). Star-P borrowed ideas from Matlab*P, but was a separate development, later becom-ing the basis for MIT spin-off Interactive Supercomputing: http://www.ll.mit.edu /HPEC/agendas/proc06/Day2/25_Edelman_Pres.pdf (or: http://bit.ly/1cB9OKd), http:// designnews.com/author.asp?doc_id=221989 (or: http://ubm.io/1cB9bQU).

1065. http://www.ll.mit.edu/mission/isr/matlabmpi/matlabmpi.html (or: http://bit.ly/1cAWllA); http://arxiv.org/abs/astro-ph/0107406 (or: http://bit.ly/1uWk2cX)

1066. http://www.ll.mit.edu/mission/isr/pmatlab/pmatlab.html (or: http://bit.ly/1cAVVM0); http://www.ll.mit.edu/HPEC/agendas/proc03/abstracts/kepner-pMatlab.pdf (or: http://bit.ly/1uWiobg). A comprehensive description can be found in Kepner's book *"Parallel MATLAB for Multicore and Multinode Computers"* (SIAM 2009, ISBN 089871673X).

1067. http://archive.osc.edu/bluecollarcomputing/applications/bcMPI/index.shtml (or: http://bit.ly/QAIwcX)

1068. http://www.ugr.es/~jfernand/mpitb_eng.html (or: http://bit.ly/1cBbFi6)

1069. http://mathworks.com/products/connections/product_detail/product_35291.html (or: http://bit.ly/1gr9zfP) — a software package that enabled MATLAB 5 to run on multiprocessor Alpha AXP RISC boards in a parallel manner.

1070. https://web.archive.org/web/20060107092058/http://www-hpc.jpl.nasa.gov/PS/MATPAR (or: http://bit.ly/1gwl2LR)

1071. http://www.cse.ohio-state.edu/presto/pubs/hips07.pdf (or: http://bit.ly/1h83ls0)

1072. http://citeseerx.ist.psu.edu/viewdoc/summary?doi=10.1.1.25.7414 (or: http://bit.ly/1cBbnbc)

1073. http://cs.utexas.edu/users/plapack (or: http://bit.ly/1gwlo59)

1074. Archived version: https://web.archive.org/web/20040603214117/http://www.tc.cornell.edu/Services/Software/CMTM (or: http://bit.ly/1cEYWeg)

1075. http://www.cs.cornell.edu/info/people/lnt/multimatlab.html (or: http://bit.ly/1cB8NSy)

1076. http://techila.fi

1077. http://walkingrandomly.com/?p=3378 (or: http://bit.ly/1cB3XVd)

1078. http://stackoverflow.com/questions/2713218/how-to-do-threading-in-matlab/12644255 (or: http://bit.ly/1cB7bIs), http://mathworks.com/matlabcentral/newsreader/view_thread/321669 (or: http://bit.ly/1cB7hzN), and http://stackoverflow.com/questions/4495000/multi-threading-in-matlab (or: http://bit.ly/18UCuJ5)

1079. http://mathworks.com/matlabcentral/fileexchange/38034-mexthread (or: http://bit.ly/MZOH8M)

1080. http://en.wikipedia.org/wiki/Threading_Building_Blocks (or: http://bit.ly/1uWyt0z). A good starting point for MATLAB-TBB integration is http://stackoverflow.com/questions/24267847/tbb-acting-strange-in-matlab-mex-file (or: http://bit.ly/1uWyT73).

1081. http://tutorialspoint.com/cplusplus/cpp_multithreading.htm (or: http://bit.ly/1c52MgQ)

1082. Additional examples of creating multi-threaded MEX code using Pthreads can be found here: http://blogs.abo.fi/alexeevpetr/2011/11/10/creating-multithreaded-code-for-simple-m-function-matlab-example (or: http://bit.ly/1cGY0zy) and here: https://web.archive.org/web/20090408024014/http://robertoostenveld.ruhosting.nl/index.php/multithreading-matlab-mex (or: http://bit.ly/1cH0vSA)

1083. http://en.wikipedia.org/wiki/Binomial_coefficient (or: http://bit.ly/1cB5NW3)

1084. http://gmplib.org ; also see §8.5.5

1085. See http://undocumentedmatlab.com/blog/explicit-multi-threading-in-matlab-part3 (or: http://bit.ly/1gzxrjm) for a discussion on how to compile on Windows

1086. http://mathworks.com/help/symbolic. See also http://walkingrandomly.com/?p=2391 (or: http://bit.ly/1cB4xCx) for recent features and §4.9.12 for performance-related tips.

1087. http://gmplib.org/manual/Binomial-Coefficients-Algorithm.html (or: http://bit.ly/1cB4Htw)

1088. http://instructables.com/id/Matlab-Multithreading-EASY (or: http://bit.ly/1cGXsK0)

1089. http://mathworks.com/matlabcentral/fileexchange/37515-mmx-multithreaded-matrix-operations-on-n-d-matrices (or: http://bit.ly/1j7WGOO); http://www.cs.washington.edu/people/postdocs/tassa/code (or: http://bit.ly/1n7O8od). Also see §4.5.2. Also see related Thomas Weibel's *MexThread* utility: http://mathworks.com/matlabcentral/fileexchange/38034-mexthread (or: http://bit.ly/MZOH8M).

1090. http://undocumentedmatlab.com/blog/explicit-multi-threading-in-matlab-part3 (or: http://bit.ly/1gzxrjm)

1091. http://openmp.org; http://en.wikipedia.org/wiki/OpenMP; http://computing.llnl.gov/tutorials/openMP (or: http://1.usa.gov/1tupYdX)

1092. See http://mathworks.com/matlabcentral/fileexchange/38372-fast-gmm-and-fisher-vectors (or: http://bit.ly/1cB5ZEO) and http://mathworks.com/matlabcentral/fileexchange/33621-fast-linear-binary-svm-classifier (or: http://bit.ly/1cB64bx) for MATLAB add-on packages that use OpenMP.

1093. http://stackoverflow.com/questions/2626230/running-multiprocess-applications-from-matlab (or: http://bit.ly/1h8nSgj)

1094. http://openmp.org/wp/openmp-specifications (or: http://bit.ly/LzfuIS)

1095. http://mathworks.com/matlabcentral/fileexchange/45501-openmp-mex-cmake (or: http://bit.ly/1lC172j)

1096. http://mathworks.com/matlabcentral/fileexchange/45505-cuda-mex-cmake (or: http://bit.ly/O6sEOq)

1097. http://mathworks.com/help/coder/ref/parfor.html#btz9_xj-9 (or: http://bit.ly/1h0tYMs)

1098. For example: http://openmp.org/forum/viewtopic.php?f=3&t=468 (or: http://bit.ly/1o5dSFl); http://openmp.org/forum/viewtopic.php?f=3&t=1279 (or: http://bit.ly/1o5dZAE)

1099. For example: http://www.arc.vt.edu/userinfo/training/2010bc/bootcamp_2010.html (or: http://bit.ly/1cB6sHd), http://walkingrandomly.com/?p=3898 (or: http://bit.ly/1gmccnf), http://walkingrandomly.com/?p=1795 (or: http://bit.ly/1cB6FtW), and https://web.archive.org/web/20130530100616/http://www.multicoreinfo.com/2009/06/parprog-part-4 (or: http://bit.ly/1cB74MS)

1100. http://developer.nvidia.com/openacc (or: http://bit.ly/1o5byhx); http://openacc-standard.org (or: http://bit.ly/1o5bQ89)

1101. http://google.com/search?q=OpenMP+Matlab (or: http://bit.ly/1o5fqz8)

1102. http://boost.org; http://en.highscore.de/cpp/boost (or: http://bit.ly/MBlmRw); https://en.wikibooks.org/wiki/C%2B%2B_Programming/Libraries/Boost (or: http://bit.ly/1kpMv8z)

1103. http://boost.org/users/download (or: http://bit.ly/1kpNSE7)

1104. http://absurdlycertain.blogspot.com/2011/09/preamble-what-follows-is-guide.html (or: http://bit.ly/MBmxAG) provides a detailed account of integrating Boost threads into MEX. Also see related http://absurdlycertain.blogspot.coml/2011/09/simpler-concurrent-matlab-programming.html (or: http://bit.ly/MBmzZc) and http://mathworks.com/matlabcentral/answers/12648 (or: http://bit.ly/MBmHrz).

1105. http://boost.org/users/history (or: http://bit.ly/1kpOPMF)

1106. http://undocumentedmatlab.com/blog/explicit-multi-threading-in-matlab-part1 (or: http://bit.ly/1c3dZZK)

1107. http://mathworks.com/help/matlab/using-java-libraries-in-matlab.html (or: http://bit.ly/1bO8D7G); http://mathworks.com/help/matlab/matlab_external/bringing-java-classes-and-methods-into-matlab-workspace.html (or: http://bit.ly/1cRDUaB)

1108. http://undocumentedmatlab.com/blog/jmi-java-to-matlab-interface (or: http://bit.ly/1kwAQou). Also see Chapter 9 of my book *Undocumented Secrets of MATLAB-Java Programming*, ISBN 9781439869031, http://UndocumentedMatlab.com/matlab-java-book (or: http://bit.ly/Zqqojt).

1109. http://mathworks.com/matlabcentral/newsreader/view_thread/148991#895929 (or: http://bit.ly/1faAFbb)

1110. http://code.google.com/p/matlabcontrol (or: http://bit.ly/1faw34N). A series of articles by Josh Kaplan explains JMI and Matlab Control: http://undocumentedmatlab.com/blog/tag/joshua-kaplan (or: http://bit.ly/1q2yW0U).

1111. http://docs.oracle.com/javase/tutorial/essential/concurrency/sync.html (or: http://bit.ly/1aQj6kD). A nice presentation of the intricacies of Java concurrency can be found in http://sourceforge.net/projects/javaconcurrenta (or: http://bit.ly/1jr8wUr), discussed here: http://www.oraclejavamagazine-digital.com/javamagazine/20131112#pg34 (or: http://bit.ly/1jr8CLD)

1112. http://mathworks.com/matlabcentral/fileexchange/27975-tcp-output-socket (or: http://bit.ly/1jracNC); http://iheartmatlab.blogspot.com (or: http://bit.ly/1jramob)

1113. http://jade.tilab.com. A MATLAB plugin for JADE (*MACSimJX*) was developed by the University of York, UK (http://agentcontrol.co.uk). Also see http://mathworks.com/matlabcentral/newsreader/view_thread/315617 (or: http://bit.ly/1faFBge), http://mathworks.com/matlabcentral/newsreader/view_thread/265152 (or: http://bit.ly/1faFPUm); http://mathworks.com/matlabcentral/newsreader/view_thread/157857 (or: http://bit.ly/1faFYay); http://mathworks.com/matlabcentral/newsreader/view_thread/166448 (or: http://bit.ly/1faG2a5)

1114. http://undocumentedmatlab.com/blog/explicit-multi-threading-in-matlab-part2 (or: http://bit.ly/1c3e5QV)

1115. http://msdn.microsoft.com/en-us/library/system.threading.thread(v=vs.100).aspx (or: http:// bit.ly/1kEmcvz)

1116. http://undocumentedmatlab.com/blog/explicit-multi-threading-in-matlab-part4 (or: http://bit.ly/1lj0Lk2)

1117. http://mathworks.com/matlabcentral/answers/94884 (or: http://bit.ly/1bWxPZN); http://mathworks.com/matlabcentral/answers/96855 (or: http://bit.ly/1bWycmX)

1118. http://undocumentedmatlab.com/blog/explicit-multi-threading-in-matlab-part4 (or: http://bit.ly/1lj0Lk2)

1119. http://technet.microsoft.com/en-us/sysinternals/bb896653 (or: http://bit.ly/IRC4TO) — see §2.4

1120. http://stackoverflow.com/questions/2626230/running-multiprocess-applications-from-matlab (or: http://bit.ly/1h8nSgj)

1121. For example, Josh Dillon's related *semaphore* utility (http://mathworks.com/matlabcentral/fileexchange/32489-semaphore or: http://bit.ly/1h83HyQ) or Andrew Smart's *semaphore_POSIX_and_Windows* utility (http://mathworks.com/matlabcentral/fileexchange/45504-semaphore_posix_and_windows or: http://bit.ly/1jDg5Hq)

1122. http://en.wikipedia.org/wiki/Load_balancing_(computing) (or: http://bit.ly/1gvm6RK)

1123. http://stackoverflow.com/a/6710779/233829 (or: http://bit.ly/1fWpggO)

1124. http://en.wikipedia.org/wiki/Inter-process_communication (or: http://bit.ly/1iMDWSe)

1125. http://stackoverflow.com/questions/872209/sharing-memory-between-processes-matlab (or: http://bit.ly/1h8i9qJ)

1126. http://mathworks.com/matlabcentral/fileexchange/28572-sharedmatrix /(or: http://bit.ly/17xIcN1); http://smlv.cc.gatech.edu/2010/08/27/shared-memory-in-matlab (or: http://b.gatech.edu/1kjnCcV)

1127. http://bengal.missouri.edu/~kes25c/SharedMemory-Windows.zip (or: http://bit.ly/1bdYhcH)

1128. http://boost.org/doc/libs/1_55_0/doc/html/interprocess/managed_memory_segments.html (or: http://bit.ly/1h8hIN2)

1129. http://infoscience.epfl.ch/record/55784/files/keleher94treadmark.pdf (or: http://bit.ly/1h8aSY1); http://research.cs.wisc.edu/areas/os/Qual/papers/treadmarks.pdf (or: http://bit.ly/1bIUsxt); https://web.archive.org/web/20071001102412/http://www.cs.rice.edu/~willy/TreadMarks/overview.html (or: http://bit.ly/1bIUWno)

1130. http://polaris.cs.uiuc.edu/matmarks (or: http://bit.ly/1h8c2Tm). Note that the current availability of *MATmarks* is unclear.

1131. http://memcached.org

1132. http://undocumentedmatlab.com/blog/inter-matlab-data-transfer-with-memcached (or: http://bit.ly/1nD58n6)

1133. http://en.wikipedia.org/wiki/Named_pipe (or: http://bit.ly/1aSnY8U)

1134. http://mathworks.com/help/matlab/import_export/share-memory-between-applications.html (or: http://bit.ly/11aH5l2)

1135. http://mathworks.com/matlabcentral/newsreader/view_thread/250214 (or: http://bit.ly/1h8cBwi)

1136. http://mathworks.com/matlabcentral/fileexchange/44077-batch-job (or: http://bit.ly/1nleJSQ)

1137. For example: http://mathworks.com/matlabcentral/fileexchange/345-tcp-udp-ip-tool-box-2-0-6 (or: http://bit.ly/1q2Ha9p); http://mathworks.com/matlabcentral/fileexchange/24524-tcp-ip-communications-in-matlab (or: http://bit.ly/1q2HkO4); http://mathworks.com/matlabcentral/fileexchange/25249-tcp-ip-socket-communications-in-matlab-using-java-classes (or: http://bit.ly/1q2HdlA).

1138. http://mathworks.com/help/matlab/matlab_external/matlab-automation-server-functions-and-properties.html (or: http://bit.ly/1kBb7LG); http://mathworks.com/help/matlab/matlab_external/introduction_brd0vd4-1.html (or: http://bit.ly/1kBbLsy); http://mathworks.com/help/matlab/ref/enableservice.html (or: http://bit.ly/1kBbrKh)

1139. http://mathworks.com/help/matlab/ref/mex.html (or: http://bit.ly/LpcnTd)

1140. http://mathworks.com/help/matlab/matlab_external/before-you-run-a-mex-file.html (or: http://bit.ly/L0XSVg); http://mathworks.com/help/matlab/matlab_external/invalid-mex-file-error.html (or: http://bit.ly/L10a6 W)

1141. http://mathworks.com/help/matlab/matlab_external/version-compatibility.html (or: http:// bit.ly/L0YOck)

1142. http://mathworks.com/help/matlab/build-cc-mex-files.html (or: http://bit.ly/1iPjr77)

1143. http://mathworks.com/help/matlab/build-`fortran-mex-files.html (or: http://bit.ly/19kK6YN)

1144. http://mathworks.com/support/compilers/current_release (or: http://bit.ly/1fvsX1o)

1145. See http://mathworks.com/support/compilers/current_release (or: http://bit.ly/1fvsX1o) for the full list.

1146. Readers interested in using Intel's C++ compiler (ICC) may benefit from two MATLAB File Exchange submissions by user Igor: http://mathworks.com/matlabcentral/fileexchange/38980-mex-setup-for-windows-x64-intel-c-compiler-13-xe (or: http://bit.ly/1fUiOL7) and http://mathworks.com/matlabcentral/fileexchange/38981-icc-mex-tools (or: http://bit.ly/1fUiT1i).

1147. http://blogs.msdn.com/b/nativeconcurrency/archive/2012/04/12/auto-vectorizer-in-visual-studio-11-overview.aspx (or: http://bit.ly/1ojtBxz)

1148. For example, Tom Minka of the LightSpeed toolbox (§8.5.6.1) believes that lcc does not produce good code: http://research.microsoft.com/en-us/um/people/minka/software/matlab.html (or: http://bit.ly/1iAqbt1).

1149. http://mathworks.com/matlabcentral/newsreader/view_thread/327545 (or: http://bit.ly/1kAC4Qu)

1150. http://mathworks.com/help/matlab/matlab_external/selecting-a-compiler.html (or: http://bit.ly/1kACEh9)

1151. http://mathworks.com/help/matlab/ref/mex.getcompilerconfigurations.html (or: http://bit.ly/L1633R); http://mathworks.com/help/matlab/matlab_external/selecting-a-compiler.html (or: http://bit.ly/1kACEh9)

1152. For example: http://cs.ubc.ca/~pcarbo/mex-tutorial (or: http://bit.ly/1iYQ7Og), http://walkingrandomly.com/?p=1959 (or: http://bit.ly/1ijYbGL)

1153. http://mathworks.com/help/matlab/ref/mex.html#btw193g-1 (or: http://bit.ly/1aHsZfb)

1154. A description of these examples can be found here: http://mathworks.com/help/matlab/matlab_external/table-of-mex-examples.html (or: http://bit.ly/1iPlnN4)

1155. http://mathworks.com/help/matlab/matlab_external/debugging-c-c-language-mex-files.html (or: http://bit.ly/L13pLB); Technical solution 1-17Z0R: http://mathworks.com/matlabcentral/answers/91741-How-do-I-debug-C-MEX-files-under-UNIX (or: http://bit.ly/LbCnRq)

1156. http://mathworks.com/help/matlab/matlab_prog/function-precedence-order.html (or: http://bit.ly/1e1p53P)

1157. http://mathworks.com/help/matlab/matlab_external/using-mex-files-to-call-c-c-and-fortran-programs.html#f26590 (or: http://bit.ly/L52RnX); http://mathworks.com/help/matlab/matlab_prog/add-help-for-your-program.html (or: http://bit.ly/L51VQg)

1158. The following discussion is based on investigations by James Tursa. Prior investigations by Peter Boetcher and Peter Li were reported here: http://undocumentedmatlab.com/blog /matlabs-internal-memory-representation (or: http://bit.ly/14Qjjhk)

1159. http://mathworks.com/help/matlab/cc-mx-matrix-library.html (or: http://bit.ly/1fVtV77); http://mathworks.com/help/matlab/fortran-mx-matrix-library.html (or: http://bit.ly /KA31DT)

1160. http://mathworks.com/help/matlab/apiref/mwsize.html (or: http://bit.ly/1fVvYbq)

1161. http://mathworks.com/help/matlab/apiref/mwindex.html (or: http://bit.ly/1fVwdDm)

1162. http://mathworks.com/help/matlab/apiref/mwsignedindex.html (or: http://bit.ly/1fVwSVp)

1163. http://mathworks.com/help/matlab/apiref/mxchar.html (or: http://bit.ly/1fVyDlq)

1164. http://mathworks.com/help/matlab/apiref/mxlogical.html (or: http://bit.ly/1fVywXd)

1165. http://mathworks.com/help/matlab/apiref/mxclassid.html (or: http://bit.ly/1fVzkeG)

1166. http://mathworks.com/help/matlab/apiref/mxcomplexity.html (or: http://bit.ly/1fVzK4J)

1167. http://mathworks.com/help/matlab/access-data.html (or: http://bit.ly/1fVEPK8). All API functions have a dedicated documentation page, for example: http://mathworks.com/help /matlab/apiref/mxgetpr.html (or: http://bit.ly/1hezESd).

1168. http://mathworks.com/help/matlab/validate-data.html (or: http://bit.ly/1heDhHR)

1169. http://mathworks.com/help/matlab/create-or-delete-array.html (or: http://bit. ly/1hePufw). Also see §8.1.6.

1170. http://stackoverflow.com/questions/1708433/matlab-avoiding-memory-allocation-in-mex (or: http://bit.ly/1eSLp0v); http://blogs.mathworks.com/loren/2007/03/22/in-place-operations-on-data#comment-16202 (or: http://bit.ly/1eSLyRC)

1171. http://stackoverflow.com/questions/19813718/mex-files-how-to-return-an-already-allocated-matlab-array (or: http://bit.ly/1fjuMvB)

1172. http://undocumentedmatlab.com/blog/matlab-mex-in-place-editing (or: http://bit.ly /1mBFFte); http://www.mk.tu-berlin.de/Members/Benjamin/mex_sharedArrays (or: http://bit.ly/1mBFNsU); http://www.mk.tu-berlin.de/Members/Benjamin /mex_SharedArray_extension (or: http://bit.ly/1mBGi6a); https://groups.google.com /forum/?hl=en#!topic/comp.soft-sys.matlab/wkHYgh-5AnU (or: http://bit.ly/1mBGank); http://mathworks.com/matlabcentral/answers/77048 (or: http://bit.ly/1fjuDsa); http:// mathworks.com/matlabcentral/newsreader/view_thread/252587 (or: http://bit.ly/1eSOs8I); http://stackoverflow.com/questions/19813718/mex-files-how-to-return-an-already-allocated-matlab-array (or: http://bit.ly/1fjuMvB)

1173. http://undocumentedmatlab.com/blog/undocumented-matlab-mex-api (or: http://bit.ly /1mBGzGc)

1174. In addition to the references listed in the previous endnote: http://mathworks.com /matlabcentral/newsreader/view_thread/21631#52838 (or: http://bit.ly/1eSO2PX); http:// undocumentedmatlab.com/blog/matlabs-internal-memory-representation#COW (or: http:// bit.ly/1eSOVYA)

1175. http://mathworks.com/matlabcentral/fileexchange/24576-inplacearray-a-semi-pointer-package-for-matlab (or: http://bit.ly/1dB7T7w); http://mathworks.com/matlabcentral /fileexchange/17476-typecast-and-typecastx-c-mex-functions (or: http://bit.ly/1e0Zrh1); and additional submissions by Bruno Luong and James Tursa.

1176. http://mathworks.com/matlabcentral/newsreader/view_thread/323578#888889 (or: http:// bit.ly/1cHLeWd)

1177. http://en.wikipedia.org/wiki/Memory_debugger (or: http://bit.ly/19UMLZv)

1178. Ken A., private communication.

1179. http://mathworks.com/help/matlab/matlab_external/memory-management.html (or: http://bit.ly/19UOzlj)

1180. http://mathworks.com/help/matlab/matlab_external/memory-management.html#f26136 (or: http://bit.ly/19UP2nA); http://mathworks.com/help/matlab/apiref/mexmakearray-persistent.html (or: http://bit.ly/19UP6Up); http://mathworks.com/help/matlab/apiref /mexmakememorypersistent.html (or: http://bit.ly/19UPemB)

1181. http://mathworks.com/help/matlab/apiref/mxdestroyarray.html (or: http://bit.ly /1eSolyM); includes pointers to various MEX examples that deallocate memory using this function.

1182. http://mathworks.com/help/matlab/apiref/mxfree.html (or: http://bit.ly/1eSoKBm); includes pointers to various MEX examples that deallocate memory using this function.

1183. http://mathworks.com/help/matlab/matlab_external/using-mex-files-to-call-c-c-and-for-tran-programs.html#brgxbfa-1 (or: http://bit.ly/1eSuew7)

1184. http://mathworks.com/matlabcentral/newsreader/view_thread/162021#410330 (or: http:// bit.ly/1lmUc0C)

1185. http://mathworks.com/matlabcentral/newsreader/view_thread/241020#617015 (or: http:// bit.ly/1eSBLLd)

1186. http://mathworks.com/matlabcentral/newsreader/view_thread/241020#617024 (or: http:// bit.ly/1eSD38U). See related: http://mathworks.com/matlabcentral/fileexchange/24349-use-c++-new-and-delete-in-matlab-mex-files (or: http://bit.ly/1eSPFgu)

1187. http://undocumentedmatlab.com/blog/accessing-private-object-properties#Set (or: http:// bit.ly/1eSWrTh)

1188. http://stackoverflow.com/questions/12496549/memory-leak-in-matlab-mex-file-managed-dll (or: http://bit.ly/1eSWyyq)

1189. http://mathworks.com/help/matlab/apiref/mexlock.html (or: http://bit.ly/1eSpFBS)

1190. http://mathworks.com/help/matlab/apiref/mexatexit.html (or: http://bit.ly/1eSp57c); http://mathworks.com/help/matlab/matlab_external/memory-management.html#f26136 (or: http://bit.ly/19UP2nA)

1191. http://mathworks.com/matlabcentral/newsreader/view_thread/241020#641536 (or: http:// bit.ly/1eSEILQ)

1192. http://mathworks.com/help/matlab/matlab_external/memory-management-issues.html (or: http://bit.ly/1eSt2J6)

1193. http://mathworks.com/matlabcentral/newsreader/view_thread/316593#866015 (or: http:// bit.ly/1epB10L); http://mathworks.com/matlabcentral/newsreader/view_thread/285063 (or: http://bit.ly/Z3UvUT)

1194. See §9.4.3 and http://mathworks.com/matlabcentral/newsreader/view_thread/257830# 670994 (or: http://bit.ly/1lmT4u1)

1195. http://mathworks.com/help/matlab/matlab_external/handling-large-mxarrays.html (or: http://bit.ly/1eSzbVJ)

1196. http://mathworks.com/help/matlab/apiref/mexevalstring.html (or: http://bit.ly/1hO8gee)

1197. http://mathworks.com/help/matlab/apiref/mexgetvariable.html (or: http://bit.ly/1hO8qSC)

1198. http://mathworks.com/help/matlab/apiref/mexputvariable.html (or: http://bit.ly /1hO8uC3)

1199. http://stackoverflow.com/questions/5248915/execution-time-of-c-program (or: http://bit. ly/1dSxyaN)

1200. http://mathworks.com/help/matlab/apiref/mexcallmatlab.html (or: http://bit.ly /1eSXXFc); http://mathworks.com/help/matlab/matlab_external/calling-functions-from-cc-mex-files.html (or: http://bit.ly/KAd9fR)

1201. http://mathworks.com/matlabcentral/fileexchange/16076-waitmex (or: http://bit.ly /17PQ0ff). Also see §10.4.2.5

1202. http://mathworks.com/help/matlab/apiref/mexcallmatlabwithtrap.html (or: http://bit.ly /1eSXLpl)

1203. http://mathworks.com/matlabcentral/newsreader/view_thread/335873#922590 (or: http:// bit.ly/Z3VeFD). Although originally intended to address a MATLAB COM API issue, it applies equally well in this case.

1204. http://mathworks.com/help/matlab/matlab_external/calling-lapack-and-blas-functions-from-mex-files.html (or: http://bit.ly/19y1m5Y)

1205. http://mathworks.com/matlabcentral/newsreader/view_thread/327906#901259 (or: http:// bit.ly/1fbbi9E)

1206. http://software.intel.com/en-us/intel-ipp#pid-3425-869 (or: http://intel.ly/1fbbt4I)

1207. http://mathworks.com/help/matlab/matlab_external/mex-file-dependent-libraries.html (or: http://bit.ly/1eeItfh)

1208. http://mathworks.com/help/matlab/matlab_external/before-you-run-a-mex-file.html (or: http://bit.ly/L0XSVg); http://mathworks.com/help/matlab/matlab_external/invalid-mex-file-error.html (or: http://bit.ly/L10a6W)

1209. http://mathworks.com/matlabcentral/newsreader/view_thread/324129#890533 (or: http://bit.ly/1dM9gLm)

1210. http://mathworks.com/help/matlab/matlab_external/upgrading-mex-files-to-use-64-bit-api.html (or: http://bit.ly/1eeFvHI)

1211. http://mathworks.com/matlabcentral/newsreader/view_thread/261651 (or: http://bit.ly/1lmTIHY)

1212. http://mathworks.com/matlabcentral/fileexchange/25977-mtimesx (or: http://bit.ly/1dB7Icd). Also see §4.5.2

1213. http://mathworks.com/matlabcentral/fileexchange/37515-mmx-multithreaded-matrix-operations-on-n-d-matrices (or: http://bit.ly/1j7WGOO). Also see §4.5.2.

1214. http://mathworks.com/matlabcentral/fileexchange/16777-lapack (or: http://bit.ly/1fXGzlN)

1215. http://mathworks.com/matlabcentral/fileexchange/index?term=tag:blas+OR+tag:lapack (or: http://bit.ly/1fXHcfd). Also see http://mathworks.com/matlabcentral/newsreader/view_thread/328184 (or: http://bit.ly/LB8k6t)

1216. http://mathworks.com/matlabcentral/fileexchange/27151-writing-matlab-cmex-code (or: http://bit.ly/15qLF39)

1217. http://cs.ubc.ca/~pcarbo/mex-tutorial (or: http://bit.ly/1iYQ7Og)

1218. http://gnu.org/software/gsl (or: http://bit.ly/1o8UXXf)

1219. http://mathworks.com/matlabcentral/fileexchange/31437-windowapi (or: http://bit.ly/1n7VWqf)

1220. http://mathworks.com/matlabcentral/fileexchange/10957-get-screen-size-dynamic (or: http://bit.ly/1n7VQ1H)

1221. http://mathworks.com/matlabcentral/answers/1115-why-does-emlmex-generate-a-slow-mex-file (or: http://bit.ly/1lmYpBt)

1222. http://mathworks.com/matlabcentral/fileexchange/30963-minmaxelem (or: http://bit.ly/1etJj7O)

1223. https://en.wikipedia.org/wiki/Loop_nest_optimization (or: http://bit.ly/1lHruCv)

1224. http://www-h.eng.cam.ac.uk/help/tpl/programs/Matlab/mex.html (or: http://bit.ly/1fJmbAK). See additional discussion of the performance-accuracy tradeoff in §3.9.2.

1225. http://mathworks.com/help/matlab/matlab_external/using-mex-files-to-call-c-c-and-fortran-programs.html (or: http://bit.ly/KALaN1)

1226. http://omegacomputer.com/staff/tadonki/mex (or: http://bit.ly/1iYMVlM)

1227. http://omegacomputer.com/staff/tadonki/automatic_%20mexfile_matlab.htm (or: http://bit.ly/1iYNagL)

1228. http://mathworks.com/help/distcomp/run-mex-functions-containing-cuda-code.html (or: http://bit.ly/1dKl3x1)

1229. https://mathworks.com/matlabcentral/answers/79046-Mex-API-wish-list (or: http://bit.ly/1hO6dqj)

1230. http://mathworks.com/matlabcentral/fileexchange/authors/29734 (or: http://bit.ly/1dYYdyn); http://mathworks.com/matlabcentral/newsreader/author/116893 (or: http://bit.ly/1j7Q4jt); http://mathworks.com/matlabcentral/answers/contributors/756104-james-tursa (or: http://bit.ly/1j7QPJ6)

1231. http://mathworks.com/matlabcentral/fileexchange/authors/29906 (or: http://bit.ly/1dYYhOC); http://mathworks.com/matlabcentral/newsreader/author/104755 (or: http://bit.ly/1j7QYfO); http://mathworks.com/matlabcentral/answers/contributors/390839-bruno-luong (or: http://bit.ly/1j7RSZH)

1232. http://mathworks.com/matlabcentral/fileexchange/authors/15233 (or: http://bit.ly/1j7SbDV); http://mathworks.com/matlabcentral/newsreader/author/106526 (or: http://bit.ly/1dYY7XB); http://mathworks.com/matlabcentral/answers/contributors/869888-jan-simon (or: http://bit.ly/1j7QCWx); http://n-simon.de/mex (or: http://bit.ly/1n7OMCh)

1233. http://mathworks.com/matlabcentral/fileexchange/authors/13308 (or: http://bit.ly/1dYYq4z)

1234. http://mathworks.com/matlabcentral/fileexchange/44163-c-mex-programming-tutorial-examples (or: http://bit.ly/LatE1V)

1235. http://mathworks.com/matlabcentral/newsreader/view_thread/173460 (or: http://bit.ly/1hOaFWd)

1236. http://mathworks.com/matlabcentral/newsreader/view_thread/323929#890047 (or: http://bit.ly/1n7UfsX)

1237. http://mathworks.com/matlabcentral/fileexchange/45522-mex-cmake (or: http://bit.ly/1lC1KsJ)

1238. http://cmake.org; http://cmake.org/Wiki/CMake/MatlabMex (or: http://bit.ly/1lC2aPZ). Also see http://cmake.org/pipermail/cmake/2011-October/046555.html (or: http://bit.ly/1lC2iz1).

1239. http://mathworks.com/matlabcentral/fileexchange/45501-openmp-mex-cmake (or: http://bit.ly/1lC172j)

1240. http://mathworks.com/matlabcentral/fileexchange/45505-cuda-mex-cmake (or: http://bit.ly/O6sEOq)

1241. http://mathworks.com/matlabcentral/newsreader/view_thread/278243 (or: http://bit.ly/1p2VQVd)

1242. http://mathworks.com/matlabcentral/fileexchange/38964-example-matlab-class-wrapper-for-a-c++-class (or: http://bit.ly/1p2VVZ9)

1243. http://stackoverflow.com/questions/11220250/how-do-i-profile-a-mex-function-in-matlab (or: http://bit.ly/1ogcTnG)

1244. http://mathworks.com/products/matlab-coder (or: http://bit.ly/14i366T); http://mathworks.com/help/coder (or: http://bit.ly/1bsFSxD)

1245. http://blogs.mathworks.com/seth/2011/04/08/welcome-to-the-coders (or: http://bit.ly/1lmY1Tz); http://blogs.mathworks.com/loren/2011/11/14/generating-c-code-from-your-matlab-algorithms (or: http://bit.ly/1lmYi93). The enhancement may actually be due to an incorporation of the Agility MCS algorithms into the Coder, although this is pure speculation.

1246. http://mathworks.com/matlabcentral/newsreader/view_thread/287119 (or: http://bit.ly/LBcmMa). The Agility website (agilityds.com) is currently defunct; here is an archived version from 2010: https://web.archive.org/web/20100122231128/http://www.agilityds.com/products/matlab_based_products/default.aspx (or: http://bit.ly/LBeX8T)

1247. A short video by Gonzalo Lezma shows how easy the entire process is: http://youtube.com/watch?v=IZJ-IlI3QR0 (or: http://bit.ly/1epwwU7).

1248. http://mathworks.com/help/coder/ug/matlab-language-features-supported-for-code-generation.html (or: http://bit.ly/1jLYLx4); http://mathworks.com/products/matlab-coder/description2.html (or: http://bit.ly/14i5IBy); http://mathworks.com/products/matlab-coder/files/matlab-coder-supported-functions-and-system-objects.pdf (or: http://bit.ly/14i3pi7)

1249. http://mathworks.com/help/fixedpoint/ug/how-working-with-matlab-classes-is-different-for-code-generation.html (or: http://bit.ly/1bsEv2d)

1250. http://mathworks.com/help/coder/release-notes.html (or: http://bit.ly/18q7Qdo); http://mathworks.com/products/matlab-coder/whatsnew.html (or: http://bit.ly/18q7TWp)

1251. http://mathworks.com/matlabcentral/answers/58305#answer_70702 (or: http://bit.ly/1bshzzX)

1252. Technical solution 1-HHLD4T: http://mathworks.com/matlabcentral/answers/93748 (or: http://bit.ly/J0LWl2). There are variants of these documents for each of the Coder releases since MATLAB R2011b (7.13), directly available here: https://mathworks.com/programs/products/download_matlab-coder_conf.html (or: http://bit.ly/J0Mqrw)

1253. https://mathworks.com/videos/matlab-to-c-made-easy-81870.html (or: http://bit.ly/J0MSpv)

1254. http://stackoverflow.com/questions/20513071/performance-tradeoff-when-is-matlab-better-slower-than-c-c#comment30666409_20513071 (or: http://bit.ly/1fk588x). The MathWorks website contains a detailed example of using MATLAB Coder to speed up a Kalman filter implementation by 63x: http://mathworks.com/help/coder/examples/c-code-generation-for-a-matlab-kalman-filtering-algorithm.html (or: http://bit.ly/1kMw72f).

1255. http://en.wikipedia.org/wiki/Dijkstra's_algorithm (or: http://bit.ly/15VBeo1)

1256. http://mathworks.com/matlabcentral/fileexchange/20025-advanced-dijkstras-minimum-path-algorithm (or: http://bit.ly/15VBqDI)

1257. http://mathworks.com/help/fixedpoint/ug/define-input-properties-programmatically-in-the-matlab-file.html (or: http://bit.ly/136xrjw)

1258. http://mathworks.com/help/fixedpoint/ref/coder.varsize.html (or: http://bit.ly/136xxYt)

1259. http://mathworks.com/matlabcentral/fileexchange/17385-dijsktra-path-finder (or: http://bit.ly/1n7TZdt)

1260. http://mathworks.com/matlabcentral/answers/1115-why-does-emlmex-generate-a-slow-mex-file (or: http://bit.ly/1lmYpBt); http://blogs.mathworks.com/loren/2011/11/14/generating-c-code-from-your-matlab-algorithms#comment-32668 (or: http://bit.ly/1lmYncK)

1261. http://mathworks.com/help/fixedpoint/ref/coder.inline.html (or: http://bit.ly/15btv44)

1262. http://mathworks.com/help/fixedpoint/ref/coder.unroll.html (or: http://bit.ly/15buvVR)

1263. http://mathworks.com/help/fixedpoint/ref/coder.nullcopy.html (or: http://bit.ly/15btUUj)

1264. http://mathworks.com/help/coder/ug/acceleration-of-matlab-algorithms-using-parallel-for-loops-parfor.html (or: http://bit.ly/1bXnoEr); http://mathworks.com/help/coder/ug/classification-of-variables-in-parfor-loops.html (or: http://bit.ly/1h0voqa)

1265. http://mathworks.com/help/coder/ref/parfor.html#btz9_xj-9 (or: http://bit.ly/1h0tYMs)

1266. http://www.microsoft.com/en-us/download/details.aspx?id=8442 (or: http://bit.ly/1bWRO9G)

1267. http://www.microsoft.com/visualstudio (or: http://bit.ly/12cgetW)

1268. http://software.intel.com/en-us/intel-compilers (or: http://intel.ly/1bWSlZd)

1269. http://mathworks.com/products/hdl-coder (or: http://bit.ly/1oktxhl); http://blogs.mathworks.com/loren/2013/04/11/matlab-to-fpga-using-hdl-codertm (or: http://bit.ly/1oktBxs); http://mathworks.com/videos/hdl-coder-overview-62492.html (or: http://bit.ly/1okuqpS)

1270. http://mathworks.com/products/hdl-verifier (or: http://bit.ly/1oku0Qo); http://mathworks.com/videos/hdl-verifier-overview-62493.html (or: http://bit.ly/1oku6aM)

1271. http://mathworks.com/products/availability#HD (or: http://bit.ly/1c1Qp4J); http://blogs.mathworks.com/loren/2013/04/11/matlab-to-fpga-using-hdl-codertm#comment-36253 (or: http://bit.ly/1c1QBRJ)

1272. http://mathworks.com/help/hdlcoder/matlab-language-support.html (or: http://bit.ly/1okw62P)

1273. http://mathworks.com/help/comm/sdrxilinx/handling-non-real-time-host-performance.html (or: http://bit.ly/1c1OTQe) Note that burst mode speedup may also be applicable for other domains, for example in the Communications System Toolbox: http://mathworks.com/help/supportpkg/usrpradio/ug/burst-mode-buffering.html (or: http://bit.ly/1wopSVQ).

1274. http://mathworks.com/products/hdl-coder (or: http://bit.ly/1oktxhl); http://blogs.mathworks.com/loren/2013/04/11/matlab-to-fpga-using-hdl-codertm (or: http://bit.ly/1oktBxs)

1275. http://mathworks.com/products/hdl-verifier (or: http://bit.ly/1oku0Qo)

1276. http://mathworks.com/videos/search.html?q=FPGA (or: http://bit.ly/1okQoJF)

1277. http://mathworks.com/matlabcentral/fileexchange/?term=FPGA (or: http://bit.ly/1okR55J)

1278. http://mathworks.com/matlabcentral/newsreader/search_results.html?dur=all&search_string=FPGA (or: http://bit.ly/1okRoNP)

1279. http://mathworks.com/matlabcentral/answers/?term=FPGA (or: http://bit.ly/1okRuVM)

1280. http://mathworks.com/matlabcentral/fileexchange/14317-findjobj-find-java-handles-of-matlab-graphic-objects (or: http://bit.ly/1dkzFAi); http://UndocumentedMatlab.com/blog/findjobj-find-underlying-java-object (or: http://bit.ly/15Thp1m)

1281. Technical solution 1-AO2EOM: http://mathworks.com/matlabcentral/answers/98613 (or: http://bit.ly/1b0OeaA)

1282. Technical solution 1-T6Y3P: http://mathworks.com/matlabcentral/answers/100607 (or: http://bit.ly/1b0H4TJ)

1283. Technical solution 1-T6Y3P: http://mathworks.com/matlabcentral/answers/100607 (or: http://bit.ly/1b0H4TJ)

1284. Technical solution 1-OV2IV: http://mathworks.com/matlabcentral/answers/99056 (or: http://bit.ly/1b0OkPA)

1285. http://mathworks.com/help/compiler/working-with-the-mcr.html#bsjag_q (or: http://bit.ly/135eL7q)

1286. http://mathworks.com/help/compiler/mcc.html#br2jars-67 (or: http://bit.ly/169tywm)

1287. http://mathworks.com/help/compiler/mcc.html#br2jars-80 (or: http://bit.ly/169ueSc)

1288. http://stackoverflow.com/questions/15967157/matlab-executable-too-slow/16059554#16059554 (or: http://bit.ly/169Ke6Y)

1289. http://undocumentedmatlab.com/blog/speeding-up-compiled-apps-startup (or: http://bit.ly/169SUtV)

1290. http://web.archive.org/web/20110830115928/http://mathworks.com/help/toolbox/compiler/brl4_f1-1.html (or: http://bit.ly/1bO1fcV)

1291. Technical solution 1-7RH0IV: http://mathworks.com/matlabcentral/answers/98048 (or: http://bit.ly/1b0Ot5s)

1292. http://matlab4compbio.blogspot.com/2010/04/how-to-solve-mcr-cache-access-problems.html (or: http://bit.ly/169YGfd)

1293. http://mathworks.com/help/compiler/mcr-component-cache-and-ctf-archive-embedding.html (or: http://bit.ly/16a99ah)

1294. Technical solution 1-6M21RQ: http://mathworks.com/matlabcentral/answers/92901 (or: http://bit.ly/1b0OxlX)

1295. MessageBox's parameters are explained here: http://msdn.microsoft.com/en-us/library/ms645505.aspx (or: http://bit.ly/169z0za)

1296. http://www.cplusplus.com/forum/windows/43982 (or: http://bit.ly/169AlWE); http://msdn.microsoft.com/en-us/library/ms645434.aspx (or: http://bit.ly/169AucU)

1297. http://mathworks.com/products/matlab-production-server (or: http://bit.ly/H9fBIM)

1298. http://stackoverflow.com/questions/16139033/how-to-make-the-mcr-starting-time-fast#16140157 (or: http://bit.ly/1gULx2U)

1299. http://mathworks.com/matlabcentral/newsreader/view_thread/293841#925800 (or: http://bit.ly/1nQnikk)

1300. http://www.matlabtips.com/professional-deployment-of-matlab-code (or: http://bit.ly/169NC1v)

1301. http://undocumentedmatlab.com/blog/splash-window-for-deployed-applications (or: http://bit.ly/169C9is)

1302. http://mathworks.com/help/matlab/matlab_external/calling-functions-in-shared-libraries.html (or: http://bit.ly/1cRbaPj)

1303. http://mathworks.com/help/matlab/matlab_external/using-net-from-matlab-an-overview.html (or: http://bit.ly/1cRbWf0)

1304. http://mathworks.com/help/matlab/matlab_external/introducing-matlab-com-integration.html (or: http://bit.ly/1cRcEcg)

1305. http://mathworks.com/help/matlab/matlab_external/how-you-can-use-web-services-with-matlab.html (or: http://bit.ly/1cRcYrm)

1306. http://mathworks.com/help/matlab/using-java-libraries-in-matlab.html (or: http://bit.ly/1bO8D7G); http://mathworks.com/help/matlab/matlab_external/bringing-java-classes-and-methods-into-matlab-workspace.html (or: http://bit.ly/1cRDUaB)

1307. For example: http://mathworks.com/matlabcentral/fileexchange/26778 (or: http://bit.ly/bagjrG)

1308. For example: the entire comp.lang.java.* Usenet tree; several dedicated forums on groups. google.com; http://community.oracle.com/community/developer/english/java; java-forums.org; forums.java.net; javakb.com; http://forums.devshed.com/java-help-9/; http://forums.codeguru.com/forumdisplay.php?5-Java-Programming

1309. blogsearch.google.com/blogsearch?q=java+code; http://onjava.com

1310. For example: java.sun.com; devx.com/Java; jguru.com; developer.com/java; javaworld.com/feature; java2s.com/Code/Java/CatalogJava.htm; oreilly.com/pub/q/all_onjava_articles

1311. For example: freejavaguide.com; java2s.com/Tutorial/Java/CatalogJava.htm; docs.oracle.com/javase/tutorial/; math.hws.edu/javanotes; javacoffeebreak.com/tutorials

1312. For example: java2s.com/Open-Source/Java/CatalogJava.htm; thefreecountry.com/sourcecode/java.shtml

1313. http://nag.com

1314. http://nag.co.uk/about_nag.asp

1315. http://nag.co.uk/numeric/MB/start.asp (or: http://bit.ly/1nyePme)

1316. http://walkingrandomly.com/?p=160 (or: http://bit.ly/1gEBcp9)

1317. http://nag.co.uk/downloads/mbdownloads.asp (or: http://bit.ly/1f0l6WZ); http://nag.co.uk/numeric/MB/start.asp#tabset-tab-4 (or: http://bit.ly/1gmaaU8)

1318. http://mathworks.com/help/optim/ug/fsolve.html (or: http://bit.ly/1gzmY9b). Also see §4.5.5.

1319. http://nag.co.uk/numeric/MB/manual64_24_1/html/C05/c05qbf.html (or: http://bit.ly/1pDxdAJ)

1320. http://walkingrandomly.com/?p=1488 (or: http://bit.ly/1gEpxqk) and a followup article here: http://walkingrandomly.com/?p=2907 (or: http://bit.ly/1gEpDyk). Mike actually reviewed *c05nb*, which was the predecessor of *c05qb* prior to NAG Mark 23 (see http://nag.co.uk/numeric/MB/manual64_24_1/html/GENINT/replace.html#C05NBF or: http://bit.ly/1bv8gBp). The *c05nb* and *c05nc* functions have similar runtimes and interface to the corresponding *c05qb* and *c05qc*, except for the "user" input/output parameter which was added in the newer functions.

1321. The corresponding speedup in 2014 for the followup article (http://walkingrandomly.com/?p=2907 or: http://bit.ly/1gEpDyk) was 3x. Note that Matt McDonnell showed that the problem presented in the followup article could be solved with pure backslash vectorization, without requiring either *fsolve* or NAG, providing a 300x (!) speedup (http://walkingrandomly.com/?p=2907#comment-34781 or: http://bit.ly/1gEAdpa)

1322. http://nag.co.uk/numeric/MB/manual64_24_1/html/C05/c05qcf.html (or: http://bit.ly/1pDyxn9)

1323. http://nag.co.uk/numeric/MB/manual64_24_1/html/X02/x02ajf.html (or: http://bit.ly/1gEwlnU)

1324. http://walkingrandomly.com/?p=1552 (or: http://bit.ly/1gm0GZ4). With regard to linear interpolation in MATLAB, see this interesting discussion: http://mathworks.com/matlabcentral/newsreader/view_thread/324342 (or: http://bit.ly/1o91pgR).

1325. http://nag.co.uk/numeric/MB/manual64_24_1/html/GENINT/threads.html (or: http://bit.ly/1gzp9to)

1326. http://nag.co.uk/numeric/MB/manual64_24_1/html/GENINT/product.html (or: http://bit.ly/1gzugK0)

1327. For example, http://nag.co.uk/numeric/MB/manual64_24_1/pdf/C05/c05intro.pdf (or: http://bit.ly/1kJAGu7), which corresponds to the online doc page http://nag.co.uk/numeric/MB/manual64_24_1/html/C05/c05intro.html (or: http://bit.ly/1kJAOKb).

1328. http://nag.co.uk/numeric/MB/calling.asp (or: http://bit.ly/1gm9ga3)

1329. http://nag.co.uk/numeric/MB/manual64_24_1/html/GENINT/threads.html (or: http://bit.ly/1gzp9to)

1330. http://nag.co.uk/numeric/MB/manual64_24_1/html/GENINT/replace.html (or: http://bit.ly/1gmatOL)

1331. http://nag.co.uk/IndustryArticles/usingtoolboxmatlab.asp (or: http://bit.ly/1dU8zjh); http://nag.co.uk/IndustryArticles/usingtoolboxmatlabpart2.asp (or: http://bit.ly/1dU8BYF); http://nag.co.uk/IndustryArticles/usingtoolboxmatlabpart3.asp (or: http://bit.ly/1dU8In1)

1332. http://nag.co.uk/downloads/chdownloads.asp (or: http://bit.ly/1gm5VI0) for Marks 22/23; The Fortran or C download page for Mark 24.

1333. http://nag.co.uk/numeric/FLOLCH/chgen23da/techdoc.html (or: http://bit.ly/1gm6G42)

1334. http://nag.co.uk/numeric/flolch/ch-faq (or: http://bit.ly/1gm6SjC)

1335. http://walkingrandomly.com/?p=3898 (or: http://bit.ly/1gmccnf). Additional NAG-MATLAB reviews by Mike Croucher: http://walkingrandomly.com/?cat=28 (or: http://bit.ly/1gmeHG7) and http://walkingrandomly.com/?p=5196 (or: http://bit.ly/1dhH7Cq).

1336. http://helsinki.fi/~mjlaine/genmex/, last updated in 1999, currently (2014) inaccessible. The latest archived version is from 2008: https://web.archive.org/web/20080506120005/http://www.helsinki.fi/~mjlaine/genmex (or: http://bit.ly/1fwCD8I)

1337. http://documents.epfl.ch/users/l/le/leuteneg/www/MATLABToolbox (or: http://bit.ly/1l9bq0w)

1338. http://documents.epfl.ch/users/l/le/leuteneg/www/MATLABToolbox/SingleClass.html (or: http://bit.ly/1nWDqS7)

1339. http://documents.epfl.ch/users/l/le/leuteneg/www/MATLABToolbox/DoubleClass.html (or: http://bit.ly/1axQ4Bh)

1340. http://documents.epfl.ch/users/l/le/leuteneg/www/MATLABToolbox/ExtendedClass.html (or: http://bit.ly/1nWEnd5)

1341. http://documents.epfl.ch/users/l/le/leuteneg/www/MATLABToolbox/Files/ComplexFunctions.pdf (or: http://bit.ly/1nWDJfD)

1342. http://documents.epfl.ch/users/l/le/leuteneg/www/MATLABToolbox/FPUasmHints.html (or: http://bit.ly/1nWDW2t). The toolbox uses the open-source Netwide Assembler (http://nasm.us) for the assembler functionality

1343. http://documents.epfl.ch/users/l/le/leuteneg/www/MATLABToolbox/DoubleClass.html (or: http://bit.ly/1axQ4Bh)

1344. Marcel Leutenegger can be contacted at marcel.leutenegger (at) a3.epfl.ch

1345. Code: http://agner.org/optimize/vectorclass.zip (or: http://bit.ly/1cBN5LL); Documentation: http://agner.org/optimize/vectorclass.pdf (or: http://bit.ly/1nWH1Q5), http://agner.org/optimize/vectorclass (or: http://bit.ly/1nWHf9V)

1346. http://advanpix.com

1347. http://advanpix.com/download/

1348. http://advanpix.com/documentation/users-manual (or: http://bit.ly/O59AAn)

1349. http://advanpix.com/documentation/function-reference (or: http://bit.ly/O59t7Z)

1350. http://advanpix.com/2013/11/03/advanpix-vs-vpa-vs-maple-dense-solvers-and-factorization (or: http://bit.ly/N6Euav). Speedups of 40x-200x are reported.

1351. http://mathworks.com/matlabcentral/fileexchange/22725-variable-precision-integer-arithmetic (or: http://bit.ly/1bJihuM)

1352. http://mathworks.com/matlabcentral/fileexchange/36534-hpf-a-big-decimal-class (or: http://bit.ly/1bJitdx)

1353. http://mathworks.com/matlabcentral/answers/116949-big-integer-speed-vpi-and-symbolic#answer_125056 (or: http://bit.ly/1bRc7Zx)

1354. http://mathworks.com/matlabcentral/fileexchange/6446-multiple-precision-toolbox-for-matlab (or: http://bit.ly/1bJiWMO). A version of this toolbox was placed on SourceForge (SF): http://sourceforge.net/projects/mptoolbox (or: http://bit.ly/1bReITb). The FEX version is newer (2008 vs. 2006) and also includes GMP and MPFR binary libraries (DLLs), which are not included in SF. The FEX version is released under the BSD license; SF was released under GPL.

1355. http://stackoverflow.com/questions/9491907/quadruple-precision-eigenvalues-eigenvectors-and-matrix-logarithms/9594783#9594783 (or: http://bit.ly/1bRlvMG)

1356. http://gmplib.org; MCT does not use GMP itself but rather GMP variants/derivatives: MPIR, MPFR and MPC

1357. http://mpfr.org

1358. Code: https://tu-ilmenau.de/fileadmin/public/at/Software/PPL4Matlab/Polymex_final.zip (or: http://bit.ly/1f28Ocr); Documentation: http://tu-ilmenau.de/fileadmin/public /at/Software/PPL4Matlab/PPL4Matlab_documentation_18Feb2013.pdf (or: http://bit.ly /1bRgunA)

1359. http://multiprecision.org

1360. http://mpir.org

1361. http://research.microsoft.com/en-us/um/people/minka/software/lightspeed (or: http:// bit.ly/1axSccc)

1362. http://research.microsoft.com/en-us/um/people/minka/software/fastfit (or: http://bit.ly /1d Uc9dm)

1363. http://spiral.net; http://spiral.ece.cmu.edu:8080/pub-spiral/pubfile/paper_146.pdf (or: http://bit.ly/1bhh9hi)

1364. http://en.wikipedia.org/wiki/Symmetric_multiprocessing (or: http://bit.ly/N6wwOF)

1365. http://en.wikipedia.org/wiki/SIMD (or: http://bit.ly/N6wC8Y)

1366. http://www.sandia.gov/~tgkolda/TensorToolbox (or: http://1.usa.gov/15p3tIz)

1367. http://pub.ist.ac.at/~schloegl/matlab/NaN (or: http://bit.ly/1jkeGDe). Also see http:// pub.ist.ac.at/~schloegl/publications/TR_OpenMP_OctaveMatlabPerformance.pdf (or: http://bit.ly/1jkfFTV)

1368. http://pub.ist.ac.at/~schloegl/matlab/tsa (or: http://bit.ly/1bsOPsp)

1369. http://pub.ist.ac.at/~schloegl/matlab/tsa/tsaperf.html (or: http://bit.ly/1bsPqdO)

1370. http://statweb.stanford.edu/~wavelab (or: http://stanford.io/1bsR92x)

1371. http://csie.ntu.edu.tw/~cjlin/liblinear (or: http://bit.ly/1jkco6W)

1372. http://tomopt.com/tomlab; http://tomwiki.com/TomSym_compared_to_the_Symbolic_ Toolbox (or: http://bit.ly/1fE0HZP)

1373. http://farfieldtechnology.com/products/toolbox (or: http://bit.ly/1jkc2NN)

1374. http://gnu.org/software/gsl (or: http://bit.ly/1o8UXXf)

1375. http://cs.ubc.ca/~pcarbo/mex-tutorial (or: http://bit.ly/1iYQ7Og)

1376. http://walkingrandomly.com/?p=2323 (or: http://bit.ly/1bsSmXI)

1377. http://dlmf.nist.gov

1378. Frank Olver, Daniel Lozier, Ronald Boisvert, Charles Clark: *NIST Handbook of Mathematical Functions, Cambridge University Press* 2010, ISBN 0521140633.

1379. http://dlmf.nist.gov/software (or: http://1.usa.gov/1o8XHny)

1380. http://en.wikipedia.org/wiki/Paging (or: http://bit.ly/13VKkBm)

1381. http://en.wikipedia.org/wiki/Virtual_memory (or: http://bit.ly/13VJeW8)

1382. http://en.wikipedia.org/wiki/Page_(computing) (or: http://bit.ly/13VK5X3)

1383. http://en.wikipedia.org/wiki/Thrashing_(computer_science) (or: http://bit.ly/13VJ8Op). The term *"churning"* is also sometimes used.

1384. http://mathworks.com/company/newsletters/articles/programming-patterns-maximizing- code-performance-by-optimizing-memory-access.html (or: http://bit.ly/1jrjibl)

1385. http://faqs.org/docs/artu/ch12s03.html (or: http://bit.ly/1evHj15) – Eric Raymond, *The Art of Unix Programming*, ISBN 0131429019, §12.3

1386. *"What Every Programmer Should Know About Memory"* (http://akkadia.org/drepper /cpumemory.pdf or: http://bit.ly/1kMxjAt) by Ulrich Drepper; *"Memory Optimization"* (http://research.scea.com/research/pdfs/GDC2003_Memory_Optimization_18Mar03.pdf or: http://bit.ly/OAKtFz) by Christer Ericson. Also see Agner Fog's *"Optimizing software in C++"* (http://agner.org/optimize/optimizing_cpp.pdf or: http://bit.ly/1fJFAQd)

1387. http://UndocumentedMatlab.com/blog/profiling-matlab-memory-usage#Java (or: http:// bit.ly/XsuL2v)

1388. http://mathworks.com/help/matlab/matlab_env/statistical-calculations-in-the-workspace- browser.html (or: http://bit.ly/VbZ60T)

1389. http://mathworks.com/matlabcentral/newsreader/view_thread/164972 (or: http://bit.ly /Mo4mHw)

1390. http://mathworks.com/support/tech-notes/1100/1106.html#2 (or: http://bit.ly/JqFvTj); http://mathworks.com/support/tech-notes/1100/1107.html##2 (or: http://bit.ly/KCkpzU)

1391. http://mathworks.com/matlabcentral/fileexchange/13548-chkmem (or: http://bit.ly /JWPRKv); Additional details in: http://blogs.mathworks.com/pick/2009/05/08/is-your-memory-fragmented-what-can-you-do-about-it (or: http://bit.ly/JWPLmb)

1392. http://mathworks.com/matlabcentral/fileexchange/21827-featurememstats (or: http://bit. ly/IR880w)

1393. http://UndocumentedMatlab.com/blog/profiling-matlab-memory-usage (or: http://bit. ly/LosnlA). The mtic/mtoc features were first reported, to the best of my knowledge, by MathWorker Bill McKeeman in August 2008, in his extensive whitepaper *"Measuring MATLAB Performance"*: http://mathworks.com/matlabcentral/fileexchange/18510-matlab-performance-measurement (or: http://bit.ly/IDpc9b).

1394. http://mathworks.com/support/tech-notes/1100/1106.html (or: http://bit.ly/cimo2V)

1395. http://UndocumentedMatlab.com/blog/undocumented-profiler-options (or: http://bit.ly /J2BkOC). This functionality is detailed in MathWorks' official guide for *"Avoiding Out of Memory Errors"*: http://mathworks.com/support/tech-notes/1100/1107.html##8, section 8.6 (or: http://bit.ly/JbgoTy).

1396. http://UndocumentedMatlab.com/blog/undocumented-profiler-options#comment-32852 (or: http://bit.ly/QaWsH6)

1397. http://blogs.mathworks.com/desktop/2010/04/26/controlling-the-java-heap-size#comment-6991 (or: http://bit.ly/9QecaC); http://stackoverflow.com/questions/2388409/how-can-i-tell-how-much-memory-a-handle-object-uses-in-matlab (or: http://bit.ly/c7lFjg); http:// www.javaworld.com/javaworld/javaqa/2003-12/02-qa-1226-sizeof.html (or: http://bit.ly/ aumv0S)

1398. Technical solution 1-3L4JU7: http://mathworks.com/matlabcentral/answers/95990 (or: http://bit.ly/1e0Po9B)

1399. http://docs.oracle.com/javase/7/docs/technotes/tools (or: http://bit.ly/1duD1C9) for Java 1.7; http://docs.oracle.com/javase/6/docs/technotes/tools (or: http://bit.ly/1duD5So) for Java 1.6; http://docs.oracle.com/javase/1.5.0/docs/tooldocs (or: http://bit.ly/1duD4hh) for Java 1.5. Some of these utilities are not be supported on Macs:http://sourceforge.net /apps/wordpress/drjava/2009/03/10/offline-memory-profiling/, http://blog.emptyway. com/2007/04/02/finding-memory-leaks-in-java-apps#comment-38741

1400. http://visualvm.java.net

1401. http://khelekore.org/jmp

1402. http://khelekore.org/jmp/tijmp

1403. http://UndocumentedMatlab.com/blog/profiling-matlab-memory-usage (or: http://bit.ly /1e0YFhV); also see related http://UndocumentedMatlab.com/blog/matlab-java-memory-leaks-performance (or: http://bit.ly/1e0ZlDS). A list of Java-centric resources is available in the Java SE Troubleshooting guide: http://oracle.com/technetwork/java/javase/index-138283.html (or: http://bit.ly/1e10j34), http://java.sun.com/javase/6/webnotes/trouble (or: http://bit.ly/1biXDdN).

1404. http://UndocumentedMatlab.com/blog/assessing-java-object-size-in-matlab (or: http://bit. ly/K023kc)

1405. http://docs.oracle.com/javase/7/docs/api/java/lang/Runtime.html (or: http://bit.ly /JTDpSu); http://blogs.mathworks.com/community/2009/08/17/calling-java-from-matlab-memory-issues (or: http://bit.ly/197xzSZ)

1406. http://mathworks.com/matlabcentral/newsreader/view_thread/296813#797410 (or: http:// bit.ly/JTBL36). Note that the suggested computation mentioned in the thread post is incorrect and uses non-existing method *availMemory()*.

1407. http://mathworks.com/matlabcentral/fileexchange/8169-matlab-memory-monitor (or: http:// bit.ly/KHhTJz)

1408. http://mathworks.com/matlabcentral/fileexchange/25579-watch-memory-allocation (or: http:// bit.ly/K5TYrw)

1409. http://mathworks.com/matlabcentral/fileexchange/27883-system-resource-monitor-for-windows (or: http://bit.ly/Jr2VYI)

1410. http://mathworks.com/matlabcentral/fileexchange/26662-system-information-class-for-windows (or: http://bit.ly/K5TwJS)

1411. http://mathworks.com/matlabcentral/fileexchange/30534-whos (or: http://bit.ly/J9g2yw)

1412. http://mathworks.com/matlabcentral/fileexchange/18708-totalmemed-m (or: http://bit.ly/IR8QLp)

1413. http://technet.microsoft.com/en-us/sysinternals/bb896653 (or: http://bit.ly/IRC4TO) — see §2.4.

1414. http://technet.microsoft.com/en-us/sysinternals/dd535533 (or: http://bit.ly/IRC8mn)

1415. http://technet.microsoft.com/en-us/library/bb490957.aspx (or: http://bit.ly/1dK8Axh). See MATLAB usage in: http://mathworks.com/matlabcentral/fileexchange/18510-matlab-performance-measurement (or: http://bit.ly/IDpc9b)

1416. http://www.nirsoft.net/utils/gdi_handles.html (or: http://bit.ly/1eaJvXf)

1417. http://mathworks.com/matlabcentral/newsreader/view_thread/15485#34519 (or: http:// bit.ly/JBfSQT); http://mathworks.com/matlabcentral/newsreader/view_thread/15988 (or: http://bit.ly/JBfWQO)

1418. http://UndocumentedMatlab.com/blog/matlabs-internal-memory-representation (or: http://bit.ly/JBfJx4)

1419. http://en.wikipedia.org/wiki/Column-major_order (or: http://bit.ly/IPX4PW)

1420. http://en.wikipedia.org/wiki/Locality_of_reference (or: http://bit.ly/13VJwMT)

1421. http://en.wikipedia.org/wiki/CPU_cache (or: http://bit.ly/JwAf1A)

1422. http://martin-thoma.com/matrix-multiplication-python-java-cpp (or: http:// bit.ly/17Fh9Bx); http://research.scea.com/research/pdfs/GDC2003_Memory_Optimization_18Mar03.pdf (or: http://bit.ly/OAKtFz); http://stackoverflow.com/questions/12264970/why-is-my-program-slow-when-looping-over-exactly-8192-elements (or: http://bit.ly/1dWQZw4)

1423. http://en.wikipedia.org/wiki/Page_%28computer_memory%29 (or: http://bit.ly/IyQ2vC)

1424. http://mathworks.com/company/newsletters/articles/programming-patterns-maximizing-code-performance-by-optimizing-memory-access.html (or: http://bit.ly/1jrjibl)

1425. http://www.mathworks.com/matlabcentral/answers/74379-function-performance-same-functions-has-very-different-speed (or: http://bit.ly/1kuswWC)

1426. http://mathworks.com/matlabcentral/newsreader/view_thread/336966 (or: http://bit.ly/Y8v2cp)

1427. http://stackoverflow.com/questions/11413855/why-is-transposing-a-matrix-of-512x512-much-slower-than-transposing-a-matrix-of-513x513 (or: http://bit.ly/1g78ydl); http://stackoverflow.com/questions/7905760/matrix-multiplication-small-difference-in-matrix-size-large-difference-in-timings (or: http://bit.ly/1g793Ek). A detailed technical explanation can be found in Agner Fog's C++ performance optimization manual: http:// agner.org/optimize/optimizing_cpp.pdf#page=87 (or: http://bit.ly/1kupstx) vis. §9.2 (page 87) and §9.10 (pages 96-98). Also see Christer Ericson's detailed presentation: http://research. scea.com/research/pdfs/GDC2003_Memory_Optimization_18Mar03.pdf (or: http://bit.ly/OAKtFz).

1428. http://UndocumentedMatlab.com/blog/matrix-processing-performance (or: http://bit.ly/J9uqut); http://mathworks.com/matlabcentral/newsreader/view_thread/330782 (or: http://bit.ly/1dB8Uw9)

1429. http://blogs.mathworks.com/loren/2008/06/25/speeding-up-matlab-applications#comment-29544 (or: http://bit.ly/1jrresZ); this conclusion seems to be as accurate in R2013b as it was in 2008 when this was posted, and as accurate for small N values as it is for very large N values.

1430. http://en.wikipedia.org/wiki/Programming_language#Static_versus_dynamic_typing (or: http://bit.ly/JZhBya)

1431. http://mathworks.com/help/matlab/matlab_prog/memory-allocation.html (or: http://bit.ly/191b2Y6)

1432. http://mathworks.com/matlabcentral/newsreader/view_thread/102704 (or: http://bit.ly/KJUtD6); presented and analyzed in http://undocumentedmatlab.com/blog/array-resizing-performance (or: http://bit.ly/1fOZ6jU).

1433. http://stackoverflow.com/questions/1548116/matrix-of-unknown-length-in-matlab/1549094#1549094 (or: http://bit.ly/15fdk6F)

1434. http://stackoverflow.com/questions/1100311/what-is-the-ideal-growth-rate-for-a-dynamically-allocated-array (or: http://bit.ly/1fOY3AG); http://en.wikipedia.org/wiki/Dynamic_array (or: http://bit.ly/1cMXBNf).

1435. http://mathworks.com/matlabcentral/fileexchange/28916-cell2vec (or: http://bit.ly/1n7OX0q)

1436. http://mathworks.com/help/matlab/matlab_prog/strategies-for-efficient-use-of-memory.html#brh72ex-38 (or: http://bit.ly/StZNxr)

1437. http://mathworks.com/matlabcentral/fileexchange/8334-incremental-growth-of-an-array-revisited (or: http://bit.ly/K05iBK); see related http://mathworks.com/matlabcentral/newsreader/view_thread/323986 (or: http://bit.ly/SpbpWw)

1438. http://mathworks.com/matlabcentral/fileexchange/34766-the-fibonacci-sequence (or: http://bit.ly/K066Xb)

1439. http://undocumentedmatlab.com/blog/array-resizing-performance#JIT (or: http://bit.ly/1fOZLlv)

1440. MathWorker Steve Eddins elaborated a bit on this new feature here: http://blogs.mathworks.com/steve/2011/05/16/automatic-array-growth-gets-a-lot-faster-in-r2011a (or: http://bit.ly/WVHSJd) and here: http://blogs.mathworks.com/steve/2011/05/20/more-about-automatic-array-growth-improvements-in-matlab-r2011a (or: http://bit.ly/YYNRKT), but he was careful not to disclose any information on the internal mechanism.

1441. See related: http://mathworks.com/matlabcentral/newsreader/view_thread/324278 (or: http://bit.ly/QMQmQb)

1442. http://blogs.mathworks.com/loren/2008/06/25/speeding-up-matlab-applications#comment-29607 (or: http://bit.ly/K95ank); http://mathworks.com/matlabcentral/newsreader/view_thread/284759#784131 (or: http://bit.ly/K95d2t); http://UndocumentedMatlab.com/blog/undocumented-profiler-options#comment-64 (or: http://bit.ly/6PURhP); http://mathworks.com/matlabcentral/newsreader/view_thread/334975#920245 (or: http://bit.ly/1jmYNiT); http://mathworks.com/matlabcentral/newsreader/view_thread/325638#895031 (or: http://bit.ly/1mqQD9l) and elsewhere

1443. http://blogs.mathworks.com/steve/2011/05/20/more-about-automatic-array-growth-improvements-in-matlab-r2011a (or: http://bit.ly/YYNRKT)

1444. http://mathworks.com/help/matlab/matlab_prog/techniques-for-improving-performance.html (or: http://bit.ly/Vc0uRl); http://undocumentedmatlab.com/blog/preallocation-performance (or: http://bit.ly/16D8Cmq); http://blogs.mathworks.com/loren/2012/11/29/understanding-array-preallocation (or: http://bit.ly/1itczA0)

1445. http://stackoverflow.com/questions/1548116/matrix-of-unknown-length-in-matlab/1548128#1548128 (or: http://bit.ly/15feZsW). Also see related: http://mathworks.com/matlabcentral/answers/68476-analysis-and-prediction-of-optimal-array-size-of-pre-allocation (or: http://bit.ly/1fOYUB7).

1446. http://blogs.mathworks.com/loren/2008/06/25/speeding-up-matlab-applications#6 (or: http://bit.ly/1jrlDD5)

1447. http://stackoverflow.com/questions/591495/matlab-preallocate-a-non-numeric-vector (or: http://bit.ly/LNUgUA) has an interesting discussion about this and related alternatives

1448. http://undocumentedmatlab.com/blog/preallocation-performance#comment-139501 (or: http://bit.ly/1iqXc6I); also read interesting discussions on StackOverflow (http://stackoverflow.com/questions/14169222/faster-way-to-initialize-arrays-via-empty-matrix-multiplication-matlab or: http://bit.ly/1kHfBR9) and the Answers forum (http://mathworks.com

/matlabcentral/answers/58055-faster-way-to-initialize-arrays-via-empty-matrix-multiplication or: http://bit.ly/1iqXTNn)

1449. http://mathworks.com/matlabcentral/answers/77109#answer_86791 (or: http://bit.ly /16D85AX). The mechanism may possibly depend on the low-level C *calloc* function, which allocates memory but postpones its actual initialization until the memory is actually used (compare MATLAB's COW mechanism, §9.5.1).

1450. http://mathworks.com/matlabcentral/newsreader/view_thread/257830 (or: http://bit.ly /1iIOwMb) includes a very interesting analysis. Also see http://UndocumentedMatlab.com/ blog/preallocation-performance#comment-130828 (or: http://bit.ly/1iIQNa8)

1451. http://undocumentedmatlab.com/blog/preallocation-performance#non-default (or: http:// bit.ly/11hzwvp); also see http://stackoverflow.com/questions/14169222/faster-way-to-initialize-arrays (or: http://bit.ly/1kHfBR9)

1452. See the vectorization guide by MathWorker Drea Thomas: http://www.ee.columbia. edu/~marios/matlab/Vectorization.pdf (or: http://bit.ly/11hijlO), origin and date unknown (probably 1995). Tony's trick has been promoted by Loren Shure (for elegance if not for performance) as recently as 2006: http://blogs.mathworks.com/loren/2006/02/22/scalar-expansion-and-more-take-2#6 (or: http://bit.ly/17GvcGZ); it was reported to be less effective as early as R14: http://mathworks.com/matlabcentral/newsreader/view_thread/261716 (or: http://bit.ly/11hAwQ8)

1453. See http://mathworks.com/matlabcentral/fileexchange/5685-writing-fast-matlab-code (or: http://bit.ly/KhGDeo) — comment made by Nemo Managna on Oct 14, 2005

1454. http://walkingrandomly.com/?p=5043#comment-615190 (or: http://bit.ly/1euk8ig)

1455. http://undocumentedmatlab.com/blog/effect-of-clear-on-performance (or: http://bit.ly /13R93To)

1456. http://undocumentedmatlab.com/blog/allocation-performance-take-2 (or: http://bit.ly /19gCd6f)

1457. See §5.1.1. Also see http://undocumentedmatlab.com/blog/allocation-performance-take-2#comment-239592 (or: http://bit.ly/1iqWhTR)

1458. http://mathworks.com/matlabcentral/fileexchange/31362-uninit-create-an-uninitialized-variable-like-zeros-but-faster (or: http://bit.ly/1fjCuSR)

1459. http://mathworks.com/matlabcentral/fileexchange/34276-vvar-class-a-fast-virtual-variable-class-for-matlab (or: http://bit.ly/N72UAg)

1460. http://sourceforge.net/projects/waterloo (or: http://bit.ly/1kM1BWu); http://waterloo. sourceforge.net (or: http://bit.ly/1kM1Ocd)

1461. http://mathworks.com/help/matlab/matlab_prog/strategies-for-efficient-use-of-memory. html#brh72ex-43 (or: http://bit.ly/S5lTwN)

1462. http://mathworks.com/matlabcentral/newsreader/view_thread/328919 (or: http://bit.ly /1fjOrYL)

1463. http://mathworks.com/help/matlab/matlab_prog/strategies-for-efficient-use-of-memory. html#brh72ex-39 (or: http://bit.ly/UghAvS)

1464. Technical solution 1-7S1YKO: http://mathworks.com/matlabcentral/answers/95640 (or: http://bit.ly/1b0OyWX); http://mathworks.com/help/matlab/matlab_prog/memory-requirements-for-a-structure-array.html (or: http://bit.ly/1dInLEB)

1465. http://stackoverflow.com/questions/591495/matlab-preallocate-a-non-numeric-vector# 603012 (or: http://bit.ly/Js7pRr)

1466. http://undocumentedmatlab.com/blog/matrix-processing-performance#comment-184418 (or: http://bit.ly/1dIowgI)

1467. http://mathworks.com/help/matlab/matlab_oop/creating-object-arrays.html#bru6o00 (or: http://bit.ly/1cYFNze)

1468. http://mathworks.com/help/matlab/matlab_oop/creating-object-arrays.html#brd4nrh (or: http://bit.ly/QASiZM)

1469. Also see: http://stackoverflow.com/questions/2510427/how-to-preallocate-an-array-of-class-in-matlab (or: http://bit.ly/IXtRgX)

1470. http://mathworks.com/help/matlab/matlab_prog/copying-objects.html (or: http://bit.ly/1h2CnOT)

1471. http://stackoverflow.com/questions/276198/matlab-class-array (or: http://bit.ly/K07HPr); http://stackoverflow.com/questions/591495/matlab-preallocate-a-non-numeric-vector#591788 (or: http://bit.ly/LNYEmq)

1472. http://mathworks.com/help/matlab/ref/matlab.mixin.copyableclass.html (or: http://bit.ly/1aYleAl). Also read the interesting discussion (with some additional copy variants) here: http://mathworks.com/matlabcentral/newsreader/view_thread/257925 (or: http://bit.ly/19jHdFU)

1473. http://undocumentedmatlab.com/blog/array-resizing-performance#Variants (or: http://bit.ly/1fOZD5z)

1474. http://mathworks.com/matlabcentral/fileexchange/28916-cell2vec (or: http://bit.ly/1n7OX0q)

1475. http://stackoverflow.com/questions/276198/matlab-class-array (or: http://bit.ly/K07HPr)

1476. http://undocumentedmatlab.com/blog/internal-matlab-memory-optimizations (or: http://bit.ly/1f3TBhH)

1477. http://mathworks.com/help/matlab/matlab_prog/memory-allocation.html#brh72ex-8 (or: http://bit.ly/191bjdx); http://blogs.mathworks.com/loren/2006/05/10/memory-management-for-functions-and-variables (or: http://bit.ly/LFmTR8); http://mathworks.com/matlabcentral/answers/152-can-matlab-pass-by-reference (or: http://bit.ly/LFnRwP); http://mathworks.com/matlabcentral/newsreader/view_thread/151196 (or: http://bit.ly/LFnD8P); http://UndocumentedMatlab.com/blog/matlab-mex-in-place-editing#COW (or: http://bit.ly/LFozu1). For a generic description of the technology, see http://en.wikipedia.org/wiki/Copy-on-write (or: http://bit.ly/J4JvfE)

1478. Also see: http://www.matlabtips.com/copy-on-write (or: http://bit.ly/JZsksC)

1479. Technical solution 1-15SO4: http://mathworks.com/matlabcentral/answers/96960 (or: http://bit.ly/InztrY)

1480. Also see: http://www.matlabtips.com/copy-on-write-in-subfunctions (or: http://bit.ly/JgUB0e)

1481. http://undocumentedmatlab.com/blog/internal-matlab-memory-optimizations#inplace (or: http://bit.ly/1eSKIUQ)

1482. http://blogs.mathworks.com/loren/2007/03/22/in-place-operations-on-data (or: http://bit.ly/J8HLBp)

1483. http://mathworks.com/company/newsletters/articles/programming-patterns-maximizing-code-performance-by-optimizing-memory-access.html (or: http://bit.ly/1jrjibl)

1484. http://blogs.mathworks.com/loren/2007/03/22/in-place-operations-on-data#comment-16224 (or: http://bit.ly/IQN1qF)

1485. http://blogs.mathworks.com/loren/2007/03/22/in-place-operations-on-data#comment-16225 (or: http://bit.ly/IQOvB0); http://www.matlabtips.com/some-places-some-rules (or: http://bit.ly/KWJ9DO)

1486. http://blogs.mathworks.com/loren/2007/03/22/in-place-operations-on-data#comment-28807 (or: http://bit.ly/IQOT2u)

1487. Adapted from http://mathworks.com/matlabcentral/newsreader/view_thread/321445 (or: http://bit.ly/O8vvmw)

1488. http://mathworks.com/matlabcentral/newsreader/view_thread/326013 (or: http://bit.ly/XyL07a)

1489. http://stackoverflow.com/questions/19566787/preparing-a-matrix-in-c-for-matlab#19568451 (or: http://bit.ly/1romp60)

1490. http://blogs.mathworks.com/loren/2007/03/22/in-place-operations-on-data#13 (or: http://bit.ly/Jxt35s)

1491. http://blogs.mathworks.com/loren/2007/03/22/in-place-operations-on-data#comment-32833 (or: http://bit.ly/JuXA4p)

1492. http://blogs.mathworks.com/loren/2007/03/22/in-place-operations-on-data#comment-16158 (or: http://bit.ly/JxybXo)

1493. e.g., http://stackoverflow.com/questions/19035371/unexpected-slowdown-of-function-that-modifies-array-in-place (or: http://bit.ly/1flfRm4)

1494. http://UndocumentedMatlab.com/blog/matlab-mex-in-place-editing (or: http://bit.ly/KgYxhA); http://www.bic.mni.mcgill.ca/users/wolforth/Matlab_Help/matlab_memory.html (or: http://bit.ly/KgQrWc)

1495. http://stackoverflow.com/questions/1708433/matlab-avoiding-memory-allocation-in-mex (or: http://bit.ly/KHRcHj)

1496. Technical solution 1-6NU359: http://mathworks.com/matlabcentral/answers/95084 (or: http://bit.ly/1b0OG8I); http://blogs.mathworks.com/loren/2007/03/22/in-place-operations-on-data#comment-16202 (or: http://bit.ly/KHQxWo)

1497. Technical solution 1-6NU359: http://mathworks.com/matlabcentral/answers/95084 (or: http://bit.ly/1b0OG8I)

1498. http://undocumentedmatlab.com/files/mxsharedcopy.c (or: http://bit.ly/GHHoiF)

1499. http://www.mk.tu-berlin.de/Members/Benjamin/mex_sharedArrays (or: http://bit.ly/JZrFaD); http://mathworks.com/matlabcentral/newsreader/view_thread/21631 (or: http://bit.ly/JZrmfY); http://undocumentedmatlab.com/blog/matlab-mex-in-place-editing#mxUnshareArray (or: http://bit.ly/1flkBbh); http://stackoverflow.com/questions/15851718 (or: http://bit.ly/GHHUgB). Also see related: http://mathworks.com/matlabcentral/newsreader/view_thread/252587 (or: http://bit.ly/JZrsV0)

1500. http://stackoverflow.com/questions/9845097/working-with-preallocated-arrays-in-matlabs-mex-function (or: http://bit.ly/15WNQxW)

1501. http://www.mk.tu-berlin.de/Members/Benjamin/mex_sharedArrays (or: http://bit.ly/JZrFaD)

1502. http://mathworks.com/matlabcentral/fileexchange/30672-mxgetpropertyptr-c-mex-function (or: http://bit.ly/Su78jQ); See related http://undocumentedmatlab.com/blog/accessing-private-object-properties (or: http://bit.ly/1papcg6).

1503. http://blogs.mathworks.com/loren/2012/01/13/best-practices-for-programming-matlab#comment-32959 (or: http://bit.ly/JxxqgV)

1504. http://mathworks.com/matlabcentral/fileexchange/24576-inplacearray (or: http://bit.ly/1dB7T7w)

1505. http://mathworks.com/matlabcentral/fileexchange/24576-inplacearray (or: http://bit.ly/1dB7T7w), see comment posted on Aug 6, 2010

1506. http://almostsure.com or: http://mathworks.com/matlabcentral/fileexchange/authors/90073. Note: it appears that Joshua stopped working with MATLAB, at least publicly, when he got his PhD and joined Google.

1507. http://mathworks.com/matlabcentral/fileexchange/28572-sharedmatrix/ (or: http://bit.ly/17xIcN1); http://smlv.cc.gatech.edu/2010/08/27/shared-memory-in-matlab (or: http://b.gatech.edu/1kjnCcV)

1508. http://mathworks.com/matlabcentral/fileexchange/28572-sharedmatrix#feedback (or: http://bit.ly/1kjwf7u)

1509. http://bengal.missouri.edu/~kes25c/SharedMemory-Windows.zip (or: http://bit.ly/1bdYhcH)

1510. http://boost.org/doc/libs/1_55_0/doc/html/interprocess/managed_memory_segments.html (or: http://bit.ly/1h8hIN2)

1511. http://mathworks.com/matlabcentral/newsreader/view_thread/248268#640233 (or: http://bit.ly/1dIsO83)

1512. http://www.matlabtips.com/what-happen-stay (or: http://bit.ly/13aCpxg)

1513. http://mathworks.com/help/matlab/matlab_oop/properties-with-constant-values.html (or: http://bit.ly/QC03yc)

1514. http://mathworks.com/help/matlab/matlab_oop/enumerations.html (or: http://bit.ly/QC1yfR)

1515. http://en.wikipedia.org/wiki/Singleton_pattern (or: http://bit.ly/13aE1Hd); http://www.oodesign.com/singleton-pattern.html (or: http://bit.ly/13aE3z1)

1516. http://mathworks.com/matlabcentral/fileexchange/24911-design-pattern-singleton (or: http://bit.ly/13aCP6Q) presents an alternative implementation.

1517. http://en.wikipedia.org/wiki/Garbage_collection_%28computer_science%29 (or: http://bit.ly/KZ3Q7U)

1518. For example: http://mathworks.com/matlabcentral/fileexchange/181-keep (or: http://bit.ly/KZ6vyo); http://mathworks.com/matlabcentral/fileexchange/182-keep2 (or: http://bit.ly/KZ6ESk); http://mathworks.com/matlabcentral/fileexchange/4259-keep3 (or: http://bit.ly/KZ6NW7); http://mathworks.com/matlabcentral/fileexchange/7129-keep4 (or: http://bit.ly/KZ6SJh); http://mathworks.com/matlabcentral/fileexchange/19494-clearbut (or: http://bit.ly/KZ7crj); http://mathworks.com/matlabcentral/fileexchange/19548-clear-except (or: http://bit.ly/KZ7sXl)

1519. http://mathworks.com/matlabcentral/fileexchange/3967-gcoll (or: http://bit.ly/JJgsKD)

1520. http://mathworks.com/help/matlab/matlab_prog/strategies-for-efficient-use-of-memory.html#brh72ex-44 (or: http://bit.ly/S5oOFJ)

1521. http://mathworks.com/help/matlab/ref/clear.html#inputarg_ItemType (or: http://bit.ly/15vdFnF); http://undocumentedmatlab.com/blog/datestr-performance#comment-60515 (or: http://bit.ly/15vdmJz)

1522. http://blogs.mathworks.com/loren/2012/01/13/best-practices-for-programming-matlab#comment-32934 (or: http://bit.ly/KWhFOO)

1523. An example of such a dilemma: http://mathworks.com/matlabcentral/answers/28332 (or: http://bit.ly/KHONMN)

1524. http://stackoverflow.com/questions/1258761/do-i-conserve-memory-in-matlab-by-declaring-variables-global (or: http://bit.ly/Jw7QcJ)

1525. For more information on stack vs. heap allocation, see any text on computer software memory management; http://stackoverflow.com/questions/79923/what-and-where-are-the-stack-and-heap (or: http://bit.ly/JzuJLG)

1526. http://en.wikipedia.org/wiki/Global_variable (or: http://bit.ly/L0Va0E)

1527. For example: http://matlab.wikia.com/wiki/FAQ#Are_global_variables_bad.3F (or: http://bit.ly/L0Uf0e); https://web.archive.org/web/20090223012054/http://www.mathworks.com/company/newsletters/news_notes/oct02/patterns.html (or: http://bit.ly/1hda1k7); http://en.wikipedia.org/wiki/Global_variable (or: http://bit.ly/L0Va0E)

1528. http://en.wikipedia.org/wiki/Action_at_a_distance_(computer_science) (or: http://bit.ly/KlnNo4)

1529. http://en.wikipedia.org/wiki/Variable_shadowing (or: http://bit.ly/JzFihS)

1530. http://stackoverflow.com/questions/3117410/global-in-matlab#3117495 (or: http://bit.ly/KW7GsD)

1531. Technical solution 1-1AQEH: http://mathworks.com/matlabcentral/answers/94145 (or: http://bit.ly/1b0OUwK)

1532. http://matlab.wikia.com/wiki/FAQ#How_can_I_share_data_between_callback_functions_in_my_GUI.28s.29.3F (or: http://bit.ly/KW7IRu)

1533. http://mathworks.com/help/matlab/ref/persistent.html (or: http://bit.ly/1i0NyZm); http://blogs.mathworks.com/videos/2013/03/14/persistent-variables-in-matlab (or: http://bit.ly/1i0N6KQ); http://blogs.mathworks.com/loren/2006/03/29/understanding-persistence (or: http://bit.ly/1i0NNUv)

1534. http://mathworks.com/help/matlab/matlab_prog/nested-functions.html#f4-73993 (or: http://bit.ly/QASwzX); also see: http://bit.ly/R3cTDx

1535. http://stackoverflow.com/questions/1258761/do-i-conserve-memory-in-matlab-by-declaring-variables-global-instead-of-passing#1261429 (or: http://bit.ly/NvdFq0)

1536. http://mathworks.com/matlabcentral/newsreader/view_thread/150145#900413 (or: http://bit.ly/11dQWqK)

1537. http://mathworks.com/help/matlab_oop/comparing-handle-and-value-classes.html (or: http://bit.ly/R3d6X5)

1538. http://stackoverflow.com/questions/4268113/matlab-takes-a-long-time-after-last-line-of-a-function (or: http://bit.ly/1lHpIBt)

1539. http://mathworks.com/matlabcentral/fileexchange/6586-packunpack-logicals (or: http://bit. ly/R1ArpT)

1540. http://mathworks.com/matlabcentral/newsreader/view_thread/321490 (or: http://bit.ly/Q538pM)

1541. http://ubcmatlabguide.github.io/html/speedup.html#60 (or: http://bit.ly/1ggXtGo); http://people.cs.ubc.ca/~murphyk/Software/matlabTutorial/html/speedup.html#37 (or: http://bit.ly/1diI1tj)

1542. http://mathworks.com/matlabcentral/newsreader/view_thread/330878#909134 (or: http://bit.ly/1hYXogq)

1543. http://mathworks.com/help/images/ref/blockproc.html (or: http://bit.ly/1dlUbG4). Also read: http://mathworks.com/matlabcentral/newsreader/view_thread/333420 (or: http://bit.ly/1dlU6Ch)

1544. http://mathworks.com/company/newsletters/articles/new-features-for-high-performance-image-processing-in-matlab.html (or: http://bit.ly/1kCL6Z8)

1545. http://mathworks.com/matlabcentral/fileexchange/24812-blockmean (or: http://bit.ly/1n7ROGM); http://mathworks.com/matlabcentral/newsreader/view_thread/330878# 909134 (or: http://bit.ly/1n7SRXe)

1546. Here is one additional example: http://mathworks.com/matlabcentral/newsreader/view_thread/320118 (or: http://bit.ly/1gQNgDh)

1547. http://mathworks.com/help/matlab/matlab_prog/anonymous-functions.html#f4-71621 (or: http://bit.ly/VY6peO)

1548. http://mathworks.com/matlabcentral/newsreader/view_thread/324323 (or: http://bit.ly/Zht7yr)

1549. http://homepages.inf.ed.ac.uk/imurray2/compnotes/octave_matlab.html#capture (or: http:// bit.ly/ZhxCcj)

1550. http://mathworks.com/support/tech-notes/1100/1106.html#3 (or: http://bit.ly/PRUISW); This webpage now leads to the newer version, but the old version of this page can still be found in http://bit.ly/PRUV8N. The memory section of this guide has recently been expanded in http://mathworks.com/help/matlab/memory.html (or: http://bit.ly/PRV8sL).

1551. Technical solution 1-1HE4G5: http://mathworks.com/matlabcentral/answers/99399 (or: http://bit.ly/1b0P5rN)

1552. http://mathworks.com/matlabcentral/answers/17171-failure-to-release-memory-when-releasing-java-objects (or: http://bit.ly/1e0LjSJ)

1553. http://mathworks.com/help/matlab/matlab_prog/strategies-for-efficient-use-of-memory. html (or: http://bit.ly/UgkbpI)

1554. http://mathworks.com/matlabcentral/newsreader/view_thread/248268 (or: http://bit.ly/KqGVDf)

1555. http://blogs.mathworks.com/loren/2009/09/11/matlab-release-2009b-best-new-feature-or (or: http://bit.ly/1dkCQry)

1556. http://msdn.microsoft.com/en-us/library/windows/desktop/aa366778(v=vs.85).aspx (or: http://bit.ly/TEwWIh)

1557. http://technet.microsoft.com/en-us/library/cc786709(WS.10).aspx (or: http://bit.ly/19HK5yG); http://support.microsoft.com/kb/291988 (or: http://bit.ly/1dkK3b4)

1558. http://superuser.com/questions/253132/what-are-the-dangers-of-manually-setting-increaseuserva (or: http://bit.ly/1dkL6rD)

1559. http://mathworks.com/products/matlab/preparing-for-64-bit-windows.html#7 (or: http://bit.ly/TBj9pv); http://mathworks.com/matlabcentral/newsreader/view_thread/278488 (or: http://bit.ly/TBjqZq)

1560. http://mathworks.com/help/matlab/ref/memory.html#zmw57dd0e308274 (or: http://bit.ly/1f3YeIx)

1561. http://mathworks.com/help/matlab/matlab_prog/strategies-for-efficient-use-of-memory. html#brh72ex-43 (or: http://bit.ly/1lHp3ju)

1562. http://stackoverflow.com/questions/3300161/matlab-free-memory-is-lost-after-calling-a-function (or: http://bit.ly/19ZsFb6); http://stackoverflow.com/questions/2149241/matlab-block-size-and-memory-management (or: http://bit.ly/19Zt9OD)

1563. http://mathworks.com/help/matlab/ref/matlabwindows.html (or: http://bit.ly/17FftYL)

1564. For example: http://mathworks.com/matlabcentral/newsreader/view_thread/295516 (or: http://bit.ly/QUAnM8); http://mathworks.com/matlabcentral/newsreader/view_thread/298192 (or: http://bit.ly/QUAt6y); http://mathworks.com/matlabcentral/newsreader/view_thread/310071#845674 (or: http://bit.ly/QUACXu); http://UndocumentedMatlab.com/blog/file-deletion-memory-leaks-performance (or: http://bit.ly/QUAEPe); http://www.mathworks.com/support/bugreports/574807 (or: http://bit.ly/1iIYzRo)

1565. http://mathworks.com/matlabcentral/newsreader/search_results?search_string=%22Memory+leak%22&dur=all (or: http://bit.ly/1iIZ9yK)

1566. http://mathworks.com/matlabcentral/newsreader/view_thread/332385#913421 (or: http://bit.ly/1bKwDUO)

1567. http://mathworks.com/matlabcentral/newsreader/view_thread/161419 (or: http://bit.ly/KqFOUl)

1568. http://blogs.mathworks.com/loren/2008/07/29/understanding-object-cleanup (or: http://bit.ly/1iIUVH7)

1569. http://stackoverflow.com/questions/424949/matlab-java-referencing-problem (or: http://bit.ly/1fBRhuo)

1570. http://blog.emptyway.com/2007/04/02/finding-memory-leaks-in-java-apps — unfortunately the blog went off the air sometime in 2012. An archived version can be found here: https://web.archive.org/web/20111126233331/http://blog.emptyway.com/2007/04/02/finding-memory-leaks-in-java-apps (or: http://bit.ly/1kAkklV).

1571. http://mathworks.com/matlabcentral/newsreader/view_thread/259528 (or: http://bit.ly/KqHHQU)

1572. http://mathforum.org/kb/message.jspa?messageID=5950839 (or: http://bit.ly/dsjsga), http://mathworks.com/matlabcentral/newsreader/view_thread/156388#399260 (or: http://bit.ly/aoEmXW) and a few others, including MATLAB's official doc: http://mathworks.com/help/techdoc/ref/set.html#f67-433534 (or: http://bit.ly/cJe0SP) and bug report: http://www.mathworks.com/support/bugreports/331335 (or: http://bit.ly/1lUthWr).

1573. http://mathworks.com/help/matlab/release-notes-older.html#brrxpv8-1 (or: http://bit.ly/JWXeaQ)

1574. http://UndocumentedMatlab.com/blog/performance-accessing-handle-properties#Java (or: http://bit.ly/JWYpHk)

1575. http://undocumentedmatlab.com/blog/converting-java-vectors-to-matlab-arrays#Performance (or: http://bit.ly/1iq6UqI)

1576. http://mathworks.com/matlabcentral/fileexchange/39081-data-compression-by-removing-redundant-points (or: http://bit.ly/ZdRzye)

1577. http://UndocumentedMatlab.com/blog/undocumented-scatter-plot-behavior (or: http://bit.ly/I4vQ6r)

1578. http://mathworks.com/matlabcentral/newsreader/view_thread/323899 (or: http://bit.ly/1oIK4f0)

1579. http://mathworks.com/help/matlab/creating_plots/optimizing-graphics-performance.html#f7-60425 (or: http://bit.ly/R3dh4V)

1580. http://UndocumentedMatlab.com/blog/plot-performance (or: http://bit.ly/Hm7eDY)

1581. http://UndocumentedMatlab.com/blog/plot-liminclude-properties (or: http://bit.ly/Hm7vXl)

1582. http://www.mathworks.com/support/bugreports/819115 (or: http://bit.ly/1e0VTcn); a patch is available for MATLAB R2011a (7.12) through R2012b (8.0).

1583. http://mathworks.com/help/matlab/creating_plots/core-graphics-objects.html#f7-45349 (or: http://bit.ly/QASY1d)

1584. http://UndocumentedMatlab.com/blog/performance-scatter-vs-line (or: http://bit.ly/Ie92gp)

1585. http://blogs.mathworks.com/videos/2013/11/18/knowing-when-to-optimize-your-graph-ics-in-matlab (or: http://bit.ly/1biHVVN)

1586. http://mathworks.com/matlabcentral/newsreader/view_thread/332947 (or: http://bit.ly/1gJsYeU)

1587. http://undocumentedmatlab.com/blog/fig-files-format (or: http://bit.ly/1geOycM)

1588. http://mathworks.com/matlabcentral/fileexchange/40790-plot-big (or: http://bit.ly/1gJz7Yx); http://blogs.mathworks.com/pick/2013/06/07/plot-real-big (or: http://bit.ly/1gJzaDG)

1589. http://mathworks.com/help/matlab/creating_plots/optimizing-graphics-performance.html#brat0ap (or: http://bit.ly/StZFxS)

1590. http://mathworks.com/matlabcentral/newsreader/view_thread/292111#782427 (or: http://bit.ly/1gJBURA)

1591. http://mathworks.com/matlabcentral/answers/68421#answer_79707 (or: http://bit.ly/18d1H4j)

1592. http://irfanview.com

1593. http://smushit.com/ysmush.it

1594. http://mathworks.com/matlabcentral/fileexchange/24812-blockmean (or: http://bit.ly/1n7ROGM); http://mathworks.com/matlabcentral/newsreader/view_thread/330878#909134 (or: http://bit.ly/1n7SRXe)

1595. www.mathworks.com/help/matlab/creating_plots/working-with-8-bit-and-16-bit-images.html (or: http://bit.ly/QAT9JK)

1596. http://mathworks.com/matlabcentral/newsreader/view_thread/317612 (or: http://bit.ly/1bOXDbf)

1597. http://stackoverflow.com/questions/19085677/matlab-efficient-image-patch-extraction#19085901 (or: http://bit.ly/1eapRLJ)

1598. http://mathworks.com/help/matlab/creating_plots/optimizing-graphics-performance.html#f7-60603 (or: http://bit.ly/1cRkcrW)

1599. http://mathworks.com/help/matlab/ref/reducepatch.html (or: http://bit.ly/1cRmXcO)

1600. http://mathworks.com/help/matlab/ref/reducevolume.html (or: http://bit.ly/1cRniMK)

1601. http://mathworks.com/help/matlab/creating_plots/optimizing-graphics-performance.html#f7-60621 (or: http://bit.ly/1cRjI5p)

1602. http://mathworks.com/matlabcentral/fileexchange/25071-matlab-offscreen-rendering-tool-box (or: http://bit.ly/18d0dam); https://github.com/tianli/matlab_offscreen (or: http://bit.ly/18d0gmr); http://tianliresearch.blogspot.com/2006_06_01_archive.html (or: http://bit.ly/18d0mKI)

1603. http://undocumentedmatlab.com/blog/modifying-default-toolbar-menubar-actions (or: http://bit.ly/1na04Yg)

1604. http://mathworks.com/matlabcentral/fileexchange/37885-mycolorbar (or: http://bit.ly/UeDY7I)

1605. http://undocumentedmatlab.com/blog/modifying-default-toolbar-menubar-actions (or: http://bit.ly/1na04Yg)

1606. http://mathworks.com/matlabcentral/answers/49868-slowing-of-program-when-adding-legend#answer_60908 (or: http://bit.ly/TpKo6G)

1607. http://mathworks.com/matlabcentral/answers/89675-plot-legend-bad-performance#answer_99523 (or: http://bit.ly/1elfkx8)

1608. http://mathworks.com/help/matlab/creating_plots/controlling-legends.html (or: http://bit.ly/1p9fSPO). For the undocumented *hasbehavior* usage, see http://UndocumentedMatlab.com/blog/handle-graphics-behavior#hasbehavior (or: http://bit.ly/1biLX0f).

1609. http://mathworks.com/matlabcentral/newsreader/view_thread/326578 (or: http://bit.ly/YXmBPA)

1610. http://mathworks.com/matlabcentral/newsreader/view_thread/315472#915507 (or: http://bit.ly/1fjbHIp). Note the related speedup suggested for *saveas*: http://mathworks.com/matlabcentral/answers/13085-line-129-in-saveas-function-is-slow (or: http://bit.ly/1mxpsH7)

1611. http://mathworks.com/matlabcentral/newsreader/view_thread/323737 (or: http://bit.ly /106oNFu)

1612. http://blogs.mathworks.com/loren/2009/06/03/hold-everything (or: http://bit.ly/LijaJC)

1613. http://undocumentedmatlab.com/blog/plot-performance#comment-167710 (or: http://bit. ly/XVoKFW)

1614. http://mathworks.com/help/matlab/creating_plots/optimizing-graphics-performance. html#f7-60621 (or: http://bit.ly/StZz9E)

1615. http://stackoverflow.com/questions/20163191/accessing-multiple-properties-with-one-handle-in-matlab#20165547 (or: http://bit.ly/IfRTuL); http://undocumentedmatlab.com /blog/performance-accessing-handle-properties#comment-299379 (or: http://bit.ly/IfSv3t)

1616. http://UndocumentedMatlab.com/blog/plot-performance#Legend (or: http://bit.ly/W84pLP). Also see §10.1.15.

1617. http://undocumentedmatlab.com/blog/performance-accessing-handle-properties (or: http://bit.ly/19EpRig) provides additional details.

1618. http://www.mathworks.com/help/matlab/matlab_oop/learning-to-use-events-and-listeners. html (or: http://bit.ly/OPinG2); http://undocumentedmatlab.com/blog/udd-events-and-listeners (or: http://bit.ly/OPivFC); http://undocumentedmatlab.com/blog /continuous-slider-callback (or: http://bit.ly/OPiBgq)

1619. http://stackoverflow.com/questions/22637069/matlab-performance-of-gaussian-filtering-using-imfilter-on-a-binary-image (or: http://bit.ly/1p4ZYCq)

1620. http://matlab.wikia.com/wiki/FAQ#How_can_I_share_data_between_callback_functions_ in_my_GUI.28s.29.3F (or: http://bit.ly/KW7IRu)

1621. http://mathworks.com/help/techdoc/ref/opengl.html (or: http://bit.ly/PnNqWo)

1622. http://mathworks.com/matlabcentral/newsreader/view_thread/328141#901832 (or: http:// bit.ly/YHhMZz)

1623. http://mathworks.com/matlabcentral/fileexchange/25071-matlab-offscreen-rendering-toolbox (or: http://bit.ly/18d0dam); https://github.com/tianli/matlab_offscreen (or: http:// bit.ly/18d0gmr); http://tianliresearch.blogspot.com/2006_06_01_archive.html (or: http://bit. ly/18d0mKI)

1624. http://psychtoolbox.org

1625. http://docs.psychtoolbox.org/MOGL (or: http://bit.ly/19lxjj3)

1626. http://mathworks.com/help/techdoc/ref/figure_props.html#WVisual (or: http://bit.ly /NGhxXk)

1627. http://UndocumentedMatlab.com/blog/figure-window-always-on-top (or: http://bit.ly /1wTNoNY)

1628. http://mathworks.com/help/techdoc/ref/figure_props.html#WVisual (or: http://bit.ly /NGhxXk)

1629. http://mathworks.com/help/matlab/creating_guis/ways-to-manage-data-in-a-guide-gui. html (or: http://bit.ly/19lxE6g); http://blogs.mathworks.com/videos/2011/11/23 /passing-data-between-guide-callbacks-without-globals-in-matlab (or: http://bit.ly/19lxLPt); http://mathworks.com/matlabcentral/fileexchange/8616-video-guide-advanced-techniques (or: http://bit.ly/15qPllJ); http://matlab.wikia.com/wiki/FAQ#How_can_I_share_data_ between_callback_functions_in_my_GUI.28s.29.3F (or: http://bit.ly/19lwU1a)

1630. http://undocumentedmatlab.com/blog/matlab-and-the-event-dispatch-thread-edt (or: http:// bit.ly/1gZUlja)

1631. http://undocumentedmatlab.com/blog/solving-a-matlab-hang-problem (or: http://bit.ly /1eAJmMw)

1632. *Undocumented Secrets of MATLAB-Java Programming*, CRC Press 2011; http://Undocumented Matlab.com/matlab-java-book (or: http://bit.ly/HNpg5X)

1633. http://docs.oracle.com/javase/tutorial/uiswing/components/html.html (or: http://bit. ly/VFTj2 k)

1634. http://UndocumentedMatlab.com/blog/html-support-in-matlab-uicomponents (or: http:// bit.ly/VFTaMs)

1635. http://docs.oracle.com/javase/tutorial/uiswing/components/table.html#renderer (or: http:// bit.ly/ZqpwLJ). I describe cell renderers and their usage in MATLAB in my ***uitable*** customization report (http://UndocumentedMatlab.com/blog/uitable-customization-report or: http:// bit.ly/ZqqfMU) and in my book *Undocumented Secrets of MATLAB-Java Programming*, ISBN 9781439869031, http://UndocumentedMatlab.com/matlab-java-book (or: http://bit.ly/Zqqojt).

1636. http://UndocumentedMatlab.com/blog/uitable-cell-colors#colors (or: http://bit.ly/ZqoCi8)

1637. http://www.youtube.com/watch?v=7gtf47D_bu0#t=341 (or: http://bit.ly/18MYgPB) and associated slides http://www.igvita.com/slides/2013/fluent-perfcourse.pdf (or: http:// www.igvita.com/slides/2013/fluent-perfcourse.pdf)

1638. http://UndocumentedMatlab.com/blog/continuous-slider-callback (or: http://bit.ly /bexwI9)

1639. http://UndocumentedMatlab.com/blog/disable-entire-figure-window (or: http://bit.ly /12Cng8r)

1640. http://mathworks.com/matlabcentral/fileexchange/15895-enabledisable-figure (or: http:// bit.ly/12Cnrkl)

1641. http://mathworks.com/matlabcentral/fileexchange/30666-blurfigure (or: http://bit.ly /VYOQ9I)

1642. See for example, http://mathworks.com/matlabcentral/newsreader/view_thread/319790 (or: http://bit.ly/J7r5Xf)

1643. http://UndocumentedMatlab.com/blog/command-window-text-manipulation (or: http:// bit.ly/JoeCni); http://UndocumentedMatlab.com/blog/cprintf-display-formatted-color-text-in-command-window#comment-56187 (or: http://bit.ly/JoeOD5)

1644. http://mathworks.com/matlabcentral/fileexchange/30297-consoleprogressbar (or: http:// bit.ly/I4bb4n); Alternatives: http://mathworks.com/matlabcentral/fileexchange/28067-text-progress-bar (or: http://bit.ly/K183Ba), http://mathworks.com/matlabcentral/ fileexchange/16213-another-text-waitbar (or: http://bit.ly/Ig7u61), http://mathworks.com /matlabcentral/fileexchange/1436-wdisp-a-pedestrian-command-window-waitbar (or: http:// bit.ly/Iq6ApV), http://mathworks.com/matlabcentral/fileexchange/8564-progress (or: http://bit.ly/Ig7Byk), http://mathworks.com/matlabcentral/fileexchange/6891-fast-progress-display (or: http://bit.ly/Iq64bj); http://mathworks.com/matlabcentral/fileexchange/24099-percent-done (or: http://bit.ly/Iq88jG)

1645. For example: http://mathworks.com/matlabcentral/fileexchange/?term=tag:waitbar (or: http://bit.ly/I3rECP); http://mathworks.com/matlabcentral/fileexchange/6922-progress-bar (or: http://bit.ly/I3rImd); http://mathworks.com/matlabcentral/fileexchange/26589-multi-progress-bar (or: http://bit.ly/I3rQlE); http://mathworks.com/matlabcentral /fileexchange/28179-progress-bars (or: http://bit.ly/I3rMlZ); http://mathworks.com /matlabcentral/fileexchange/22161-waitbar-with-time-estimation (or: http://bit.ly/I3rvj0); etc.

1646. For example: http://mathworks.com/matlabcentral/fileexchange/23838-ledbar (or: http:// bit.ly/Igacs8), http://mathworks.com/matlabcentral/fileexchange/16663-simple-waitbar (or: http://bit.ly/Igaf7s), http://mathworks.com/matlabcentral/fileexchange/26284-tooltip-waitbar (or: http://bit.ly/IgaHSX)

1647. http://mathworks.com/matlabcentral/fileexchange/14773-statusbar (or: http://bit.ly/Iga1gB)

1648. http://mathworks.com/matlabcentral/fileexchange/16076-waitmex (or: http://bit.ly/17 PQ0ff)

1649. http://undocumentedmatlab.com/blog/animated-busy-spinning-icon (or: http://bit.ly /1eKoFU4)

1650. http://docs.oracle.com/javase/tutorial/uiswing/components/progress.html (or: http:// bit.ly/IiRu6N); used in http://mathworks.com/matlabcentral/fileexchange/file_infos /14773-statusbar (or: http://bit.ly/IiRl2Y)

1651. Refer for example to the issue of the speed of MATLAB's ***comet*** demo animation, http:// blogs.mathworks.com/pick/2013/11/01/comet3-with-speed-control (or: http://bit.ly /1bivTeT)

1652. http://matlabtips.com/waiting-for-the-waitbar (or: http://bit.ly/1gR5cuA)

1653. http://mathworks.com/matlabcentral/fileexchange/26589-multi-progress-bar (or: http://bit.ly/I3rQlE)

1654. http://mathworks.com/matlabcentral/fileexchange/25634-easy-n-fast-smoothing-for-1-d-to-n-d-data (or: http://bit.ly/JfS6iq)

1655. http://mathworks.com/matlabcentral/fileexchange/37878-quick-spline-smoothing-for-1-d-data (or: http://bit.ly/JfS9ux)

1656. http://mathworks.com/matlabcentral/fileexchange/24443-slm-shape-language-modeling (or: http://bit.ly/JfU1Ub); requires the Optimization Toolbox

1657. http://mathworks.com/matlabcentral/fileexchange/?term=tag:smoothing (or: http://bit.ly/JfSvRV). Patrick Mineault's spline utility (http://mathworks.com/matlabcentral/fileexchange/32509-fast-b-spline-class or: http://bit.ly/JfTYYf) deserves special mention in this regard.

1658. http://mathworks.com/help/matlab/creating_guis/callback-sequencing-and-interruption.html (or: http://bit.ly/WArhq0)

1659. http://mathworks.com/help/matlab/creating_guis/callback-sequencing-and-interruption.html (or: http://bit.ly/WArhq0)

1660. http://mathworks.com/help/matlab/creating_guis/callback-sequencing-and-interruption.html (or: http://bit.ly/WArhq0); http://mathworks.com/help/techdoc/ref/uicontrol_props.html#Interruptible (or: http://bit.ly/WAsDRy)

1661. http://UndocumentedMatlab.com/blog/controlling-callback-re-entrancy (or: http://bit.ly/qmFwYw)

1662. http://UndocumentedMatlab.com/blog/continuous-slider-callback#Property_Listener (or: http://bit.ly/b49NwE)

1663. http://www.luxford.com/high-performance-windows-timers (or: http://bit.ly/116KnWD)

1664. http://undocumentedmatlab.com/blog/pause-for-the-better (or: http://bit.ly/116KHoh)

1665. http://UndocumentedMatlab.com/blog/waiting-for-asynchronous-events#Polling (or: http://bit.ly/YBfAAz)

1666. http://UndocumentedMatlab.com/blog/pause-for-the-better (or: http://bit.ly/NRMUSw)

1667. http://mathworks.com/matlabcentral/newsreader/view_thread/250288 (or: http://bit.ly/17vC21h)

1668. http://www.eevblog.com/forum/chat/does-anyone-know-how-to-improve-matlab-performance-with-2d-figures (or: http://bit.ly/IZGEat); http://mathworks.com/matlabcentral/newsreader/view_thread/328388 (or: http://bit.ly/IZGO1A)

1669. http://mathworks.com/matlabcentral/newsreader/view_thread/328388 (or: http://bit.ly/1ninlKR); http://stackoverflow.com/a/21559178/233829 (or: http://bit.ly/1nino9u)

1670. http://mathworks.com/matlabcentral/fileexchange/40790-plot-big (or: http://bit.ly/1gJz7Yx); http://blogs.mathworks.com/pick/2013/06/07/plot-real-big (or: http://bit.ly/1gJzaDG)

1671. http://mathworks.com/matlabcentral/fileexchange/15850-dsplot-downsampled-plot (or: http://bit.ly/1eXIVNw); http://blogs.mathworks.com/pick/2007/08/06/downsampling-data-for-faster-plotting (or: http://bit.ly/1eXIXF6)

1672. http://mathworks.com/matlabcentral/fileexchange/42191-jplot (or: http://bit.ly/1eXJmr1)

1673. http://mathworks.com/matlabcentral/fileexchange/27359-turbo-plot (or: http://bit.ly/1eXKryY)

1674. http://java.sun.com/javase/6/docs/api/javax/swing/ToolTipManager.html (or: http://bit.ly/9fG6sV)

1675. http://UndocumentedMatlab.com/blog/additional-uicontrol-tooltip-hacks#tooltip_timing (or: http://bit.ly/IZwzdC)

1676. http://UndocumentedMatlab.com/blog/matlab-and-the-event-dispatch-thread-edt (or: http://bit.ly/1gZUlja)

1677. http://undocumentedmatlab.com/blog/solving-a-matlab-hang-problem (or: http://bit.ly/1eAJmMw)

1678. http://undocumentedmatlab.com/blog/uicontextmenu-performance (or: http://bit.ly/1eARf4r)

1679. http://blogs.mathworks.com/loren/2006/05/10/memory-management-for-functions-and-variables#comment-31065 (or: http://bit.ly/19mCdCq). Note that I have not independently confirmed this report.

1680. http://icl.cs.utk.edu/~mucci/latest/pubs/IDC-Pres-2012.pdf#page=19 (or: http://bit.ly/1jAueTn)

1681. https://youtube.com/watch?v=IsGbEkbSivE (or: http://bit.ly/1iXoH9r) shows a 75x speedup by simply adding semi-colons.

1682. http://mathworks.com/matlabcentral/answers/110294#answer_118945 (or: http://bit.ly/1hTmKcR)

1683. http://mathworks.com/matlabcentral/newsreader/view_thread/268923#704048 (or: http://bit.ly/15f5ATz)

1684. http://mathworks.com/matlabcentral/newsreader/view_thread/320272#877503 (or: http://bit.ly/YpeMxf)

1685. http://mathworks.com/matlabcentral/newsreader/view_thread/328228#902075 (or: http://bit.ly/YpitTN); we could also use *wc –l* on Linux platforms.

1686. http://en.wikipedia.org/wiki/Memory-mapped_file (or: http://bit.ly/ZkZRaE). A detailed technical explanation can be found here: http://msdn.microsoft.com/en-us/library/ms810613.aspx (or: http://bit.ly/11axlXY)

1687. http://mathworks.com/help/matlab/import_export/share-memory-between-applications.html (or: http://bit.ly/11aH5l2)

1688. http://mathworks.com/help/matlab/ref/memmapfile.html (or: http://bit.ly/ZW2bkc)

1689. http://mathworks.com/help/matlab/import_export/overview-of-memory-mapping.html#braidws-6 (or: http://bit.ly/11ayKha)

1690. http://mathworks.com/help/matlab/import_export/deleting-a-memory-map.html#braidws-55 (or: http://bit.ly/ZoCzRn)

1691. http://mathworks.com/help/matlab/import_export/overview-of-memory-mapping.html#braidws-14 (or: http://bit.ly/11aD8gi)

1692. http://lists.freebsd.org/pipermail/freebsd-questions/2004-June/050371.html (or: http://bit.ly/Zl2Wr9)

1693. http://mathworks.com/help/matlab/import_export/constructing-a-memmapfile-object.html#braidws-28 (or: http://bit.ly/11aG5NY)

1694. http://mathworks.com/matlabcentral/answers/125185-read-csv-file-fast-convert-csv-to-mat (or: http://bit.ly/1kkIGAn)

1695. http://mathworks.com/matlabcentral/newsreader/view_thread/300706 (or: http://bit.ly/17vyLyU)

1696. http://mathworks.com/help/matlab/ref/textread.html (or: http://bit.ly/1kkIAZp)

1697. http://stackoverflow.com/questions/19159975/remove-reoccuring-lines-from-text-file-with-enhanced-performance (or: http://bit.ly/19lgBkX)

1698. http://mathworks.com/matlabcentral/newsreader/view_thread/329256 (or: http://bit.ly/1epdpGp)

1699. http://en.wikipedia.org/wiki/Data_buffer (or: http://bit.ly/YVglHU)

1700. http://mathworks.com/matlabcentral/newsreader/view_thread/330753#908695 (or: http://bit.ly/1epfiTD); http://stackoverflow.com/questions/16632964/matlab-speed-up-reading-of-ascii-file (or: http://bit.ly/1ysgEqs)

1701. http://mathworks.com/help/matlab/matlab_prog/strategies-for-efficient-use-of-memory.html#brh72ex-39 (or: http://bit.ly/UghAvS)

1702. http://mathworks.com/matlabcentral/newsreader/view_thread/334934 (or: http://bit.ly/1kGipeP)

1703. http://mathworks.com/matlabcentral/newsreader/view_thread/336506#924126 (or: http://bit.ly/1tewS54)

1704. http://mathworks.com/matlabcentral/newsreader/view_thread/299353 (or: http://bit.ly/YVv0Tk)

1705. Technical solution 1-PV371: http://mathworks.com/matlabcentral/answers/96284 (or: http://bit.ly/1b0Pbjg)

1706. http://UndocumentedMatlab.com/blog/improving-fwrite-performance (or: http://bit.ly/XJfWGH); http://blogs.mathworks.com/loren/2006/04/19/high-performance-file-io (or: http://bit.ly/11vR304)

1707. See, for example, http://mathworks.com/matlabcentral/newsreader/view_thread/325950 (or: http://bit.ly/Z6qwbJ)

1708. http://mathworks.com/matlabcentral/newsreader/view_thread/329704#906473 (or: http://bit.ly/1ysuU2r)

1709. http://en.wikipedia.org/wiki/Data_compression (or: http://bit.ly/102FAa0)

1710. Proprietary MAT format: http://mathworks.com/help/pdf_doc/matlab/matfile_format.pdf (or: http://bit.ly/Z5lPOW); HDF5: http://www.hdfgroup.org/HDF5 (or: http://bit.ly/Z5m6RO)

1711. For the proprietary MAT format we can use, for example:C: http://mathworks.com/matlabcentral/fileexchange/26731-portable-matfile-exporter-in-c (or: http://bit.ly/Z5mOP4); C#: http://mathworks.com/matlabcentral/fileexchange/16319-csmatio-mat-file-io-api-for-net-2-0 (or: http://bit.ly/Z5mAHF); Java: http://mathworks.com/matlabcentral/fileexchange/10759-jmatio-matlabs-mat-file-io-in-java (or: http://bit.ly/Z5mCPT) or: http://sourceforge.net/projects/jmatio/ For the HDF5 format there are multiple public adaptors, but MATLAB's implementation is not pure HDF5 so their usage is highly questionable. For both formats we could use MATLAB's *matOpen* adaptor for C and Fortran: http://mathworks.com/help/matlab/apiref/matopen.html (or: http://bit.ly/Z5sxEB).

1712. http://www.hdfgroup.org/HDF5 (or: http://bit.ly/Z5m6RO); http://mathworks.com/help/matlab/hdf5-files.html (or: http://bit.ly/10K77eh)

1713. http://www.hdfgroup.org/doc_resource/SZIP (or: http://bit.ly/Z5nFiM)

1714. http://www.hdfgroup.org/ftp/HDF5/examples/examples-by-api/matlab/HDF5_M_Examples/h5ex_d_gzip.m (or: http://bit.ly/10K7vcR)

1715. http://mathworks.com/matlabcentral/answers/15521#answer_23983 (or: http://bit.ly/ZmUaoB)

1716. http://mathworks.com/help/matlab/ref/save.html#inputarg_version (or: http://bit.ly/Z5mks6)

1717. http://mathworks.com/matlabcentral/newsreader/view_thread/255244#663437 (or: http://bit.ly/11Q3rKW)

1718. http://www.mathworks.com/support/bugreports/331250 (or: http://bit.ly/18GI52I)

1719. http://www.mathworks.com/support/bugreports/784028 (or: http://bit.ly/108ELtv)

1720. http://www.mathworks.com/support/bugreports/938249 (or: http://bit.ly/108xKJa); http://www.mathworks.com/support/bugreports/736830 (or: http://bit.ly/1575aPC)

1721. http://mathworks.com/matlabcentral/newsreader/view_thread/255244#663448 (or: http://bit.ly/11Q4smf)

1722. http://stackoverflow.com/questions/4950630/matlab-differences-between-mat-versions (or: http://bit.ly/Z5uuki)

1723. http://mathworks.com/matlabcentral/fileexchange/39721-save-mat-files-more-quickly (or: http://bit.ly/Z5DNRa)

1724. http://mathworks.com/matlabcentral/answers/15521 (or: http://bit.ly/11c8zGO)

1725. http://www.hdfgroup.org/ftp/HDF5/examples/examples-by-api/matlab/HDF5_M_Examples/h5ex_d_gzip.m (or: http://bit.ly/10K7vcR); additional examples are provided in http://www.hdfgroup.org/ftp/HDF5/examples/examples-by-api/api18-m.html (or: http://bit.ly/10RdL1i)

1726. http://mathworks.com/matlabcentral/newsreader/view_thread/268923 (or: http://bit.ly/152qXaN)

1727. http://mathworks.com/matlabcentral/fileexchange/34564-fast-serialize-deserialize (or: http://bit.ly/14H4roy). We have already encountered another utility by Cristian, *microcache*, in the Memoization section (§3.2.4.2).

1728. http://undocumentedmatlab.com/blog/serializing-deserializing-matlab-data (or: http://bit.ly/1ilyapo). Also see §3.2.6. Note that these functions are apparently limited to ~2GB data: http://undocumentedmatlab.com/blog/improving-save-performance#comment-200289 (or: http://bit.ly/N70ilV). These blog articles were the source for Jan Berling's *Bytestream Save Toolbox* on the MATLAB File Exchange: http://mathworks.com/matlabcentral/fileexchange/45743-bytestream-save-toolbox (or: http://bit.ly/1ppG8C5), which enables fast saving and loading of workspace variables to/from MAT files.

1729. http://mathworks.com/matlabcentral/newsreader/view_thread/299353#830958 (or: http://bit.ly/11sm7iY)

1730. http://mathworks.com/matlabcentral/fileexchange/29457-serialize-deserialize (or: http://bit.ly/11smslI)

1731. http://www.mathworks.com/support/bugreports/857319 (or: http://bit.ly/10wgoER)

1732. See related: http://stackoverflow.com/questions/4814569/what-is-the-fastest-way-to-load-data-in-matlab (or: http://bit.ly/Z6Wi7Z)

1733. http://www.mathworks.com/support/bugreports/331250 (or: http://bit.ly/18GI52I); http://mathworks.com/help/matlab/hdf5-files.html (or: http://bit.ly/10K77eh). Also run *doc('hdf5')* in MATLAB.

1734. http://mathworks.com/help/matlab/low-level-functions.html (or: http://bit.ly/N6YvgE)

1735. http://mathworks.com/help/matlab/hdf4.html (or: http://bit.ly/1r89FAW), note specifically the low-level functions. An older version of this documentation section, which may provide a different viewpoint, can be found in: http://web.archive.org/web/20121221183052/http://www.mathworks.com/help/matlab/ref/hdf.html (or: http://bit.ly/1r88Bgd).

1736. See for example: http://mathworks.com/matlabcentral/newsreader/view_thread/129639 (or: http://bit.ly/ZATFOB)

1737. http://www.mathworks.com/support/bugreports/784028 (or: http://bit.ly/108ELtv)

1738. http://www.javaworld.com/community/node/8362 (or: http://bit.ly/10BfYAw); http://stackoverflow.com/questions/4146402/how-to-read-and-write-a-zip-file-in-java (or: http://bit.ly/10BfGJT); http://www.mkyong.com/java/how-to-compress-files-in-zip-format (or: http://bit.ly/10Bg7UP)

1739. http://docs.oracle.com/javase/1.5.0/docs/api/java/util/zip/ZipEntry.html (or: http://bit.ly/ZdOdLv)

1740. http://docs.oracle.com/javase/6/docs/api/java/util/zip/ZipOutputStream.html#write (byte[], int, int) (or: http://bit.ly/XzSX2Q)

1741. http://docs.oracle.com/javase/1.5.0/docs/api/java/io/ByteArrayOutputStream.html (or: http://bit.ly/ZdOuxZ)

1742. http://docs.oracle.com/javase/1.5.0/docs/api/java/io/FileOutputStream.html (or: http://bit.ly/ZdOxdi)

1743. http://UndocumentedMatlab.com/blog/savezip-utility (or: http://bit.ly/1ppmoSl); http://mathworks.com/matlabcentral/fileexchange/47698-savezip (or: http://bit.ly/XY6lj2)

1744. http://undocumentedmatlab.com/blog/fixing-matlabs-actxserver (or: http://bit.ly/XAtxQx)

1745. http://mathworks.com/matlabcentral/fileexchange/10465-xlswrite1 (or: http://bit.ly/XAxjt8)

1746. http://mathworks.com/matlabcentral/fileexchange/15192-officedoc-read-write-format-ms-office-docs-xls-doc-ppt (or: http://bit.ly/XAxe8N)

1747. http://mathworks.com/matlabcentral/fileexchange/340-saveppt (or: http://bit.ly/17zCX2O); http://mathworks.com/matlabcentral/fileexchange/19322-saveppt2 (or: http://bit.ly/17zCWfh); and again http://mathworks.com/matlabcentral/fileexchange/15192-officedoc-read-write-format-ms-office-docs-xls-doc-ppt (or: http://bit.ly/XAxe8N)

1748. http://undocumentedmatlab.com/blog/running-vb-code-in-matlab#Macros (or: http://bit.ly/ZqVCHx)

1749. http://mathworks.com/help/matlab/import_export/supported-file-formats.html (or: http://bit.ly/Z8kbuG). Switchyard functions are described in §4.6.1.4.
1750. http://en.wikipedia.org/wiki/Graphics_Interchange_Format (or: http://bit.ly/YFDSbc)
1751. http://mathworks.com/help/matlab/ref/imread.html#f25-722074 (or: http://bit.ly/15zymhw)
1752. http://www.mathworks.com/support/bugreports/793155 (or: http://bit.ly/156ASN9)
1753. http://mathworks.com/help/matlab/release-notes.html#bs31l21-1 (or: http://bit.ly/YGo3Bj)
1754. http://www.mathworks.com/support/bugreports/863941 (or: http://bit.ly/108v6TB); http://www.mathworks.com/support/bugreports/914792 (or: http://bit.ly/108wc1K); http://www.mathworks.com/support/bugreports/872110 (or: http://bit.ly/108wdTm); http://www.mathworks.com/support/bugreports/861335 (or: http://bit.ly/108wiqf)
1755. http://www.mathworks.com/support/bugreports/863941 (or: http://bit.ly/108v6TB)
1756. http://mathworks.com/help/matlab/ref/imread.html#f25-721031 (or: http://bit.ly/YGwEni); http://stackoverflow.com/questions/6157606/what-is-the-fastest-way-to-load-multiple-image-tiff-file-in-matlab (or: http://bit.ly/YGENIC)
1757. http://www.matlabtips.com/how-to-load-tiff-stacks-fast-really-fast (or: http://bit.ly/Zk8jT4)
1758. http://www.matlabtips.com/how-to-load-tiff-stacks-fast-really-fast#comment-270 (or: http://bit.ly/Zk8jT4)
1759. http://www.mathworks.com/support/bugreports/799101 (or: http://bit.ly/18GFJkE)
1760. http://stackoverflow.com/a/24910562/233829 (or: http://bit.ly/Y8SO8h)
1761. http://docs.oracle.com/javase/tutorial/essential/io (or: http://bit.ly/1fzcUfp)
1762. http://docs.oracle.com/javase/7/docs/api/java/nio/file/Files.html (or: http://bit.ly/1fz8DbX)
1763. http://en.wikipedia.org/wiki/New_I/O (or: http://bit.ly/IWcPXo); http://docs.oracle.com/javase/tutorial/essential/io/fileio.html (or: http://bit.ly/IWcZ0Z); http://docs.oracle.com/javase/1.5.0/docs/guide/nio (or: http://bit.ly/IWd2Ka); http://tutorials.jenkov.com/java-nio (or: http://bit.ly/IWd4BP); https://blogs.oracle.com/slc/entry/javanio_vs_javaio (or: http://bit.ly/IWdlEB); https://www.ibm.com/developerworks/java/tutorials/j-nio/section2.html (or: http://ibm.co/IWduIa)
1764. http://docs.oracle.com/javase/7/docs/api/java/io/File.html (or: http://bit.ly/1fz8Z2p)
1765. http://en.wikipedia.org/wiki/Zero-copy (or: http://bit.ly/IWdA2u)
1766. http://sourceforge.net/projects/jmatio (or: http://bit.ly/MEY3qq)
1767. http://mathworks.com/help/matlab/matlab_external/file-handling-with-c.html (or: http://bit.ly/1kftB1J)
1768. http://undocumentedmatlab.com/blog/explicit-multi-threading-in-matlab-part3 (or: http://bit.ly/1gzxrjm)
1769. http://mathworks.com/matlabcentral/newsreader/view_thread/324838#892755 (or: http://bit.ly/LbqhYC)
1770. http://mathworks.com/matlabcentral/fileexchange/40016-recursive-directory-searching-for-multiple-file-specs (or: http://bit.ly/1bbrR2t); Selected as a MATLAB-Central Pick-of-the-Week (POTW): http://blogs.mathworks.com/pick/2013/02/15/recursive-directory-searching-for-multiple-file-specs-revisited (or: http://bit.ly/1ka5Ngv)
1771. http://mathworks.com/matlabcentral/answers/52705-test-existence-of-files-with-exist (or: http://bit.ly/Pd5fuU) discusses various issues with *exist*'s performance and possible alternatives
1772. http://mathworks.com/matlabcentral/fileexchange/28249-getfullpath (or: http://bit.ly/1fU9irm); http://blogs.mathworks.com/pick/2011/04/01/be-absolute-about-your-relative-path-with-getfullpath (or: http://bit.ly/1fU9pDk)
1773. http://mathworks.com/matlabcentral/fileexchange/24671-filetime (or: http://bit.ly/1n7Q6VN)
1774. http://mathworks.com/matlabcentral/fileexchange/17291-fdep (or: http://bit.ly/ZG55zv)
1775. http://mathworks.com/matlabcentral/fileexchange/17291-fdep/content/hfdep/fdephtml.html (or: http://bit.ly/ZG5WAi)

1776. http://blogs.mathworks.com/pick/2011/09/30/determine-file-dependencies (or: http://bit.ly/ZG6u9m)

1777. http://mathworks.com/matlabcentral/fileexchange/10702-exporttozip (or: http://bit.ly/ZG3GsK)

1778. http://blogs.mathworks.com/pick/2009/09/18/easier-and-less-error-prone-creation-of-zip-files (or: http://bit.ly/ZG6KVE)

1779. http://mathworks.com/matlabcentral/fileexchange/30724-exporttozip (or: http://bit.ly/ZG4SMB)

1780. http://mathworks.com/matlabcentral/fileexchange/15924-farg (or: http://bit.ly/ZG8r5y)

1781. http://en.wikipedia.org/wiki/grep (or: http://bit.ly/ZGcpv3)

1782. http://mathworks.com/matlabcentral/fileexchange/9647-grep (or: http://bit.ly/105lrwx)

1783. http://blogs.mathworks.com/pick/2012/05/25/grep-text-searching-utility (or: http://bit.ly/105lt7I)

1784. http://winmerge.org

1785. http://mathworks.com/matlabcentral/fileexchange/14596-cmp (or: http://bit.ly/ZH6INi)

1786. http://blogs.mathworks.com/pick/2013/11/15/whats-in-that-mat-file (or: http://bit.ly/1bix6D4); http://mathworks.com/matlabcentral/fileexchange/42159-matwho (or: http://bit.ly/1bix7GZ). http://mathworks.com/matlabcentral/newsreader/view_thread/316593 (or: http://bit.ly/1epy7Jt) explains the underlying mechanism, using MEX's *matGetDir()* function.

1787. http://en.wikipedia.org/wiki/Bytecode (or: http://bit.ly/11PDBVq); http://en.wikipedia.org/wiki/P-code_machine (or: http://bit.ly/1eAwLsI)

1788. http://en.wikipedia.org/wiki/Machine_code (or: http://bit.ly/ZQ5XjC)

1789. http://mathworks.com/matlabcentral/answers/142465-hdfread-nfs-and-i-o-performance (or: http://bit.ly/1AgYjiP)

1790. http://undocumentedmatlab.com/blog/images-in-matlab-uicontrols-and-labels (or: http://bit.ly/11PdQEF)

1791. http://mathworks.com/matlabcentral/newsreader/view_thread/268923#704048 (or: http://bit.ly/15f5ATz)

1792. http://faqs.org/docs/artu/ch12s04.html (or: http://bit.ly/ZGkFVU) — Eric Raymond, *The Art of Unix Programming*, ISBN 0131429019, §12.4.

1793. http://www.mathworks.com/support/bugreports/905821 (or: http://bit.ly/156PhZJ). This bug was fixed in the R2013b (MATLAB 8.2) release.

1794. The following user report may possibly be due to this issue: http://mathworks.com/matlabcentral/answers/120907-memory-usage-and-speed (or: http://bit.ly/1ppOiuo).

1795. http://mathworks.com/matlabcentral/answers/13085-line-129-in-saveas-function-is-slow (or: http://bit.ly/1mxpsH7)

1796. http://mathworks.com/matlabcentral/fileexchange/29569-filerename (or: http://bit.ly/1mxrffk) – up to 50x faster than *movefile*!

1797. http://mathworks.com/matlabcentral/fileexchange/37384-fileresize (or: http://bit.ly/1mxsR8R); http://mathworks.com/matlabcentral/fileexchange/24671-filetime (or: http://bit.ly/1mxtjE2); http://mathworks.com/matlabcentral/fileexchange/30395-filerealcase (or: http://bit.ly/1iXk8fc); http://mathworks.com/matlabcentral/fileexchange/28249-getfullpath (or: http://bit.ly/1fU9irm)

1798. http://mathworks.com/matlabcentral/newsreader/view_thread/323717#889416 (or: http://bit.ly/1lmXp06)

1799. http://mathworks.com/matlabcentral/newsreader/view_thread/250214 (or: http://bit.ly/1h8cBwi)

Index

Milton Keynes UK
Ingram Content Group UK Ltd.
UKHW020828141024
449569UK00008B/588

9 781482 211290